Mathematical Methods

Second Edition

Mathematical Methods

MERLE C. POTTER

Michigan State University

JACK GOLDBERG

University of Michigan

PRENTICE-HALL, INC., *Englewood Cliffs, New Jersey* 07632

Library of Congress Cataloging-in-Publication Data

Potter, Merle C.
    Mathematical methods.

    Rev. ed. of: Mathematical methods in the
physical sciences. c1978.
    Bibliography: p.
    Includes index.
    1. Mathematics—1961–   2. Engineering
mathematics.   3. Science—Methodology.   I. Goldberg,
Jack L. (Jack Leonard)   II. Potter,
Merle C. Mathematical methods in the physical
sciences.   III. Title.
QA37.2.P668   1987   515′.024′53   86–12221
ISBN 0–13–561184–9

© 1987, 1978 by Prentice-Hall, Inc.
A Division of Simon & Schuster
Englewood Cliffs, New Jersey 07632

Editorial/production: Nicholas Romanelli
Manufacturing buyer: Rhett Conklin
Cover design: Ben Santora

Printed in the United States of America

10 9 8 7 6 5 4 3 2 1

ISBN 0-13-561184-9

Prentice-Hall International (UK) Limited, *London*
Prentice-Hall of Australia Pty. Limited, *Sydney*
Prentice-Hall Canada Inc., *Toronto*
Prentice-Hall Hispanoamericana, S.A., *Mexico*
Prentice-Hall of India Private Limited, *New Delhi*
Prentice-Hall of Japan, Inc., *Tokyo*
Prentice-Hall of Southeast Asia Pte. Ltd., *Singapore*
Editora Prentice-Hall do Brasil, Ltda., *Rio de Janeiro*

In memory of
a dear friend, colleague, and gifted educator,
Roger Low

Contents

Preface

The purpose of this book is to introduce students of the physical sciences to several mathematical methods often essential to the successful solution of real problems. The methods chosen are those most frequently used in typical physics and engineering applications. The treatment is not intended to be exhaustive; the subject of each chapter can be found as the title of a book which treats the material in much greater depth. The reader is encouraged to consult such a book should more study be desired in any of the areas introduced.

Perhaps it would be helpful to discuss the motivation that led to the writing of this text. Undergraduate education in the physical sciences has become more advanced and sophisticated with the advent of the space age and computers, with their demand for the solution of very difficult problems. During the recent past, mathematical topics usually reserved for graduate study have become part of the undergraduate program. It is now common to find an applied mathematics course, usually covering one topic, that follows differential equations in engineering and physical science curricula. Choosing the content of this mathematics course is often difficult. In each of the physical science disciplines, different phenomena are investigated which result in a variety of mathematical models. To be sure, a number of outstanding textbooks exist that present advanced and comprehensive treatment of these methods. However, these texts are usually written at a level too advanced for the undergraduate student, and the material is so exhaustive that it inhibits the effective presentation of the mathematical techniques as a tool for the analysis of some of the simpler problems encountered by the undergraduate. This book was written to provide for an additional course, or two, after a course in differential equations, to permit more than one topic to be introduced in a term or semester, and to make the material comprehensive to the undergraduate. However, rather than assume a knowledge of differential equations, we have included all of the essential material usually found in a course on that subject, so that this text can also be used in an introductory course on differential equations or in a second applied course on differential equations. Selected sections from several of the chapters would constitute such courses.

Ordinary differential equations, including a number of physical applications, are reviewed in Chapter One. The use of series methods are presented in Chapter Two. Subsequent chapters present Laplace transforms, matrix theory and applications, vector analysis, Fourier series, partial differential equations, numerical methods using finite differences, and complex variables. The material is presented so that more than one subject, perhaps four subjects, can be covered in a single course, depending on the topics chosen and the completeness of coverage. The style of presentation is such that the step-by-step derivations may be followed by the reader with a minimum of assistance from the instructor. Liberal use of examples and homework problems should aid the student in the study of the mathematical methods presented.

The first edition of this text was written solely by the first author, an engineering professor at Michigan State University. When it became time to write a second edition, it was decided to add a mathematician coauthor who would add additional material to several of the chapters and increase the mathematical exactness and rigor of the concepts presented. The presentation was to remain at the undergraduate level. Consequently, the second author, a mathematician from The University of Michigan, who is a gifted teacher at both the undergraduate and the graduate levels, agreed to become the coauthor. It is hoped that this second edition has more effectively accomplished the objectives stated above.

Merle C. Potter / Jack Goldberg

1

Ordinary Differential Equations

1.1   Introduction

Differential equations play a vital role in the solution of many problems encountered when modeling physical phenomena. All the disciplines in the physical sciences, each with its own unique physical situations, require that the student be able to derive the necessary mathematical equations (often differential equations) and then solve the equations to obtain the desired solutions. We shall consider a variety of physical situations that lead to differential equations, examine representative problems from several disciplines, and develop the standard methods to solve these equations.

An equation relating an unknown function to its various derivatives is a differential equation; thus

$$\frac{du}{dx} = u \qquad (1.1.1)$$

$$\frac{d^2 f}{dx^2} + 2xf = e^x \qquad (1.1.2)$$

$$\frac{\partial^2 u}{\partial x^2} + \frac{\partial^2 u}{\partial y^2} = 0 \qquad (1.1.3)$$

are examples of differential equations. A *solution* of a differential equation is a function defined and differentiable sufficiently often so that when the function and its derivatives are substituted into the equation, the resulting expression is an identity. Thus $u(x) = e^x$ is a solution of Eq. 1.1.1 because*

---

*Primes or dots will often be used to denote differentiation. Hence $u'(x) = du/dx$ and $\dot{u}(t) = du/dt$.

$u'(x) = e^x = u(x)$. The function $e^x \sin y$ is a solution of Eq. 1.1.3 because

$$\frac{\partial^2}{\partial x^2}(e^x \sin y) = e^x \sin y \tag{1.1.4}$$

$$\frac{\partial^2}{\partial y^2}(e^x \sin y) = -e^x \sin y \tag{1.1.5}$$

and hence, for all x and y,

$$\frac{\partial^2}{\partial x^2}(e^x \sin y) + \frac{\partial^2}{\partial y^2}(e^x \sin y) = 0. \tag{1.1.6}$$

Often it is not possible to express a solution in terms of combinations of "elementary" functions. Such is the case with Eq. 1.1.2. In these circumstances we must turn to alternative methods for describing the solutions. Under this category we list numerical methods, power series, asymptotic series, iteration methods, and phase-plane analysis. In this chapter we confine ourselves primarily to equations for which elementary functions, or their integrals, suffice to represent the solution.

1.2 Definitions

An *ordinary differential equation* is one in which only the derivatives with respect to one variable appear. A *partial differential equation* contains partial derivatives with respect to more than one independent variable. Equations 1.1.1 and 1.1.2 are ordinary differential equations, while Eq. 1.1.3 is a partial differential equation. Using our convention,

$$\frac{\partial F(x, t)}{\partial t} = xt$$

is an ordinary differential equation but

$$\frac{\partial^2 F(x, t)}{\partial t\, \partial x} = t$$

is a partial differential equation.*

The dependent variable usually models the unknown quantity sought after in some physical problem, or some quantity closely related to it. For example, if the lift on an airfoil is the quantity desired, we would solve a partial differential equation to find the unknown velocity $v(x, y)$—the dependent variable—from which we can calculate the pressure and consequently the lift.

---

*Some authors would consider both equations as partial differential equations. The techniques for solution do not depend on so arbitrary a matter as a name.

The *order* of a differential equation is the order of the highest derivative occurring in the equation. The order of both Eqs. 1.1.2 and 1.1.3 is 2; the order of Eq. 1.1.1 is 1 and the order of the equation

$$\frac{d^3u}{dx^3} + x^4 u^5 \frac{d^2u}{dx^2} - \sin u = 0 \tag{1.2.1}$$

is 3. The most general first-order equation that we* consider is

$$u' = f(x, u). \tag{1.2.2}$$

Similarly, the most general second-order equation is

$$u'' = f(x, u, u'). \tag{1.2.3}$$

In the nth-order case

$$u^{(n)} = f(x, u, u', \dots, u^{(n-1)}). \tag{1.2.4}$$

The nth-order equation is called *linear* if f has the special form

$$u^{(n)} = q(x) - P_{n-1}(x)u - P_{n-2}(x)u' - \cdots - P_0(x)u^{(n-1)}. \tag{1.2.5}$$

Rewriting this expression gives us the *standard* form for the nth-order linear equation:

$$u^{(n)} + P_0(x)u^{(n-1)} + \cdots + P_{n-2}(x)u' + P_{n-1}(x)u = g(x). \tag{1.2.6}$$

If $g(x) = 0$, the linear equation is called *homogeneous*; otherwise, it is *nonhomogeneous*. An equation that is not linear is *nonlinear*. The equation

$$u'' + \frac{1}{x}u' + \left(1 - \frac{n^2}{x^2}\right)u = 0 \tag{1.2.7}$$

is a homogeneous, second-order linear differential equation. The equation

$$u'' + 4uu' = 0 \tag{1.2.8}$$

is nonlinear but also of second order. (We do not distinguish between homogeneous and nonhomogeneous equations in the nonlinear case.)

Some differential equations are particularly easy to solve. For example, the linear differential equation

$$\frac{du}{dx} = g(x) \tag{1.2.9}$$

has the solution

$$u(x) = \int g(x)\, dx + C, \tag{1.2.10}$$

---

*Some authors allow the more general representation $F(x, u, u') = 0$.

where C is an arbitrary constant. This follows from the Fundamental Theorem of Calculus, which implies that

$$\frac{du}{dx} = \frac{d}{dx} \left[\int g(x) \, dx + C \right] = \frac{d}{dx} \int g(x) \, dx = g(x). \quad (1.2.11)$$

Unless $g(x)$ is one of a relatively sparse family of functions, it will not be possible to express $u(x)$ in any simpler form than the indefinite integral of $g(x)$.

Equation 1.2.10 raises a notational issue. Writing $u(x)$ in the form

$$u(x) = \int g(x) \, dx + C, \quad (1.2.12)$$

even when C is specified, does not readily suggest a means for expressing or computing individual values of u, such as $u(0)$. An alternative form for u is

$$u(x) = \int_{x_0}^{x} g(s) \, ds + C. \quad (1.2.13)$$

Note carefully that u is a function of x, the upper limit of integration, not s, the "dummy" variable. Indeed, u is also expressible as

$$u(x) = \int_{x_0}^{x} g(t) \, dt + C. \quad (1.2.14)$$

It is not advisable to write

$$u(x) = \int_{x_0}^{x} g(x) \, dx + C. \quad (1.2.15)$$

This will often lead to errors, especially when attempting to differentiate the equation.

For certain rather special equations, repeated integrations provide a means for obtaining solutions. For example,

$$\frac{d^n u}{dx^n} = 0 \quad (1.2.16)$$

has the family of solutions

$$u(x) = c_0 + c_1 x + \cdots + c_{n-1} x^{n-1}, \quad (1.2.17)$$

obtained by integrating Eq. 1.2.16 n times; the n arbitrary constants, c_0, $c_1, \ldots, c_{n-1}$, are constants of integration. The differential equations considered in this chapter possess solutions that will be obtained with more difficulty than the above; however, there will be times when simple equations such as $u^{(n)} = g(x)$ do model phenomena of interest.

A *general solution* of an nth-order, linear equation is a family of solutions containing n "essential" arbitrary constants. The family of solutions given by

Eq. 1.2.17 is one example. Another is the family

$$f(x) = Ax + B(x^3 + 1), \tag{1.2.18}$$

a general solution of the linear equation

$$(2x^3 - 1)f'' - 6x^2 f' + 6xf = 0. \tag{1.2.19}$$

In contrast, $f(x) = Ae^{x+B}$ is a solution of $y'' - y = 0$ for each choice of A and B, but this family of solutions is not a general solution since both A and B are not essential, for

$$Ae^{x+B} = Ae^B e^x = Ce^x. \tag{1.2.20}$$

Thus, in spite of the appearance of the two arbitrary constants A and B, the family of solutions described by the set of functions Ae^{x+B} is the same family described by Ae^x. (A precise definition of a general solution—and hence of essential arbitrary constants—will be given in Section 1.5; until then we make do with the intuitive ideas suggested above.)

We do not define general solutions* for nonlinear equations because experience has shown that this notion plays a minor role in the theory of these equations. Part of the reason for this is the sparsity of interesting nonlinear equations for which general solutions are known. In most applications it is the specific solutions that are of most importance anyway. A *specific solution* is a single function[†] that solves the given differential equation. When a general solution is known, specific solutions are obtained by assigning values to each of its arbitrary constants. The nonlinear equation[‡]

$$y'^2 + xy' - y = 0 \tag{1.2.21}$$

has a family of solutions $y_g(x) = cx + c^2$. [Since the differential equation is nonlinear, we refrain from calling $y_g(x)$ a general solution.] Each choice of c results in a specific solution of Eq. 1.2.21; the function $y(x) = -x^2/4$ is also a specific solution, but not one that can be obtained from the family $y_g(x)$.

Under certain reasonable assumptions, a unique specific solution to a first-order equation is determined by demanding that the solution meet an *initial condition*, a condition that specifies the value of the solution at some

---

*This viewpoint is not taken by all authors. The student will find many texts in which general solutions are defined for some nonlinear equations.

[†]A function is a rule that assigns to each x in some domain a unique value denoted by $f(x)$; there are no arbitrary constants in $f(x)$. The domains for ordinary differential equations are one of the following types: $-\infty < x < \infty$, $-\infty < x < b$, $a < x < \infty$, and $a < x < b$, where a and b are finite.

[‡]This equation is one member of a family of nonlinear equations known collectively as *Clairaut's equation*.

$x = x_0$. The differential equation together with the initial condition is an *initial-value problem*. The equations

$$u' = f(x, u)$$
$$u(x_0) = u_0$$

(1.2.22)

form the most general, first-order, initial-value problem. Two conditions must be given for second-order equations; the initial-value problem is

$$u'' = f(x, u, u')$$
$$u(x_0) = u_0, \qquad u'(x_0) = u_0'$$

(1.2.23)

Here both conditions are obtained at the same point, $x = x_0$. If the conditions are given at different points, a *boundary-value problem* results. A very common boundary-value problem is

$$u'' = f(x, u, u')$$
$$u(x_0) = u_0, \qquad u(x_1) = u_1.$$

(1.2.24)

Other boundary-value problems are possible; for example, a derivative may be specified at one of the points.

PROBLEMS

In each case decide whether the equation is linear or nonlinear, homogeneous or nonhomogeneous, and state its order.

1. $u'/u = 1 + x$ **2.** $uu' = 1 + x$

3. $\sin u' = u$ **4.** $u'' - 2u' + u = \cos x$

5. $u'' = x^2$ **6.** $u'' = u$

7. $u'' = u^2$ **8.** $(u^2)' = -u$

Find families of solutions to each differential equation.

9. $u' = x^2 + 2$ **10.** $u' = \sin x + e^x$

11. $u' = x + \cos^2 x$ **12.** $u'' = 2x$

13. $u''' = x^2$ **14.** $u^{iv} = x - 2$

15. Verify that each member of the family of functions given by Eq. 1.2.17 solves Eq. 1.2.16.

16. Verify that $Ax + B(x^3 + 1)$ satsifies Eq. 1.2.19 for each choice of A and B.

17. Show that $A(x - c_1)(x - c_2) + B(x - c_3) + C$ has only three essential arbitrary constants.

Verify that the given function satisfies the differential equation.

18. $u = \cos 2x$, $u'' + 4u = 0$

19. $u = e^{2x}$, $u'' - 4u = 0$

20. $u^2 + x^2 = 10$, $uu' + x = 0$

21. $u = e^{-3x} + 12e^{-2x}$, $u'' + 5u' + 6u = 0$

22. The acceleration of an object is given by $a = d^2s/dt^2$, where s is the displacement. For a constant deceleration of 20 m/s^2, find the distance an object travels before coming to rest if the initial velocity is 100 m/s.

23. An object is dropped from a house roof 8 m above the ground. How long does it take to hit the ground? Use $a = -9.81$ m/s^2 in the differential equation $a = d^2y/dt^2$, y being positive upward.

24. Verify that $y(x) = cx + c^2$ is a solution of Eq. 1.2.21 for each c. Verify that $y(x) = -x^2/4$ is also a solution of the same equation.

25. Verify that the initial-value problem

$$y'^2 + xy' - y = 0$$

$$y(2) = -1$$

has two specific solutions

$$y_1(x) = 1 - x \quad \text{and} \quad y_2(x) = \frac{-x^2}{4}.$$

26. Verify that the initial-value problem

$$y'^2 + xy' - y = 0$$

$$y(-1) = 2$$

has specific solutions

$$y_1(x) = 1 - x \quad \text{and} \quad y_2(x) = 2(x + 2).$$

27. Verify that $u(x) = A \sin(ax) + B \cos(ax)$ is a solution to $u'' + a^2u = 0$. If $u(0) = 10$ and $u(\pi/2a) = 20$, determine the specific solution.

28. The deflection of a 10-m-long cantilever beam with constant loading is found by solving $u^{iv} = 0.006$. Find the maximum deflection of the beam. Each cantilever end requires both deflection and slope to be zero.

1.3   Differential Equations of First Order

1.3.1   SEPARABLE EQUATIONS

Some first-order equations can be reduced to

$$h(u)\frac{du}{dx} = g(x), \tag{1.3.1}$$

which is equivalent to

$$h(u) \, du = g(x) \, dx. \tag{1.3.2}$$

This first-order equation is *separable* because the variables and their corresponding differentials appear on different sides of the equation. Hence, the solution is obtained by integration:

$$\int h(u) \, du = \int g(x) \, dx + C. \tag{1.3.3}$$

Unless the indefinite integral on the left-hand side of Eq. 1.3.3 is a particularly simple function of u, Eq. 1.3.3 is not an improvement over Eq. 1.3.2. To illustrate this point, let $h(u) = \sin \sqrt{u}$ and $g(x) = e^{x^2}$. Then Eq. 1.3.3 becomes

$$\int \sin \sqrt{u} \, du = \int e^{x^2} \, dx + C, \tag{1.3.4}$$

which is an expression that defines u with no more clarity than its differential form,

$$\sin \sqrt{u} \, du = e^{x^2} \, dx. \tag{1.3.5}$$

The following example is more to our liking.

Example 1.3.1:   Find the solutions to the nonlinear equation

$$x \frac{du}{dx} + u^2 = 4.$$

SOLUTION: The equation is separable and may be written as

$$\frac{du}{4 - u^2} = \frac{dx}{x}.$$

To aid in the integration we write

$$\frac{1}{4 - u^2} = \frac{1/4}{2 - u} + \frac{1/4}{2 + u}.$$

Our equation becomes

$$\frac{1}{4} \frac{du}{2 - u} + \frac{1}{4} \frac{du}{2 + u} = \frac{dx}{x}.$$

This is integrated to give

$$-\tfrac{1}{4} \ln (2 - u) + \tfrac{1}{4} \ln (2 + u) = \ln x + \tfrac{1}{4} \ln C$$

where $\tfrac{1}{4} \ln C$ is constant, included because of the indefinite integration. In this last equation u, x, and C are restricted so that each logarithm is defined (i.e., $|u| < 2$, $x > 0$, and $C > 0$). After some algebra this is put in the equivalent form

$$\frac{2 + u}{2 - u} = x^4 C,$$

which can be written as

$$u(x) = \frac{2(Cx^4 - 1)}{Cx^4 + 1}.$$

If the constant of integration had been chosen as just plain C, an equivalent but more complicated expression would have resulted. We chose $\frac{1}{4} \ln C$ to provide a simpler appearing solution. The restrictions on u, x, and C, introduced earlier to ensure the existence of the logarithms, are seen to be superfluous. An easy differentiation verifies that for each C, the solution is defined for all x, $-\infty < x < \infty$. ■

The linear, homogeneous equation

$$\frac{du}{dx} + p(x)u = 0 \tag{1.3.6}$$

is separable. It can be written as

$$\frac{du}{u} = -p(x)\,dx \tag{1.3.7}$$

and hence the solution is

$$\ln u = -\int p(x)\,dx + C. \tag{1.3.8}$$

If we write $F(x) = e^{\int p(x)\,dx}$, then the solution takes the form

$$|u(x)| = \frac{e^C}{F(x)}. \tag{1.3.9}$$

This last form suggests examining

$$u(x) = \frac{K}{F(x)}. \tag{1.3.10}$$

In fact, for each K this represents a solution of Eq. 1.3.6. Therefore, Eq. 1.3.10 represents a family of solutions of Eq. 1.3.6.

Certain equations that are not separable can be made separable by a change of variables. An important class of such equations may be described by the formula

$$\frac{du}{dx} = f\left(\frac{u}{x}\right). \tag{1.3.11}$$

Then, setting $u = vx$ to define the new dependent variable v, we obtain

$$\frac{du}{dx} = x\frac{dv}{dx} + v. \tag{1.3.12}$$

Substituting into Eq. 1.3.11 results in

$$x\frac{dv}{dx} + v = f(v), \tag{1.3.13}$$

which, in turn, leads to the equation

$$\frac{dv}{f(v) - v} = \frac{dx}{x},$$
(1.3.14)

which can be solved by integration.

Example 1.3.2:   Determine a family of solutions to the differential equation

$$xu \frac{du}{dx} - u^2 = x^2.$$

SOLUTION: The equation in the given form is not separable and it is nonlinear. However, the equation can be put in the form

$$\frac{u}{x} \frac{du}{dx} - \frac{u^2}{x^2} = 1$$

by dividing by x^2. This is in the form of Eq. 1.3.11, since we can write

$$\frac{du}{dx} = \frac{1 + (u/x)^2}{u/x}.$$

Define a new dependent variable to be $v = u/x$, so that

$$\frac{du}{dx} = x \frac{dv}{dx} + v.$$

Substitute back into the given differential equation and obtain

$$v\left(x \frac{dv}{dx} + v\right) - v^2 = 1.$$

This can be put in the separable form

$$v \, dv = \frac{dx}{x}.$$

Integration of this equation yields

$$\frac{v^2}{2} = \ln |x| + C.$$

Substitute $v = u/x$ and obtain $u(x)$ to be

$$u(x) = \sqrt{2}x(C + \ln |x|)^{1/2}.$$

This represents a solution for each C such that $C + \ln |x| > 0$, as is proved by differentiation and substitution into the given equation.   ∎

PROBLEMS

Find a family of solutions to each differential equation.

1. $u' = 10 u$
        2. $u' = 10 u^2$

3. $u' = 2u + 3$

4. $u' = u \sin x$

5. $u' = \cot u \sin x$

6. $x^2 u' + u^2 = 1$

7. $x(x + 2)u' = u^2$

8. $5x \, du + x^2 u \, dx = 0$

Find a family of solutions to each equation.

9. $xu' + 2x = u$

10. $x^2 u' = xu + u^2$

11. $x^3 + u^3 - xu^2 u' = 0$

12. $3u + (u + x)u' = 0$

13. $xu' = (x - u)^2 + u$ (let $x - u = y$)

14. $(x + 2u + 1)u' = x + 2u + 4$ (*Hint*: Let $x + 2u = y$.)

Solve each initial value problem.

15. $u' = 2u - 1,$ $u(0) = 2$

16. $u' \tan x = u + 1,$ $u(2) = 0$

17. $xu' + u = 2x,$ $u(1) = 10$

18. $xu' = (u - x)^3 + u,$ $u(0) = 0$ (*Hint*: Let $v = u - x$.)

1.3.2   EXACT EQUATIONS

The equation $du/dx = f(x,u)$ can be written in many different forms. For instance, given any $N(x, u)$, define $M(x, u)$ by the equation

$$M(x, u) = -f(x, u)N(x, u). \tag{1.3.15}$$

Then

$$\frac{du}{dx} = f(x, u) = -\frac{M(x, u)}{N(x, u)} \tag{1.3.16}$$

leads to

$$M(x, u) \, dx + N(x, u) \, du = 0. \tag{1.3.17}$$

In this form the differential equation suggests the question: Does there exist $\phi(x, u)$ such that $d\phi = M \, dx + N \, du$? The *total differential* of $\phi(x, u)$ is defined

$$d\phi = \frac{\partial \phi}{\partial x} \, dx + \frac{\partial \phi}{\partial u} \, du. \tag{1.3.18}$$

Note that if $\phi(x, u) = K$, then $d\phi = 0$.

The equation $M \, dx + N \, du = 0$ is *exact* if there exists $\phi(x, u)$ such that

$$d\phi = M(x, u) \, dx + N(x, u) \, du \tag{1.3.19}$$

or, equivalently,

$$\frac{\partial \phi}{\partial x} = M \quad \text{and} \quad \frac{\partial \phi}{\partial u} = N, \tag{1.3.20}$$

a consequence of Eq. 1.3.18. If $M\,dx + N\,du = 0$ is known to be exact, then it follows that

$$d\phi = M\,dx + N\,du = 0, \tag{1.3.21}$$

so that

$$\phi(x, u) = K. \tag{1.3.22}$$

If Eq. 1.3.22 is simple enough, it may be possible to solve for u as a function of x and then verify that this u is a solution of Eq. 1.3.17. This is the tack we take.

Example 1.3.3: Verify, by finding ϕ, that

$$-\frac{u}{x^2}\,dx + \frac{1}{x}\,du = 0$$

is exact. Find u from $\phi(x, u) = K$ and verify that u solves the given differential equation.

SOLUTION: We determine all possible ϕ by solving (see Eq. 1.3.20),

$$\frac{\partial \phi}{\partial x} = M = -\frac{u}{x^2}.$$

Integration implies that $\phi(x, u) = u/x + h(u)$ if the given differential equation is exact. The function $h(u)$ is an arbitrary differentiable function of u, analogous to an arbitrary constant of integration. The second equation in Eq. 1.3.20 yields

$$\frac{\partial \phi}{\partial u} = \frac{\partial}{\partial u}\left[\frac{u}{x} + h(u)\right]$$

$$= \frac{1}{x} + h'(u) = N = \frac{1}{x}.$$

Therefore,

$$\phi(x, u) = \frac{u}{x} + C$$

which, for any C, satisfies both parts of Eq. 1.3.20. Hence, the given differential equation is exact. Moreover, we determine, using Eq. 1.3.22,

$$u(x) = Ax,$$

where $A = K - C$, and verify that

$$-\frac{u}{x^2}\,dx + \frac{1}{x}\,du = -\frac{Ax}{x^2}\,dx + \frac{1}{x}(A\,dx) \equiv 0. \quad\blacksquare$$

If it had been the case that our given differential equation was not exact, it would have been impossible to solve Eq. 1.3.20. The next example illustrates this point.

Example 1.3.4:   Show that

$$-u \, dx + x \, du = 0$$

is not exact.

SOLUTION: We find that

$$\frac{\partial \phi}{\partial x} = M = -u$$

requires $\phi(x, u) = -xu + h(u)$. However,

$$\frac{\partial \phi}{\partial u} = \frac{\partial}{\partial u}[-xu + h(u)] = -x + h'(u) = N = x$$

requires $h'(u) = 2x$, an obvious contradiction.

The student should note that

$$-u \, dx + x \, du = x^2\left(-\frac{u}{x^2} \, dx + \frac{1}{x} \, du\right) = x^2 d\left(\frac{u}{x}\right) = 0,$$

and thus the apparently trivial modification of multiplying an exact equation by x^2 destroys its exactness.   ■

The pair of equations in Eq. 1.3.20 imply by differentiation that

$$\frac{\partial^2 \phi}{\partial x \, \partial u} = \frac{\partial M}{\partial u} \quad \text{and} \quad \frac{\partial^2 \phi}{\partial u \, \partial x} = \frac{\partial N}{\partial x}.$$

Hence, assuming that the order of differentiation can be interchanged, a situation that is assumed in all our work in this text, we have

$$\frac{\partial M}{\partial u} = \frac{\partial N}{\partial x}. \tag{1.3.23}$$

We use Eq. 1.3.23 as a negative test. If it fails to hold, then $M \, dx + N \, du = 0$ is not exact* and we need not attempt to solve for ϕ. In Example 1.3.2, $M = -u$ and $N = x$ and hence

$$\frac{\partial M}{\partial u} = -1 \neq \frac{\partial N}{\partial x} = 1. \tag{1.3.24}$$

This saves much useless labor.

Example 1.3.5:   Find the specific solution of the differential equation

$$(2 + x^2 u)\frac{du}{dx} + xu^2 = 0 \quad \text{if} \quad u(1) = 2.$$

---

*We do not prove that $\partial M/\partial u = \partial N/\partial x$ implies that $M \, dx + N \, du = 0$ is exact because such a proof would take us far afield. Moveover, in any particular case, knowing that $M \, dx + N \, du = 0$ is exact does not circumvent the need to solve Eq. 1.3.20 for ϕ. Once ϕ is found, $M \, dx + N \, du = 0$ is exact by construction.

SOLUTION: The differential equation is found to be exact by identifying

$$N = 2 + x^2u, \qquad M = xu^2.$$

Appropriate differentiation results in

$$\frac{\partial N}{\partial x} = 2xu, \qquad \frac{\partial M}{\partial u} = 2xu$$

From

$$\frac{\partial \phi}{\partial x} = M = xu^2$$

we deduce that

$$\phi(x, u) = \frac{x^2u^2}{2} + h(u).$$

We continue as follows:

$$\frac{\partial \phi}{\partial u} = x^2u + h'(u) = N = 2 + x^2u.$$

We deduce that $h'(u) = 2$ and hence that

$$\phi(x, u) = \frac{x^2u^2}{2} + 2u.$$

Using Eq. 1.3.22, we can write

$$\frac{x^2u^2}{2} + 2u = K,$$

which defines $u(x)$. Given that $u(1) = 2$, we find $K = 6$. Finally, using the quadratic formula,

$$u(x) = -\frac{2}{x^2} + \frac{2}{x^2}\sqrt{1 + 3x^2}.$$

We use the plus sign so that $u(1) = 2$. Implicit differentiation of $x^2u^2/2 + 2u = 6$ is the easiest way of verifying that u is a solution.   ∎

PROBLEMS

Verify that each exact equation is linear and solve by separating variables.

1. $-\dfrac{u}{x^2} \, dx + \dfrac{1}{x} \, du = 0$
    2. $2xu \, dx + x^2 \, du = 0$

Show that each equation is exact and find a solution.

3. $(2 + x^2)\dfrac{du}{dx} + 2xu = 0$
    4. $x^2 + 3u^2 \dfrac{du}{dx} = 0$

5. $\sin 2x \dfrac{du}{dx} + 2u \cos 2x = 0$
    6. $e^x\left(\dfrac{du}{dx} + u\right) = 0$

7. Show that the equation $u' = f(x, u)$, written as $f(x, u)\, dx - du = 0$, is exact if and only if f is a function of x alone. What is ϕ when this equation is exact?

8. The separable equation $h(u)\, dx + g(x)\, du = 0$ is exact. Find ϕ and thus verify that it is exact.

9. Find ϕ and thus verify that

$$e^{\int p_0(x)\, dx}[p_0(x)u - g(x)]\, dx + e^{\int p_0(x)\, dx}\, du = 0$$

is exact.

Solve each initial-value problem.

10. $(1 + x^2)u' + 2xu = 0, \quad u(0) = 1$

11. $(x + u)u' + u = x, \qquad u(1) = 0$

12. $(u' + u)e^x = 0, \qquad\quad u(0) = 0$

13. If

$$M(x, u)\, dx + N(x, u)\, du = 0$$

is exact, then so is

$$kM(x, u)\, dx + kN(x, u)\, du = 0$$

for any constant k. Why? Under the same assumptions show that

$$f(x)M(x, u)\, dx + f(x)N(x, u)\, du = 0$$

is not exact unless $f(x) = k$, a constant.

14. Show that

$$M(x, u)\, dx + N(x, u)\, du = 0$$

is exact if and only if

$$[M(x, u) + g(x)]\, dx + [N(x, u) + h(u)]\, du = 0$$

is exact.

1.3.3   INTEGRATING FACTORS

The equation $M\, dx + N\, du = 0$ is rarely exact. This is not surprising since exactness depends so intimately on the forms of M and N. As we have seen in Example 1.3.4, even a relatively insignificant modification of M and N can destroy exactness. On the other hand, this raises the question of whether an inexact equation can be altered to make it exact. The function $I(x, u)$ is an *integrating factor* if

$$I(x, u)[M(x, u)\, dx + N(x, u)\, du] = 0 \qquad (1.3.25)$$

is exact. To find I, we solve

$$\frac{\partial(IM)}{\partial u} = \frac{\partial(IN)}{\partial x}, \qquad (1.3.26)$$

a prospect not likely to be easier than solving $M\,dx + N\,du = 0$. In at least one case, however, we can find $I(x, u)$. Consider the general, linear equation

$$u' + p(x)u = g(x), \tag{1.3.27}$$

which can be put in the form

$$du + [p(x)u - g(x)]\,dx = 0. \tag{1.3.28}$$

We search for an integrating factor which is a function of x alone, that is, $I(x, u) = F(x)$. Then, from Eq. 1.3.26, noting that $M(x, u) = p(x)u - g(x)$ and $N(x, u) = 1$,

$$\frac{\partial}{\partial u}\{F(x)[p(x)u - g(x)]\} = \frac{\partial}{\partial x}F(x) \tag{1.3.29}$$

is the required condition on $F(x)$. Hence,

$$F(x)p(x) = F'(x). \tag{1.3.30}$$

This is a homogeneous first-order equation for $F(x)$. By inspection we find

$$F(x) = e^{\int p(x)\,dx}. \tag{1.3.31}$$

Using this expression for $F(x)$ we can form the differential

$$\begin{aligned} d(Fu) &= F\,du + u\,dF \\ &= F\,du + up(x)F(x)\,dx = F(x)g(x)\,dx \end{aligned} \tag{1.3.32}$$

using* Eq. 1.3.28. Integrating the above gives us

$$F(x)u(x) = \int F(x)g(x)\,dx + K. \tag{1.3.33}$$

Solving for u gives

$$u(x) = \frac{1}{F(x)}\int F(x)g(x)\,dx + \frac{K}{F(x)}. \tag{1.3.34}$$

This formula is the standard form of the general solution of the linear, first-order, homogeneous equation

$$\frac{du}{dx} + p(x)u = g(x). \tag{1.3.35}$$

If $g(x) = 0$ then Eq. 1.3.35 is homogeneous and Eq. 1.3.34 reduces to $u(x) = K/F(x)$; compare this with Eq. 1.3.10.

---

*Using the expression for $F(x)$ given in Eq. 1.3.31 yields

$$\frac{dF}{dx} = \left(\frac{d}{dx}\int p(x)\,dx\right)e^{\int p(x)\,dx} = p(x)F(x).$$

Example 1.3.6:   Solve the linear equation

$$x^2 \frac{du}{dx} + 2u = 5x$$

for the standard form of the general solution.

SOLUTION: The differential equation is first order and linear but is not separable. Thus, let us use an integrating factor to aid in the solution.

Following Eq. 1.3.35, the equation is written in the form

$$\frac{du}{dx} + \frac{2}{x^2}u = \frac{5}{x}.$$

The integrating factor is provided by Eq. 1.3.31 and is

$$F(x) = e^{\int (2/x^2)\,dx} = e^{-2/x}.$$

Equation 1.3.34 then provides the solution

$$u(x) = e^{2/x}\left[\int \frac{5}{x} e^{-2/x}\,dx + K\right].$$

This is left in integral form because the integration cannot be written in terms of elementary functions. If the integrals that arise in these formulas can be evaluated in terms of elementary functions, this should be done.   ■

Equation 1.3.34 does not readily lend itself to solving the initial-value problem

$$\frac{du}{dx} + p(x)u = g(x)$$

$$u(x_0) = u_0 \tag{1.3.36}$$

since, as we have remarked earlier, we cannot conveniently express $u(x_0)$ when u is defined by indefinite integrals. To remedy this deficiency, let $F(x)$ be expressed by

$$F(x) = \exp\left[\int_{x_0}^{x} p(s)\,ds\right] \tag{1.3.37}$$

so that $F(x_0) = 1$. Then an alternative to Eq. 1.3.34 is

$$u(x) = \frac{1}{F(x)} \int_{x_0}^{x} F(t)g(t)\,dt + \frac{K}{F(x)}. \tag{1.3.38}$$

At x_0,

$$u(x_0) = \frac{1}{F(x_0)} \int_{x_0}^{x_0} F(t)g(t)\,dt + \frac{K}{F(x_0)} = K. \tag{1.3.39}$$

Hence,

$$u(x) = \frac{1}{F(x)} \int_{x_0}^{x} F(t)g(t) \, dt + \frac{u_0}{F(x)} \tag{1.3.40}$$

solves the initial-value problem (Eq. 1.3.36).

Example 1.3.7:   Solve the initial-value problem

$$\frac{du}{dx} + 2u = 2, \qquad u(0) = 2.$$

SOLUTION: Here $p(x) = 2$, so

$$F(x) = \exp\left(\int_0^x 2 \, dx\right) = e^{2x}.$$

Thus,

$$u(x) = e^{-2x} \int_0^x e^{2t} \cdot 2 \, dt + \frac{2}{e^{2x}} = e^{-2x}(e^{2x} - 1) + 2e^{-2x} = 1 + e^{-2x}. \quad \blacksquare$$

Example 1.3.8:   Solve the initial-value problem

$$\frac{du}{dx} + 2u = 2, \qquad u(0) = 0.$$

SOLUTION: Since only the initial condition has been changed, we can utilize the work in Example 1.3.7 to obtain

$$u(x) = e^{-2x}(e^{2x} - 1) + \frac{0}{e^{2x}} = 1 - e^{-2x}. \quad \blacksquare$$

PROBLEMS

1. It is not necessary to include a constant of integration in the expression for the integrating factor $F(x) = \exp \int p(x) \, dx$. Include an integration constant and show that the solution (Eq.1.3.34), is unaltered.

2. If $g(x) = 0$ in Eq. 1.3.35, show that $u(x) = K/F(x)$ by solving the resulting separable equation.

Find the general solution to each differential equation.

3. $u' + u = 2$

4. $u' + 2u = 2x$

5. $u' + xu = 10$

6. $u' - 2u = e^x$

7. $u' + u = xe^{-x}$

8. $u' - u = \cos x$

9. $xu' - 2u = xe^x$

10. $x^2 u' - u = 2 \sin (1/x)$

Solve each initial-value problem.

11. $u' + 2u = 2e^{-2x}$, $u(0) = 2$

12. $u' + xu = e^{-x^2}$, $u(1) = 0$

13. $u' - u = x$, $u(0) = 1$

14. $u' - 2u = 4$, $u(0) = 0$

1.4   Physical Applications

There are abundant physical phenomena that can be modeled with first-order differential equations that fall into one of the classes of the previous section. We shall consider several such phenomena, derive the appropriate describing equations, and provide the correct solutions. Other applications will be included in the Problems.

1.4.1   SIMPLE ELECTRICAL CIRCUITS

Consider the circuit in Fig. 1.1, containing a resistance R, inductance L, and capacitance C in series. A known electromotive force $v(t)$ is impressed across the terminals. The differential equation relating the current i to the electromotive force may be found by applying Kirchhoff's first law,* which states that the voltage impressed on a closed loop is equal to the sum of the voltage drops in the rest of the loop. Letting q be the electric charge on the capacitor and recalling that the current i flowing through the capacitor is related to the charge by

$$i = \frac{dq}{dt}, \tag{1.4.1}$$

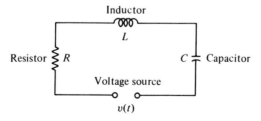

FIGURE 1.1.   *RLC* circuit.

---

*Kirchhoff's second law states that the current flowing into any point in an electrical circuit must equal the current flowing out from that point.

we can write

$$v(t) = L\frac{d^2q}{dt^2} + R\frac{dq}{dt} + \frac{1}{C}q, \tag{1.4.2}$$

where the values, q, v, L, R, and C are in physically consistent units—coulombs, volts, henrys, ohms, and farads, respectively. In the equation above we have used the following experimental observations:

$$\text{voltage drop across a resistor} = iR$$

$$\text{voltage drop across a capacitor} = \frac{q}{C} \tag{1.4.3}$$

$$\text{voltage drop across an inductor} = L\frac{di}{dt}.$$

Differentiating Eq. 1.4.2 with respect to time and using Eq. 1.4.1, where i is measured in amperes, we have

$$\frac{dv}{dt} = L\frac{d^2i}{dt^2} + R\frac{di}{dt} + \frac{1}{C}i. \tag{1.4.4}$$

If dv/dt is nonzero, Eq. 1.4.4 is a linear, nonhomogeneous, second-order differential equation.

If there is no capacitor in the circuit, Eq. 1.4.4 reduces to

$$\frac{dv}{dt} = L\frac{d^2i}{dt^2} + R\frac{di}{dt}. \tag{1.4.5}$$

Integrating, we have (Kirchhoff's first law requires that the constant of integration be zero)

$$L\frac{di}{dt} + Ri = v(t). \tag{1.4.6}$$

The solution to this equation will be provided in the following example.

Example 1.4.1:   Using the integrating factor, solve Eq. 1.4.6 for the case where the electromotive force is given by $v = V \sin \omega t$.

SOLUTION: First, put Eq. 1.4.6 in the standard form

$$\frac{di}{dt} + \frac{R}{L}i = \frac{V}{L}\sin \omega t.$$

Using Eq. 1.3.31 we find that the integrating factor is

$$F(t) = e^{(R/L)t}.$$

According to Eq. 1.3.34 the solution is

$$i(t) = e^{-(R/L)t} \left[\int \frac{V}{L} \sin \omega t \, e^{(R/L)t} \, dt + K \right],$$

where K is the constant of integration. Simplification of this equation yields, after integrating by parts,

$$i(t) = V \left[\frac{R \sin \omega t - \omega L \cos \omega t}{R^2 + \omega^2 L^2} \right] + K e^{-(R/L)t}.$$

If the current $i = i_0$ at $t = 0$, we calculate the constant K to be given by

$$K = i_0 + \frac{V \omega L}{R^2 + \omega^2 L^2}$$

and finally that

$$i(t) = V \left[\frac{R \sin \omega t - \omega L \cos \omega t}{R^2 + \omega^2 L^2} \right] + \left[i_0 + \frac{V \omega L}{R^2 + \omega^2 L^2} \right] e^{-(R/L)t}.$$

In this example we simplified the problem by removing the capacitor. We can also consider a similar problem where the capacitor is retained but the inductor is removed; we would then obtain a solution for the voltage. In Section 1.7 we consider the solution of the general second-order equation 1.4.4 where all components are included. ■

1.4.2   THE RATE EQUATION

A number of phenomena can be modeled by a first-order equation called a *rate equation*. It has the general form

$$\frac{du}{dt} = f(u, t), \tag{1.4.7}$$

indicating that the rate of change of the dependent quantity u may be dependent on both time and u. We shall derive the appropriate rate equation for the concentration of salt in a solution. Other rate equations will be included in the Problems.

Consider a tank with volume V (in cubic meters, m^3), containing a salt solution of concentration $C(t)$. The initial concentration is C_0 (in kilograms per cubic meter, kg/m^3). A brine containing a concentration C_1 is flowing into the tank at the rate q (in cubic meters per second, m^3/s), and an equal flow of the mixture is issuing from the tank. The salt concentration is kept uniform throughout by continual stirring. Let us develop a differential equation that can be solved to give the concentration C as a function of time. The equation is derived by writing a balance equation on the amount (in kilograms) of salt contained in the tank:

$$\text{amount in} - \text{amount out} = \text{amount of increase} \tag{1.4.8}$$

For a small time increment Δt this becomes

$$C_1 q\,\Delta t - Cq\,\Delta t = C(t + \Delta t)V - C(t)V, \qquad (1.4.9)$$

assuming that the concentration of the solution leaving is equal to the concentration $C(t)$ in the tank. The volume V of solution is maintained at a constant volume since the outgoing flow rate is equal to the incoming flow rate. The equation above may be rearranged to give

$$q(C_1 - C) = V\frac{C(t + \Delta t) - C(t)}{\Delta t}. \qquad (1.4.10)$$

Now, if we let the time increment Δt shrink to zero and recognize that

$$\lim_{\Delta t \to 0} \frac{C(t + \Delta t) - C(t)}{\Delta t} = \frac{dC}{dt}, \qquad (1.4.11)$$

we arrive at the rate equation for the concentration of salt in a solution,

$$\frac{dC}{dt} + \frac{q}{V}C = \frac{qC_1}{V}. \qquad (1.4.12)$$

The solution is provided in the following example.

Example 1.4.2:   The initial concentration of salt in a 10-m^3 tank is 0.02 g/m^3. A brine flows into the tank at 2 m^3/s with a concentration of 0.01 g/m^3. Determine the time necessary to reach a concentration of 0.011 g/m^3 in the tank if the outflow equals the inflow.

SOLUTION: Equation 1.4.12 is the equation to be solved. Using $q = 2$, $V = 10$ and $C_1 = 0.01$, we have

$$\frac{dC}{dt} + \frac{2}{10}C = \frac{2 \times 0.01}{10}.$$

The integrating factor is

$$F(t) = e^{\int (1/5)\,dt} = e^{t/5}.$$

The solution, referring to Eq. 1.3.34, is then

$$C(t) = e^{-t/5}\left[\int 0.002\,e^{t/5}\,dt + A\right] = 0.01 + Ae^{-t/5},$$

where A is the arbitrary constant. Using the initial condition there results

$$0.02 = 0.01 + A,$$

so that

$$A = 0.01.$$

The solution is then

$$C(t) = 0.01[1 + e^{-t/5}].$$

Setting $C(t) = 0.011$, we have

$$0.011 = 0.01[1 + e^{-t/5}].$$

Solving for the time, we have

$$0.1 = e^{-t/5}$$

or

$$t = 11.51 \text{ s.} \quad \blacksquare$$

1.4.3   FLUID FLOW

In the absence of viscous effects it has been observed that a liquid (water, for example) will flow from a hole with a velocity

$$v = \sqrt{2gh} \qquad \text{m/s,} \tag{1.4.13}$$

where h is the height of the free surface of the liquid above the hole and g is the local acceleration of gravity (usually assumed to be 9.81 m/s^2). Bernoulli's equation, which may have been presented in a physics course, will yield the result above. Let us develop a differential equation that will relate the height of the free surface and time, thereby allowing us to determine how long it will take to empty a particular reservoir. Assume the hole of diameter d to be in the bottom of a cylindrical tank of diameter D with the initial water height h_0 meters above the hole. The incremental volume ΔV of liquid escaping from the hole during the time increment Δt is

$$\Delta V = vA\,\Delta t = \sqrt{2gh}\,\frac{\pi d^2}{4}\,\Delta t. \tag{1.4.14}$$

This small volume change must equal the volume lost in the tank due to the decrease in liquid level Δh. This is expressed as

$$\Delta V = -\frac{\pi D^2}{4}\,\Delta h. \tag{1.4.15}$$

Equating the two expressions above and taking the limit as $\Delta t \to 0$, we have

$$\frac{dh}{dt} = -\sqrt{2gh}\,\frac{d^2}{D^2}. \tag{1.4.16}$$

This equation is immediately separable and is put in the form

$$h^{-1/2}\,dh = -\sqrt{2g}\,\frac{d^2}{D^2}\,dt, \tag{1.4.17}$$

which is integrated to provide the solution, using $h = h_0$ at $t = 0$,

$$h(t) = \left[-\sqrt{\frac{g}{2}}\,\frac{d^2}{D^2}t + \sqrt{h_0}\right]^2. \tag{1.4.18}$$

The time t_e necessary to drain the tank completely would be (set $h = 0$)

$$t_e = \frac{D^2}{d^2} \sqrt{\frac{2h_0}{g}} \qquad \text{seconds.} \qquad (1.4.19)$$

Additional examples of physical phenomena are included in the Problems.

PROBLEMS

1. A constant voltage of 12 V is impressed on a series circuit composed of a 10-Ω resistor and a 10^{-4}-H inductor. Determine the current after 2 μs if the current is zero at $t = 0$.

2. An exponentially increasing voltage of $0.2e^{2t}$ is impressed on a series circuit containing a 20-Ω resistor and a 10^{-3}-H inductor. Calculate the resulting current as a function of time using $i = 0$ at $t = 0$.

3. A series circuit composed of a 50-Ω resistor and a 10^{-7}-F capacitor is excited with the voltage 12 sin 2t. What is the general expression for the charge on the capacitor? For the current?

4. A constant voltage of 12 V is impressed on a series circuit containing a 200-Ω resistor and a 10^{-6}-F capacitor. Determine the general expression for the charge. How long will it take before the capacitor is half-charged?

5. The initial concentration of salt in 10 m³ of solution 0.2 kg/m³. Fresh water flows into the tank at the rate of 0.1 m³/s until the volume is 20 m³, at which time t_f the solution flows out at the same rate as it flows into the tank. Express the concentration C as a function of time. One function will express $C(t)$ for $t < t_f$ and another for $t > t_f$.

6. An average person takes 18 breaths per minute and each breath exhales 0.0016 m³ of air containing 4% CO_2. At the start of a seminar with 300 participants, the room air contains 0.4% CO_2. The ventilation system delivers 10 m³ of air per minute to the 1500-m³ room. Find an expression for the concentration level of CO_2 in the room.

7. Determine an expression for the height of water in the funnel shown. What time is necessary to drain the funnel?

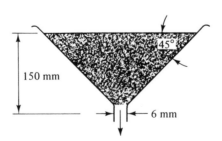

150 mm

45°

6 mm

8. A square tank, 3 m on a side, is filled with water to a depth of 2 m. A vertical slot 6 mm wide from the top to the bottom allows the water to drain out. Determine the height h as a function of time and the time necessary for one half of the water to drain out.

9. A body falls from rest and is resisted by air drag. Determine the time necessary to reach a velocity of 50 m/s if the 100-kg body is resisted by a force equal to (a) 0.01 V and (b) 0.004 V^2. Check if the equation $M(dV/dt) = Mg - D$, where D is the drag force, describes the motion.

10. Calculate the velocity of escape from the earth for a rocket fired radially outward on the surface ($R \cong 6400$ km) of the earth. Use Newton's law of gravitation, which states that $dv/dt = k/r^2$, where, for the present problem, $k = -gR^2$. Also, to eliminate t, use $dt = dr/v$.

11. The rate in kilowatts (kW) at which heat is conducted in a solid is proportional to the area and the temperature gradient with the constant of proportionality being the thermal conductivity k (kW/m $\cdot$ °C). For a long, laterally insulated rod this takes the form $q = -kA(dT/dx)$. At the left end heat is transferred at the rate of 10 kW. Determine the temperature distribution in the rod if the right end at $x = 2$ m is held constant at 50°C. The cross-sectional area is 1200 mm² and $k = 100$ kW/m $\cdot$ °C.

12. An object at a temperature of 80°C to be cooled is placed in a refrigerator maintained at 5°C. It has been observed that the rate of temperature change of such an object is proportional to the surface area A and the difference between its temperature T and the temperature of the surrounding medium. Determine the time for the temperature to reach 8°C if the constant of proportionality $\alpha = 0.02$ (s·m²)$^{-1}$ and $A = 0.2$ m².

13. The evaporation rate of moisture from a sheet hung on a clothesline is proportional to the moisture content. If one half of the moisture is lost in the first 20 minutes, calculate the time necessary to evaporate 95% of the moisture.

1.5   Linear Differential Equations

1.5.1   INTRODUCTION AND A FUNDAMENTAL THEOREM

Many of the differential equations that describe physical phenomena are linear differential equations and among these, the second-order equation is the most common and the most important special case. In this section we present certain aspects of the general theory of the second-order equation; the theory for the nth-order equation is often a straightforward extension of these ideas.

In the general, the second-order equation is

$$\frac{d^2u}{dx^2} + p_0(x)\frac{du}{dx} + p_1(x)u = g(x). \qquad a < x < b \qquad (1.5.1)$$

The function $g(x)$ is the *forcing function* and $p_0(x)$, $p_1(x)$ are coefficient functions; in many applications the forcing function has jump discontinuities. Figure 1.2 illustrates three common forcing functions, each with jump discontinuities. The graphs in Fig. 1.2 suggest the following definition of a jump discontinuity for $g(x)$ at $x = x_0$:

$$\textit{limit from the left} = \lim_{\substack{x \to x_0 \\ x < x_0}} g(x) \equiv g_0^-$$

$$\textit{limit from the right} = \lim_{\substack{x \to x_0 \\ x > x_0}} g(x) \equiv g_0^+.$$

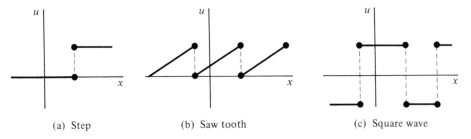

(a) Step   (b) Saw tooth   (c) Square wave

FIGURE 1.2.   Some common forcing functions with jump discontinuities.

The jump is $|g_0^+ - g_0^-|$ and is always finite. Although we do not require $g(x)$ to have a value at x_0, we usually define $g(x_0)$ as the average of the limits from the left and the right:

$$g(x_0) = \frac{1}{2}(g^+ + g^-). \qquad (1.5.2)$$

Figure 1.3 illustrates this point. Two ways in which a function can have a discontinuity that is not a jump are illustrated in Fig. 1.4.

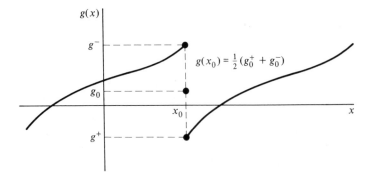

FIGURE 1.3.   The definition of $g(x_0)$ where a jump discontinuity exists at x_0.

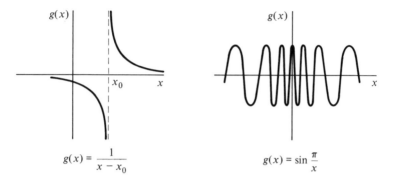

FIGURE 1.4.   Functions with a discontinuity that is not a jump discontinuity.

We study Eq. 1.5.1 in the interval I: $a < x < b$, an interval in which $a = -\infty$ or $b = +\infty$ or both are possible. We assume:

1. $p_0(x)$ and $p_1(x)$ are continuous in I.
2. $g(x)$ is sectionally continuous in I.

A *sectionally* continuous function $g(x)$ in $a < x < b$ is a function with only a finite number of jump discontinuities in each finite subinterval of I and in which, for a and b finite, g_a^+ and g_b^- exist.* Thus, the unit-step, the sawtooth, and the squarewave are sectionally continuous functions (see Fig. 1.2). The function graphed in Fig. 1.3 is sectionally continuous. If $p_0(x)$ and $p_1(x)$ are continuous and $g(x)$ sectionally continuous in I, then Eq. 1.5.1 and its corresponding initial-value problem are called *standard*.

Theorem 1.1:   (The Fundamental Theorem):   *The standard initial-value problem*

$$\frac{d^2u}{dx^2} + p_0(x)\frac{du}{dx} + p_1(x)u = g(x), \qquad a < x < b$$

$$u(x_o) = u_0, \qquad u'(x_0) = u_0', \qquad a < x_0 < b \tag{1.5.3}$$

has one and only one solution in I.

For a proof of this existence and uniqueness theorem, we refer the reader to a textbook on ordinary differential equations.

---

*It is unreasonable to expect g_a^- to exist since g is, presumably, undefined for $x < a$. Similarly, g_b^- is the only reasonable limit at the right end point of I.

It follows immediately from this theorem that $u(x) \equiv 0$ is the only solution of

$$\frac{d^2u}{dx^2} + p_0(x)\frac{du}{dx} + p_1(x)u = 0$$

$$\text{(1.5.4)}$$

$$u(x_0) = 0, \qquad u'(x_0) = 0$$

in I. This corollary has a physical analog: A system at rest and in equilibrium and undisturbed by external forces remains at rest and in equilibrium.

PROBLEMS

Which of the following functions have a discontinuity at $x = 0$? Indicate whether the discontinuity is a jump discontinuity.

1. $g(x) = \ln x$

2. $g(x) = \ln |x|$

3. $g(x) = |x|$

4. $g(x) = 1/x^2$

5. $g(x) = e^{-x}$

6. $g(x) = \begin{cases} (\sin x)/x, & x \neq 0 \\ 0, & x = 0 \end{cases}$

7. $g(x) = \begin{cases} x \sin \pi/x, & x \neq 0 \\ 0, & x = 0 \end{cases}$

Which of the following functions are sectionally continuous?

8. $g(x) = \ln x, \quad x > 0$

9. $g(x) = \ln x, \quad x > 1$

10. $g(x) = \begin{cases} 1/x, & x \neq 0 \\ 0, & x = 0 \end{cases} \qquad \text{for} \quad -1 < x < 1$

11. $g(x) = \text{sgn } (x) = \begin{cases} +1 & \text{if} \quad x > 0 \\ -1 & \text{if} \quad x < 0 \\ 0 & \text{if} \quad x = 0 \end{cases}$

12. $g(x) = \begin{cases} 0, & x < 0 \\ |\sin x|, & x \geq 0 \end{cases}$

13. $g(x) = \begin{cases} 1, & 0 \leq x \leq 1 \\ 0, & \text{otherwise} \end{cases}$

14. What is the analog of the Fundamental Theorem for the following first-order initial-value problem?

$$\frac{du}{dx} + p(x)u = g(x), \qquad a < x < b$$

$$u(x_0) = u_0, \qquad a < x_0 < b.$$

15. In view of Problem 14, consider this paradox: The initial-value problem

$$x\frac{du}{dx} - 2u = 0, \qquad u(0) = 0$$

has the two solutions $u_1(x) = x^2$ and $u_2(x) = -x^2$. Resolve the dilemma.

1.5.2   LINEAR DIFFERENTIAL OPERATORS

Given any twice differentiable function $u(x)$, the expression

$$L[u] \equiv \frac{d^2u}{dx^2} + p_0(u)\frac{du}{dx} + p_1(x)u \qquad (1.5.5)$$

defines a function, $r(x)$. For example, set

$$L[u] = \frac{d^2u}{dx^2} + 3\frac{du}{dx} + 2u. \qquad (1.5.6)$$

Then, for $u(x) = 1 - x$, e^x, e^{-x}, and K, we have

$$r(x) = L[-x + 1] = 3(-1) + 2(-x + 1) = -2x - 1$$

$$r(x) = L[e^x] = e^x + 3e^x + 2e^x = 6e^x$$

$$r(x) = L[e^{-x}] = e^x - 3e^x + 2e^x = 0$$

$$r(x) = L[K] = 2K.$$

The formula $L[u] = r(x)$ may be viewed in an operator context: For each allowable input $u(x)$, L produces the *output* $r(x)$. We call L a *differential operator*. It is linear because, as we may easily verify,

$$L[c_1u_1 + c_2u_2] = c_1L[u_1] + c_2L[u_2] \qquad (1.5.7)$$

for each pair of constants c_1 and c_2* (see Problem 1 at the end of this section). Three consequences of Eq. 1.5.7 are

$$\begin{array}{lll} (1) & L[0] = 0 & \\ (2) & L[cu] = cL[u] & (1.5.8) \\ (3) & L[u + v] = L[u] + L[v] & \end{array}$$

Item (1) follows by choosing $c_1 = c_2 = 0$ in Eq. 1.5.7; the other two are equally obvious.

We may now interpret the differential equation

$$L[u] = \frac{d^2u}{dx^2} + p_0(x)\frac{du}{dx} + p_1(x)u = g(x) \qquad (1.5.9)$$

in this manner: Given L and $g(x)$, find $u(x)$ such that

$$L[u] \equiv g(x), \qquad a < x < b. \qquad (1.5.10)$$

Theorem 1.2:   (The Superposition Principle):   *If u_1 and u_2 are solutions of*

$$\frac{d^2u}{dx^2} + p_0(x)\frac{du}{dx} + p_1(x)u = 0, \qquad (1.5.11)$$

---

*There are other linear operators besides differential operators. In Chapter 3 we study an important class of linear integral operators called the Laplace transform.

then so is

$$u(x) = c_1u_1(x) + c_2u_2(x) \tag{1.5.12}$$

for every constant c_1 and c_2.

PROOF: In terms of operators, $L[u_1] = L[u_2] = 0$ by hypothesis. But

$$L[c_1u_1 + c_2u_2] = c_1L[u_i] + c_2L[u_2] = c_1 \cdot 0 + c_2 \cdot 0 = 0 \tag{1.5.13}$$

by linearity (Eq. 1.5.7).

PROBLEMS

1. Let $L_1[u] = du/dx + p_0(x)u$. Verify that L_1 is a linear operator by using Eq. 1.5.7.

2. Let L_n be defined as follows:

$$L_n[u] \equiv \frac{d^n u}{dx^n} + p_0(x)\frac{d^{n-1}u}{dx^{n-1}} + \cdots + p_{n-1}(x)u.$$

Verify that this operator is linear by showing that

$$L_n[c_1u_1 + c_2u_2] = c_1L_n[u_1] + c_2L_n[u_2].$$

3. Prove that $L[cu] = cL[u]$ for every c and $L[u + v] = L[u] + L[v]$ implies Eq. 1.5.7. (*Note:* it is unnecessary to know anything at all about the structure of L except what is given in this problem.)

4. A *second Principle of Superposition:* Suppose that L is a linear operator and $L[u] = g_1$, $L[v] = g_2$. Prove that $L[u + v] = g_1 + g_2$.

5. Suppose that $c_1 + c_2 = 1$ and $L[u] = L[v] = g$. Prove that $L[c_1u + c_2v] = g$.

6. Let $L[u] \equiv \dfrac{d^2u}{dx^2} + 2b\dfrac{du}{dx} + cu = 0$, where b and c are constants. Verify that

$$L[e^{\lambda x}] = e^{\lambda x}(\lambda^2 + 2b\lambda + c).$$

7. Suppose that L is a linear operator. Suppose that $L[u_n] = 0$ and $L[u_p] = g(x)$. Show that $L[cu_n + u_p] = g(x)$ for every scalar c.

8. Suppose that L is a linear operator and $L[u_1] = L[u_2] = g(x)$. Show that $L[u_1 - u_2] = 0$.

1.5.3 WRONSKIANS AND GENERAL SOLUTIONS

If $u_1(x)$ and $u_2(x)$ are solutions of

$$L[u] = \frac{d^2u}{dx^2} + p_0(x)\frac{du}{dx} + p_1(x)u = 0, \tag{1.5.14}$$

then we define the *Wronskian*, $W(x; u_1, u_2)$, as follows:

$$W(x) = W(x; u_1, u_2) = \begin{vmatrix} u_1 & u_2 \\ u_1' & u_2' \end{vmatrix} = u_1 u_2' - u_2 u_1'. \qquad (1.5.15)$$

Now,

$$\begin{aligned} W'(x) &= u_1 u_2'' + u_1' u_2' - u_2' u_1' - u_2 u_1'' \\ &= u_1 u_2'' - u_2 u_1'' \\ &= u_1[-p_0 u_2' - p_1 u_2] - u_2[-p_0 u_1' - p_1 u_1] \\ &= -p_0(u_1 u_2' - u_2 u_1') \\ &= -p_0 W(x). \qquad (1.5.16) \end{aligned}$$

This is a first-order equation whose general solution may be written

$$W(x) = Ke^{-\int p_0(x)\,dx}. \qquad (1.5.17)$$

A critical fact follows directly from this equation.

Theorem 1.3: *On the interval I: $a < x < b$, either*

$$W(x) \equiv 0$$

or

$$W(x) > 0$$

or

$$W(x) < 0.$$

PROOF: Since $p_0(x)$ is continuous on I, so is $\int p_0(x)\,dx$. Therefore,

$$e^{\int p_0(x)\,dx} > 0 \qquad \text{on } I.$$

Hence, $W = 0$ if and only if $K = 0$, it is positive if and only if $K > 0$ and is negative if and only if $K < 0$.

We say that $u_1(x)$ and $u_2(x)$ are *independent* if $W(x)$ is not zero on $a < x < b$. (see Problem 11). A pair of independent solutions is called a *basic solution set*.

Theorem 1.4: *If $u_1(x)$ and $u_2(x)$ is a basic solution set of*

$$\frac{d^2 u}{dx^2} + p_0(x)\frac{du}{dx} + p_1(x)u = 0 \qquad (1.5.18)$$

and $\hat{u}$ is any solution, then there are constants c_1 and c_2 such that

$$\hat{u}(x) = c_1 u_1(x) + c_2 u_2(x). \qquad (1.5.19)$$

PROOF: Define the numbers r_1 and r_2 by

$$r_1 = \hat{u}(x_0), \qquad r_2 = \hat{u}'(x_0). \tag{1.5.20}$$

Now consider the initial-value problem

$$\frac{d^2u}{dx^2} + p_0(x)\frac{du}{dx} + p_1(x)u = 0$$

$$u(x_0) = r_1, \qquad u'(x_0) = r_2. \tag{1.5.21}$$

By construction u solves this problem. By the Fundamental Theorem (Theorem 1.1) u is the only solution. However, we can always find c_1 and c_2 so that $c_1u_1 + c_2u_2$ will also solve the initial-value problem 1.5.21. Hence, the theorem is proved.

To find the values of c_1 and c_2 we set

$$r_1 = c_1u_1(x_0) + c_2u_2(x_0)$$

$$r_2 = c_1u_1'(x_0) + c_2u_2'(x_0). \tag{1.5.22}$$

But this system has a solution (unique at that) if

$$\begin{vmatrix} u_1(x_0) & u_2(x_0) \\ u_1'(x_0) & u_2'(x_0) \end{vmatrix} \neq 0. \tag{1.5.23}$$

It is a well-known theorem of linear algebra (see Chapter 4) that a system of n equations in n unknowns has a unique solution if and only if the determinant of its coefficients is not zero. This determinant is $W(x_0)$ and by hypothesis $W(x) \neq 0$ on I [recall that u_1 and u_2 is a basic set, which means that $W(x)$ cannot vanish on $a < x < b$].

Because of this theorem, the family of all functions $\{c_1u_1 + c_2u_2\}$ is the set of all solutions of Eq. 1.5.14 in I. For this reason, we often call

$$u(x) = c_1u_1(x) + c_2u_2(x) \tag{1.5.24}$$

the *general solution* of Eq. 1.5.14.

Note that the proof is constructive. It provides a precise computation for c_1 and c_2 given the initial values and basic solution set. From this point of view, the constants c_1 and c_2 in Eq. 1.5.24 are "essential" if $W(x) \neq 0$.

One last point. There is no unique basic solution set. Every pair of solutions from the set $\{c_1u_1 + c_2u_2\}$ for which $W(x) \neq 0$ provides a satisfactory basic solution pair (see Problems 8–10).

PROBLEMS

Find the Wronskians of each equation.

1. $u'' + \alpha u' + \beta u = 0$, α, β constants

2. $u'' + \dfrac{1}{x}u' + p_1(x)u = 0, \quad 0 < x < \infty$

3. $u'' - \dfrac{1}{x}u' + p_1(x)u = 0, \quad 0 < x < \infty$

4. $u'' + I(x)u = 0, \qquad\qquad -\infty < x < \infty$

5. One solution of $u'' + 2\alpha u' + \alpha^2 u = 0$, α constant, is $e^{-\alpha x}$. Find the Wronskian and, by using the definition $W(x) = u_1 u_2' - u_2 u_1'$, find a second independent solution.

6. Use Cramer's rule to solve the system (1.5.22) and thereby find

$$c_1 = \frac{\begin{vmatrix} r_1 & u_2(x_0) \\ r_2 & u_2'(x_0) \end{vmatrix}}{W(x_0)}, \qquad c_2 = \frac{\begin{vmatrix} u_1(x_0) & r_1 \\ u_1'(x_0) & r_2 \end{vmatrix}}{W(x_0)}.$$

7. Suppose that u_1 and u_2 is a basic solution set of Eq. 1.5.14. Suppose also that $u_1(x_0) = 0$, $a < x < b$. Use the definition of the Wronskian and Theorem 1.3 to prove that $u_2(x_0) \neq 0$.

Show that if u_1 and u_2 is a basic solution set of $L[u] = 0$ so are these:

8. $u_1 + u_2, \; u_1 - u_2$.

9. $u_1, \; u_1 + u_2$.

10. $\alpha u_1 + \beta u_2, \; \gamma u_1 + \delta u_2$. For which choice of α, β, γ, δ?

11. If $u_2 = ku_1$, show that $W(x; u_1, u_2) = 0$.

1.5.4 THE GENERAL SOLUTION OF THE NONHOMOGENEOUS EQUATION

Suppose that $u_1(x)$ and $u_2(x)$ form a basic solution set for the associated homogeneous equation of

$$L[u] = \frac{d^2 u}{dx^2} + p_0(x)\frac{du}{dx} + p_1(x)u = g(x). \tag{1.5.25}$$

Then $L[u_1] = L[u_2] = 0$ and $W(x) \neq 0$ for all x, $a < x < b$. Let $u_p(x)$ be a particular solution of Eq. 1.5.25. Then, for every choice of c_1 and c_2,

$$u(x) = u_p(x) + c_1 u_1(x) + c_2 u_2(x) \tag{1.5.26}$$

also solves Eq. 1.5.25. This is true since

$$\begin{aligned} L[u] &= L[u_p + c_1 u_1 + c_2 u_2] \\ &= L[u_p] + c_1 L[u_1] + c_2 L[u_2] \\ &= g(x) \end{aligned} \tag{1.5.27}$$

by linearity and the definitions of u_p, u_1 and u_2. We call $u(x)$ the general solution of Eq. 1.5.25 for this reason.

Theorem 1.5: *If û is a solution of Eq. 1.5.25, then there exists constants c_1 and c_2 such that*

$$\hat{u}(x) = u_p(x) + c_1 u_1(x) + c_2 u_2(x). \qquad (1.5.28)$$

PROOF: We leave the details to the student. A significant simplification in the argument occurs, by observing that $\hat{u} - u_p$ solves the associated homogeneous equation and then relying on Theorem 1.4.

Example 1.5.1:   Verify that $u(x) = x^2$ and $v(x) = x - 1$ is a basic solution set of

$$L[u] = x(x - 2)u'' - 2(x - 1)u' + 2u = 0, \qquad 0 < x < 2.$$

SOLUTION: It is easy to differentiate and show that

$$L[x^2] = L[x - 1] = 0.$$

The Wronskian, $W(x) = uv' - u'v$, is also easy to check:

$$W(x) = \begin{vmatrix} x^2 & x - 1 \\ 2x & 1 \end{vmatrix} = x^2 - 2x(x - 1) = x(x - 2) \neq 0, \qquad 0 < x < 2.$$

Thus, $u(x)$ and $v(x)$ is a basic solution set of $L[u]$. ∎

Example 1.5.2:   Let L be defined as in Example 1.5.1. Show that the particular solution $u_p = x^3$ is a solution of $L[u] = 2x^2(x - 3)$ and find a specific solution of the initial-value problem

$$L[u] = 2x^2(x - 3), \qquad u(1) = 0, \qquad u'(1) = 0.$$

SOLUTION: To show that $u_p = x^3$ solves $L[u] = 2x^2(x - 3)$, we substitute into the given $L[u]$ and find

$$L[x^3] = x(x - 2)(6x) - 2(x - 1)(3x^2) + 2(x^3)$$

$$= 6x^3 - 12x^2 - 6x^3 + 6x^2 + 2x^3 = 2x^2(x - 3).$$

The general solution,* by Theorem 1.5, using $u(x)$ and $v(x)$ from Example 1.5.1, is

$$u(x) = x^3 + c_1 x^2 + c_2(x - 1);$$

but

$$u(1) = 1 + c_1, \qquad u'(1) = 3 + 2c_1 + c_2.$$

We determine c_1 and c_2 by setting $u(1) = u'(1) = 0$; that is,

$$1 + c_1 = 0, \qquad 3 + 2c_1 + c_2 = 0,$$

which leads to the unique solution $c_1 = c_2 = -1$, and therefore

$$u(x) = x^3 - x^2 - (x - 1) = x^3 - x^2 - x + 1$$

is the unique solution to the given initial-value problem. ∎

---

*Technically, we ought to put the equation in standard form by dividing by $x(x - 2)$; but all steps would be essentially the same.

1.6 Homogeneous, Second-Order, Linear Equations with Constant Coefficients

We will focus out attention on second-order differential equations with constant coefficients. The homogeneous equation is written in standard form as

$$\frac{d^2 u}{dx^2} + a\frac{du}{dx} + bu = 0. \tag{1.6.1}$$

We seek solutions of the form

$$u = e^{mx}. \tag{1.6.2}$$

When this function is substituted into Eq. 1.6.1, we find that

$$e^{mx}(m^2 + am + b) = 0. \tag{1.6.3}$$

Thus, $u = e^{mx}$ is a solution of Eq 1.6.1 if and only if

$$m^2 + am + b = 0. \tag{1.6.4}$$

This is the *characteristic equation*. It has the two roots

$$m_1 = -\frac{a}{2} + \frac{1}{2}\sqrt{a^2 - 4b}, \qquad m_2 = -\frac{a}{2} - \frac{1}{2}\sqrt{a^2 - 4b}. \tag{1.6.5}$$

It then follows that

$$u_1 = e^{m_1 x}, \qquad u_2 = e^{m_2 x} \tag{1.6.6}$$

are solutions of Eq. 1.6.1. The Wronskian of these solutions is

$$W(x) = \begin{vmatrix} e^{m_1 x} & e^{m_2 x} \\ m_1 e^{m_1 x} & m_2 e^{m_2 x} \end{vmatrix} = (m_2 - m_1)e^{(m_1 + m_2)x}, \tag{1.6.7}$$

which is zero if and only if $m_1 = m_2$. Hence, if $m_1 \neq m_2$, then a general solution is

$$u(x) = c_1 e^{m_1 x} + c_2 e^{m_2 x}. \tag{1.6.8}$$

Let us consider two cases $(a^2 - 4b) > 0$ and $(a^2 - 4b) < 0$. If $(a^2 - 4b) > 0$, the solution takes the form

$$u(x) = e^{-ax/2}(c_1 e^{1/2\sqrt{a^2 - 4b}\,x} + c_2 e^{-1/2\sqrt{a^2 - 4b}\,x}). \tag{1.6.9}$$

Using the appropriate identities,* this solution can be put in the following two equivalent forms:

$$u(x) = e^{-ax/2}[A \sinh\left(\tfrac{1}{2}\sqrt{a^2 - 4b}\,x\right) + B \cosh\left(\tfrac{1}{2}\sqrt{a^2 - 4b}\,x\right)] \tag{1.6.10}$$

$$u(x) = c_3 e^{-ax/2} \sinh\left(\tfrac{1}{2}\sqrt{a^2 - 4b}\,x + c_4\right). \tag{1.6.11}$$

---

*The appropriate identities are

$$e^x = \cosh x + \sinh x$$

$$\sinh(x + y) = \sinh x \cosh y + \sinh y \cosh x$$

$$\cosh^2 x - \sinh^2 x = 1.$$

If $(a^2 - 4b) < 0$, the general solution takes the form, using $i = \sqrt{-1}$,

$$u(x) = e^{-ax/2}(c_1 e^{i/2\sqrt{a^2 - 4b}\,x} + c_2 e^{i/2\sqrt{a^2 - 4b}\,x}), \qquad (1.6.12)$$

which, with the appropriate identities,* can be put in the following two equivalent forms:

$$u(x) = e^{-ax/2}[A \sin\left(\tfrac{1}{2}\sqrt{a^2 - 4b}\,x\right) + B \cos\left(\tfrac{1}{2}\sqrt{a^2 - 4b}\,x\right)] \quad (1.6.13)$$

$$u(x) = c_3 e^{-ax/2} \cos\left(\tfrac{1}{2}\sqrt{a^2 - 4b}\,x + c_4\right). \qquad (1.6.14)$$

If a particular form of the solution is not requested, the form of Eq 1.6.9 is used if $(a^2 - 4b) > 0$ and the form of Eq. 1.6.13 is used if $(a^2 - 4b) < 0$.

If $(a^2 - 4b) = 0$, $m_1 = m_2$ and a double root occurs. For this case the solution 1.6.8 no longer is a general solution. What this means is that the assumption that there are two linearly independent solutions of Eq 1.6.1 of the form e^{mx} is false, an obvious conclusion since $W(x) = 0$. To find a second solution we make the assumption that it is of the form

$$u_2(x) = v(x)e^{mx}, \qquad (1.6.15)$$

where $m^2 + am + b = 0$. Substitute into Eq. 1.6.1 and we have

$$(m^2 e^{mx} + ame^{mx} + be^{mx})v + (2me^{mx} + ae^{mx})\frac{dv}{dx} + e^{mx}\frac{d^2v}{dx^2} = 0. (1.6.16)$$

The coefficient of v is zero since $m^2 + am + b = 0$. The coefficient of dv/dx is zero since we are assuming that $m^2 + am + b = 0$ has equal roots, that is, $m = -a/2$. Hence,

$$\frac{d^2v}{dx^2} = 0. \qquad (1.6.17)$$

Therefore, $v(x) = x$ suffices; the second solution is then

$$u_2(x) = xe^{mx} = xe^{-ax/2}. \qquad (1.6.18)$$

A general solution is

$$u(x) = c_1 e^{-ax/2} + c_2 xe^{-ax/2} = (c_1 + c_2 x)e^{-ax/2}. \qquad (1.6.19)$$

Note that, using $u_1 = e^{-ax/2}$ and $u_2 = xe^{-ax/2}$, the Wronskian is

$$W(x) = e^{-ax} > 0 \qquad \text{for all } x. \qquad (1.6.20)$$

The arbitrary constants in the solutions above are used to find specific solutions to initial or boundary-value problems.

---

*The appropriate identities are

$$e^{i\theta} = \cos\theta + i\sin\theta$$

$$\cos(\alpha + \beta) = \cos\alpha \cos\beta - \sin\alpha \sin\beta$$

$$\cos^2\beta + \sin^2\beta = 1.$$

The technique above can also be used for solving differential equations with constant coefficients of order greater than 2. The substitution $u = e^{mx}$ leads to a characteristic equation which is solved for the various roots. The solution follows as above.

Example 1.6.1:   Determine a general solution of the differential equation

$$\frac{d^2u}{dx^2} + 5\frac{du}{dx} + 6u = 0.$$

Express the solution in terms of exponentials.

SOLUTION: We assume that the solution has the form $u(x) = e^{mx}$. Substitute this into the differential equation and find the characteristic equation to be

$$m^2 + 5m + 6 = 0.$$

This is factored into

$$(m + 3)(m + 2) = 0.$$

The roots are obviously

$$m_1 = -3, \qquad m_2 = -2.$$

The two independent solutions are then

$$u_1(x) = e^{-3x}, \qquad u_2(x) = e^{-2x}.$$

These solutions are superimposed to yield the general solution

$$u(x) = c_1e^{-3x} + c_2e^{-2x}. \quad \blacksquare$$

Example 1.6.2:   Find the solution of the initial-value problem

$$\frac{d^2u}{dx^2} + 6\frac{du}{dx} + 9u = 0, \qquad u(0) = 2, \qquad \frac{du}{dx}(0) = 0.$$

SOLUTION: Assume a solution of the form $u(x) = e^{mx}$. The characteristic equation

$$m^2 + 6m + 9 = 0$$

yields the roots

$$m_1 = -3, \qquad m_2 = -3.$$

The roots are identical; therefore, the general solution is (see Eq. 1.6.19)

$$u(x) = c_1e^{-3x} + c_2xe^{-3x}.$$

This solution must satisfy the initial conditions. Using $u(0) = 2$, we have

$$2 = c_1.$$

Differentiating the expression for $u(x)$ gives

$$\frac{du}{dx} = (c_1 + c_2x)(-3e^{-3x}) + c_2e^{-3x}$$

and therefore

$$\frac{du}{dx}(0) = -3c_1 + c_2 = 0.$$

Hence,

$$c_2 = 6.$$

The specific solution is then

$$u(x) = 2(1 + 3x)e^{-3x}. \quad \blacksquare$$

Example 1.6.3:   Find a general solution of the differential equation

$$\frac{d^2u}{dx^2} + 2\frac{du}{dx} + 5u = 0.$$

SOLUTION: The assumed solution $u(x) = e^{mx}$ leads to the characteristic equation

$$m^2 + 2m + 5 = 0.$$

The roots to this equation are

$$m_1 = -1 + 2i, \qquad m_2 = -1 - 2i.$$

A general solution is then

$$u(x) = c_1 e^{(-1+2i)x} + c_2 e^{(-1-2i)x}.$$

Alternate general solutions, having the virtue that the functions involved are real, can be written as (see Eqs. 1.6.13 and 1.6.14)

$$u(x) = e^{-x}(A \cos 2x + B \sin 2x)$$

or

$$u(x) = c_3 e^{-x} \sin(2x + c_4).$$

Note that the second of these forms is equivalent to Eq. 1.6.14 since $\cos(2x + a) = \sin(2x + b)$ for the appropriate choices of a and b. Also, note that

$$W(x) = \begin{vmatrix} e^{-x}\cos 2x & e^{-x}\sin 2x \\ \dfrac{d}{dx}(e^{-x}\cos 2x) & \dfrac{d}{dx}(e^{-x}\sin 2x) \end{vmatrix} = 2e^{-2x} > 0$$

for all real x. $\blacksquare$

PROBLEMS

Find a general solution in terms of exponentials for each differential equation.

1. $u'' - u' - 6u = 0$ **2.** $u'' - 9u = 0$

3. $u'' + 9u = 0$ **4.** $4u'' + u = 0$

5. $u'' - 4u' + 4u = 0$ **6.** $u'' + 4u' + 4u = 0$

7. $u'' - 4u' - 4u = 0$ **8.** $u'' + 4u' - 4u = 0$

9. $u'' - 4u = 0$ **10.** $u'' - 4u' + 8u = 0$

11. $u'' + 2u' + 10u = 0$ **12.** $2u'' + 6u' + 5u = 0$

Write a general solution, in the form of Eq. 1.6.9 or 1.6.13, for each equation.

13. $u'' - u' - 6u = 0$ **14.** $u'' - 9u = 0$

15. $u'' + 9u = 0$ **16.** $4u'' + u = 0$

17. $u'' - 4u' - 4u = 0$ **18.** $u'' + 4u' + 4u = 0$

19. $u'' - 4u = 0$ **20.** $u'' - 4u' + 8u = 0$

21. $u'' + 2u' + 10u = 0$ **22.** $u'' + 5u' + 3u = 0$

Find a general solution, in the form of Eq. 1.6.11 or 1.6.14, for each equation.

23. $u'' - u' - 6u = 0$ **24.** $u'' - 9u = 0$

25. $u'' + 9u = 0$ **26.** $4u'' + u = 0$

27. $u'' - 4u' - 4u = 0$ **28.** $u'' + 4u' - 4u = 0$

29. $u'' - 4u = 0$ **30.** $u'' - 4u' + 8u = 0$

31. $u'' + 2u' + 5u = 0$ **32.** $u'' + 5u' + 3u = 0$

Solve each initial-value problem. Express each answer in the form of Eq. 1.6.10, 1.6.13, or 1.6.19.

33. $u'' + 9u = 0,$ $u(0) = 0,$ $u'(0) = 1$

34. $u'' + 5u' + 6u = 0,$ $u(0) = 2,$ $u'(0) = 0$

35. $u'' + 4u' + 4u = 0,$ $u(0) = 0,$ $u'(0) = 2$

36. $u'' - 4u = 0,$ $u(0) = 2,$ $u'(0) = 1$

Find the answer to each initial-value problem. Express each answer in the form of Eq. 1.6.9 or 1.6.13.

37. $u'' + 9u = 0,$ $u(0) = 0,$ $u'(0) = 1$

38. $u'' + 5u' + 6u = 0,$ $u(0) = 2,$ $u'(0) = 0$

39. $u'' - 4u = 0,$ $u(0) = 2,$ $u'(0) = 1$

Determine the solution to each initial-value problem. Express each answer in the form Eq. 1.6.10 or 1.6.13.

40. $u'' + 9u = 0,$ $u(0) = 0,$ $u'(0) = 1$

41. $u'' + 5u' + 6u = 0,$ $u(0) = 2,$ $u'(0) = 0$

42. $u'' - 4u = 0,$ $u(0) = 2,$ $u'(0) = 1$

43. Consider the differential equation

$$u^{(n)} + a_1 u^{(n-1)} + \cdots + a_{n-1}u' + a_n u = 0.$$

The characteristic equation for this differential equation is

$$m^n + a_1 m^{n-1} + \cdots + a_{n-1} m + a_n = 0.$$

Let $m_1, m_2, \ldots, m_n$ be the roots of this algebraic equation. Explain why $e^{m_1 x}$, $e^{m_2 x}, \ldots, e^{m_n x}$ are solutions of the differential equation.

Use the result of Problem 43 to solve each differential equation.

44. $u^{(3)} - u = 0$ **45.** $u^{(3)} + 2u^{(2)} - u^{(1)} + 2u = 0$

46. $u^{(4)} - u^{(2)} = 0$ **47.** $u^{(4)} - u = 0$

48. $u^{(4)} - u^{(3)} = 0$

1.7   Spring-Mass System: Free Motion

There are many physical phenomena that are described with linear, second-order, homogeneous differential equations. We wish to discuss one such phenomenon, the free motion of a spring-mass system, as an illustrative example. We shall restrict ourselves to systems with *1 degree of freedom;* that is, only one independent variable is needed to describe the motion. Systems requiring more than one independent variable, such as a system with several masses and springs, lead to simultaneous ordinary differential equations and will not be considered in this section. However, see Chapter 5.

Consider the simple spring-mass system shown in Fig. 1.5. We shall make the following assumptions:

1. The mass M, measured in kilograms, is constrained to move in the vertical directions only.
2. The viscous damping C, with units of kilograms per second, is proportional to the velocity dy/dt. For relatively small velocities this is usually acceptable; however, for large velocities the damping is more nearly proportional to the square of the velocity.
3. The force in the spring is Kd, where d is the distance measured in meters from the unstretched position. The spring modulus K, with units of newtons per meter (N/m), is assumed constant.
4. The mass of the spring is negligible compared with the mass M.
5. No external forces act on the system.

Newton's second law is used to describe the motion of the lumped mass. It states that the sum of the forces acting on a body in any particular direction equals the mass of the body multiplied by the acceleration of the body in that direction. This is written as

$$\sum F_y = M a_y \qquad (1.7.1)$$

for the y direction. Consider that the mass is suspended from an unstretched spring, as shown in Fig. 1.5a. The spring will then deflect a distance h, where

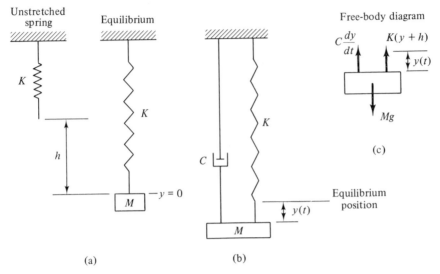

FIGURE 1.5. Spring–mass system.

h is found from the relationship

$$Mg = hK, \qquad (1.7.2)$$

which is a simple statement that for static equilibrium the weight must equal the force in the spring. The weight is the mass times the local acceleration of gravity. At this stretched position we attach a viscous damper, a dashpot, and allow the mass to undergo motion about the equilibrium position. A free-body diagram of the mass is shown in Fig. 1.5c. Applying Newton's second law, we have, with the positive direction downward,

$$Mg - C\frac{dy}{dt} - K(y + h) = M\frac{d^2y}{dt^2} \qquad (1.7.3)$$

Using Eq. 1.7.2, this simplifies to

$$M\frac{d^2y}{dt^2} + C\frac{dy}{dt} + Ky = 0. \qquad (1.7.4)$$

This is a second-order, linear, homogeneous, ordinary differential equation. Let us first consider the situation where the viscous damping coefficient C is sufficiently small that the viscous damping term may be neglected.

1.7.1 UNDAMPED MOTION

For the case where C is small, it may be acceptable, especially for small time spans, to neglect the damping. If this is done, the differential equation that describes the motion is

$$M\frac{d^2y}{dt^2} + Ky = 0. \tag{1.7.5}$$

We assume a solution of the form e^{mt}, which leads to the characteristic equation

$$m^2 + \frac{K}{M} = 0. \tag{1.7.6}$$

The two roots are

$$m_1 = \sqrt{\frac{K}{M}}\, i, \qquad m_2 = -\sqrt{\frac{K}{M}}\, i. \tag{1.7.7}$$

The solution is then

$$y(t) = Ae^{\sqrt{K/M}\, it} + Be^{-\sqrt{K/M}\, it} \tag{1.7.8}$$

or equivalently (see Eq. 1.6.13),

$$y(t) = c_1 \cos \sqrt{\frac{K}{M}}\, t + c_2 \sin \sqrt{\frac{K}{M}}\, t, \tag{1.7.9}$$

where $A + B = c_1$ and $i(A - B) = c_2$. The mass will undergo its first complete cycle as t goes from zero to $2\pi/\sqrt{K/M}$. Thus, one cycle is completed in $2\pi/\sqrt{K/M}$ seconds, the *period*. The number of cycles per second, the *frequency*, is then $\sqrt{K/M}/2\pi$. The *angular frequency* ω_0 is given by

$$\omega_0 = \sqrt{\frac{K}{M}}. \tag{1.7.10}$$

The solution is then written in the preferred form,

$$y(t) = c_1 \cos \omega_0 t + c_2 \sin \omega_0 t. \tag{1.7.11}$$

This is the motion of the undamped mass. It is often referred to as a *harmonic oscillator*. It is important to note that the sum of the sine and cosine terms in Eq. 1.7.11 can be written as (see Eq. 1.6.14)

$$y(t) = \Delta \cos (\omega_0 t - \delta) \tag{1.7.12}$$

where the *amplitude* Δ is related to c_1 and c_2 by $\Delta = \sqrt{c_1^2 + c_2^2}$ and $\tan \delta = c_2/c_1$. In this form Δ and δ are the arbitrary constants.

Two initial conditions, the initial displacement and velocity, are necessary to determine the two arbitrary constants. For a zero initial velocity the motion is sketched in Fig. 1.6.

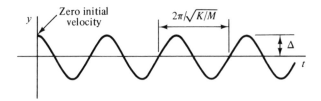

FIGURE 1.6.   Harmonic oscillation.

PROBLEMS

1. Derive the differential equation that describes the motion of a mass M swinging from the end of a string of length L. Assume small angles. Find the general solution of the differential equation.

2. Determine the motion of a mass moving toward the origin with a force of attraction proportional to the distance from the origin. Assume that the 10-kg mass starts at rest at a distance of 10 m and that the constant of proportionality is 10 N/m. What will the speed of the mass be 5 m from the origin?

3. A spring-mass system has zero damping. Find the general solution and determine the frequency of oscillation if $M = 4$ kg and $K = 100$ N/m.

4. Calculate the time necessary for a 0.03-kg mass hanging from a spring with spring constant 0.5 N/m to undergo one complete oscillation.

5. A 4-kg mass is hanging from a spring with $K = 100$ N/m. Sketch, on the same plot, the two specific solutions found from (a) $y(0) = 0.5$ m, $\dot{y}(0) = 0$, and (b) $y(0) = 0$, $\dot{y}(0) = 10$ m/s. The coordinate y is measured from the equilibrium position.

6. Solve the initial-value problem resulting from the undamped motion of a 2-kg mass suspended by a 50-N/m spring if $y(0) = 2$ m and $\dot{y}(0) = -10$ m/s.

7. Sketch, on the same plot, the motion of a 2-kg mass and that of a 10-kg mass suspended by a 50-N/m spring if motion starts from the equilibrium position with $\dot{y}(0) = 10$ m/s.

1.7.2   DAMPED MOTION

Let us now include the viscous damping term in the equation. This is necessary for long time spans, since viscous damping is always present, however small, or for short time periods, in which the damping coefficient C is not small. The describing equation is

$$M\frac{d^2y}{dt^2} + C\frac{dy}{dt} + Ky = 0. \qquad (1.7.13)$$

Assuming a solution of the form e^{mt}, the characteristic equation,

$$Mm^2 + Cm + K = 0, \tag{1.7.14}$$

results. The roots of this equation are

$$m_1 = -\frac{C}{2M} + \frac{1}{2M}\sqrt{C^2 - 4MK}, \qquad m_2 = -\frac{C}{2M} - \frac{1}{2M}\sqrt{C^2 - 4MK}.$$

$$\tag{1.7.15}$$

Let $\Omega = \sqrt{C^2 - 4KM}/2M$. The solution for $m_1 \neq m_2$ is then written as

$$y(t) = Ae^{-(C/2M)t + \Omega t} + Be^{-(C/2M)t - \Omega t} \tag{1.7.16}$$

or, equivalently,

$$y(t) = e^{-(C/2M)t}[Ae^{\Omega t} + Be^{-\Omega t}]. \tag{1.7.17}$$

The solution obviously takes on three different forms depending on the magnitude of the damping. The three cases are:

Case 1. *Overdamping* $\qquad C^2 - 4KM > 0.$ $\qquad m_1$ and m_2 are real.

Case 2. *Critical damping* $\qquad C^2 - 4KM = 0.$ $\qquad m_1 = m_2.$

Case 3. *Underdamping* $\qquad C^2 - 4KM < 0.$ $\qquad m_1$ and m_2 are complex.

Let us investigate each case separately.

Case 1. *Overdamping.* For this case the damping is so large that $C^2 > 4KM$. The roots m_1 and m_2 are real and the solution is best presented as in Eq. 1.7.17. Several overdamped motions are shown in Fig. 1.7. For large time the solution approaches $y = 0$.

Case 2. *Critical damping.* For this case the damping is just equal to $4KM$. There is a double root of the characteristic equation, so the solution is (see Eq. 1.6.19)

$$y(t) = Ae^{mt} + Bte^{mt}. \tag{1.7.18}$$

For the spring-mass system this becomes

$$y(t) = e^{-(C/2M)t}[A + Bt]. \tag{1.7.19}$$

A sketch of the solution is not unlike that of the overdamped case. It is shown in Fig. 1.8.

Case 3. *Underdamping.* The most interesting of the three cases is that of underdamped motion. If $C^2 - 4KM$ is negative, we may write Eq. 1.7.17

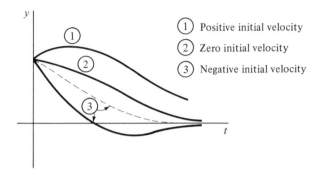

1 Positive initial velocity
2 Zero initial velocity
3 Negative initial velocity

(a) Positive initial displacement

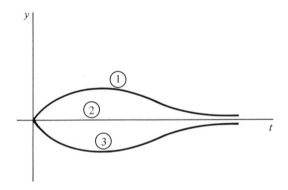

(b) Zero initial displacement

FIGURE 1.7.   Overdamped motion.

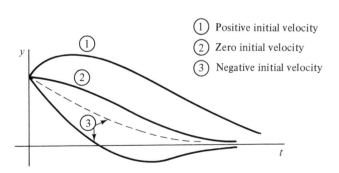

1 Positive initial velocity
2 Zero initial velocity
3 Negative initial velocity

(a) Positive initial displacement

FIGURE 1.8.   Critically damped motion.

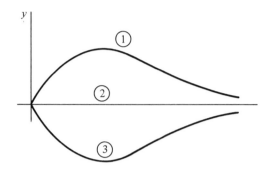

(b) Zero initial displacement

FIGURE 1.8.   (*cont.*)

as, using $\Omega = \sqrt{4KM - C^2}/2M$,

$$y(t) = e^{-(C/2M)t}[Ae^{i\Omega t} + Be^{-i\Omega t}]. \qquad (1.7.20)$$

This is expressed in the equivalent form (see Eq. 1.6.13)

$$y(t) = e^{-(C/2M)t}[c_1 \cos \Omega t + c_2 \sin \Omega t]. \qquad (1.7.21)$$

The motion is an oscillating motion with a decreasing amplitude with time. The frequency of oscillation is $\sqrt{4KM - C^2}/4\pi M$ and approaches that of the undamped case as $C \to 0$. Equation 1.7.21 can be written in a form from which a sketch can more easily be made. It is (see Eq. 1.6.14)

$$y(t) = Ae^{-(C/2M)t} \cos (\Omega t - \delta), \qquad (1.7.22)$$

where

$$\tan \delta = \frac{c_2}{c_1}, \qquad A = \sqrt{c_1^2 + c_2^2}. \qquad (1.7.23)$$

The underdamped motion is sketched in Fig. 1.9 for an initial zero velocity. The motion damps out for large time.

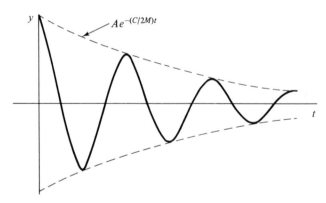

FIGURE 1.9.   Underdamped motion.

The ratio of successive maximum amplitudes is a quantity of particular interest for underdamped oscillations. We will show in Example 1.7.1 that this ratio is given by

$$\frac{y_n}{y_{n+2}} = e^{\pi C / \Omega M}. \tag{1.7.24}$$

It is constant for a particular underdamped motion for all time. The logarithm of this ratio is called the *logarithmic decrement D*:

$$D = \ln \frac{y_n}{y_{n+2}} = \frac{\pi C}{\Omega M}. \tag{1.7.25}$$

Returning to the definition of Ω, this is expressed as

$$D = \frac{2\pi C}{\sqrt{4KM - C^2}} \tag{1.7.26}$$

In terms of the *critical damping, $C_c = 2\sqrt{KM}$*, this is

$$D = \frac{2\pi C}{\sqrt{C_c^2 - C^2}} \tag{1.7.27}$$

or, alternatively,

$$\frac{C}{C_c} = \frac{D}{\sqrt{D^2 + 4\pi^2}} \tag{1.7.28}$$

Since y_n and y_{n+2} are easily measured, the logarithmic decrement D can be evaluated quite simply. This allows a quick method for determining the fraction of the critical damping that exists in a particular system.

Example 1.7.1: Determine the ratio of successive maximum amplitudes for the free motion of an underdamped oscillation.

SOLUTION: The displacement function for an underdamped spring-mass system can be written as

$$y(t) = Ae^{-(C/2M)t} \cos (\Omega t - \delta).$$

To find the maximum amplitude we set $dy/dt = 0$ and solve for the particular t that yields this condition. Differentiating, we have

$$\frac{dy}{dt} = -\left[\frac{C}{2M} \cos (\Omega t - \delta) + \Omega \sin (\Omega t - \delta) \right] Ae^{-(C/2M)t} = 0.$$

This gives

$$\tan (\Omega t - \delta) = -\frac{C}{2M\Omega}$$

or, more generally,

$$\tan^{-1} \left(-\frac{C}{2M\Omega} \right) + n\pi = \Omega t - \delta.$$

The time at which a maximum occurs in the amplitude is given by

$$t = \frac{\delta}{\Omega} - \frac{1}{\Omega} \tan^{-1} \frac{C}{2M\Omega} + \frac{n\pi}{\Omega}.$$

where $n = 0$ represents the first maximum, $n = 2$ the second maximum, and so on. For $n = 1$, a minimum results. We are interested in the ratio y_n/y_{n+2}. If we let

$$B = \frac{\delta}{\Omega} - \frac{1}{\Omega} \tan^{-1} \frac{C}{2M\Omega}$$

this ratio becomes

$$\frac{y_n}{y_{n+2}} = \frac{Ae^{-(C/2M)[B+(n\pi/\Omega)]} \cos\left[\Omega\left(B + \frac{n\pi}{\Omega}\right) - \delta\right]}{Ae^{-(C/2M)[B+(n+2/\Omega)\pi]} \cos\left[\Omega\left(B + \frac{n+2}{\Omega}\pi\right) - \delta\right]}$$

$$= e^{\pi C/\Omega M} \frac{\cos\left[B\Omega + \cancel{n\pi}^{1} - \delta\right]}{\cos\left[B\Omega + \cancel{n\pi} - \delta + 2\pi\right]} = e^{\pi C/\Omega M}.$$

Hence, we see that the ratio of successive maximum amplitudes is dependent only on M, K, and C and is independent of time. It is constant for a particular spring-mass system. ◾

1.7.3   THE ELECTRICAL CIRCUIT ANALOG

We now consider the solution to Eq. 1.4.4 for the case $dv/dt = 0$. By comparing Eq. 1.4.4 with Eq. 1.7.4, we see that we can interchange the spring-mass system parameters with the circuit parameters as follows:

Spring-Mass		Series Circuit
M	$\rightarrow$	L
C	$\rightarrow$	R
K	$\rightarrow$	$1/C$

The solutions that we have just considered for $y(t)$ may then be taken as solutions for $i(t)$.

Thus, for the undamped circuit, we have $R = 0$, and there is no dissipation of electrical energy. The current in this case is given by (see Eq. 1.7.11)

$$i(t) = c_1 \cos \omega_0 t + c_2 \sin \omega_0 t, \tag{1.7.29}$$

where

$$\omega_0 = \sqrt{\frac{1}{LC}}. \tag{1.7.30}$$

This value is typically very large for electrical circuits, since both L and C are usually quite small.

For the damped circuit the solution for $i(t)$ may be deduced from Eq. 1.7.17 to be

$$i(t) = e^{-(R/2L)t}[Ae^{\sqrt{R^2-4L/C}\,(t/2L)} + Be^{-\sqrt{R^2-4L/C}\,(t/2L)}]. \qquad (1.7.31)$$

Now the damping criteria become

Case 1.   *Overdamped* $R^2 - \dfrac{4L}{C} > 0$

Case 2.   *Critically damped* $R^2 - \dfrac{4L}{C} = 0$

Case 3.   *Underdamped* $R^2 - \dfrac{4L}{C} < 0$

Example 1.7.2: Use Kirchhoff's second law to establish the differential equation for the parallel electrical circuit shown. Give the appropriate analogies with the spring-mass system and write the solution to the resulting differential equation.

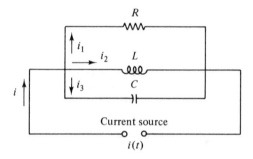

SOLUTION: Kirchhoff's second law states that the current flowing to a point in a circuit must equal the current flowing away from the point. This demands that

$$i(t) = i_1 + i_2 + i_3$$

Use the observed relationships of current to impressed voltage for the components of our circuit,

$$\text{current flowing through a resistor} = \frac{v}{R}$$

$$\text{current flowing through a capacitor} = C\frac{dv}{dt}$$

$$\text{current flowing through an inductor} = \frac{1}{L}\int v\,dt.$$

The equation above becomes

$$i(t) = \frac{v}{R} + C\frac{dv}{dt} + \frac{1}{L}\int v\,dt.$$

If we assume the current source to be a constant and differentiate our expression for $i(t)$, we find the differential equation to be

$$C\frac{d^2v}{dt^2} + \frac{1}{R}\frac{dv}{dt} + \frac{v}{L} = 0.$$

The analogy with the spring-mass system is

$$M \longrightarrow C$$

$$C \longrightarrow \frac{1}{R}$$

$$K \longrightarrow \frac{1}{L}$$

The solution to the homogeneous equation above is

$$v(t) = e^{-(t/2CR)}[Ae^{\sqrt{(1/R^2)-(4C/L)}(t/2C)} + Be^{-\sqrt{(1/R^2)-(4C/L)}(t/2C)}]. \quad \blacksquare$$

PROBLEMS

1. A damped spring-mass system involves a mass of 4 kg, a spring with $K = 64$ N/m, and a dashpot with $C = 32$ kg/s. The mass is displaced 1 m from its equilibrium position and released from rest. Sketch $y(t)$ for the first 2 s.

2. A damped spring-mass system is given an initial velocity of 50 m/s from the equilibrium position. Find $y(t)$ if $M = 4$ kg, $K = 64$ N/m, and $C = 40$ kg/s.

3. A body weighs 50 N and hangs from a spring with spring constant of 50 N/m. A dashpot is attached to the body. If the body is raised 2 m from its equilibrium position and released from rest, determine the solution if (a) $C = 17.7$ kg/s and (b) $C = 40$ kg/s.

4. After a period of time a dashpot deteriorates, so the damping coefficient decreases. For Problem 2 sketch $y(t)$ if the damping coefficient is reduced to 20 kg/s.

5. Solve the overdamped motion of a spring–mass system with $M = 2$ kg, $C = 32$ kg/s, $K = 100$ N/m if $y(0) = 0$ and $\dot{y}(0) = 10$ m/s. Express your answer in the form of Eq. 1.7.17.

6. Show that the general solution of the overdamped motion of a spring–mass system can be written as

$$y(t) = c, e^{-(C/2M)t} \sinh\frac{\sqrt{C^2 - 4KM}\,t}{2M + C_2}.$$

7. A maximum occurs for the overdamped motion of curve 1 of Fig. 1.7. For Problem 5 determine the time at which this maximum occurs.

8. For the overdamped motion of curve 1 of Fig. 1.7, show that the maximum occurs when

$$t = \frac{2M}{\sqrt{C^2 - 4KM}} \tanh^{-1} \frac{\sqrt{C^2 - 4KM}}{C}.$$

9. Find y_{max} for the motion of Problem 5.

10. Using the results of Problems 6 and 8 find an expression for y_{max} of curve 1 of Fig. 1.7 if v_0 is the initial velocity.

11. Determine the time between consecutive maximum amplitudes for a spring-mass system in which $M = 30$ kg, $K = 2000$ N/m, and $C = 300$ kg/s.

12. Find the damping as a percentage of critical damping for the motion $y(t) = 2e^{-t} \sin t$. Also find the time for the first maximum and sketch the curve.

13. Find the displacement $y(t)$ for a mass of 5 kg hanging from a spring with $K = 100$ N/m if there is a dashpot attached having $C = 30$ kg/s. The initial conditions are $y(0) = 1$ m and $dy/dt \,(0) = 0$. Express the solution in all three forms. Refer to Eqs. (1.7.20), (1.7.21), (1.7.22).

14. An electrical circuit is composed of an inductor with $L = 10^{-3}$ H, a capacitor with $C = 2 \times 10^{-5}$ F, and a resistor. Determine the critical resistance that will just lead to an oscillatory current if the elements are connected (a) in series and (b) in parallel.

15. The amplitudes of two successive maximum currents in a series circuit containing an inductor with $L = 10^{-4}$ H and a capacitor with $C = 10^{-6}$ F are measured to be 0.02 A and 0.01 A. Determine the resistance and write the solution for $i(t)$ in the form of Eq. (1.7.22).

16. Determine the current $i(t)$ in a series circuit containing a resistor with $R = 20 \,\Omega$, a capacitor with $C = 10^{-6}/2$ F, and an inductor with $L = 10^{-3}$ H. The initial conditions are $i(0) = 10$ A and $(di/dt)(0) = 0$.

17. An input torque on a circular shaft is $T(t)$. It is resisted by a clamping torque proportional to the rate of angle change $d\theta/dt$ and an elastic torque proportional to the angle itself, the constants of proportionality being c and k, respectively. We have observed that the moment of inertia I times the angular acceleration $d^2\theta/dt^2$ equals the net torque. Write the appropriate differential equation and note the analogy with the spring-mass system.

1.8   Nonhomogeneous, Second-Order, Linear Equations with Constant Coefficients

A general solution of the second-order equation of the form

$$\frac{d^2u}{dx^2} + a\frac{du}{dx} + bu = g(x) \qquad\qquad (1.8.1)$$

is found by adding any particular solution $u_p(x)$ to a general solution $u_h(x)$ of the homogeneous equation

$$\frac{d^2u}{dx} + a\frac{du}{dx} + bu = 0. \tag{1.8.2}$$

The solution of the homogeneous equation was presented in Section 1.6; therefore, we must only find $u_p(x)$. One approach that may be taken is called the *method of undetermined coefficients*. Three common types of functions, which are terms often found in $g(x)$, are listed below. Let us present the form of $u_p(x)$ for each.

1. $g(x)$ is a polynomial of degree n and $k = 0$ is not a root of the characteristic equation. Choose

$$u_p(x) = A_0 + A_1x + \cdots + A_nx^n, \tag{1.8.3}$$

where $A_0, A_1, \ldots, A_n$ are undetermined coefficients. If $k = 0$ is a single root of the characteristic equation, choose

$$u_p(x) = x(A_0 + A_1x + \cdots + A_nx^n). \tag{1.8.4}$$

If $k = 0$ is a double root, choose

$$u_p(x) = x^2(A_0 + A_1x + \cdots + A_nx^n). \tag{1.8.5}$$

2. $g(x)$ is an exponential function Ce^{kx}, and k is not a root of the characteristic equation. Choose

$$u_p(x) = Ae^{kx}. \tag{1.8.6}$$

If k is a single root of the characteristic equation,

$$u_p(x) = Axe^{kx} \tag{1.8.7}$$

and if k is a double root,

$$u_p(x) = Ax^2e^{kx}. \tag{1.8.8}$$

3. $g(x)$ is a sine or cosine function (e.g., $C \cos kx$), and ik is not a root of the characteristic equation. Choose

$$u_p(x) = A \cos kx + B \sin kx. \tag{1.8.9}$$

If ik is a single root of the characteristic equation,

$$u_p(x) = Ax \cos kx + Bx \sin kx. \tag{1.8.10}$$

(*Note:* ik cannot be a double root, since a and b are real. The real equation $m^2 + am + b$ has ik and $-ik$ as roots)

Should $g(x)$ include a combination of the above functions, the particular solution would be found by superimposing the appropriate particular solutions listed above. For functions $g(x)$ that are not listed above, the particular solution must be found using some other technique. Variation of parameters, presented in Section 1.11, will always yield a particular solution.

Example 1.8.1:   Find a general solution of the differential equation

$$\frac{d^2u}{dx^2} + u = x^2.$$

SOLUTION: The solution of the homogeneous equation

$$\frac{d^2u}{dx^2} + u = 0$$

is found to be

$$u_h(x) = c_1 \cos x + c_2 \sin x.$$

A particular solution is assumed to have the form

$$u_p(x) = Ax^2 + Bx + C.$$

This is substituted into the original differential equation to give

$$2A + Ax^2 + Bx + C = x^2$$

Equating coefficients of the various powers of x, we have

$$
\begin{aligned}
x^0: & \quad 2A + C = 0 \\
x^1: & \quad B = 0 \\
x^2: & \quad A = 1.
\end{aligned}
$$

These equations are solved simultaneously to give the particular solution

$$u_p(x) = x^2 - 2.$$

Finally, a general solution is

$$
\begin{aligned}
u(x) &= u_h(x) + u_p(x) \\
&= c_1 \cos x + c_2 \sin x + x^2 - 2. \quad \blacksquare
\end{aligned}
$$

Example 1.8.2:   Find the general solution of the differential equation

$$\frac{d^2u}{dx^2} + 4u = 2 \sin 2x.$$

SOLUTION: The solution of the homogeneous equation is

$$u_h(x) = c_1 \cos 2x + c_2 \sin 2x.$$

One root of the characteristic equation is $2i$; hence, we assume a solution

$$u_p(x) = Ax \cos 2x + Bx \sin 2x.$$

Substitute this into the original differential equation:

$$
\begin{aligned}
-2A \sin 2x + 2B \cos 2x - 2A \sin 2x + 2B \cos 2x - 4A \cos 2x \\
- 4Bx \sin 2x + 4Ax \cos 2x + 4Bx \sin 2x = 2 \sin 2x.
\end{aligned}
$$

Equating coefficients yields

$$\sin 2x: \quad -2A - 2A = 2$$
$$\cos 2x: \quad 2B + 2B = 0$$
$$x \sin 2x: \quad -4B + 4B = 0$$
$$x \cos 2x: \quad -4A + 4A = 0.$$

These equations require that $A = -\frac{1}{2}$ and $B = 0$. Thus,

$$u_p(x) = -\frac{1}{2}x \cos 2x.$$

A general solution is then

$$u(x) = u_h(x) + u_p(x)$$

$$= c_1 \cos 2x + c_2 \sin 2x - \frac{1}{2}x \cos 2x. \quad \blacksquare$$

Example 1.8.3:   Find a particular solution of the differential equation

$$\frac{d^2u}{dx^2} + \frac{du}{dx} + 2u = 4e^x + 2x^2.$$

SOLUTION: Assume the particular solution to have the form

$$u_p(x) = Ae^x + Bx^2 + Cx + D.$$

Substitute this into the given differential equation and there results

$$Ae^x + 2B + Ae^x + 2Bx + C + 2Ae^x + 2Bx^2 + 2Cx + 2D = 4e^x + 2x^2$$

Equating the various coefficients yields

$$e^x: \quad A + A + 2A = 4$$
$$x^0: \quad 2B + C + 2D = 0$$
$$x^1: \quad 2B + 2C = 0$$
$$x^2: \quad 2B = 2.$$

From the equations above we find $A = 1$, $B = 1$, $C = -1$, and $D = -\frac{1}{2}$. Thus,

$$u_p(x) = e^x + x^2 - x - \frac{1}{2}. \quad \blacksquare$$

We conclude this section with one more illustration. The equation

$$\frac{d^2u}{dx^2} - u = -1 \tag{1.8.11}$$

has general solutions

$$\hat{u}(x) = Ae^x + Be^{-x} + 1 \tag{1.8.12}$$

and

$$\bar{u}(x) = c_1 e^x + c_2 e^{-x} + 2 \cosh^2 \frac{x}{2}, \qquad (1.8.13)$$

which explains why we refer to "a" general solution rather than "the" general solution. We leave it to the student to show that the family of solutions described by $\hat{u}$ and $\bar{u}$ are identical, despite the radical difference in appearance between $\hat{u}$ and $\bar{u}$.

PROBLEMS

Find a particular solution for each differential equation.

1. $\dfrac{d^2u}{dx^2} + 2u = 2x$

2. $\dfrac{d^2u}{dx^2} + \dfrac{du}{dx} + 2u = 2x$

3. $\dfrac{d^2u}{dx^2} + u = e^{-x}$

4. $\dfrac{d^2u}{dx^2} - u = e^x$

5. $\dfrac{d^2u}{dx^2} + 10u = 5 \sin x$

6. $\dfrac{d^2u}{dx^2} + 9u = \cos 3x$

7. $\dfrac{d^2u}{dx^2} + 4\dfrac{du}{dx} + 4u = e^{-2x}$

8. $\dfrac{d^2u}{dx^2} + 9u = x^2 + \sin 3x$

Find a general solution for each differential equation

9. $\dfrac{d^2u}{dx^2} + u = e^{2x}$

10. $\dfrac{d^2u}{dx^2} + 4\dfrac{du}{dx} + 4u = x^2 + x + 4$

11. $\dfrac{d^2u}{dx^2} + 9u = x^2 + \sin 2x$

12. $\dfrac{d^2u}{dx^2} + 4u = \sin 2x$

13. $\dfrac{d^2u}{dx^2} - 16u = e^{4x}$

14. $\dfrac{d^2u}{dx^2} + 5\dfrac{du}{dx} + 6u = 3 \sin 2x$

Find the solution for each initial-value problem.

15. $\dfrac{d^2u}{dx^2} + 4\dfrac{du}{dx} + 4u = x^2, \qquad u(0) = 0, \qquad \dfrac{du}{dx}(0) = \dfrac{1}{2}$

16. $\dfrac{d^2u}{dx^2} + 4u = 2 \sin x, \qquad u(0) = 1, \qquad \dfrac{du}{dx}(0) = 0$

17. $\dfrac{d^2u}{dx^2} + 4\dfrac{du}{dx} + 5u = x^2 + 5, \qquad u(0) = 0, \qquad \dfrac{du}{dx}(0) = 0$

18. $\dfrac{d^2u}{dx^2} + 4u = 2 \sin 2x, \qquad u(0) = 0, \qquad \dfrac{du}{dx}(0) = 0$

19. $\dfrac{d^2u}{dx^2} + 6\dfrac{du}{dx} + 10u = \cos 2x, \quad u(0) = 0, \quad \dfrac{du}{dx}(0) = 0$

20. $\dfrac{d^2u}{dx^2} - 16u = 2e^{4x}, \qquad\qquad u(0) = 0, \quad \dfrac{du}{dx}(0) = 0$

21. Suppose that k is a root (real or complex) of the characteristic equation of $u'' + au' + bu = 0$. Explain why Ae^{kx} cannot be a solution of $u'' + au' + bu = g(x)$, for any $g(x) \neq 0$ regardless of the choice of A.

22. If $k = 0$ is a root of the characteristic equation of $u'' + au' + bu = 0$, show that $b = 0$.

23. Use the result of Problem 22 to show that no choice of undetermined constants, $A_0, A_1, \ldots, A_n$ will yield a solution of the form $A_0 + A_1x + \cdots + A_nx^n$ for the equation

$$u'' + au' + bu = c_0 + c_1x + \cdots + c_nx^n$$

if $k = 0$ is a root of the characteristic equation of $u'' + au' + bu = 0$.

1.9 Spring–Mass System: Forced Motion

The spring–mass system shown in Fig. 1.5 is acted upon by a force $F(t)$, a *forcing function*, as shown in Fig. 1.10. The equation describing this motion is again found by applying Newton's second law to the mass M. We have

$$F(t) + Mg - K(y + h) - C\frac{dy}{dt} = M\frac{d^2y}{dt^2}, \qquad (1.9.1)$$

where h is as defined in Fig. 1.5, so that $Mg = Kh$. The equation above becomes

$$M\frac{d^2y}{dt^2} + C\frac{dy}{dt} + Ky = F(t). \qquad (1.9.2)$$

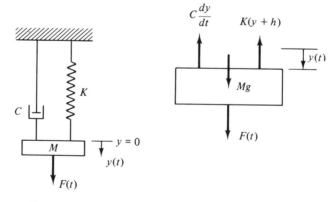

FIGURE 1.10. Spring–mass system with a forcing function.

It is a nonhomogeneous equation and can be solved by the techniques introduced in Section 1.8.

We shall discuss the form of the solution for a sinusoidal forcing function,

$$F(t) = F_0 \cos \omega t. \tag{1.9.3}$$

The particular solution has the form

$$y_p(t) = A \cos \omega t + B \sin \omega t. \tag{1.9.4}$$

Substitute into Eq. 1.9.2 to obtain

$$[(K - M\omega^2)A + \omega CB] \cos \omega t + [(K - M\omega^2)B - \omega CA] \sin \omega t$$
$$= F_0 \cos \omega t. \tag{1.9.5}$$

Equating coefficients of $\cos \omega t$ and $\sin \omega t$ results in

$$(K - M\omega^2)A + \omega CB = F_0$$
$$-\omega CA + (K - M\omega^2)B = 0. \tag{1.9.6}$$

A simultaneous solution yields

$$A = F_0 \frac{K - M\omega^2}{(K - M\omega^2)^2 + \omega^2 C^2}$$

$$B = F_0 \frac{\omega C}{(K - M\omega^2)^2 + \omega^2 C^2}. \tag{1.9.7}$$

The particular solution is then

$$y_p(t) = \frac{(K - M\omega^2)F_0}{(K - M\omega^2)^2 + \omega^2 C^2} \left[\cos \omega t + \frac{\omega C}{K - M\omega^2} \sin \omega t \right]. \tag{1.9.8}$$

This is added to the homogeneous solution presented in Section 1.7 to form the general solution

$$y(t) = e^{-(C/2M)t}[Ae^{\sqrt{C^2-4MK}\,(t/2M)} + Be^{-\sqrt{C^2-4MK}\,(t/2M)}]$$

$$+ \frac{(K - M\omega^2)F_o}{(K - M\omega^2)^2 + \omega^2 C^2} \left[\cos \omega t + \frac{\omega C}{K - M\omega^2} \sin \omega t \right]. \tag{1.9.9}$$

Let us now discuss this solution in some detail.

1.9.1   RESONANCE

An interesting and very important phenomenon is observed in the solution above if we let the damping coefficient C, which is often very small, be zero. The general solution is then (see Eq. 1.7.11 and let $C = 0$ in Eq. 1.9.8)

$$y(t) = c_1 \cos \omega_0 t + c_2 \sin \omega_0 t + \frac{F_0}{M(\omega_0^2 - \omega^2)} \cos \omega t, \tag{1.9.10}$$

where $\omega_0 = \sqrt{K/M}$ and $\omega_0/2\pi$ is the natural frequency of the free oscillation. Consider the condition $\omega \to \omega_0$; that is, the input frequency approaches the natural frequency. We observe from Eq. 1.9.10 that the amplitude of the particular solution becomes unbounded as $\omega \to \omega_0$. This condition is referred to as *resonance*. The amplitude, of course, does not become unbounded in a physical situation; the damping term may limit the amplitude, the physical situation may change for large amplitude, or failure may occur. The latter must be guarded against in the design of oscillating systems. Soldiers break step on bridges so that resonance will not occur. The spectacular failure of the Tacoma Narrows bridge provided a very impressive example of resonant failure. One must be extremely careful to make the natural frequency of oscillating systems different, if at all possible, from the frequency of any probable forcing function.

If $\omega = \omega_0$, Eq. 1.9.10 is, of course, not a solution to the differential equation with no damping. For that case $i\omega_0$ is a root of the characteristic equation

$$m^2 + \omega_0^2 = 0 \tag{1.9.11}$$

of the undamped spring–mass system. The particular solution takes the form

$$y_p(t) = t(A \cos \omega_0 t + B \sin \omega_0 t). \tag{1.9.12}$$

By substituting into the differential equation

$$\frac{d^2 y}{dt^2} + \omega_0^2 y = \frac{F_0}{M} \cos \omega_0 t, \tag{1.9.13}$$

we find the particular solution to be

$$y_p(t) = \frac{F_0}{2M\omega_0} t \sin \omega_0 t. \tag{1.9.14}$$

As time t becomes large, the amplitude becomes large and will be limited by either damping, a changed physical condition, or failure. The particular solution $y_p(t)$ for resonance is shown in Fig. 1.11.

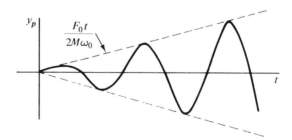

FIGURE 1.11.   The particular solution for resonance.

1.9.2   NEAR RESONANCE

Another phenomenon occurs when the forcing frequency is approximately equal to the natural frequency; that is, the quantity $\omega_0 - \omega$ is small. Let us consider a particular situation for which $dy/dt\,(0) = 0$ and $y(0) = 0$. The arbitrary constants in Eq. 1.9.10 are then

$$c_2 = 0, \qquad c_1 = -\frac{F_0}{M(\omega_0^2 - \omega^2)} \qquad (1.9.15)$$

The solution then becomes

$$y(t) = \frac{F_0}{M(\omega_0^2 - \omega^2)}[\cos \omega t - \cos \omega_0 t]. \qquad (1.9.16)$$

With the use of a trigonometric identity, this can be put in the form*

$$y(t) = \frac{2F_0}{M(\omega_0^2 - \omega^2)} \sin\left[(\omega_0 + \omega)\frac{t}{2}\right] \sin\left[(\omega_0 - \omega)\frac{t}{2}\right]. \qquad (1.9.17)$$

The quantity $\omega_0 - \omega$ is small; thus, the period of the sine wave $\sin[(\omega_0 - \omega)(t/2)]$ is large compared to the period of $\sin[(\omega_0 + \omega)(t/2)]$. For $\omega_0 \cong \omega$, we can write

$$\frac{\omega_0 + \omega}{2} \cong \omega, \qquad \frac{\omega_0 - \omega}{2} = \epsilon, \qquad (1.9.18)$$

where ϵ is small. Then the near-resonance equation 1.9.17 is expressed as

$$y(t) = \left[\frac{2F_0 \sin \epsilon t}{M(\omega_0^2 - \omega^2)}\right] \sin \omega t, \qquad (1.9.19)$$

where the quantity in brackets is the slowly varying amplitude. A plot of $y(t)$ is sketched in Fig. 1.12. The larger wavelength wave appears as a "beat" and can often be heard when two sound waves are of approximately the same frequency.

---

*This is accomplished by writing

$$\cos \omega t = \cos\left[\left(\frac{\omega + \omega_0}{2}\right)t + \left(\frac{\omega - \omega_0}{2}\right)t\right]$$

and

$$\cos \omega_0 t = \cos\left[\left(\frac{\omega + \omega_0}{2}\right)\right]t - \left(\frac{\omega - \omega_0}{2}\right)t\right]$$

and then using the trigonometric identity

$$\cos(\alpha + \beta) = \cos \alpha \cos \beta - \sin \alpha \sin \beta.$$

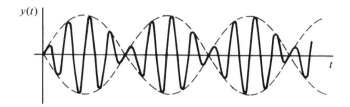

FIGURE 1.12.   Near resonance—"beats."

PROBLEMS

1. Find the solution for $M(d^2y/dt^2) + C(dy/dt) + Mg = 0$. Show that this represents the motion of a body rising with drag proportional to velocity.

2. For Problem 1 assume that the initial velocity is 100 m/s upward, $C = 0.4$ kg/s, and $M = 2$ kg. How high will the body rise?

3. For the body of Problem 1 calculate the time required for the body to rise to the maximum height and compare this to the time it takes for the body to fall back to the original position.

4. A body weighing 100 N is dropped from rest. The drag is assumed to be proportional to the first power of the velocity with the constant of proportionality being 0.5. Approximate the time necessary for the body to attain terminal velocity. Define terminal velocity to be equal to $0.99V_\infty$, where V_∞ is the velocity attained as $t \rightarrow \infty$.

5. Find a general solution to the equation $M(d^2y/dt^2) + Ky = F_0 \cos \omega t$ and verify Eq. 1.9.10 by letting $\omega_0 = \sqrt{K/M}$.

6. A 2-kg mass is suspended by a spring with $K = 32$ N/m. A force of $0.1 \sin 4t$ is applied to the mass. Calculate the time required for failure to occur if the spring breaks when the amplitude of the oscillation exceeds 0.5 m. The motion starts from rest and damping is neglected.

Solve each initial-value problem.

7. $\ddot{y} + 9y = 8 \cos 2t$, $y(0) = 0$, $\dot{y}(0) = 0$

8. $\ddot{y} + 9y = 8 \cos 3t$ $y(0) = 0$, $\dot{y}(0) = 0$

9. $\ddot{y} + 16y = -2 \sin 4t$, $y(0) = 2$, $\dot{y}(0) = 0$

10. $\ddot{y} + 16y = 2 \sin t$ $y(0) = 0$, $\dot{y}(0) = 10$

11. $\ddot{y} + 25y = t - 2$ $y(0) = 1$, $\dot{y}(0) = 4$

12. $\ddot{y} + y = 2e^{-t}$ $y(0) = 0$, $\dot{y}(0) = 2$

For each simple series circuit (see Fig. 1.1), find the current $i(t)$ if $i(0) = q(0) = 0$.

13. $C = 0.02$ F, $L = 0.5$ H, $R = 0$,   and $v = 10 \sin 10t$

14. $C = 10^{-4}$ F, $L = 1.0$ H, $R = 0$ and $v = 120 \sin 100t$

15. $C = 10^{-3}$ F, $L = 0.1$ H, $R = 0$,   and $v = 240 \cos 10t$

16. A 20-N weight is suspended by a frictionless spring with $k = 98$ N/m. A force of $2 \cos 7t$ acts on the weight. Calculate the frequency of the "beat" and find the maximum amplitude of the motion, which starts from rest.

17. A simple series circuit, containing a 10^{-3} F capacitor and a 0.1 H inductor, has an imposed voltage of $120 \cos 101t$. Determine the frequency of the "beat" and find the maximum current.

1.9.3   FORCED OSCILLATIONS WITH DAMPING

The homogeneous solution

$$y_h(t) = e^{-(C/2M)t}[Ae^{\sqrt{C^2-4Mk}\,(t/2M)} + Be^{-\sqrt{C^2-4MK}\,(t/2M)}] \qquad (1.9.20)$$

for damped oscillations includes a factor $e^{-(C/2M)t}$ which is approximately zero after a sufficiently long time. Thus, the general solution $y(t)$ tends to the particular solution $y_p(t)$ after a long time; hence, $y_p(t)$ is called the *steady-state solution*. For short times the homogeneous solution must be included and $y(t) = y_h(t) + y_p(t)$ is the *transient solution*.

With damping included, the amplitude of the particular solution is not unbounded as $\omega \to \omega_0$, but it can still become large. The condition of resonance can be approached, for the case of small damping. Hence, even with some damping, the condition $\omega = \omega_0$ is to be avoided, if at all possible.

We are normally interested in the amplitude. To better display the amplitude for the input $F_0 \cos \omega t$, write Eq. 1.9.8 in the equivalent form

$$y_p(t) = \frac{F_0}{\sqrt{M^2(\omega_0^2 - \omega^2)^2 + \omega^2 C^2}} \cos(\omega t - \alpha), \qquad (1.9.21)$$

where we have used $\omega_0^2 = K/M$. The angle α is called the *phase angle* or *phase lag*. The amplitude Δ of the oscillation is

$$\Delta = \frac{F_0}{\sqrt{M^2(\omega_0^2 - \omega^2)^2 + \omega^2 C^2}} \qquad (1.9.22)$$

We can find the maximum amplitude by setting $d\Delta/d\omega = 0$. Do this and find that the maximum amplitude occurs when

$$\omega^2 = \omega_0^2 - \frac{C^2}{2M^2} \qquad (1.9.23)$$

Note that for sufficiently large damping, $C^2 > 2M^2\omega_0^2$, there is no value of ω that represents a maximum for the amplitude. However, if $C^2 < 2M^2\omega_0^2$, then the maximum occurs at the value of ω as given by Eq. 1.9.23. Substituting this into Eq. 1.9.22 gives the maximum amplitude as

$$\Delta_{max} = \frac{2F_0 M}{C\sqrt{4M^2\omega_0^2 - C^2}} \tag{1.9.24}$$

The amplitude given by Eq. 1.9.22 is sketched in Fig. 1.13 as a function of ω. Large relative amplitudes can thus be avoided by a sufficient amount of damping, or by making sure $|\omega - \omega_0|$ is relatively large.

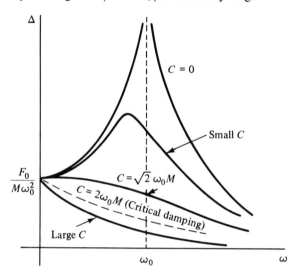

FIGURE 1.13. Amplitude as a function of ω for various degrees of damping.

Example 1.9.1: The ratio of successive maximum amplitudes for a particular spring–mass system for which $K = 100$ N/m and $M = 4$ kg is found to be 0.8 when the system undergoes free motion. If a forcing function $F = 10 \cos 4t$ is imposed on the system, determine the maximum amplitude of the steady-state motion.

SOLUTION: Damping causes the amplitude of the free motion to decrease with time. The logarithmic decrement is found to be (see Eq. 1.7.25).

$$D = \ln\frac{y_n}{y_{n+2}} = \ln\frac{1}{0.8} = 0.223.$$

The damping is then calculated from Eq. 1.7.28. it is

$$C = C_c\frac{D}{\sqrt{D^2 + 4\pi^2}} = 2\sqrt{KM}\,\frac{D}{\sqrt{D^2 + 4\pi^2}}$$

$$= 2\sqrt{100 \times 4}\,\frac{0.223}{\sqrt{0.223^2 + 4\pi^2}} = 1.42 \text{ kg/s}.$$

The natural frequency of the undamped system is

$$\omega_0 = \sqrt{\frac{K}{M}} = \sqrt{\frac{100}{4}} = 5 \text{ rad/s}.$$

The maximum deflection has been expressed by Eq. 1.9.24. It is now calculated to be

$$\Delta_{max} = \frac{2F_0 M}{C\sqrt{4M^2\omega_0^2 - C^2}}$$

$$= \frac{2 \times 10 \times 4}{1.42\sqrt{4 \times 4^2 \times 5^2 - 1.42^2}} = 1.41 \text{ m}. \quad \blacksquare$$

Example 1.9.2:   For the network shown, using Kirchhoff's laws, determine the currents $i_1(t)$ and $i_2(t)$, assuming all currents to be zero at $t = 0$.

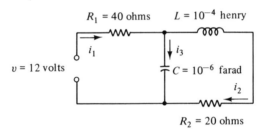

SOLUTION: Using Kirchhoff's first law on the circuit on the left, we find that (see Eqs. 1.4.3)

$$40i_1 + \frac{q}{10^{-6}} = 12, \tag{1}$$

where q is the charge on the capacitor. For the circuit around the outside of the network, we have

$$40i_1 + 10^{-4}\frac{di_2}{dt} + 20i_2 = 12. \tag{2}$$

Kirchhoff's second law requires that

$$i_1 = i_2 + i_3 \tag{3}$$

Using the relationship

$$i_3 = \frac{dq}{dt} \tag{4}$$

and the initial conditions, that $i_1 = i_2 = i_3 = 0$ at $t = 0$, we can solve the set of equations above. To do this, substitute (4) and (1) into (3). This gives

$$\frac{1}{40}(12 - 10^6 q) - \frac{dq}{dt} = i_2.$$

Substituting this and (1) into (2) results in

$$10^{-4}\frac{d^2q}{dt^2} + 22.5\frac{dq}{dt} + 1.5 \times 10^6 q = 6.$$

The appropriate initial conditions can be found from (1) and (4) to be $q = 12 \times 10^{-6}$ and $dq/dt = 0$ at $t = 0$. Solving the equation above, using the methods of this chapter, gives the charge as

$$q(t) = e^{-1.12 \times 10^5 t}[c_1 \cos 48{,}200t + c_2 \sin 48{,}200t] + 4 \times 10^{-6}.$$

The initial conditions allow the constants to be evaluated. They are

$$c_1 = 8 \times 10^{-6}, \qquad c_2 = 0.468 \times 10^{-6}.$$

The current $i_1(t)$ is found using (1) to be

$$i_1(t) = 0.2 - e^{-1.12 \times 10^5 t}[0.2 \cos 48{,}200t + 0.468 \sin 48{,}200t].$$

The current $i_2(t)$ is found by using (4) and (3). It is

$$i_2(t) = 0.2 + e^{-1.12 \times 10^5 t}[-0.2 \cos 48{,}200t + 2.02 \sin 48{,}200t].$$

Note the high frequency and rapid decay rate, which is typical of electrical circuits. ■

PROBLEMS

1. Using the sinusoidal forcing function as $F_0 \sin \omega t$, and with $C = 0$ show that the amplitude $F_0/M(\omega_0^2 - \omega^2)$ of the particular solution remains unchanged for the spring–mass system.

2. Show that the particular solution given by Eq. 1.9.14 for $\omega = \omega_0$ follows from the appropriate equations for $F(t) = F_0 \cos \omega_0 t$.

Find the steady-state solution for each differential equation.

3. $\dfrac{d^2y}{dt^2} + \dfrac{dy}{dt} + 4y = 2 \sin 2t$

4. $\dfrac{d^2y}{dt^2} + 2\dfrac{dy}{dt} + y = \cos 3t$

5. $\dfrac{d^2y}{dt^2} + \dfrac{dy}{dt} + y = 2 \sin t + \cos t$

6. $\dfrac{d^2y}{dt^2} + 0.1\dfrac{dy}{dt} + 2y = 2 \sin 2t$

7. $\dfrac{d^2y}{dt^2} + 2\dfrac{dy}{dt} + 5y = \sin t - 2 \cos 3t$

8. $\dfrac{d^2y}{dt^2} + \dfrac{dy}{dt} + 2y = \cos t - \sin 2t$

Determine the transient solution for each differential equations.

9. $\dfrac{d^2y}{dt^2} + 5\dfrac{dy}{dt} + 4y = \cos 2t$

10. $\dfrac{d^2y}{dt^2} + 7\dfrac{dy}{dt} + 10y = 2 \sin t - \cos 2t$

11. $\dfrac{d^2y}{dt^2} + 4\dfrac{dy}{dt} + 4y = 4 \sin t$

12. $\dfrac{d^2y}{dt^2} + 0.1\dfrac{dy}{dt} + 2y = \cos 2t$

Solve for the specific solution to each initial-value problem.

13. $\dfrac{d^2y}{dt^2} + 5\dfrac{dy}{dt} + 6y = 52 \cos 2t,$ $y(0) = 0, \dfrac{dy}{dt}(0) = 0$

14. $\dfrac{d^2y}{dt^2} + 2\dfrac{dy}{dt} + y = 2 \sin t,$ $y(0) = 0, \dfrac{dy}{dt}(0) = 0$

15. $\dfrac{d^2y}{dt^2} + 2\dfrac{dy}{dt} + 10y = 26 \sin 2t,$ $y(0) = 1, \dfrac{dy}{dt}(0) = 0$

16. $\dfrac{d^2y}{dt^2} + 0.1\dfrac{dy}{dt} + 2y = 20.2 \cos t,$ $y(0) = 0, \dfrac{dy}{dt}(0) = 10$

17. $\dfrac{d^2y}{dt^2} + 3\dfrac{dy}{dt} + 2y = 10 \sin t,$ $y(0) = 0, \dfrac{dy}{dt}(0) = 0$

18. $\dfrac{d^2y}{dt^2} + 0.02\dfrac{dy}{dt} + 16y = 2 \sin 4t,$ $y(0) = 0, \dfrac{dy}{dt}(0) = 0$

19. The motion of a 3-kg mass, hanging from a spring with $K = 12$ N/m, is damped with a dashpot with $C = 5$ kg/s. (a) Show that Eq. 1.9.22 gives the amplitude of the steady-state solution if $F(t) = F_0 \sin \omega t$. (b) Determine the phase lag and amplitude of the steady-state solution if a force $F = 20 \sin 2t$ acts on the mass.

20. For Problem 19 let the forcing function be $F(t) = 20 \sin \omega t$. Calculate the maximum possible amplitude of the steady-state solution and the associated forcing-function frequency.

21. A forcing function $F = 10 \sin 2t$ is to be imposed on a spring-mass system with $M = 2$ kg and $K = 8$ N/m. Determine the damping coefficient necessary to limit the amplitude of the resulting motion to 2 m.

22. A constant voltage of 12 V is impressed on a series circuit containing elements with $R = 30$ Ω, $L = 10^{-4}$ H, and $C = 10^{-6}$ F. Determine expressions for both the charge on the capacitor and the current if $q = i = 0$ at $t = 0$.

23. A series circuit is composed of elements with $R = 60$ Ω, $L = 10^{-3}$ H, and $C = 10^{-5}$ F. Find an expression for the steady-state current if a voltage of $120 \cos 120\pi t$ is applied at $t = 0$.

24. A circuit is composed of elements with $R = 80 \, \Omega$, $L = 10^{-4}$ H, and $C = 10^{-6}$ F connected in parallel. The capacitor has an initial charge of 10^{-4} C. There is no current flowing through the capacitor at $t = 0$. What is the current flowing through the resistor at $t = 10^{-4}$ s?

25. The circuit of Problem 24 is suddenly subjected to a current source of $2 \cos 200t$. Find the steady-state voltage across the elements.

26. The inductor and the capacitor are interchanged in Example 1.9.2. Determine the resulting current $i_2(t)$ flowing through R_2. Also, find the steady-state charge on the capacitor.

1.10   The Cauchy–Euler Equation

In the preceding sections we have discussed differential equations with constant coefficients. In this section we present the solution to a class of second-order differential equations with variable coefficients. Such a class of equations is called the *Cauchy–Euler equation* of order 2. (We abbreviate Cauchy–Euler by C–E.) It is

$$x^2\frac{d^2u}{dx^2} + ax\frac{du}{dx} + bu = 0 \qquad (1.10.1)$$

We search for solutions of the form

$$u(x) = x^m. \qquad (1.10.2)$$

This function is substituted into Eq. 1.10.1 to obtain

$$x^2m(m - 1)x^{m-2} + ax\,mx^{m-1} + bx^m = 0 \qquad (1.10.3)$$

or, equivalently,

$$[m(m - 1) + am + b]x^m = 0. \qquad (1.10.4)$$

By setting the quantity in brackets equal to zero, we can find two roots for m. This *characteristic equation,* written as

$$m^2 + (a - 1)m + b = 0, \qquad (1.10.5)$$

yields the two distinct roots m_1 and m_2 with corresponding independent solutions

$$u_1 = |x|^{m_1} \quad \text{and} \quad u_2 = |x|^{m_2} \qquad (1.10.6)$$

The general solution, for distinct roots, is then

$$u(x) = c_1|x|^{m_1} + c_2|x|^{m_2} \qquad (1.10.7)$$

valid in every interval* not containing $x = 0$.

---

*The Wronskian is $W(x) = (m_1 - m_2)x^a$.

If a double root results from the characteristic equation, that is, $m_1 = m_2$ then u_1 and u_2 are not independent and Eq. 1.10.7 is not a general solution. To find a second independent solution, assuming that $u_1 = x^m$ is one solution, we assume, as in Eq. 1.6.15, that

$$u_2 = v(x)u_1. \tag{1.10.8}$$

Following the steps outlined in the equations following Eq. 1.6.15, we find that

$$v(x) = \ln|x|. \tag{1.10.9}$$

A general solution, for double roots, is then

$$u(x) = (c_1 + c_2 \ln|x|)|x|^m \tag{1.10.10}$$

valid in every interval (a, b) provided that $x = 0$ is not in (a, b).
   We note in passing that $m_1 = m_2$ can occur only if

$$b = \left(\frac{a-1}{2}\right)^2, \tag{1.10.11}$$

so that $m = -(a - 1)/2$ and Eq. 1.10.8 becomes

$$u(x) = (c_1 + c_2 \ln|x|)|x|^{-(a-1)/2}. \tag{1.10.12}$$

Example 1.10.1:   Find a general solution to the differential equation

$$x^2\frac{d^2u}{dx^2} - 5x\frac{du}{dx} + 8u = 0.$$

SOLUTION: The characteristic equation is

$$m^2 - 6m + 8 = 0.$$

The two roots are

$$m_1 = 4, \qquad m_2 = 2,$$

with corresponding independent solutions

$$u_1 = x^4, \qquad u_2 = x^2.$$

A general solution is then

$$u(x) = c_1x^4 + c_2x^2. \quad \blacksquare$$

Example 1.10.2:   Determine the solution to the initial-value problem

$$x^2\frac{d^2u}{dx^2} - 3x\frac{du}{dx} + 4u = 0$$

$u(1) = 2, u'(1) = 8.$

   SOLUTION: The characteristic equation is

$$m^2 - 4m + 4 = 0.$$

A double root $m = 2$ occurs; thus, the general solution is (see Eq. 1.10.10)

$$u(x) = (c_1 + c_2 \ln |x|)x^2.$$

To use the initial conditions, we must have du/dx. We find

$$\frac{du}{dx} = \left(\frac{c_2}{x}\right)x^2 + (c_1 + c_2 \ln |x|)2x.$$

The initial conditions then give

$$2 = (c_1 + c_2 \overbrace{\ln 1}^{0})1^2$$

$$8 = \left(\frac{c_2}{1}\right)1^2 + (c_1 + c_2 \overbrace{\ln 1}^{0})2.$$

These two equations result in

$$c_1 = 2, \qquad c_2 = 4.$$

Finally, the solution is

$$u(x) = 2(1 + 2 \ln |x|)x^2 = 2(1 + \ln x^2)x^2. \quad \blacksquare$$

PROBLEMS

Determine a general solution for each differential equation.

1. $x^2 u'' + 7xu' + 8u = 0$ **2.** $x^2 u'' + 9xu' + 12u = 0$

3. $x^2 u'' - 12u = 24x$ **4.** $x^2 u'' + 2xu' - 12u = 24$

Solve each initial-value problem.

5. $x^2 u'' + 9xu' + 12u = 0$ $u(1) = 2,$ $u'(1) = 0$

6. $x^2 u'' + 2xu' - 12u = 12,$ $u(1) = 0,$ $u'(1) = 0$

7. Show that $v(x) = \ln |x|$ does, in fact, follow by using Eq. 1.10.8.

8. The C–E equation of nth order is

$$x^n u^{(n)} + a_1 x^{n-1} u^{(n-1)} + \cdots + a_{n-1}xu' + a_n u = 0.$$

Find a general solution for the C–E equation of order $n = 1$; assume that $u(x) = x^m$ is a solution.

9. Find the characteristic equation for the C–E equation of order $n = 3$.

1.11 *Variation of Parameters*

In Section 1.8 we discussed particular solutions arising from forcing functions of very special types. In this section we present a method applicable to any sectionally continuous input function.

Consider the equation

$$\frac{d^2u}{dx^2} + P_0(x)\frac{du}{dx} + P_1(x)u = g(x). \qquad (1.11.1)$$

A general solution $u(x)$ is found by adding a particular solution $u_p(x)$ to a general solution of the homogeneous equation, to obtain

$$u(x) = c_1u_1(x) + c_2u_2(x) + u_p(x), \qquad (1.11.2)$$

where $u_1(x)$ and $u_2(x)$ are solutions to the homogeneous equation

$$\frac{d^2u}{dx^2} + P_0(x)\frac{du}{dx} + P_1(x)u = 0. \qquad (1.11.3)$$

To find a particular solution, assume that the solution has the form

$$u_p(x) = v_1(x)u_1(x) + v_2(x)u_2(x). \qquad (1.11.4)$$

Differentiate and obtain

$$\frac{du_p}{dx} = v_1\frac{du_1}{dx} + v_2\frac{du_2}{dx} + u_1\frac{dv_1}{dx} + u_2\frac{dv_2}{dx}. \qquad (1.11.5)$$

We seek a solution such that

$$u_1\frac{dv_1}{dx} + u_2\frac{dv_2}{dx} = 0. \qquad (1.11.6)$$

We are free to impose this one restriction on $v_1(x)$ and $v_2(x)$ without loss of generality, as the following analysis shows. We have

$$\frac{du_p}{dx} = v_1\frac{du_1}{dx} + v_2\frac{du_2}{dx}. \qquad (1.11.7)$$

Differentiating this equation again results in

$$\frac{d^2u_p}{dx^2} = v_1\frac{d^2u_1}{dx^2} + v_2\frac{d^2u_2}{dx^2} + \frac{dv_1}{dx}\frac{du_1}{dx} + \frac{dv_2}{dx}\frac{du_2}{dx}. \qquad (1.11.8)$$

Substituting into Eq. 1.11.1, we find that

$$v_1\left(\frac{d^2u_1}{dx^2} + P_0\frac{du_1}{dx} + P_1u_1\right) + v_2\left(\frac{d^2u_2}{dx^2} + P_0\frac{du_2}{dx} + P_1u_2\right) + \frac{dv_1}{dx}\frac{du_1}{dx}$$

$$+ \frac{dv_2}{dx}\frac{du_2}{dx} = g(x). \qquad (1.11.9)$$

The quantities in parentheses are both zero since u_1 and u_2 are solutions of the homogeneous equation. Hence,

$$\frac{dv_1}{dx}\frac{du_1}{dx} + \frac{dv_2}{dx}\frac{du_2}{dx} = g(x). \qquad (1.11.10)$$

This equation and Eq. 1.11.6 are solved simultaneously to find

$$\frac{dv_1}{dx} = -\frac{u_2 g(x)}{u_1 \dfrac{du_2}{dx} - u_2 \dfrac{du_1}{dx}}, \qquad \frac{dv_2}{dx} = \frac{u_1 g(x)}{u_1 \dfrac{du_2}{dx} - u_2 \dfrac{du_1}{dx}}. \qquad (1.11.11)$$

The quantity in the denominator is the *Wronskian W* of $u_1(x)$ and $u_2(x)$,

$$W(x) = u_1 \frac{du_2}{dx} - u_2 \frac{du_1}{dx} \qquad (1.11.12)$$

We can now integrate Eqs. 1.11.11 and obtain

$$v_1(x) = -\int \frac{u_2 g}{W} dx, \qquad v_2(x) = \int \frac{u_1 g}{W} dx. \qquad (1.11.13)$$

A particular solution is then

$$u_p(x) = -u_1 \int \frac{u_2 g}{W} dx + u_2 \int \frac{u_1 g}{W} dx. \qquad (1.11.14)$$

A general solution of the nonhomogeneous equation follows by using this expression for $u_p(x)$ in Eq. 1.11.2. This technique is referred to as the *method of variation of parameters*.

Example 1.11.1:   A general solution of $(d^2 u/dx^2) + u = x^2$ was found in Example 1.8.1. Find a particular solution to this equation using the method of variation of parameters.

SOLUTION: Two independent solutions are

$$u_1(x) = \sin x, \qquad u_2(x) = \cos x.$$

The Wronskian is then

$$W(x) = u_1 \frac{du_2}{dx} - u_2 \frac{du_1}{dx}$$

$$= -\sin^2 x - \cos^2 x = -1.$$

A particular solution is then found from Eq. 1.11.14 to be

$$u_p(x) = \sin x \int x^2 \cos x \, dx - \cos x \int x^2 \sin x \, dx.$$

This is integrated by parts twice to give

$$u_p(x) = x^2 - 2.$$

The particular solution derived in the preceding example is the same as that found in Example 1.8.1. The student should not make too much of this coincidence. There are, after all, infinitely many particular solutions; for instance, two others are

$$\bar{u}_p(x) = \sin x + x^2 - 2$$

$$\hat{u}_p(x) = \cos x + x^2 - 2. \qquad (1.11.15)$$

The reason no trigonometric term appeared in $u_p(x)$ was due to our implicit choice of zero for the arbitrary constants of integration in Eq. 1.11.14.

One way to obtain a unique particular solution is to require that the particular solution satisfy the initial conditions, $u_p(x_0) = u'_p(x_0) = 0$ for some convenient x_0. In this example $x_0 = 0$ seems reasonable and convenient. Let

$$u(x) = c_1 \sin x + c_2 \cos x + x^2 - 2. \qquad (1.11.16)$$

Then, imposing the initial conditions,

$$u(0) = c_2 - 2 = 0$$

$$u'(0) = c_1 = 0. \qquad (1.11.17)$$

Hence,

$$u_p(x) = 2 \cos x + x^2 - 2 \qquad (1.11.18)$$

is the required particular solution. Note that this method does not yield the "intuitively obvious" best choice, $u_p(x) = x^2 - 2$. ■

PROBLEMS

Find a general solution for each differential equation.

1. $u'' + u = x \sin x$

2. $u'' + 5u' + 4u = xe^x$

3. $u'' + 4u' + 4u = xe^{-2x}$

4. $u'' + u = \sec x$

5. $u'' - 2u' + 2u = x^2 e^x$

6. $u'' - 4u' + 4u = x^{-1}e^x$

7. $x^2 u'' + xu' - u = 9$

8. $x^2 u'' + xu' - u = 2x^2$

8. $x^2 u'' - 2xu' + 4u = x \cos x$

10. $xu'' - u' = (1 + x)x$

Find a particular solution for each differential equation.

11. $\ddot{y} + y = t \sin t$

12. $\ddot{y} + 5\dot{y} + 4y = te^t$

13. $\ddot{y} + 4\dot{y} + 4y = te^{-2t}$

14. (a) Show that Eq. 1.11.14 may be rewritten as

$$u_p(x) = \int_{x_0}^x \frac{g(s)}{W(s)} \begin{vmatrix} u_1(s) & u_2(s) \\ u_1(x) & u_2(x) \end{vmatrix} ds.$$

(b) Use the result in part (a) to show that the solution of Eq. 1.11.1 with initial conditions $u(x_0) = 0$ and $u'(x_0) = 0$ is

$$u(x) = \int_{x_0}^x g(s) \frac{u_1(s)u_2(x) - u_1(x)u_2(s)}{u_1(s)u'_2(s) - u'_1(s)u_2(s)} ds.$$

Hint: If $F(x) = \int_a^x g(x, s)\, ds$ then

$$F'(x) = g(x, x) + \int_a^x \frac{\partial}{\partial x} g(x, s)\, ds.$$

15. Use the results in Problem 14 to obtain $u_p(x) = 2\cos x + x^2 - 2$ as the solution of Example 1.11.1 satisfying $u(0) = 0$, $u'(0) = 0$.

16. Use the results of Problem 14 to write a particular solution of $u'' + b^2 u = g(x)$ in the form

$$u_p(x) = \frac{1}{b} \int_0^x g(s) \sin b(x - s) \, ds.$$

17. Use the results of Problem 14 to write a particular solution of $u'' - b^2 u = g(x)$ in the form

$$u_p(x) = \frac{1}{b} \int_0^x g(s) \sinh b(x - s) \, ds.$$

18. Verify that the functions $u_p(x)$ in Problems 16 and 17 satisfy $u_p(0) = u_p'(0) = 0$.

19. Use the results of Problem 14 to show that

$$u_p(x) = \int_0^x g(s)(x - s) e^{-a(x-s)} \, ds$$

is a particular solution of $u'' + 2au' + a^2 u = 0$.

1.12   Miscellania

1.12.1   CHANGE OF DEPENDENT VARIABLES

By carefully choosing $f(x)$, the change of dependent variables from u to y via the transformation $u(x) = f(x)y(x)$ can be extremely useful, as we now illustrate. This change of variables,

$$u(x) = f(x)y(x) \tag{1.12.1}$$

converts

$$u'' + P_0(x)u' + P_1(x)u = g(x) \tag{1.12.2}$$

into a second-order equation in $y(x)$. There results

$$u = fy$$
$$u' = fy' + f'y \tag{1.12.3}$$
$$u'' = f''y + 2f'y' + fy'',$$

which, when substituted into Eq. 1.12.2 and rearranged, gives

$$fy'' + (2f' + P_0 f)y' + (f'' + P_0 f' + P_1 f)y = g(x). \tag{1.12.4}$$

Equation 1.12.4 takes on different forms and serves a variety of purposes depending on the choice of $f(x)$. As an illustration, suppose that $g(x) = 0$ and

$f(x)$ is a solution of Eq. 1.12.2. Then

$$f'' + p_0 f' + p_1 f = 0 \tag{1.12.5}$$

and hence, Eq. 1.12.4 reduces to

$$f y'' + (2f' + p_0 f) y' = 0. \tag{1.12.6}$$

The latter equation is a linear, first order equation in y' and its general solution is easy to get. This is precisely the method used in Section 1.6, Eq. 1.6.15, and in Section 1.10, Eq. 1.10.8, to obtain a second solution to the equations

$$u'' + au' + bu = 0 \tag{1.12.7}$$

and

$$x^2 u'' + axu' + bu = 0 \tag{1.12.8}$$

when $m_1 = m_2$. (See Problems 5 and 6).

The substitution 1.12.1 is useful in the nth-order case. Given a solution $f(x)$ of the associated homogeneous equation the change of variables $u = f(x)y$ leads to an $(n - 1)^{\text{st}}$-order equation in y'.

PROBLEMS

1. Use the fact that $f(x) = e^x$ solves $(x - 1)u'' - xu' + u = 0$ to find the general solution of

$$(x - 1)u'' - xu' + u = 1.$$

   Use $u(x) = e^x y(x)$ and Eq. 1.12.4.

2. Find the general solution of

$$u'' + 4u' + 4u = e^{-2x}$$

   using the ideas of this section.

3. The function $f(x) = \sin x$ solves

$$\tan^2 x u'' - 2 \tan x u' + (2 + \tan^2 x)u = 0.$$

   Find its general solution.

4. Find an "elementary" method that yields the general solution of

$$u'' - xp(x)u' + p(x)u = 0.$$

   Hint: Change dependent variables and try to find a clever choice for $f(x)$.

5. Suppose that the characteristic equation of $u'' + au' + bu = 0$ has equal roots, $m_1 = m_2 = -a/2$. Use the method of this section to obtain a second solution.

6. Suppose that the characteristic equation of $x^2 u'' + axu' + bu = 0$ has roots $m_1 = m_2 = -a/2$. It is assumed that one solution of the form $|x|^{m_1}$ exists; find the second solution using the method of this section.

7. Suppose that $f(x)$ is a solution of $u'' + p_0(x)u' + p_1(x)u = 0$. Show that a particular solution of $u'' + p_0(x)u' + p_1(x)u = g(x)$ can always be found in the form $u_p(x) = f(x)y(x)$.

8. Use the result of Problem 7 to find a general solution of $u'' + p_0(x)u' + xp_1(x)u = g(x)$, given that $f(x)$ is a solution of the corresponding homogeneous equation.

1.12.2   THE NORMAL FORM

In Section 1.12.1 we chose $f(x)$ so that the coefficient of y is zero. We may choose f so that the coefficient of y' is zero. Let $f(x)$ be any solution of

$$2f' + p_0(x)f = 0. \tag{1.12.9}$$

That is,

$$f(x) = e^{-(1/2)\int p_0(x)\,dx}. \tag{1.12.10}$$

From the hypothesis that the coefficient of the y' term in Eq. 1.12.4 is zero, we have

$$fy'' + [p_1(x)f + p_0(x)f' + f'']y = 0. \tag{1.12.11}$$

By differentiating Eq. 1.12.9 we have

$$2f'' = -p_0'f - p_0f'. \tag{1.12.12}$$

Substituting the expressions for f' and f'' in Eqs. 1.12.9 and 1.12.12 into Eq. 1.12.11, we obtain

$$y'' + [p_1(x) - \tfrac{1}{4}p_0^2(x) - \tfrac{1}{2}p_0'(x)]y = 0. \tag{1.12.13}$$

This is the *normal form* of

$$u'' + p_0(x)u' + p_1(x)u = 0. \tag{1.12.14}$$

The coefficient of y,

$$I_u(x) = p_1(x) - \tfrac{1}{4}p_0^2(x) - \tfrac{1}{2}p_0'(x), \tag{1.12.15}$$

is the *invariant* of Eq. 1.12.14. This terminology is motivated by the following rather surprising theorem.

Theorem 1.6: *Suppose that*

$$w'' + \hat{p}_0(x)w' + \hat{p}_1(x)w = 0 \tag{1.12.16}$$

results from the change of variables, $u = h(x)w$, applied to Eq. 1.12.14. Then the normal forms of Eqs. 1.12.14 and 1.12.16 are identical.

PROOF: The invariant for Eq. 1.12.16 is

$$I_w(x) = \hat{p}_1(x) - \tfrac{1}{4}\hat{p}_0^2(x) - \tfrac{1}{2}\hat{p}_0'(x). \tag{1.12.17}$$

In view of Eq. 1.12.4, the relationships between $\hat{p}_1$ and $\hat{p}_0$ and p_0 and p_1 are these:

$$\frac{2h'}{h} + p_0 = \hat{p}_0$$

$$\frac{h''}{h} + p_0\frac{h'}{h} + p_1 = \hat{p}_1. \tag{1.12.18}$$

Thus,

$$\hat{p}_0^2 = 4\left(\frac{h'}{h}\right)^2 + p_0^2 + 4\frac{h'}{h}p_0 \tag{1.12.19}$$

and

$$\hat{p}_0' = 2\frac{h''}{h} - 2\left(\frac{h'}{h}\right)^2 + p_0'. \tag{1.12.20}$$

Substituting these two expressions into Eq. 1.12.17 results in

$$I_w(x) = p_1 - \tfrac{1}{4}p_0^2 - \tfrac{1}{2}p_0' = I_u(x), \tag{1.12.21}$$

which completes the proof.

Example 1.12.1:   Find the normal form of

$$u'' + au' + bu = 0.$$

SOLUTION: The invariant is

$$I_u = b - \frac{a^2}{4}.$$

The normal form of the equation is then

$$y'' + (b - \tfrac{1}{4}a^2)y = 0. \quad \blacksquare$$

Example 1.12.2:   Find the normal form of

$$x^2u'' - 2xu' + (a^2x^2 + 2)u = 0$$

and thus find its general solution.

SOLUTION: Here, $p_0(x) = -2/x$ and $p_1(x) = a^2 + 2/x^2$. Thus

$$I_u(x) = a^2 + \frac{2}{x^2} - \frac{1}{4}\left(\frac{4}{x^2}\right) - \frac{1}{2}\left(\frac{2}{x^2}\right) = a^2.$$

Therefore, the normal form is

$$y'' + a^2y = 0.$$

Now, using Eq. 1.12.10, we have

$$f(x) = e^{-(1/2)\int p_0(x)\,dx}$$
$$= e^{\int (1/x)\,dx} = e^{\ln x} = x,$$

so that

$$u(x) = f(x)y(x) = x(c_1 \cos ax + c_2 \sin ax). \quad \blacksquare$$

PROBLEMS

Find the normal form of each differential equation.

1. $x(1 - x)u'' + [\gamma - (\alpha + \beta + 1)x]u' - \alpha\beta u = 0$

2. $(1 - x^2)u'' - 2xu' + n(n + 1)u = 0$

3. $xu'' + (\gamma - x)u' - \alpha u = 0$

4. $x^2u'' + xu' + (x^2 - n^2)u = 0$

5. $xu'' + (1 - \gamma)u' + u = 0$

6. $u'' - 2xu' + 2nu = 0$

The above equations are "classical": (1) the hypergeometric equation, (2) the Legendre equation, (3) the confluent hypergeometric equation, (4) the Bessel equation, (5) the Bessel–Clifford equation, and (6) the Hermite equation.

7. Find a general solution of

$$xu'' - 2(x - 1)u' + 2(x - 1)u = 0$$

by showing that its normal form has constant coefficients.

8. Prove that $u'' + p_0(x)u' + p_1(x)u = 0$ can be transformed into an equation with constant coefficients using $u = fy$ if and only if $I_u(x) = \text{const}$.

9. Suppose that $u = fy$ transforms $u'' + p_0(x)u' + p_1(x)u = 0$ into its normal form. Suppose that $u = hw$ transforms $u'' + p_0(x)u' + p_1(x)u = 0$ into $w'' + \hat{p}_0(x)w' + \hat{p}_1(x)w = 0$. Find $r(x)$ so that $w = r(x)y$ transforms $w'' + \hat{p}_0(x)w' + \hat{p}_1(x)w = 0$ into its normal form. What is the relationship, if any, between $r(x)$, $h(x)$, and $f(x)$?

1.12.3   CHANGE OF INDEPENDENT VARIABLE

It is sometimes useful to change the independent variable; to change from the independent variable x to the independent variable z, we define

$$z = h(x) \qquad\qquad (1.12.22)$$

Then

$$\frac{dz}{dx} = h'(x) \qquad\qquad (1.12.23)$$

and, using the chain rule, we have

$$\frac{du}{dx} = \frac{du}{dz}\frac{dz}{dx} = h'(x)\frac{du}{dz}.$$

(1.12.24)

Also, by the chain rule and the product rule,

$$\frac{d^2u}{dx^2} = \frac{d}{dx}\left(\frac{du}{dx}\right) = \frac{d}{dx}\left(h'(x)\frac{du}{dz}\right)$$

$$= h''(x)\frac{du}{dz} + h'(x)\frac{d}{dx}\left(\frac{du}{dz}\right)$$

$$= h''(x)\frac{du}{dz} + h'(x)\frac{d}{dz}\left(\frac{du}{dz}\right)\frac{dz}{dx}$$

$$= h''\frac{du}{dz} + h'^2\frac{d^2u}{dz^2}.$$

(1.12.25)

We substitute into

$$\frac{d^2u}{dx^2} + p_0(x)\frac{du}{dx} + p_1(x)u = 0$$

(1.12.26)

to obtain

$$h'^2\frac{d^2u}{dz^2} + (h'' + h'p_0)\frac{du}{dz} + p_1u = 0.$$

(1.12.27)

In Eq. 1.12.27, it is understood that h'', h', p_0, and p_1 must be written as functions of z using $z = h(x)$ and $x = h^{-1}(z)$. Therefore, the efficacy of this substitution depends, in part, on the simplicity of the inverse, $h^{-1}(z)$. An example will illustrate.

Example 1.12.3:   Introduce the change of variables $z = \ln x$ $(x = e^z)$ in the Cauchy–Euler equation

$$x^2\frac{d^2u}{dx^2} + ax\frac{du}{dx} + bu = 0.$$

Solve the resulting equation and thereby solve the Cauchy–Euler equation.

SOLUTION: We have $dz/dx = 1/x$ and $d^2z/dx^2 = -1/x^2$. Therefore, using Eqs. 1.12.24 and 1.12.25, there results

$$x^2\left(-\frac{1}{x^2}\frac{du}{dz} + \frac{1}{x^2}\frac{d^2u}{dz^2}\right) + ax\left(\frac{1}{x}\frac{du}{dz}\right) + bu = 0,$$

from which

$$\frac{d^2u}{dz^2} + (a - 1)\frac{du}{dz} + bu = 0.$$

This constant-coefficient equation has the following solution:

$$u(z) = c_1 e^{m_1 z} + c_2 e^{m_2 z} \qquad \text{for real } m_1 \neq m_2$$

or

$$u(z) = (c_1 + c_2 z)e^{mz} \qquad \text{for } m_1 = m_2 = m$$

or

$$u(z) = e^{\alpha z}(c_1 \cos \beta z + c_2 \sin \beta z) \qquad \text{for complex } m = \alpha \pm i\beta.$$

In terms of x we get, using $e^{mz} = e^{m \ln |x|} = |x|^m$,

$$u(x) = c_1 |x|^{m_1} + c_2 |x|^{m_2}$$

or

$$u(x) = (c_1 + c_2 \ln |x|)|x|^m$$

or

$$u(x) = |x|^\alpha [c_1 \cos (\beta \ln |x|) + c_2 \cos (\beta \ln |x|)],$$

respectively. ∎

Example 1.12.4: Introduce the change of variables $x = \cos \phi$ in

$$(1 - x^2)\frac{d^2 u}{dx^2} - 2x\frac{du}{dx} + \lambda(\lambda + 1)u = 0.$$

SOLUTION: The new variable ϕ can be written as

$$\phi = \cos^{-1} x.$$

It then follows that

$$1 - x^2 = 1 - \cos^2 \phi = \sin^2 \phi$$

$$\frac{d\phi}{dx} = -\frac{1}{\sin \phi}.$$

Also,

$$\frac{d^2 \phi}{dx^2} = -\frac{d}{d\phi}\left(\frac{1}{\sin \phi}\right)\frac{d\phi}{dx} = \frac{\cos \phi}{\sin^2 \phi}\left(-\frac{1}{\sin \phi}\right) = -\frac{\cos \phi}{\sin^3 \phi}.$$

The coefficients are

$$p_0(x) = \frac{-2x}{1 - x^2} = -2\frac{\cos \phi}{\sin^2 \phi}, \qquad p_1(x) = \frac{\lambda(\lambda + 1)}{1 - x^2} = \frac{\lambda(\lambda + 1)}{\sin^2 \phi}.$$

Therefore, in terms of ϕ (see Eq. 1.12.27),

$$\frac{1}{\sin^2 \phi}\frac{d^2 u}{d\phi^2} + \left(-\frac{\cos \phi}{\sin^3 \phi} + \frac{2 \cos \phi}{\sin^3 \phi}\right)\frac{du}{d\phi} + \frac{\lambda(\lambda + 1)}{\sin^2 \phi}u = 0.$$

Simplifying yields

$$\frac{d^2 u}{d\phi^2} + \cot \phi\frac{du}{d\phi} + \lambda(\lambda + 1)u = 0.$$

This can be put in the alternative form

$$\frac{1}{\sin\phi}\frac{d}{d\phi}\left(\sin\phi\frac{du}{d\phi}\right) + \lambda(\lambda + 1)u = 0.$$

Either is acceptable, although the latter form is more useful.   ∎

PROBLEMS

1. Show that $z = h(x) = \int \sqrt{p_1(x)}\, dx$ changes

$$\frac{d^2u}{dx^2} + p_0(x)\frac{du}{dx} + p_1(x)u = 0$$

into an equation with constant coefficients if

$$\frac{p_1'(x) + 2p_0(x)p_1(x)}{p_1^{3/2}(x)}$$

is constant.

Use the result of Problem 1 to find a general solution to each equation.

2. $u'' + \tan x u' + \cos^2 xu = 0$ **3.** $xu'' - 3u' + 16x^7u = 0$

4. $x^4u'' + x^2(2x - 3)u' + 2u = 0$ **5.** $2xu'' + (5x^2 - 2)u' + 2x^3u = 0$

6. $xu'' + (8x^2 - 1)u' + 20x^3u = 0$

7. Consider the change of independent variable $z = \int x^{k/2}\, dx$, $k \neq -2$, for the equation

$$u'' + \frac{1}{x}u' + x^ku = 0.$$

Show that the equation in z is

$$z^2u'' + zu' + z^2u = 0.$$

TABLE 1.1. Differential Equations

Differential Equation	Method of Solution
Separable equation: $f_1(x)g_1(u)\, dx + f_2(x)g_2(u)\, du = 0$	$\displaystyle\int \frac{f_1(x)}{f_2(x)}\, dx + \int \frac{g_2(u)}{g_1(u)}\, du = C$
Exact equation: $M(x, u)\, dx + N(x, u)\, du = 0,$ where $\partial M/\partial u = \partial N/\partial x.$	$\displaystyle\int M\, \partial x + \int \left(N - \frac{\partial}{\partial u}\int M\, \partial x\right) du = C,$ where ∂x indicates that the integration is to be performed with respect to x keeping u constant.

TABLE 1.1. Differential Equations (cont.)

Differential Equation	Method of Solution
Linear first-order equation: $$\frac{du}{dx} + p(x)u = g(x)$$	$$ue^{\int p\,dx} = \int ge^{\int p\,dx}\,dx + C$$
Bernoulli's equation: $$\frac{du}{dx} + p(x)u = g(x)u^n$$	$$te^{(1-n)\int p\,dx}$$ $$= (1-n)\int ge^{(1-n)\int p\,dx}\,dx + C,$$ where $t = u^{1-n}$. If $n = 1$, the solution is $$\ln u = \int (g - p)\,dx + C.$$
Homogeneous equation: $$\frac{du}{dx} = F\left(\frac{u}{x}\right)$$	$$\ln x = \int \frac{dt}{F(t)} - t + C,$$ where $t = u/x$. If $F(t) = t$, the solution is $u = Cx$.
Reducible to homogeneous: $$(a_1x + b_1u + c_1)\,dx$$ $$+ (a_2x + b_2u + c_2)\,du = 0$$ $$\frac{a_1}{a_2} \neq \frac{b_1}{b_2}$$	Set $y = a_1x + b_1u + c_1$ $t = a_2x + b_2u + c_2$. Eliminate x and u and the equation becomes homogeneous.
Reducible to separable: $$(a_1x + b_1u + c_1)\,dx$$ $$+ (a_2x + b_2u + c_2)\,du = 0$$ $$\frac{a_1}{a_2} = \frac{b_1}{b_2}$$	Set $y = a_1x + b_1u$. Eliminate x or u and the equation becomes separable.
First-order equation: $$uF(xu)\,dx + xG(xu)\,du = 0$$	$$\ln x = \int \frac{G(t)\,dt}{t[G(t) - F(t)]} + C,$$ where $t = xu$. If $G(t) = F(t)$, the solution is $xu = C$.
Linear, homogeneous second-order equation: $$\frac{d^2u}{dx^2} + a\frac{du}{dx} + bu = 0$$ a, b are real constants	Let m_1, m_2 be roots of $m^2 + am + b = 0$. Then there are three cases; Case 1. m_1, m_2 real and distinct: $$u = c_1e^{m_1x} + c_2e^{m_2x}$$ Case 2. m_1, m_2 real and equal: $$u = c_1e^{m_1x} + c_2xe^{m_2x}$$ Case 3. $m_1 = p + qi$, $m_2 = p - qi$: $$u = e^{px}(c_1 \cos qx + c_2 \sin qx),$$ where $p = -a/2$, $q = \sqrt{4b - a^2}/2$.

TABLE 1.1. *Differential Equations (cont.)*

Differential Equation	**Method of Solution**
Linear, nonhomogeneous *second-order equation:* $$\frac{d^2u}{dx^2} + a\frac{du}{dx} + bu = g(x)$$ *a, b* real constants	There are three cases corresponding to those immediately above: Case 1. $u = c_1 e^{m_1 x} + c_2 e^{m_2 x}$ $$+ \frac{e^{m_1 x}}{m_1 - m_2} \int e^{-m_1 x} g(x)\, dx$$ $$+ \frac{e^{m_2 x}}{m_2 - m_1} \int e^{-m_2 x} g(x)\, dx$$ Case 2. $u = c_1 e^{m_1 x} + c_2 x e^{m_1 x}$ $$+ x e^{m_1 x} \int e^{-m_1 x} g(x)\, dx$$ $$- e^{m_1 x} \int e^{-m_1 x} g(x)\, dx$$ Case 3. $u = e^{px}(c_1 \cos qx + c_2 \sin qx)$ $$+ \frac{e^{px} \sin qx}{q} \int e^{-px} g(x) \cos qx\, dx$$ $$- \frac{e^{px} \cos qx}{q} \int e^{-px} g(x) \sin qx\, dx$$
Cauchy–Euler equation: $$x^2\frac{d^2u}{dx^2} + ax\frac{du}{dx} + bu = g(x)$$ *a, b* real constants	Putting $x = e^t$, the equation becomes $$\frac{d^2u}{dt^2} + (a - 1)\frac{du}{dt} + bu = g(e^t)$$ and can then be solved as a linear second-order equation.
Bessel's equation: $$x^2\frac{d^2u}{dx^2} + x\frac{du}{dx} + (n^2x^2 - \lambda^2)u = 0$$	$u = c_1 J_\lambda(nx) + c_2 Y_\lambda(nx)$
Transformed Bessel's equation: $$x^2\frac{d^2u}{dx^2} + (2p + 1)x\frac{du}{dx}$$ $$+ (\alpha^2 x^{2r} + \beta^2)u = 0$$	$u = x^{-p}\left[c_1 J_{q/r}\left(\frac{\alpha}{r}x^r\right) + c_2 Y_{q/r}\left(\frac{\alpha}{r}x^r\right)\right],$ where $q = \sqrt{p^2 - \beta^2}$.
Legendre's equation: $$(1 - x^2)\frac{d^2u}{dx^2} - 2x\frac{du}{dx} + \lambda(\lambda + 1)u = 0$$	$u = c_1 P_\lambda(x) + c_2 Q_\lambda(x)$
Riccati's equation: $$\frac{du}{dx} + p(x)u + q(x)u^2 = g(x)$$	Set $u = y'/(qy)$, where $y' = dy/dx$. There results $$\frac{d^2y}{dx^2} + \left(p - \frac{q'}{q}\right)\frac{dy}{dx} - gqy = 0.$$ This second-order, linear equation is then solved.

TABLE 1.1. *Differential Equations* (*cont.*)

Differential Equation	Method of Solution
Error function equation:	
$\dfrac{d^2u}{dx^2} + 2x\,\dfrac{du}{dx} - 2n^u = 0$	$u = i^n \text{ erfc } x$
n integer	where $i^n \text{ erfc } x = \displaystyle\int_x^\infty i^{n-1} \text{ erfc } t \, dt$
	$i^0 \text{ erfc } x = \text{erfc } x$
	$\text{efrc } x = 1 - \text{erf } x$
	$\quad = \dfrac{2}{\sqrt{\pi}} \displaystyle\int_x^\infty e^{-t^2} \, dt$

2

Series Method

2.1 Introduction

We have studied linear differential equations with constant coefficients and have solved such equations using exponential functions. In general, a linear differential equation with variable coefficients cannot be solved in terms of exponential functions. We did, however, solve a special equation with variable coefficients, the Cauchy–Euler equation, by assuming a solution of the form x^n. A more general method will be presented that utilizes infinite sums of powers to obtain a solution.

2.2 *Properties of Power Series*

A *power series* is the sum of the infinite number of terms of the form $b_k(x - a)^k$ and is written

$$b_0 + b_1(x - a) + b_2(x - a)^2 + \cdots = \sum_{n=0}^{\infty} b_n(k - a)^n, \quad (2.2.1)$$

where $a, b_0, b_1, \ldots$ are constants. A power series does not include terms with negative or fractional powers. None of the following are power series:

(a) $1 + (x - 1) + (x - 1)(x - 2) + (x - 1)(x - 2)(x - 3) + \cdots$

(b) $1 + (x^2 - 1) + (x^2 - 1)^2 + (x^2 - 1)^3 + \cdots$

(c) $\dfrac{1}{x} + 1 + x + x^2 + \cdots$

(d) $x^{1/2} + x^{3/2} + x^{5/2} + \cdots$

There are several properties of power series that we will consider before we look at some examples illustrating their use. The sum s_m of the first m terms is

$$s_m = b_0 + b_1(x - a) + \cdots + b_m(x - a)^m \qquad (2.2.2)$$

and is called the mth *partial sum* of the series. The series *converges* at $x = x_0$ if

$$\lim_{m \to \infty} s_m(x_o) = \lim_{m \to \infty} [b_0 + b_1(x_0 - a) + \cdots + b_m(x_0 - a)^m] \qquad (2.2.3)$$

exists; otherwise, it diverges at $x = x_0$. Clearly, every power series converges at $x = a$. Usually, there is an interval over which the power series converges with midpoint at $x = a$. That is, the series converges for those x for which

$$|x - a| < R, \qquad (2.2.4)$$

where R is the *radius of convergence*. This radius is given by

$$\frac{1}{R} = \lim_{n \to \infty} \left| \frac{b_{n+1}}{b_n} \right| \qquad (2.2.5)$$

when this limit exists. This formula will not be developed here.

A function $f(x)$ is *analytic** at the point $x = a$ if it can be expressed as a power series $\Sigma_{n=0}^{\infty} b_n(x - a)^n$ with $R > 0$. (We use the terms "expressed," "expanded," and "represented" interchangeably.) It follows from techniques of elementary calculus that the coefficients in the series 2.2.1 are related to the derivatives of $f(x)$ at $x = a$ by the formula

$$b_n = \frac{1}{n!} f^{(n)}(a) \qquad (2.2.6)$$

for each n, and $\Sigma_{n=0}^{\infty} b_n(x - a)^n$ converges to $f(x)$ in $|x - a| < R$. We write

$$f(x) = \sum_{n=0}^{\infty} b_n(x - a)^n, \qquad |x - a| < R \qquad (2.2.7)$$

This power series is called the *Taylor series* of $f(x)$, expanded about the *center* $x = a$. If expanded about the special point $x = 0$, it may then be referred to as a *Maclaurin series*. Taylor series expansions of some well-known functions expanded about $x = 0$ are tabulated in Table 2.1.

The symbol $\binom{n}{k}$ is a convenient notation for the *binomial coefficient*

$$\binom{n}{k} = \frac{n!}{k! \, (n - k)!} = \frac{n(n - 1) \cdots (n - k + 1)}{k!}. \qquad (2.2.8)$$

It can be used whenever the expressions above occur.

---

*The term "regular" is often used synonymously with "analytic."

TABLE 2.1 Taylor Series Expansions of Some Simple Functions

$$\frac{1}{1-x} = 1 + x + x^2 + \cdots, \qquad\qquad |x| < 1$$

$$e^x = 1 + x + \frac{x^2}{2!} + \cdots, \qquad\qquad |x| < \infty$$

$$\sin x = x - \frac{x^3}{3!} + \frac{x^5}{5!} - \cdots, \qquad\qquad |x| < \infty$$

$$\cos x = 1 - \frac{x^2}{2!} + \frac{x^4}{4!} - \cdots, \qquad\qquad |x| < \infty$$

$$\ln(1+x) = x - \frac{x^2}{2} + \frac{x^3}{3} - \cdots, \qquad\qquad -1 \le x < 1$$

$$\sinh x = x + \frac{x^3}{3!} + \frac{x^5}{5!} + \cdots, \qquad\qquad |x| < \infty$$

$$\cosh x = 1 + \frac{x^2}{2!} + \frac{x^4}{4!} + \cdots, \qquad\qquad |x| < \infty$$

$$(1+x)^\alpha = 1 + \alpha x + \frac{1}{2!}\alpha(\alpha-1)x^2 + \cdots, \qquad\qquad |x| < 1$$

Two important properties of a power series are contained in the following theorem:

Theorem 2.1: *If*

$$f(x) = \sum_{n=0}^{\infty} b_n (x - a)^n, \tag{2.2.9}$$

then

$$f'(x) = \sum_{n=1}^{\infty} n b_n (x - a)^{n-1} \tag{2.2.10}$$

and

$$\int_a^x f(t)\, dt = \sum_{n=0}^{\infty} \frac{b_n}{n+1} (x - a)^{n+1}. \tag{2.2.11}$$

In words, if $f(x)$ is analytic at $x = a$ then $f'(x)$ and $\int f\, dx$ are also analytic at $x = a$ and their power-series expansions about the center $x = a$ may be obtained by term-by-term differentiation and integration, respectively. Note that R does not change.

Example 2.2.1:   Derive the Taylor series expansion of $\sin x$ about the center $x = 0$.

SOLUTION: For the function $f(x) = \sin x$, Eq. 2.2.6 yields the b_n so that

$$\sin x = \overset{0}{\cancel{\sin 0}} + x \cos 0 - \frac{x^2}{2!} \overset{0}{\cancel{\sin 0}} - \frac{x^3}{3!} \cos 0 + \cdots$$

$$= x - \frac{x^3}{3!} + \frac{x^3}{5!} - \cdots. \quad \blacksquare$$

Example 2.2.2:   By using the expansion in Table 2.1, find a series expansion for $1/(x^2 - 4)$.

SOLUTION: First we factor the given function,

$$\frac{1}{x^2 - 4} = \frac{1}{x - 2}\frac{1}{x + 2}.$$

Next, we write the fractions in the form

$$\frac{1}{x - 2} = -\frac{1}{2 - x} = -\frac{1}{2}\left(\frac{1}{1 - x/2}\right)$$

$$\frac{1}{x + 2} = \frac{1}{2 + x} = \frac{1}{2}\left(\frac{1}{1 + x/2}\right).$$

Now, we use the first expansion in Table 2.1, replacing x with $x/2$ for the first fraction and x with $(-x/2)$ for the second fraction. There results

$$\frac{1}{x - 2} = -\frac{1}{2}\left[1 + \frac{x}{2} + \left(\frac{x}{2}\right)^2 + \left(\frac{x}{2}\right)^3 + \cdots\right]$$

$$= -\frac{1}{2} - \frac{x}{4} - \frac{x^2}{8} - \frac{x^3}{16} - \cdots$$

$$\frac{1}{x + 2} = \frac{1}{2}\left[1 + \left(-\frac{x}{2}\right) + \left(-\frac{x}{2}\right)^2 + \left(-\frac{x}{2}\right)^3 + \cdots\right]$$

$$= \frac{1}{2} - \frac{x}{4} + \frac{x^2}{8} - \frac{x^3}{16} + \cdots.$$

Finally, we add these two series to obtain

$$\frac{1}{x - 2} + \frac{1}{x + 2} = \frac{2x}{x^2 - 4}$$

$$= -\frac{x}{2} - \frac{x^3}{8} - \frac{x^5}{32} - \cdots$$

$$\therefore \quad \frac{1}{x^2 - 4} = -\frac{1}{4}\left(1 + \frac{x^2}{4} + \frac{x^4}{16} + \cdots\right).$$

We could also have multiplied the two series to obtain the desired result.   ■

Example 2.2.3:   Find the Taylor series expansions for the functions $(1 - x)^{-k}$ for $k = 2, 3, \cdots$.

SOLUTION: We use the first expansion in Table 2.1 and repeated differentiation:

$$\frac{1}{1 - x} = 1 + x + x^2 + \cdots$$

$$\frac{d}{dx}\left(\frac{1}{1 - x}\right) = \frac{1}{(1 - x)^2} = 1 + 2x + 3x^2 + \cdots$$

$$\frac{d^2}{dx^2}\left(\frac{1}{1-x}\right) = \frac{2}{(1-x)^3} = 2 + 6x + 12x^2 + \cdots$$

$$\vdots$$

$$\frac{d^k}{dx^k}\left(\frac{1}{1-x}\right) = \frac{k!}{(1-x)^{k+1}} = \sum_{n=k}^{\infty} n(n-1)\cdots(n-k+1)x^{n-k}.$$

Therefore, for each $k = 1, 2, \cdots,$

$$\frac{1}{(1-x)^{k+1}} = \frac{1}{k!} \sum_{n=k}^{\infty} n(n-1)\cdots(n-k+1)x^{n-k}$$

or, using our special notation (Eq. 2.2.8),

$$\frac{1}{(1-x)^{k+1}} = \sum_{n=k}^{\infty} \binom{n}{k} x^{n-k} = \sum_{n=0}^{\infty} \binom{n+k}{k} x^n.$$

Now, replace k by $k - 1$ to obtain

$$\frac{1}{(1-x)^k} = \sum_{n=0}^{\infty} \binom{n+k-1}{k-1} x^n, \qquad k \geq 1$$

where, by convention,

$$\binom{n}{0} = 1. \quad \blacksquare$$

Example 2.2.4:   Find the Taylor series expansion of $\tan^{-1} x$ about $x = 0$.

SOLUTION: We know that

$$\tan^{-1} x = \int_0^x \frac{dt}{1 + t^2}.$$

Using Table 2.1, the function $1/(1 + t^2)$ is expanded about $x = 0$, obtaining

$$\frac{1}{1 + t^2} = 1 - t^2 + t^4 - \cdots.$$

Hence, we have

$$\tan^{-1} x = \int_0^x \frac{dt}{1 + t^2} = \int_0^x (1 - t^2 + t^4 - \cdots)\, dt = x - \frac{x^3}{3} + \frac{x^5}{5} - \cdots$$

by integrating. In fact, the series for $\tan^{-1} x$ converges at $x = 1$ by the alternating series test, whch states that a series converges if the signs of the terms of the series alternate, and the nth term tends monotonically to zero. We then have the interesting result that

$$\tan^{-1} 1 = \frac{\pi}{4} = 1 - \frac{1}{3} + \frac{1}{5} - \frac{1}{7} + \cdots. \quad \blacksquare$$

We will have occasion to need the first few coefficients of a Taylor series in cases in which $f^{(n)}(x)$ cannot be readily computed. Consider the following examples.

Example 2.2.5: Find the first three nonzero coefficients in the Taylor series expansion for $1/\cos x$ about $x = 0$.

SOLUTION: The function $\cos x$ is expanded about $x = 0$ as follows:

$$\frac{1}{\cos x} = \frac{1}{1 - x^2/2! + x^4/4! - \cdots}$$

$$= \frac{1}{1 - (x^2/2! - x^4/4! + \cdots)}.$$

Using the first series in Table 2.1 and replacing x which appears there with $(x^2/2! - x^4/4! + \cdots)$, there results

$$\frac{1}{\cos x} = 1 + \left(\frac{x^2}{2!} - \frac{x^4}{4!} + \cdots\right) + \left(\frac{x^2}{2!} - \frac{x^4}{4!} + \cdots\right)^2 + \cdots$$

$$= 1 + \frac{x^2}{2!} + \left(\frac{1}{(2!)^2} - \frac{1}{4!}\right)x^4 + \cdots$$

$$= 1 + \frac{x^2}{2} + \frac{5x^4}{24} + \cdots.$$

Note: To obtain the first three terms, we can ignore all but the square of the first term in $(x^2/2! - x^4/4! + \cdots)^2$ and all the terms in the higher powers of this series because they generate the coefficients of x^{2k}, $k > 2$. ∎

Example 2.2.6: Find the first three coefficients of the expansion of $e^{\sin x}$ about $x = 0$.

SOLUTION: We have

$$e^x = 1 + x + \frac{x^2}{2!} + \cdots$$

so that

$$e^{\sin x} = 1 + \sin x + \frac{\sin^2 x}{2} + \cdots$$

$$= 1 + \left(x - \frac{x^3}{3!} + \cdots\right) + \frac{1}{2}\left(x - \frac{x^3}{3!} + \cdots\right)^2 + \cdots$$

$$= 1 + x + \frac{x^2}{2} + \cdots.$$

Note that these are the same first terms for the series expansion of e^x. This is not surprising since for small x we know that x approximates $\sin x$. If we were to compute additional terms in the series, they would, of course, differ from those of the expansion of e^x. ∎

We conclude this section by studying a more convenient and somewhat simpler method for determining the radius of convergence than determining the limit in Eq. 2.2.5. A point is a *singularity* of $f(x)$ if $f(x)$ is not analytic there. For instance, $x = 0$ is a singularity of each of the functions $1/x$, $\ln x$, $\sqrt{x}$, and $|x|$. Locate all the singularities of a proposed $f(x)$ in the complex plane. In so doing, consider x to be a complex variable with real and imaginary parts. As an example, consider the function $x/[(x^2 + 9)(x - 6)]$. It has singularities at the following points: $x = 6$, $3i$, $-3i$. The singular points are plotted in Fig. 2.1a. If we expand about the origin, the radius of convergence is established by drawing a circle, with center at the origin, passing through the nearest singular point, as shown in Fig. 2.1b. This gives $R = 3$, a rather surprising result, since the first singular point on the x axis is at $x = 6$. The singularity at $x = 3i$ prevents the series from converging for $x \geq 3$. If we expand about $x = 5$, that is, in powers of $(x - 5)$, the nearest singularity would be located at $(6, 0)$. This would give a radius of convergence of $R = 1$ and the series would converge for $6 > x > 4$. It is for this reason that $\sin x$, $\cos x$, $\sinh x$, $\cosh x$, and e^x have $R = \infty$; no singularities exist for these functions (technically, we should say no singularities in the finite plane).

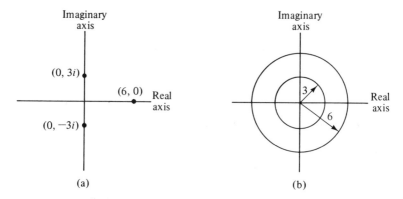

FIGURE 2.1.   Singular points and convergence regions of the function $x/[(x^2 + 9)(x - 6)]$.

If the functions $p_0(x)$ and $p_1(x)$ in

$$\frac{d^2u}{dx^2} + p_0(x)\frac{du}{dx} + p_1(x)u = 0 \qquad (2.2.12)$$

are analytic at $x = a$, then $x = a$ is an *ordinary point* of the equation; otherwise, it is a *singular point* of the equation. Thus, $x = a$ is a singular point of the equation if it is a singular point of either $p_0(x)$ or $p_1(x)$.

Theorem 2.2:   *If $x = 0$ is an ordinary point of Eq. 2.2.12, then there exists a pair of basic solutions*

$$u_1(x) = \sum_{n=0}^{\infty} a_n x^n, \qquad u_2(x) = \sum_{n=0}^{\infty} b_n x^n \qquad (2.2.13)$$

in which the series converges in $|x| < R$. *The radius of convergence is at least as large as the distance from the origin to the singularity of* $p_0(x)$ *or* $p_1(x)$ *closest to the origin.*

It then follows immediately that if $p_0(x)$ and $p_1(x)$ are polynomials, the series representations of the solutions converge for all x.

Suppose that $p_0(x)$ or $p_1(x)$ is the function of Fig. 2.1. And suppose that we are interested in the series solution in the interval $3 < x < 6$. For that situation we could expand about the point $x_0 = 4.5$, halfway between 3 and 6, and express the basic solutions as

$$u_1(x) = \sum_{n=0}^{\infty} a_n(x - x_0)^n, \qquad u_2(x) = \sum_{n=0}^{\infty} b_n(x - x_0)^n. \quad (2.2.14)$$

Or, we could transform the independent variable from x to t using $t = x - x_0$. Then, the solutions are

$$u_1(t) = \sum_{n=0}^{\infty} a_n t^n, \qquad u_2(t) = \sum_{n=0}^{\infty} b_n t^n. \quad (2.2.15)$$

The final result could then be expressed in terms of x by letting $t = x - x_0$.

PROBLEMS

Derive a power series expansion of each function by expanding in a Taylor series about $a = 0$.

1. $\dfrac{1}{1 - x}$ **2.** e^x **3.** $\sin x$

4. $\cos x$ **5.** $\ln x$ **6.** $\ln (1 + x)$

7. $\dfrac{1}{1 + x}$ **8.** $\dfrac{1}{x + 2}$ **9.** $\dfrac{1}{x^2 + 3x + 2}$

10. $\dfrac{7}{x^2 - x - 12}$ **11.** e^{2x+1} **12.** e^{-x^2}

13. $\sin x^2$ **14.** $\tan x$ **15.** $\ln \dfrac{x + 1}{2}$

16. $\ln \dfrac{4 - x^2}{4}$ **17.** $\dfrac{e^x}{x + 4}$ **18.** $e^{-x} \sin x$

Find a power series expansion for each integral by first expanding the integrand about $a = 0$.

19. $\displaystyle\int_0^x \dfrac{dt}{1 + t}$ **20.** $\displaystyle\int_0^x \dfrac{dt}{4 - t^2}$ **21.** $\displaystyle\int_0^x \dfrac{t\,dt}{1 + t^2}$

22. $\displaystyle\int \sin^2 x\,dx$ **23.** $\displaystyle\int \tan x\,dx$ **24.** $\displaystyle\int \sin x \cos x\,dx$

25. The function $(x^2 - 1)/[(x - 4)(x^2 + 1)]$ is to be expanded in a power series about (a) the origin, (b) the point $a = 1$, and (c) the point $a = 2$. Determine the radius of convergence for each expansion.

For each equation, list all singular points and determine the radius of convergence if we expand about the origin.

26. $\dfrac{d^2u}{dx^2} + (x^2 - 1)u = x^2$

27. $(x^2 - 1)\dfrac{d^2u}{dx^2} + u = x^2$

28. $x(x^2 + 4)\dfrac{d^2u}{dx^2} + x\dfrac{du}{dx} = 0$

29. $\dfrac{d^2u}{dx^2} + xu = \dfrac{1}{1 - x}$

30. $\dfrac{d^2u}{dx^2} + \dfrac{x - 1}{x + 1}\dfrac{du}{dx} + u = 0$

31. $\cos x\dfrac{d^2u}{dx^2} + u = \sin x$

Determine the radius of convergence for each series.

32. $\displaystyle\sum_{n=0}^{\infty} x^n$

33. $\displaystyle\sum_{n=0}^{\infty} \dfrac{1}{n!}x^n$

34. $\displaystyle\sum_{n=0}^{\infty} \dfrac{n(n - 1)}{2^n}x^n$

35. $\displaystyle\sum_{n=0}^{\infty} 2^n x^n$

36. $\displaystyle\sum_{n=0}^{\infty} \dfrac{1}{n!}(x - 2)^n$

37. $\displaystyle\sum_{n=0}^{\infty} \dfrac{(-1)^n}{(2n)!}(x - 1)^n$

Find a series expansion about $a = 1$ for each function.

38. $\dfrac{1}{x}$

39. $\dfrac{1}{x(x - 2)}$

40. $\dfrac{1}{x^2 - 4}$

41. $\dfrac{1}{x(x^2 + 4x + 4)}$

2.3   *Solutions of Ordinary Differential Equations*

The existence of the power series solutions of

$$\frac{d^2u}{dx^2} + p_0(x)\frac{du}{dx} + p_1(x)u = 0, \qquad |x| < R \qquad (2.3.1)$$

is guaranteed by Theorem 2.2. In this section we show how to obtain the coefficients of the series. The method is best explained by using a specific example. Let us solve the differential equation

$$\frac{d^2u}{dx^2} + x^2u = 0 \qquad (2.3.2)$$

Using power series, assume that

$$u(x) = \sum_{n=0}^{\infty} b_n x^n. \qquad (2.3.3)$$

Substitute into the given differential equation and find

$$\sum_{n=2}^{\infty} n(n-1)b_n x^{n-2} + x^2 \sum_{n=0}^{\infty} b_n x^n = 0. \tag{2.3.4}$$

Let $n - 2 = m$ in the first series and multiply the x^2 into the second series. Then

$$\sum_{m=0}^{\infty} (m+2)(m+1)b_{m+2} x^m + \sum_{n=0}^{\infty} b_n x^{2+n} = 0. \tag{2.3.5}$$

Now let $n + 2 = m$ in the second series. We have

$$\sum_{m=0}^{\infty} (m+2)(m+1)b_{m+2} x^m + \sum_{m=2}^{\infty} b_{m-2} x^m = 0. \tag{2.3.6}$$

The first series starts at $m = 0$, but the second starts at $m = 2$. Thus, in order to add the series, we must extract the first two terms from the first series. There results, letting $m = n$,

$$2b_2 + 6b_3 x + \sum_{n=2}^{\infty} (n+2)(n+1)b_{n+2} x^n + \sum_{n=2}^{\infty} b_{n-2} x^n = 0. \tag{2.3.7}$$

Now we can combine the two series, resulting in

$$2b_2 + 6b_3 x + \sum_{n=2}^{\infty} [(n+2)(n+1)b_{n+2} + b_{n-2}]x^n = 0. \tag{2.3.8}$$

Equating coefficients of the various powers of x gives

$$x^0: \qquad 2b_2 = 0 \qquad\qquad \therefore b_2 = 0$$

$$x^1: \qquad 6b_3 = 0 \qquad\qquad \therefore b_3 = 0$$

$$x^n: \qquad (n+2)(n+1)b_{n+2} + b_{n-2} = 0$$

$$\therefore b_{n+2} = -\frac{b_{n-2}}{(n+2)(n+1)}, \qquad n \geq 2. \tag{2.3.9}$$

With $b_2 = 0$, Eq. 2.3.9 implies that $b_6 = b_{10} = b_{14} = \cdots = 0$; with $b_3 = 0$, we have $b_7 = b_{11} = b_{15} = \cdots = 0$. Equation 2.3.9 also implies a relationship between $b_0, b_4, b_8, \cdots$ and $b_1, b_5, b_9, \cdots$. We call Eq. 2.3.9 a *two-term recursion* and digress to explore a simple technique for obtaining its solution.

Consider the following tabulation:

1	$n = 2$	$b_4 = -\dfrac{b_0}{4 \cdot 3}$
2	$n = 6$	$b_8 = -\dfrac{b_4}{8 \cdot 7}$

$$
\begin{array}{ccc}
\vdots & \vdots & \vdots \\
k & n = 4k - 2 & b_{4k} = -\dfrac{b_{4k-4}}{4k(4k - 1)}.
\end{array}
$$

The kth line provides a general formula for computing any given line. The first line in the table is obtained by setting $k = 1$ in the kth line. In fact, the kth line is obtained by generalizing from the first two (or three) lines. Line 2 is constructed so that Eq. 2.3.9 has b_4 on its right-hand side. Therefore, in line 2, $n = 6$. If the pattern is obvious, we jump to the kth line; if not, we try line 3 and continue until the general line can be written. Once the table is completed we multiply all the equations in the third column:

$$
b_4 \cdot b_8 \cdots b_{4k-4} \cdot b_{4k} = (-1)^k \frac{b_0 \cdot b_4 \cdot b_8 \cdots b_{4k-4}}{3 \cdot 4 \cdot 7 \cdot 8 \cdots (4k - 1)4k}. \tag{2.3.10}
$$

We cancel $b_4, b_8, \cdots, b_{4k-4}$ from both sides to obtain

$$
b_{4k} = (-1)^k \frac{b_0}{3 \cdot 4 \cdot 7 \cdot 8 \cdots (4k - 1)4k} \tag{2.3.11}
$$

for $k = 1, 2, 3, \cdots$. Since Eq. 2.3.11 expresses each coefficient $b_4, b_8, b_{12}, \cdots$ as a function of b_0, we call Eq. 2.3.11 a *solution of recursion* 2.3.9.

The solution of the recursion leads directly to a solution of Eq. 2.3.3. We choose $b_0 = 1$ without loss of generality and find

$$
u_1(x) = 1 + \sum_{n=1}^{\infty} b_{4k} x^{4k} = 1 + \sum_{k=1}^{\infty} \frac{(-1)^k x^{4k}}{3 \cdot 4 \cdot 7 \cdot 8 \cdots (4k - 1)4k}. \tag{2.3.12}
$$

Having taken care of b_0 we now explore how $b_5, b_9, b_{13}, \cdots$ are related to b_1. This provides us with the first line in our table so that the table takes the form

$$
\begin{array}{ccc}
1 & n = 3 & b_5 = -\dfrac{b_1}{5 \cdot 4} \\[2mm]
2 & n = 7 & b_9 = -\dfrac{b_5}{9 \cdot 8} \\[2mm]
\vdots & \vdots & \vdots \\[2mm]
k & n = 4k - 1 & b_{4k+1} = -\dfrac{b_{4k-3}}{(4k + 1)4k}.
\end{array}
$$

Again, a multiplication and cancellation yield a second solution of the recursion:

$$
b_{4k+1} = (-1)^k \frac{b_1}{4 \cdot 5 \cdot 8 \cdot 9 \cdots 4k(4k + 1)}, \tag{2.3.13}
$$

$k = 1, 2, 3, \cdots$. Now set $b_1 = 1$ and from Eq. 2.3.3

$$u_2(x) = x + \sum_{k=1}^{\infty} \frac{(-1)^k x^{4k+1}}{4 \cdot 5 \cdot 8 \cdot 9 \cdots 14k(4k + 1)}. \qquad (2.3.14)$$

The solutions $u_1(x)$ and $u_2(x)$ form a basic set because

$$W(x; u_1, u_2) = \begin{vmatrix} u_1(x) & u_2(x) \\ u_1'(x) & u_2'(x) \end{vmatrix}$$

implies that

$$W(0; u_1(0), u_2(0)) = \begin{vmatrix} 1 & 0 \\ 0 & 1 \end{vmatrix} = 1 \qquad (2.3.15)$$

and, by Theorem 1.5, that $W(x) > 0$. The general solution is

$$u(x) = c_1 u_1(x) + c_2 u_2(x)$$

$$= c_1 \left[1 + \sum_{k=1}^{\infty} \frac{(-1)^k x^{4k}}{3 \cdot 4 \cdot 7 \cdot 8 \cdots (4k - 1)4k} \right]$$

$$+ c_2 \left[x + \sum_{k=1}^{\infty} \frac{(-1)^k x^{4k+1}}{4 \cdot 5 \cdot 8 \cdot 9 \cdots 4k(4k + 1)} \right]. \qquad (2.3.16)$$

We often expand the above, showing three terms in each series, as

$$u(x) = c_1 \left(1 - \frac{x^4}{12} + \frac{x^8}{672} + \cdots \right) + c_2 \left(x - \frac{x^5}{20} + \frac{x^9}{1440} + \cdots \right).$$

$$(2.3.17)$$

Since $p_0(x) = 0$ and $p_1(x) = x^2$, the two series converge for all x by Theorem 2.2. The ratio* test also establishes this conclusion.

We could have found the first three terms in the expressions for $u_1(x)$ and $u_2(x)$, as shown in Eq. 2.3.17, without actually solving the recursion. The advantage of having the general term in the expansion of the solution is both theoretical and practical.

Example 2.3.1:   Find a power series solution to the initial-value problem

$$\frac{d^2 u}{dx^2} + 9u = 0, \qquad u(0) = 1, \quad u'(0) = 0.$$

SOLUTION: Assume the solution to be the power series.

$$u(x) = \sum_{n=0}^{\infty} b_n x^n.$$

---

*The ratio test for convergence states that if the absolute value of the ratio of the $(n + 1)$th term to the nth term approaches a limit r, then the series converges if $r < 1$.

The second derivative is

$$\frac{d^2u}{dx^2} = \sum_{n=2}^{\infty} n(n-1)b_n x^{n-2}.$$

Substitute these back into the given differential equation to get

$$\sum_{n=2}^{\infty} n(n-1)b_n x^{n-2} + \sum_{n=0}^{\infty} 9b_n x^n = 0.$$

In the leftmost series replace n by $n + 2$:

$$\sum_{n=0}^{\infty} (n+2)(n+1)b_{n+2} x^n + \sum_{n=0}^{\infty} 9b_n x^n = \sum_{n=0}^{\infty} [(n+2)(n+1)b_{n+2} + 9b_n]x^n = 0.$$

Now, for this equation to be satisfied for all x we demand that every coefficient of each power of x be zero. That is,

$$x^0: \qquad 2b_2 + 9b_0 = 0$$

$$x^1: \qquad 6b_3 + 9b_1 = 0$$

$$\vdots$$

$$x^n: \qquad (n+2)(n+1)b_{n+2} + 9b_n = 0.$$

Since $u(0) = b_0 = 1$ and $u'(0) = b_1 = 0$, we deduce that

$$b_1 = b_3 = b_5 = \cdots = 0$$

and that $b_2, b_4, b_6, \cdots$ are determined by $b_0 = 1$. The two-term recursion forces $n = 0$ in line 1 and $n = 2$ in line 2. Specifically, our table takes the following form:

1	$n = 0$	$b_2 = \dfrac{-9}{2 \cdot 1}$
2	$n = 2$	$b_4 = \dfrac{-9b_2}{4 \cdot 3}$
$\vdots$	$\vdots$	$\vdots$
k	$n = 2k - 2$	$b_{2k} = \dfrac{-9b_{2k-2}}{(2k-1)2k}.$

Hence, multiplying all equations in the third column and simplifying, we obtain

$$b_{2k} = \frac{(-9)^k}{(2k)!}.$$

Using $n = 2k$ and $k = 1, 2, 3, \cdots$, in our original expansion, we obtain the result

$$u(x) = 1 + \sum_{k=1}^{\infty} \frac{(-9)^k x^{2k}}{(2k)!}.$$

This can be written in the equivalent form

$$u(x) = 1 + \sum_{k=1}^{\infty} \frac{(-1)^k (3x)^{2k}}{(2k)!}.$$

In expanded form

$$u(x) = 1 - \frac{(3x)^2}{2!} + \frac{(3x)^4}{4!} - \cdots .$$

This is recognized as

$$u(x) = \cos 3x,$$

which is the solution we would expect using the methods of Chapter 1. It is not always possible, however, to put the power-series solution in a form that is recognizable as a well-known function. The solution is usually left in series form with the first few terms written explicitly. ■

PROBLEMS

Solve each differential equation for a general solution using the power-series method by expanding about $a = 0$. Note the radius of convergence for each solution.

1. $\dfrac{du}{dx} + u = 0$

2. $\dfrac{du}{dx} + ku = 0$

3. $(1 - x)\dfrac{du}{dx} + u = 0$

4. $\dfrac{du}{dx} + xu = 0$

5. $\dfrac{d^2 u}{dx^2} - 4u = 0$

6. $(x^2 - 1)\dfrac{d^2 u}{dx^2} - 4u = 0$

7. $\dfrac{d^2 u}{dx^2} + 2\dfrac{du}{dx} + u = 0$

8. $\dfrac{d^2 u}{dx^2} + 6\dfrac{du}{dx} + 5u = 0$

Find a specific solution to each differential equation by expanding about $a = 0$. State the limits of convergence for each series.

9. $x\dfrac{du}{dx} + u \sin x = 0,$ \qquad\qquad $u(0) = 1$

10. $(4 - x^2)\dfrac{d^2 u}{dx^2} + 2u = 0$ \qquad\qquad $u(0) = 0, \quad \dfrac{du}{dx}(0) = 1$

11. $\dfrac{d^2 u}{dx^2} + (1 - x)u = 0$ \qquad\qquad $u(0) = 1, \quad \dfrac{du}{dx}(0) = 0$

12. $\dfrac{d^2 u}{dx^2} - x^2\dfrac{du}{dx} + u \sin x = 0$ \qquad\qquad $u(0) = 0, \quad \dfrac{du}{dx}(0) = 1$

13. Solve $(1 - x)\, df/dx - f = 0$ using a power series expansion. Let $f = 6$ for $x = 0$, and expand about $x = 0$. Obtain five terms in the series and compare with the exact solution for values of $x = 0, \frac{1}{4}, \frac{1}{2}, 1,$ and 2.

14. The solution to $(1 - x)\, df/dx - f = 0$ is desired in the interval from $x = 1$ to $x = 2$. Expand about $a = 2$ and determine the value of $f(x)$ at $x = 1.9$ if $f(2) = 1$. Compare with the exact solution.

Find a general solution of each differential equation by expanding about the point specified.

15. $(x - 2)\dfrac{d^2u}{dx^2} + u = 0$ about $a = 1$

16. $x^2\dfrac{d^2u}{dx^2} + u = 0$ about $a = 1$

17. $\dfrac{d^2u}{dx^2} + xu = 0$ about $a = 2$

18. Solve the differential equation $(d^2u/dx^2) + x^2u = 0$ using the power-series method if $u(0) = 4$ and $u'(0) = -2$. Find an approximate value for $u(x)$ at $a = 2$.

19. Solve the differential equation $x^2(d^2u/dx^2) + 4u = 0$ by expanding about the point $x = 2$. Find an approximate value for $u(3)$ if $u(2) = 2$ and $u'(2) = 4$.

20. If $x(d^2u/dx^2) + (x - 1)u = 0$ find approximate values for $u(x)$ at $x = 1$ and at $x = 3$. We know that $u(2) = 10$ and $u'(2) = 0$.

2.3.1   LEGENDRE'S EQUATION

A differential equation that attracts much attention in the solution of a number of physical problems is *Legendre's equation,*

$$(1 - x^2)\frac{d^2u}{dx^2} - 2x\frac{du}{dx} + \lambda(\lambda + 1)u = 0. \qquad (2.3.18)$$

It is encountered most often when modeling a phenomenon in spherical coordinates. The parameter λ is a nonnegative, real constant.* Legendre's equation is written in standard form as

$$\frac{d^2u}{dx^2} - \frac{2x}{1 - x^2}\frac{du}{dx} + \frac{\lambda(\lambda + 1)}{1 - x^2}u = 0. \qquad (2.3.19)$$

The variable coefficients can be expressed as a power series about the origin and thus are analytic at $x = 0$. They are not analytic at $x = \pm 1$. Let us find the power-series solution of Legendre's equation valid for $-1 < x < 1$.

Assume a power-series solution

$$u(x) = \sum_{n=0}^{\infty} b_n x^n. \qquad (2.3.20)$$

Substitute into Eq. (2.3.18) and let $\lambda(\lambda + 1) = \alpha$. Then

---

*The solution for a negative value of λ, say λ_n, is the same as that for $\lambda = -(\lambda_n + 1)$; hence, it is sufficient to consider only nonnegative values.

$$(1 - x^2)\sum_{n=2}^{\infty} n(n - 1)b_n x^{n-2} - 2x\sum_{n=1}^{\infty} nb_n x^{n-1} + \alpha \sum_{n=0}^{\infty} b_n x^n = 0. \quad (2.3.21)$$

This can be written as

$$\sum_{n=2}^{\infty} n(n - 1)b_n x^{n-2} - \sum_{n=2}^{\infty} n(n - 1)b_n x^n - \sum_{n=1}^{\infty} 2nb_n x^n + \alpha \sum_{n=0}^{\infty} \alpha b_n x^n = 0.$$

$$(2.3.22)$$

The first sum can be rewritten as

$$\sum_{n=2}^{\infty} n(n - 1)b_n x^{n-2} = \sum_{n=0}^{\infty} (n + 2)(n + 1)b_{n+2} x^n. \quad (2.3.23)$$

Then, extracting the terms for $n = 0$ and $n = 1$, Eq. 2.3.22 becomes

$$\sum_{n=2}^{\infty} \{(n + 2)(n + 1)b_{n+2} - [n(n - 1) + 2n - \alpha]b_n\}x^n$$

$$+ 2b_2 + \alpha b_0 + (6b_3 - 2b_1 + \alpha b_1)x = 0. \quad (2.3.24)$$

Equating coefficients of like powers of x to zero, we find that

$$b_2 = -\frac{\alpha}{2}b_0$$

$$b_3 = \frac{2 - \alpha}{6}b_1 \quad (2.3.25)$$

$$b_{n+2} = \frac{n^2 + n - \alpha}{(n + 2)(n + 1)}b_n, \qquad n = 2, 3, 4, \cdots .$$

Substituting $\lambda(\lambda + 1) = \alpha$ back into the coefficients, we have

$$b_{n+2} = \frac{(n - \lambda)(n + \lambda + 1)}{(n + 2)(n + 1)}b_n, \qquad n = 2, 3, 4 \cdots . \quad (2.3.26)$$

There are two arbitrary coefficients b_0 and b_1. The coefficients with even subscripts can be expressed in terms of b_0 and those with odd subscripts in terms of b_1. The solution can then be written as

$$u(x) = b_0 u_1(x) + b_1 u_2(x), \quad (2.3.27)$$

where

$$u_1(x) = 1 - \frac{\lambda(\lambda + 1)}{2!}x^2 + \frac{(\lambda - 2)\lambda(\lambda + 1)(\lambda + 3)}{4!}x^4 + \cdots$$

$$(2.3.28)$$

$$u_2(x) = x - \frac{(\lambda - 1)(\lambda + 2)}{3!}x^3 + \frac{(\lambda - 3)(\lambda - 1)(\lambda + 2)(\lambda + 4)}{5!}x^5 + \cdots$$

are the two independent solutions.

We can solve the two-term recursion in Eq. 2.3.25 by the technique of Section 2.2—and the student is asked to do so in the homework problems but the resulting expression is not convenient. Temporarily, we are content to leave the answer in the forms of Eqs. 2.3.28.

2.3.2   LEGENDRE POLYNOMIALS AND FUNCTIONS

Let us investigate the solution $u_1(x)$ and $u_2(x)$ (Eqs. 2.3.28), for various positive values of λ. If λ is an integer,

$$\lambda = 0, \qquad u_1(x) = 1$$
$$\lambda = 2, \qquad u_1(x) = 1 - 3x^2 \qquad\qquad\qquad (2.3.29)$$
$$\lambda = 4, \qquad u_1(x) = 1 - 10x^2 + \tfrac{35}{3}x^4, \qquad \text{etc.}$$

All the higher-power terms contain factors that are zero. Thus, only polynomials result. For odd integers,

$$\lambda = 1, \qquad u_2(x) = x$$
$$\lambda = 3, \qquad u_2(x) = x - \tfrac{5}{3}x^3 \qquad\qquad\qquad (2.3.30)$$
$$\lambda = 5, \qquad u_2(x) = x - \tfrac{14}{3}x^3 + \tfrac{21}{5}x^5, \qquad \text{etc.}$$

The polynomials above represent independent solutions to Legendre's equation for the various λ's indicated; that is, if $\lambda = 5$, one independent solution is $x - \tfrac{14}{3}x^3 + \tfrac{21}{5}x^5$. Obviously, if $u_1(x)$ is a solution to the differential equation, then $Cu_1(x)$, where C is a constant, is also a solution. We shall choose the constant C such that the polynomials above all have the value unity at $x = 1$. If we do that, the polynomials are called *Legendre polynomials.* Several are

$$P_0(x) = 1, \qquad\qquad\qquad\qquad P_1(x) = x$$
$$P_2(x) = \tfrac{1}{2}(3x^2 - 1), \qquad\qquad P_3(x) = \tfrac{1}{2}(5x^3 - 3x) \qquad (2.3.31)$$
$$P_4(x) = \tfrac{1}{8}(35x^4 - 30x^2 + 3), \qquad P_5(x) = \tfrac{1}{8}(63x^5 - 70x^3 + 15x).$$

We can write Legendre polynomials in the general form

$$P_\lambda(x) = \sum_{n=0}^{N} (-1)^n \frac{(2\lambda - 2n)!}{2^\lambda n!(\lambda - n)!(\lambda - 2n)!} x^{\lambda - 2n}, \qquad (2.3.32)$$

where $N = \lambda/2$ if λ is even and $N = (\lambda - 1)/2$ if λ is odd. Some Legendre polynomials are sketched in Fig. 2.2.

When λ is an even integer, $u_2(x)$ has the form of an infinite series, and when λ is an odd integer, $u_1(x)$ is expressed as an infinite series. *Legendre's functions of the second kind* are multiples of the infinite series defined by

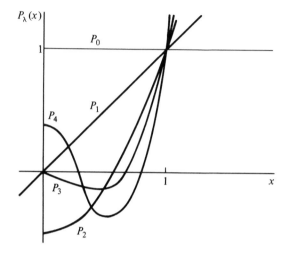

FIGURE 2.2. Legendre polynomials.

$$Q_\lambda(x) = \begin{cases} u_1(1)u_2(x), & \lambda \text{ even} \\ -u_2(1)u_1(x), & \lambda \text{ odd.} \end{cases} \tag{2.3.33}$$

The general solution of Legendre's equation is now written as

$$u(x) = c_1 P_\lambda(x) + c_2 Q_\lambda(x). \tag{2.3.34}$$

Several Legendre functions of the second kind can be shown, by involved manipulation, to be

$$\begin{aligned}
Q_0(x) &= \frac{1}{2} \ln \frac{1+x}{1-x} \\
Q_1(x) &= xQ_0(x) - 1 \\
Q_2(x) &= P_2(x)Q_0(x) - \tfrac{3}{2}x \\
Q_3(x) &= P_3(x)Q_0(x) - \tfrac{5}{2}x^2 + \tfrac{2}{3} \\
Q_4(x) &= P_4(x)Q_0(x) - \tfrac{35}{8}x^3 + \tfrac{55}{24}x \\
Q_5(x) &= P_5(x)Q_0(x) - \tfrac{63}{8}x^4 + \tfrac{49}{8}x^2 - \tfrac{8}{15}.
\end{aligned} \tag{2.3.35}$$

Note that all the functions are singular at the point $x = 1$, since $Q_0(x) \to \infty$ as $x \to 1$, and thus the functions above are valid only for $|x| < 1$.

If we make the change of variables $x = \cos \phi$, we transform Legendre's equation 2.3.18 into*

---

*This is a change of independent variable (see Section 1.12.3, Example 1.12.4).

$$\frac{d^2u}{d\phi^2} + \cot \phi \frac{du}{d\phi} + \lambda(\lambda + 1)u = 0 \qquad (2.3.36)$$

or, equivalently,

$$\frac{1}{\sin \phi} \frac{d}{d\phi}\left(\sin \phi \frac{du}{d\phi}\right) + \lambda(\lambda + 1)u = 0. \qquad (2.3.37)$$

Legendre's equations of this form arise in various physical problems in which spherical coordinates are used.

Example 2.3.2: Find the specific solution to the differential equation

$$(1 - x^2)\frac{d^2u}{dx^2} - 2x\frac{du}{dx} + 12u = 0 \qquad \text{if} \quad u'(0) = 4$$

and the function $u(x)$ is well behaved at $x = 1$. The latter condition is often imposed in physical situations.

SOLUTION: We note that the given differential equation is Legendre's equation with λ determined from

$$\lambda(\lambda + 1) = 12.$$

This can be written as

$$(\lambda + 4)(\lambda - 3) = 0,$$

giving

$$\lambda = -4, 3.$$

We choose the positive root and write the general solution as

$$u(x) = c_1 P_3(x) + c_2 Q_3(x).$$

If the function is to be well behaved at $x = 1$, we must let $c_2 = 0$, since $Q_3(1)$ is not defined. The other condition gives

$$4 = c_1 P_3'(0) = -\tfrac{3}{2}c_1 \quad \text{or} \quad c_1 = -\tfrac{8}{3}.$$

The solution is then

$$u(x) = -\tfrac{4}{3}(5x^3 - 3x). \quad \blacksquare$$

PROBLEMS

1. Verify by substitution that the Legendre polynomials of Eqs. 2.3.31 satisfy Legendre's equation.

2. Write an expression for $P_8(x)$.

Show that

3. $P_\lambda(-x) = (-1)^\lambda P_\lambda(x)$

4. $\dfrac{dP_\lambda}{dx}(-x) = (-1)^{\lambda+1}\dfrac{dP_\lambda}{dx}(x)$

5. Verify that the formula

$$P_\lambda(x) = \frac{1}{2^\lambda \lambda!}\frac{d^\lambda}{dx^\lambda}(x^2 - 1)^\lambda$$

yields the first four Legendre polynomials. This is known as the *Rodrigues formula* and can be used for all Legendre polynomials with λ a positive integer.

6. Verify the formulas

$$\frac{dP_{\lambda+1}}{dx} - \frac{dP_{\lambda-1}}{dx} = (2\lambda + 1)P_\lambda$$

$$\int_x^1 P_\lambda(x)\, dx = \frac{1}{2\lambda + 1}[P_{\lambda-1}(x) - P_{\lambda+1}(x)]$$

for $\lambda = 2$ and $\lambda = 4$.

Determine the general solution for each differential equation valid near the origin.

7. $(1 - x^2)\dfrac{d^2u}{dx^2} - 2x\dfrac{du}{dx} + 12u = 0$

8. $(1 - x^2)\dfrac{d^2u}{dx^2} - 2x\dfrac{du}{dx} + 6u = x$

9. $4(1 - x^2)\dfrac{d^2u}{dx^2} - 8x\dfrac{du}{dx} + 3u = 0$

10. $\dfrac{1}{\sin\phi}\dfrac{d}{d\phi}\left(\sin\phi\dfrac{du}{d\phi}\right) + 6u = 0$ (*Hint:* Let $x = \cos\phi$.)

11. Find the specific solution to the differential equation

$$(1 - x^2)\frac{d^2u}{dx^2} - 2x\frac{du}{dx} + 20u = 14x^2.$$

At $x = 0$, $u = 3$ and the function has a finite value at $x = 1$.

12. Expand $(1 - 2xt + t^2)^{-1/2}$ in powers of t. Set

$$(1 - 2xt + t^2)^{-1/2} = \sum_{n=0}^{\infty} P_n(x)t^n.$$

Show that $P_n(x)$ is the Legendre polynomial of degree n. (*Hint:* Use Table 2.1.)

Use the result in Problem 12 to show that

13. $P_n(-x) = (-1)^n P_n(x)$

14. $P_n(1) = 1,$ $P_n(-1) = (-1)^n$

15. $P_{2n+1}(0) = 0, \quad P_{2n}(0) = \dfrac{(-1)^n 1 \cdot 3 \cdot 5 \cdots (2n-1)}{2^n n!}$

16. Use the Rodrigues formula in Problem 5 and integration by parts to show that

(a) $\displaystyle\int_{-1}^{1} P_n(x)P_m(x)\, dx = 0, \quad n \neq m$ (b) $\displaystyle\int_{-1}^{1} P_n^2(x)\, dx = \dfrac{2}{2n+1}$

2.3.3   HERMITE POLYNOMIALS

The equation

$$\frac{d^2u}{dx^2} - 2x\frac{du}{dx} + 2\lambda u = 0 \tag{2.3.38}$$

provides another "classical" set of solutions known as *Hermite polynomials* when λ is a nonnegative integer. These polynomials play a significant role in statistics. To find a solution, set

$$u(x) = \sum_{n=0}^{\infty} b_n x^n. \tag{2.3.39}$$

Then

$$u'(x) = \sum_{n=1}^{\infty} nb_n x^{n-1}, \qquad u''(x) = \sum_{n=2}^{\infty} n(n-1)b_n x^{n-2}. \tag{2.3.40}$$

Substituting in Eq. 2.3.38 and making the usual adjustments in the indices of summation, there results

$$2b_2 + 2\lambda b_0 + \sum_{n=1}^{\infty} [(n+2)(n+1)b_{n+2} - 2(n-\lambda)b_n]x^n = 0. \tag{2.3.41}$$

Hence,

$$
\begin{aligned}
x^0: &\qquad b_2 = \frac{2(-\lambda)}{2 \cdot 1}b_0 \\[2mm]
x^1: &\qquad b_3 = \frac{2(1-\lambda)}{3 \cdot 2}b_1 \\[2mm]
x^n: &\qquad b_{n+2} = \frac{2(n-\lambda)}{(n+2)(n+1)}b_n.
\end{aligned}
\tag{2.3.42}
$$

This two-term recursion is easy to solve and we find

$$b_{2k} = \frac{2^k(-\lambda)(2-\lambda)(4-\lambda)\cdots(2k-2-\lambda)}{(2k)!}b_0 \tag{2.3.43}$$

and

$$b_{2k+1} = \frac{2^k(1 - \lambda)(3 - \lambda) \cdots (2k - 1 - \lambda)}{(2k + 1)!} b_1. \qquad (2.3.44)$$

Choose $b_0 = b_1 = 1$ and these relationships lead to the basic solution pair

$$u_1(x) = 1 + \sum_{k=1}^{\infty} \frac{2^k(-\lambda)(2 - \lambda) \cdots (2k - 2 - \lambda)}{(2k)!} x^{2k}. \qquad (2.3.45)$$

$$u_2(x) = x + \sum_{k=1}^{\infty} \frac{2^k(1 - \lambda)(3 - \lambda) \cdots (2k - 1 - \lambda)}{(2k + 1)!} x^{2k+1}. \qquad (2.3.46)$$

It is now apparent that Eq. 2.3.38 will have polynomial solutions when λ is a nonnegative integer. For λ even,

$$\begin{aligned}
\lambda = 0: \quad & u_1(x) = 1 \\
\lambda = 2: \quad & u_1(x) = 1 - 2x^2 \qquad (2.3.47) \\
\lambda = 4: \quad & u_1(x) = 1 - 4x^2 + \tfrac{4}{3}x^4.
\end{aligned}$$

For λ odd,

$$\begin{aligned}
\lambda = 1: \quad & u_2(x) = x \\
\lambda = 3: \quad & u_2(x) = x - \tfrac{2}{3}x^3 \qquad (2.3.48) \\
\lambda = 5: \quad & u_2(x) = x - \tfrac{4}{3}x^3 + \tfrac{4}{15}x^5.
\end{aligned}$$

Certain multiples of these polynomials are called *Hermite polynomials*. They are

$$\begin{aligned}
H_0(x) &= 1 \\
H_1(x) &= 2x \\
H_2(x) &= -2 + 4x^2 \qquad (2.3.49) \\
H_3(x) &= -12x + 8x^3
\end{aligned}$$

and, in general,

$$H_n(x) = n! \sum_{k=0}^{N} \frac{(-1)^k(2x)^{n-2k}}{k! (n - 2k)!}, \qquad (2.3.50)$$

where $N = n/2$ if n is even and $N = (n - 1)/2$ if n is odd.

PROBLEMS

1. Expand $e^{2xt - t^2}$ in powers of t. Set

$$e^{2xt - t^2} = \sum_{n=0}^{\infty} \frac{H_n(x)}{n!} t^n.$$

Show that $H_n(x)$ is given by Eq. 2.3.50.

Use the result of Problem 1 to show that

2. $H_n(-x) = (-1)^n H_n(x)$

3. $H_{2n+1}(0) = 0;$ $H_{2n}(0) = (-1)^n 2^{2n} 1 \cdot 3 \cdot 5 \cdots (2n-1)$

4. $H_n(x) = (-1)^n e^{x^2} \dfrac{d^n}{dx^n}(e^{-x^2})$

5. Use the result in Problem 4 to show that

$$\int_{-\infty}^{\infty} e^{-x^2} H_n(x)H_m(x) \, dx = 0, \qquad m \neq n$$

(*Hint:* Try integration by parts.)

6. Verify that Eq. 2.3.50 yields Eq. 2.3.49 for $n = 1, 2, 3$. Find $H_4(x)$.

7. Compute $H_n(x)$ and evaluate $H_n(0)$ by

(a) Differentiating Eq. 2.3.50.

(b) Differentiating the expansion in Problem 1 with respect to x.

Which method is easier?

2.4   The Method of Frobenius:
Solutions about Regular Singular Points

There are second-order differential equations that appear in physical applications which have coefficients that cannot be expressed in power series about the center $a = 0$; the origin is a singular point of such equations. Even so, the method described in Section 2.3 may yield a solution valid about the origin for such equations. The Cauchy–Euler equation

$$x^2 \frac{d^2 u}{dx^2} + \frac{3}{2}x \frac{du}{dx} - \frac{1}{2}u = 0 \tag{2.4.1}$$

is an example in which the power-series method fails. A pair of basic solutions is $u_1(x) = x^{-1}$ and $u_2(x) = \sqrt{x}$; neither function is analytic* at $x = 0$.
Consider the equation

$$L[u] = x^2 \frac{d^2 u}{dx^2} + xp_0(x) \frac{du}{dx} + p_1(x)u = 0, \tag{2.4.2}$$

where $p_0(x)$ and $p_1(x)$ are analytic at $x = 0$. As we have seen above, we cannot

---

*In the sections that follow, x^r ($r =$ an integer) and $\ln x$ appear repeatedly. In each case we assume that $x > 0$ to avoid writing $|x|^r$ and $\ln|x|$. For $x < 0$ we make the transformation $t = -x$ and solve the resulting equation for $t > 0$. Often the equation in t is identical to the equation in x.

expect a power series solution for this equation. The following more general series always provides at least one solution:

$$u(x) = x^r \sum_{n=0}^{\infty} a_n x^n, \qquad a_0 \neq 0. \tag{2.4.3}$$

Such a series is a *Frobenius series*. It reduces to a power series if r is a nonnegative integer. If $x = 0$ is not an ordinary point of Eq. 2.4.2, it is a *regular singular point*. The power-series part of the Frobenius series will converge in $|x| < R$, where R is at least as great as the distance from the origin to the nearest of the singular points of $p_0(x)$ and $p_1(x)$.

To solve Eq. 2.4.2, we assume a solution in Frobenius series and expand $p_0(x)$ and $p_1(x)$ in the series forms

$$p_0(x) = \sum_{n=0}^{\infty} b_n x^n \tag{2.4.4}$$

$$p_1(x) = \sum_{n=0}^{\infty} c_n x^n. \tag{2.4.5}$$

If it happens that $b_0 = c_0 = c_1 = 0$, then $x = 0$ is an ordinary point—not a regular singular point—because Eq. 2.4.2 would have a factor of x^2 and, after division by this factor, the resulting equation would have an ordinary point at the origin.

Differentiating the series expansion 2.4.3 yields

$$\frac{du}{dx} = \sum_{n=0}^{\infty} (n + r)a_n x^{n+r-1} \tag{2.4.6}$$

and

$$\frac{d^2u}{dx^2} = \sum_{n=0}^{\infty} (n + r - 1)(n + r)a_n x^{n+r-2}. \tag{2.4.7}$$

Substitution of the series expressions above into Eq. 2.4.2 gives, in expanded form,

$$L[u] = [r(r - 1)a_0 x^r + (r + 1)ra_1 x^{r+1} + \cdots]$$
$$+ [(b_0 + b_1 x + \cdots)(ra_0 x^r + (r + 1)a_1 x^{r+1} + \cdots]$$
$$+ [(c_0 + c_1 x + \cdots)(a_0 x^r + a_1 x^{r+1} + \cdots)]. \tag{2.4.8}$$

If we collect like powers of x, we get

$$L[u] = [r(r - 1) + b_0 r + c_0]a_0 x^r$$
$$+ \sum_{n=1}^{\infty} \{[(n + r)(n + r - 1) + b_0(n + r) + c_0]a_n$$
$$+ \sum_{k=1}^{n} [(n - k + r)b_k + c_k]a_{n-k}\}x^{n+r}. \tag{2.4.9}$$

We call the bracketed coefficient of the x^r term, $F(r)$; that is, since $a_0 \neq 0$, then $F(r) = 0$ is necessary for $L[u] = 0$:

$$F(r) = r(r - 1) + b_0 r + c_0 = 0. \qquad (2.4.10)$$

This equation is known as the *indicial equation*. It has roots r_1 and r_2. Now, note that

$$F(n + r) = (n + r)(n + r - 1) + b_0(n + r) + c_0. \qquad (2.4.11)$$

Hence, $L[u]$ may be written as

$$L[u] = F(r)a_0 x^r + \sum_{n=1}^{\infty} \{a_n F(n + r)$$

$$+ \sum_{k=1}^{n} [(n - k + r)b_k + c_k]a_{n-k}\}x^{n+r}. \qquad (2.4.12)$$

If we write

$$G_n(1) = (n - 1 + r)b_1 + c_1$$

$$G_n(2) = (n - 2 + r)b_2 + c_2 \qquad (2.4.13)$$

$$\vdots$$

$$G_n(k) = (n - k + r)b_k + c_k,$$

then

$$L[u] = F(r)a_0 x^r + \sum_{n=1}^{\infty} \{a_n F(n + r) + \sum_{k=1}^{n} G_n(k)a_{n-k}\}x^{n+r}. \qquad (2.4.14)$$

Since $L[u] = 0$, each coefficient must vanish:

$$F(r)a_0 = 0$$

$$F(n + r)a_n = -\sum_{k=1}^{n} G_n(k)a_{n-k} \qquad (2.4.15)$$

for $n = 1, 2, 3, \cdots$. [As we remarked above, $a_0 \neq 0$ implies that $F(r) = 0$.]

It is common practice in this method to use the substitution

$$s = r_1 - r_2 \qquad (2.4.16)$$

in which $r_1 \geq r_2$ if r_1 is real. Then

$$F(r) = (r - r_1)(r - r_2) \qquad (2.4.17)$$

so that

$$F(n + r_1) = (n + r_1 - r_1)(n + r_1 - r_2) = n(n + s) \qquad (2.4.18)$$

while

$$F(n + r_2) = (n + r_2 - r_1)(n + r_2 - r_2) = n(n - s). \qquad (2.4.19)$$

Therefore, if we set $r = r_1$ in the recursion 2.4.15, we obtain

$$0 \cdot a_0 = 0$$

$$n(n + s)a_n = -\sum_{k=1}^{n} G_n(k)a_{n-k} \qquad (2.4.20)$$

and if $r = r_2$,

$$0 \cdot a_0 = 0$$

$$n(n - s)a_n = -\sum_{k=1}^{n} G_n(k)a_{n-k}. \qquad (2.4.21)$$

We may always solve recursion 2.4.20 since $n(n + s)$ is never zero. Hence, $a_1, a_2, a_3, \cdots$ can all be determined as multiples of a_0 and

$$u_1(x) = x^{r_1} \sum_{n=0}^{\infty} a_n x^n, \qquad a_0 = 1 \qquad (2.4.22)$$

is a Frobenius representation of one of the solutions of Eq. 2.4.2. If s is not an integer, recursion 2.4.21 generates coefficients, say $d_1, d_2, d_3, \cdots$, which are all multiples of a_0. Hence,

$$u_2(x) = x^{r_2} \sum_{n=0}^{\infty} d_n x^n, \qquad d_0 = 1 \qquad (2.4.23)$$

represents a second, independent solution.

However, if $s = N$, a positive integer, then,

$$N \cdot 0 \cdot a_N = -\sum_{k=1}^{N} G_N(k)a_{N-k}. \qquad (2.4.24)$$

This equation does not determine a_N and the method becomes vastly more complicated when finding the second solution. This case will be treated in Section 2.9.2.

If $s = 0$, then recursions 2.4.20 and 2.4.21 are the same and only one solution of the form 2.4.22 exists. The technique for obtaining the second independent solution will be presented in Section 2.9.1.

We illustrate the use of these ideas in the following sections by applications to some important differential equations. In order to do so in a convenient fashion, we digress to study the gamma function, $\Gamma(\lambda)$. Before this excursion, let us consider an example.

Example 2.4.1: Find a general solution, valid near the origin, of the differential equation

$$8x^2 \frac{d^2u}{dx^2} + 6x \frac{du}{dx} + (x - 1)u = 0.$$

SOLUTION: For this equation $p_0(x) = \frac{6}{8}$ and $p_1(x) = (-1 + x)/8$. Thus, $b_0 = \frac{3}{4}$, $b_1 = b_{2} = \cdots = 0$, $c_0 = -\frac{1}{8}$, $c_1 = \frac{1}{8}$, $c_2 = c_3 = \cdots = 0$. The indicial equation is then

$$F(r) = r^2 + (\tfrac{3}{4} - 1)r - \tfrac{1}{8}$$

$$= r^2 - \tfrac{1}{4}r - \tfrac{1}{8}$$

$$= (r + \tfrac{1}{4})(r - \tfrac{1}{2}) = 0$$

$$\therefore \quad r_1 = \tfrac{1}{2}, \qquad r_2 = -\tfrac{1}{4}.$$

Equation 2.4.12 yields

$$L[u] = (r + \tfrac{1}{4})(r - \tfrac{1}{2})a_0 x^r + [a_1(r + 1 + \tfrac{1}{4})(r + 1 - \tfrac{1}{2}) + \tfrac{1}{8}a_0]x^{r+1}$$

$$+ \sum_{n=2}^{\infty} [a_n(r + n + \tfrac{1}{4})(r + n - \tfrac{1}{2}) + \tfrac{1}{8}a_{n-1}]x^{r+n} = 0.$$

The above then demands that

$$a_1 = -\frac{a_0}{8(r + 5/4)(r + 1/2)}$$

$$a_n = -\frac{a_{n-1}}{8(r + n + 1/4)(r + n - 1/2)}, \qquad n = 2, 3, \cdots.$$

The recursion above gives

$$a_k = \frac{(-1)^k a_0}{8^k[(r + \tfrac{5}{4})(r + \tfrac{9}{4}) \cdots (r + k + \tfrac{1}{4})][(r + \tfrac{1}{2})(r + \tfrac{3}{2}) \cdots (r + k - \tfrac{1}{2})]}.$$

We can let $a_0 = 1$ without loss of generality, and, referring to Eq. 2.4.3, we have

$$u_r(x) = x^r + \sum_{k=1}^{\infty} \frac{(-1)^k x^{k+r}}{8^k[(r + \tfrac{5}{4})(r + \tfrac{9}{4}) \cdots (r + k + \tfrac{1}{4})][(r + \tfrac{1}{2})(r + \tfrac{3}{2}) \cdots (r + k - \tfrac{1}{2})]}.$$

Setting $r_1 = \frac{1}{2}$ and $r_2 = -\frac{1}{4}$, respectively, the two independent solutions are

$$u_1(x) = x^{1/2} + \sum_{k=1}^{\infty} \frac{(-1)^k x^{k+1/2}}{8^k \cdot \tfrac{7}{4} \cdot \tfrac{11}{4} \cdots (k + \tfrac{3}{4})k!}$$

$$u_2(x) = x^{-1/4} + \sum_{k=1}^{\infty} \frac{(-1)^k x^{k-1/4}}{8^k \cdot \tfrac{1}{4} \cdot \tfrac{5}{4} \cdots (k - \tfrac{3}{4})k!}.$$

A general solution is then

$$u(x) = Au_1(x) + Bu_2(x).$$

We can put the solution in the alternative form by expanding for the first three terms in each series:

$$u(x) = A[x^{1/2} - \tfrac{1}{14}x^{3/2} + \tfrac{1}{616}x^{5/2} + \cdots]$$

$$+ B[x^{-1/4} - \tfrac{1}{2}x^{3/4} + \tfrac{1}{40}x^{7/4} + \cdots]. \quad \blacksquare$$

PROBLEMS

Find a general solution, valid in the vicinity of the origin, for each differential equation (roots not differing by an integer).

1. $2x\dfrac{d^2u}{dx^2} + (1 - x)\dfrac{du}{dx} + u = 0$

2. $16x\dfrac{d^2u}{dx^2} + 3(1 + 1/x)u = 0$

3. $2x(1 - x)\dfrac{d^2u}{dx^2} + \dfrac{du}{dx} - u = 0$

4. $2x\dfrac{d^2u}{dx^2} + (1 + 4x)\dfrac{du}{dx} + u = 0$

5. $4x^2(1 - x)\dfrac{d^2u}{dx^2} - x\dfrac{du}{dx} + (1 - x)u = 0$

6. $2x^2\dfrac{d^2u}{dx^2} - 7x\dfrac{du}{dx} + (x - 10) = 0$

7. $2x^2\dfrac{d^2u}{dx^2} + x(x - 1)\dfrac{du}{dx} + u = 0$

8. $2x^2\dfrac{d^2u}{dx^2} + x^2\dfrac{du}{dx} - u = 0$

9. $2x^2u'' + x(2x + 3)u' + (3x - 1)u = 0$

10. $2x(x - 1)u'' + 3(x - 1)u' - u = 0$

11. $2xu'' + (1 - x)u' - u = 0$

12. $3xu'' + 2(1 - x)u' - 4u = 0$

13. $3x^2u'' - xu' - 4u = 0$

14. $2x^2u'' + x(4x - 1)u' + 2(3x - 1)u = 0$

Use the Frobenius method to solve each differential equation about $a = 0$, an ordinary point of the equation.

15. $u'' + u = 0$

16. $u'' + (1 - x)u = 0$

17. The equation of Example 2.3.1.

18. $u'' + au' + bu = 0$

19. In the expansions assumed for $p_0(x)$ and $p_1(x)$ given by Eqs. 2.4.4 and 2.4.5, let $c_0 = c_1 = b_0 = 0$. Show that Eq. 2.4.2 has an ordinary point at $x = 0$.

2.5 The Gamma Function

Because of its importance and common use, we present this introduction to the gamma function $\Gamma(\lambda)$. The gamma function is defined by the improper integral

$$\Gamma(\lambda + 1) = \int_0^\infty e^{-t}t^\lambda \, dt \qquad (2.5.1)$$

which converges for all $\lambda > -1$.

To deduce some of the properties of the gamma function, let us integrate Eq. 2.5.1 by parts:

$$\int_0^\infty e^{-t}t^\lambda \, dt = -e^{-t}t^\lambda \, \Big|_0^\infty + \lambda \int_0^\infty e^{-t}t^{\lambda-1} \, dt \qquad (2.5.2)$$

$$\left(\begin{matrix} u = t^\lambda & dv = e^{-t} \, dt \\ du = \lambda t^{\lambda-1} \, dt & v = -e^{-t} \end{matrix} \right).$$

The quantity $e^{-t}t^\lambda$ vanishes at $t = \infty$ and $t = 0$. Thus, we have

$$\Gamma(\lambda + 1) = \lambda \int_0^\infty e^{-t}t^{\lambda-1} \, dt. \qquad (2.5.3)$$

The last integral is simply $\Gamma(\lambda)$. Thus, we have the important property

$$\Gamma(\lambda + 1) = \lambda \Gamma(\lambda). \qquad (2.5.4)$$

If we let $\lambda = 0$ in Eq. 2.5.1, there results

$$\Gamma(1) = \int_0^\infty e^{-t} \, dt$$

$$= -e^{-t} \Big|_0^\infty = 1. \qquad (2.5.5)$$

Using Eq. 2.5.4, there follows

$$\Gamma(2) = 1 \cdot \Gamma(1) = 1$$
$$\Gamma(3) = 2 \cdot \Gamma(2) = 2! \qquad (2.5.6)$$
$$\Gamma(4) = 3 \cdot \Gamma(3) = 3!$$

The equations above represent another important property of the gamma function. If λ is a positive integer,

$$\Gamma(\lambda + 1) = \lambda! \qquad (2.5.7)$$

It is interesting to note that $\Gamma(\lambda)$ is defined for all real λ except $\lambda = 0$,

$-1, -2, \cdots$, by the functional equation $\Gamma(\lambda + 1) = \lambda\Gamma(\lambda)$; in fact, we need to know $\Gamma(\lambda)$ only for $1 \leq \Gamma \leq 2$ to compute $\Gamma(\lambda)$ for all real values of λ. This tabulation is given in Table A2 of the Appendix. Figure 2.3 illustrates the graph of $\Gamma(\lambda)$.

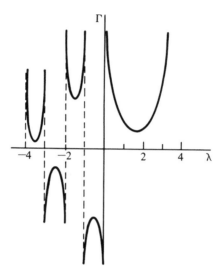

FIGURE 2.3.   The gamma function.

The gamma function at half-integer values are multiples of $\sqrt{\pi}$ (see Eq. 2.5.13 below). To see this, we begin with the known result (see Table A3 in the Appendix),

$$\int_0^\infty e^{-x^2}\, dx = \frac{\sqrt{\pi}}{2}. \tag{2.5.8}$$

Set $x^2 = t$ so that $dx = t^{-1/2}\, dt/2$ and therefore

$$\int_0^\infty e^{-x^2}\, dx = \frac{1}{2}\int_0^\infty e^{-t}\, t^{-1/2}\, dt$$

$$= \frac{1}{2}\Gamma\!\left(\frac{1}{2}\right) = \Gamma\!\left(\frac{3}{2}\right). \tag{2.5.9}$$

From the tabulated results in the Appendix (Table A.2), we verify the foregoing result, namely, that $\sqrt{\pi}/2 = \Gamma(3/2) = 0.886227$.

Example 2.5.1:   Evaluate the integral $\int_0^\infty x^{5/4} e^{-\sqrt{x}}\, dx$.

SOLUTION: The gamma functions are quite useful in evaluating integrals of this type. To make the exponent of the exponential function equal to $-t$, we let

$$x = t^2, \qquad dx = 2t\, dt.$$

Then the integral becomes (the limits remain unchanged)

$$\int_0^\infty x^{5/4} e^{-\sqrt{x}} \, dx = 2 \int_0^\infty t^{7/2} e^{-t} \, dt.$$

By using Eq. 2.5.1, we have

$$2 \int_0^\infty e^{-t} t^{7/2} \, dt = 2\Gamma(\tfrac{9}{2}).$$

The recursion 2.5.4 gives

$$\Gamma(\tfrac{9}{2}) = \tfrac{7}{2} \cdot \tfrac{5}{2} \cdot \tfrac{3}{2} \Gamma(\tfrac{3}{2}) = \tfrac{105}{8} \Gamma(\tfrac{3}{2}).$$

From Eq. 2.5.9

$$\Gamma(\tfrac{9}{2}) = \tfrac{105}{8} \tfrac{\sqrt{\pi}}{2} = \tfrac{105}{8} \cdot 0.8862.$$

Finally, the value of the integral is

$$\int_0^\infty x^{5/4} e^{-\sqrt{x}} \, dx = 2 \times \tfrac{105}{8} \times 0.8862 = 23.27. \quad \blacksquare$$

Example 2.5.2:   Express the product

$$f(r) = r(r + h)(r + 2h) \cdots [r + (n - 1)h]$$

as a quotient of gamma functions.

SOLUTION: We have

$$f(r) = (r/h)(r/h + 1)(r/h + 2) \cdots (r/h + n - 1)h^n$$

$$= h^n \frac{\Gamma(r/h + 1)}{\Gamma(r/h)} \frac{\Gamma(r/h + 2)}{\Gamma(r/h + 1)} \cdots \frac{\Gamma(r/h + r)}{\Gamma(r/h + n - 1)}$$

$$= h^n \frac{\Gamma(r/h + n)}{\Gamma(r/h)}$$

obtained by using the recursion 2.5.4 with $\lambda = r/h$. $\blacksquare$

Some special cases of the result of Example 2.5.2 are interesting. For a particular case, set $r = 1$ and $h = 2$. Then

$$1 \cdot 3 \cdot 5 \cdots (2n - 1) = 2^n \frac{\Gamma(n + \tfrac{1}{2})}{\Gamma(\tfrac{1}{2})} \tag{2.5.10}$$

But $\tfrac{1}{2}\Gamma(\tfrac{1}{2}) = \Gamma(\tfrac{3}{2}) = \sqrt{\pi}/2$. Hence,

$$1 \cdot 3 \cdot 5 \cdots (2n - 1) = \frac{2^n}{\sqrt{\pi}} \Gamma\left(n + \frac{1}{2}\right). \tag{2.5.11}$$

However,

$$1 \cdot 3 \cdot 5 \cdot \cdots \cdot (2n - 1) = 1 \cdot 3 \cdot 5 \cdots (2n - 1)\frac{2 \cdot 4 \cdot 6 \cdots 2n}{2 \cdot 4 \cdot 6 \cdots 2n}$$

$$= \frac{(2n)!}{2^n n!} . \qquad (2.5.12)$$

So combining the two equations above, we get

$$\Gamma\left(n + \frac{1}{2}\right) = \frac{(2n)!}{2^{2n} n!} \sqrt{\pi} \qquad (2.5.13)$$

for $n = 1, 2, \cdots$.

PROBLEMS

Evaluate each integral.

1. $\displaystyle\int_0^\infty \sqrt{x}\, e^{-x}\, dx$ **2.** $\displaystyle\int_0^\infty x^2 e^{-x^2}\, dx$ **3.** $\displaystyle\int_0^\infty x^3 e^{-x}\, dx$

4. $\displaystyle\int_0^\infty x^{-4} e^{-\sqrt{x}}\, dx$ **5.** $\displaystyle\int_0^\infty \frac{1}{\sqrt{x}} e^{-x^2} dx$ **6.** $\displaystyle\int_0^\infty (1 - x)^3 e^{\sqrt{x}}\, dx$

7. $\displaystyle\int_1^\infty x^2 e^{1-x}\, dx$ **8.** $\displaystyle\int_0^\infty x^3 e^{-x^{1/2}}\, dx$

Use the results in Example 2.5.2 to write each product as a quotient of the gamma function.

9. $2 \cdot 4 \cdot 6 \cdots (2n)$

10. $1 \cdot 4 \cdot 7 \cdots (3n - 2)$

11. $a(a + 1) \cdots (a + n - 1)$

12. Use Eq. 2.5.4 to explain why either $\Gamma(0)$ is meaningless or this equation is invalid for $\lambda = 0$.

13. Show that the improper integral $\int_0^\infty e^{-t} t^\lambda\, dt$ converges for $\lambda > -1$ and diverges for $\lambda \le -1$. (*Hint:* Write the integral as

$$\int_0^\infty e^{-t} t^\lambda\, dt = \int_0^1 e^{-t} t^\lambda\, dt + \int_1^\infty e^{-t}\, t^\lambda\, dt$$

and note that $e^{-t} \le 1$ for $0 \le t \le 1$ and $e^{-t/2} t^\lambda \to 0$ as $t \to +\infty$.)

14. Let $\Psi(\lambda) = \dfrac{d}{d\lambda} \ln \Gamma(\lambda)$. Show that

$$\Psi(\lambda + n + 1) = \frac{1}{\lambda + 1} + \frac{1}{\lambda + 2} + \cdots + \frac{1}{\lambda + n} + \Psi(\lambda + 1)$$

where n is a postive integer. Hence,

$$\Psi(n + 1) = 1 + \frac{1}{2} + \cdots + \frac{1}{n} + \Psi(1)$$

$[\Psi(1) = -\gamma = -0.57721566490$, approximately. This constant is the *Euler constant*.]

15. Show that

$$\frac{d}{d\lambda}\left[\frac{1}{\Gamma(\lambda)}\right] = -\frac{\Psi(\lambda)}{\Gamma(\lambda)}.$$

2.6   The Bessel–Clifford Equation

As an example of the method of Frobenius, consider the *Bessel–Clifford equation*,

$$x\frac{d^2u}{dx^2} + (1 - \gamma)\frac{du}{dx} + u = 0 \tag{2.6.1}$$

where the parameter $\gamma \geq 0$. Since $p_0(x) = 1 - \gamma$ and $p_1(x) = x$ (see Eq. 2.4.2), we have

$$b_0 = 1 - \gamma, \qquad\qquad c_1 = 1$$
$$b_1 = b_2 = \cdots = 0, \qquad c_0 = c_2 = c_3 = \cdots = 0. \tag{2.6.2}$$

The indicial equation is

$$F(r) = r(r - 1) + (1 - \gamma)r = r(r - \gamma) = 0. \tag{2.6.3}$$

Thus, $r_1 = \gamma$, $r_2 = 0$, and $s = r_1 - r_2 = \gamma$. Since $b_k = 0$ for all $k > 0$ and $c_k = 0$ for all k except $k = 1$ and

$$G_n(k) = (n - k + r)b_k + c_k, \qquad 1 \leq k \leq n, \tag{2.6.4}$$

we have

$$G_n(1) = c_1 = 1 \quad \text{and} \quad G_n(k) = 0, \qquad k > 1. \tag{2.6.5}$$

The recursions 2.4.20 and 2.4.21 reduce to the convenient forms

$$n(n + \gamma)a_n = -a_{n-1}, \qquad n = 1, 2, \cdots \tag{2.6.6}$$

and

$$n(n - \gamma)a_n = -a_{n-1} \qquad n = 1, 2, \cdots \tag{2.6.7}$$

For γ neither zero nor a positive integer, we have

$$u_1(x) = x^\gamma\left[1 + \sum_{k=1}^{\infty} \frac{(-1)^k x^k}{k!(1 + \gamma)(2 + \gamma) \cdots (k + \gamma)}\right] \tag{2.6.8}$$

and

$$u_2(x) = 1 + \sum_{k=1}^{\infty} \frac{(-1)^k x^k}{k!(1 - \gamma)(2 - \gamma) \cdots (k - \gamma)}, \tag{2.6.9}$$

so that $u(x) = Au_1(x) + Bu_2(x)$ is the general solution. If $\gamma = 0$, then $u_1 = u_2$ and this method fails to yield a second, independent solution. If γ is a positive integer, then $n(n - \gamma)a_n = -a_{n-1}$ breaks down when $n = \gamma$, and again, the method does not yield a second, independent solution. Methods for obtaining the second, independent solution for $\gamma = 0$ or for γ a positive integer will be presented in a subsequent section.

PROBLEMS

1. Set $\gamma = 0$ in the Bessel–Clifford equation and verify that

$$\sum_{k=0}^{\infty} \frac{(-1)^k x^k}{(k!)^2} = 1 - x + \frac{x^2}{4} - \frac{x^3}{36} + \cdots$$

is a solution by direct substitution.

2. Set $\gamma = 1$ in the Bessel–Clifford equation and verify that

$$\sum_{k=0}^{\infty} \frac{(-1)^k x^k}{k!(k + 1)!} = 1 - \frac{x}{2} + \frac{x^2}{12} - \cdots$$

is a solution by direct substitution.

2.7 Laguerre Polynomials

The equation

$$x \frac{d^2 u}{dx^2} + (1 - x) \frac{du}{dx} + Nu = 0 \tag{2.7.1}$$

has a polynomial solution for each positive integer N. To see this, we begin by identifying $p_0(x) = 1 - x$ and $p_1(x) = Nx$ (see Eq. 2.4.2) so that

$$\begin{array}{llll} b_0 = 1, & b_1 = -1, & b_k = 0, & k > 1 \\ c_0 = 0, & c_1 = N, & c_k = 0, & k > 1. \end{array} \tag{2.7.2}$$

Hence, the indicial equation is

$$F(r) = r(r - 1) + r = r^2 = 0. \tag{2.7.3}$$

Therefore, $r_1 = r_2 = 0$ and $s = 0$. For each $n \geq 1$,

$$\begin{array}{lll} G_n(k) = (n - k)b_k + c_k, & 1 \leq k \leq n \\ \quad\quad\quad = 0 & \text{if } k > 1. \end{array} \tag{2.7.4}$$

Thus,

$$G_n(1) = 1 - n + N$$

$$G_n(k) = 0, \qquad k = 2, 3, \cdots, n \qquad (2.7.5)$$

and the recursions 2.4.20 and 2.4.21 reduce to the single recursion

$$n^2 a_n = -(-n + 1 + N)a_{n-1}, \qquad n = 1, 2, \cdots. \qquad (2.7.6)$$

Set $a_0 = 1$. Then, for each $k \geq 1$,*

$$a_k = \frac{(-1)^k N(N - 1) \cdots (N - k + 1)}{(k!)^2}. \qquad (2.7.7)$$

The binomial coefficient

$$\binom{N}{k} = \frac{N(N - 1) \cdots (N - k + 1)}{k!} = \frac{N!}{k!(N - k)!} \qquad (2.7.8)$$

leads to a neater solution:

$$a_k = \frac{(-1)^k \binom{N}{k}}{k!}, \qquad k = 1, 2, \cdots \qquad (2.7.9)$$

and therefore

$$u(x) = L_n(x) = 1 + \sum_{k=1}^{N} \frac{(-1)^k \binom{N}{k}}{k!} x^k = \sum_{k=0}^{N} \frac{(-1)^k \binom{N}{k}}{k!} x^k \qquad (2.7.10)$$

is a polynomial solution of Eq. 2.7.1. The family of polynomials

$$L_0(x) = 1$$

$$L_1(x) = 1 - x$$

$$L_2(x) = 1 - 2x + \frac{x^2}{2!} \qquad (2.7.11)$$

$$L_n(x) = \sum_{k=0}^{N} \frac{(-1)^k \binom{N}{k}}{k!} x^k$$

are the *Laguerre polynomials*.

PROBLEMS

1. Prove the validity of the Rodrigues formula,

---

*When $k = N + 1, N + 2, \cdots, a_k = 0$.

$$L_n(x) = \frac{e^x}{n!} \frac{d^n}{dx^n} (x^n e^{-x}).$$

2. Use the Rodrigues formula in Problem 1 and integration by parts to establish that

$$\int_0^\infty e^{-x} L_n(x) L_m(x)\, dx = 0, \qquad n \neq m$$

$$\int_0^\infty e^{-x} L_n^2(x)\, dx = 1.$$

2.8   Roots Differing by an Integer: The Wronskian Method

We have seen in Section 2.7 that

$$x^2 \frac{d^2 u}{dx^2} + x p_0(x) \frac{du}{dx} + p_1(x) u = 0 \qquad (2.8.1)$$

always has a solution of the form

$$u_1(x) = x^{r_1} \sum_{n=0}^\infty a_n x^n, \qquad a_0 \neq 0 \qquad (2.8.2)$$

for some r_1. We know from general principles that there exists an independent solution $u_2(x)$ in some interval $0 < x < b$. The Wronskian $W(x)$ of $u_1(x)$ and $u_2(x)$ is defined as

$$W(x) = u_1(x) u_2'(x) - u_1'(x) u_2(x), \qquad (2.8.3)$$

and we know from Eq. 1.5.17 that

$$W(x) = K e^{-\int [p_0(x)/x]\, dx}. \qquad (2.8.4)$$

Equating the two relationships above, we can write

$$\frac{u_1(x) u_2'(x) - u_1'(x) u_2(x)}{u_1^2(x)} = \frac{K}{u_1^2(x)} e^{-\int [p_0(x)/x]\, dx}. \qquad (2.8.5)$$

This is

$$\frac{d}{dx}\left(\frac{u_2(x)}{u_1(x)} \right),$$

so Eq. 2.8.5 can be integrated to yield

$$\frac{u_2(x)}{u_1(x)} = K \int \frac{1}{u_1^2(x)} e^{-\int [p_0(x)/x]\, dx}\, dx. \qquad (2.8.6)$$

This last relationship yields $u_2(x)$ for any choice of $u_1(x)$. With no loss of generality we can pick* $K = a_0 = 1$ and substitute the known series expansion 2.8.2 for $u_1(x)$ in Eq. 2.8.6. First,

---

*We take $K = 1$ and $a_0 = 1$ throughout this section.

$$p_0(x) = b_0 + b_1 x + b_2 x^2 + \cdots \tag{2.8.7}$$

so that

$$-\int \frac{p_0(x)}{x}\, dx = -b_0 \ln x - b_1 x - \frac{1}{2} b_2 x^2 - \cdots . \tag{2.8.8}$$

Hence,

$$
\begin{aligned}
e^{-\int [p_0(x)/x]\, dx} &= e^{-b_0 \ln x - b_1 x - \cdots} \\
&= e^{-b_0 \ln x} e^{-b_1 x - b_2 x^2/2 - \cdots} \\
&= x^{-b_0}\left[1 - (b_1 x + \tfrac{1}{2} b_2 x^2 + \cdots) \right. \\
&\qquad \left. + \frac{(b_1 x + \tfrac{1}{2} b_2 x^2 + \cdots)^2}{2!} + \cdots \right] \\
&= x^{-b_0}(1 - b_1 x + k_2 x^2 + k_3 x^3 + \cdots),
\end{aligned}
\tag{2.8.9}
$$

where k_i are functions of $b_1, b_2 \cdots , b_i$. Next,

$$\frac{1}{u_1^2(x)} = \frac{1}{x^{2r_1}} \frac{1}{(1 + a_1 x + \cdots)^2} = \frac{1}{x^{2r_1}}(1 - 2a_1 x + \cdots). \tag{2.8.10}$$

and we find that

$$
\begin{aligned}
\frac{1}{u_1^2(x)} e^{-\int [p_0(x)/(x)]\, dx} &= \frac{1}{x^{2r_1 + b_0}}(1 - b_1 x + \cdots)(1 - 2a_1 x + \cdots) \\
&= \frac{1}{x^{2r_1 + b_0}}[1 - (b_1 + 2a_1)x + \cdots].
\end{aligned}
\tag{2.8.11}
$$

But since $F(r) = (r - r_1)(r - r_2) = r(r - 1) + b_0 r + c_0$, we see that $r_1 + r_2 = 1 - b_0$. By definition, we know that $s = r_1 - r_2$, so that, by adding these equations, we get

$$2r_1 + b_0 = 1 + s \tag{2.8.12}$$

and therefore

$$\frac{1}{u_1^2(x)} e^{-\int [p_0(x)/x]\, dx} = \frac{1}{x^{1+s}}[1 - (b_1 + 2a_1)x + \cdots]. \tag{2.8.13}$$

Let us now consider two special cases.

Case 1.   *Equal roots so that $s = 0$. Substituting Eq. 2.8.13 into Eq. 2.8.6 gives us*

$$
\begin{aligned}
u_2(x) &= u_1(x)\left[\int \frac{1}{x}\, dx - \int (b_1 + 2a_1)\, dx + \cdots \right] \\
&= u_1(x) \ln x + u_1(x)[-(b_1 + 2a_1)x + \cdots].
\end{aligned}
\tag{2.8.14}
$$

We can go one step further. Because $a_0 = 1$,

$$u_2(x) = u_1(x) \ln x + x^{r_1}[-(b_1 + 2a_1)x + \cdots]. \qquad (2.8.15)$$

[*Note:* If $s = 0$, the expansion for $u_2(x)$ will always include a term containing $\ln x$.]

Case 2. *Roots differing by an integer, $s = N$, N positive.* As in case 1, substitute Eq. 2.8.13 into Eq. 2.8.6 and obtain

$$u_2(x) = u_1(x) \left[\int \frac{dx}{x^{N+1}} - \int \frac{b_1 + 2a_1}{x^N} dx + \cdots + \int \frac{c}{x} dx + \cdots \right]$$

$$= u_1(x) \left[-\frac{1}{N} x^{-N} + \frac{b_1 + 2a_1}{N-1} x^{-N+1} + c \ln x + \cdots \right]$$

$$= cu_1(x) \ln x + u_1(x) \left[-\frac{1}{N} + d_1 x + \cdots \right] x^{-N}, \qquad (2.8.16)$$

where c represents a constant. Since $u_1(x) = x^{r_1}(1 + a_1x + \cdots)$ and $r_1 - N = r_1 - s = r_2$, we can express this second independent solution as

$$u_2(x) = cu_1(x) \ln x + x^{r_2} \left[-\frac{1}{N} + \left(d_1 - \frac{a_1}{N} \right) x + \cdots \right] \qquad (2.8.17)$$

If $c = 0$, the $\ln x$ term will not appear in the solution $u_2(x)$.

Although a few of the coefficients in Eq. 2.8.15 or 2.8.17 can be computed in this manner, no pattern in the formation of the coefficients can usually be discerned. In the next section we explore an alternative method which often yields formulas for the coefficients in the second solution. The main importance of the technique illustrated here resides in these conclusions.

1. If the roots of the indicial equation are equal, then a second solution always contains a $\ln x$ term and is of the form

$$u_2(x) = u_1(x) \ln x + x^{r_1} \sum_{n=0}^{\infty} d_n x^n. \qquad (2.8.18)$$

2. If the roots of the indicial equation differ by an integer, either

$$u_2(x) = x^{r_2} \sum_{n=0}^{\infty} d_n x^n \qquad (2.8.19)$$

or

$$u_2(x) = cu_1(x) \ln x + x^{r_2} \sum_{n=0}^{\infty} d_n x^n \qquad (2.8.20)$$

represents a second solution.

3. If $p_0(x) = 0$ or $p_0(x) = b_0$, the integral in Eq. 2.8.6 simplifies. If $p_0(x) = 0$, then

$$u_2(x) = u_1(x) \int \frac{dx}{u_1^2(x)}. \qquad (2.8.21)$$

If $p_0(x) = b_0$, then

$$u_2(x) = u_1(x) \int \frac{dx}{x^{b_0} u_1^2(x)}. \qquad (2.8.22)$$

PROBLEMS

1. Verify that $k_2 = \frac{1}{2}(b_1^2 - b_2)$ in Eq. 2.8.9.

2. Extend Eq. 2.8.10 to

$$\frac{1}{u_1^2(x)} = \frac{1}{x^{2r_1}}[1 - 2a_1 x + (3a_1^2 - 2a_2)x^2 + \cdots].$$

3. Use the results of Problems 1 and 2 to obtain the following extended expression for Eq. 2.8.13:

$$\frac{1}{u_1^2(x)} e^{-\int [p_0(x)/x]\,dx} = \frac{1}{x^{1+s}}[1 - (b_1 + 2a_1)x$$

$$+ (2a_1 b_1 + \tfrac{1}{2}b_1^2 - \tfrac{1}{2}b_2 + 3a_1^2 - 2a_2)x^2 + \cdots].$$

4. Let $s = 0$ in the result of Problem 3 and thus obtain the following extended form for Eq. 2.8.15:

$$u_2(x) = u_1(x) \ln x + x^{r_1}[-(b_1 + 2a_1)x + \tfrac{1}{2}(-\tfrac{1}{2}b_2 + \tfrac{1}{2}b_1^2 - a_1^2 - 2a_2)x^2 + \cdots].$$

5. In the Bessel–Clifford equation (Eq. 2.6.1), with $\gamma = 0$, show that

$$u_2(x) = u_1(x) \ln x + 2x - \tfrac{3}{4}x^2 + \cdots,$$

where

$$u_1(x) = \sum_{k=0}^{\infty} \frac{(-1)^k x^k}{k!\, k!}.$$

6. In the Laguerre equation (Eq. 2.7.1) with $N = 0$, show that

$$u_2(x) = \int \frac{e^x}{x}\, dx$$

using Eq. 2.8.6.

7. In the Laguerre equation (Eq. 2.7.1) with $N = 1$, we have $L_1(x) = 1 - x = u_1(x)$. Show that

$$u_2(x) = (1 - x) \ln x + 3x + \cdots$$

using Eq. 2.8.15, and

$$u_2(x) = (1 - x) \ln x + 3x - \tfrac{1}{4}x^2 + \cdots$$

using the results of Problem 4.

8. Show that

$$\frac{1}{u_1^2(x)} e^{-\int [p_0(x)/x] \, dx} = \frac{1}{x^{1+s}}$$

for the Cauchy–Euler equation, $x^2u'' + b_0xu' + c_0u = 0$. Let $u_1(x) = x^{r_1}$.

(a) Suppose that $s = 0$. Use Eq. 2.8.6 to show that

$$u_2(x) = x^{r_1} \ln x.$$

(b) Suppose that $s \neq 0$. Use Eq. 2.8.6 to show that

$$u_2(x) = -\frac{1}{s}x^{r_1-s}.$$

Reconcile this with the expected second solution $u_2(x) = x^{r_2}$.

9. One solution of $x^2u'' - x(1 + x)u' + u = 0$ is $u_1(x) = xe^x$. Verify this. Show that a second solution can be written

$$u_2(x) = u_1(x) \ln x - x^2 + \cdots.$$

10. Show that

$$u_1(x) = \sum_{n=0}^{\infty} \frac{x^{n+1/2}}{2^n(n!)^2}$$

and

$$u_2(x) = u_1(x) \ln x + x^{1/2}[-x - \tfrac{3}{16}x^2 - \cdots]$$

are solutions of $4x^2u'' + (1 - 2x)u = 0$.

11. Verify that $u_1(x) = x$ is a solution of $x^2u'' - x(1 - x)u' + (1 - x)u = 0$. Use Eq. 2.8.6 to show that

$$u_2(x) = x \ln x + x \sum_{n=1}^{\infty} \frac{(-1)^n x^n}{n \cdot n!}$$

is a second solution.

2.9 Roots Differing by an Integer: Series Method

Once again denote

$$L[u] = x^2\frac{d^2u}{dx^2} + xp_0(x)\frac{du}{dx} + p_1(x)u \qquad (2.9.1)$$

and let

$$u(x) = x^r \sum_{n=0}^{\infty} a_n x^n, \qquad a_0 \neq 0. \qquad (2.9.2)$$

Following the analysis in Section 2.4, we find

$$L[u] = F(r)a_0 x^r + \sum_{n=1}^{\infty} \left[F(n + r)a_n + \sum_{k=1}^{n} G_n(k)a_{n-k} \right] x^{n+r}. \quad (2.9.3)$$

We want $L[u(x)] = 0$. For the purposes of this section we show that for any r, we can get

$$L[u] = F(r)a_0 x^r. \qquad (2.9.4)$$

That is, we can find $a_1, a_2, \cdots$, so that the infinite series part of Eq. 2.9.3 is identically zero. To do this we solve the recursion

$$F(n + 1)a_n = - \sum_{k=1}^{n} G_n(k)a_{n-k}, \qquad n = 1, 2, \cdots \qquad (2.9.5)$$

without choosing r. Obviously, in order that $L[u] = 0$, Eq. 2.9.4 demands that $F(r) = 0$, but we ignore this for the time being. A specific equation will be used to illustrate the procedure. The equation is

$$\frac{d^2 u}{dx^2} - u = 0. \qquad (2.9.6)$$

For this equation $F(r) = r^2 - r$, so that $r_1 = 0$, and $r_2 = 1$; all b_n are zero since $p_0(x) = 0$, and all $c_n = 0$ except $c_2 = -1$ since $p_1(x) = -x^2$. Referring to Eq. 2.4.12, we have

$$L[u] = (r^2 - r)a_0 x^r + (r^2 + r)a_1 x^{r+1}$$

$$+ \sum_{n=2}^{\infty} \{[(r + n)^2 - (r + n)]a_n - a_{n-2}\} x^{n+r}, \qquad (2.9.7)$$

So we set

$$(r^2 + r)a_1 = 0$$

$$(r + n)(r + n - 1)a_n = a_{n-2}, \qquad n \geq 2. \qquad (2.9.8)$$

Thus, $a_1 = 0$ and the recursion gives $a_3 = a_5 = \cdots = 0$. For the coefficients with even subscripts the recursion gives

$$a_{2k} = \frac{a_0}{(r + 1)(r + 2) \cdots (r + 2k)}, \qquad k \geq 1. \qquad (2.9.9)$$

We can arbitrarily set $a_0 = 1$, so that $u_r(x)$ is defined as (see Eq. 2.4.3),

$$u_r(x) = x^r + \sum_{k=1}^{\infty} \frac{x^{2k+r}}{(r + 1)(r + 2) \cdots (r + 2k)}. \qquad (2.9.10)$$

We can readily verify that $L[u_r] = (r^2 - r)x^r$, thus confirming Eq. 2.9.4. Setting $r_1 = 0$ and $r_2 = 1$, respectively, gives

$$u_1(x) = 1 + \sum_{k=1}^{\infty} \frac{x^{2k}}{(2k)!} \tag{2.9.11}$$

$$u_2(x) = x + \sum_{k=1}^{\infty} \frac{x^{2k+1}}{(2k+1)!} \tag{2.9.12}$$

The general solution of Eq. 2.9.6 is then

$$u(x) = Au_1(x) + Bu_2(x). \tag{2.9.13}$$

A homework problem at the end of this section will show that this is equivalent to the solution found using the methods of Chapter 1, that is,

$$u(x) = c_1 e^x + c_2 e^{-x}. \tag{2.9.14}$$

2.9.1 $s = 0$.

In this case $F(r) = (r - r_1)^2$ and Eq. 2.9.4 becomes

$$L[u_r(x)] = (r - r_1)^2 a_0 x^r. \tag{2.9.15}$$

Set $a_0 = 1$. In the present notation, $u_r(x)|_{r=r_1}$ is a solution. Consider

$$\frac{\partial}{\partial r} L[u_r(x)] = L\left[\frac{\partial}{\partial r} u_r(x)\right]$$

$$= \frac{\partial}{\partial r}[(r - r_1)^2 x^r]$$

$$= 2(r - r_1)x^r + (r - r_1)^2 x^r \ln x. \tag{2.9.16}$$

If we set $r = r_1$, the equation above shows us that

$$L\left[\frac{\partial}{\partial r} u_r(x)\Big|_{r=r_1}\right] = 0. \tag{2.9.17}$$

Hence,

$$\frac{\partial}{dr} u_r(x)\Big|_{r=r_1}$$

is also a solution. Thus,

$$u_r(x)\Big|_{r=r_1} \quad \text{and} \quad \frac{\partial}{\partial r} u_r(x)\Big|_{r=r_1}$$

are, as we shall see, a basic solution set. An example will illustrate the details.

Example 2.9.1: Find two independent solutions of

$$x\frac{d^2u}{dx^2} + \frac{du}{dx} - u = 0$$

SOLUTION: For this equation $b_0 = 1$, $c_1 = -1$, and $F(r) = r^2$, so that $r_1 = r_2 = 0$. It then follows that (see Eq. 2.4.12),

$$L[u] = r^2 a_0 x^r + [(r + 1)^2 a_1 - a_0]x^{r+1} + \sum_{n=2}^{\infty} [a_n(r + n)^2 - a_{n-1}]x^{n+r},$$

so that

$$(r + 1)^2 a_1 - a_0 = 0$$

$$(r + n)^2 a_n - a_{n-1} = 0, \qquad n \geq 2$$

We then have

$$a_1 = \frac{a_0}{(r + 1)^2}$$

$$a_k = \frac{a_0}{(r + 1)^2 (r + 2)^2 \cdots (r + k)^2}, \qquad k \geq 2$$

Thus, letting $a_0 = 1$,

$$u_r(x) = x^r \left[1 + \sum_{k=1}^{\infty} \frac{x^k}{(r + 1)^2 (r + 2)^2 \cdots (r + k)^2}\right].$$

The first independent solutin is then, setting $r = 0$,

$$u_r(x)\big|_{r=0} = u_1(x) = 1 + \sum_{k=1}^{\infty} \frac{x^k}{(k!)^2}.$$

Using the result of this section, we find the second independent solution to be*

$$u_2(x) = \frac{\partial}{\partial r} u_r(x)\big|_{r=0}$$

$$= \left\{x^r \ln x \left[1 + \sum_{k=1}^{\infty} \frac{x^k}{(r + 1)^2 (r + 2)^2 \cdots (r + k)^2}\right] \right.$$

$$\left. + x^r \sum_{k=1}^{\infty} x^k \frac{\partial}{\partial r}\left[(r + 1)^{-2}(r + 2)^{-2} \cdots (r + k)^{-2}\right]\right\}_{r=0}.$$

The easiest way to compute the derivative of the product appearing in the second series is to use logarithmic differentiation, as follows. Set

$$f(r) = (r + 1)^{-2}(r + 2)^{-2} \cdots (r + k)^{-2}$$

so that

$$f(0) = 2^{-2} \cdot 3^{-2} \cdots k^{-2} = \frac{1}{(k!)^2}.$$

Then,

$$\ln [f(r)] = -2[\ln (r + 1) + \ln (r + 2) + \cdots + \ln (r + k)]$$

and thus

---

*Recall that $da^x/dx = a^x \ln a$.

$$\frac{d}{dr} \ln \left[f(r) \right] = \frac{f'(r)}{f(r)} = -2 \left[\frac{1}{r+1} + \frac{1}{r+2} + \cdots + \frac{1}{r+k} \right].$$

Finally,

$$f'(r) = -2f(r) \left[\frac{1}{r+1} + \frac{1}{r+2} + \cdots + \frac{1}{r+k} \right].$$

Substitute this into the series above and obtain, setting $r = 0$,

$$u_2(x) = \frac{\partial}{\partial r} u_r(x) \Big|_{r=0} = \ln x \left[1 + \sum_{k=1}^{\infty} \frac{x^k}{(k!)^2} \right]$$

$$+ \sum_{k=1}^{\infty} x^k \left[-2f(0) \left(1 + \frac{1}{2} + \frac{1}{3} + \cdots + \frac{1}{k} \right) \right],$$

$$= u_1(x) \ln x - \sum_{k=1}^{\infty} \frac{2x^k}{(k!)^2} h_k$$

where the kth partial sum of the harmonic series $(1 + \frac{1}{2} + \frac{1}{3} + \cdots + \frac{1}{m} + \cdots)$ is written as

$$h_k = 1 + \frac{1}{2} + \frac{1}{3} + \cdots + \frac{1}{k}. \quad \blacksquare$$

PROBLEMS

Determine the general solution for each differential equation by expanding in a series about the origin (equal roots).

1. $x \dfrac{d^2 u}{dx^2} + \dfrac{du}{dx} + u = 0$

2. $x(1 - x) \dfrac{d^2 u}{dx^2} + \dfrac{du}{dx} + u = 0$

3. $x^2 \dfrac{d^2 u}{dx^2} - 3x \dfrac{du}{dx} + (4 - x)u = 0$

4. Solve the Bessel–Clifford equation (Eq. 2.6.1), with $\gamma = 0$. Compare with the answer given in Problem 5 of Section 2.8.

5. Find a general solution of the Laguerre equation (Eq. 2.7.1) and compare with the answer to Problem 7 of Section 2.8.

6. Show that the general solution given by Eq. 2.9.13 is the same family of functions as the general solution given by Eq. 2.9.14.

7. Show that

$$1 + \frac{1}{3} + \frac{1}{5} + \cdots + \frac{1}{2n-1} = h_{2n} - \frac{1}{2} h_n,$$

where

$$h_n = 1 + \frac{1}{2} + \cdots + \frac{1}{n}.$$

8. Use the method of this section to show that

$$u_1(x) = x \sum_{n=0}^{\infty} \frac{x^n}{n!}, \qquad u_2(x) = u_1(x) \ln x - x \sum_{n=1}^{\infty} \frac{h_n x^n}{n!}$$

are solutions of $x^2 u'' - x(1 + x)u' + u = 0$.

9. Solve $4x^2 u'' + (1 - 2x)u = 0$.

10. Solve $xu'' + (1 - x)u' - u = 0$.

2.9.2 $s = N$, N A POSITIVE INTEGER

We suppose that

$$s = r_1 - r_2 = N, \tag{2.9.18}$$

where N is a postive integer. Two very different possibilities exist. In some circumstances, we get two independent Frobenius series solutions. The Cauchy–Euler equation is an illustration of this situation. A second alternative is one Frobenius series and a solution involving $\ln x$. (See Section 2.8 for a discussion of how these disparate cases arise.) The indicial equation is $F(r) = (r - r_1)(r - r_2)$, and therefore we have (see Eq. 2.9.4)

$$L[u_r(x)] = (r - r_1)(r - r_2)a_0 x^r. \tag{2.9.19}$$

In Section 2.9.1 we obtained a second solution by computing $\partial u_r / \partial r$ and noting that

$$L\left[\frac{\partial u}{\partial r}\bigg|_{r=r_1}\right] = \frac{\partial}{\partial r} L[u_r]|_{r=r_1} = 0. \tag{2.9.20}$$

But this fortunate circumstance is due entirely to the fact that r_1 is a double root; that is, $F(r) = (r - r_1)^2$. This method fails if $r_1 \neq r_2$. We can force a double root into the coefficient of x^r in Eq. 2.9.19 by choosing

$$a_0 = (r - r_2)c \tag{2.9.21}$$

where c is an arbitrary constant. Now

$$\frac{\partial}{\partial r} L[u_r] = L\left[\frac{\partial u_r}{\partial r}\right] = \frac{\partial}{\partial r}\{(r - r_1)(r - r_2)^2 c x^r\}$$

$$= (r - r_2)^2 c x^r + 2(r - r_1)(r - r_2)c x^r$$

$$+ (r - r_1)(r - r_2)^2 c x^r \ln x \tag{2.9.22}$$

which vanishes identically for $r = r_2$. Thus, our second solution will be

$$\frac{\partial u_r}{\partial r}\bigg|_{r=r_2} = u_2(x) \tag{2.9.23}$$

where u_r is defined by using Eq. 2.9.21. An example will make this clearer.

Example 2.9.2:   Using Eq. 2.9.23 find the second solution of the differential equation

$$x^2 \frac{d^2 u}{dx^2} + x \frac{du}{dx} + (x^2 - 1)u = 0.$$

SOLUTION: We have $p_0(x) = 1$ and $p_1(x) = -1 + x^2$, so that $b_0 = 1$, $c_0 = -1$, and $c_2 = 1$. The remaining coefficients are zero. Since $r_1 = 1$ and $r_2 = -1$, we have $s = 2$ and

$$F(r) = r^2 - 1$$

$$= (r - 1)(r + 1)$$

$$\therefore \quad F(n + r) = (n + r - 1)(n + r + 1).$$

From Eq. 2.4.12, with $b_1 = b_2 = c_1 = 0$ and $c_2 = 1$, we have

$$L[u] = (r - 1)(r + 1)a_0 x^r + a_1 r(r + 2)x^{r+1}$$

$$+ \sum_{n=2}^{\infty} [(n + r - 1)(n + r + 1)a_n + a_{n-2}]x^{n+r}.$$

Thus, we are forced to set

$$a_1 r(r + 2) = 0$$

$$(n + r - 1)(n + r + 1)a_n = -a_{n-2}, \qquad n = 2, 3, \ldots$$

to ensure that $L[u] = (r - 1)(r + 1)a_0 x^r$. Since we do not wish to specify r, we must select $a_1 = 0$. Thus, the recursion implies that $a_3 = a_5 = \cdots = 0$. For $k \geq 1$, the recursion yields

$$a_{2k} = \frac{(-1)^k a_0}{(r + 1)(r + 3) \cdots (r + 2k - 1)} \frac{1}{(r + 3)(r + 5) \cdots (r + 2k + 1)}.$$

Define $f(r) = (r + 3)^{-2}(r + 5)^{-2} \cdots (r + 2k - 1)^{-2}(r + 2k + 1)^{-1}$ and set $a_0 = r + 1$. Then it follows that

$$a_{2k} = (-1)^k f(r), \qquad k = 1, 2, \ldots$$

and

$$L[u] = (r - 1)(r + 1)^2 x^r$$

and

$$u_r(x) = x^r \left[(r + 1) + \sum_{k=1}^{\infty} (-1)^k f(r)x^{2k} \right].$$

Hence, with $r = 1$,

$$u_1(x) = x \left[2 + \sum_{k=1}^{\infty} \frac{(-1)^k x^{2k}}{2^{2k-1} k!(k + 1)!} \right]$$

since

$$f(1) = 4^{-2} \cdot 6^{-2} \cdots (2k)^{-2}(2k + 2)^{-1} = \frac{1}{2^{2k-1} k!(k + 1)!}.$$

We now differentiate $u_r(x)$ and obtain

$$\frac{\partial u_r}{\partial r} = u_r(x) \ln x + x^r \left[1 + \sum_{k=1}^{\infty} (-1)^k f'(r) x^{2k} \right].$$

Set $r = -1$ and then

$$u_2(x) = \frac{\partial u_r}{\partial r} \bigg|_{r=-1}$$

$$= u_r(x)|_{r=-1} \ln x + x^{-1} \left[1 + \sum_{k=1}^{\infty} (-1)^k f'(-1) x^{2k} \right].$$

Now,

$$f(-1) = 2^{-2} \cdot 4^{-2} \cdots (2k - 2)^{-2} (2k)^{-1}$$

$$= \frac{1}{2^{2k-1}(k - 1)!\, k!}$$

and

$$f'(r) = f(r) \left[\frac{-2}{r + 3} + \frac{-2}{r + 5} + \cdots + \frac{-2}{r + 2k - 1} + \frac{-1}{r + 2k + 1} \right]$$

so that

$$f'(-1) = \frac{-1}{2^{2k-1}(k - 1)!\, k!} \left(1 + \frac{1}{2} + \cdots + \frac{1}{k - 1} + \frac{1}{2k} \right).$$

Since $h_k - h_{k-1} = 1/k$, we may write

$$f'(-1) = \frac{-1}{2^{2k-1}(k - 1)!\, k!} [h_{k-1} + \tfrac{1}{2}(h_k - h_{k-1})]$$

$$= \frac{-1}{2^{2k-1}(k - 1)!\, k!} \frac{h_k + h_{k-1}}{2}.$$

We can now express the second solution $u_2(x)$ as follows:

$$u_2(x) = x^{-1} \left[\sum_{k=1}^{\infty} \frac{(-1)^k x^{2k}}{2^{2k-1}(k - 1)!\, k!} \right] \ln x$$

$$+ x^{-1} \left[1 - \sum_{k=1}^{\infty} \frac{(-1)^k x^{2k}}{2^{2k}(k - 1)!\, k!} (h_k + h_{k-1}) \right].$$

Finally, we adjust the form of $u_2(x)$ by altering the index of summation:

$$u_2(x) = \sum_{k=0}^{\infty} \frac{(-1)^{k+1} x^{2k+1}}{2^{2k+1} k!\, (k + 1)!} \ln x + \frac{1}{x} - \sum_{k=0}^{\infty} \frac{(-1)^{k+1} x^{2k+1}}{2^{2k+2} k!\, (k + 1)!} (h_{k+1} + h_k).$$

This is the preferred form of the second solution. Note that the function multiplying $\ln x$ is related to $u_1(x)$ and that $u_1(x)$ can be written as

$$u_1(x) = x\left[2 + \sum_{k=1}^{\infty} \frac{(-1)^k x^{2k}}{2^{2k-1} k! (k+1)!}\right]$$

$$= x\sum_{k=0}^{\infty} \frac{(-1)^k x^{2k}}{2^{2k-1} k! (k+1)!} = \sum_{k=0}^{\infty} \frac{(-1)^k x^{2k+1}}{2^{2k-1} k! (k+1)!}. \quad \blacksquare$$

PROBLEMS

Solve each differential equation for the general solution valid about $x = 0$ (roots differing by an integer).

1. $x\dfrac{d^2u}{dx^2} - u = 0$

2. $x^2\dfrac{d^2u}{dx^2} + x\dfrac{du}{dx} + (x^2 - 1)u = 0$

3. $4x^2\dfrac{d^2u}{dx^2} - 4x(1 - x)\dfrac{du}{dx} + 3u = 0$

4. Solve the Bessel–Clifford equation (Eq. 2.6.1), with $\gamma = N$ a positive integer.

5. Solve $x^2u'' + x^2u' - 2xu = 0$.

6. Solve $xu'' + (3 + 2x)u' + 4u = 0$.

7. Solve $xu'' + u = 0$

8. Solve $xu'' + (4 + 3x)u' + 3u = 0$.

2.10 Bessel's Equation

The family of linear equations

$$x^2\frac{d^2u}{dx^2} + x\frac{du}{dx} + (x^2 - \lambda^2)u = 0, \qquad (2.10.1)$$

known collectively as *Bessel's equation*, is perhaps the single most important nonelementary equation in mathematical physics. Its solutions have been studied in literally thousands of research papers. It often makes its appearance in solving partial differential equations in cylindrical coordinates.

The parameter λ is real and nonnegative and, as we shall see, affects the nature of the solution sets in a very dramatic way. Since the Bessel equation has a regular singular point at $x = 0$, we apply the Frobenius method by assuming a solution of the form

$$u(x) = \sum_{n=0}^{\infty} a_n x^{n+r}. \qquad (2.10.2)$$

2.10.1   ROOTS NOT DIFFERING BY AN INTEGER

Instead of applying the formulas* developed in Section 2.4, it is simpler and more instructive to substitute $u(x)$ and its derivatives into Eq. 2.10.1. This yields

$$\sum_{1=0}^{\infty} (n - r)(n + r - 1)a_n x^{n+r} + \sum_{n=0}^{\infty} (n + r)a_n x^{n+r}$$

$$+ \sum_{n=0}^{\infty} a_n x^{n+r+z} - \sum_{n=0}^{\infty} \lambda^2 a_n x^{n+r} = 0 \qquad (2.10.3)$$

Changing the third summation so that the exponent on x is $n + r$, we have

$$\sum_{n=0}^{\infty} [(n + r)(n + r - 1) + (n + r) - \lambda^2]a_n x^{n+r} + \sum_{n=2}^{\infty} a_{n-2} x^{n+r} = 0.$$

$$(2.10.4)$$

Writing out the first two terms on the first summation gives

$$[r(r - 1) + r - \lambda^2]a_0 x^r + [(1 + r)r + (1 + r) - \lambda^2]a_1 x^{1+r}$$

$$+ \sum_{n=2}^{\infty} \{[(n + r)(n + r - 1) + (n + r) - \lambda^2]a_n + a_{n-2}\}x^{n+r} = 0$$

$$(2.10.5)$$

Equating coefficients of like powers of x to zero gives

$$(r^2 - \lambda^2)a_0 = 0 \qquad (2.10.6)$$

$$(r^2 + 2r + 1 - \lambda^2)a_1 = 0 \qquad (2.10.7)$$

$$[(n + r)^2 - \lambda^2]a_n + a_{n-2} = 0 \qquad (2.10.8)$$

Equation 2.10.6 requires that

$$r^2 - \lambda^2 = 0 \qquad (2.10.9)$$

since $a_0 \neq 0$ according to the method of Frobenius. The indicial equation above has roots $r_1 = \lambda$ and $r_2 = -\lambda$.

Next, we shall find $u_1(x)$ corresponding to $r_1 = \lambda$. Equation 2.10.7 gives $a_1 = 0$, since the quantity in parentheses is not zero. From the recursion relation 2.10.8, we find that $a_3 = a_5 = a_7 = \cdots = 0$. All the coefficients with an odd subscript vanish. For the coefficients with an even subscript, we find that

---

*The student is asked to use these formulas in the Problems following this section.

$$a_2 = -\frac{a_0}{2^2(\lambda + 1)}$$

$$a_4 = -\frac{a_2}{2^2 \cdot 2(\lambda + 2)} = \frac{a_0}{2^4 \cdot 2(\lambda + 1)(\lambda + 2)} \tag{2.10.10}$$

$$a_6 = -\frac{a_4}{2^2 \cdot 3(\lambda + 3)} = -\frac{a_0}{2^6 \cdot 3 \cdot 2(\lambda + 1)(\lambda + 2)(\lambda + 3)}, \quad \text{etc.}$$

In general, we can relate the coefficients with even subscripts to the arbitrary coefficient a_0 by the equation

$$a_{2n} = \frac{(-1)^n a_0}{2^{2n} n!(\lambda + 1)(\lambda + 2) \cdots (\lambda + n)}, \quad n = 0, 1, 2, \cdots. \tag{2.10.11}$$

Because a_0 is arbitrary, it is customary to normalize that a_n's by letting

$$a_0 = \frac{1}{2^\lambda \Gamma(\lambda + 1)}. \tag{2.10.12}$$

With the introduction of the normalizing factor 2.10.12, we have

$$a_{2n} = \frac{(-1)^n}{2^{2n+\lambda} n! \Gamma(\lambda + n + 1)}, \quad n = 0, 1, 2, \cdots, \tag{2.10.13}$$

where we have used

$$\Gamma(\lambda + n + 1) = (\lambda + n)(\lambda + n - 1) \cdots (\lambda + 1)\Gamma(\lambda + 1). \tag{2.10.14}$$

By substituting the coefficients above into our series solution 2.10.2 (replace n with $2n$), we have found one independent solution of Bessel's equation to be

$$J_\lambda(x) = \sum_{n=0}^\infty \frac{(-1)^n x^{2n+\lambda}}{2^{2n+\lambda} n! \Gamma(\lambda + n + 1)}, \tag{2.10.15}$$

where $J_\lambda(x)$ is called the *Bessel function of the first kind* of order λ. The series converges for all values of x, since there are no singular points other than $x = 0$; this results in an infinite radius of convergence. Sketches of $J_0(x)$ and $J_1(x)$ are shown in Fig. 2.4. Table A4 in the Appendix gives the numerical values of $J_0(x)$ and $J_1(x)$ for $0 < x < 15$.

The solution corresponding to $r_2 = -\lambda$ is found simply by replacing λ with $(-\lambda)$. This can be verified by following the steps leading to the expression for $J_\lambda(x)$. Hence, if λ is not an integer, the solution

$$J_{-\lambda}(x) = \sum_{n=0}^\infty \frac{(-1)^n x^{2n-\lambda}}{2^{2n-\lambda} \Gamma(n - \lambda + 1)} \tag{2.10.16}$$

is a second independent solution. It is singular at $x = 0$. The general solution is then

$$u(x) = A J_\lambda(x) + B J_{-\lambda}(x). \tag{2.10.17}$$

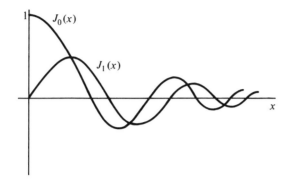

FIGURE 2.4.   Bessel functions of the first kind.

If λ is zero or an integer, $J_{-\lambda}(x)$ is not independent but can be shown to be related to $J_{\lambda}(x)$ by the relation (see Problem 10)

$$J_{-\lambda}(x) = (-1)^n J_{\lambda}(x). \qquad (2.10.18)$$

A second, independent solution for λ zero or a positive integer is given in subsequent sections.

PROBLEMS

1. Apply the formulas developed in Section 2.4 and find expressions for $J_{\lambda}(x)$ and $J_{-\lambda}(x)$.

2. Write out the first four terms in the expansion for (a) $J_0(x)$ and (b) $J_1(x)$.

3. From the expansions in Problem 1, calculate $J_0(2)$ and $J_1(2)$ to four decimal places. Compare with the tabulated values in the Appendix.

4. If we were interested in $J_0(x)$ and $J_1(x)$ for small x only (say, for $x < 0.1$), what algebraic expressions could be used to approximate $J_0(x)$ and $J_1(x)$?

5. Using the expressions from Problem 4, find $J_0(0.1)$ and $J_1(0.1)$ and compare with the tabulated values in the Appendix.

Write the general solution for each equation.

6. $x^2 u'' + xu' + (x^2 - \frac{1}{16})u = 0$

7. $xu'' + u' + (x - \frac{1}{9}x)u = 0$

8. $4x^2 u'' + 4xu' + (4x^2 - 1)u = 0$

9. Solve $u'' + u = 0$ by substituting $u = \sqrt{x}\, v$ and solving the resulting Bessel's equation. Then show that the solution is equivalent to $A \sin x + B \cos x$.

10. Suppose that $\lambda \geq 0$ is an integer. Use the infinite series expansion of $J_{-\lambda}(x)$ to show that

$$J_{-\lambda}(x) = (-1)^{\lambda} J_{\lambda}(x).$$

(*Hint*: When λ is an integer, the series for $J_{-\lambda}$ begins with $n = \lambda$.)

11. Let $I_n(x)$ be a solution of the *modified Bessel equation*,

$$x^2 \frac{d^2 u}{dx^2} + x \frac{du}{dx} - (x^2 + \lambda^2)u = 0.$$

Find $I_n(x)$ by the Frobenius method. (Assume that the roots do not differ by an integer.)

12. Use the results of Problem 11 to verify that

$$I_n(x) = i^{-n} J_n(ix)$$

where $i = \sqrt{-1}$ and $\lambda = n$.

13. Show that

$$J_{1/2}(x) = \left(\frac{2}{\pi x}\right)^{1/2} \sin x$$

and

$$J_{-1/2}(x) = \left(\frac{2}{\pi x}\right)^{1/2} \cos x$$

by examining the respective power series.

2.10.2   EQUAL ROOTS

If $\lambda = 0$, Bessel's equation takes the form

$$x^2 \frac{d^2 u}{dx^2} + x \frac{du}{dx} + x^2 u = 0. \tag{2.10.19}$$

In this case, $b_0 = 1$, $c_2 = 1$, and $F(r) = r^2$, the case of equal roots; hence,

$$L[u] = r^2 a_0 x^r + (r + 1)^2 a_1 x^{r+1} + \sum_{n=2}^{\infty} [(n + r)^2 a_n + a_{n-2}]x^{n+r}.$$
$$\tag{2.10.20}$$

So we set

$$(r + 1)^2 a_1 = 0 \tag{2.10.21}$$

and

$$(n + r)^2 a_n = -a_{n-2}, \qquad n = 2, 3, \ldots \tag{2.10.22}$$

Thus, we have $a_1 = 0$ and the recursion gives $a_3 = a_5 = \cdots = 0$. For the coefficients with an even subscript

$$a_{2k} = \frac{(-1)^k a_0}{(r + 2)^2 (r + 4)^2 \cdots (r + 2k)^2}. \tag{2.10.23}$$

Set $a_0 = 1$ and define u_r as follows:

$$u_r(x) = x^r + \sum_{k=1}^{\infty} \frac{(-1)^k x^{2k+r}}{(r+2)^2(r+4)^2 \cdots (r+2k)^2}. \quad (2.10.24)$$

Note that setting $r = 0$ gives $L[u_r(x)|_{r=0}] = 0$ and

$$u_1(x) = 1 + \sum_{k=1}^{\infty} \frac{(-1)^k x^{2k}}{2^2 \cdot 4^2 \cdots (2k)^2}$$

$$= 1 + \sum_{k=1}^{\infty} \frac{(-1)^k x^{2k}}{2^{2k}(k!)^2} = J_0(x), \quad (2.10.25)$$

the same solution as before. It is the second independent solution that is now needed. It is found by expressing $u_r(x)$ as

$$u_r(x) = x^r \left[1 + \sum_{k=1}^{\infty} \frac{(-1)^k x^{2k}}{(r+2)^2(r+4)^2 \cdots (r+2k)^2} \right]. \quad (2.10.26)$$

Differentiating the expression above gives

$$\frac{\partial u_r}{\partial r} = x^r \ln x \left[1 + \sum_{k=1}^{\infty} \frac{(-1)^k x^{2k}}{(r+2)^2(r+4)^2 \cdots (r+2k)^2} \right]$$

$$+ x^r \left\{ \sum_{k=1}^{\infty} (-1)^k x^{2k} \frac{\partial}{\partial r} [(r+2)^{-2} \cdots (r+2k)^{-2}] \right\} \quad (2.10.27)$$

The derivative of the product appearing in the second sum is computed using logarithmic differentiation. Set

$$f(r) = (r+2)^{-2}(r+4)^{-2} \cdots (r+2k)^{-2}, \quad (2.10.28)$$

so that

$$f(0) = 2^{-2} \cdot 4^{-2} \cdots (2k)^{-2}$$

$$= \frac{1}{2^{2k}(k!)^2} \quad (2.10.29)$$

and

$$f'(r) = -2f(r)\left(\frac{1}{r+2} + \frac{1}{r+4} + \cdots + \frac{1}{r+2k} \right). \quad (2.10.30)$$

We substitute $f'(r)$ into Eq. 2.10.27 and obtain

$$\frac{\partial u_r}{\partial r} = x^r \ln x \left[1 + \sum_{k=1}^{\infty} (-1)^k x^{2k} f(r) \right]$$

$$+ x^r \sum_{k=1}^{\infty} (-1)^k x^{2k}(-2)f(r)\left(\frac{1}{r+2} + \cdots + \frac{1}{r+2k} \right) \quad (2.10.31)$$

$$= u_r(x) \ln x + x^r \sum_{k=1}^{\infty} (-1)^k x^{2k}(-2)f(r)\left(\frac{1}{r+2} + \cdots + \frac{1}{r+2k} \right).$$

Setting $r = 0$ yields

$$u_2(x) = \left. \frac{\partial u_r}{\partial r} \right|_{r=0}$$

$$= J_0(x) \ln x + \sum_{k=1}^{\infty} \frac{(-1)^{k+1} x^{2k}}{2^{2k}(k!)^2} h_k. \qquad (2.10.32)$$

The *Bessel function of the second kind of order zero* is a linear combination of $J_0(x)$ and $u_2(x)$. Specifically,

$$Y_0(x) = \frac{2}{\pi} u_2(x) + \frac{2}{\pi} (\gamma - \ln 2) J_0(x), \qquad (2.10.33)$$

where γ is the *Euler constant*, $\gamma = 0.57721566490$. Finally, we write

$$Y_0(x) = \frac{2}{\pi} J_0(x) \left[\ln \frac{x}{2} + \gamma \right] + \sum_{k=1}^{\infty} \frac{(-1)^{k+1} x^{2k}}{2^{2k}(k!)^2} h_k \qquad (2.10.34)$$

and the general solution is

$$u(x) = A J_0(x) + B Y_0(x). \qquad (2.10.35)$$

PROBLEMS

1. Find the "logarithmic" solution of the modified Bessel equation, $\lambda = 0$,

$$x^2 \frac{d^2 u}{dx^2} + x \frac{du}{dx} - x^2 u = 0.$$

2. Use Eq. 2.10.34 and Table A4 for $J_0(x)$ to compute $Y_0(x)$ for $x = 0.1, 0.2, \ldots,$ 1.0. Approximate the infinite series part of the definition of $Y_0(x)$ with its first two nonzero terms.

3. Explain how Eq. 2.10.34 arises from Eq. 2.10.32.

2.10.3   ROOTS DIFFERING BY AN INTEGER

The Bessel equation with $\lambda = 1$ was solved in Example 2.9.2. The expression for the second solution was found to be

$$u_2(x) = \sum_{k=0}^{\infty} \frac{(-1)^{k+1} x^{2k+1}}{2^{2k+1} k!(k+1)!} \ln x + \frac{1}{x} - \sum_{k=0}^{\infty} \frac{(-1)^{k+1} x^{2k+1}}{2^{2k+2} k!(k+1)!} (h_{k+1} + h_k).$$

$$(2.10.36)$$

Using $-J_1(x)$ to represent the series multiplying $\ln x$,

$$u_2(x) = -J_1(x) \ln x + \frac{1}{x} - \frac{1}{2} \sum_{k=0}^{\infty} \frac{(-1)^k x^{2k+1}}{2^{2k+1} k! (k+1)!} (h_{k+1} + h_k). \quad (2.10.37)$$

A standard form for this "second" solution is denoted by $Y_1(x)$ and is called *Bessel's function of the second kind of order* 1. It is defined as

$$Y_1(x) = -\frac{2}{\pi} u_2(x) + \frac{2}{\pi} (\gamma - \ln 2) J_1(x), \quad (2.10.38)$$

where γ is Euler's constant. (Compare with Eq. 2.10.34.) In neater form,

$$Y_1(x) = \frac{2}{\pi} J_1(x) \left[\ln \frac{x}{2} + \gamma \right] - \frac{2}{\pi x} + \frac{x}{\pi} \sum_{k=0}^{\infty} \frac{(-1)^{k-1} x^{2k} (h_{k+1} + h_k)}{2^{2k+1} k! (k+1)!}.$$

$$(2.10.39)$$

Hence, the general solution of Bessel's equation of order 1 is

$$u(x) = A J_1(x) + B Y_1(x). \quad (2.10.40)$$

A similar analysis holds for Bessel's equation of order N, where N is a positive integer. The equation is

$$x^2 \frac{d^2 u}{dx^2} + x \frac{du}{dx} + (x^2 - N^2) u = 0 \quad (2.10.41)$$

and a pair of basic solutions is (see Eq. 2.10.15)

$$J_N(x) = \sum_{n=0}^{\infty} \frac{(-1)^n x^{2n+N}}{2^{2n+N} n! (n+N)!} \quad (2.10.42)$$

and

$$Y_N(x) = \frac{2}{\pi} J_N(x) \left[\ln \frac{x}{2} + \gamma \right]$$

$$- \frac{1}{\pi} \sum_{n=0}^{\infty} \frac{(-1)^n x^{2n+N}}{2^{2n+N} n! (n+N)!} (h_{n+N} + h_n) - \frac{1}{\pi} \sum_{n=0}^{N-1} \frac{(N-n-1)! x^{2n-N}}{2^{2n-N} n!}$$

$$(2.10.43)$$

with the general solution

$$u(x) = A J_N(x) + B Y_N(x). \quad (2.10.44)$$

Graphs of $Y_0(x)$ and $Y_1(x)$ are shown in Fig. 2.5. Since $Y_0(x)$ and $Y_1(x)$ are not defined at $x = 0$, that is, $Y_0(0) = Y_1(0) = -\infty$, the solution 2.10.44 for a problem with a finite boundary condition at $x = 0$ requires that $B = 0$; the solution would then only involve Bessel functions of the first kind.

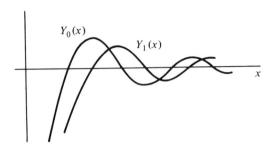

FIGURE 2.5.   Bessel functions of the second kind.

PROBLEMS

1. Derive Eq. 2.10.39 from Eq. 2.10.38.

2. Derive Eq. 2.10.43 by obtaining the second solution of Eq. 2.10.41 by the method of Frobenius.

Show that Y_n satisfies each relationship.

3. $xY_n'(x) = xY_{n-1}(x) - nY_n(x)$

4. $xY_n'(x) = xY_{n+1}(x) + nY_n(x)$

5. $2Y_n'(x) = Y_{n-1}(x) + Y_{n+1}(x)$

6. $2nY_n(x) = x[Y_{n-1}(x) - Y_{n+1}(x)]$

7. Use Eq. 2.10.15 to show that

$$\frac{\partial}{\partial \lambda} J_\lambda(x) = J_\lambda(x) \ln \frac{x}{2} - \left(\frac{x}{2}\right)^\lambda \sum_{k=0}^{\infty} \frac{(-1)^k \Psi(\lambda + k + 1)}{2^k \Gamma(\lambda + k + 1)} \frac{x^{2k}}{k!}$$

and hence

$$\frac{\partial}{\partial \lambda} J_\lambda(x) \bigg|_{\lambda=0} = \frac{\pi}{2} Y_0(x).$$

[*Hint*: Use Problems 14 and 15 of Section 2.5 and the fact that $h_n = \Psi(n + 1) + \gamma$.]

8. Show that the Wronskian,

$$W(J_n, Y_n) = J_{n+1} Y_n - J_n Y_{n+1} = \frac{2}{\pi x}.$$

9. Show that

$$\frac{d}{dx} Y_0(x) = -Y_1(x), \qquad \frac{d}{dx} J_0(x) = -J_1(x).$$

10. Prove that Wronskian, for x not an integer, is

$$W(J_n, J_{-n}) = k/x, \qquad k \neq 0.$$

11. Use the results of Problem 13, Section 2.10.1 and Problem 9 above to evaluate $W(J_{1/2}, J_{-1/2})$.

2.10.4   BASIC IDENTITIES

In the manipulation of Bessel functions a number of helpful identities are used. This section is devoted to presenting some of the most important of these identities. Let use first show that

$$\frac{d}{dx}[x^{\lambda+1}J_{\lambda+1}(x)] = x^{\lambda+1}J_\lambda(x). \tag{2.10.45}$$

The series expansion (2.10.15) gives

$$x^{\lambda+1}J_{\lambda+1}(x) = \sum_{n=0}^{\infty} \frac{(-1)^n x^{2n+2\lambda+2}}{2^{2n+\lambda+1}n!\Gamma(\lambda + n + 2)}. \tag{2.10.46}$$

This is differentiated to yield

$$\begin{aligned}
\frac{d}{dx}[x^{\lambda+1}J_{\lambda+1}(x)] &= \sum_{n=0}^{\infty} \frac{(-1)^n(2n + 2\lambda + 2)x^{2n+2\lambda+1}}{2^{2n+\lambda+1}n!\Gamma(\lambda + n + 2)} \\
&= \sum_{n=0}^{\infty} \frac{(-1)^n 2(n + \lambda + 1)x^{2n+2\lambda+1}}{2 \cdot 2^{2n+\lambda}n!(\lambda + n + 1)\Gamma(\lambda + n + 1)} \\
&= x^{\lambda+1}\sum_{n=0}^{\infty} \frac{(-1)^n x^{2n+\lambda}}{2^{2n+\lambda}n!\Gamma(\lambda + n + 1)} \\
&= x^{\lambda+1}J_\lambda(x). \tag{2.10.47}
\end{aligned}$$

This proves the relationship (2.10.45). Following this procedure, we can show that

$$\frac{d}{dx}[x^{-\lambda}J_\lambda(x)] = -x^{-\lambda}J_{\lambda+1}(x). \tag{2.10.48}$$

From the two identities above we can perform the indicated differentiation on the left-hand sides and arrive at

$$x^{\lambda+1}\frac{dJ_{\lambda+1}}{dx} + (\lambda + 1)x^\lambda J_{\lambda+1} = x^{\lambda+1}J_\lambda$$

$$\tag{2.10.49}$$

$$x^{-\lambda}\frac{dJ_\lambda}{dx} - \lambda x^{-\lambda-1}J_\lambda = -x^{-\lambda}J_{\lambda+1}.$$

Let us multiply the first equation above by $x^{-\lambda-1}$ and the second by x^λ. There results

$$\frac{dJ_{\lambda+1}}{dx} + \frac{\lambda+1}{x}J_{\lambda+1} = J_\lambda$$

$$\frac{dJ_\lambda}{dx} - \frac{\lambda}{x}J_\lambda = -J_{\lambda+1}.$$

(2.10.50)

If we now replace $\lambda + 1$ with λ in the first equation, we have

$$\frac{dJ_\lambda}{dx} + \frac{\lambda}{x}J_\lambda = J_{\lambda-1}.$$

(2.10.51)

This equation can be added to the second equation of 2.10.50 to obtain

$$\frac{dJ_\lambda}{dx} = \tfrac{1}{2}(J_{\lambda-1} - J_{\lambda+1}).$$

(2.10.52)

Equation 2.10.51 can also be subtracted from the second equation of (2.10.50) to obtain the important recurrence relation

$$J_{\lambda+1}(x) = \frac{2\lambda}{x}J_\lambda(x) - J_{\lambda-1}(x).$$

(2.10.53)

This allows us to express Bessel functions of higher order in terms of Bessel functions of lower order. This is the reason that tables only give $J_0(x)$ and $J_1(x)$ as entries. All higher-order Bessel functions can be related to $J_0(x)$ and $J_1(x)$. By rewriting Eq. 2.10.53, we can also relate Bessel functions of higher negative order to $J_0(x)$ and $J_1(x)$. We would use

$$J_{\lambda-1}(x) = \frac{2\lambda}{x}J_\lambda(x) - J_{\lambda+1}(x).$$

(2.10.54)

In concluding this section, let us express the differentiation identities 2.10.45 and 2.10.48 as integration identities. By integrating once we have

$$\int x^{\lambda+1}J_\lambda(x)\,dx = x^{\lambda+1}J_{\lambda+1}(x) + C$$

(2.10.55)

$$\int x^{-\lambda}J_{\lambda+1}(x)\,dx = -x^{-\lambda}J_\lambda(x) + C.$$

These formulas are used when integrating Bessel functions.

Example 2.10.1:   Find numerical values for the quantities $J_4(3)$ and $J_{-4}(3)$ using the recurrence relations.

SOLUTION: We use the recurrence relation 2.10.53 to find a value for $J_4(3)$. It gives

$$J_4(3) = \frac{2 \cdot 3}{3} J_3(3) - J_2(3)$$

$$= 2\left[\frac{2 \cdot 2}{3} J_2(3) - J_1(3)\right] - J_2(3)$$

$$= \frac{5}{3}\left[\frac{2}{3} J_1(3) - J_0(3)\right] - 2J_1(3)$$

$$= -\frac{8}{9} J_1(3) - \frac{5}{3} J_0(3)$$

$$= -\frac{8}{9} \times 0.339 - \frac{5}{3} \times (-0.260) = 0.132.$$

Now, to find a value for $J_{-4}(3)$ we use Eq. 2.10.54 to get

$$J_{-4}(3) = \frac{2(-3)}{3} J_{-3}(3) - J_{-2}(3)$$

$$= -2\left[\frac{2(-2)}{3} J_{-2}(3) - J_{-1}(3)\right] - J_{-2}(3)$$

$$= \frac{5}{3}\left[\frac{2(-1)}{3} J_{-1}(3) - J_0(3)\right] + 2J_{-1}(3)$$

$$= \frac{8}{9}[-J_1(3)] - \frac{5}{3} J_0(3) = 0.132.$$

We see that $J_4(x) = J_{-4}(x)$.   ■

Example 2.10.2:   Integrals involving Bessel functions are often encountered in the solution of physically motivated problems. Determine an expression for

$$\int x^2 J_2(x) \, dx.$$

SOLUTION: To use the second integration formula of 2.10.55, we put the integral in the form

$$\int x^2 J_2(x) \, dx = \int x^3[x^{-1} J_2(x)] \, dx.$$

$$u = x^3 \qquad\qquad dv = x^{-1} J_2(x) \, dx$$

$$du = 3x^2 \qquad\qquad v = -x^{-1} J_1(x).$$

Then

$$\int x^2 J_2(x) \, dx = -x^2 J_1(x) + 3 \int x J_1(x) \, dx.$$

Again we integrate by parts:

$$u = x \qquad\qquad dv = J_1(x) \, dx$$

$$du = dx \qquad\qquad v = -J_0(x).$$

There results

$$\int x^2 J_2(x) \, dx = -x^2 J_1(x) - 3x J_0(x) + 3 \int J_0(x) \, dx.$$

The last integral, $\int J_0(x) \, dx$, cannot be evaluated using our integration formulas. Because it often appears when integrating Bessel functions, it has been tabulated, although we will not include it in this work. However, we must recognize when we arrive at $\int J_0(x) \, dx$, our integration is complete. In general, whenever we integrate $\int x^n J_m(x) \, dx$ and $n + m$ is even and positive, the integral $\int J_0(x) \, dx$ will appear.   ∎

PROBLEMS

Evaluate each term.

1. $J_3(2)$ **2.** $J_5(5)$ **3.** $\dfrac{dJ_0}{dx}$ at $x = 2$

4. $\dfrac{dJ_2}{dx}$ at $x = 4$ **5.** $\dfrac{dJ_1}{dx}$ at $x = 1$ **6.** $\dfrac{dJ_3}{dx}$ at $x = 1$

Find an expression in terms of $J_1(x)$ and $J_0(x)$ for each integral.

7. $\displaystyle\int x^3 J_2(x) \, dx$ **8.** $\displaystyle\int x J_2(x) \, dx$ **9.** $\displaystyle\int \frac{J_4(x)}{x} \, dx$

10. $\displaystyle\int x J_1(x) \, dx$ **11.** $\displaystyle\int x^3 J_1(x) \, dx$ **12.** $\displaystyle\int \frac{J_3(x)}{x} \, dx$

13. We know that

$$J_{1/2}(x) = \left(\frac{2}{\pi x}\right)^{1/2} \sin x, \qquad J_{-1/2}(x) = \left(\frac{2}{\pi x}\right)^{1/2} \cos x.$$

Use Eq. 2.10.53 to find expressions for $J_{3/2}(x)$ and $J_{5/2}(x)$.

Prove each identity for the modified Bessel functions of Problem 11 of Section 2.10.1.

14. $x I'_n(x) = x I_{n-1}(x) - n I_n(x)$

15. $x I'_n(x) = x I_{n+1}(x) + n I_n(x)$

16. $2 I'_n(x) = I_{n-1}(x) + I_{n+1}(x)$

17. $2 n I_n(x) = x[I_{n-1}(x) - I_{n+1}(x)]$

18. Express $I_{1/2}(x)$ and $I_{-1/2}(x)$ in terms of elementary functions analogous to the expressions for $J_{1/2}(x)$ and $J_{-1/2}(x)$ in Problem 13.

2.11 Nonhomogeneous Equations

A general solution of

$$\frac{d^2u}{dx^2} + p_0(x)\frac{du}{dx} + p_1(x)u = f(x) \tag{2.11.1}$$

is the sum of a general solution of

$$\frac{d^2u}{dx^2} + p_0(x)\frac{du}{dx} + p_1(x)u = 0 \tag{2.11.2}$$

and any particular solution of Eq. 2.11.1. If a general solution of Eq. 2.11.2 is known, the method of variation of parameters will always generate a particular solution of Eq. 2.11.1. When Eq. 2.11.2 has its solution expressed as a power or Frobenius series, the general solution of Eq. 2.11.1 will also be in series form. An example will illustrate.

Example 2.11.1: Find a particular solution of

$$\frac{d^2u}{dx^2} + x^2u = x.$$

SOLUTION: The point $x = 0$ is an ordinary point of the given equation and the technique of Section 2.2 provides the following pair of independent solutions:

$$u_1(x) = 1 - \frac{x^4}{3\cdot 4} + \frac{x^8}{3\cdot 4\cdot 7\cdot 8} - \cdots$$

$$u_2(x) = x - \frac{x^5}{4\cdot 5} + \frac{x^9}{4\cdot 5\cdot 8\cdot 9} - \cdots.$$

In the method of variation of parameters, we assume a solution in the form

$$u_p(x) = v_1u_1 + v_2u_2$$

and determine $v_1(x)$ and $v_2(x)$. The equations for v_1 and v_2 are (see Section 1.11)

$$v_1'u_1 + v_2'u_2 = 0$$

$$v_1'u_1' + v_2'u_2' = x,$$

which have the unique solution

$$v_1'(x) = -\frac{xu_2(x)}{W}$$

$$v_2'(x) = \frac{xu_1(x)}{W(x)}$$

where $W(x)$ is the Wronskian of u_1 and u_2. However,

$$W(x) = Ke^{-\int [p_0(x)/x]\,dx} = Ke^0 = K$$

since $p_0(x) = 0$. We pick $K = 1$, since retaining K simply generates a multiple of $u_p(x)$. Hence, the expressions above be integrated to give

$$v_1(x) = -\int xu_2(x)\,dx$$

$$= -\int \left(x^2 - \frac{x^6}{4\cdot 5} + \frac{x^{10}}{4\cdot 5\cdot 8\cdot 9} - \cdots \right) dx$$

$$= -\frac{x^3}{3} + \frac{x^7}{4\cdot 5\cdot 7} - \frac{x^{11}}{4\cdot 5\cdot 8\cdot 9\cdot 11} + \cdots$$

and

$$v_2(x) = \int xu_1(x)\,dx$$

$$= \int \left(x - \frac{x^5}{3\cdot 4} + \frac{x^9}{3\cdot 4\cdot 7\cdot 8} - \cdots \right) dx$$

$$= \frac{x^2}{2} - \frac{x^6}{3\cdot 4\cdot 6} + \frac{x^{10}}{3\cdot 4\cdot 7\cdot 8\cdot 10} - \cdots .$$

The regular pattern of the coefficients is obscured when we substitute these series into the expression for $u_p(x)$:

$$u_p(x) = \left(-\frac{x^3}{3} + \frac{x^7}{4\cdot 5\cdot 7} - \cdots \right)\left(1 - \frac{x^4}{3\cdot 4} + \cdots \right)$$

$$+ \left(\frac{x^2}{2} - \frac{x^6}{3\cdot 4\cdot 6} + \cdots \right)\left(x - \frac{x^5}{4\cdot 5} + \cdots \right)$$

$$= \frac{x^3}{6} - \frac{x^7}{6\cdot 6\cdot 7} + \cdots . \quad \blacksquare$$

An alternative method sometimes yields the pattern of the coefficients. We replace the nonhomogeneous term $f(x)$ by its power series and substitute

$$u_p(x) = \sum_{n=0}^{\infty} b_n x^n \qquad (2.11.3)$$

into the left-hand side of Eq. 2.11.1 It is usually convenient to add the conditions that $u(0) = 0 = u'(0)$, which means that $b_0 = b_1 = 0$ in Eq. 2.11.3. The following example illustrates this method.

Example 2.11.2:   Find the solution to the initial-value problem

$$\frac{d^2 u}{dx^2} + x^2 u = x, \qquad u(0) = u'(0) = 0.$$

SOLUTION: We ignore $u_1(x)$ and $u_2(x)$ and proceed by assuming that $u_p(x)$ can be expressed as the power series of Eq. 2.11.3. We have

$$x^2 u_p(x) = \sum_{n=0}^{\infty} b_n x^{n+2} = \sum_{n=2}^{\infty} b_{n-2} x^n$$

and

$$u_p''(x) = \sum_{n=2}^{\infty} (n-1)n b_n x^{n-2} = \sum_{n=0}^{\infty} (n+1)(n+2) b_{n+2} x^n.$$

Hence,

$$u_p'' + x^2 u_p = 2b_2 + 6b_3 x + \sum_{n=2}^{\infty} [(n+1)(n+2)b_{n+2} + b_{n-2}] x^n = x.$$

Identifying the coefficients of like powers of x leads to

$$x^0: \qquad\qquad\qquad b_2 = 0$$

$$x^1: \qquad\qquad\qquad 6b_3 = 1$$

$$x^n: \qquad (n+2)(n+1)b_{n+2} + b_{n-2} = 0, \qquad n \geq 2.$$

In view of the fact that $b_0 = b_1 = b_2 = 0$ and that the subscripts in the two-term recursion differ by four, we solve the recursion by starting with $n = 5$. The table

1	$n = 5$	$b_7 = -\dfrac{b_3}{7 \cdot 6}$
2	$n = 9$	$b_{11} = -\dfrac{b_7}{11 \cdot 10}$
$\vdots$	$\vdots$	$\vdots$
k	$n = 4k + 1$	$b_{4k+3} = -\dfrac{b_{4k-1}}{(4k+3)(4k+2)}$

leads to the solution

$$b_{4k+3} = \frac{(-1)^k b_3}{6 \cdot 7 \cdot 10 \cdot 11 \cdots (4k+2)(4k+3)}.$$

Thus, noting that $b_3 = \frac{1}{6}$,

$$u_p(x) = \frac{1}{6}\left[x^3 + \sum_{k=1}^{\infty} \frac{(-1)^k x^{4k+1}}{6 \cdot 7 \cdots (4k+2)(4k+3)} \right].$$

This method is preferred whenever it can be effected, since it generates either the pattern of the coefficients or, failing that, as many coefficients as desired. The first two terms in the expansion in Example 2.11.1 are verified using the series above.

The general solution is, then, using $u_1(x)$ and $u_2(x)$ from Example 2.11.1,

$$u(x) = A u_1(x) + B u_2(x) + u_p(x).$$

Using $u(0) = 0$ requires that $A = 0$; using $u'(0) = 0$ requires that $B = 0$. The solution is then

$$u(x) = \frac{1}{6}\left[x^3 + \sum_{k=1}^{\infty} \frac{(-1)^k x^{4k+1}}{6 \cdot 7 \cdots (4k + 2)(4k + 3)} \right].$$

Any variation in the initial conditions from $u(0) = u'(0) = 0$ brings in the series for $u_1(x)$ or $u_2(x)$. For instance, since

$$u(x) = Au_1(x) + Bu_2(x) + u_p(x),$$

the initial conditions, $u(0) = -1$, $u'(0) = 2$, lead to the solution

$$u(x) = -u_1(x) + 2u_2(x) + \frac{1}{6}\left[x^3 - \frac{x^5}{6 \cdot 7} + \cdots \right]. \quad \blacksquare$$

PROBLEMS

Solve each initial-value problem by expanding about $x = 0$.

1. $(4 - x^2)u'' + 2u = x^2 + 2x,$ $u(0) = 0,$ $u'(0) = 0$

2. $u'' + (1 - x)u = 4x,$ $u(0) = 1,$ $u'(0) = 0$

3. $u'' - x^2 u' + u \sin x = 4 \cos x,$ $u(0) = 0,$ $u'(0) = 1$

4. Solve $(1 - x)f' - f = 2x$ using a power-series expansion. Let $f = 6$ at $x = 0$, and expand about $x = 0$. Obtain five terms in the series and compare with the exact solution for values of $x = 0, \frac{1}{4}, \frac{1}{2}, 1$, and 2.

5. Solve the differential equation $u'' + x^2 u = 2x$ using a power-series expansion if $u(0) = 4$ and $u'(0) = -2$. Find an approximate value for $u(x)$ at $x = 2$.

Solve each differential equation for a general solution using the power-series method by expanding about $x = 0$. Note the radius of convergence for each solution.

6. $\dfrac{du}{dx} + u = x^2$

7. $(1 - x)\dfrac{du}{dx} + u = x$

8. $x\dfrac{du}{dx} + x^2 u = \sin x$

9. $\dfrac{d^2 u}{dx^2} + 2\dfrac{du}{dx} + u = x^2$

10. $\dfrac{d^2 u}{dx^2} + 5\dfrac{du}{dx} + 6u = x^2 + 2 \sin x$

11. The solution to $(1 - x)\, df/dx - f = 2x$ is desired in the interval from $x = 1$ to $x = 2$. Expand about $x = 2$ and determine the value of $f(x)$ at $x = 1.9$ if $f(2) = 1$. Compare with the exact solution.

3

Laplace Transforms

3.1 Introduction

The solution of a linear, ordinary differential equation with constant coefficients may be obtained by using the Laplace transformation. It is particularly useful in solving nonhomogeneous equations that result when modeling systems involving discontinuous, periodic input functions. It is not necessary, however, when using Laplace transforms that a homogeneous solution and a particular solution be added together to form the general solution. In fact, we do not find a general solution when using Laplace transforms. The initial conditions must be given and with them we obtain the specific solution to the nonhomogeneous equation directly, with no additional steps. This makes the technique quite attractive.

Another attractive feature of using Laplace transforms to solve a differential equation is that the transformed equation is an algebraic equation. The algebraic equation is then used to determine the solution to the differential equation.

The general technique of solving a differential equation using Laplace transforms involves finding the transform of each term in the equation, solving the resulting algebraic equation in terms of the new transformed variable, then finally solving the inverse problem to retrieve the original variables. We shall follow that order in this chapter. Let us first find the Laplace transform of the various quantities that occur in our differential equations.

3.2   The Laplace Transform

Let the function $f(t)$ be the dependent variable of an ordinary differential equation that we wish to solve. Multiply $f(t)$ by e^{-st} and integrate with respect to t from 0 to infinity. The independent variable t integrates out and there remains a function of s, say $F(s)$. This is expressed as

$$F(s) = \int_0^\infty f(t)e^{-st}\,dt. \qquad (3.2.1)$$

The function $F(s)$ is called the *Laplace transform* of the function $f(t)$. We will return often to this definition of the Laplace transform. It is usually written as

$$\mathscr{L}(f) = F(s) = \int_0^\infty f(t)e^{-st}\,dt, \qquad (3.2.2)$$

where the script $\mathscr{L}$ denotes the Laplace transform operator. We shall consistently use a lowercase letter to represent a function and its capital to denote its Laplace transform; that is, $Y(s)$ denotes the Laplace transform of $y(t)$. The *inverse Laplace transform* will be denoted by $\mathscr{L}^{-1}$, resulting in

$$f(t) = \mathscr{L}^{-1}(F). \qquad (3.2.3)$$

There are two threats to the existence of the Laplace transform; the first is that

$$\int_0^{t_0} f(t)e^{-st}\,dt$$

may not exist because, for instance, $\lim_{t \to t_0} f(t) = +\infty$. The second is that, as an improper integral

$$\int_0^\infty f(t)e^{-st}\,dt$$

diverges. We avoid the first pitfall by requiring $f(t)$ to be sectionally continuous (see Section 3.3 and Fig. 3.1). The second problem is avoided by assuming that there exists a constant M such that

$$|f(t)| \le Me^{bt} \qquad \text{for all} \quad t \ge 0. \qquad (3.2.4)$$

Functions that satisfy Eq. 3.2.4 are of *exponential order* as $t \to \infty$. If $f(t)$ is of exponential order

$$\left| \int_0^\infty f(t)e^{-st}\,dt \right| \le \int_0^\infty |f(t)|e^{-st}\,dt \le M \int_0^\infty e^{(b-s)t}\,dt = \frac{M}{s-b}. \qquad (3.2.5)$$

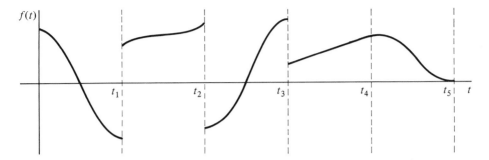

FIGURE 3.1.   Sectionally continuous function.

The function e^{t^2} does not possess a Laplace transform; note that it is not of exponential order. It is an unusual function not often encountered in the solution of real problems. By far the majority of functions representing physical quantities will possess Laplace transforms. Thus, if $f(t)$ is sectionally continuous in every finite interval and of exponential order as $t \rightarrow \infty$, then

$$\mathcal{L}[f(t)] = \int_0^\infty f(t)e^{-st}\, dt \qquad (3.2.6)$$

exists. Moreover, the inequality proved above leads to

(i) $sF(s)$ is bounded as $s \rightarrow \infty$, from which it follows that

(ii) $\lim\limits_{s \to \infty} F(s) = 0$.

Thus, $F(s) = 1$, for instance, is not a Laplace transform of any function in our class. We will assume that any $f(t)$ is sectionally continuous on every interval $0 < t < t_0$ and is of exponential order as $t \rightarrow \infty$. Thus $\mathcal{L}(f)$ exists.

Before considering some examples that demonstrate how the Laplace transforms of various functions are found, let us consider some important properties of the Laplace transform. First, the Laplace transform operator $\mathcal{L}$ is a *linear operator*. This is expressed as

$$\mathcal{L}[af(t) + bg(t)] = a\mathcal{L}(f) + b\mathcal{L}(g), \qquad (3.2.7)$$

where a and b are constants. To verify that this is true, we simply substitute the quantity $[af(t) + bg(t)]$ into the definition for the Laplace transform, obtaining

$$\mathcal{L}[af(t) + bg(t)] = \int_0^\infty [af(t) + bg(t)]e^{-st}\, dt$$

$$= a \int_0^\infty f(t)e^{-st}\, dt + b \int_0^\infty g(t)e^{-st}\, dt$$

$$= a\mathcal{L}(f) + b\mathcal{L}(g). \qquad (3.2.8)$$

The second property is often called the *first shifting property*. It is expressed as

$$\mathcal{L}[e^{at}f(t)] = F(s - a), \tag{3.2.9}$$

where $F(s)$ is the Laplace transform of $f(t)$. This is proved by using $e^{at}f(t)$ in place of $f(t)$ in Eq. 3.2.2; there results

$$\mathcal{L}[e^{at}f(t)] = \int_0^\infty e^{at}f(t)e^{-st}\,dt = \int_0^\infty f(t)e^{-(s-a)t}\,dt. \tag{3.2.10}$$

Now, let $s - a = \lambda$. Then we have

$$\mathcal{L}[e^{at}f(t)] = \int_0^\infty f(t)e^{-\lambda t}\,dt$$

$$= F(\lambda) = F(s - a). \tag{3.2.11}$$

The third property is the *second shifting property*. It is stated as follows: If the Laplace transform of $f(t)$ is known to be

$$\mathcal{L}(f) = F(s), \tag{3.2.12}$$

and if

$$g(t) = \begin{cases} f(t - a), & t > a \\ 0, & t < a, \end{cases} \tag{3.2.13}$$

then the Laplace transform of $g(t)$ is

$$\mathcal{L}(g) = e^{-as}F(s) \tag{3.2.14}$$

To show this result, the Laplace transform of $g(t)$ given by Eq. 3.2.13 is

$$\mathcal{L}(g) = \int_0^\infty g(t)e^{-st}\,dt = \int_0^a 0 \cdot e^{-st}\,dt + \int_a^\infty f(t - a)e^{-st}\,dt. \tag{3.2.15}$$

Make the substitution $\tau = t - a$. Then $d\tau = dt$ and we have

$$\mathcal{L}(g) = \int_0^\infty f(\tau)e^{-s(\tau+a)}\,d\tau = e^{-as}\int_0^\infty f(\tau)e^{-s\tau}\,d\tau = e^{-as}F(s) \tag{3.2.16}$$

and the second shifting property is verified.

The fourth property follows from a change of variables. Set $\tau/a = t$ in

$$F(s) = \int_0^\infty e^{-st}f(t)\,dt. \tag{3.2.17}$$

Then, since $dt = d\tau/a$,

$$F(s) = \frac{1}{a}\int_0^\infty e^{-(s/a)\tau} f\left(\frac{\tau}{a}\right) d\tau \tag{3.2.18}$$

which may be written, using $a\delta = s$,

$$aF(a\delta) = \mathscr{L}\left[f\left(\frac{t}{a}\right)\right]. \tag{3.2.19}$$

The four properties above simplify the task of finding the Laplace transform of a particular function $f(t)$, or the inverse transform of $F(s)$. This will be illustrated in the following examples. Table 3.1, which gives the Laplace transform of a variety of functions, is found at the end of this chapter.

Example 3.2.1:   Find the Laplace transform of the unit step function

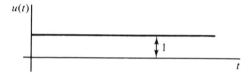

$$u_0(t) = \begin{cases} 1, & t > 0 \\ 0, & t < 0. \end{cases}$$

SOLUTION: Using the definition of the Laplace transform, we have

$$\mathscr{L}(u_0) = \int_0^\infty u_0(t)e^{-st}\, dt = \int_0^\infty e^{-st}\, dt = -\frac{1}{s}e^{-st}\bigg|_0^\infty = \frac{1}{s}.$$

This will also be used as the Laplace transform of unity, that is $\mathscr{L}(1) = 1/s$, since the integration occurs between zero and infinity, as above.   ∎

Example 3.2.2:   Use the first shifting property to find the Laplace transform of e^{at}.

SOLUTION: Equation 3.2.9 provides us with

$$\mathscr{L}(e^{at}) = F(s - a),$$

where the transform of unity is, from Example 3.2.1,

$$F(s) = \frac{1}{s}.$$

We simply substitute $s - a$ for s and obtain

$$\mathscr{L}(e^{at}) = \frac{1}{s - a}.\quad ∎$$

Example 3.2.3:   Use the second shifting property and find the Laplace transform of the unit step function $u_a(t)$ defined by

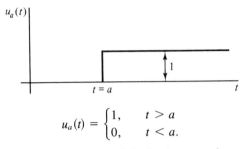

$$u_a(t) = \begin{cases} 1, & t > a \\ 0, & t < a. \end{cases}$$

Check the result by using the definition of the Laplace transform.

SOLUTION: Using the second shifting theorem given by Eq. 3.2.14, there results

$$\mathcal{L}(u_a) = e^{-as} F(s)$$

$$= \frac{1}{s} e^{-as},$$

where $F(s)$ is the Laplace transform of the unit step function.

To check the result above, we use the definition of the Laplace transform:

$$\mathcal{L}(u_a) = \int_0^\infty u_a(t) e^{-st} \, dt = \int_0^a 0 \cdot e^{-st} \, dt + \int_a^\infty e^{-st} \, dt = -\frac{1}{s} e^{-st} \Big|_a^\infty = \frac{1}{s} e^{-as}.$$

This, of course, checks the result obtained with the second shifting theorem. ∎

Example 3.2.4:   Determine the Laplace transform of $\sin \omega t$ and $\cos \omega t$ by using

$$e^{i\theta} = \cos \theta + i \sin \theta,$$

the first shifting property, and the linearity property.

SOLUTION: The first shifting property allows us to write (see Example 3.2.2)

$$\mathcal{L}(e^{i\omega t}) = \frac{1}{s - i\omega}$$

$$= \frac{1}{s - i\omega} \frac{s + i\omega}{s + i\omega} = \frac{s + i\omega}{s^2 + \omega^2} = \frac{s}{s^2 + \omega^2} + i \frac{\omega}{s^2 + \omega^2}.$$

Using the linearity property expressed by Eq. 3.2.7, we have

$$\mathcal{L}(e^{i\omega t}) = \mathcal{L}(\cos \omega t + i \sin \omega t) = \mathcal{L}(\cos \omega t) + i\mathcal{L}(\sin \omega t).$$

Equating the real and imaginary parts of the two equations above results in

$$\mathcal{L}(\sin \omega t) = \frac{\omega}{s^2 + \omega^2}$$

$$\mathcal{L}(\cos \omega t) = \frac{s}{s^2 + \omega^2}.$$

These two Laplace transforms could have been obtained by substituting directly into Eq. 3.2.2, each of which would have required integrating by parts twice. ■

Example 3.2.5:   Find the Laplace transform of t^k.

SOLUTION: The Laplace transform of t^k is given by

$$\mathcal{L}(t^k) = \int_a^\infty t^k e^{-st}\, dt.$$

To integrate this, we make the substitution

$$\xi = st, \qquad dt = \frac{d\xi}{s}.$$

There then results

$$\mathcal{L}(t^k) = \int_0^\infty t^k e^{-st}\, dt = \frac{1}{s^{k+1}} \int_0^\infty \xi^k e^{-\xi}\, d\xi = \frac{1}{s^{k+1}} \Gamma(k + 1),$$

where the gamma function $\Gamma(k + 1)$ is as defined by Eq. 2.5.1. If k is an integer, say $k = n$, then

$$\Gamma(n + 1) = n!$$

and we obtain

$$\mathcal{L}(t^n) = \frac{n!}{s^{n+1}}.\quad ■$$

Example 3.2.6:   Use the linearity property and find the Laplace transform of cosh ωt.

SOLUTION: The cosh ωt can be written as

$$\cosh \omega t = \tfrac{1}{2}(e^{\omega t} + e^{-\omega t}).$$

The Laplace transform is then

$$\mathcal{L}(\cosh \omega t) = \mathcal{L}(\tfrac{1}{2}e^{\omega t} + \tfrac{1}{2}e^{-\omega t}) = \tfrac{1}{2}\mathcal{L}(e^{\omega t}) + \tfrac{1}{2}\mathcal{L}(e^{-\omega t}).$$

Using the results of Example 3.2.2, we have

$$\mathcal{L}(\cosh \omega t) = \frac{1}{2(s - \omega)} + \frac{1}{2(s + \omega)} = \frac{s}{s^2 - \omega^2}.\quad ■$$

Example 3.2.7:   Find the Laplace transform of the function

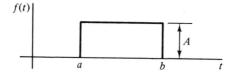

$$f(t) = \begin{cases} 0, & t < a \\ A, & a < t < b \\ 0, & b < t. \end{cases}$$

Use the result of Example 3.2.3.

SOLUTION: The function $f(t)$ can be written in terms of the unit step function as

$$f(t) = Au_a(t) - Au_b(t).$$

The Laplace transform is, from Example 3.2.3,

$$\mathscr{L}(f) = \frac{A}{s}e^{-as} - \frac{A}{s}e^{-bs} = \frac{A}{s}[e^{-as} - e^{-bs}]. \quad \blacksquare$$

Example 3.2.8: An extension of the function shown in Example 3.2.7 is the function shown. If $\epsilon \to 0$, the *unit impulse function* results. It is often denoted by $\delta_0(t)$. It* has an area of unity, its height approaches ∞ as its base approaches zero. Find $\mathscr{L}(f)$ for the unit impulse function if it occurs (a) at $t = 0$ as shown, and (b) at $t = a$.

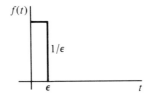

SOLUTION: Let us use the results of Example 3.2.7. With the function $f(t)$ shown, the Laplace transform is, using $A = 1/\epsilon$,

$$\mathscr{L}(f) = \frac{1}{\epsilon s}[1 - e^{-\epsilon s}].$$

To find the limit as $\epsilon \to 0$, expand $e^{-\epsilon s}$ in a series. This gives

$$e^{-\epsilon s} = 1 - \epsilon s + \frac{\epsilon^2 s^2}{2!} - \frac{\epsilon^3 s^3}{3!} + \cdots.$$

Hence,

$$\frac{1 - e^{-\epsilon s}}{\epsilon s} = 1 - \frac{\epsilon s}{2!} + \frac{\epsilon^2 s^2}{3!} - \cdots.$$

As $\epsilon \to 0$, the expression above approaches unity. Thus,

$$\mathscr{L}(\delta_0) = 1.$$

---

*Strictly speaking, $\delta_0(t)$ is not a function. Moreover, $\lim_{\epsilon \to 0} \mathscr{L}(f) = \mathscr{L}(\delta_0)$ does not make sense; witness the fact that $\mathscr{L}(\delta_0) = 1$ contradicts $\lim_{s \to \infty} \mathscr{L}[f(t)] = 0$. The resolution of these logical difficulties requires the theory of distributions, a subject we do not explore in this text.

If the impulse function occurs at a time $t = a$, it is denoted by $\delta_a(t)$. Then, using the second shifting property, we have

$$\mathcal{L}(\delta_a) = e^{-as}. \quad \blacksquare$$

Examples of the use of the impulse function are a concentrated load $P\delta_a(x)$ located at $x = a$, or an electrical potential $V\delta_a(t)$ applied instantaneously to a circuit at $t = a$.

Example 3.2.9:   Find the Laplace transform of

$$f(t) = \begin{cases} 0, & 0 < t < 1 \\ t^2, & 1 < t < 2 \\ 0, & 2 < t. \end{cases}$$

SOLUTION: The function $f(t)$ is written in terms of the unit step function as

$$f(t) = u_1(t)t^2 - u_2(t)t^2.$$

We cannot apply the second shifting property with $f(t)$ in this form since, according to Eq. 3.2.13, we must have for the first term a function of $(t - 1)$ and for the second term a function of $(t - 2)$. The function $f(t)$ is thus rewritten as follows:

$$f(t) = u_1(t)[(t - 1)^2 + 2(t - 1) + 1] - u_2(t)[(t - 2)^2 + 4(t - 2) + 4].$$

Now, we can apply the second shifting property to the result above, to obtain,

$$\mathcal{L}(f) = \mathcal{L}\{u_1(t)[(t - 1)^2 + 2(t - 1) + 1]\} - \mathcal{L}\{u_2(t)[(t - 2)^2 + 4(t - 2) + 4]\}.$$

For the first set of braces $f(t) = t^2 + 2t + 1$, and for the second set of braces $f(t) = t^2 + 4t + 4$. The result is

$$\mathcal{L}(f) = e^{-s}\left[\frac{2}{s^3} + \frac{2}{s^2} + \frac{1}{s}\right] - e^{-2s}\left[\frac{2}{s^3} + \frac{4}{s^2} + \frac{4}{s}\right].$$

Note that, in general, $f(t)$ is not the function given in the statement of the problem, which in this case was $f(t) = t^2$. $\quad \blacksquare$

Example 3.2.10:   The square-wave function is as shown. Determine its Laplace transform.

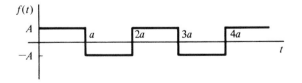

SOLUTION: The function $f(t)$ can be represented using the unit step function. It is

$$f(t) = Au_0(t) - 2Au_a(t) + 2Au_{2a}(t) - 2Au_{3a}(t) + \cdots.$$

The Laplace transform of the above is, referring to Example 3.2.3,

$$\mathcal{L}(f) = A\left[\frac{1}{s} - \frac{2}{s}e^{-as} + \frac{2}{s}e^{-2as} - \frac{2}{s}e^{-3as} + \cdots\right]$$

$$= \frac{A}{s}[1 - 2e^{-as}(1 - e^{-as} + e^{-2as} - \cdots)]$$

Letting $e^{-as} = \xi$, we have

$$\mathcal{L}(f) = \frac{A}{s}[1 - 2\xi(1 - \xi + \xi^2 - \xi^3 + \cdots)].$$

The quantity in parentheses is recognized as the series expansion for $1/(1 + \xi)$. Hence, we can write

$$\mathcal{L}(f) = \frac{A}{s}\left[1 - \frac{2e^{-as}}{1 + e^{-as}}\right].$$

This can be put in the form

$$\mathcal{L}(f) = \frac{A}{s}\frac{1 - e^{-as}}{1 + e^{-as}} = \frac{A}{s}\frac{e^{as/2} - e^{-as/2}}{e^{as/2} + e^{-as/2}}\frac{e^{-as/2}}{e^{-as/2}} = \frac{A}{s}\frac{e^{as/2} - e^{-as/2}}{e^{as/2} + e^{-as/2}}.$$

This form is recognized as

$$\mathcal{L}(f) = \frac{A}{s}\tanh\frac{as}{2}. \quad \blacksquare$$

Example 3.2.11:   Use the Laplace transforms from Table 3.1 and find $f(t)$ when $F(s)$ is given by

$$(a) \ \frac{2s}{s^2 + 4} \qquad (b) \ \frac{6s}{s^2 + 4s + 13} \qquad (c) \ \frac{4e^{-2s}}{s^2 - 16}$$

SOLUTION: (a) The Laplace transform of $\cos \omega t$ is

$$\mathcal{L}(\cos \omega t) = \frac{s}{s^2 + \omega^2}.$$

Then,

$$\mathcal{L}(2 \cos 2t) = 2\mathcal{L}(\cos 2t) = \frac{2s}{s^2 + 2^2}.$$

Thus, if $F(s) = 2s/(s^2 + 4)$, then $f(t)$ is given by

$$f(t) = \mathcal{L}^{-1}\left(\frac{2s}{s^2 + 4}\right) = 2 \cos 2t.$$

(b) Let us write the given $F(s)$ as (this is suggested by the term $4s$ in the denominator)

$$F(s) = \frac{6s}{s^2 + 4s + 13} = \frac{6(s + 2) - 12}{(s + 2)^2 + 9} = \frac{6(s + 2)}{(s + 2)^2 + 9} - \frac{12}{(s + 2)^2 + 9}.$$

Using the first shifting property, Eq. 3.2.9, we can write

$$\mathcal{L}(e^{-2t} \cos 3t) = \frac{s + 2}{(s + 2)^2 + 9}$$

$$\mathcal{L}(e^{-2t} \sin 3t) = \frac{3}{(s + 2)^2 + 9}.$$

It then follows that

$$\mathcal{L}^{-1}\left(\frac{6s}{s^2 + 4s + 13}\right) = 6e^{-2t} \cos 3t - 4e^{-2t} \sin 3t,$$

or we have

$$f(t) = 2e^{-2t}(3 \cos 3t - 2 \sin 3t).$$

(c) The second shifting property suggests that we write

$$\frac{4e^{-2s}}{s^2 - 16} = e^{-2s}\left[\frac{4}{s^2 - 16}\right]$$

and find the $f(t)$ associated with the quantity in brackets; that is

$$\mathcal{L}(\sinh 4t) = \frac{4}{s^2 - 16}$$

or

$$\mathcal{L}^{-1}\left(\frac{4}{s^2 - 16}\right) = \sinh 4t.$$

Finally, there results, using Eq. 3.2.13,

$$f(t) = \begin{cases} \sinh 4(t - 2), & t > 2 \\ 0, & t < 2. \end{cases}$$

In terms of the unit step function, this can be written as

$$f(t) = u_2(t) \sinh 4(t - 2). \quad \blacksquare$$

PROBLEMS

Find the Laplace transform of each function by direct integration.

1. $2t$ 2. $t - 3$

3. e^{3t} 4. $2 \sin t$

5. $\cos 4t$ 6. $t^{1/2}$

7. $2t^{3/2}$ 8. $4t^2 - 3$

9. $\sinh 2t$ 10. $(t - 2)^2$

11. $\cosh 4t$ 12. e^{2t-1}

13.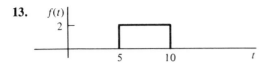

14.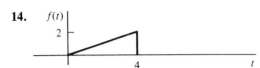

15.

Use the first shifting property and Table 3.1 to find the Laplace transform of each function.

16. $3te^{3t}$

17. $t^2 e^{-t}$

18. $e^{-2t} \cos 4t$

19. $e^{2t} \sinh 2t$

20. $3^{-t} \sin 2t$

21. $4e^{-2t} \cosh t$

22. $e^{-t}(\cos 4t - 2 \sin 4t)$

23. $e^{-2t}(\sinh 2t + 3 \cosh 2t)$

24. $e^{-2t}(t^2 + 4t + 5)$

Use the second shifting property and Table 3.1 to find the Laplace transform of each function. Sketch each function.

25. $u_2(t)$

26. $u_4(t) \sin \pi t$

27. $f(t) = \begin{cases} 0 & 0 < t < 2 \\ 2t & 2 < t < 4 \\ 0 & 4 < t \end{cases}$

28. $\dfrac{t}{2} - u_4(t)\dfrac{t}{2}$

29. $u_4(t)(6 - t) - u_6(t)(6 - t)$

30. $f(t) = \begin{cases} t & 0 < t < 2 \\ 2 & 2 < t \end{cases}$

31. $f(t) = \begin{cases} \sin t & 0 < t < 2\pi \\ 0 & 2\pi < t \end{cases}$

32. $f(t) = \begin{cases} \sin t & 0 < t < \pi \\ \sin 2t & \pi < t \end{cases}$

Express each hyperbolic function in terms of exponential functions and, with the use of Table 3.1, find the Laplace transform.

33. $2 \cosh 2t \sin 2t$

34. $2 \sinh 3t \cos 2t$

35. $4 \cosh 2t \sinh 3t$

36. $6 \sinh t \cos t$

37. $4 \sinh 2t \sinh 4t$

38. $2 \cosh t \cos 2t$

Use Table 3.1 to find the function $f(t)$ corresponding to each Laplace transform.

39. $\dfrac{1}{s}\left(\dfrac{2}{s^2} + \dfrac{1}{s} - 2\right)$

40. $\dfrac{1}{s^2}\left(\dfrac{3}{s} + 2\right)$

41. $\dfrac{2s}{(s + 3)^2}$

42. $\dfrac{s}{(s+1)^3}$

43. $\dfrac{1}{s(s+1)}$

44. $\dfrac{1}{s^2(s-2)}$

45. $\dfrac{1}{(s-2)(s+1)}$

46. $\dfrac{1}{(s-1)(s+2)}$

47. $\dfrac{2s}{(s-1)^2(s+1)}$

48. $\dfrac{e^{-s}}{s+1}$

49. $\dfrac{e^{-2s}}{s(s+1)^2}$

50. $\dfrac{4}{s^2+2s+5}$

51. $\dfrac{4s+3}{s^2+4s+13}$

52. $\dfrac{2}{s^2-2s-3}$

53. $\dfrac{3s+1}{s^2-4s-5}$

54. $\dfrac{4se^{-2\pi s}}{s^2+2s+5}$

55. $\dfrac{2}{(s^2-1)(s^2+1)}$

56. $\dfrac{2s+3}{(s^2+4s+13)^2}$

57. If $\mathscr{L}[f(t)] = F(s)$, show that $\mathscr{L}[f(at)] = (1/a)F(s/a)$; use the definition of a Laplace transform. Then, if $\mathscr{L}(\cos t) = s/(s^2+1)$, find $\mathscr{L}(\cos 4t)$.

3.3 Laplace Transforms of Derivatives and Integrals

The operations of differentiation and integration are significantly simplified when using Laplace transforms. Differentiation results when the Laplace transform of a function is multiplied by the transformed variable s and integration corresponds to dividing by s, as we shall see.

Let us consider a function $f(t)$ that is continuous and possesses a derivative $f'(t)$ that is sectionally continuous. An example of such a function is sketched in Fig. 3.2. We shall not allow discontinuities in the function $f(t)$, although we will discuss such a function subsequently. The Laplace transform of a derivative is defined to be

$$\mathscr{L}(f') = \int_0^\infty f'(t)e^{-st}\,dt. \tag{3.3.1}$$

This can be integrated by parts if we let

$$u = e^{-st} \qquad\qquad dv = f'(t)\,dt = df$$
$$du = -se^{-st}\,dt \qquad v = f. \tag{3.3.2}$$

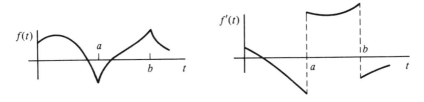

FIGURE 3.2.   Continuous function possessing a sectionally continuous derivative.

Then

$$\mathcal{L}(f') = fe^{-st}\Big|_0^\infty + s\int_0^\infty f(t)e^{-st}\,dt. \tag{3.3.3}$$

Assuming that the quantity fe^{-st} vanishes at the upper limit, this is written as

$$\mathcal{L}(f') = -f(0) + s\int_0^\infty f(t)e^{-st}\,dt = s\mathcal{L}(f) - f(0). \tag{3.3.4}$$

This result can be easily extended to the second-order derivative; however, we must demand that the first derivative $f(t)$ be continuous. Then, with the use of Eq. 3.3.4, we have

$$\mathcal{L}(f'') = s\mathcal{L}(f') - f'(0)$$
$$= s[s\mathcal{L}(f) - f(0)] - f'(0)$$
$$= s^2\mathcal{L}(f) - sf(0) - f'(0). \tag{3.3.5}$$

Note that the values of f and its derivatives must be known at $t = 0$ when finding the Laplace transforms of the derivatives.

Higher-order derivatives naturally follow giving us the relationship,

$$\mathcal{L}(f^{(n)}) = s^n\mathcal{L}(f) - s^{n-1}f(0) - s^{n-2}f'(0) - \cdots - f^{(n-1)}(0), \tag{3.3.6}$$

where all the functions $f(t), f'(t), \ldots, f^{(n-1)}(t)$ are continuous, with the quantities $f^{(n-1)}e^{-st}$ vanishing at infinity; the quantity $f^{(n)}(t)$ is sectionally continuous.

Now, let us find the Laplace transform of a function possessing a discontinuity. Consider the function $f(t)$ to have one discontinuity at $t = a$, with $f(a^+)$ the right-hand limit and $f(a^-)$ the left-hand limit as shown in Fig. 3.3. The Laplace transform of the first derivative is then

$$\mathcal{L}(f') = \int_0^a f'(t)e^{-st}\,dt + \int_a^\infty f'(t)e^{-st}\,dt. \tag{3.3.7}$$

Integrating by parts allows us to write

$$\mathcal{L}(f') = fe^{-st}\Big|_0^{a^-} + s\int_0^{a^-} f(t)e^{-st}\,dt + fe^{-st}\Big|_{a^+}^\infty + s\int_{a^+}^\infty f(t)e^{-st}\,dt$$

$$= f(a^-)e^{-as} - f(0) + s\int_0^{a^-} f(t)e^{-st}\,dt - f(a^+)e^{-as}$$

$$+ s\int_{a^+}^\infty f(t)e^{-st}\,dt. \tag{3.3.8}$$

The two integrals above can be combined, since there is no contribution to the integral between $t = a^-$ and $t = a^+$. We then have

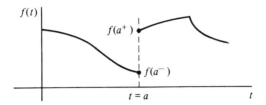

FIGURE 3.3.   Function $f(t)$ with one discontinuity.

$$\mathcal{L}(f') = s \int_0^\infty f(t)e^{-st}\, dt - f(0) - [f(a^+) - f(a^-)]e^{-as}$$

$$= s\mathcal{L}(f) - f(0) - [f(a^+) - f(a^-)]e^{-as}. \qquad (3.3.9)$$

If two discontinuities exist in $f(t)$, the second discontinuity would be accounted for by adding the appropriate terms to the equation above.

We shall now find the Laplace transform of a function expressed as an integral. Let the integral be given by

$$g(t) = \int_0^t f(\tau)\, d\tau, \qquad (3.3.10)$$

where the dummy variable of integration is arbitrarily chosen as τ; the variable t occurs only as the upper limit. The first derivative is then*

$$g'(t) = f(t). \qquad (3.3.11)$$

We also note that $g(0) = 0$. Now, applying Eq. 3.3.4, we have

$$\mathcal{L}(g') = s\mathcal{L}(g) - \overset{0}{\cancel{g(0)}} \qquad (3.3.12)$$

or, using Eq. 3.3.11, this can be written as

$$\mathcal{L}(g) = \frac{\mathcal{L}(g')}{s} = \frac{1}{s}\mathcal{L}(f). \qquad (3.3.13)$$

Written explicitly in terms of the integral, this is

$$\mathcal{L}\left(\int_0^t f(\tau)\, d\tau\right) = \frac{1}{s}\mathcal{L}(f). \qquad (3.3.14)$$

These transforms of derivatives and integrals are obviously necessary

---

*Liebnitz's rule of differentiating an integral is

$$\frac{d}{dt}\int_{a(t)}^{b(t)} f(\tau,\, t)\, d\tau = \frac{db}{dt}f(b,\, t) - \frac{da}{dt}f(a,\, t) + \int_a^b \frac{\partial f}{\partial t}\, d\tau.$$

when solving differential equations or integro-differential equations. They will also, however, find application in obtaining the Laplace transforms of various functions and the inverse transforms. Before we turn to the solution of differential equations, let us illustrate the latter use.

Example 3.3.1:   Find the Laplace transform of $f(t) = t^2$. Use the transform of a derivative.

SOLUTION: We can use the Laplace transform of the third derivative obtained in Eq. 3.3.6. From the given function we have $f(0) = 0$, $f'(0) = 0$, and $f''(0) = 2$; this allows us to write

$$\mathcal{L}(f''') = s^3 \mathcal{L}(f) - s^2 f(0) - s f'(0) - f''(0)$$

and, recognizing that $f''' = 0$,

$$\mathcal{L}(0) = s^3 \mathcal{L}(f) - 2 = 0,$$

since $\mathcal{L}(0) = 0$. This results in

$$\mathcal{L}(t^2) = \frac{2}{s^3}. \quad \blacksquare$$

Example 3.3.2:   Use the transform of a derivative and find the Laplace transform of $f(t) = t \sin t$ assuming that $\mathcal{L}(\cos t)$ is known.

$$f'(t) = t \cos t + \sin t$$

$$f''(t) = 2 \cos t - t \sin t.$$

The transform of a second derivative is

$$\mathcal{L}(f'') = s^2 \mathcal{L}(f) - s f(0) - f'(0)$$

where we have used $f(0) = 0$ and $f'(0) = 0$. Thus, Eq. 3.3.5 gives

$$\mathcal{L}(2 \cos t - t \sin t) = s^2 \mathcal{L}(t \sin t).$$

This can be written as

$$2\mathcal{L}(\cos t) - \mathcal{L}(t \sin t) = s^2 \mathcal{L}(t \sin t)$$

or

$$(s^2 + 1)\mathcal{L}(t \sin t) = 2\mathcal{L}(\cos t) = \frac{2s}{s^2 + 1}.$$

Finally, we have

$$\mathcal{L}(t \sin t) = \frac{2s}{(s^2 + 1)^2}. \quad \blacksquare$$

Example 3.3.3:   Find $f(t)$ if $F(s) = 8/(s^2 + 4)^2$ by using $\mathcal{L}(t \sin 2t) = 2s/(s^2 + 4)^2$.

SOLUTION: We write the given transform as

$$\mathcal{L}(f) = F(s) = \frac{1}{s} \frac{8s}{(s^2 + 4)^2}.$$

Equation 3.3.14 allows us to write

$$\mathcal{L}\left(\int_0^t 4\tau \sin 2\tau \, d\tau \right) = \frac{1}{s} \frac{8s}{(s^2 + 4)^2}.$$

Hence,

$$f(t) = 4 \int_0^t \tau \sin 2\tau \, d\tau.$$

This is integrated by parts. If we let

$$u = \tau \qquad dv = \sin 2\tau \, d\tau$$
$$du = d\tau \qquad v = -\tfrac{1}{2} \cos 2\tau.$$

there results

$$f(t) = -2t \cos 2t + 2 \int_0^t \cos 2\tau \, d\tau = -2t \cos 2t + \sin 2t. \quad \blacksquare$$

PROBLEMS

1. Write an expression for $\mathcal{L}(f^{(iv)})$.

2. Write an expression for $\mathcal{L}(f')$ if two discontinuities occur in $f(t)$, one at $t = a$ and the other at $t = b$.

3. Use Eq. 3.3.4 to find the Laplace transform of $f(t) = e^t$.

Use Eq. 3.3.5 to find the Laplace transform of each function.

4. $\sin \omega t$ 5. $\cos \omega t$ 6. $\sinh at$

7. $\cosh at$ 8. e^{2t} 9. t

10. If $f(t) = \begin{cases} t, & 0 < t < 1 \\ 1, & 1 < t, \end{cases}$ find $\mathcal{L}(f)$. Also, find $\mathcal{L}(f')$. Is Eq. 3.3.4 verified for this $f(t)$?

11. If $f(t) = \begin{cases} t, & 0 < t < 1 \\ 0, & 1 < t, \end{cases}$ find $\mathcal{L}(f)$. Also, find $\mathcal{L}(f')$. Does Eq. 3.3.4 hold for this $f(t)$? Verify that Eq. 3.3.9 holds for this $f(t)$.

Using the equations for the Laplace transforms of derivatives from Section 3.3 and Table 3.1, find the transform of each function.

12. te^t | **13.** $t \sin 2t$ | **14.** $t \cos t$

15. $t^2 \sin t$ | **16.** $te^t \sin t$ | **17.** $(t^2 + 1) \cos 2t$

18. $t \cos 2t$ | **19.** $t^2 e^t$ | **20.** $t \sinh 2t$

Find the functon $f(t)$ corresponding to each Laplace transform.

21. $\dfrac{1}{s^2 + 2s}$ **22.** $\dfrac{2}{s^2 - s}$ **23.** $\dfrac{4}{s^3 + 4s}$

24. $\dfrac{4}{s^4 + 4s^2}$ **25.** $\dfrac{6}{s^3 - 9s}$ **26.** $\dfrac{6}{s^4 - 9s^2}$

27. $\dfrac{2}{s^4 + 2s^2}$ **28.** $\dfrac{1}{s}\dfrac{s-1}{s+1}$ **29.** $\dfrac{1}{s^2}\dfrac{s-1}{s^2+1}$

3.4   Derivatives and Integrals of Laplace Transforms

The problem of determining the Laplace transform of a particular function or the function corresponding to a particular transform can often be simplified by either differentiating or integrating a Laplace transform. First, let us find the Laplace transform of the quantity $tf(t)$. It is, by definition,

$$\mathcal{L}(tf) = \int_0^\infty tf(t)e^{-st}\,dt. \tag{3.4.1}$$

Using Liebnitz's rule of differentiating an integral (see footnote on p. 161) we can differentiate Eq. 3.2.2 and obtain

$$F'(s) = \frac{d}{ds}\int_0^\infty f(t)e^{-st}\,dt = \int_0^\infty f(t)\frac{\partial}{\partial s}(e^{-st})\,dt$$

$$= -\int_0^\infty tf(t)e^{-st}\,dt. \tag{3.4.2}$$

Comparing this with Eq. 3.4.1, there follows

$$\mathcal{L}(tf) = -F'(s). \tag{3.4.3}$$

The second derivative is

$$F''(s) = \int_0^\infty f(t)\frac{\partial^2}{\partial s^2}(e^{-st})\,dt = \int_0^\infty t^2 f(t)e^{-st}\,dt \tag{3.4.4}$$

or

$$\mathcal{L}(t^2 f) = F''(s). \tag{3.4.5}$$

In general, this is written as

$$\mathcal{L}(t^n f) = (-1)^n F^{(n)}(s). \tag{3.4.6}$$

Next, we will find the Laplace transform of $f(t)/t$. Let

$$f(t) = tg(t). \qquad (3.4.7)$$

Then, using Eq. 3.4.3, the Laplace transform of the equation above is

$$F(s) = \mathcal{L}(f) = \mathcal{L}(tg) = -G'(s). \qquad (3.4.8)$$

This is written as

$$-dG = F(s) \, ds. \qquad (3.4.9)$$

Thus

$$-G(s) = \int_{\infty}^{s} F(\lambda) \, d\lambda, \qquad (3.4.10)$$

where* $G(s) \to 0$ as $s \to \infty$. The dummy variable of integration is written arbitrarily as λ. We then have

$$G(s) = \int_{s}^{\infty} F(\lambda) \, d\lambda, \qquad (3.4.11)$$

where the limits of integration have been interchanged to remove the negative sign. Finally, referring to Eq. 3.4.7, we see that

$$\mathcal{L}(f/t) = \mathcal{L}(g) = G(s) = \int_{s}^{\infty} F(\lambda) \, d\lambda. \qquad (3.4.12)$$

The use of the expressions above for the derivatives and integral of a Laplace transform will be demonstrated in the following examples.

Example 3.4.1:   Differentiate the Laplace transform of $f(t) = \sin \omega t$, thereby determining $\mathcal{L}(t \sin \omega t)$. Use $\mathcal{L}(\sin \omega t) = \omega/(s^2 + \omega^2)$.

SOLUTION: Equation 3.4.3 allows us to write

$$\mathcal{L}(t \sin \omega t) = -\frac{d}{ds}\mathcal{L}(\sin \omega t) = -\frac{d}{ds}\left(\frac{\omega}{s^2 + \omega^2}\right) = \frac{2\omega s}{(s^2 + \omega^2)^2}.$$

This transform was obviously much easier to obtain using Eq. 3.4.3 than the technique used in Example 3.3.2.   ∎

Example 3.4.2:   Find the Laplace transform of $(e^{-t} - 1)/t$ using the transforms

$$\mathcal{L}(e^{-t}) = \frac{1}{s + 1} \quad \text{and} \quad \mathcal{L}(1) = \frac{1}{s}.$$

---

*This limit is a consequence of the assumption that $g(t)$ is sectionally continuous, and item (ii) following Eq. 3.2.6, where $\mathcal{L}[g(t)] = G(s)$.

SOLUTION: The Laplace transform of the function $f(t) = e^{-t} - 1$ is

$$\mathcal{L}(f) = \frac{1}{s+1} - \frac{1}{s}.$$

Equation 3.4.12 gives us

$$\mathcal{L}(f/t) = \int_s^\infty \left(\frac{1}{\Delta + 1} - \frac{1}{\Delta} \right) d\Delta$$

$$= \left[\ln(\Delta + 1) - \ln \Delta \right]_s^\infty = \ln \frac{\Delta + 1}{\Delta} \bigg|_s^\infty = \ln \frac{s}{s+1}.$$

This problem could be reformulated to illustrate a function that had no Laplace transform. Consider the function $(e^{-t} - 2)/t$. The solution would have resulted in $\ln(s+1)/s^2\big|_s^\infty$. At the upper limit this quantity is not defined and thus $\mathcal{L}(f/t)$ does not exist. ∎

Example 3.4.3: Determine the inverse Laplace transform of $\ln[s^2/(s^2+4)]$.

SOLUTION: We know that if we differentiate $\ln[s^2/(s^2+4)]$ we will arrive at a recognizable function. Letting $G(s) = \ln[s^2/(s^2+4)] = \ln s^2 - \ln(s^2+4)$, we have (see Eq. 3.4.8)

$$F(s) = -G'(s) = -\frac{2s}{s^2} + \frac{2s}{s^2+4}$$

$$= -\frac{2}{s} + \frac{2s}{s^2+4}.$$

Now, the inverse transform of $F(s)$ is, referring to Table 3.1,

$$f(t) = -2 + 2\cos 2t.$$

Finally, the desired inverse transform is ,

$$\mathcal{L}^{-1}\left(\ln \frac{s^2}{s^2+4} \right) = \frac{f(t)}{t} = -\frac{2}{t}(1 - \cos 2t). \quad ∎$$

PROBLEMS

Determine the Laplace transform of each function using Table 3.1 and the equations of Section 3.4.

1. $2t \sin 3t$

2. $t \cos 2t$

3. $t^2 \sin 2t$

4. $t^2 \sinh t$

5. $te^t \cos 2t$

6. $t(e^t - e^{-t})$

7. $t(e^t - e^{-2t})$

8. $te^{-t} \sin t$

9. $t^2 e^{-t} \sin t$

10. $t \cosh t$

11. $\frac{2}{t}(1 - \cos 2t)$

12. $\frac{2}{t}(1 - \cosh 2t)$

13. $\dfrac{1}{t}(e^{2t} - e^{-2t})$ **14.** $\dfrac{1}{t}(e^{2t} - 1)$

15. Use Eq. 3.4.3 to find an expression for the Laplace transform of $f(t) = t^n e^{at}$ using $\mathcal{L}(e^{at}) = 1/(s - a)$.

Find the function $f(t)$ that corresponds to each Laplace transform using the equations of Section 3.4.

16. $\dfrac{1}{(s + 2)^2}$ **17.** $\dfrac{4s}{(s^2 + 4)^2}$ **18.** $\dfrac{s}{(s^2 - 4)^2}$

19. $\ln \dfrac{s}{s - 2}$ **20.** $\ln \dfrac{s - 2}{s + 3}$ **21.** $\ln \dfrac{s^2 - 4}{s^2 + 4}$

22. $\ln \dfrac{s^2 + 1}{s^2 + 4}$ **23.** $\ln \dfrac{s^2}{s^2 + 4}$ **24.** $\ln \dfrac{s^2 + 4s + 5}{s^2 + 2s + 5}$

3.5   Laplace Transforms of Periodic Functions

Before we turn to the solution of differential equations using Laplace transforms, we shall consider the problem of finding the transform of periodic functions. The nonhomogeneous part of differential equations often involve such periodic functions. A periodic function is one that is sectionally continuous and for some a satisfies

$$f(t) = f(t + a) = f(t + 2a) = f(t + 3a)$$
$$= \cdots = f(t + na) = \cdots . \qquad (3.5.1)$$

This is illustrated in Fig. 3.4. We can write the transform of these $f(t)$ as the series of integrals

$$\mathcal{L}(f) = \int_0^\infty f(t)e^{-st}\, dt$$

$$= \int_0^a f(t)e^{-st}\, dt + \int_a^{2a} f(t)e^{-st}\, dt + \int_{2a}^{3a} f(t)e^{-st}\, dt + \cdots . \qquad (3.5.2)$$

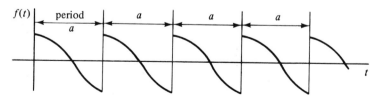

FIGURE 3.4.   Periodic function.

In the second integral, let $t = \tau + a$; in the third integral, let $t = \tau + 2a$; in the fourth, let $t = \tau + 3a$; etc; then the limits on each integral are 0 and a. There results

$$\mathcal{L}(f) = \int_0^a f(t)e^{-st}\, dt + \int_0^a f(\tau + a)e^{-s(\tau + a)}\, d\tau$$

$$+ \int_0^a f(\tau + 2a)e^{-s(\tau + 2a)}\, d\tau + \cdots. \qquad (3.5.3)$$

The dummy variable of integration τ can be set equal to t, and with the use of Eq. 3.5.1 we have

$$\mathcal{L}(f) = \int_0^a f(t)e^{-st}\, dt + e^{-as}\int_0^a f(t)e^{-st}\, dt + e^{-2as}\int_0^a f(t)e^{-st}\, dt + \cdots$$

$$= [1 + e^{-as} + e^{-2as} + \cdots] \int_0^a f(t)e^{-st}\, dt. \qquad (3.5.4)$$

Using the series expansion, $1/(1 - x) = 1 + x + x^2 + \cdots$, we can write the equation above as

$$\mathcal{L}(f) = \frac{1}{1 - e^{-as}} \int_0^a f(t)e^{-st}\, dt. \qquad (3.5.5)$$

Example 3.5.1:   Determine the Laplace transform of the square-wave function shown. Compare with Example 3.2.10.

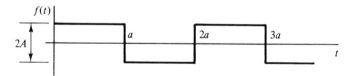

SOLUTION: The function $f(t)$ is periodic with period $2a$. Using Eq. 3.5.5, we have

$$\mathcal{L}(f) = \frac{1}{1 - e^{-2as}} \int_0^{2a} f(t)e^{-st}\, dt$$

$$= \frac{1}{1 - e^{-2as}} \left[\int_0^a Ae^{-st}\, dt + \int_a^{2a} (-A)e^{-st}\, dt \right]$$

$$= \frac{1}{1 - e^{-2as}} \left[-\frac{A}{s}e^{-st}\Big|_0^a + \frac{A}{s}e^{-st}\Big|_a^{2a} \right]$$

$$= \frac{1}{1 - e^{-2as}} \left[\frac{A}{s}(-e^{-as} + 1 + e^{-2as} - e^{-as}) \right]$$

$$= \frac{A}{s} \frac{1 - 2e^{-as} + e^{-2as}}{1 - e^{-2as}} = \frac{A}{s} \frac{(1 - e^{-as})(1 - e^{-as})}{(1 - e^{-as})(1 + e^{-as})}$$

$$= \frac{A}{s} \frac{1 - e^{-as}}{1 + e^{-as}}.$$

This is the same result obtained in Example 3.2.10. It can be put in the more desired form, as in Example 3.2.10,

$$\mathcal{L}(f) = \frac{A}{s} \tanh \frac{as}{2}. \quad \blacksquare$$

Example 3.5.2: Find the Laplace transform of the half-wave-rectified sine wave shown with period 2π and amplitude 1.

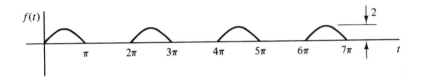

SOLUTION: The function $f(t)$ is given by

$$f(t) = \begin{cases} \sin t, & 0 < t < \pi \\ 0, & \pi < t < 2\pi. \end{cases}$$

Equation 3.5.5 provides us with

$$\mathcal{L}(f) = \frac{1}{1 - e^{-2\pi s}} \int_0^{2\pi} f(t)e^{-st}\, dt = \frac{1}{1 - e^{-2\pi s}} \int_0^{\pi} \sin t e^{-st}\, dt.$$

Integrate by parts:

$$u = \sin t \qquad dv = e^{-st}\, dt$$

$$du = \cos t\, dt \qquad v = -\frac{1}{s}e^{-st}$$

Then

$$\int_0^{\pi} \sin t e^{-st}\, dt = -\frac{1}{s}e^{-st} \underset{0}{\overset{0}{\sin t}}\Big|_0^{\pi} + \frac{1}{s}\int_0^{\pi} \cos t e^{-st}\, dt.$$

Integrate by parts again:

$$u = \cos t \qquad dv = e^{-st}\, dt$$

$$du = -\sin t\, dt \qquad v = -\frac{1}{s}e^{-st}.$$

Again,

$$\int_0^\pi \sin t e^{-st}\, dt = \frac{1}{s}\left[-\frac{1}{s}e^{-st} \cos t \Big|_0^\pi - \frac{1}{s}\int_0^\pi \sin t e^{-st}\, dt \right].$$

This is rearranged to give

$$\int_0^\pi \sin t e^{-st}\, dt = \frac{s^2}{s^2+1}\left[\frac{1}{s^2}(e^{-s\pi} + 1) \right] = \frac{1 + e^{-\pi s}}{s^2 + 1}.$$

Finally,

$$\mathcal{L}(f) = \frac{1 + e^{-\pi s}}{(1 - e^{-2\pi s})(s^2 + 1)} = \frac{1}{(1 - e^{-\pi s})(s^2 + 1)}. \quad ■$$

Example 3.5.3:   Find the Laplace transform of the periodic function shown.

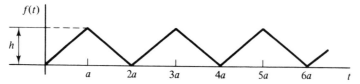

SOLUTION: We can find the transform for the function by using Eq. 3.5.5 with

$$f(t) = \begin{cases} \dfrac{h}{a}t, & 0 < t < a \\[2mm] \dfrac{h}{a}(2a - t), & a < t < 2a. \end{cases}$$

This is not too difficult a task; but if we recognize that the $f(t)$ of this example is simply the integral of the square wave of Example 3.5.1 if we let $h = Aa$, then we can use Eq. 3.3.14 in the form

$$\mathcal{L}(f) = \mathcal{L}\left(\int_0^t g(\tau)\, d\tau \right) = \frac{1}{s}\mathcal{L}(g),$$

where $\mathcal{L}(g)$ is given in Example 3.5.1. There results

$$\mathcal{L}(f) = \frac{h}{as^2}\tanh\frac{as}{2}.$$

This example illustrates that some transforms may be easier to find using the results of the preceding sections.   ■

PROBLEMS

Determine the Laplace transform for each periodic function. The first period is stated. Also, sketch several periods of each function.

1. $f(t) = \sin t$, $0 < t < \pi$ **2.** $f(t) = t$, $0 < t < 2$

3. $f(t) = 2 - t, \quad 0 < t < 2$

4. $f(t) = t - 2, \quad 0 < t < 4$

5. $f(t) = t^2, \quad 0 < t < \pi$

6. $f(t) = \begin{cases} 1, & 0 < t < 2 \\ 0 & 2 < t < 4 \end{cases}$

7. $f(t) = \begin{cases} t, & 2 < t < 4 \\ 0, & 0 < t < 2 \end{cases}$

8. $f(t) = \begin{cases} 2 - t, & 0 < t < 1 \\ 0, & 1 < t < 2 \end{cases}$

9. $f(t) = \begin{cases} 2, & 0 < t < 1 \\ 0, & 1 < t < 2 \\ -2, & 2 < t < 3 \\ 0, & 3 < t < 4 \end{cases}$

3.6   *Inverse Laplace Transforms: Partial Fractions*

When solving differential equations using Laplace transforms we must frequently make use of partial fractions in finding the inverse of a transform. In this section we shall present a technique that will organize this procedure.

3.6.1   UNREPEATED LINEAR FACTOR $(s - a)$

Consider the ratio of two polynomials $P(s)$ and $Q(s)$ such that the degree of $Q(s)$ is greater than the degree of $P(s)$. Then a theorem of algebra allows us to write the partial-fraction expansion as

$$F(s) = \frac{P(s)}{Q(s)} = \frac{A_1}{s - a_1} + \frac{A_2}{s - a_2} + \frac{A_3}{s - a_3} + \cdots + \frac{A_n}{s - a_n},$$

$$(3.6.1)$$

where it is assumed that $Q(s)$ can be factored into n factors with distinct roots $a_1, a_2, a_3, \ldots, a_n$. Let us attempt to find one of the coefficients, for example A_3. Multiply Eq. 3.6.1 by $(s - a_3)$ and let $s \to a_3$; there results

$$\lim_{s \to a_3} \frac{P(s)}{Q(s)} (s - a_3) = A_3, \qquad (3.6.2)$$

since all other terms are multiplied by $(s - a_3)$, which goes to zero as $s \to a_3$.

Now, we may find the limit shown above. It is found as follows:

$$\lim_{s \to a_3} \frac{P(s)}{Q(s)} (s - a_3) = \lim_{s \to a_3} \left[P(s) \frac{s - a_3}{Q(s)} \right] = P(a_3) \frac{0}{0}. \qquad (3.6.3)$$

Because the quotient $0/0$ appears, we use L'Hospital's rule and differentiate both numerator and denominator with respect to s and then let $s \to a_3$. This yields

$$A_3 = P(a_3) \lim_{s \to a_3} \frac{1}{Q'(s)} = \frac{P(a_3)}{Q'(a_3)}. \tag{3.6.4}$$

We could, of course, have chosen any coefficient; so, in general,

$$A_i = \frac{P(a_i)}{Q'(a_i)} \quad \text{or} \quad A_i = \frac{P(a_i)}{[Q(s)/(s - a_i)]_{s=a_i}}. \tag{3.6.5}$$

This second formula is obtained from the limit in Eq. 3.6.2. With either of these formulas, the coefficients of the partial fractions are quite easily obtained.

Example 3.6.1:   The Laplace transform $F(s)$ of $f(t)$ is given as

$$\frac{s^3 + 3s^2 - 2s + 4}{s(s - 1)(s - 2)(s^2 + 4s + 3)}.$$

Find $f(t)$.

SOLUTION: The partial-fraction representation of $F(s)$ is

$$F(s) = \frac{A_1}{s} + \frac{A_2}{s - 1} + \frac{A_3}{s - 2} + \frac{A_4}{s + 1} + \frac{A_5}{s + 3}.$$

The A_i will be found using the second formula of Eq. 3.6.5. For the given $F(s)$, we have

$$P(s) = s^3 + 3s^2 - 2s + 4$$

$$Q(s) = s(s - 1)(s - 2)(s^2 + 4s + 3)$$

$$= s(s - 1)(s - 2)(s + 3)(s + 1).$$

For the first root, $a_1 = 0$. Letting $s = 0$ in the expressions for $P(s)$ and $Q(s)/s$, there results

$$A_1 = \frac{P(0)}{[Q(s)/s]_{s=0}} = \frac{4}{6}.$$

Similarly, we have, with $a_2 = 1$, $a_3 = 2$, $a_4 = -1$, and $a_5 = -3$,

$$A_2 = \frac{P(1)}{[Q(s)(s - 1)]_{s=1}} = \frac{6}{-8} \qquad A_3 = \frac{P(2)}{[Q(s)/(s - 2)]_{s=2}} = \frac{20}{30}$$

$$A_4 = \frac{P(-1)}{[Q(s)/(s + 1)]_{s=-1}} = \frac{8}{-12} \qquad A_5 = \frac{P(-3)}{[Q(s)/(s + 5)]_{s=-5}} = \frac{10}{120}.$$

The partial-fraction representation is then

$$F(s) = \frac{\frac{2}{3}}{s} - \frac{\frac{3}{4}}{s - 1} + \frac{\frac{2}{3}}{s - 2} - \frac{\frac{2}{3}}{s + 1} + \frac{\frac{1}{12}}{s + 3}.$$

Table 3.1 is consulted to finding $f(t)$. There results

$$f(t) = \frac{2}{3} - \frac{3}{4}e^t + \frac{2}{3}e^{2t} - \frac{2}{3}e^{-t} + \frac{1}{12}e^{-3t}. \quad \blacksquare$$

3.6.2   REPEATED LINEAR FACTOR $(s - a)^m$

If there are repeated roots in $Q(s)$, such as $(s - a)^m$, we have the sum of partial fractions

$$F(s) = \frac{P(s)}{Q(s)} = \frac{B_m}{(s - a_1)^m} + \cdots + \frac{B_2}{(s - a_1)^2} + \frac{B_1}{s - a_1}$$

$$+ \frac{A_2}{s - a_2} + \frac{A_3}{s - a_3} + \cdots . \qquad (3.6.6)$$

The A_i, the coefficients of the terms resulting from distinct roots, are given in Eq. 3.6.5. But the B_i are given by

$$B_i = \frac{1}{(m - i)!} \frac{d^{m-i}}{ds^{m-1}} \left[\frac{P(s)}{Q(s)/(s - a_1)^m} \right]_{s=a_1}$$

$$B_m = \frac{P(a_1)}{[Q(s)/(s - a_1)^m]_{s=a_1}}. \quad \blacksquare \qquad (3.6.7)$$

Example 3.6.2:   Find the inverse Laplace transform of

$$F(s) = \frac{s^2 - 1}{(s - 2)^2(s^2 + s - 6)}.$$

SOLUTION: The denominator $Q(s)$ can be written as

$$Q(s) = (s - 2)^3(s + 3)$$

Thus, a triple root occurs and we use the partial-fraction expansion given by Eq. 3.6.6, that is,

$$F(s) = \frac{B_3}{(s - 2)^3} + \frac{B_2}{(s - 2)^2} + \frac{B_1}{s - 2} + \frac{A_2}{s + 3}.$$

The constants B_i are determined from Eq. 3.6.7 to be

$$B_3 = \left[\frac{s^2 - 1}{s + 3} \right]_{s=2} = \frac{3}{5}$$

$$B_2 = \frac{1}{1!} \frac{d}{ds} \left[\frac{s^2 - 1}{s + 3} \right]_{s=2} = \left[\frac{s^2 + 6s + 1}{(s + 3)^2} \right]_{s=2} = \frac{17}{25}$$

$$B_1 = \frac{1}{2!} \frac{d^2}{ds^2} \left[\frac{s^2 - 1}{s + 3} \right]_{s=2} = \frac{82}{125}.$$

The constant A_2 is, using $a_2 = -3$,

$$A_2 = \frac{P(a_2)}{[Q(s)/(s + 3)]_{s=a_2}} = \frac{8}{125}.$$

Hence, we have

$$F(s) = \frac{\frac{3}{5}}{(s - 2)^3} + \frac{\frac{17}{25}}{(s - 2)^2} + \frac{\frac{82}{125}}{s - 2} + \frac{\frac{8}{125}}{s + 3}.$$

Table 3.1, at the end of the chapter, allows us to write $f(t)$ as

$$f(t) = \frac{3}{10}t^2 e^{2t} + \frac{17}{25}te^{2t} + \frac{82}{125}e^{2t} + \frac{8}{125}e^{-3t}. \quad \blacksquare$$

3.6.3 UNREPEATED QUADRATIC FACTOR $[(s - a)^2 + b^2]$

Suppose that a quadratic factor appears in $Q(s)$, such as $[(s - a)^2 + b^2]$. The transform $F(s)$ written in partial fractions is

$$F(s) = \frac{P(s)}{Q(s)} = \frac{B_1 s + B_2}{(s - a)^2 + b^2} + \frac{A_1}{s - a_1} + \frac{A_2}{s - a_2} + \cdots,$$

$$(3.6.8)$$

where B_1 and B_2 are real constants. Now, multiply by the quadratic factor and let $s \to (a + ib)$. There results

$$B_1(a + ib) + B_2 = \left\{\frac{P(s)}{Q(s)/[(s - a)^2 + b^2]}\right\}_{s=a+ib} \quad (3.6.9)$$

The equation above involves complex numbers. The real part and the imaginary part allow both B_1 and B_2 to be calculated. The A_i are given by Eq. 3.6.5.

Example 3.6.3: The Laplace transform of the displacement function $y(t)$ for a forced, frictionless, spring-mass system is found to be

$$Y(s) = \frac{\omega F_0/M}{(s^2 + \omega_0^2)(s^2 + \omega^2)}$$

for a particular set of initial conditions. Find $y(t)$.

SOLUTION: The function $Y(s)$ can be written as

$$Y(s) = \frac{A_1 s + A_2}{s^2 + \omega_0^2} + \frac{B_1 s + B_2}{s^2 + \omega^2}.$$

The functions $P(s)$ and $Q(s)$ are

$$P(s) = \frac{\omega F_0}{M}$$

$$Q(s) = (s^2 + \omega_0^2)(s^2 + \omega^2).$$

With the use of Eq. 3.6.9, we have

$$A_1(i\omega_0) + A_2 = \frac{\omega F_0/M}{(i\omega_0)^2 + \omega^2} = \frac{\omega F_0/M}{\omega^2 - \omega_0^2}$$

$$B_1(i\omega) + B_2 = \frac{\omega F_0/M}{(i\omega)^2 + \omega_0^2} = -\frac{\omega F_0/M}{\omega^2 - \omega_0^2}$$

where $a = 0$ and $b = \omega_0$ in the first equation; in the second equation $a = 0$ and $b = \omega$. Equating real and imaginary parts:

$$A_1 = 0, \qquad A_2 = \frac{\omega F_0/M}{\omega^2 - \omega_0^2}$$

$$B_1 = 0, \qquad B_2 = -\frac{\omega F_0/M}{\omega^2 - \omega_0^2}.$$

The partial-fraction representation is then

$$Y(s) = \frac{\omega F_0/M}{\omega^2 - \omega_0^2}\left[\frac{1}{s^2 + \omega_0^2} - \frac{1}{s^2 + \omega^2}\right].$$

Finally, using Table 3.1, we have

$$y(t) = \frac{\omega F_0/M}{\omega^2 - \omega_0^2}\left[\frac{1}{\omega_0}\sin \omega_0 t - \frac{1}{\omega}\sin \omega t\right]. \quad \blacksquare$$

3.6.4 REPEATED QUADRATIC FACTOR $[(s - a)^2 + b^2]^m$

If the square of a quadratic factor appears in $Q(s)$, the transform $F(s)$ is expanded in partial fractions as

$$F(s) = \frac{P(s)}{Q(s)} = \frac{C_1 s + C_2}{[(s - a)^2 + b^2]^2} + \frac{B_1 s + B_2}{(s - a)^2 + b^2}$$

$$+ \frac{A_1}{s - a_1} + \frac{A_2}{s - a_2} + \cdots \qquad (3.6.10)$$

The undetermined constants are obtained from the equations

$$C_1(a + ib) + C_2 = \left\{\frac{P(s)}{Q(s)/[(s - a)^2 + b^2]^2}\right\}_{s=a+ib} \qquad (3.6.11)$$

and

$$B_1(a + ib) + B_2 = \frac{d}{ds}\left\{\frac{P(s)}{Q(s)/[(s - a)^2 + b^2]^2}\right\}_{s=a+ib}. \qquad (3.6.12)$$

The A_i are again given by Eq. 3.6.5.

PROBLEMS

Find the function $f(t)$ corresponding to each Laplace transform.

1. $\dfrac{120s}{(s - 1)(s + 2)(s^2 - 2s - 3)}$

2. $\dfrac{5s^2 + 20}{s(s - 1)(s^2 + 5s + 4)}$

3. $\dfrac{s^3 + 2s}{(s^2 + 3s + 2)(s^2 + s - 6)}$

4. $\dfrac{s^2 + 2s + 1}{(s - 1)(s^2 + 2s - 3)}$

5. $\dfrac{8}{s^2(s-2)(s^2-4s+4)}$

6. $\dfrac{s^2-3s+2}{s^2(s-1)^2(s-5s+4)}$

7. $\dfrac{s^2-1}{(s^2+4)(s^2+1)}$

8. $\dfrac{5}{(s^2+400)(s^2+441)}$

9. $\dfrac{s-1}{(s+1)(s^2+4)}$

10. $\dfrac{s^2+1}{(s+1)^2(s^2+4)}$

11. $\dfrac{50}{(s^2+4)^2(s^2+1)}$

12. $\dfrac{10}{(s^2+4)^2(s^2+1)^2}$

3.7 A Convolution Theorem

The question arises whether we can express $\mathcal{L}^{-1}[F(s)G(s)]$ in terms of

$$\mathcal{L}^{-1}[F(s)] = f(t) \quad \text{and} \quad \mathcal{L}^{-1}[G(s)] = g(t). \qquad (3.7.1)$$

To see how this may be done, we first note that

$$F(s)G(s) = \int_0^\infty e^{-s\tau} F(s)g(\tau)\, d\tau \qquad (3.7.2)$$

since

$$G(s) = \int_0^\infty e^{-s\tau} g(\tau)\, d\tau. \qquad (3.7.3)$$

Also,

$$\mathcal{L}^{-1}[e^{-s\tau} F(s)] = f(t-\tau)u_\tau(t), \qquad (3.7.4)$$

where $u_\tau(t)$ is the unit step function of Example 3.2.3. (See Eq. 3.2.16 and the discussion of the second shifting property.)
    Equation 3.7.4 implies that

$$e^{-s\tau} F(s) = \int_0^\infty e^{-st} f(t-\tau)u_\tau(t)\, dt. \qquad (3.7.5)$$

This latter expression can be substituted into Eq. 3.7.2 to obtain

$$F(s)G(s) = \int_0^\infty \int_0^\infty e^{-st} f(t-\tau)g(\tau)u_\tau(t)\, dt\, d\tau$$

$$= \int_0^\infty \int_\tau^\infty e^{-st} f(t-\tau)g(\tau)\, dt\, d\tau \qquad (3.7.6)$$

since $u_\tau(t) = 0$ for $0 < t < \tau$ and $u_\tau(t) = 1$ for $t \geq \tau$. Now, consider the $\tau = t$ line shown in Fig. 3.5. The double integral may be viewed as an

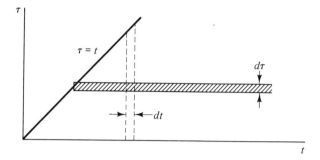

FIGURE 3.5.   The integration strips.

integration using horizontal strips: first integrate in dt from τ to ∞ followed by an integration in $d\tau$ from 0 to ∞. Alternatively, we may integrate using vertical strips: first integrate in $d\tau$ from 0 to t followed by an integration in dt from 0 to ∞; this results in the expression

$$F(s)G(s) = \int_0^\infty \int_0^t e^{-st} f(t - \tau)g(\tau) \, d\tau \, dt$$

$$= \int_0^\infty e^{-st} \left[\int_0^t f(t - \tau)g(\tau) \, d\tau \right] dt. \qquad (3.7.7)$$

But referring to Eq. 3.2.2, the equation above says that

$$\mathcal{L}\left[\int_0^t f(t - \tau)g(\tau) \, d\tau \right] = F(s)G(s) \qquad (3.7.8)$$

or, equivalently,

$$\mathcal{L}^{-1}\left[F(s)G(s) \right] = \int_0^t f(t - \tau)g(\tau) \, d\tau$$

$$= \int_0^t g(t - \tau)f(\tau) \, d\tau. \qquad (3.7.9)$$

The second integral in Eq. 3.7.9 follows from a simple change of variables (see Problem 1). Equation 3.7.9 is called a *convolution theorem* and the integrals on the right-hand sides are *convolution integrals*.

We often adopt a simplified notation. We write

$$f * g = \int_0^t f(t - \tau)g(\tau) \, d\tau \qquad (3.7.10)$$

and hence the convolution theorem (Eq. 3.7.9) may be expressed as

$$\mathcal{L}[f(t) * g(t)] = F(s)G(s). \qquad (3.7.11)$$

(See Problem 1.)

Example 3.7.1:   Suppose that $\mathcal{L}^{-1}[F(s)] = f(t)$. Find

$$\mathcal{L}^{-1}\left[\frac{1}{s}F(s)\right].$$

SOLUTION: Since $\mathcal{L}^{-1}[1/s] = 1$ we have, from Eq. 3.7.9,

$$\mathcal{L}^{-1}\left[\frac{1}{s}F(s)\right] = \int_0^t f(\tau)\,d\tau.$$

Compare with Eq. 3.3.14.   ∎

3.7.1   THE ERROR FUNCTION

A function that plays an important role in statistics is the *error function*,

$$\text{erf}\,(x) = \frac{2}{\sqrt{\pi}}\int_0^x e^{-t^2}\,dt. \tag{3.7.12}$$

In Table 3.1, with $f(t) = t^{k-1}\,e^{at}$, let $a = -1$ and $k = \frac{1}{2}$. Then

$$\frac{e^{-t}}{\sqrt{t}} = \mathcal{L}^{-1}\left[\frac{\Gamma(\frac{1}{2})}{\sqrt{s+1}}\right]. \tag{3.7.13}$$

But $\Gamma(\frac{1}{2}) = \sqrt{\pi}$ (see Section 2.5), so

$$\frac{e^{-t}}{\sqrt{t\pi}} = \mathcal{L}^{-1}\left[\frac{1}{\sqrt{s+1}}\right]. \tag{3.7.14}$$

From the convolution theorem,

$$\mathcal{L}^{-1}\left[\frac{1}{s\sqrt{s+1}}\right] = \int_0^t (1)\,\frac{e^{-\tau}}{\sqrt{\pi\tau}}\,d\tau. \tag{3.7.15}$$

Set $\sqrt{\tau} = x$. Then $\frac{1}{2}\tau^{-1/2}\,d\tau = dx$ and hence

$$\mathcal{L}^{-1}\left[\frac{1}{s\sqrt{s+1}}\right] = \frac{2}{\sqrt{\pi}}\int_0^{\sqrt{t}} e^{-x^2}\,dx = \text{erf}\,(\sqrt{t}). \tag{3.7.16}$$

Example 3.7.2:   Show that

$$\mathcal{L}^{-1}\left[\frac{1}{1+\sqrt{1+s}}\right] = -1 + \text{erf}(\sqrt{t}) + \frac{e^{-t}}{\sqrt{\pi t}}.$$

SOLUTION: We have

$$\frac{1}{1+\sqrt{1+s}} = \frac{1-\sqrt{1+s}}{1-1-s} = -\frac{1}{s} + \frac{\sqrt{1+s}}{s}$$

$$= -\frac{1}{s} + \frac{1+s}{s\sqrt{1+s}} = -\frac{1}{s} + \frac{1}{s\sqrt{1+s}} + \frac{1}{\sqrt{1+s}}.$$

Using Eqs. 3.7.9 and 3.7.13, the inverse transform is

$$\mathcal{L}^{-1}\left[\frac{1}{1 + \sqrt{1 + s}}\right] = \mathcal{L}^{-1}\left(-\frac{1}{s}\right) + \mathcal{L}^{-1}\left(\frac{1}{s\sqrt{1 + s}}\right) + \mathcal{L}^{-1}\left(\frac{1}{\sqrt{1 + s}}\right)$$

$$= -1 + \text{erf}\,(\sqrt{t}) + \frac{e^{-t}}{\sqrt{\pi t}}$$

as proposed.   ■

PROBLEMS

Use the definition of $f * g$ as given in Eq. 3.7.10 to show each of the following.

1. $f * g = g * f$ by use of the change of variable $\beta = t - \tau$.

2. $f * (g * h) = (f * g) * h$

3. $f * (g + h) = f * g + f * h$

4. $f * kg = k(f * g)$ for any scalar k

5. $1 * f = \int_0^t f(\tau)\,d\tau$

6. $1 * f'(t) = f(t) - f(0)$

7. $(f * g)' = g(0)f(t) + g'(t) * f(t) = f(0)g(t) + g(t) * f'(t)$

Compute $f * g$ given the following.

8. $f(t) = g(t) = e^{at}$

9. $f(t) = \sin t$, $g(t) = e^{-at}$

10. $f(t) = g(t) = \sin \omega t$

Use the convolution theorem to find each inverse Laplace transform.

11. $\mathcal{L}^{-1}\left[\dfrac{1}{(s^2 + a^2)^2}\right]$

12. $\mathcal{L}^{-1}\left[\dfrac{s^2}{s^2 - a^2}\right]$

13. $\mathcal{L}^{-1}\left[\dfrac{1}{s^2(s^2 + \omega^2)}\right]$

14. Let $F(s) = \mathcal{L}[f(t)]$. Use the convolution theorem to show that

$$\mathcal{L}^{-1}\left[\frac{F(s)}{(s + a)^2 + b^2}\right] = \frac{1}{b}e^{-at}\int_0^t f(\tau)e^{a\tau}\sin b(t - \tau)\,d\tau.$$

15. Let $F(s) = \mathcal{L}[f(t)]$. Use the convolution theorem to show that

$$\mathcal{L}^{-1}\left[\frac{F(s)}{(s + a)^2}\right] = \int_0^t \tau e^{-a\tau}f(t - \tau)\,d\tau.$$

3.8 Solution of Differential Equations

We are now in a position to solve linear ordinary differential equations with constant coefficients. The technique will be demonstrated with second-order equations, as was done in Chapter 1. The method is, however, applicable to any linear, differential equation. To solve a differential equation, we shall find the Laplace transform of each term of the differential equation, using the techniques presented in Sections 3.2 through 3.5. The resulting algebraic equation will then be organized into a form for which the inverse can be readily found. For nonhomogeneous equations this usually involves partial fractions as discussed in Section 3.6. Let us demonstrate the procedure for the equation

$$\frac{d^2y}{dt^2} + a\frac{dy}{dt} + by = r(t). \tag{3.8.1}$$

Equations 3.3.4 and 3.3.5 allow us to write this differential equation as the algebraic equation

$$s^2Y(s) - sy(0) - y'(0) + a[sY(s) - y(0)] + bY(s) = R(s), \tag{3.8.2}$$

where $Y(s) = \mathcal{L}(y)$ and $R(s) = \mathcal{L}(r)$. This algebraic equation is referred to as the *subsidiary equation* of the given differential equation. It can be rearranged in the form

$$Y(s) = \frac{(s + a)y(0) + y'(0)}{s^2 + as + b} + \frac{R(s)}{s^2 + as + b}. \tag{3.8.3}$$

Note that the initial conditions are responsible for the first term on the right and the nonhomogeneous part of the differential equation is responsible for the second term. To find the desired solution, our task is simply to find the inverse Laplace transform

$$y(t) = \mathcal{L}^{-1}(Y). \tag{3.8.4}$$

Let us illustrate with several examples.

Example 3.8.1: Find the solution of the differential equation that represents the damped harmonic motion of the spring-mass system shown,

$$\frac{d^2y}{dt^2} + 4\frac{dy}{dt} + 8y = 0,$$

with initial conditions $y(0) = 2$, $\dot{y}(0) = 0$. For the derivation of this equation, see Section 1.7.

SOLUTION: The subsidiary equation is found by taking the Laplace transform of the given differential equation:

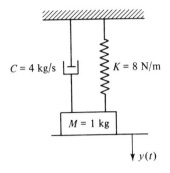

$$s^2Y - sy(0) - \cancel{y'(0)}^{\,0} + 4[sY - y(0)] + 8Y = 0.$$

This is rearranged and put in the form

$$Y(s) = \frac{2s + 8}{s^2 + 4s + 8}.$$

To use Table 3.1 we write this as

$$Y(s) = \frac{2(s + 2) + 4}{(s + 2)^2 + 4} = \frac{2(s + 2)}{(s + 2)^2 + 4} + \frac{4}{(s + 2)^2 + 4}.$$

The inverse transform is then found to be

$$y(t) = e^{-2t}\, 2 \cos 2t + e^{-2t}\, 2 \sin 2t = 2e^{-2t}(\cos 2t + \sin 2t). \quad \blacksquare$$

Example 3.8.2:   An inductor of 2 H and a capacitor of 0.02 F is connected in series with an imposed voltage of 100 sin ωt volts. Determine the charge $q(t)$ on the capacitor as a function of ω if the initial charge on the capacitor and current in the circuit are zero.

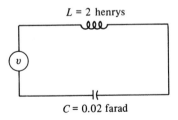

$L = 2$ henrys

$C = 0.02$ farad

SOLUTION: Kirchhoff's laws allow us to write (see Section 1.4)

$$2\frac{di}{dt} + \frac{q}{0.02} = 100 \sin \omega t,$$

where $i(t)$ is the current in the circuit. Using $i = dq/dt$, we have

$$2\frac{d^2q}{dt^2} + 50q = 100 \sin \omega t.$$

The Laplace transform of this equation is

$$2[s^2 Q - \cancel{2q(0)}^{\,0} - \cancel{q'(0)}^{\,0}] + 50Q = \frac{100\omega}{s^2 + \omega^2},$$

using $i(0) = q'(0) = 0$. The transform of $q(t)$ is then

$$Q(s) = \frac{50\omega}{(s^2 + \omega^2)(s^2 + 25)}.$$

The appropriate partial fractions are

$$Q(s) = \frac{A_1 s + A_2}{s^2 + \omega^2} + \frac{B_1 s + B_2}{s^2 + 25}.$$

The constants are found from (see Eq. 3.6.9)

$$A_1(i\omega) + A_2 = \frac{50\omega}{-\omega^2 + 25}, \quad B_1(5i) + B_2 = \frac{50\omega}{-25 + \omega^2}.$$

They are

$$A_1 = 0, \quad A_2 = \frac{50\omega}{25 - \omega^2}, \quad B_1 = 0, \quad B_2 = \frac{50\omega}{\omega^2 - 25}.$$

Hence,

$$Q(s) = \frac{50}{25 - \omega^2}\left[\frac{\omega}{s^2 + \omega^2} - \frac{\omega}{s^2 + 25}\right].$$

The inverse Laplace transform is

$$q(t) = \frac{50}{25 - \omega^2}[\sin \omega t - \sin 5t].$$

This solution is acceptable if $\omega \neq 5$ rad/s, and we observe that the amplitude becomes unbounded as $\omega \to 5$ rad/s. If $\omega = 5$ rad/s, the Laplace transform becomes

$$Q(s) = \frac{250}{(s^2 + 25)^2} = \frac{A_1 s + A_2}{(s^2 + 25)^2} + \frac{B_1 s + B_2}{s^2 + 25}.$$

Using Eqs. 3.6.11 and 3.6.12, we have

$$A_1(5i) + A_2 = 250, \quad B_1(5i) + B_2 = \frac{d}{ds}(250) = 0.$$

We have

$$A_1 = 0, \quad A_2 = 250, \quad B_1 = 0, \quad B_2 = 0.$$

Hence,

$$Q(s) = \frac{250}{(s^2 + 25)^2}.$$

The inverse is

$$q(t) = 250\left[\frac{1}{2 \times 5^3}(\sin 5t - 5t \cos 5t)\right]$$

$$= \sin 5t - 5t \cos 5t. \quad \blacksquare$$

Observe, in the example above, that the amplitude becomes unbounded as t gets large. This is *resonance*, a phenomenon that occurs in undamped oscillatory systems with input frequency equal to the natural frequency of the system. See Section 1.9.1 for a discussion of resonance.

Example 3.8.3:   As an example of a differential equation that has boundary conditions at two locations, consider a beam loaded as shown. The differential equation that describes the deflection $y(x)$ is

$$\frac{d^4y}{dx^4} = \frac{w}{EI}$$

with boundary conditions $y(0) = y''(0) = y(L) = y''(L) = 0$. Find $y(x)$.

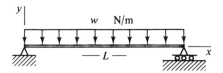

SOLUTION: The Laplace transform of the differential equation is, according to Eq. 3.3.6,

$$s^4 Y - s^3 \cancel{y(0)}^{\,0} - s^2 y'(0) - s\cancel{y''(0)}^{\,0} - y'''(0) = \frac{w}{EIs}.$$

The two unknown initial conditions are replaced with

$$y'(0) = c_1 \quad \text{and} \quad y'''(0) = c_2.$$

We then have

$$Y(s) = \frac{c_1}{s^2} + \frac{c_2}{s^4} + \frac{w}{EIs^5}.$$

The inverse Laplace transform is

$$y(x) = c_1 x + c_2\frac{x^3}{6} + \frac{w}{EI}\frac{x^4}{24}.$$

The boundary conditions on the right end are now satisfied:

$$y(L) = c_1 L + c_2\frac{L^3}{6} + \frac{w}{EI}\frac{L^4}{24} = 0$$

$$y''(L) = c_2 L + \frac{w}{EI}\frac{L^2}{2} = 0.$$

Hence,

$$c_2 = -\frac{wL}{2EI}, \qquad c_1 = \frac{wL^3}{EI}.$$

Finally, the desired solution for the deflection of the beam is

$$y(x) = \frac{w}{24EI}[xL^3 - 2x^3L + x^4]. \quad \blacksquare$$

Example 3.8.4:   Solve the differential equation

$$\frac{d^2y}{dt^2} + 0.02\frac{dy}{dt} + 25y = f(t),$$

which describes a slightly damped oscillating system where $f(t)$ is as shown. Assume that the system starts from rest.

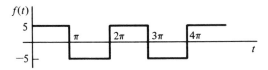

SOLUTION: The subsidiary equation is found by taking the Laplace transform of the given differential equation:

$$s^2Y - sy(0) - y'(0) + 0.02[sY - y(0)] + 25Y = F(s),$$

where $F(s)$ is given by (see Example 3.2.10)

$$F(s) = \frac{5}{s}[1 - 2e^{-\pi s} + 2e^{-2\pi s} - 2e^{-3\pi s} + \cdots].$$

Since the system starts from rest, $y(0) = y'(0) = 0$. The subsidiary equation becomes

$$Y(s) = \frac{F(s)}{s^2 + 0.02s + 25}$$

$$= \frac{5}{s(s^2 + 0.02s + 25)}[1 - 2e^{-\pi s} + 2e^{-2\pi s} - \cdots].$$

Now let us find the inverse term by term. We must use*

$$\mathcal{L}^{-1}\left[\frac{5}{(s + 0.01)^2 + 5^2}\right] = e^{-0.01t}\sin 5t.$$

With the use of Eq. 3.3.14, we have, for the first term (see Example 3.5.2 for integration by parts),

$$y_0(t) = \mathcal{L}^{-1}\left[\frac{5}{s[(s + 0.01)^2 + 5^2]}\right]$$

$$= \int_0^t e^{-0.01\tau}\sin 5\tau\, d\tau$$

---

*We write $s^2 + 0.02s + 25 = (s + 0.01)^2 + 24.999 \cong (s + 0.01)^2 + 5^5$.

$$= 1 - \frac{1}{5}e^{-0.01t}[\cos 5t + 0.002 \sin 5t].$$

The inverse of the next term is found using the second shifting property (see Eq. 3.2.14):

$$y_1(t) = \mathcal{L}^{-1}\left[e^{-\pi s}\frac{5}{s(s^2 + 0.02s + 25)}\right]$$

$$= -2u_n(t)\left\{1 - \frac{1}{5}e^{-0.01(t-\pi)}[\cos 5(t - \pi) + 0.002 \sin 5(t - \pi)]\right\}$$

$$= -2u_n(t)\{1 + [1 - y_0(t)]e^{0.01\pi}\},$$

where $u_n(t)$ is the unit step function and we have used $\cos(t - \pi) = -\cos t$ and $\sin(t - \pi) = -\sin t$. The third term provides us with

$$y_2(t) = \mathcal{L}^{-1}\left[e^{-2\pi s}\frac{5}{s(s^2 + 0.02s + 25)}\right]$$

$$= -2u_{2\pi}(t)\left\{1 - \frac{1}{5}e^{-0.01(t-2\pi)}[\cos 5(t - 2\pi) + 0.002 \sin 5(t - 2\pi)]\right\}$$

$$= -2u_{2\pi}(t)\{1 - [1 - y_0(t)]e^{0.02\pi}\},$$

and so on. The solution $y(t)$ is

$$y(t) = y_0(t) = 1 - \frac{1}{5}e^{-0.01t}[\cos 5t + 0.002 \sin 5t], \qquad 0 < t < \pi$$

$$y(t) = y_0(t) + y_1(t)$$

$$= -1 - \frac{1}{5}e^{-0.01t}[\cos 5t + 0.002 \sin 5t][1 + 2e^{0.01\pi}],$$

$$\pi < t < 2\pi$$

$$y(t) = y_0(t) + y_1(t) + y_2(t)$$

$$= 1 - \frac{1}{5}e^{-0.01t}[\cos 5t + 0.002 \sin 5t][1 + 2e^{0.01\pi} + 2e^{0.02\pi}],$$

$$2\pi < t < 3\pi.$$

Now let us find the solution for large t, that is, for $n\pi < t < (n + 1)\pi$, with n large. Generalize the results above and obtain*

---

*In the manipulations we will use

$$\frac{1}{1 + x} = 1 + x + x^2 + x^3 + \cdots + x^n + x^{n+1} + \cdots$$

$$= 1 + x + x^2 + \cdots + x^n + x^{n+1}(1 + x + x^2 + \cdots)$$

$$= 1 + x + x^2 + \cdots + x^n + x^{n+1}/(1 - x).$$

Hence,

$$\frac{1 - x^{n+1}}{1 - x} = 1 + x + x^2 + \cdots + x^n.$$

$$y(t) = y_0(t) + y_1(t) + \cdots + y_n(t), \qquad n\pi < t < (n+1)\pi$$

$$= (-1)^n - \frac{1}{5}e^{-0.01t}[\cos 5t + 0.002 \sin 5t][1 + 2e^{0.01\pi}$$

$$+ 2e^{0.002\pi} + \cdots + 2e^{0.01n\pi}]$$

$$= (-1)^n + \frac{1}{5}e^{-0.01t}[\cos 5t + 0.002 \sin 5t]$$

$$- \frac{2}{5}e^{-0.01t}[\cos 5t + 0.002 \sin 5t][1 + 2e^{0.01\pi} + \cdots + 2e^{0.01n\pi}]$$

$$= (-1)^n + \frac{1}{5}e^{-0.01t}[\cos 5t + 0.002 \sin 5t]$$

$$- \frac{2}{5}e^{-0.01t}[\cos 5t + 0.002 \sin 5t]\frac{1 - e^{(n+1)0.01\pi}}{1 - e^{0.01\pi}}$$

$$= (-1)^n + \left[\frac{1}{5} - \frac{2}{5(1 - e^{0.01\pi})}\right]e^{-0.01t}[\cos 5t + 0.002 \; \sin 5t]$$

$$+ \frac{2e^{(n+1)0.01\pi - 0.01t}}{5(1 - e^{0.01\pi})}[\cos 5t + 0.002 \sin 5t].$$

Then, letting t be large and in the interval $n\pi < t < (n+1)\pi$, $e^{-0.01t} \to 0$ and we have

$$y(t) = (-1)^n + \frac{2e^{(n+1)0.01\pi - 0.01t}}{5(1 - e^{0.01\pi})}(\cos 5t + 0.002 \sin 5t)$$

$$\cong (-1)^n - 12.5e^{0.01[(n+1)\pi - t]} \cos 5t.$$

This is the steady-state response due to the square-wave input function shown in the example. One period is sketched here. The second half of the period is obtained by

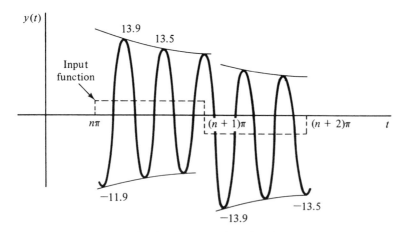

(For this sketch, n is considered even)

replacing n with $n + 1$. Note that the input frequency of 1 rad/s results in a periodic response of 5 rad/s. Note also the large amplitude of the response, a resonance-type behavior. This is surprising, since the natural frequency of the system with no damping is 5 rad/s. This phenomenon occurs quite often when systems with little damping are subjected to nonsinusoidal periodic input functions.   ∎

PROBLEMS

Determine the solution for each initial-value problem.

1. $\dfrac{d^2y}{dt^2} + 4y = 0,$ $y(0) = 0, \quad y'(0) = 10$

2. $\dfrac{d^2y}{dt^2} - 4y = 0,$ $y(0) = 2, \quad y'(0) = 0$

3. $\dfrac{d^2y}{dt^2} + y = 2,$ $y(0) = 0, \quad y'(0) = 2$

4. $\dfrac{d^2y}{dt^2} + 4y = 2\cos t,$ $y(0) = 0, \quad y'(0) = 0$

5. $\dfrac{d^2y}{dt^2} + 4y = 2\cos 2t,$ $y(0) = 0, \quad y'(0) = 0$

6. $\dfrac{d^2y}{dt^2} + y = e^t + 2,$ $y(0) = 0, \quad y'(0) = 0$

7. $\dfrac{d^2y}{dt^2} + 5\dfrac{dy}{dt} + 6y = 0,$ $y(0) = 0, \quad y'(0) = 20$

8. $\dfrac{d^2y}{dt^2} + 4\dfrac{dy}{dt} + 4y = 0,$ $y(0) = 1, \quad y'(0) = 0$

9. $\dfrac{d^2y}{dt^2} - 2\dfrac{dy}{dt} - 8y = 0,$ $y(0) = 1, \quad y'(0) = 0$

10. $\dfrac{d^2y}{dt^2} + 5\dfrac{dy}{dt} + 6y = 12,$ $y(0) = 0, \quad y'(0) = 10$

11. $\dfrac{d^2y}{dt^2} + 2\dfrac{dy}{dt} + y = 2t,$ $y(0) = 0, \quad y'(0) = 0$

12. $\dfrac{d^2y}{dt^2} + 4\dfrac{dy}{dt} + 4y = 4\sin 2t,$ $y(0) = 1, \quad y'(0) = 0$

13. $\dfrac{d^2y}{dt^2} + 4\dfrac{dy}{dt} + 104y = 2\cos 10t,$ $y(0) = 0, \quad y'(0) = 0$

14. $\dfrac{d^2y}{dt^2} + 2\dfrac{dy}{dt} + 101y = 5\sin 10t,$ $y(0) = 0, \quad y'(0) = 20$

Solve for the displacement $y(t)$ if $y(0) = 0$, $y'(0) = 0$. Use a combination of the following friction coefficients and forcing functions.

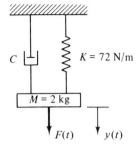

15. $C = 0$ kg/s   (a) $F(t) = 2$ N

16. $C = 2$ kg/s   (b) $F(t) = 10 \sin 2t$

17. $C = 24$ kg/s   (c) $F(t) = 10 \sin 6t$

18. $C = 40$ kg/s   (d) $F(t) = 10[u_0(t) - u_{4n}(t)]$

(e) $F(t) = 10e^{-0.2t}$

(f) $F(t) = 100\delta_0(t)$

For a particular combination of the following resistances and input voltages, calculate the current $i(t)$ if the circuit is quiescent at $t = 0$, that is, the initial charge on the capacitor $q(0) = 0$ and $i(0) = 0$. Sketch the solution.

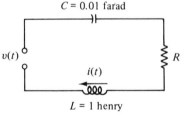

19. $R = 0 \ \Omega$ (a) $v(t) = 10$ V

20. $R = 16 \ \Omega$ (b) $v(t) = 10 \sin 10t$

21. $R = 20 \ \Omega$ (c) $v(t) = 5 \sin 10t$

22. $R = 25 \ \Omega$ (d) $v(t) = 10[u_0(t) - u_{2\pi}(t)]$

(e) $v(t) = 10\delta_0(t)$

(f) $v(t) = 20e^{-t}$

Calculate the response due to the input function $f(t)$ for one of the systems shown. Assume each system to be quiescent at $t = 0$. The function $f(t)$ is given as sketched.

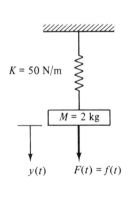

(a)

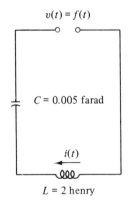

(b)

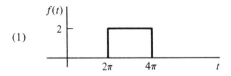

(1)

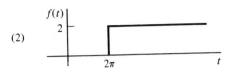

(2)

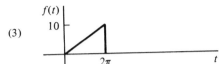

(3)

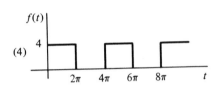

(4)

(6)

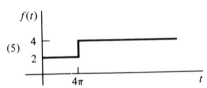

(5)

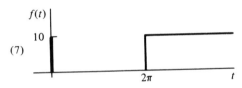

(7)

Determine the response function due to the input function for one of the systems shown. Each system is quiescent at $t = 0$. Use an input function $f(t)$ from Problems 23 and 24.

25.

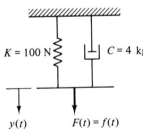

$K = 100\ \text{N}$ $C = 4\ \text{kg/s}$

$y(t)$ $F(t) = f(t)$

26.

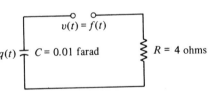

$v(t) = f(t)$

$q(t)$ $C = 0.01\ \text{farad}$ $R = 4\ \text{ohms}$

27. Find the deflection $y(x)$ of the beam shown. The differential equation that describes the deflection is

$$\frac{d^4 y}{dx^4} = \frac{w(x)}{EI}, \qquad w(x) = P\delta_{L/2}(x) + w[u_0(x) - u_{L/2}(x)].$$

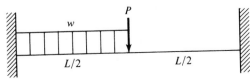

P

w

$L/2$ $L/2$

3.9 Special Techniques

3.9.1 POWER SERIES

If the power series for $f(t)$, written as

$$f(t) = \sum_{n=0}^{\infty} a_n t^n, \tag{3.9.1}$$

has an infinite radius of convergence—or equivalently, if $f(t)$ has no singularities—and if $f(t)$ has exponential order as $t \to \infty$, then

$$F(s) = \mathcal{L}(f) = \sum_{n=0}^{\infty} a_n \mathcal{L}(t^n) = \sum_{n=0}^{\infty} n! \, a_n \frac{1}{s^{n+1}}. \tag{3.9.2}$$

If the series 3.9.2 is easily recognized as combinations of known functions, this technique can be quite useful.

Example 3.9.1: Find the Laplace transform of $f(t) = (e^{-t} - 1)/t$ (see Example 3.4.2).

SOLUTION: Since the given $f(t)$ can be expanded in a power series (see Section 2.2), we can write

$$\frac{e^{-t} - 1}{t} = \sum_{n=1}^{\infty} \frac{(-1)^n}{n!} t^{n-1}.$$

It then follows that

$$\mathcal{L}\left[\frac{e^{-t} - 1}{t}\right] = \sum_{n=1}^{\infty} \frac{(-1)^n}{n!} \frac{(n-1)!}{s^n}$$

$$= \sum_{n=1}^{\infty} \frac{(-1)^n}{ns^n}$$

$$= -\frac{1}{s} + \frac{1}{2s^2} - \frac{1}{3s^3} + \cdots$$

$$= -\ln\left(1 + \frac{1}{s}\right). \quad \blacksquare$$

Example 3.9.2: Find the Laplace transform of $f(t) = t^{-1/2} \operatorname{erf}(\sqrt{t})$.

SOLUTION: By definition (see Eq. 3.7.12),

$$t^{-1/2} \operatorname{erf}(\sqrt{t}) = \frac{2t^{-1/2}}{\sqrt{\pi}} \int_0^{\sqrt{t}} e^{-x^2} \, dx$$

$$= \frac{2t^{-1/2}}{\sqrt{\pi}} \int_0^{\sqrt{t}} \sum_{n=0}^{\infty} \frac{(-1)^n x^{2n}}{n!} \, dx$$

$$= \frac{2t^{-1/2}}{\sqrt{\pi}} \sum_{n=0}^{\infty} \frac{(-1)^n (\sqrt{t})^{2n+1}}{(2n+1)n!}$$

$$= \frac{2}{\sqrt{\pi}} \sum_{n=0}^{\infty} \frac{(-1)^n t^n}{(2n+1)n!}.$$

Therefore,

$$\mathscr{L}[t^{-1/2} \, \mathrm{erf}\,(\sqrt{t})] = \frac{2}{\sqrt{\pi}} \sum_{n=0}^{\infty} \frac{(-1)^n}{2n+1} \frac{1}{s^{n+1}}.$$

The power series in $1/s$ can be recognized as $(2/\sqrt{\pi s}) \tan^{-1}(1/\sqrt{s})$. So

$$\mathscr{L}[t^{-1/2} \, \mathrm{erf}\,(\sqrt{t})] = \frac{2}{\sqrt{\pi s}} \tan^{-1} \frac{1}{\sqrt{s}}. \quad \blacksquare$$

Example 3.9.3:   Show that

$$\mathscr{L}[J_0(t)] = \frac{1}{\sqrt{s^2+1}},$$

where $J_0(t)$ is the Bessel function of index zero.

SOLUTION: The Taylor series for $J_0(t)$ is given in Eq. 2.4.13:

$$J_0(t) = \sum_{k=0}^{\infty} \frac{(-1)^k t^{2k}}{2^{2k} k! \, k!}.$$

Hence,

$$\mathscr{L}[J_0(t)] = \sum_{k=0}^{\infty} \frac{(-1)^k (2k)!}{2^{2k} k! \, k!} \frac{1}{s^{2k+1}},$$

but

$$(2k)! = 2 \cdot 4 \cdot 6 \cdots 2k \cdot 1 \cdot 3 \cdots (2k-1)$$
$$= 2^k k! \cdot 1 \cdot 3 \cdots (2k-1).$$

Thus,

$$\mathscr{L}[J_0(t)] = \frac{1}{s} \left[1 + \sum_{k=1}^{\infty} \frac{(-1)^k \cdot 1 \cdot 3 \cdots (2k-1)}{2^k k! \, s^{2k}} \right].$$

Use of the binomial theorem is one way of establishing that

$$\left(1 + \frac{1}{s^2} \right)^{-1/2} = 1 + \frac{(-1)(\frac{1}{2})}{1!} \left(\frac{1}{s^2} \right) + \frac{(-\frac{1}{2})(-\frac{3}{2})(-\frac{5}{2})}{2!} \left(\frac{1}{s} \right)^2 + \cdots$$

$$+ \frac{(-1)^k (1)(3) \cdots (2k-1)}{2^k k!} \frac{1}{s^{2k}} + \cdots. \tag{2}$$

Finally, using Eq. (2) in Eq. (1) we have

$$\mathscr{L}[J_0(t)] = \frac{1}{s} \left(1 + \frac{1}{s^2} \right)^{-1/2} = \frac{1}{\sqrt{s^2+1}}. \quad \blacksquare$$

PROBLEMS

1. Expand $\sin \sqrt{t}$ in an infinite series and show that $\mathcal{L}(\sin \sqrt{t}) = (\sqrt{\pi}/2s^{3/2})e^{-1/4s}$.

2. Use the identity

$$\frac{d}{dt}J_0(t) = -J_1(t)$$

and the Laplace transform of J_0 to derive

$$\mathcal{L}[J_1(t)] = 1 - \frac{s}{\sqrt{s^2 + 1}}.$$

3. Show that

$$\mathcal{L}^{-1}\left[\frac{1}{s^{n+1}}e^{-1/s}\right] = t^{n/2}J_n(2\sqrt{t}).$$

4. Expand $1/t \sin (tk)$ in powers of t to prove that

$$\mathcal{L}\left[\frac{1}{t}\sin (kt)\right] = \tan^{-1}\frac{k}{s}.$$

5. Find $\mathcal{L}[J_0(2t)]$.

TABLE 3.1. Laplace Transforms

	$f(t)$	$F(s) = \mathcal{L}\{f(t)\}$
1	1	$\dfrac{1}{s}$
2	t	$\dfrac{1}{s^2}$
3	t^{n-1}	$\dfrac{(n-1)!}{s^n}$ $(n = 1, 2, \cdots)$
4	$t^{-1/2}$	$\dfrac{\sqrt{\pi}}{s}$
5	$t^{1/2}$	$\dfrac{\sqrt{\pi}}{2s^{3/2}}$
6	t^{k-1}	$\dfrac{\Gamma(k)}{s^k}$ $(k > 0)$
7	e^{at}	$\dfrac{1}{s-a}$
8	te^{at}	$\dfrac{1}{(s-a)^2}$
9	$t^{n-1}e^{at}$	$\dfrac{(n-1)!}{(s-a)^n}$ $(n = 1, 2, \cdots)$
10	$t^{k-1}e^{at}$	$\dfrac{\Gamma(k)}{(s-a)^k}$ $(k > 0)$
11	$e^{at} - e^{bt}$	$\dfrac{a-b}{(s-a)(s-b)}$ $(a \neq b)$

TABLE 3.1. *Laplace Transforms* (*Cont.*)

	$f(t)$	$F(s) = \mathcal{L}\{f(t)\}$
12	$ae^{at} - be^{bt}$	$\dfrac{(a-b)s}{(s-a)(s-b)}$ $\quad (a \neq b)$
13	$\delta_0(t)$	1
14	$\delta_a(t)$	e^{-as}
15	$u_a(t)$	$e^{-as/s}$
16	$\ln t$	$\dfrac{1}{s}\left(\ln\dfrac{1}{s} - 0.5772156\cdots\right)$
17	$\sin \omega t$	$\dfrac{\omega}{s^2 + \omega^2}$
18	$\cos \omega t$	$\dfrac{s}{s^2 + \omega^2}$
19	$\sinh at$	$\dfrac{a}{s^2 - a^2}$
20	$\cosh at$	$\dfrac{s}{s^2 - a^2}$
21	$e^{at} \sin \omega t$	$\dfrac{\omega}{(s-a)^2 + \omega^2}$
22	$e^{at} \cos \omega t$	$\dfrac{s-a}{(s-a)^2 + \omega^2}$
23	$1 - \cos \omega t$	$\dfrac{\omega^2}{s(s^2 + \omega^2)}$
24	$\omega t - \sin \omega t$	$\dfrac{\omega^3}{s^2(s^2 + \omega^2)}$
25	$\sin \omega t - \omega t \cos \omega t$	$\dfrac{2\omega^3}{(s^2 + \omega^2)^2}$
26	$t \sin \omega t$	$\dfrac{\omega s}{(s^2 + \omega^2)^2}$
27	$\sin \omega t + \omega t \cos \omega t$	$\dfrac{2\omega s^2}{(s^2 + \omega^2)^2}$
28	$\cos at - \cos bt$	$\dfrac{(b^2 - a^2)s}{(s^2 + a^2)(s^2 + b^2)}$ $\quad (a^2 \neq b^2)$
29	$\sin at \cosh at - \cos at \sinh at$	$\dfrac{4a^3}{s^4 + 4a^4}$
30	$\sin at \sinh at$	$\dfrac{2a^2 s}{s^4 + 4a^4}$
31	$\sinh at - \sin at$	$\dfrac{2a^3}{s^4 - a^4}$
32	$\cosh at - \cos at$	$\dfrac{2a^2 s}{s^4 - a^4}$
33	$e^{at} f(t)$	$F(s-a)$
34	$\dfrac{1}{a}f\left(\dfrac{t}{a}\right)$	$F(as)$
35	$\dfrac{1}{a}e^{-(b/a)t}f\left(\dfrac{t}{a}\right)$	$F(as + b)$
36	$f(t - c)u_c(t)$	$e^{-cs}F(s)$
37	$\displaystyle\int_0^t f(\tau)g(t - \tau)\, d\tau$	$F(s)G(s)$

4

The Theory of Matrices

4.1 Introduction

The theory of matrices arose as a means to solve simultaneous, linear, algebraic equations. Its present uses span the entire spectrum of mathematical ideas, including numerical analysis, statistics, differential equations, and optimization theory, to mention a few of its applications. In this chapter we develop notation, terminology, and the central ideas most closely allied to the physical sciences.

4.2 Notation and Terminology

A *matrix* is a rectangular array of numbers; its *order* is the number of rows and columns that define the array. Thus, the matrices

(a) $\begin{bmatrix} 1 & 0 & -1 \\ 2 & 5 & 7 \end{bmatrix}$ (b) $\begin{bmatrix} -2 & 1 & 1 \\ 0 & 0 & 0 \\ 1 & 1 & 1 \end{bmatrix}$ (c) $\begin{bmatrix} x \\ y \\ z \end{bmatrix}$

(d) $\begin{bmatrix} 1 & 1 - i & 1 + i \end{bmatrix}$ (e) $[0]$

have orders 2×3, 3×3, 3×1, 1×3, and 1×1, respectively. (The order 2×3 is read "two by three.")

In general, the matrix **A**, defined by

$$\mathbf{A} = \begin{bmatrix} a_{11} & a_{12} & \cdots & a_{1q} \\ a_{21} & a_{22} & \cdots & a_{2q} \\ \vdots & & & \\ a_{p1} & a_{p2} & \cdots & a_{pq} \end{bmatrix}, \tag{4.2.1}$$

is order $p \times q$. The numbers a_{ij} are called the *entries* or *elements* of **A**; the first subscript defines its row position, the second its column position.

In general, we will use uppercase bold letters to represent matrices, but sometimes it is convenient to explicitly mention the order of **A** or display a typical element, by use of the notations $\mathbf{A}_{p \times q}$ and (a_{ij}),

$$\mathbf{A}_{3 \times 3} = (i^j) = \begin{bmatrix} 1 & 1 & 1 \\ 2 & 2^2 & 2^3 \\ 3 & 3^2 & 3^3 \end{bmatrix}$$

(4.2.2)

$$\mathbf{A}_{2 \times 4} = (i - j) = \begin{bmatrix} 0 & -1 & -2 & -3 \\ 1 & 0 & -1 & -2 \end{bmatrix}$$

In the system of simultaneous equations

$$
\begin{aligned}
2x_1 - x_2 + x_3 - x_4 &= 1 \\
x_1 \quad\quad - x_3 \quad\quad &= 1 \\
x_2 + x_3 + x_4 &= -1
\end{aligned}
$$

(4.2.3)

the matrix

$$\mathbf{A} = \begin{bmatrix} 2 & -1 & 1 & -1 \\ 1 & 0 & -1 & 0 \\ 0 & 1 & 1 & 1 \end{bmatrix}$$

(4.2.4)

is the *coefficient matrix* and

$$\mathbf{B} = \begin{bmatrix} 2 & -1 & 1 & -1 & 1 \\ 1 & 0 & -1 & 0 & 1 \\ 0 & 1 & 1 & 1 & -1 \end{bmatrix}$$

(4.2.5)

is the *augmented matrix*. The augmented matrix is the coefficient matrix with an extra column containing the right-hand-side constants.

The ith row of the general matrix 4.2.1 is denoted by $\mathbf{A}_{i*}$, the jth column by $\mathbf{A}_{*j}$. Thus, in the matrix 4.2.4,

$$
\begin{aligned}
\mathbf{A}_{1*} &= \begin{bmatrix} 2 & -1 & 1 & -1 \end{bmatrix} \\
\mathbf{A}_{2*} &= \begin{bmatrix} 1 & 0 & -1 & 0 \end{bmatrix} \\
\mathbf{A}_{3*} &= \begin{bmatrix} 0 & 1 & 1 & 1 \end{bmatrix},
\end{aligned}
$$

(4.2.6)

while

$$\mathbf{A}_{*1} = \begin{bmatrix} 2 \\ 1 \\ 0 \end{bmatrix}, \qquad \mathbf{A}_{*2} = \begin{bmatrix} -1 \\ 0 \\ 1 \end{bmatrix}, \qquad \mathbf{A}_{*3} = \begin{bmatrix} 1 \\ -1 \\ 1 \end{bmatrix}, \qquad \mathbf{A}_{*4} = \begin{bmatrix} -1 \\ 0 \\ 1 \end{bmatrix}$$

(4.2.7)

Square matrices have the same number of rows and columns. The *diagonal* entries of the $\mathbf{A}_{n \times n}$ matrix are $a_{11}, a_{22}, \ldots, a_{nn}$; the *off-diagonal* entries are a_{ij}, $i \neq j$. Matrices with off-diagonal entries of zero are *diagonal matrices*. The following are diagonal matrices:

$$A = \begin{bmatrix} 1 & 0 \\ 0 & -1 \end{bmatrix}, \qquad B = \begin{bmatrix} 0 & 0 & 0 \\ 0 & 0 & 0 \\ 0 & 0 & 0 \end{bmatrix},$$

$$C = \begin{bmatrix} 1 & 0 & 0 \\ 0 & 2 & 0 \\ 0 & 0 & 1 \end{bmatrix}, \qquad D = [-2] \tag{4.2.8}$$

The *identity matrix* I_n is the $n \times n$ diagonal matrix in which $a_{ii} = 1$ for all i. So

$$I_1 = [1], \qquad I_2 = \begin{bmatrix} 1 & 0 \\ 0 & 1 \end{bmatrix}, \qquad I_3 = \begin{bmatrix} 1 & 0 & 0 \\ 0 & 1 & 0 \\ 0 & 0 & 1 \end{bmatrix}. \tag{4.2.9}$$

When the context makes the order of I_n clear, we drop the subscript n.

Upper triangular matrices are square matrices whose entries below the diagonal are all zero. *Lower triangular matrices* are square matrices whose off-diagonal entries lying above the diagonal are zero. We use U and L as generic names for these matrices. For example,

$$L_1 = \begin{bmatrix} 1 & 0 \\ 2 & -1 \end{bmatrix}, \qquad L_2 = \begin{bmatrix} 0 & 0 \\ 0 & 0 \end{bmatrix}, \qquad L_3 = I \tag{4.2.10}$$

are all lower triangular—the subscript here simply distinguishes different lower triangular matrices. Similarly,

$$U_1 = \begin{bmatrix} 0 & 1 \\ 0 & -1 \end{bmatrix}, \qquad U_2 = [7], \qquad U_3 = I \tag{4.2.11}$$

are all upper triangular. Note that diagonal matrices are both upper and lower triangular and every matrix that is both upper and lower triangular is a diagonal matrix.

Finally, we define the O matrix to have all entries equal to zero; that is, the entries of the square matrix O_n are $a_{ij} = 0$. Thus, for instance,

$$O_2 = \begin{bmatrix} 0 & 0 \\ 0 & 0 \end{bmatrix}. \tag{4.2.12}$$

PROBLEMS

Write each matrix in full.

1. $A_{3 \times 3} = [j^i]$

2. $A_{2 \times 4} = [i + j]$

3. $A_{3 \times 3} = [i]$

4. $A_{3 \times 3} = [j]$

What are the coefficient and augmented matrices for each of the following?

5. $x_1 \qquad\quad = 0$
$\qquad x_2 \qquad = 0$
$\qquad\qquad x_3 = 0$

6. $x_1 + x_2 + x_3 = 0$

7. $x_1 = x_2$
$\quad x_2 = x_3$
$\quad x_3 = 1$

8. $x_1 = 0, \quad x_2 = 1, \quad x_3 = 1$

9. Is $\mathbf{O}_n$ upper triangular? Lower triangular? Diagonal?

10. Identify which of the following groups of numbers are matrices.

(a) $[0 \quad 2]$ (b) $\begin{bmatrix} 0 \\ 2 \end{bmatrix}$ (c) $\begin{bmatrix} 0 \\ 1 & 2 \end{bmatrix}$ (d) $\begin{bmatrix} 0 & 1 \\ 2 & 3 \end{bmatrix}$ (e) $\begin{bmatrix} 1 & 0 \\ & 3 & 2 \end{bmatrix}$

(f) $\begin{bmatrix} 1 & 2 \\ 3 \end{bmatrix}$ (g) $\begin{bmatrix} 2x & x^2 \\ 2 & 0 \end{bmatrix}$ (h) $[2x \quad x^2]$ (i) $\begin{bmatrix} x \\ xy \\ z \end{bmatrix}$

(j) $\begin{bmatrix} x & y & z \\ 1 & 0 & 1 \\ x^2 & y^2 & z^2 \end{bmatrix}$ (k) $\begin{bmatrix} 2 - i & i \\ 2 & 3 + i \end{bmatrix}$

11. Given

$$\mathbf{A} = \begin{bmatrix} 5 & 2 & 0 & -3 \\ 4 & -2 & 7 & 0 \\ 1 & 0 & 6 & 8 \\ -2 & 4 & 0 & 9 \end{bmatrix},$$

identify the following elements:

(a) a_{22} (b) a_{32} (c) a_{23} (d) a_{11} (e) a_{14}

4.3 The Solution of Simultaneous Equations by Gaussian Elimination

Our first application of matrix theory is connected with its oldest use—the solution of a system of algebraic equations. Consider, for example, the following equations:

$$x \qquad\quad + \ z = 1$$
$$2x + y + \ z = 0 \qquad\qquad (4.3.1)$$
$$x + y + 2z = 1.$$

We can solve these equations by elimination. Proceed systematically, eliminating the first unknown, x, from the second and third equations using the first equation. This results in the system

$$x \quad + z = \quad 1$$
$$y - z = -2 \qquad (4.3.2)$$
$$y + z = \quad 0.$$

Next, eliminate y from the third equation and get

$$x \quad + z = \quad 1$$
$$y - z = -2 \qquad (4.3.3)$$
$$2z = \quad 2.$$

We now have $z = 1$ from the third equation, and from this deduce $y = -1$ from the second equation and $x = 0$ from the first equation.

It should be clear that the elimination process depends on the coefficients of the equations and not the unknowns. We could have collected all the coefficients in Eqs. 4.3.1 in a rectangular array, the augmented matrix,

$$\begin{bmatrix} 1 & 0 & 1 & 1 \\ 2 & 1 & 1 & 0 \\ 1 & 1 & 2 & 1 \end{bmatrix}, \qquad (4.3.4)$$

and eliminated x and y using the rows of the array as though they were equations. For instance, (-2) times each entry in the first row added, entry by entry, to the second row and (-1) times each entry in the first row added to the third row yields the array

$$\begin{bmatrix} 1 & 0 & 1 & 1 \\ 0 & 1 & -1 & -2 \\ 0 & 1 & 1 & 0 \end{bmatrix}, \qquad (4.3.5)$$

which exhibits the coefficients of Eqs. 4.3.2. The zeros in the first column of Eq. 4.3.5 refer to the fact that x no longer appears in any equation but the first. The elimination of y from the third equation requires the replacement of the 1 in the third row, second column, by 0. We do this by subtracting the second row from the third and thus obtain the coefficients of Eqs. 4.3.3 displayed in the array

$$\begin{bmatrix} 1 & 0 & 1 & 1 \\ 0 & 1 & -1 & -2 \\ 0 & 0 & 2 & 2 \end{bmatrix}. \qquad (4.3.6)$$

Once the equations have been manipulated this far, it is not essential to perform any further simplifications. For the sake of completeness we observe that dividing the third row by 2 (which amounts to dividing the equation $2z = 2$ by 2), then adding the new third row to the second and subtracting it from the first, leads to the array

$$\begin{bmatrix} 1 & 0 & 0 & 0 \\ 0 & 1 & 0 & -1 \\ 0 & 0 & 1 & 1 \end{bmatrix},$$
(4.3.7)

This corresponds to the equations

$$x = \quad 0$$

$$y = -1$$
(4.3.8)

$$z = \quad 1.$$

The equations used to simplify Eqs. 4.3.1 to 4.3.8 are *elementary row operations*. They are of three types:

1. Interchange any two rows.
2. Add the multiple of one row to another.
3. Multiply a row by a nonzero constant.

The crucial point here is that an elementary row operation replaces a system of equations by another system, the latter having exactly the same solution as the former. So $x = 0$, $y = 1$, and $z = 1$ is the unique solution of Eq. 4.3.1.

The foregoing reduction of several variables in each equation of a system to one variable in each equation is referred to as *Gaussian elimination*.

Example 4.3.1: Solve the equations

$$x \quad\quad + z = 1$$

$$2x \quad\quad + z = 0$$

$$x + y + z = 1.$$

SOLUTION: We begin with the augmented matrix

$$\begin{vmatrix} 1 & 0 & 1 & 1 \\ 2 & 0 & 1 & 0 \\ 1 & 1 & 1 & 1 \end{vmatrix}.$$

Then proceed to manipulate the matrix in the following manner:

$$\begin{bmatrix} 1 & 0 & 1 & 1 \\ 2 & 0 & 1 & 0 \\ 1 & 1 & 1 & 1 \end{bmatrix} \longrightarrow \begin{bmatrix} 1 & 0 & 1 & 1 \\ 0 & 0 & -1 & -2 \\ 0 & 1 & 0 & 0 \end{bmatrix} \longrightarrow \begin{bmatrix} 1 & 0 & 0 & -1 \\ 0 & 0 & -1 & -2 \\ 0 & 1 & 0 & 0 \end{bmatrix}.$$

The arrows denote the application of one or more elementary row operations. The rightmost matrix in this *arrow diagram* represents the system

$$x = -1$$

$$-z = -2$$

$$y = \quad 0.$$

Thus, the solution of the given system is $x = -1$, $y = 0$, and $z = 2$. ∎

Example 4.3.2:   Solve the system

$$-x \quad\quad + z = -1$$
$$x + y \quad\quad = 0$$
$$z = 0.$$

SOLUTION: We apply elementary row operations to the augmented matrix of this system, so

$$\begin{bmatrix} -1 & 0 & 1 & -1 \\ 1 & 1 & 0 & 0 \\ 0 & 0 & 1 & 0 \end{bmatrix} \longrightarrow \begin{bmatrix} 1 & 0 & -1 & 1 \\ 0 & 1 & 1 & -1 \\ 0 & 0 & 1 & 0 \end{bmatrix} \longrightarrow \begin{bmatrix} 1 & 0 & 0 & 1 \\ 0 & 1 & 0 & -1 \\ 0 & 0 & 1 & 0 \end{bmatrix}.$$

Hence $x = 1$, $y = -1$, and $z = 0$ is the unique solution.   ■

Example 4.3.3:   Solve the system

$$x + y + z = 1$$
$$x - y + z = 3$$
$$x \quad\quad + z = 2.$$

SOLUTION: We have

$$\begin{bmatrix} 1 & 1 & 1 & 1 \\ 1 & -1 & 1 & 3 \\ 1 & 0 & 1 & 2 \end{bmatrix} \longrightarrow \begin{bmatrix} 1 & 1 & 1 & 1 \\ 0 & -2 & 0 & 2 \\ 0 & -1 & 0 & 1 \end{bmatrix}$$

$$\longrightarrow \begin{bmatrix} 1 & 1 & 1 & 1 \\ 0 & 1 & 0 & -1 \\ 0 & 1 & 0 & -1 \end{bmatrix} \longrightarrow \begin{bmatrix} 1 & 0 & 1 & 2 \\ 0 & 1 & 0 & -1 \\ 0 & 0 & 0 & 0 \end{bmatrix}.$$

Hence,

$$x + z = 2$$
$$y = -1$$
$$0 = 0$$

and there are infinitely many solutions of the given system. Let $z = c$. Then $x = 2 - c$, $y = -1$, and $z = c$ is a solution for every choice of c.   ■

The system in Example 4.3.3 is inconsistent if any constant but 2 appears on the right-hand side of the last equation. If we attempt to solve

$$x + y + z = 1$$
$$x - y + z = 3 \quad\quad\quad\quad (4.3.9)$$
$$x \quad\quad + z = K,$$

we get

$$\begin{bmatrix} 1 & 1 & 1 & 1 \\ 1 & -1 & 1 & 3 \\ 1 & 0 & 1 & K \end{bmatrix} \longrightarrow \begin{bmatrix} 1 & 1 & 1 & 1 \\ 0 & -2 & 0 & 2 \\ 0 & -1 & 0 & K-1 \end{bmatrix}$$ (4.3.10)

$$\longrightarrow \begin{bmatrix} 1 & 1 & 1 & 1 \\ 0 & 1 & 0 & -1 \\ 0 & 1 & 0 & 1-K \end{bmatrix} \longrightarrow \begin{bmatrix} 1 & 1 & 1 & 1 \\ 0 & 1 & 0 & -1 \\ 0 & 0 & 0 & 2-K \end{bmatrix}.$$

This represents the system

$$x + y + z = 1$$
$$y = -1$$ (4.3.11)
$$0 = 2 - K$$

and the last equation is contradictory unless $K = 2$. This conclusion holds for Eqs. 4.3.9 as well.

The number of equations need not be the same as the number of unknowns. The method outlined above is still the method of choice. Two examples will illustrate this point.

Example 4.3.4:   Find all the solutions of

$$t + x + y + z = 1$$
$$t - x - y + z = 0$$
$$2t + x + y - z = 2.$$

SOLUTION: The augmented matrix is

$$\begin{bmatrix} 1 & 1 & 1 & 1 & 1 \\ 1 & -1 & -1 & 1 & 0 \\ 2 & 1 & 1 & -1 & 2 \end{bmatrix}.$$

Hence, with elementary row operations the matrix above becomes

$$\begin{bmatrix} 1 & 1 & 1 & 1 & 1 \\ 0 & -2 & -2 & 0 & -1 \\ 0 & -1 & -1 & -3 & 0 \end{bmatrix} \longrightarrow \begin{bmatrix} 1 & 1 & 1 & 1 & 1 \\ 0 & 1 & 1 & 0 & \frac{1}{2} \\ 0 & -1 & -1 & -3 & 0 \end{bmatrix}$$

$$\longrightarrow \begin{bmatrix} 1 & 1 & 1 & 1 & 1 \\ 0 & 1 & 1 & 0 & \frac{1}{2} \\ 0 & 0 & 0 & -3 & \frac{1}{2} \end{bmatrix} \longrightarrow \begin{bmatrix} 1 & 0 & 0 & 1 & \frac{1}{2} \\ 0 & 1 & 1 & 0 & \frac{1}{2} \\ 0 & 0 & 0 & 1 & -\frac{1}{6} \end{bmatrix}.$$

This last matrix is the augmented matrix for the system

$$t + z = \tfrac{1}{2}$$
$$x + y = \tfrac{1}{2}$$
$$z = -\tfrac{1}{6}.$$

Starting with the last equation, $z = -\frac{1}{6}$. From the second equation, we find $x = \frac{1}{2} - c$ if $y = c$. From the first equation, $t = \frac{2}{3}$. Thus, the family of all solutions is the set $t = \frac{2}{3}$, $x = \frac{1}{2} - c$, $y = c$, $z = -\frac{1}{6}$, and c arbitrary.   ■

Example 4.3.5:   Find all solutions to

$$t - x + 2y - z = 1$$
$$t \quad\quad + y + z = 0$$
$$-x + y - 2z = -1.$$

SOLUTION: We have

$$\begin{bmatrix} 1 & -1 & 2 & -1 & 1 \\ 1 & 0 & 1 & 1 & 0 \\ 0 & -1 & 1 & -2 & -1 \end{bmatrix} \longrightarrow \begin{bmatrix} 1 & -1 & 2 & -1 & 1 \\ 0 & 1 & -1 & 2 & -1 \\ 0 & -1 & 1 & -2 & -1 \end{bmatrix}$$

$$\longrightarrow \begin{bmatrix} 1 & -1 & 2 & -1 & 1 \\ 0 & 1 & -1 & 2 & -1 \\ 0 & 0 & 0 & 0 & -2 \end{bmatrix}$$

The third row of the last matrix in this arrow diagram stands for the contradictory equation

$$0t + 0x + 0y + 0z = -2.$$

Hence, the given system has no solutions.   ■

PROBLEMS

Reduce each matrix to upper triangular form by repeated use of row operation 2.

1. $\begin{bmatrix} 1 & 0 & 0 \\ -2 & 2 & 0 \\ 1 & 3 & -1 \end{bmatrix}$ **2.** $\begin{bmatrix} 0 & 1 & 0 \\ 1 & 0 & 0 \\ 0 & 0 & 1 \end{bmatrix}$ **3.** $\begin{bmatrix} a & b \\ c & d \end{bmatrix}$, $a \neq 0$

4. $\begin{bmatrix} a & b \\ c & d \end{bmatrix}$, $a = 0$ **5.** $\begin{bmatrix} 1 & 2 & 1 \\ 2 & 4 & -2 \\ 0 & 0 & 1 \end{bmatrix}$

6. Explain why a system of equations whose matrix of coefficients is upper triangular can be solved without further simplification, provided that there is a solution. Write an example illustrating the case with no solutions.

7. Find all solutions of

$$x_1 + x_2 - x_3 = 1.$$

8. Find all solutions of

$$x_1 + x_2 - x_3 = 1$$
$$x_1 + x_2 + x_3 = -1.$$

9. Relate the set of solutions of Problem 8 to that of Problem 7.

Solve each system of linear, algebraic equations.

10. $x - y = 6$
   $x + y = 0$

11. $2x - 2y = 4$
   $2x + y = 3$

12. $3x + 4y = 7$
   $2x - 5y = 2$

13. $3x + 2y - 6z = 0$
   $x - y + z = 4$
   $y + z = 3$

14. $x - 3y + z = -2$
   $x - 3y - z = 0$
   $-3y + z = 0$

15. $x_1 + x_2 + x_3 = 4$
   $x_1 - x_2 - x_3 = 2$
   $x_1 - 2x_2 = 0$

Find the column vector representing the solution to each set of algebraic equations.

16. $x - y = 2$
   $x + y = 0$

17. $x + z = 4$
   $2x + 3z = 8$

18. $x + 2y + z = -2$
   $x + y = 3$
   $x + z = 4$

19. $x_1 - x_2 + x_3 = 5$
   $2x_1 - 4x_2 + 3x_3 = 0$
   $x_1 - 6x_2 + 2x_3 = 3$

4.4   Rank and the Row-Echelon Normal Form

Suppose that a sequence of elementary row operations is applied to $\mathbf{A}$, resulting in the arrow diagram

$$\mathbf{A} \longrightarrow \mathbf{A}_1 \longrightarrow \mathbf{A}_2 \longrightarrow \cdots \longrightarrow \mathbf{A}_R. \qquad (4.4.1)$$

For any $\mathbf{A}$, it is always possible to arrange the row operations so that $\mathbf{A}_R$ has these four properties:

1. All the zero rows of $\mathbf{A}_R$ are its last rows.
2. The first nonzero entry in a nonzero row is 1. This is called the *leading one* of a nonzero row.
3. The leading one is the only nonzero entry in its column.
4. The leading one in row i is to the left of the leading one in row j if $i < j$.

Any matrix with these four properties is said to be in *row-echelon normal form*, RENF for short. A crucial theorem follows:

Theorem 4.1:   *Every matrix has a unique RENF which can be attained by a finite sequence of row operations.*

The existence of the RENF, $\mathbf{A}_R$, is not difficult to prove—the uniqueness provides something of a challenge. We invite the reader to construct both arguments!

Here are some matrices* in RENF:

(a) $\mathbf{I}_n$ (b) $\mathbf{O}_{m \times n}$ (c) $\begin{bmatrix} 1 & * & * & * \\ 0 & 0 & 0 & 0 \\ 0 & 0 & 0 & 0 \end{bmatrix}$ (d) $\begin{bmatrix} 1 & * & * \end{bmatrix}$

(e) $\begin{bmatrix} 1 \\ 0 \\ 0 \end{bmatrix}$ (f) $\begin{bmatrix} 1 & 0 & * & \cdots & * \\ 0 & 1 & * & \cdots & * \end{bmatrix}$ (g) $\begin{bmatrix} 1 & 0 \\ 0 & 1 \\ 0 & 0 \\ 0 & 0 \end{bmatrix}$

Note that $\mathbf{O}_{m \times n}$ satisfies the last three criteria in the definition of RENF vacuously; there are no leading ones.

If $\mathbf{A}_R$ is the RENF of $\mathbf{A}$, then the *rank* of $\mathbf{A}$, written rank $\mathbf{A}$, is the number of nonzero rows of $\mathbf{A}_R$. The matrices (a)–(g) above have ranks of n, 0, 1, 1, 1, 2, and 2, respectively. The following theorem regards the rank.

Theorem 4.2: *For each* $\mathbf{A}_{m \times n}$

$$\text{rank } \mathbf{A} \leq m \quad and \quad \text{rank } \mathbf{A} \leq n. \tag{4.4.2}$$

PROOF: By definition, rank $\mathbf{A}$ is a count of a subset of the number of rows, so rank $\mathbf{A} \leq m$ is obvious. But rank $\mathbf{A}$ is also the number of leading ones. There cannot be more leading ones than columns, so rank $\mathbf{A} \leq n$ is also trivial.

Consider the system

$$a_{11}x_1 + a_{12}x_2 + \cdots + a_{1n}x_n = r_1$$
$$a_{21}x_1 + a_{22}x_2 + \cdots + a_{2n}x_n = r_2 \tag{4.4.3}$$
$$\vdots$$
$$a_{m1}x_1 + a_{m2}x_2 + \cdots + a_{mn}x_n = r_m.$$

with coefficient matrix $\mathbf{A}$ and augmented matrix $\mathbf{B}$:

$$\mathbf{A} = \begin{bmatrix} a_{11} & a_{12} & \cdots & a_{1n} \\ a_{21} & a_{22} & \cdots & a_{2n} \\ \vdots & & & \\ a_{m1} & a_{m2} & \cdots & a_{mn} \end{bmatrix}, \quad \mathbf{B} = \begin{bmatrix} a_{11} & a_{12} & \cdots & a_{1n} & r_1 \\ a_{21} & a_{22} & \cdots & a_{2n} & r_2 \\ \vdots & & & \\ a_{m1} & a_{m2} & \cdots & a_{mn} & r_m \end{bmatrix}.$$

$$\tag{4.4.4}$$

Theorem 4.3: *System 4.4.3 is consistent*[†] *if and only if*

$$\text{rank } \mathbf{A} = \text{rank } \mathbf{B}. \tag{4.4.5}$$

---

*The entries designated with $*$ in a matrix represent any number.
[†]A consistent system is a set of simultaneous equations with at least one solution. An inconsistent system has no solutions.

PROOF: Let $\mathbf{B}_R$ be the RENF of $\mathbf{B}$. The RENF of $\mathbf{A}$ can be obtained from $\mathbf{B}_R$ by striking out the last column* of $\mathbf{B}_R$. Then rank $\mathbf{B}$ = rank $\mathbf{A}$ or rank $\mathbf{B}$ = rank $\mathbf{A}$ + 1 because either $\mathbf{B}_R$ contains the same number of leading ones or one more leading one. In the latter case, the last nonzero row of $\mathbf{B}_R$ is

$$[0, 0, \ldots, 0, 1], \tag{4.4.6}$$

which, as in Example 4.3.5, signals no solutions to the given system. In the former case, the system always has at least one solution.

Corollary 4.4:   *If $r_i = 0$ for $i = 1, 2, \ldots, m$, then system 4.4.3 is consistent.*

A row such as 4.4.6 is impossible in this case, so rank $\mathbf{A}$ = rank $\mathbf{B}$. The corollary is trivial for another reason: $x_1 = x_2 = \cdots = x_n = 0$ is always a solution when $r_i = 0$.

In a row-echelon normal form matrix, columns containing a leading one are *leading columns*; the remaining columns are *free columns*. The number of leading columns is equal to the rank of $\mathbf{A}$. If we set rank $\mathbf{A} = r$ and η = number of free columns then r is the number of leading columns, and

$$n = r + \eta \tag{4.4.7}$$

PROBLEMS

1. Suppose that $\mathbf{B}$ is in row-echelon form and $\mathbf{B}$ is $m \times n$. Explain why the matrix obtained by striking out the last columns of $\mathbf{B}$ is a matrix also in row-echelon form. [Examine the matrices (a)–(g) above Eq. 4.4.2.]

2. Which of the following matrices are in RENF?

(a) $\begin{bmatrix} 1 & 0 & 1 \\ 0 & 0 & 1 \\ 0 & 0 & 0 \end{bmatrix}$ (b) $[2]$ (c) $[1]$

(d) $[0]$ (e) $\begin{bmatrix} 0 & 0 \\ 0 & 1 \\ 0 & 0 \end{bmatrix}$ (f) $\begin{bmatrix} 0 & 0 & 1 \\ 0 & 0 & 0 \end{bmatrix}$

(g) $\begin{bmatrix} 1 & * & 0 \\ 0 & 0 & 1 \end{bmatrix}$ (h) $\begin{bmatrix} 1 & * & 0 \\ 0 & 0 & 0 \end{bmatrix}$ (i) $\begin{bmatrix} 0 & * & 0 \\ 0 & 0 & 1 \end{bmatrix}$

(j) $\begin{bmatrix} 0 & 1 & 0 \\ 0 & 0 & 1 \end{bmatrix}$ (k) $\begin{bmatrix} 0 & 1 & 2 & 3 \\ 0 & 0 & 0 & 1 \end{bmatrix}$ (l) $\begin{bmatrix} 0 & 1 & 2 & 0 \\ 0 & 0 & 0 & 1 \end{bmatrix}$

3. Find the ranks of the matrices (a)–(g) in the text above Eq. 4.4.2. For each matrix determine the leading columns.

---

*See Problem 1.

4. Find the ranks of the matrices in Problem 2. For each matrix determine the leading columns.

5. Explain why the number of leading columns of $\mathbf{A}$ is the rank of $\mathbf{A}$.

4.5 The Arithmetic of Matrices

We have seen the convenience afforded by simply operating on the array of coefficients of a system of equations rather than on the equations themselves. Further work along these lines will support this view. Ultimately, mathematicians thought of giving these arrays an existence of their own apart from their connection with simultaneous equations. It is this aspect we now explore.

Let the $m \times n$ matrices $\mathbf{A}$ and $\mathbf{B}$ be given by

$$
\mathbf{A} = \begin{bmatrix} a_{11} & a_{12} & \cdots & a_{1n} \\ a_{21} & a_{22} & \cdots & a_{2n} \\ \vdots & & & \\ a_{m1} & a_{m2} & \cdots & a_{mn} \end{bmatrix}, \qquad \mathbf{B} = \begin{bmatrix} b_{11} & b_{12} & \cdots & b_{1n} \\ b_{21} & b_{22} & \cdots & b_{2n} \\ \vdots & & & \\ b_{m1} & b_{m2} & \cdots & b_{mn} \end{bmatrix}
$$

$$(4.5.1)$$

Then $\mathbf{A} = \mathbf{B}$ if $a_{ij} = b_{ij}$ for each $i = 1, 2, \ldots, m$ and for each $j = 1, 2, \ldots, n$. Implicit in the definition of equality is the assumption that the orders of $\mathbf{A}$ and $\mathbf{B}$ are the same. The equality of two matrices implies the equality of m times n numbers, the corresponding entries of the matrices.

In addition to matrix equality, we define addition of matrices and multiplication of matrices by a constant. For the matrices $\mathbf{A}$ and $\mathbf{B}$ of Eq. 4.5.1 and any scalar k, we define $\mathbf{A} + \mathbf{B}$ and $k\mathbf{A}$ by the expressions

$$
\mathbf{A} + \mathbf{B} = \begin{bmatrix} a_{11} + b_{11} & a_{12} + b_{12} & \cdots & a_{1n} + b_{1n} \\ a_{21} + b_{21} & a_{22} + b_{22} & \cdots & a_{2n} + b_{2n} \\ \vdots & & & \\ a_{m1} + b_{m1} & a_{m2} + b_{m2} & & a_{mn} + b_{mn} \end{bmatrix},
$$

$$
k\mathbf{A} = \begin{bmatrix} ka_{11} & ka_{12} & \cdots & ka_{1n} \\ ka_{21} & ka_{22} & \cdots & ka_{2n} \\ \vdots & & & \\ ka_{m1} & ka_{m2} & \cdots & ka_{mn} \end{bmatrix} \tag{4.5.2}
$$

The definitions of Eq. 4.5.2 easily imply that

(a) $\mathbf{A} + \mathbf{B} = \mathbf{B} + \mathbf{A}$ (b) $\mathbf{A} + (\mathbf{B} + \mathbf{C}) = (\mathbf{A} + \mathbf{B}) + \mathbf{C}$

(c) $\mathbf{A} + \mathbf{O} = \mathbf{A}$ (d) $\mathbf{A} + (-1)\mathbf{A} = \mathbf{O}$

(e) $0\mathbf{A} = \mathbf{O}$ (f) $k(h\mathbf{A}) = (kh)\mathbf{A}$

(g) $k(\mathbf{A} + \mathbf{B}) = k\mathbf{A} + k\mathbf{B}$ (h) $(k + h)\mathbf{A} = k\mathbf{A} + h\mathbf{A}$

$$(4.5.3)$$

If we understand by $\mathbf{B} - \mathbf{A}$, a matrix such that $(\mathbf{B} - \mathbf{A}) + \mathbf{A} = \mathbf{B}$, then (d) enables us to find such a matrix and provides a definition of subtraction, for

$$
\begin{aligned}
[\mathbf{B} + (-1)\mathbf{A}] + \mathbf{A} &= \mathbf{B} + [(-1)\mathbf{A} + \mathbf{A}] \\
&= \mathbf{B} + [\mathbf{A} + (-1)\mathbf{A}] \\
&= \mathbf{B} + \mathbf{O} \\
&= \mathbf{B}.
\end{aligned}
\tag{4.5.4}
$$

Thus, $\mathbf{B} - \mathbf{A}$ is defined as

$$\mathbf{B} - \mathbf{A} = \mathbf{B} + (-1)\mathbf{A}. \tag{4.5.5}$$

Matrices having a single column are so important that an exception is made to our convention that matrices are always written in boldface uppercase letters. We call the matrix

$$
\mathbf{r} = \begin{bmatrix} r_1 \\ r_2 \\ \vdots \\ r_m \end{bmatrix}
\tag{4.5.6}
$$

a *vector* and use a boldface lowercase letter.

We shall find it helpful occasionally to interchange the rows with the columns of a matrix. The new matrix that results is called the *transpose* of the original matrix. The transpose $\mathbf{A}^T$ of the matrix displayed by Eq. 4.5.1 is

$$
\mathbf{A}^T = \begin{bmatrix}
a_{11} & a_{21} & \cdots & a_{m1} \\
a_{12} & a_{22} & \cdots & a_{m2} \\
\vdots & & & \\
a_{1n} & a_{2n} & \cdots & a_{mn}
\end{bmatrix}
\tag{4.5.7}
$$

Note that if a matrix is square, its transpose is also square; however, if a matrix is $m \times n$, its transpose is $n \times m$. An example of a matrix and its transpose is

$$
\mathbf{A} = \begin{bmatrix} 2 & 0 \\ 3 & -1 \\ 1 & 1 \\ 0 & 0 \end{bmatrix}, \qquad
\mathbf{A}^T = \begin{bmatrix} 2 & 3 & 1 & 0 \\ 0 & -1 & 1 & 0 \end{bmatrix}.
\tag{4.5.8}
$$

The transpose of a vector is a matrix with a single row, a *row vector*. So

$$\mathbf{r}^T = [r_1, r_2, \ldots, r_n]. \tag{4.5.9}$$

The commas in a row vector are omitted if the meaning is clear. If $\mathbf{C} = \mathbf{A} + \mathbf{B}$, then $\mathbf{C}^T = \mathbf{A}^T + \mathbf{B}^T$ follows from the definitions.

A matrix $\mathbf{A}$ is *symmetric* if $\mathbf{A}^T = \mathbf{A}$; it is *antisymmetric* (or *skew-symmetric*) if $\mathbf{A}^T = -\mathbf{A}$. Note that symmetric and antisymmetric matrices must be square. The matrix

$$\begin{bmatrix} 2 & 1 & 3 & 4 \\ 1 & 0 & -2 & 0 \\ 3 & -2 & 1 & -1 \\ 4 & 0 & -1 & 0 \end{bmatrix}$$

is symmetric, and the matrix

$$\begin{bmatrix} 0 & -1 & 2 \\ 1 & 0 & -3 \\ -2 & 3 & 0 \end{bmatrix}$$

is skew-symmetric.

Any square matrix can be written as the sum of a symmetric matrix and a skew-symmetric matrix. This is done as follows:

$$\mathbf{A} = \left(\frac{\mathbf{A}}{2} + \frac{\mathbf{A}^T}{2}\right) + \left(\frac{\mathbf{A}}{2} - \frac{\mathbf{A}^T}{2}\right), \tag{4.5.10}$$

where the symmetric matrix $\mathbf{A}_s$ and the antisymmetric matrix $\mathbf{A}_a$ are given by

$$\mathbf{A}_s = \frac{\mathbf{A}}{2} + \frac{\mathbf{A}^T}{2}, \qquad \mathbf{A}_a = \frac{\mathbf{A}}{2} - \frac{\mathbf{A}^T}{2}. \tag{4.5.11}$$

Note that $(\mathbf{A}^T)^T = \mathbf{A}$ is needed in establishing this result.

Example 4.5.1:   Given the two matrices

$$\mathbf{A} = \begin{bmatrix} 0 & 2 & 5 \\ 1 & -2 & 1 \\ 2 & 3 & 1 \end{bmatrix}, \qquad \mathbf{B} = \begin{bmatrix} -1 & 2 & 0 \\ 0 & 2 & 1 \\ 6 & -6 & 0 \end{bmatrix},$$

find $\mathbf{A} + \mathbf{B}$, $5\mathbf{A}$, and $\mathbf{B} - 5\mathbf{A}$.

SOLUTION: To find the sum $\mathbf{A} + \mathbf{B}$, we simply add corresponding elements $a_{ij} + b_{ij}$ and obtain

$$\mathbf{A} + \mathbf{B} = \begin{bmatrix} -1 & 4 & 5 \\ 1 & 0 & 2 \\ 8 & -3 & 1 \end{bmatrix}.$$

Following Eq. 4.5.2, the product $5\mathbf{A}$ is

$$5A = \begin{bmatrix} 0 & 10 & 25 \\ 5 & -10 & 5 \\ 10 & 15 & 5 \end{bmatrix}.$$

Now we subtract each element of the preceding matrix from the corresponding element of **B**, that is, $b_{ij} - 5a_{ij}$, and find

$$B - 5A = \begin{bmatrix} -1 & -8 & -25 \\ -5 & 12 & -4 \\ -4 & -21 & -5 \end{bmatrix}. \quad \blacksquare$$

Example 4.5.2:   Express the matrix

$$A = \begin{bmatrix} 2 & 0 & 3 \\ 2 & 0 & 2 \\ -3 & 4 & 2 \end{bmatrix}$$

as the sum of a symmetric matrix and a skew-symmetric matrix.

SOLUTION: First, let us write the transpose A^T. It is

$$A^T = \begin{bmatrix} 2 & 2 & -3 \\ 0 & 0 & 4 \\ 3 & 2 & 2 \end{bmatrix}.$$

Now, using Eq. 4.5.11, the symmetric part of **A** is

$$A_s = \frac{1}{2}(A + A^T) = \frac{1}{2}\begin{bmatrix} 4 & 2 & 0 \\ 2 & 0 & 6 \\ 0 & 6 & 4 \end{bmatrix} = \begin{bmatrix} 2 & 1 & 0 \\ 1 & 0 & 3 \\ 0 & 3 & 2 \end{bmatrix}.$$

The skew-symmetric part is given by

$$A_a = \frac{1}{2}(A - A^T) = \frac{1}{2}\begin{bmatrix} 0 & -2 & 6 \\ 2 & 0 & -2 \\ -6 & 2 & 0 \end{bmatrix} = \begin{bmatrix} 0 & -1 & 3 \\ 1 & 0 & -1 \\ -3 & 1 & 0 \end{bmatrix}.$$

Obviously, the given matrix **A** is the sum

$$A = A_s + A_a$$

$$= \begin{bmatrix} 2 & 1 & 0 \\ 1 & 0 & 3 \\ 0 & 3 & 2 \end{bmatrix} + \begin{bmatrix} 0 & -1 & 3 \\ 1 & 0 & -1 \\ -3 & 1 & 0 \end{bmatrix} = \begin{bmatrix} 2 & 0 & 3 \\ 2 & 0 & 2 \\ -3 & 4 & 2 \end{bmatrix}.$$

This provides us with a check on the manipulations above. $\blacksquare$

PROBLEMS

1. For arbitrary **A**, prove $(kA)^T = kA^T$.

2. Prove that a symmetric (or skew-symmetric) matrix must be a square matrix.

3. What matrices are simultaneously symmetric and skew-symmetric?

4. Prove that $(A^T)^T = A$.   5. Prove that $(A + B)^T = A^T + B^T$.

6. Show that $A/2 + A^T/2$ is symmetric and $A/2 - A^T/2$ is skew-symmetric.

7. Explain why the diagonal entries of a skew-symmetric matrix are all zero.

8. Show that an upper triangular symmetric matrix is a diagonal matrix and that an upper triangular skew-symmetric matrix is **O**.

Let

$$A = \begin{bmatrix} 2 & 1 & 0 \\ 1 & -1 & -2 \\ 4 & 2 & 0 \end{bmatrix}, \quad B = \begin{bmatrix} 1 & 1 & 1 \\ 0 & 0 & 0 \\ 2 & 1 & -3 \end{bmatrix}, \quad C = \begin{bmatrix} 2 & 3 & 1 \\ 0 & 2 & 0 \\ -1 & 2 & -1 \end{bmatrix}.$$

Determine the following.

9. $A + B$
   $B + A$

10. $A - B$
   $B - A$

11. $A + (B - C)$
   $(A + B) - C$

12. $4A + 4B$
   $4(A + B)$

13. $2A - 4C$
   $2(A - 2C)$

14. A^T

15. For the matrices **A**, **B**, and **C** above, show that $(A + B)^T = A^T + B^T$.

16. For the matrix **A** above, show that $A + A^T$ is symmetric and $A - A^T$ is skew-symmetric.

Let

$$A = \begin{bmatrix} 2 & 4 & 6 \\ 0 & 4 & -2 \\ -4 & 2 & 2 \end{bmatrix}, \quad B = \begin{bmatrix} 0 & -8 & 6 \\ -2 & 0 & 2 \\ 2 & -4 & 4 \end{bmatrix}.$$

Find the following.

17. A_s and A_a (see Eq. 4.5.11)

18. B_s and B_a.

19. $(A + B)_s$ and $(A - B)_s$

4.6   *Matrix Multiplication: Definition*

There are several ways that matrix multiplication could be defined. We shall motivate our definition by considering the simultaneous set of equations

$$a_{11}x_1 + a_{12}x_2 + a_{13}x_3 = r_1$$
$$a_{21}x_1 + a_{22}x_2 + a_{23}x_3 = r_2 \qquad (4.6.1)$$
$$a_{31}x_1 + a_{32}x_2 + a_{33}x_3 = r_3.$$

These equations could be written, using the summation symbol, as

$$\sum_{j=1}^{3} a_{ij}x_j = r_i, \qquad (4.6.2)$$

where the first equation is formed by choosing $i = 1$, the second equation letting $i = 2$, and the third equation with $i = 3$. The quantity a_{ij} contains the nine elements $a_{11}, a_{12}, a_{13}, \ldots, a_{33}$; it is a 3×3 matrix. The quantities x_j and r_i each contain three elements and are treated as vectors. Hence, we write Eqs. 4.6.1 in matrix notation as

$$\mathbf{Ax} = \mathbf{r}. \qquad (4.6.3)$$

We must define the product of the matrix $\mathbf{A}$ and the vector $\mathbf{x}$ so that Eqs. 4.6.1 result. This, of course, demands that the number of rows in the vector $\mathbf{x}$ equal the number of columns in the matrix $\mathbf{A}$. Matrix multiplication is generalized as follows: The matrix product of the matrix $\mathbf{A}$ and the matrix $\mathbf{B}$ is the matrix $\mathbf{C}$ whose elements are computed from

$$c_{ij} = \sum_{k=1}^{r} a_{ik}b_{kj}. \qquad (4.6.4)$$

For the definition above to be meaningful, the number of columns in $\mathbf{A}$ must be equal to the number of rows in $\mathbf{B}$. If $\mathbf{A}$ is an $m \times r$ matrix and $\mathbf{B}$ an $r \times n$ matrix, then $\mathbf{C}$ is an $m \times n$ matrix. Note that the matrix multiplication $\mathbf{AB}$ would not be defined if both $\mathbf{A}$ and $\mathbf{B}$ were 2×3 matrices. $\mathbf{AB}$ is defined, however, if $\mathbf{A}$ is 2×3 and $\mathbf{B}$ is 3×2; the product $\mathbf{AB}$ would then be a 2×2 matrix and the product $\mathbf{BA}$ is a 3×3 matrix. Obviously, matrix multiplication is not, in general, commutative; that is,

$$\mathbf{AB} \neq \mathbf{BA} \qquad (4.6.5)$$

must be assumed unless we have reason to believe the contrary. In fact, the product $\mathbf{BA}$ may not even be defined, even if $\mathbf{AB}$ exists.

The multiplication of two matrices $\mathbf{A}$ and $\mathbf{B}$ to form the matrix $\mathbf{C}$ is displayed as

$$
\begin{bmatrix}
c_{11} & c_{12} & \cdots & c_{1n} \\
c_{21} & c_{22} & \cdots & c_{2n} \\
\vdots & \vdots & & \vdots \\
c_{i1} & \cdots & \boxed{c_{ij}} & \cdots & c_{in} \\
\vdots & \vdots & & \vdots \\
c_{m1} & c_{m2} & \cdots & c_{mn}
\end{bmatrix}
$$

$$
=
\begin{bmatrix}
a_{11} & a_{12} & \cdots & a_{1r} \\
\vdots & \vdots & & \vdots \\
\boxed{a_{i1} \quad a_{i2} \quad \cdots \quad a_{ir}} \\
\vdots & \vdots & & \vdots \\
a_{m1} & a_{m2} & \cdots & a_{mr}
\end{bmatrix}
\begin{bmatrix}
b_{11} & b_{12} & \cdots & \boxed{b_{1j}} & \cdots & b_{1n} \\
b_{21} & b_{22} & \cdots & \boxed{b_{2j}} & \cdots & b_{2n} \\
\vdots & \vdots & & \vdots & & \vdots \\
b_{r1} & b_{r2} & \cdots & \boxed{b_{rj}} & \cdots & b_{rn}
\end{bmatrix}. \qquad (4.6.6)
$$

Observe that the element c_{ij} depends on the elements in row i of **A** and the elements in column j of **B**. If the elements of row i of **A** and the elements of column j of **B** are considered to be the components of vectors, then the element c_{ij} is simply the scalar (dot) product of the two vectors. Written out we have

$$c_{ij} = a_{i1}b_{1j} + a_{i2}b_{2j} + a_{i3}b_{3j} + \cdots + a_{ir}b_{rj}. \qquad (4.6.7)$$

This is, of course, the same equation as Eq. 4.6.4.

In the matrix product **AB** the matrix **A** is referred to as the *premultiplier* and the matrix **B** as the *postmultiplier*. The matrix **A** is postmultiplied by **B**, or **B** is premultiplied by **A**.

It is now an easier task to manipulate matrix equations such as Eq. 4.6.3. For example, suppose that the unknown vector **x** were related to another unknown vector **y** by the matrix equation

$$\mathbf{x} = \mathbf{B}\mathbf{y} \qquad (4.6.8)$$

where **B** is a known coefficient matrix. We could then substitute Eq. 4.6.8 into Eq. 4.6.3 and obtain

$$\mathbf{A}\mathbf{B}\mathbf{y} = \mathbf{r}. \qquad (4.6.9)$$

The matrix product **AB** is determined following the multiplication rules outlined above.

Example 4.6.1: Several examples of the multiplication of two matrices will be given here using the following:

$$\mathbf{A} = \begin{bmatrix} 2 \\ 3 \\ -4 \end{bmatrix}, \quad \mathbf{B} = [2, \ -1, \ 0], \quad \mathbf{C} = \begin{bmatrix} 2 & 3 & -1 \\ 0 & 1 & 4 \end{bmatrix}$$

$$\mathbf{D} = \begin{bmatrix} 3 & 0 & 1 \\ 2 & -2 & 1 \\ 0 & 2 & 0 \end{bmatrix}, \quad \mathbf{E} = \begin{bmatrix} 2 & -1 & 1 \\ 1 & 0 & 0 \\ 2 & 0 & 1 \end{bmatrix}.$$

SOLUTION: **A** is a 3×1 matrix and **B** is a 1×3 matrix. The product matrices **BA** and **AB** are

$$\mathbf{BA} = [2, \ -1, \ 0] \begin{bmatrix} 2 \\ 3 \\ -4 \end{bmatrix} = [2 \cdot 2 + (-1)(3) + 0(-4)] = [1]$$

$$\mathbf{AB} = \begin{bmatrix} 2 \\ 3 \\ -4 \end{bmatrix} [2, \ -1, \ 0] = \begin{bmatrix} 2 \cdot 2 & 2 \cdot (-1) & 2 \cdot 0 \\ 3 \cdot 2 & 3 \cdot (-1) & 3 \cdot 0 \\ -4 \cdot 2 & -4 \cdot (-1) & -4 \cdot 0 \end{bmatrix}$$

$$= \begin{bmatrix} 4 & -2 & 0 \\ 6 & -3 & 0 \\ -8 & 4 & 0 \end{bmatrix}.$$

From these expressions it is obvious that $\mathbf{AB} \neq \mathbf{BA}$. In fact, the rows and columns of the product matrix are even different. The first product is often called a *scalar product*, since the product yields a matrix with only one scalar element.

Now consider the product of a 2×3 matrix and a 3×1 matrix, $\mathbf{CA}$. The product matrix is

$$\mathbf{CA} = \begin{bmatrix} 2 & 3 & -1 \\ 0 & 1 & 4 \end{bmatrix} \begin{bmatrix} 2 \\ 3 \\ -4 \end{bmatrix} = \begin{bmatrix} 2 \cdot 2 + 3 \cdot 3 + -1 \cdot (-4) \\ 0 \cdot 2 + 1 \cdot 3 + 4 \cdot (-4) \end{bmatrix} = \begin{bmatrix} 17 \\ -13 \end{bmatrix}.$$

The product $\mathbf{AC}$ does not exist since matrix multiplication of a 3×1 matrix with a 2×3 matrix is not defined.

The product of two 3×3 matrices will now be attempted. We have

$$\mathbf{DE} = \begin{bmatrix} 3 & 0 & 1 \\ 2 & -2 & 1 \\ 0 & 2 & 0 \end{bmatrix} \begin{bmatrix} 2 & -1 & 1 \\ 1 & 0 & 0 \\ 2 & 0 & 1 \end{bmatrix} = \begin{bmatrix} 8 & -3 & 4 \\ 4 & -2 & 3 \\ 2 & 0 & 0 \end{bmatrix}.$$

Check this result using the procedure discussed preceding Eq. 4.6.7. Then verify that

$$\mathbf{ED} = \begin{bmatrix} 2 & -1 & 1 \\ 1 & 0 & 0 \\ 2 & 0 & 1 \end{bmatrix} \begin{bmatrix} 3 & 0 & 1 \\ 2 & -2 & 1 \\ 0 & 2 & 0 \end{bmatrix} = \begin{bmatrix} 4 & 4 & 1 \\ 3 & 0 & 1 \\ 6 & 2 & 2 \end{bmatrix}. \quad \blacksquare$$

In certain special circumstances, $\mathbf{AB} = \mathbf{BA}$. Two simple illustrations are

(1)
$$\mathbf{AI} = \mathbf{IA} = \mathbf{A} \tag{4.6.10}$$

and

(2)
$$\mathbf{OA} = \mathbf{AO} = \mathbf{O}. \tag{4.6.11}$$

It is true, that for all $\mathbf{A}$, $\mathbf{B}$, and $\mathbf{C}$,

$$\mathbf{A(BC)} = \mathbf{(AB)C}$$
$$\mathbf{A(B + C)} = \mathbf{AB} + \mathbf{AC} \tag{4.6.12}$$
$$\mathbf{(B + C)A} = \mathbf{BA} + \mathbf{CA}$$

provided that the orders in each multiplication are correct.

A striking example of the peculiarity of matrix multiplication is the product

$$\begin{bmatrix} 1 & 1 \\ -1 & -1 \end{bmatrix} \begin{bmatrix} 1 & 1 \\ -1 & -1 \end{bmatrix} = \begin{bmatrix} 0 & 0 \\ 0 & 0 \end{bmatrix}. \tag{4.6.13}$$

Thus, $\mathbf{AB} = \mathbf{O}$ does not imply that either $\mathbf{A}$ or $\mathbf{B}$ are zero. Also,

$$\begin{bmatrix} 1 & 1 \\ -1 & -1 \end{bmatrix}\begin{bmatrix} 1 & 1 \\ 0 & 0 \end{bmatrix} = \begin{bmatrix} 1 & 1 \\ -1 & -1 \end{bmatrix}\begin{bmatrix} 0 & 0 \\ 1 & 1 \end{bmatrix} = \begin{bmatrix} 1 & 1 \\ -1 & -1 \end{bmatrix} \qquad (4.6.14)$$

shows that $\mathbf{AB} = \mathbf{AC}$ does not imply that $\mathbf{B} = \mathbf{C}$ even though $\mathbf{A} \neq \mathbf{O}$. That is, there is no "law of cancellation," at least without more restrictive conditions than $\mathbf{A} \neq \mathbf{O}$.

The failure of the commutivity of multiplication complicates the rules of algebra. For example,

$$\begin{aligned}(\mathbf{A} + \mathbf{B})^2 &= (\mathbf{A} + \mathbf{B})(\mathbf{A} + \mathbf{B}) \\ &= (\mathbf{A} + \mathbf{B})\mathbf{A} + (\mathbf{A} + \mathbf{B})\mathbf{B} \\ &= \mathbf{A}^2 + \mathbf{BA} + \mathbf{AB} + \mathbf{B}^2 \neq \mathbf{A}^2 + 2\mathbf{AB} + \mathbf{B}^2 \qquad (4.6.15)\end{aligned}$$

unless $\mathbf{A}$ and $\mathbf{B}$ commute. However, it is true that

$$(\mathbf{A} + \mathbf{I})^2 = \mathbf{A}^2 + 2\mathbf{A} + \mathbf{I}. \qquad (4.6.16)$$

The transpose of the product of two matrices equals the product of the transposes taken in reverse order; that is,

$$(\mathbf{AB})^T = \mathbf{B}^T \mathbf{A}^T \qquad (4.6.17)$$

This is most readily verified by writing the equation in index form. Let $\mathbf{C} = \mathbf{AB}$. Then $\mathbf{C}^T = (\mathbf{AB})^T$ and is given by

$$c_{ij}^T = c_{ji} = \sum_{k=1}^{n} a_{jk}b_{ki} = \sum_{k=1}^{n} a_{kj}^T b_{ik}^T = \sum_{k=1}^{n} b_{ik}^T a_{kj}^T. \qquad (4.6.18)$$

This expression is observed to be the index form of the product $\mathbf{B}^T \mathbf{A}^T$, thereby verifying Eq. 4.6.17. The preceding should also be verified using some particular examples.

Example 4.6.2:   Verify the statement expressed in Eq. 4.6.17 if

$$\mathbf{A} = \begin{bmatrix} 3 \\ 0 \\ -1 \end{bmatrix} \quad \text{and} \quad \mathbf{B} = [2, \quad -1, \quad 1].$$

SOLUTION: The matrix product $\mathbf{AB}$ is found to be

$$\mathbf{AB} = \begin{bmatrix} 6 & -3 & 3 \\ 0 & 0 & 0 \\ -2 & 1 & -1 \end{bmatrix}.$$

The transpose matrices are

$$\mathbf{A}^T = [3, \quad 0, \quad -1], \qquad \mathbf{B}^T = \begin{bmatrix} 2 \\ -1 \\ 1 \end{bmatrix}.$$

The product $\mathbf{B}^T\mathbf{A}^T$ is found to be

$$\mathbf{B}^T\mathbf{A}^T = \begin{bmatrix} 6 & 0 & -2 \\ -3 & 0 & 1 \\ 3 & 0 & -1 \end{bmatrix}.$$

This, obviously, is the transpose of the matrix product $\mathbf{AB}$. ■

PROBLEMS

Find each product.

1. $\begin{bmatrix} 1 & 3 \\ 3 & 1 \end{bmatrix}\begin{bmatrix} 1 & 4 \\ 4 & 1 \end{bmatrix}$

2. $\begin{bmatrix} 1 & 1 & 0 \\ 0 & 1 & 2 \\ 0 & 0 & 1 \end{bmatrix}\begin{bmatrix} 1 & -2 & 2 \\ 0 & 2 & 2 \\ 0 & 0 & 1 \end{bmatrix}$

3. $\begin{bmatrix} 2 & 0 & 0 \\ 0 & 1 & 0 \\ 0 & 1 & -1 \end{bmatrix}\begin{bmatrix} x \\ y \\ z \end{bmatrix}$

4. $[a, \quad b, \quad c]\begin{bmatrix} a \\ b \\ c \end{bmatrix}$

5. $\begin{bmatrix} -6 & 7 \\ 7 & -8 \end{bmatrix}\begin{bmatrix} 8 & 7 \\ 7 & 6 \end{bmatrix}$

6. $\begin{bmatrix} 2 & 5 \\ 1 & 3 \end{bmatrix}\begin{bmatrix} 11 & 30 \\ -4 & -11 \end{bmatrix}\begin{bmatrix} 3 & -5 \\ -1 & 2 \end{bmatrix}$

7. $[0][1, \quad 7, \quad -2]$

8. Verify $\mathbf{A}(\mathbf{B} + \mathbf{C}) = \mathbf{AB} + \mathbf{AC}$ when

$$\mathbf{A} = \begin{bmatrix} 1 & 2 \\ 3 & -1 \end{bmatrix}, \qquad \mathbf{B} = \begin{bmatrix} 0 & 1 \\ 2 & 3 \end{bmatrix}, \qquad \mathbf{C} = \begin{bmatrix} 1 & 1 \\ 0 & 1 \end{bmatrix}.$$

9. Find $\mathbf{A}^2$, $\mathbf{A}^3$, $\mathbf{A}^4$ where

$$\mathbf{A} = \begin{bmatrix} 0 & 1 & 1 \\ 0 & 0 & 1 \\ 0 & 0 & 0 \end{bmatrix}.$$

10. Find a formula for $\mathbf{A}^n$ where

$$\mathbf{A} = \begin{bmatrix} 1 & 1 \\ 0 & 1 \end{bmatrix}.$$

Let $\mathbf{A} = \begin{bmatrix} 1 & 2 \\ 0 & 1 \end{bmatrix}$. Compute

11. $3\mathbf{A}^2 - 9\mathbf{A} + 6\mathbf{I}$

12. $3(\mathbf{A} - \mathbf{I})(\mathbf{A} - 2\mathbf{I})$

13. $3(\mathbf{A} - 2\mathbf{I})(\mathbf{A} - \mathbf{I})$

Verify $(\mathbf{A} + \mathbf{I})^3 = \mathbf{A}^3 + 3\mathbf{A}^2 + 3\mathbf{A} + \mathbf{I}$ for each equation.

14. $\mathbf{A} = \mathbf{I}$

15. $\mathbf{A} = \mathbf{O}$

16. $\mathbf{A} = \begin{bmatrix} 1 & 1 & 1 \\ 1 & 1 & 1 \\ 1 & 1 & 1 \end{bmatrix}$

17. $\mathbf{A} = \begin{bmatrix} 1 & 0 & -1 \\ 1 & 2 & 2 \\ -1 & 1 & 0 \end{bmatrix}$

Expand each expression.

18. $[x_1, \quad x_2, \quad x_3] \begin{bmatrix} 1 & -1 & 0 \\ 0 & 1 & 1 \\ 0 & 1 & -1 \end{bmatrix} \begin{bmatrix} x_1 \\ x_2 \\ x_3 \end{bmatrix}$ **19.** $[x_1, \quad x_2, \quad x_3] \begin{bmatrix} 2 & 0 & 0 \\ 0 & 1 & 0 \\ 1 & 1 & 1 \end{bmatrix} \begin{bmatrix} x_1 \\ x_2 \\ x_3 \end{bmatrix}$

20. Prove $(\mathbf{AB})^T = \mathbf{B}^T\mathbf{A}^T$.

21. Find two examples of 3×3 matrices such that $\mathbf{AB} = \mathbf{O}$ but $\mathbf{A} \neq \mathbf{O}$ and $\mathbf{B} \neq \mathbf{O}$.

22. Use your answers to Problem 21 to construct two examples of matrices $\mathbf{A}, \mathbf{B},$ and $\mathbf{C}$ such that $\mathbf{AB} = \mathbf{AC}$ but $\mathbf{B} \neq \mathbf{C}$.

23. If $\mathbf{AB} = \mathbf{BA}$ prove that $\mathbf{A}$ and $\mathbf{B}$ are square matrices with the same dimensions.

24. Prove that $\mathbf{AB}$ is upper triangular if $\mathbf{A}$ and $\mathbf{B}$ are.

25. Using the matrices of Problem 2, show that matrix multiplication given by Eq. 4.6.6 can be written as $\mathbf{C} = [\mathbf{AB}_{*1}, \mathbf{AB}_{*2}, \ldots, \mathbf{AB}_{*n}]$.

26. Prove each equation in Eq. 4.6.12.

27. Use Eq. 4.6.17 to show that $(\mathbf{ABC})^T = \mathbf{C}^T\mathbf{B}^T\mathbf{A}^T$.

28. Show by example that $(\mathbf{AB})^2 \neq \mathbf{A}^2\mathbf{B}^2$ in general. Show that $\mathbf{AB} = \mathbf{BA}$ does imply $(\mathbf{AB})^n = \mathbf{A}^n\mathbf{B}^n$.

29. Suppose that $\mathbf{A}$ is upper triangular and $a_{ii} = 0$, $i = 1, 2, \ldots, n$. Show that $\mathbf{A}^n = \mathbf{O}$.

30. Show that $\mathbf{A}^T\mathbf{A}$ and $\mathbf{AA}^T$ are symmetric.

Let

$$\mathbf{A} = \begin{bmatrix} 1 \\ -1 \\ 2 \end{bmatrix} \quad \mathbf{B} = [2, \quad 4, \quad -1], \quad \mathbf{C} = \begin{bmatrix} 3 & 2 & 1 \\ -2 & 0 & -1 \\ 1 & 0 & 1 \end{bmatrix},$$

$$\mathbf{D} = \begin{bmatrix} -1 & 0 & 2 \\ 1 & 2 & 1 \\ 2 & -1 & -1 \end{bmatrix}.$$

Find the following.

31. AB   **32.** BA   **33.** (AB)C   **34.** A(BC)

35. CA   **36.** CD   **37.** BD   **38.** DA

39. Let $\mathbf{A}$ and $\mathbf{B}$ be 3×3 diagonal matrices. What is $\mathbf{AB}$? $\mathbf{BA}$? Does this result generalize to $n \times n$ diagonal matrices?

Let

$$\mathbf{A} = \begin{bmatrix} 0 & 3 & 1 \\ -1 & 2 & 0 \\ 0 & 0 & 1 \end{bmatrix}, \quad \mathbf{B} = \begin{bmatrix} 1 & 0 & 0 \\ -1 & 2 & 1 \\ 3 & 1 & 0 \end{bmatrix}, \quad \mathbf{C} = \begin{bmatrix} 2 \\ 0 \\ -1 \end{bmatrix}, \quad \mathbf{D} = [1, 2, 0].$$

Determine the following sums and products and identify those that are not defined.

40. $(\mathbf{A} + \mathbf{B})\mathbf{C}$ and $\mathbf{AC} + \mathbf{BC}$ **41.** $\mathbf{A}(\mathbf{BC})$ and $(\mathbf{AB})\mathbf{C}$

42. $D(A + B)$ and $DA + DB$ **43.** $(AB)^T$ and $B^T A^T$

44. $A^T A$ and AA^T **45.** $C^T C$ and CC^T

46. A^2 and A^3 **47.** C^2

48. $A + C$ **49.** $A^2 - 2B + 3I$

50. $2AC + DB - 4I$

Let

$$A = \begin{bmatrix} 2 & 0 & 0 \\ 0 & -1 & 0 \\ 0 & 0 & 3 \end{bmatrix}, \quad B = \begin{bmatrix} 2 & 1 & 3 \\ 1 & -1 & 2 \\ 1 & 3 & 2 \end{bmatrix}, \quad C = \begin{bmatrix} 2 \\ 1 \\ -1 \end{bmatrix}.$$

Find the following.

51. AB and BA. Are they equal?   **52.** AC.

53. $C^T A$.

4.7 The Inverse of a Matrix

Division is not a concept defined for matrices. In its place and to serve similar purposes, we introduce the notion of the *inverse*. The square matrix A is *nonsingular*, or has an *inverse* (or is *invertible*), if there exists a square matrix B such that

$$AB = BA = I. \tag{4.7.1}$$

It is immediately clear that not all matrices have inverses since if $A = O$ Eq. 4.7.1 is false for every B. However, if there exists a B that satisfies Eq. 4.7.1 for a given A, there is only one such B. For suppose that $AC = I$. Then

$$B(AC) = B(I) = B. \tag{4.7.2}$$

But

$$B(AC) = (BA)C = (I)C = C. \tag{4.7.3}$$

Hence, $B = C$.

Since there is never more than one matrix satisfying Eq. 4.7.1 for a given A, we call the matrix B the *inverse* of A and denote it by A^{-1}, so that Eq. 4.7.1 can be written

$$AA^{-1} = A^{-1}A = I. \tag{4.7.4}$$

A matrix that is not invertible, that is, one for which A^{-1} does not exist, is called *singular* or *noninvertible*.

The following matrices are singular:

(a) $\begin{bmatrix} 1 & 1 & 1 \\ 1 & 1 & 1 \\ 1 & 1 & 1 \end{bmatrix}$ (b) $\begin{bmatrix} 0 & 0 \\ a & b \end{bmatrix}$ (c) $\begin{bmatrix} 1 & 1 & 1 \\ 2 & 2 & 2 \\ 3 & 3 & 3 \end{bmatrix}$ (d) $[0]$

Example 4.7.1:   Verify that the matrix in (c) above is singular.

SOLUTION: To verify that the matrix in (c) is indeed singular, we attempt to find a matrix such that $\mathbf{AB} = \mathbf{I}$. Hence, we set

$$\begin{bmatrix} 1 & 1 & 1 \\ 2 & 2 & 2 \\ 3 & 3 & 3 \end{bmatrix} \begin{bmatrix} b_{11} & b_{12} & b_{13} \\ b_{21} & b_{22} & b_{23} \\ b_{31} & b_{32} & b_{33} \end{bmatrix} = \begin{bmatrix} 1 & 0 & 0 \\ 0 & 1 & 0 \\ 0 & 0 & 1 \end{bmatrix}.$$

But we arrive at a contradiction by computing the entries in the (1, 1) and (2, 1) positions:

$$b_{11} + b_{21} + b_{31} = 1$$

$$2b_{11} + 2b_{21} + 2b_{31} = 0.$$

Thus, we cannot find a matrix $\mathbf{B}$ and we conclude that $\mathbf{A}$ is singular.   ∎

Except in rather special circumstances, it is not a trivial task to discover whether $\mathbf{A}$ is singular, particularly if the order of the square matrix $\mathbf{A}$ is rather large, say 8 or more. Before discussing systematic methods for finding $\mathbf{A}^{-1}$, when it exists, it is helpful to exhibit two matrices whose inverses are easy to compute:

1. The diagonal matrix $\mathbf{D}$ with diagonal elements $(a_{11}, a_{22}, \ldots, a_{nn})$ is singular if and only if $a_{ii} \neq 0$ for some $i = 1, 2, \ldots, n$. Its inverse is a diagonal matrix with diagonal entries $(a_{11}^{-1}, a_{22}^{-1}, \ldots, a_{nn}^{-1})$.

2. If $ad - bc \neq 0$, then $\begin{bmatrix} a & b \\ c & d \end{bmatrix}$ is nonsingular and $\dfrac{1}{ad - bc} \begin{bmatrix} d & -b \\ -c & a \end{bmatrix}$ is its inverse.

Finally, let us note some properties associated with the inverse matrix.

1. The inverse of the product of two matrices is the product of the inverse in the reverse order:

$$\mathbf{(AB)}^{-1} = \mathbf{B}^{-1}\mathbf{A}^{-1}. \tag{4.7.5}$$

The argument to show this follows:

$$\mathbf{AB(B^{-1}A^{-1})} = \mathbf{A(BB^{-1})A^{-1}}$$

$$= \mathbf{A(IA^{-1})} = \mathbf{AA^{-1}} = \mathbf{I}. \tag{4.7.6}$$

2. The inverse of the transpose is the transpose of the inverse:

$$\mathbf{(A^T)}^{-1} = \mathbf{(A^{-1})}^T. \tag{4.7.7}$$

We prove this by taking transposes of $\mathbf{AA^{-1}}$ and $\mathbf{A^{-1}A}$; the details are left for the Problems.

3. The inverse of the inverse is the given matrix:

$$\mathbf{(A^{-1})}^{-1} = \mathbf{A}. \tag{4.7.8}$$

The proof is Problem 5 at the end of this section.

The existence of an inverse is a remedy for the lack of a law of cancellation. For suppose that $\mathbf{AB} = \mathbf{AC}$ and $\mathbf{A}$ is invertible. Then we can conclude that $\mathbf{B} = \mathbf{C}$. We cannot simply cancel $\mathbf{A}$ on both sides; however, because

$$\mathbf{AB} = \mathbf{AC} \tag{4.7.9}$$

we can write

$$\mathbf{A}^{-1}(\mathbf{AB}) = \mathbf{A}^{-1}(\mathbf{AC}) = (\mathbf{A}^{-1}\mathbf{A})\mathbf{C} = (\mathbf{I})\mathbf{C} = \mathbf{C}. \tag{4.7.10}$$

Also,

$$\mathbf{A}^{-1}(\mathbf{AB}) = (\mathbf{A}^{-1}\mathbf{A})\mathbf{B} = (\mathbf{I})\mathbf{B} = \mathbf{B}. \tag{4.7.11}$$

Hence, we may write

$$\mathbf{B} = \mathbf{C}. \tag{4.7.12}$$

A second illustration relates to the matrix representation for a system of n equations and n unknowns. Suppose that

$$\mathbf{Ax} = \mathbf{b} \tag{4.7.13}$$

and $\mathbf{A}^{-1}$ exists; then, since $\mathbf{A}^{-1}(\mathbf{Ax}) = \mathbf{x}$,

$$\mathbf{A}^{-1}(\mathbf{Ax}) = \mathbf{A}^{-1}\mathbf{b} \tag{4.7.14}$$

implies that

$$\mathbf{x} = \mathbf{A}^{-1}\mathbf{b}. \tag{4.7.15}$$

To determine the solution vector $\mathbf{x}$ we must compute $\mathbf{A}^{-1}$. In the next section we present an efficient algorithm for computing $\mathbf{A}^{-1}$ and several illustrative examples.

PROBLEMS

1. If $\mathbf{A}^{-1}$ exists and $\mathbf{A}$ has dimensions $n \times n$, what are the dimensions of $\mathbf{A}^{-1}$? Must $\mathbf{A}^{-1}$ be square?

2. If $\mathbf{A}$ has dimensions $n \times n$ and $\mathbf{AC} = \mathbf{B}$ and $\mathbf{A}^{-1}$ exists, then $\mathbf{A}^{-1}$ and $\mathbf{B}$ have dimensions that allow $\mathbf{A}^{-1}\mathbf{B}$. Why? Is the same true for $\mathbf{BA}^{-1}$?

3. If $\mathbf{AC} = \mathbf{B}$, $\mathbf{A}$ is $n \times n$ and $\mathbf{C}$ is a column matrix, what are the dimensions of $\mathbf{B}$?

4. Prove that $(\mathbf{A}^T)^{-1} = (\mathbf{A}^{-1})^{T.}$

5. Prove that $(\mathbf{A}^{-1})^{-1} = \mathbf{A}$.

6. Show by example that $(\mathbf{A} + \mathbf{B})^{-1} \neq \mathbf{A}^{-1} + \mathbf{B}^{-1}$, in general.

7. Verify that $\mathbf{A} = \begin{bmatrix} a & b \\ c & d \end{bmatrix}$ implies that

$$\mathbf{A}^{-1} = \frac{1}{ad - bc} \begin{bmatrix} d & -b \\ -c & a \end{bmatrix}$$

if $ad - bc \neq 0$.

8. Explain why the matrices (a)–(d) in the text after Eq. 4.7.4 are all singular.

9. Show that $(\mathbf{A}^{-1}\mathbf{B}\mathbf{A})^2 = \mathbf{A}^{-1}\mathbf{B}^2\mathbf{A}$. Generalize to $(\mathbf{A}^{-1}\mathbf{B}\mathbf{A})^n$.

10. Show that $(\mathbf{A}^n)^{-1} = (\mathbf{A}^{-1})^n$.

11. Suppose that $\mathbf{u}^T = [u_1, u_2, \ldots, u_n]$. Write out $\mathbf{u}\mathbf{u}^T$ and following the argument in Example 4.7.1, show that $\mathbf{u}\mathbf{u}^T$ is singular for $n \geq 2$. Under what conditions is $\mathbf{u}\mathbf{u}^T$ singular if $n = 1$?

12. It is possible to establish this weak form of definition 4.7.1: If $\mathbf{A}$ is square and there is a $\mathbf{B}$ such that either $\mathbf{A}\mathbf{B} = \mathbf{I}$ or $\mathbf{B}\mathbf{A} = \mathbf{I}$, then $\mathbf{B} = \mathbf{A}^{-1}$. Assume this theorem and show the following:
   (a) If $\mathbf{A}\mathbf{B}$ is invertible, then so are $\mathbf{A}$ and $\mathbf{B}$.
   (b) Use part (a) to establish: if $\mathbf{A}$ is singular, so is $\mathbf{A}\mathbf{B}$ for every $\mathbf{B}$, and likewise $\mathbf{B}\mathbf{A}$.

13. Let

$$\mathbf{A} = \begin{bmatrix} a_{11} & 0 & \cdots & 0 \\ 0 & a_{22} & \cdots & 0 \\ \vdots & \vdots & & \\ 0 & 0 & \cdots & a_{nn} \end{bmatrix}$$

Show that

$$\mathbf{A}^{-1} = \begin{bmatrix} a_{11}^{-1} & 0 & \cdots & 0 \\ 0 & a_{22}^{-1} & \cdots & 0 \\ \vdots & \vdots & & \\ 0 & 0 & \cdots & a_{nn}^{-1} \end{bmatrix}$$

by computing $\mathbf{A}\mathbf{A}^{-1}$ and $\mathbf{A}^{-1}\mathbf{A}$.

4.8   The Computation of $\mathbf{A}^{-1}$

Given $\mathbf{A}$, we wish to find $\mathbf{X}$ so that $\mathbf{A}\mathbf{X} = \mathbf{I}$. Suppose that

$$\mathbf{X} = [\mathbf{X}_{*1}, \mathbf{X}_{*2}, \ldots, \mathbf{X}_{*n}]. \tag{4.8.1}$$

Then, by definition of matrix multiplication (see Eq. 4.6.6 and Problem 25 of Section 4.6),

$$\mathbf{A}\mathbf{X} = [\mathbf{A}\mathbf{X}_{*1}, \mathbf{A}\mathbf{X}_{*2}, \ldots, \mathbf{A}\mathbf{X}_{*n}] = \mathbf{I}. \tag{4.8.2}$$

Therefore, to find $\mathbf{X}$ we need to solve the following n systems simultaneously:

$$\mathbf{AX}_{*1} = \begin{bmatrix} 1 \\ 0 \\ \vdots \\ 0 \end{bmatrix}, \quad \mathbf{AX}_{*2} = \begin{bmatrix} 0 \\ 1 \\ \vdots \\ 0 \end{bmatrix}, \quad \ldots, \quad \mathbf{AX}_{*n} = \begin{bmatrix} 0 \\ 0 \\ \vdots \\ 1 \end{bmatrix}. \quad (4.8.3)$$

We can do this by forming the augmented matrix

$$[\mathbf{A} \;\vdots\; \mathbf{I}] \quad\quad (4.8.4)$$

and using row reduction until the elements of $\mathbf{A}$ reduce to an identity matrix. An example will illustrate this point.

Example 4.8.1:   Find $\mathbf{A}^{-1}$ if $\mathbf{A} = \begin{bmatrix} 1 & 2 \\ -1 & 1 \end{bmatrix}$.

SOLUTION: Here we have

$$\begin{bmatrix} 1 & 2 \\ -1 & 1 \end{bmatrix} \mathbf{A}^{-1} = \begin{bmatrix} 1 & 0 \\ 0 & 1 \end{bmatrix},$$

where the unknown $\mathbf{A}^{-1}$ will be represented by

$$\mathbf{A}^{-1} = \begin{bmatrix} a & c \\ b & d \end{bmatrix}.$$

This yields two systems of two equations in two unknowns; namely

$$\begin{bmatrix} 1 & 2 \\ -1 & 1 \end{bmatrix} \begin{bmatrix} a \\ b \end{bmatrix} = \begin{bmatrix} 1 \\ 0 \end{bmatrix} \quad \text{and} \quad \begin{bmatrix} 1 & 2 \\ -1 & 1 \end{bmatrix} \begin{bmatrix} c \\ d \end{bmatrix} = \begin{bmatrix} 0 \\ 1 \end{bmatrix}.$$

We now augment $\mathbf{A}$ with both right-hand sides and proceed to solve both systems at once by row reduction of the augmented matrix

$$\begin{bmatrix} 1 & 2 & 1 & 0 \\ -1 & 1 & 0 & 1 \end{bmatrix}.$$

We select our operations so as to reduce $\mathbf{A}$ to an identity matrix. Thus, adding the first row to the second and then $-\frac{2}{3}$ of the second to the first yields

$$\begin{bmatrix} 1 & 0 & \frac{1}{3} & -\frac{2}{3} \\ 0 & 3 & 1 & 1 \end{bmatrix}.$$

Dividing the second row by 3 gives

$$\begin{bmatrix} 1 & 0 & \frac{1}{3} & -\frac{2}{3} \\ 0 & 1 & \frac{1}{3} & \frac{1}{3} \end{bmatrix}.$$

We have deduced that

$$\begin{bmatrix} a \\ b \end{bmatrix} = \begin{bmatrix} \frac{1}{3} \\ \frac{1}{3} \end{bmatrix} \quad \text{and} \quad \begin{bmatrix} c \\ d \end{bmatrix} = \begin{bmatrix} -\frac{2}{3} \\ \frac{1}{3} \end{bmatrix}$$

and hence,

$$\mathbf{A}^{-1} = \begin{bmatrix} a & c \\ b & d \end{bmatrix} = \begin{bmatrix} \frac{1}{3} & -\frac{2}{3} \\ \frac{1}{3} & \frac{1}{3} \end{bmatrix}.$$

We have found a matrix $\mathbf{A}^{-1}$, such that $\mathbf{A}\mathbf{A}^{-1} = \mathbf{I}$. One may verify that $\mathbf{A}^{-1}\mathbf{A} = \mathbf{I}$ also. It is true* in general that for square matrices $\mathbf{A}$ and $\mathbf{B}$, $\mathbf{A}\mathbf{B} = \mathbf{I}$ if an only if $\mathbf{B}\mathbf{A} = \mathbf{I}$. ∎

Example 4.8.2: Invert the matrix

$$\mathbf{A} = \begin{bmatrix} 1 & 0 & 1 \\ 2 & 1 & 1 \\ 1 & 1 & 2 \end{bmatrix}.$$

SOLUTION: The augmented matrix may be reduced as follows:

$$\begin{bmatrix} 1 & 0 & 1 & 1 & 0 & 0 \\ 2 & 1 & 1 & 0 & 1 & 0 \\ 1 & 1 & 2 & 0 & 0 & 1 \end{bmatrix} \longrightarrow \begin{bmatrix} 1 & 0 & 1 & 1 & 0 & 0 \\ 0 & 1 & -1 & -2 & 1 & 0 \\ 0 & 0 & 2 & 1 & -1 & 1 \end{bmatrix}$$

$$\longrightarrow \begin{bmatrix} 1 & 0 & 0 & \frac{1}{2} & \frac{1}{2} & -\frac{1}{2} \\ 0 & 1 & 0 & -\frac{3}{2} & \frac{1}{2} & \frac{1}{2} \\ 0 & 0 & 1 & \frac{1}{2} & -\frac{1}{2} & \frac{1}{2} \end{bmatrix}.$$

Thus, the inverse of $\mathbf{A}$ is

$$\mathbf{A}^{-1} = \begin{bmatrix} \frac{1}{2} & \frac{1}{2} & -\frac{1}{2} \\ -\frac{3}{2} & \frac{1}{2} & \frac{1}{2} \\ \frac{1}{2} & -\frac{1}{2} & \frac{1}{2} \end{bmatrix}.$$

In the next example we see how this method detects singular matrices. ∎

Example 4.8.3: Show that $\mathbf{A}$ is singular if

$$\mathbf{A} = \begin{bmatrix} 2 & 2 & 1 \\ 3 & 3 & -2 \\ 1 & 1 & -3 \end{bmatrix}.$$

SOLUTION: The student is invited to complete the details in the following arrow diagram:

$$\begin{bmatrix} 2 & 2 & 1 & 1 & 0 & 0 \\ 3 & 3 & -2 & 0 & 1 & 0 \\ 1 & 1 & -3 & 0 & 0 & 1 \end{bmatrix} \longrightarrow \cdots \longrightarrow \begin{bmatrix} 2 & 2 & 1 & 1 & 0 & 0 \\ 1 & 1 & -3 & -1 & 1 & 0 \\ 0 & 0 & 0 & 1 & -1 & 1 \end{bmatrix}.$$

Thus, $\mathbf{A}\mathbf{A}^{-1} = \mathbf{I}$ is equivalent to

$$\begin{bmatrix} 2 & 2 & 1 \\ 1 & 1 & -3 \\ 0 & 0 & 0 \end{bmatrix} \mathbf{A}^{-1} = \begin{bmatrix} 1 & 0 & 0 \\ -1 & 1 & 0 \\ 1 & -1 & 1 \end{bmatrix}.$$

But

$$\begin{bmatrix} 2 & 2 & 1 \\ 1 & 1 & -3 \\ 0 & 0 & 0 \end{bmatrix} \mathbf{A}^{-1} = \begin{bmatrix} * & * & * \\ * & * & * \\ 0 & 0 & 0 \end{bmatrix},$$

a contradiction of the equation preceding it. Hence, $\mathbf{A}$ is singular. ∎

---

*We do not prove this theorem here.

The preceding examples are illustrations of this principle:

Theorem 4.5:   *The RENF of a square matrix is either the identity matrix or an upper triangular matrix with at least one row of zeros.*

PROOF: Every RENF of a square matrix is upper triangular, call it $\mathbf{U}$. If the leading ones are the diagonal entries, then the RENF is $\mathbf{I}_n$. So suppose that at least one diagonal entry of the RENF is zero. Then the number of leading ones must be less than n. Hence, one row, at least, must be a row of zeros. In fact, by definition of RENF, the last row of $\mathbf{U}$ is a row of zeros.

The implication of this theorem is that row reduction on Eq. 4.8.4 yields $\mathbf{I}$ or detects that $\mathbf{A}$ is singular, whichever is the case; for

$$[\mathbf{A} \quad \vdots \quad \mathbf{I}] \longrightarrow \cdots \longrightarrow [\mathbf{U} \quad \vdots \quad \mathbf{B}] \tag{4.8.5}$$

implies that

$$\mathbf{I} \longrightarrow \cdots \longrightarrow \mathbf{B}, \tag{4.8.6}$$

where $\mathbf{U}$ is the RENF of $\mathbf{A}$. Suppose that $\mathbf{A}$ is singular. Then the last row of $\mathbf{U}$ is a row of zeros. Since $\mathbf{AX} = \mathbf{I}$ implies that $\mathbf{UX} = \mathbf{B}$, the last row of $\mathbf{B}$ is a row of zeros. The diagram 4.8.6 states* that $\mathbf{Ix} = \mathbf{0}$ has the same solution sets as $\mathbf{Bx} = \mathbf{0}$. But $\mathbf{Ix} = \mathbf{0}$ imples that $\mathbf{x} = \mathbf{0}$. Since the last row of $\mathbf{B}$ is a zero row,

$$\mathbf{B} \begin{bmatrix} 0 \\ 0 \\ \vdots \\ 0 \\ 1 \end{bmatrix} = \mathbf{0} \tag{4.8.7}$$

and $\mathbf{Ix} = \mathbf{0}$ and $\mathbf{Bx} = \mathbf{0}$ do not have the same solution sets. Hence, $\mathbf{AX} = \mathbf{I}$ is a contradiction and $\mathbf{A}$ cannot be nonsingular.

PROBLEMS

Use row operations to decide whether each matrix is singular. When the inverse does exist, find it.

1. $\begin{bmatrix} -1 & 0 & 1 \\ 1 & 0 & 0 \\ 0 & 0 & 1 \end{bmatrix}$ 　　　　　　**2.** $\begin{bmatrix} 2 & 0 & 1 \\ 0 & 3 & 4 \\ 0 & 0 & 7 \end{bmatrix}$

3. $\begin{bmatrix} 1 & 2 \\ -2 & 1 \end{bmatrix}$ 　　　　　　**4.** $\begin{bmatrix} 0 & 0 & 1 \\ 0 & 1 & 0 \\ 1 & 0 & 0 \end{bmatrix}$

---

*In the following discussion and throughout Chapters 4 and 5 we will use a boldface zero to denote a vector with all components equal to zero.

5. $\begin{bmatrix} \cos \theta & \sin \theta \\ -\sin \theta & \cos \theta \end{bmatrix}$ **6.** $\begin{bmatrix} 2 & 0 & 0 \\ 4 & -1 & 0 \\ 0 & 1 & -1 \end{bmatrix}$

7. $\begin{bmatrix} 1 & 0 & 2 \\ 0 & 1 & 0 \\ 0 & 5 & 0 \end{bmatrix}$

Verify that each pair of matrices are inverses of each other.

8. $\begin{bmatrix} 1 & 2 \\ -1 & 1 \end{bmatrix}$ $\begin{bmatrix} \frac{1}{3} & -\frac{2}{3} \\ \frac{1}{3} & \frac{1}{3} \end{bmatrix}$

9. $\begin{bmatrix} 1 & 0 & 1 \\ 2 & 1 & 1 \\ 1 & 2 & 2 \end{bmatrix}$ $\begin{bmatrix} 0 & \frac{2}{3} & -\frac{1}{3} \\ -1 & \frac{1}{3} & \frac{1}{3} \\ 1 & -\frac{2}{3} & \frac{1}{3} \end{bmatrix}$

10. Explain why the diagram $\mathbf{I} \to \cdots \to \mathbf{B}$ means that $\mathbf{Ix} = \mathbf{0}$ has the same solution as $\mathbf{Bx} = \mathbf{0}$.

11. Explain why $\mathbf{UX} = \mathbf{B}$ implies that $\mathbf{B}$ has a row of zeros if $\mathbf{U}$ has a row of zeros. What row of $\mathbf{B}$ is a row of zeros?

12. Explain why Eq. 4.8.7 is true.

Find the inverse of each symmetric matrix and conclude that the inverse of a symmetric matrix is also symmetric.

13. $\begin{bmatrix} 2 & 1 \\ 1 & 1 \end{bmatrix}$ **14.** $\begin{bmatrix} 3 & 1 & 2 \\ 1 & 0 & 1 \\ 2 & 1 & -1 \end{bmatrix}$ **15.** $\begin{bmatrix} 0 & 2 & 3 \\ 2 & 0 & 2 \\ 3 & 2 & 0 \end{bmatrix}$ **16.** $\begin{bmatrix} 2 & 1 & 1 & 1 \\ 1 & 2 & 0 & 0 \\ 1 & 0 & 0 & 1 \\ 1 & 0 & 1 & 2 \end{bmatrix}$

Find the inverse matrix $\mathbf{A}^{-1}$ (if one exists) if $\mathbf{A}$ is given by

17. $\begin{bmatrix} 1 & -1 \\ 1 & 1 \end{bmatrix}$ **18.** $\begin{bmatrix} 2 & 6 \\ 1 & 3 \end{bmatrix}$ **19.** $\begin{bmatrix} 2 & 0 \\ 0 & 1 \end{bmatrix}$

20. $\begin{bmatrix} 1 & 2 \\ 0 & 0 \end{bmatrix}$ **21.** $\begin{bmatrix} -3 & 5 \\ -2 & 6 \end{bmatrix}$ **22.** $\begin{bmatrix} 1 & 0 & 2 \\ 2 & 1 & 1 \\ 1 & 1 & 1 \end{bmatrix}$

23. $\begin{bmatrix} 1 & 2 & 2 \\ 1 & 1 & 2 \\ 1 & -2 & 2 \end{bmatrix}$ **24.** $\begin{bmatrix} 3 & 1 & 2 \\ -1 & 2 & 1 \\ 0 & 1 & 1 \end{bmatrix}$ **25.** $\begin{bmatrix} 0 & 1 & 1 & 0 \\ 0 & 0 & 1 & 1 \\ 1 & 0 & 1 & 1 \\ 1 & 1 & 1 & 1 \end{bmatrix}$

4.9   Determinants of n × n Matrices

The reader has probably encountered determinants of 2×2 and 3×3 matrices and may recall the formulas

$$\begin{vmatrix} a_{11} & a_{12} \\ a_{21} & a_{22} \end{vmatrix} = a_{11}a_{22} - a_{21}a_{12} \tag{4.9.1}$$

$$\begin{vmatrix} a_{11} & a_{12} & a_{13} \\ a_{21} & a_{22} & a_{23} \\ a_{31} & a_{32} & a_{33} \end{vmatrix} = \begin{aligned} & a_{11}a_{22}a_{33} + a_{12}a_{23}a_{31} \\ & + a_{13}a_{21}a_{32} - a_{13}a_{22}a_{31} \\ & - a_{12}a_{21}a_{33} - a_{11}a_{23}a_{32} \end{aligned} \qquad (4.9.2)$$

The determinant has many uses. We cite a few:

1. The system

$$a_{11}x + a_{12}y + a_{13}z = 0$$
$$a_{21}x + a_{22}y + a_{23}z = 0 \qquad (4.9.3)$$
$$a_{31}x + a_{32}y + a_{33}z = 0$$

has a nontrivial solution (x, y, and z not all zero) if and only if the determinant of the coefficient matrix vanishes, that is, if and only if

$$\begin{vmatrix} a_{11} & a_{12} & a_{13} \\ a_{21} & a_{22} & a_{23} \\ a_{31} & a_{32} & a_{33} \end{vmatrix} = 0. \qquad (4.9.4)$$

2. The construction of the solutions to simultaneous equations using Cramer's rule* involves the quotients of various determinants.
3. If v_1 and v_2 are vectors with two entries defining the sides of a parallelogram, as sketched in Fig. 4.1, and if $[v_1, v_2]$ denotes the matrix with these as columns, then, abbreviating absolute value by "abs,"

$$\text{abs}\,|v_1, v_2| = \text{area of the parallelogram.} \qquad (4.9.5)$$

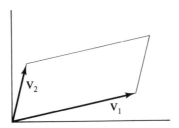

FIGURE 4.1.   A parallelogram.

4. If v_1, v_2, and v_3 are vectors with three entries defining the sides of a parallelepiped, (see Fig. 4.2), and if $[v_1, v_2, v_3]$ denotes the matrix with these as columns, then

$$\text{abs}\,|v_1, v_2, v_3| = \text{volume of the parallelepiped.} \qquad (4.9.6)$$

---

*Cramer's rule was part of your course in algebra; it is presented again as Eq. 4.9.23.

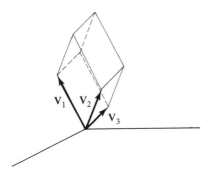

FIGURE 4.2. A parallelopiped.

The determinant can be defined for any square matrix in such a way that these applications, among many others, are preserved in higher dimensions.

Formula 4.9.2 for the value of $|\mathbf{A}|$ when $\mathbf{A}$ is 3×3 can be remembered by the following familiar device. Write the first two columns of the determinant to the right of $\mathbf{A}$ and then sum the products of the elements of the various diagonals using negative signs with the diagonals sloping upward:

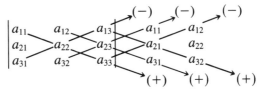

In general, the determinant of the matrix $\mathbf{A}$ is given by

$$|\mathbf{A}| = \sum (-1)^k a_{1i} a_{2j} \cdots a_{ns}, \qquad (4.9.7)$$

where the summation extends over all possible arrangements of the n second subscripts and k is the total number of inversions* in the sequence of the second subscript. We do not, however, use this definition in actual calculations. Rather, the value of the nth-order determinant is most generally found by exploiting a number of consequences of this definition. We list these without proof: the interested reader is referred to a textbook on linear algebra.

Some important properties of a determinant are:

1. If two rows or columns of $\mathbf{A}$ are interchanged to form $\mathbf{A}'$, then

$$|\mathbf{A}| = -|\mathbf{A}'|. \qquad (4.9.8)$$

---

*The number of inversions is the number of pairs of elements in which a larger number precedes a smaller one; for example, the numbers (1, 5, 2, 4, 3) form the four inversions (5, 2), (5, 4), (5, 3), and (4, 3).

2. If a row or column of **A** is multiplied by α to form **A'**, then

$$\alpha|\mathbf{A}| = |\mathbf{A}'|. \tag{4.9.9}$$

3. If a multiple of one row (or column) of **A** is added to another row (or column) of **A** to form **A'**, then

$$|\mathbf{A}| = |\mathbf{A}'|. \tag{4.9.10}$$

4. If **A** is either upper or lower triangular with diagonal entries a_{11}, $a_{22}, \ldots , a_{nn}$, then

$$|\mathbf{A}| = a_{11}a_{22} \cdots a_{nn}. \tag{4.9.11}$$

5. $|\mathbf{AB}| = |\mathbf{A}\|\mathbf{B}|.$ $\hspace{4cm}$ (4.9.12)
6. $|\mathbf{A}^T| = |\mathbf{A}|.$ $\hspace{4cm}$ (4.9.13)

Now suppose that $\mathbf{A} \to \mathbf{A}_1$ by a single elementary row operation. In view of Eqs. 4.9.8 to 4.9.10, $|\mathbf{A}| = \alpha|\mathbf{A}_1|$ where $\alpha \neq 0$. If $\mathbf{A} \to \mathbf{A}_1 \to \cdots \to \mathbf{A}_m$, then $|\mathbf{A}| = \alpha|\mathbf{A}_m|$, $\alpha \neq 0$. This leads to an easy proof of the following especially important theorem.

Theorem 4.6: **A** *is singular if and only if* $|\mathbf{A}| = 0$.

PROOF: From Theorem 4.5 and the discussion immediately following it, we know that **A** has an inverse if and only if

$$\mathbf{A} \longrightarrow \cdots \longrightarrow \mathbf{I}. \tag{4.9.14}$$

It follows from the analysis made above—after Eq. 4.9.13—that $|\mathbf{A}| = \alpha|\mathbf{I}| = \alpha \neq 0$. Hence, $|\mathbf{A}| = 0$ is a contradiction, implying that **A** is singular, and conversely.

Example 4.9.1:   Calculate $|\mathbf{A}|$, using row operations, given

$$\mathbf{A} = \begin{bmatrix} 0 & 1 & 1 \\ 1 & 0 & 0 \\ 0 & 0 & 1 \end{bmatrix}.$$

SOLUTION: We add the second row to the first, then substract the first row from the second:

$$|\mathbf{A}| = \begin{vmatrix} 1 & 1 & 1 \\ 1 & 0 & 0 \\ 0 & 0 & 1 \end{vmatrix} = \begin{vmatrix} 1 & 1 & 1 \\ 0 & -1 & -1 \\ 0 & 0 & 1 \end{vmatrix} = 1 \cdot -1 \cdot 1 = -1. \quad \blacksquare$$

Example 4.9.2:   Calculate $|\mathbf{A}|$ where

$$\mathbf{A} = \begin{bmatrix} 1 & 2 & 1 & 3 \\ -1 & 1 & 3 & 2 \\ 1 & 0 & 2 & 3 \\ -1 & 1 & 1 & 4 \end{bmatrix}.$$

SOLUTION: By various applications of elementary row operations we can express the determinant of **A** as

$$|\mathbf{A}| = \begin{vmatrix} 1 & 2 & 1 & 3 \\ 0 & 3 & 4 & 5 \\ 0 & -2 & 1 & 0 \\ 0 & 3 & 2 & 7 \end{vmatrix} = \begin{vmatrix} 1 & 2 & 1 & 3 \\ 0 & 3 & 4 & 5 \\ 0 & 0 & \frac{11}{3} & \frac{10}{3} \\ 0 & 0 & -2 & -2 \end{vmatrix}.$$

Continuing, we have

$$|\mathbf{A}| = \begin{vmatrix} 1 & 2 & 1 & 3 \\ 0 & 3 & 4 & 5 \\ 0 & 0 & \frac{11}{3} & \frac{10}{3} \\ 0 & 0 & 0 & \frac{42}{11} \end{vmatrix} = 1 \cdot 3 \cdot \frac{11}{3} \cdot \frac{42}{11} = 42. \quad \blacksquare$$

Example 4.9.3: Show that $|k\mathbf{A}| = k^n|\mathbf{A}|$, where **A** is $n \times n$.

SOLUTION: This follows from Eq. 4.9.9 after noting that $k\mathbf{A}$ has each row of **A** multiplied by k and there are n rows in **A**. $\quad \blacksquare$

Example 4.9.4: Show that $|\mathbf{A}^{-1}| = |\mathbf{A}|^{-1}$.

SOLUTION: Since

$$\mathbf{A}\mathbf{A}^{-1} = \mathbf{I} \quad \text{and} \quad |\mathbf{I}| = 1,$$

we have

$$|\mathbf{A}\mathbf{A}^{-1}| = |\mathbf{A}||\mathbf{A}^{-1}| = 1$$

from Eq. 4.9.12. Thus,

$$|\mathbf{A}^{-1}| = \frac{1}{|\mathbf{A}|} = |\mathbf{A}|^{-1} \quad \blacksquare$$

PROBLEMS

1. Show that $|\mathbf{S}^{-1}\mathbf{A}\mathbf{S}| = |\mathbf{A}|$.

2. Show that $|\mathbf{A}^n| = |\mathbf{A}|^n$.

3. Find **A** and **B** to illustrate that $|\mathbf{A} + \mathbf{B}| = |\mathbf{A}| + |\mathbf{B}|$ is false, in general.

The following problems illustrate the definition 4.9.7. We call $a_{1i}a_{2j} \cdots a_{ns}$ a *term*.

4. Show that the right-hand side of Eq. 4.9.7 is a sum of $n!$ terms.

5. Show that $a_{11}a_{22} \cdots a_{nn}$ is a term.

6. Explain why a term is the product of n entries in **A**, one from each row and each column. Hence, no term contains two entries of **A** from the same row or two from the same column.

7. Use the results of Problems 5 and 6 to explain why the determinant of an upper triangular matrix is the product of its diagonal entries.

8. Show that Eq. 4.9.9 is an immediate consequence of Eq. 4.9.7.

9. Show that Eqs. 4.9.1 and 4.9.2 are consequences of the definition 4.9.7.

Using the products of the diagonal elements, evaluate each determinant.

10. $\begin{vmatrix} 2 & 0 \\ -1 & 3 \end{vmatrix}$
11. $\begin{vmatrix} 1 & 2 \\ 1 & 3 \end{vmatrix}$
12. $\begin{vmatrix} 2 & -2 \\ -1 & 1 \end{vmatrix}$

13. $\begin{vmatrix} 3 & 1 & 0 \\ 1 & 3 & -1 \\ 2 & -1 & 0 \end{vmatrix}$
14. $\begin{vmatrix} 4 & -1 & 3 \\ 2 & 2 & 2 \\ 1 & -2 & 4 \end{vmatrix}$

Show the following, by computation.

15. $\begin{vmatrix} 3 & 2 & -1 \\ 6 & 3 & 0 \\ 3 & 1 & 2 \end{vmatrix} = 3 \begin{vmatrix} 1 & 2 & -1 \\ 2 & 3 & 0 \\ 1 & 1 & 2 \end{vmatrix}$
16. $\begin{vmatrix} 3 & 2 & -1 \\ 6 & 3 & 0 \\ 3 & 1 & 2 \end{vmatrix} = - \begin{vmatrix} 2 & 3 & -1 \\ 3 & 6 & 0 \\ 1 & 3 & 2 \end{vmatrix}$

17. $\begin{vmatrix} 3 & 2 & -1 \\ 6 & 3 & 0 \\ 3 & 1 & 2 \end{vmatrix} = 3 \begin{vmatrix} 3+2 & 2 & -1 \\ 6+3 & 3 & 0 \\ 3+1 & 1 & 2 \end{vmatrix}$
18. $\begin{vmatrix} 3 & 2 & -1 \\ 6 & 3 & 0 \\ 3 & 1 & 2 \end{vmatrix} = \begin{vmatrix} 3+10 & 2 & -1 \\ 6+15 & 3 & 0 \\ 3+5 & 1 & 2 \end{vmatrix}$

19. $\begin{vmatrix} 3 & 2 & -1 \\ 6 & 3 & 0 \\ 3 & 1 & 2 \end{vmatrix} = \begin{vmatrix} 3+3 & 2+1 & -1+2 \\ 6 & 3 & 0 \\ 3 & 1 & 2 \end{vmatrix}$

20. $\begin{vmatrix} 3 & -3 & -1 \\ 6 & -6 & 0 \\ 3 & -3 & 2 \end{vmatrix} = 0$

Evaluate each determinant by using Eq. 4.9.11.

21. $\begin{vmatrix} 3 & 1 & 3 \\ 2 & 0 & 4 \\ -1 & 2 & -2 \end{vmatrix}$
22. $\begin{vmatrix} 2 & 3 & 4 \\ -1 & 0 & 3 \\ 1 & 2 & 3 \end{vmatrix}$
23. $\begin{vmatrix} 1 & 1 & 1 \\ -1 & -2 & 2 \\ 1 & 2 & 3 \end{vmatrix}$

24. $\begin{vmatrix} 2 & 1 & 3 \\ 4 & 2 & 6 \\ -3 & 1 & 0 \end{vmatrix}$
25. $\begin{vmatrix} 4 & 3 & 1 & 4 \\ 3 & 0 & 0 & 3 \\ 1 & 2 & 2 & 1 \\ 0 & -1 & 3 & 2 \end{vmatrix}$
26. $\begin{vmatrix} 3 & 1 & -1 & 0 \\ 2 & 2 & 2 & 1 \\ -1 & 3 & 0 & 4 \\ 8 & 6 & -2 & 2 \end{vmatrix}$

27. $\begin{vmatrix} 1 & 1 & 1 & 1 \\ -2 & 3 & 1 & 0 \\ 4 & 3 & 8 & 1 \\ 7 & 5 & -2 & 0 \end{vmatrix}$
28. $\begin{vmatrix} 2 & -1 & 6 & 3 \\ -2 & 4 & 5 & -1 \\ 3 & 4 & 3 & 2 \\ 1 & -1 & 2 & 3 \end{vmatrix}$

29. Verify by computation the fifth property listed for a determinant, using

$$\mathbf{A} = \begin{bmatrix} 2 & 3 & 1 \\ -1 & 0 & 2 \\ 3 & 4 & 1 \end{bmatrix}, \quad \mathbf{B} = \begin{bmatrix} 2 & -1 & 3 \\ 6 & 7 & -1 \\ 3 & 4 & 2 \end{bmatrix}.$$

First, using matrix multiplications, find $\mathbf{C} = \mathbf{AB}$. Then show that $|\mathbf{C}| = |\mathbf{A}||\mathbf{B}|$.

4.9.1   MINORS AND COFACTORS

The determinant of the matrix formed from $\mathbf{A}$ by striking out the ith row and jth column of $\mathbf{A}$ is called the *minor* of a_{ij}. For example, if

$$\mathbf{A} = \begin{bmatrix} 1 & -1 & 2 \\ 0 & -2 & 3 \\ 4 & -4 & 6 \end{bmatrix}, \qquad (4.9.15)$$

then the three minors of the elements in the first row of $\mathbf{A}$ are

$$\begin{vmatrix} -2 & 3 \\ -4 & 6 \end{vmatrix}, \qquad \begin{vmatrix} 0 & 3 \\ 4 & 6 \end{vmatrix}, \qquad \begin{vmatrix} 0 & -2 \\ 4 & -4 \end{vmatrix},$$

respectively.

The *cofactor* of a_{ij} is A_{ij} and is $(-1)^{i+j}$ times its minor. Hence, the cofactors of the three elements above are, respectively,

$$A_{11} = (-1)^2 \begin{vmatrix} -2 & 3 \\ -4 & 6 \end{vmatrix}, \qquad A_{12} = (-1)^3 \begin{vmatrix} 0 & 3 \\ 4 & 6 \end{vmatrix},$$

$$A_{13} = (-1)^4 \begin{vmatrix} 0 & -2 \\ 4 & -4 \end{vmatrix}. \qquad (4.9.16)$$

The importance of the cofactors is due to the following:*

$$|\mathbf{A}| = \sum_{j=1}^{n} a_{ij} A_{ij} = \sum_{j=1}^{n} a_{ji} A_{ji}. \qquad (4.9.17)$$

That is, the value of the determinant of the square matrix $\mathbf{A}$ is given by the sum of the products of the elements of any row or column with their respective cofactors.

Example 4.9.5:   Using cofactors, find the determinant of

$$\mathbf{A} = \begin{bmatrix} 3 & 2 & 1 \\ -1 & 0 & 1 \\ 1 & 2 & 2 \end{bmatrix}$$

by expanding by the first row and then the first column.

SOLUTION: Expanding by the first row, we have

$$\begin{vmatrix} 3 & 2 & 1 \\ -1 & 0 & 1 \\ 1 & 2 & 2 \end{vmatrix} = 3 \begin{vmatrix} 0 & 1 \\ 2 & 2 \end{vmatrix} - 2 \begin{vmatrix} -1 & 1 \\ 1 & 2 \end{vmatrix} + 1 \begin{vmatrix} -1 & 0 \\ 1 & 2 \end{vmatrix}$$

$$= 3(-2) - 2(-2 - 1) + 1(-2) = -2.$$

---

*See any textbook on linear algebra.

Expanding by the first column, there results

$$\begin{vmatrix} 3 & 2 & 1 \\ -1 & 0 & 1 \\ 1 & 2 & 2 \end{vmatrix} = 3\begin{vmatrix} 0 & 1 \\ 2 & 2 \end{vmatrix} - (-1)\begin{vmatrix} 2 & 1 \\ 2 & 2 \end{vmatrix} + 1\begin{vmatrix} 2 & 1 \\ 0 & 1 \end{vmatrix}$$

$$= 3(-2) + 1(4 - 2) + 1(2) = -2. \quad \blacksquare$$

Example 4.9.6:   Evaluate the determinant of Example 4.9.2 by expanding, using cofactors of the first row.

SOLUTION: The determinant of the matrix is

$$|\mathbf{A}| = A_{11} + 2A_{12} + A_{13} + 3A_{14}.$$

We evaluate each 3×3 determinant by expanding, using its first row:

$$A_{11} = 1(8 - 3) - 3(0 - 3) + 2(0 - 2) = 10$$
$$A_{12} = -[(-1)(8 - 3) - 3(4 + 3) + 2(1 + 2)] = 20$$
$$A_{13} = (-1)(0 - 3) - (1)(4 + 3) + 2(1 - 0) = -2$$
$$A_{14} = -[(-1)(0 - 2) - (1)(1 + 2) + 3(1 - 0)] = -2.$$

Hence,

$$|\mathbf{A}| = 1 \cdot 10 + 2 \cdot 20 + 1 \cdot (-2) + 3 \cdot (-2) = 42. \quad \blacksquare$$

PROBLEMS

Using cofactors, evaluate

$$\begin{vmatrix} 3 & 2 & -1 \\ 3 & 0 & 3 \\ -1 & 2 & 1 \end{vmatrix}.$$

1. Expand by the first row.

2. Expand by the second row.

3. Expand by the first column.

4. Expand by the second column.

Using cofactors, evaluate

$$\begin{vmatrix} 2 & 0 & 8 & 6 \\ -1 & 4 & 2 & 0 \\ 0 & -1 & 3 & 0 \\ 3 & 5 & 7 & 3 \end{vmatrix}.$$

5. Expand by the first row.

6. Expand by the third row.

7. Expand by the first column.

8. Expand by the fourth column.

Use the method of cofactors to find the value of each determinant.

9. $\begin{vmatrix} 2 & 0 \\ 0 & 1 \end{vmatrix}$ **10.** $\begin{vmatrix} 1 & 2 \\ 0 & 0 \end{vmatrix}$

11. $\begin{vmatrix} 0 & 2 & 3 \\ 2 & 0 & 2 \\ 3 & 2 & 0 \end{vmatrix}$ **12.** $\begin{vmatrix} 1 & 0 & 2 \\ 2 & 1 & 1 \\ 1 & 1 & 1 \end{vmatrix}$

13. $\begin{vmatrix} 1 & 2 & 2 \\ 1 & 1 & 2 \\ 1 & -2 & 2 \end{vmatrix}$ **14.** $\begin{vmatrix} 3 & 1 & 2 \\ -1 & 2 & 1 \\ 0 & 1 & 1 \end{vmatrix}$

15. $\begin{vmatrix} 2 & 1 & 1 & 1 \\ 1 & 2 & 0 & 0 \\ 1 & 0 & 0 & 1 \\ 1 & 0 & 1 & 2 \end{vmatrix}$ **16.** $\begin{vmatrix} 0 & 1 & 1 & 0 \\ 0 & 0 & 1 & 1 \\ 1 & 0 & 1 & 1 \\ 1 & 1 & 1 & 1 \end{vmatrix}$

4.9.2   THE ADJOINT

We are now in a position to define a matrix closely related to $\mathbf{A}^{-1}$. The *adjoint matrix* $\mathbf{A}^+$ is the transpose of the matrix obtained from the square matrix $\mathbf{A}$ by replacing each element a_{ij} of $\mathbf{A}$ with its cofactor A_{ij}. It is displayed as

$$\mathbf{A}^+ = \begin{bmatrix} A_{11} & A_{21} & \cdots & A_{n1} \\ A_{12} & A_{22} & \cdots & A_{n2} \\ \vdots & & & \\ A_{1n} & A_{2n} & \cdots & A_{nn} \end{bmatrix}. \tag{4.9.18}$$

Note that the cofactor A_{ij} occupies the position of a_{ji}, not the position of a_{ij}.

One can now establish the relationship

$$\mathbf{A}\mathbf{A}^+ = \mathbf{A}^+\mathbf{A} = |\mathbf{A}|\mathbf{I}. \tag{4.9.19}$$

(The proof is left to the reader, see Problem 16.) Hence, if $\mathbf{A}^{-1}$ exists, we have the result

$$\mathbf{A}^{-1} = \frac{\mathbf{A}^+}{|\mathbf{A}|}. \tag{4.9.20}$$

This formula for $\mathbf{A}^{-1}$ is not convenient for computation, since it requires producing n^2 determinants of $(n-1) \times (n-1)$ matrices to find $|\mathbf{A}|$.

Equation 4.9.20 does, however, lead to the well-known *Cramer's rule*: Suppose that $\mathbf{A}$ is $n \times n$ and

$$\mathbf{A} = [\mathbf{A}_{*1}, \mathbf{A}_{*2}, \ldots, \mathbf{A}_{*n}], \quad \mathbf{x} = \begin{bmatrix} x_1 \\ x_2 \\ \vdots \\ x_n \end{bmatrix}, \quad \mathbf{r} = \begin{bmatrix} r_1 \\ r_2 \\ \vdots \\ r_n \end{bmatrix}, \tag{4.9.21}$$

then the system

$$\mathbf{Ax} = \mathbf{r} \tag{4.9.22}$$

has solution

$$x_1 = \frac{|[\mathbf{r}, \mathbf{A}_{*2}, \ldots, \mathbf{A}_{*n}]|}{|\mathbf{A}|}, \quad x_2 = \frac{|[\mathbf{A}_{*1}, \mathbf{r}, \mathbf{A}_{*3}, \ldots, \mathbf{A}_{*n}]|}{|\mathbf{A}|}, \ldots,$$

$$x_n = \frac{|[\mathbf{A}_{*1}, \mathbf{A}_{*2}, \ldots, \mathbf{r}]|}{|\mathbf{A}|}. \tag{4.9.23}$$

The proof follows: Let $\mathbf{A}_1 = [\mathbf{r}, \mathbf{A}_{*2}, \ldots, \mathbf{A}_{*n}]$. Expand $|\mathbf{A}_1|$ by cofactors of the first column. This results in

$$|\mathbf{A}_1| = r_1 A_{11} + r_2 A_{21} + \cdots + r_n A_{n1}. \tag{4.9.24}$$

Now, consider the first component of

$$\mathbf{x} = \mathbf{A}^{-1}\mathbf{r} = \frac{\mathbf{A}^+\mathbf{r}}{|\mathbf{A}|}. \tag{4.9.25}$$

The first component is

$$x_1 = \frac{r_1 A_{11} + r_2 A_{21} + \cdots + r_n A_{n1}}{|\mathbf{A}|}. \tag{4.9.26}$$

The argument is essentially the same for each component of $\mathbf{x}$.

PROBLEMS

Find the adjoint matrix $\mathbf{A}^+$ and the inverse matrix $\mathbf{A}^{-1}$ (if one exists) if $\mathbf{A}$ is given by

1. $\begin{bmatrix} 1 & -1 \\ 1 & 1 \end{bmatrix}$
   2. $\begin{bmatrix} 2 & 6 \\ 1 & 3 \end{bmatrix}$
   3. $\begin{bmatrix} 2 & 0 \\ 0 & 1 \end{bmatrix}$

4. $\begin{bmatrix} 1 & 2 \\ 0 & 0 \end{bmatrix}$
   5. $\begin{bmatrix} -3 & 5 \\ -2 & 6 \end{bmatrix}$
   6. $\begin{bmatrix} 1 & 0 & 2 \\ 2 & 1 & 1 \\ 1 & 1 & 1 \end{bmatrix}$

7. $\begin{bmatrix} 1 & 2 & 2 \\ 1 & 1 & 2 \\ 1 & -2 & 2 \end{bmatrix}$
   8. $\begin{bmatrix} 3 & 1 & 2 \\ -1 & 2 & 1 \\ 0 & 1 & 1 \end{bmatrix}$
   9. $\begin{bmatrix} 0 & 1 & 1 & 0 \\ 0 & 0 & 1 & 1 \\ 1 & 0 & 1 & 1 \\ 1 & 1 & 1 & 1 \end{bmatrix}$

Solve each system of linear, algebraic equations by Cramer's rule.

10. $\begin{aligned} x - y &= 6 \\ x + y &= 0 \end{aligned}$
   11. $\begin{aligned} x - 2y &= 4 \\ 2x + y &= 3 \end{aligned}$
   12. $\begin{aligned} 3x + 4y &= 7 \\ 2x - 5y &= 2 \end{aligned}$

13. $\begin{aligned} 3x + 2y - 6z &= 0 \\ x - y + z &= 4 \\ y + z &= 3 \end{aligned}$
   14. $\begin{aligned} x - 3y + z &= -2 \\ x - 3y - z &= 0 \\ -3y + z &= 0 \end{aligned}$
   15. $\begin{aligned} x_1 + x_2 + x_3 &= 4 \\ x_1 - x_2 - x_3 &= 2 \\ x_1 - 2x_2 &= 0 \end{aligned}$

16. The following result, analogous to Eq. 4.9.1, may be found in any text on linear algebra:

$$0 = \sum_{j=1}^{n} a_{kj} A_{ij} = \sum_{j=1}^{n} a_{jk} A_{ji} \qquad \text{if } k \ne i.$$

Now use Eq 4.9.1 and this result to establish

$$\mathbf{A}^+\mathbf{A} = \mathbf{A}\mathbf{A}^+ = |\mathbf{A}|\mathbf{I}.$$

17. Use the result of Problem 16 to prove that $\mathbf{A}^{-1}$ exists if and only if $|\mathbf{A}| \ne 0$.

4.10 Linear Independence

In Chapter 1 we saw the importance of the linear independence of a set of solutions of a differential equation. The central idea was that a sufficiently large set of independent solutions enabled us to solve any initial-value problem. An analogous situation exists for the theory of systems of algebraic equations. We explore this idea in detail in this section.

Suppose that vector $\mathbf{y}$ is defined by the sum

$$\mathbf{y} = a_1\mathbf{x}_1 + a_2\mathbf{x}_2 + \cdots + a_k\mathbf{x}_k, \qquad k \ge 1. \qquad (4.10.1)$$

Then $\mathbf{y}$ is a *linear combination* of $\{\mathbf{x}_1, \mathbf{x}_2, \ldots, \mathbf{x}_k\}$. The scalars a_1, $a_2, \ldots, a_k$ may be real or complex numbers. If all the scalars are zero, then Eq. 4.10.1 is called *trivial* and, of course, $\mathbf{y} = \mathbf{0}$. On the other hand, $\mathbf{y}$ may be the zero vector without all the scalars zero as seen in the sum, $\mathbf{0}^T = 2[1, 1] - [2, 2]$. When a linear combination is the zero vector without all the scalars being zero, we call the combination *nontrivial*.

A set of vectors $\{\mathbf{x}_1, \mathbf{x}_2, \ldots, \mathbf{x}_n\}$ is *linearly dependent* if $\mathbf{0}$ is a nontrivial combination of vectors from this set; that is,

$$\mathbf{0} = a_1\mathbf{x}_1 + a_2\mathbf{x}_2 + \cdots + a_k\mathbf{x}_k, \qquad k \ge 1 \qquad (4.10.2)$$

where at least one scalar is not zero. If the given vectors are not linearly independent, they are *linearly dependent*. It then follows that if $\{\mathbf{x}_1, \mathbf{x}_2, \ldots, \mathbf{x}_k\}$ is linearly independent and

$$\mathbf{0} = a_1\mathbf{x}_1 + a_2\mathbf{x}_2 + \cdots + a_k\mathbf{x}_k, \qquad (4.10.3)$$

then

$$a_1 = a_2 = \cdots = a_k = 0. \qquad (4.10.4)$$

Example 4.10.1: Demonstrate the linear dependence of the following vectors:

(a) $[1, 1, 0], [-1, 1, 0], [0, 1, 0]$

(b) $\mathbf{0}, \mathbf{x}_1, \mathbf{x}_2, \mathbf{x}_3$

(c) $\mathbf{x}_1, \mathbf{x}_2 - \mathbf{x}_1, 2\mathbf{x}_1 + \mathbf{x}_2$

SOLUTION: First of all, for the vectors of part (a), we attempt to select the scalar coefficients such that zero results on the right-hand side. Hence,

$$\begin{bmatrix} 1 \\ 1 \\ 0 \end{bmatrix} + \begin{bmatrix} -1 \\ 1 \\ 0 \end{bmatrix} - 2\begin{bmatrix} 0 \\ 1 \\ 0 \end{bmatrix} = \begin{bmatrix} 0 \\ 0 \\ 0 \end{bmatrix}.$$

Next,

$$1 0 + 0x_1 + 0x_2 + 0x_3 = 0$$

and

$$3x_1 + (x_2 - x_1) - (2x_1 + x_2) = 0.$$

The above shows the dependence of the vectors given in parts (a), (b), and (c), respectively.   ∎

Example 4.10.2:   Show that $[1, 1, 0, 1]$, $[1, 0, 0, 1]$, $[1, -1, 0, 1]$, and $[0, 0, 1, 0]$ is a linearly dependent set by finding the scalars so that Eq. 4.10.2 holds.

SOLUTION: We apply elementary row operations to the matrix

$$\begin{bmatrix} 1 & 1 & 0 & 1 & x_1 \\ 1 & 0 & 0 & 1 & x_2 \\ 1 & -1 & 0 & 1 & x_3 \\ 0 & 0 & 1 & 0 & x_4 \end{bmatrix}.$$

The rows of this matrix are given vectors while the last column is simply used to keep track of the various row operations used in the reduction. Note, for instance, that the last column of the matrix

$$\begin{bmatrix} 1 & 1 & 0 & 1 & x_1 \\ 0 & -1 & 0 & 0 & x_2 - x_1 \\ 0 & -2 & 0 & 0 & x_3 - x_1 \\ 0 & 0 & 1 & 0 & x_4 \end{bmatrix}$$

exhibits the row operations used. Continuing, we obtain

$$\begin{bmatrix} 1 & 1 & 0 & 1 & x_1 \\ 0 & -1 & 0 & 0 & x_2 - x_1 \\ 0 & 0 & 0 & 0 & x_3 - x_1 - 2(x_2 - x_1) \\ 0 & 0 & 1 & 0 & x_4 \end{bmatrix}.$$

The third row shows that

$$[0, 0, 0, 0] = x_3 - x_1 - 2(x_2 - x_1)$$

or, more neatly,

$$x_1 - 2x_2 + x_3 = 0$$

which is the required nontrivial sum.   ∎

The point illustrated by this example is that there is a simple algorithm to detect whether a set is linearly dependent—and if linearly dependent, evaluate the scalars in the *dependency relationship* 4.10.2.

The special case in which we have n vectors $x_1, x_2, \ldots, x_n$, each with n entries, leads to a determinant test. Set

$$X = \begin{bmatrix} x_1^T \\ x_2^T \\ \vdots \\ x_n^T \end{bmatrix}_{n \times n}. \tag{4.10.4}$$

Using an arrow diagram

$$X \longrightarrow \cdots \longrightarrow U \tag{4.10.5}$$

where U is the RENF of X and is upper triangular. By Theorem 4.6, U is either I or has a row of zeros. Using the properties of a determinant, we know that

$$|X| = k|U|, \qquad k \neq 0. \tag{4.10.6}$$

If U has a row of zeros, $|U| = 0$, which implies that $|X| = 0$. If $|X| = 0$, then U must have a row of zeros since U cannot be I. Thus, we have the following theorem.

Theorem 4.7:   *If X is a square matrix,*

$$|X| = 0 \tag{4.10.7}$$

if and only if the columns of X form a linearly dependent set.

Example 4.10.3:   Find those numbers t for which $[1 - t, 0, 0]$, $[1, 1 - t, 0]$, and $[1, 1, 1 - t]$ are linearly dependent.

SOLUTION: Consider the matrix

$$A = \begin{bmatrix} 1 - t & 0 & 0 \\ 1 & 1 - t & 0 \\ 1 & 1 & 1 - t \end{bmatrix}.$$

Then

$$|A| = (1 - t)^3 = 0$$

if and only if

$$t = 1. \quad \blacksquare$$

Example 4.10.4:   Show that the n vectors

$$e_1 = \begin{bmatrix} 1 \\ 0 \\ \vdots \\ 0 \end{bmatrix}, \quad e_2 = \begin{bmatrix} 0 \\ 1 \\ \vdots \\ 0 \end{bmatrix}, \quad \cdots, \quad e_n = \begin{bmatrix} 0 \\ 0 \\ \vdots \\ 1 \end{bmatrix}$$

are linearly independent.

SOLUTION: The appropriate matrix is **I**. We know that

$$|\mathbf{I}| = 1$$

so that the given vectors are linearly independent by Theorem 4.7. ∎

PROBLEMS

Which of the following sequences of vectors are linearly dependent?

1. $[1, 0, 1], [1, 1, -1], [-1, 1, -3]$

2. $[1, 1, 0, 1], [-1, 2, 0, 0], [0, 0, 1, 0], [1, -1, -1, 1]$

3. $[-1, 0, 0, 1], [2, -1, 1, 1], [0, -1, 1, 3]$

4. Find k so that $[k, 0, 1], [1, k, -1], [-1, 1, k]$ are linearly dependent.

5. If $\mathbf{x}_1, \mathbf{x}_2, \mathbf{x}_3, \mathbf{x}_4$ is a linearly independent sequence, is the sequence $\mathbf{x}_1, \mathbf{x}_2, \mathbf{x}_3$ linearly independent? Why? Generalize.

6. If $\mathbf{x}_1, \mathbf{x}_2, \mathbf{x}_3, \mathbf{x}_4$ is a linearly dependent sequence, is the sequence $\mathbf{x}_1, \mathbf{x}_2, \mathbf{x}_3$, linearly dependent? Explain.

For each sequence of linearly dependent vectors, express **0** as a nontrivial linear combination.

7. Problem 1

8. Problem 3

9. $[-1, 2, 0, 0], [1, 2, -1, 0], [1, 1, 0, 1], [1, 5, -1, 1]$

10. $[1, 1, 0, 1], [1, 0, 0, 1], [0, 0, 1, 0], [1, -1, 0, 1]$

11. A sequence consisting of a single vector, $\mathbf{x} \neq \mathbf{0}$, is linearly independent. Why?

12. Give a linearly dependent sequence of at least two vectors, none of which is zero, such that

$$\mathbf{0} = a_1\mathbf{x}_1 + a_2\mathbf{x}_2 + \cdots + a_k\mathbf{x}_k$$

is nontrivial. Prove that at least two of the scalars are not zero.

4.11   Homogeneous Systems

We return to the solution of m equations in n unknowns represented by

$$\mathbf{Ax} = \mathbf{r}. \tag{4.11.1}$$

In this section we assume that $\mathbf{r} = \mathbf{0}$ and call

$$\mathbf{Ax} = \mathbf{0} \tag{4.11.2}$$

homogeneous. Every homogeneous system is consistent because $\mathbf{A0} = \mathbf{0}$ shows that Eq. 4.11.2 always has the trivial solution $\mathbf{x} = \mathbf{0}$. Our interest, therefore, centers on those matrices $\mathbf{A}$, for which $\mathbf{Ax} = \mathbf{0}$ has nontrivial solutions. If Eq. 4.11.2 has the solution $\mathbf{x} = \mathbf{x}_1 \neq \mathbf{0}$, then

$$\mathbf{A}(c\mathbf{x}_1) = c\mathbf{Ax}_1 = c\mathbf{0} = \mathbf{0} \tag{4.11.3}$$

shows that $c\mathbf{x}_1$ is a solution for every choice of the scalar c. Thus, if $\mathbf{Ax} = \mathbf{0}$ has even one nontrivial solution, it has infinitely many solutions.

Suppose that $\mathbf{A}$ is $m \times n$. We say that $\mathbf{A}$ has *full rank* if rank $\mathbf{A} = n$. The matrices in (a) below have full rank; those in (b) do not:

(a) $\begin{bmatrix} 1 & 0 & 0 \\ 0 & 1 & 0 \\ 0 & 0 & 1 \end{bmatrix}$, $\begin{bmatrix} 1 & 0 \\ 0 & 1 \\ * & * \\ \vdots & \\ * & * \end{bmatrix}$

(b) $\begin{bmatrix} 1 & * & 0 \\ 0 & 0 & 1 \\ 0 & 0 & 0 \\ \vdots & \\ 0 & 0 & 0 \end{bmatrix}$, $\begin{bmatrix} 1 & 0 & * & \cdots & * \\ 0 & 1 & * & \cdots & * \end{bmatrix}$

The critical difference between the cases in which $\mathbf{Ax} = \mathbf{0}$ has the unique, trivial solution $\mathbf{x} = \mathbf{0}$ and the cases in which $\mathbf{Ax} = \mathbf{0}$ has infinitely many solutions is most easily seen by studying an example. We choose $\mathbf{A}$ in RENF but not with full rank. We shall discover that it is precisely when $\mathbf{A}$ does not have full rank that $\mathbf{Ax} = \mathbf{0}$ has nontrivial solutions.

Let

$$\mathbf{A} = \begin{bmatrix} 1 & 2 & 0 & -1 & -2 \\ 0 & 0 & 1 & 1 & 1 \\ 0 & 0 & 0 & 0 & 0 \\ 0 & 0 & 0 & 0 & 0 \end{bmatrix}. \tag{4.11.4}$$

Then, $\mathbf{Ax} = \mathbf{0}$ represents

$$x_1 + 2x_2 \quad - x_4 - 2x_5 = 0$$
$$x_3 + x_4 + \quad x_5 = 0$$
$$0 = 0$$
$$0 = 0. \tag{4.11.5}$$

The coefficients of the unknowns x_i are the entries in the ith column of $\mathbf{A}$. For instance, x_5 has the coefficients -2, 1, 0, 0, in Eq. 4.11.5, the entries in column 5 of $\mathbf{A}$. The unknowns whose coefficients are the entries in a leading column are *basic variables*; the remaining unknowns are *free variables*. In the system 4.11.5, x_1 and x_3 are basic and x_2, x_4, x_5 are free. Since, in RENF, each basic variable appears in one and only one equation, each choice of free variables leads to a unique determination of the basic variables, and therefore to a unique solution. We distinguish *basic solutions* of $\mathbf{Ax} = \mathbf{0}$ by the following definition:

> A basic solution is one in which a single free variable is assigned the value one and the remaining free variables (if any) are set zero.

For the $\mathbf{A}$ in Eq. 4.11.4 we obtain three basic solutions corresponding to the three free variables:

Solution 1. Set $x_2 = 1$, $x_4 = x_5 = 0$. Then

$$\mathbf{x}_1 = \begin{bmatrix} -2 \\ 1 \\ 0 \\ 0 \\ 0 \end{bmatrix}.$$

Solution 2. Set $x_4 = 1$, $x_2 = x_5 = 0$. Then

$$\mathbf{x}_2 = \begin{bmatrix} 1 \\ 0 \\ -1 \\ 1 \\ 0 \end{bmatrix}.$$

Solution 3. Set $x_5 = 1$, $x_2 = x_4 = 0$. Then

$$\mathbf{x}_3 = \begin{bmatrix} 2 \\ 0 \\ -1 \\ 0 \\ 1 \end{bmatrix}.$$

We call the set of all basic solutions a *basic set of solutions*. A basic set of solutions is a linearly independent set. To see why this is the case, consider

$$c_1\mathbf{x}_1 + c_2\mathbf{x}_2 + c_3\mathbf{x}_3 = c_1 \begin{bmatrix} * \\ 1 \\ * \\ 0 \\ 0 \end{bmatrix} + c_2 \begin{bmatrix} * \\ 0 \\ * \\ 1 \\ 0 \end{bmatrix} + c_3 \begin{bmatrix} * \\ 0 \\ * \\ 0 \\ 1 \end{bmatrix} = \mathbf{0}, \qquad (4.11.6)$$

where the basic variables are ignored. Equation 4.11.6 clearly shows that $c_1 = c_2 = c_3 = 0$. In the general case, assuming for convenience that the free variables are the last k variables,

$$c_1 \begin{bmatrix} * \\ \vdots \\ * \\ 1 \\ 0 \\ 0 \\ \vdots \\ 0 \end{bmatrix} + c_2 \begin{bmatrix} * \\ \vdots \\ * \\ 0 \\ 1 \\ 0 \\ \vdots \\ 0 \end{bmatrix} + \cdots + c_k \begin{bmatrix} * \\ \vdots \\ * \\ 0 \\ 0 \\ 0 \\ \vdots \\ 1 \end{bmatrix} = \mathbf{0}, \qquad (4.11.7)$$

implies that $c_1 = c_2 = \cdots = c_k = 0$.

Example 4.11.1:   Find the basic set of solutions of

$$\begin{bmatrix} 1 & 1 & 1 & 1 \\ 1 & -1 & -1 & 1 \\ 2 & 0 & 0 & 1 \end{bmatrix} \mathbf{x} = \mathbf{0}.$$

SOLUTION: We must first reduce $\mathbf{A}$ to its RENF:

$$\begin{bmatrix} 1 & 1 & 1 & 1 \\ 1 & -1 & -1 & 1 \\ 2 & 0 & 0 & 1 \end{bmatrix} \longrightarrow \begin{bmatrix} 1 & 1 & 1 & 1 \\ 0 & -2 & -2 & 0 \\ 2 & 0 & 0 & 2 \end{bmatrix} \longrightarrow \begin{bmatrix} 1 & 1 & 1 & 1 \\ 0 & 1 & 1 & 0 \\ 2 & 0 & 0 & 2 \end{bmatrix}$$

$$\longrightarrow \begin{bmatrix} 1 & 1 & 1 & 1 \\ 0 & 1 & 1 & 0 \\ 0 & -2 & -2 & 0 \end{bmatrix} \longrightarrow \begin{bmatrix} 1 & 1 & 1 & 1 \\ 0 & 1 & 1 & 0 \\ 0 & 0 & 0 & 0 \end{bmatrix} \longrightarrow \begin{bmatrix} 1 & 0 & 0 & 1 \\ 0 & 1 & 1 & 0 \\ 0 & 0 & 0 & 0 \end{bmatrix}.$$

Since the first two columns are the leading columns, the variables x_1 and x_2 are basic and x_3 and x_4 are free. There are, therefore, two basic solutions. The equations are

$$x_1 + x_4 = 0$$
$$x_2 + x_3 = 0.$$

So set $x_3 = 1$, $x_4 = 0$ and obtain

$$\mathbf{x}_1 = \begin{bmatrix} 0 \\ -1 \\ 1 \\ 0 \end{bmatrix}.$$

Set $x_4 = 1$, $x_3 = 0$ and obtain

$$\mathbf{x}_2 = \begin{bmatrix} -1 \\ 0 \\ 0 \\ 1 \end{bmatrix}.$$

Note the placement of the zeros and the ones among the free variables:

$$\begin{bmatrix} * \\ * \\ 1 \\ 0 \end{bmatrix}, \begin{bmatrix} * \\ * \\ 0 \\ 1 \end{bmatrix}. \quad \blacksquare$$

Theorem 4.8:   *The homogeneous system* $\mathbf{Ax} = \mathbf{0}$ *has the unique solution* $\mathbf{x} = \mathbf{0}$ *if and only if* $\mathbf{A}$ *has full rank.*

PROOF: Suppose that $\mathbf{A}$ is $m \times n$ and $\mathbf{A}$ has full rank. Then rank $\mathbf{A} = n$ and the RENF of $\mathbf{A}$, written $\mathbf{A}_R = \mathbf{U}$, is either

$$\mathbf{U} = \mathbf{I}_n \quad \text{or} \quad \mathbf{U} = \left[\frac{\mathbf{I}_n}{\mathbf{O}} \right]. \tag{4.11.8}$$

In either case $\mathbf{Ux} = \mathbf{0}$ implies that $\mathbf{x} = \mathbf{0}$. Hence, $\mathbf{Ax} = \mathbf{0}$ implies that $\mathbf{x} = \mathbf{0}$ when $\mathbf{A}$ has full rank. Conversely, suppose that rank $\mathbf{A} = r < n$. Then at least one variable is free and there is a basic solution. But basic solutions are never trivial, so if $\mathbf{Ax} = \mathbf{0}$ has only the trivial solution, then rank $\mathbf{A} = n$.

For square matrices, full rank is equivalent to the following:

1. $\mathbf{A}^{-1}$ exists.
2. $|\mathbf{A}| \neq 0$.
3.   The rows or columns of $\mathbf{A}$ are linearly independent.

Let $\mathbf{x}_1, \mathbf{x}_2, \ldots, \mathbf{x}_k$ be the basic set of solutions of $\mathbf{Ax} = \mathbf{0}$. Then

$$\mathbf{x}_h = c_1 \mathbf{x}_1 + c_2 \mathbf{x}_2 + \cdots + c_k \mathbf{x}_k \tag{4.11.9}$$

is a *general solution*. It is understood that Eq. 4.11.9 represents a family of solutions, one for each choice of the k scalars, $c_1, c_2, \ldots, c_k$. To warrant the name "general solution," we need to show two things: first, that $\mathbf{x}_h$ is a solution. But this is trivial, for

$$\mathbf{Ax}_h = \mathbf{A} \sum_{i=1}^{k} c_i \mathbf{x}_i$$

$$= \sum_{i=1}^{k} c_i \mathbf{Ax}_i = \mathbf{0} \tag{4.11.10}$$

since for each i, $\mathbf{Ax}_i = \mathbf{0}$. Second, that for each solution $\mathbf{x}_0$, there is a choice of constants such that

$$\mathbf{x}_0 = \sum_{i=1}^{k} c_i \mathbf{x}_i. \tag{4.11.11}$$

The argument for the second point is more subtle. Let's examine the argument for a particular example.

Example 4.11.2:   Show that

$$\mathbf{x}_0 = \begin{bmatrix} -1 \\ -2 \\ 2 \\ 1 \end{bmatrix}$$

is a solution of the system in Example 4.11.1 and find c_1, c_2 so that

$$\mathbf{x}_0 = c_1\mathbf{x}_1 + c_2\mathbf{x}_2.$$

SOLUTION: First,

$$\begin{bmatrix} 1 & 1 & 1 & 1 \\ 1 & -1 & -1 & 1 \\ -2 & 0 & 0 & 2 \end{bmatrix} \begin{bmatrix} -1 \\ -2 \\ 2 \\ 1 \end{bmatrix} = \begin{bmatrix} 0 \\ 0 \\ 0 \end{bmatrix}.$$

Second, to find c_1 and c_2 we use the following:

$$\begin{bmatrix} -1 \\ -2 \\ 2 \\ 1 \end{bmatrix} = c_1 \begin{bmatrix} * \\ * \\ 1 \\ 0 \end{bmatrix} + c_2 \begin{bmatrix} * \\ * \\ 0 \\ 1 \end{bmatrix}.$$

Clearly, $c_1 = 2$ and $c_2 = 1$ and these are the only choices. Once c_1 and c_2 are fixed, the free variables are determined; namely, $x_3 = c_1$ and $x_4 = c_2$. When the free variables are fixed, the solution is defined.   ■

In general, suppose that the variables $x_{n-k+1}, x_{n-k+2}, \ldots, x_n$ are free— these are the last k variables in $\mathbf{x}$. Let $\mathbf{x}_0$ be a solution of $\mathbf{Ax} = \mathbf{0}$. Say that

$$\mathbf{x}_0 = \begin{bmatrix} * \\ \vdots \\ * \\ \alpha_1 \\ \alpha_2 \\ \vdots \\ \alpha_k \end{bmatrix}. \qquad (4.11.12)$$

Then

$$\mathbf{x}_0 = \begin{bmatrix} * \\ \vdots \\ * \\ \alpha_1 \\ \alpha_2 \\ \vdots \\ \alpha_k \end{bmatrix} = c_1 \begin{bmatrix} * \\ \vdots \\ * \\ 1 \\ 0 \\ \vdots \\ 0 \end{bmatrix} + c_2 \begin{bmatrix} * \\ \vdots \\ * \\ 0 \\ 1 \\ \vdots \\ 0 \end{bmatrix} + \cdots + c_k \begin{bmatrix} * \\ \vdots \\ * \\ 0 \\ 0 \\ \vdots \\ 1 \end{bmatrix}. \qquad (4.11.13)$$

Thus, $c_1 = \alpha_1$, $c_2 = \alpha_2$, $\ldots$, $c_k = \alpha_k$ and the basic variables take care of themselves.

Thus, for $\mathbf{Ax} = \mathbf{0}$, we have the following principle: *The general solution is a family of solutions containing every solution of* $\mathbf{Ax} = \mathbf{0}$. In the event that $\mathbf{A}$ has full rank, the general solution is a set with a single member, $\mathbf{x} = \mathbf{0}$.

Although we do not make explicit use of the fact, in some more complete accounts, attention is paid to the number of basic solutions. This number is the *nullity* of $\mathbf{A}$, written $\eta(\mathbf{A})$, and abbreviated by η. Since η is exactly the number of free columns, and r (the rank of $\mathbf{A}$) is the number of leading columns,

$$\eta + r = n, \tag{4.11.14}$$

where $\mathbf{A}$ is $m \times n$; or

$$\eta(\mathbf{A}) + \text{rank } \mathbf{A} = \text{number of columns of } \mathbf{A}. \tag{4.11.15}$$

PROBLEMS

In the following problems, $\mathbf{A}$ is a square matrix.

1. Prove: $\mathbf{A}^{-1}$ exists if and only if $\mathbf{A}$ has full rank.

2. Prove: $|\mathbf{A}| \neq 0$ if and only if $\mathbf{A}$ has full rank.

3. Prove: The rows (or columns) of $\mathbf{A}$ are a linearly independent set if and only if $\mathbf{A}$ has full rank.

4. Matrix multiplication is a linear operator. Show that this is the case by explaining why

$$\mathbf{A}(\alpha\mathbf{x}_1 + \beta\mathbf{x}_2) = \alpha\mathbf{A}\mathbf{x}_1 + \beta\mathbf{A}\mathbf{x}_2.$$

5. Use the result of Problem 4 to establish this theorem: if $\mathbf{x}_1, \mathbf{x}_2, \ldots, \mathbf{x}_n$ are solutions of $\mathbf{A}\mathbf{x} = \mathbf{0}$, then so is every linear combination of these solutions.

6. If $\mathbf{A}$ has full rank, explain why the number of rows is not greater than the number of columns.

In each problem find a basic solution set, a general solution, and verify Eq. 4.11.5.

7. $x_1 + x_2 - x_3 + x_4 = 0$

8. $x_1 - x_2 + x_3 + x_4 = 0$
  $x_4 = 0$

9. $x_1 - x_2 + x_3 + x_4 = 0$
  $x_3 - 2x_4 = 0$

10. $x_1 = x_2 = x_3 = 0$

11. $x_1 + x_2 = 0$
  $x_1 - x_2 = 0$

12. $x_1 + x_2 + x_3 = 0$
  $x_2 + x_3 = 0$

13. $x_1 - x_2 + x_3 = 0$
  $x_1 + 2x_2 - x_3 = 0$

14. $x_1 = 0$
  $x_1 + x_2 + x_3 = 0$

15. $x_1 + x_2 - x_3 = 0$
  $3x_1 + 4x_2 - x_3 = 0$
  $x_1 + 2x_2 + x_3 = 0$

16. $[1, 1, \ldots, 1]\mathbf{x} = \mathbf{0}$

17. $\mathbf{J}\mathbf{x} = \mathbf{0}$, where $\mathbf{J}$ is an $n \times n$ matrix of all 1's.

18. $\mathbf{u}\mathbf{u}^T\mathbf{x} = \mathbf{0}$, $\mathbf{u} \neq \mathbf{0}$

19. $\mathbf{u}^T\mathbf{u}\mathbf{x} = \mathbf{0}$

20. $\mathbf{u}^T\mathbf{v}\mathbf{x} = \mathbf{0}$, $\mathbf{u}^T\mathbf{v} \neq 0$

21. $\mathbf{u}\mathbf{v}^T\mathbf{x} = \mathbf{0}$, $\mathbf{u} \neq \mathbf{0}$, $\mathbf{v} \neq \mathbf{0}$

22. If $\mathbf{A}^{-1}$ exists, explain why $\mathbf{A}$ has no free variables. Use Eq. 4.11.5.

4.12 Nonhomogeneous Equations

The theory, associated with the solution of the nonhomogeneous system

$$\mathbf{Ax} = \mathbf{r}, \qquad \mathbf{r} \neq \mathbf{0} \qquad (4.12.1)$$

parallels the corresponding theory associated with the solution of linear, nonhomogeneous differential equations. We find a general solution, $\mathbf{x}_h$, of the associated homogeneous system

$$\mathbf{Ax} = \mathbf{0} \qquad (4.12.2)$$

and add to it a particular solution of Eq. 4.12.1. We must first take care of one minor difference between these theories; system 4.12.1 may be inconsistent. We can check this by comparing the ranks of $\mathbf{A}$ and $[\mathbf{A} \vdots \mathbf{r}]$ (see Section 4.4). So assume that Eq. 4.12.1 is consistent and that $\mathbf{x}_p$ is a "particular" solution; that is,

$$\mathbf{Ax}_p = \mathbf{r}. \qquad (4.12.3)$$

Let $\mathbf{x}_h$ be the general solution of $\mathbf{Ax} = \mathbf{0}$. Then,

$$\mathbf{A}(\mathbf{x}_p + \mathbf{x}_h) = \mathbf{Ax}_p + \mathbf{Ax}_h = \mathbf{r} + \mathbf{0} = \mathbf{r}. \qquad (4.12.4)$$

So, as expected, the family $\{\mathbf{x}_h + \mathbf{x}_p\}$ is a set of solutions. We must show that it contains all solutions. This is surprisingly easy. Let $\mathbf{x}_0$ be any particular solution so that

$$\mathbf{Ax}_0 = \mathbf{r}. \qquad (4.12.5)$$

Now

$$\mathbf{A}(\mathbf{x}_0 - \mathbf{x}_p) = \mathbf{Ax}_0 - \mathbf{Ax}_p$$
$$= \mathbf{r} - \mathbf{r} = \mathbf{0}. \qquad (4.12.6)$$

Hence, $\mathbf{x}_0 - \mathbf{x}_p$ is a solution of the associated homogeneous system. By the results of Section 4.11,

$$\mathbf{x}_0 - \mathbf{x}_p = c_1\mathbf{x}_1 + c_2\mathbf{x}_2 + \cdots + c_k\mathbf{x}_k \qquad (4.12.7)$$

and hence,

$$\mathbf{x}_0 = \mathbf{x}_p + (c_1\mathbf{x}_1 + c_2\mathbf{x}_2 + \cdots + c_k\mathbf{x}_k). \qquad (4.12.8)$$

That is, $\mathbf{x}_0$ is a member of $\{\mathbf{x}_p + \mathbf{x}_h\}$.

We conclude this section with an observation on constructing particular solutions. A *fundamental solution* of $\mathbf{Ax} = \mathbf{r}$ is the solution obtained by setting all free variables of $\mathbf{Ax}$ equal to zero.

Example 4.12.1: Find the fundamental solution and the general solution of

$$x_1 + x_2 + x_3 = 1$$
$$2x_1 + 2x_2 + 2x_3 = 2$$
$$3x_1 + 3x_2 + 3x_3 = 3.$$

SOLUTION: We have

$$\begin{bmatrix} 1 & 1 & 1 & 1 \\ 2 & 2 & 2 & 2 \\ 3 & 3 & 3 & 3 \end{bmatrix} \longrightarrow \begin{bmatrix} 1 & 1 & 1 & 1 \\ 0 & 0 & 0 & 0 \\ 0 & 0 & 0 & 0 \end{bmatrix}.$$

Hence, x_1 is a basic variable and x_2, x_3 are free. The associated homogeneous system, in RENF, is

$$x_1 + x_2 + x_3 = 0.$$

Thus, there are two basic solutions:

$$\mathbf{x}_1 = \begin{bmatrix} -1 \\ 1 \\ 0 \end{bmatrix}, \quad \mathbf{x}_2 = \begin{bmatrix} -1 \\ 0 \\ 1 \end{bmatrix}.$$

The fundamental solution is obtained by setting $x_2 = x_3 = 0$ in

$$x_1 + x_2 + x_3 = 1.$$

Therefore,

$$\mathbf{x}_p = \begin{bmatrix} 1 \\ 0 \\ 0 \end{bmatrix}.$$

The general solution is

$$\mathbf{x} = \mathbf{x}_h + \mathbf{x}_p = c_1 \begin{bmatrix} -1 \\ 1 \\ 0 \end{bmatrix} + c_2 \begin{bmatrix} -1 \\ 0 \\ 1 \end{bmatrix} + \begin{bmatrix} 1 \\ 0 \\ 0 \end{bmatrix}. \quad \blacksquare$$

Example 4.12.2:   Find the fundamental solution of

$$\begin{bmatrix} 1 & 2 & 0 & -1 & 1 & 1 & 0 & 3 \\ 0 & 0 & 1 & 1 & 0 & 0 & 0 & -1 \\ 0 & 0 & 0 & 0 & 0 & 0 & 1 & 2 \\ 0 & 0 & 0 & 0 & 0 & 0 & 0 & 0 \end{bmatrix} \mathbf{x} = \begin{bmatrix} 3 \\ -5 \\ 7 \\ 0 \end{bmatrix}.$$

SOLUTION: Since the coefficient matrix is already in RENF, we can identify the free variables by inspection. They are x_2, x_4, x_5, x_6, x_8. We set these zero and see instantly that $x_1 = 3$, $x_3 = -5$, $x_7 = 7$. So

$$\mathbf{x}_p = \begin{bmatrix} 3 \\ 0 \\ -5 \\ 0 \\ 0 \\ 0 \\ 7 \\ 0 \end{bmatrix}$$

is the fundamental solution.

As an additional note: there are $\eta = 5$ basic solutions, so the general solution has the form

$$\mathbf{x} = \mathbf{x}_p + c_1\mathbf{x}_1 + c_2\mathbf{x}_2 + c_3\mathbf{x}_3 + c_4\mathbf{x}_4 + c_5\mathbf{x}_5. \quad \blacksquare$$

PROBLEMS

1. Find a general solution of the system in Example 4.12.2.

Find a particular solution of each system.

2. $x + y + z = 1$
$\quad x - y \quad = 1$

3. $x - y = 1$

4. $\quad x + y + z = -1$
$\quad x - y - z = \quad 0$
$\quad 2x \qquad\quad = \quad 0$

5. $x + y + z + t = 1$
$\qquad y + z + t = 1$
$\qquad\quad z + t = 1$

6. $x - y + \quad z - t = -1$
$\quad x - y + 2z - t = \quad 1$

7. Find a general solution for each choice of b_1, b_2 of

$$x + y + z = b_1$$
$$x - y \quad = b_2.$$

Find a general solution of each system.

8. $x_1 + x_2 - x_3 + x_4 = 1$

9. $x_1 - x_2 + x_3 + x_4 = \quad 1$
$\qquad\qquad\qquad\quad x_4 = -1$

10. $x_1 - x_2 + x_3 + \quad x_4 = 0$
$\qquad\qquad\quad x_3 - 2x_4 = 2$

11. $x_1 + x_2 + x_3 = -1$
$\qquad\quad x_2 + x_3 = \quad 0$

12. $x_1 - \quad x_2 + x_3 = 1$
$\quad x_1 + 2x_2 - x_3 = 1$

13. $x_1 + x_2 + x_3 = 0$
$\quad x_1 \qquad\qquad = 1$

14. $\mathbf{J}\mathbf{x} = \begin{bmatrix} 1 \\ 1 \\ \vdots \\ 1 \end{bmatrix}$ (see Problem 12 in Section 4.11)

15. If $\mathbf{x}_1$ and $\mathbf{x}_2$ are solutions of $\mathbf{A}\mathbf{x} = \mathbf{b}$, show that $\mathbf{x}_1 - \mathbf{x}_2$ is a solution of $\mathbf{A}\mathbf{x} = \mathbf{0}$.

16. If $\mathbf{x}_1, \mathbf{x}_2, \ldots, \mathbf{x}_k$ are solutions of $\mathbf{A}\mathbf{x} = \mathbf{b}$, show that $\alpha_1\mathbf{x}_1 + \alpha_2\mathbf{x}_2 + \cdots + \alpha_n\mathbf{x}_n$ is also a solution of $\mathbf{A}\mathbf{x} = \mathbf{b}$ if and only if $\alpha_1 + \alpha_2 + \cdots + \alpha_n = 1$.

17. If $\mathbf{A}$ is nonsingular, explain why the general solution of $\mathbf{A}\mathbf{x} = \mathbf{b}$ is $\mathbf{x}_0 = \mathbf{A}^{-1}\mathbf{b}$. *Hint:* Show that $\mathbf{A}^{-1}\mathbf{b}$ is a solution and the only solution.

5

Matrix Applications

5.1 Introduction

In this chapter we study two of the many important applications of the theory of matrices: the method of "least squares" and the solution of systems of linear differential equations. To accomplish these aims we need to introduce the notions of length and direction and solve the so-called eigenvalue problem.

Section 5.2 is essential to both problems; but the reader may skip Sections 5.3 and 5.4 if only the solutions of systems of differential equations are of interest. If neither application is of interest, one may still profit from a study of Sections 5.2, 5.5, and 5.6.

5.2 Norms and Inner Products

The vector $\mathbf{x} = \begin{bmatrix} x_1 \\ x_2 \end{bmatrix}$ is represented geometrically as the directed line segment from O: $(0, 0)$ to P: (x_1, x_2). The length of $\mathbf{x}$ is the length of this line segment; that is,

$$\text{length of } \mathbf{x} = \sqrt{x_1^2 + x_2^2}. \tag{5.2.1}$$

In n dimensions the *norm* of $\mathbf{x}$, written $\|\mathbf{x}\|$, is defined as

$$\|\mathbf{x}\| = \left(\sum_{k=1}^{n} x_k^2 \right)^{1/2} \tag{5.2.2}$$

where **x** is the n-dimensional vector

$$\mathbf{x} = \begin{bmatrix} x_1 \\ x_2 \\ \vdots \\ x_n \end{bmatrix}. \tag{5.2.3}$$

Thus, the norm of **x** is a generalization to n dimensions of the two-dimensional notion of length. In one dimension $\|\mathbf{x}\| = \text{abs}(x_1)$, the absolute value of x_1. In three dimensions $\|\mathbf{x}\|$ is the length of the directed line segment form O: $(0, 0, 0)$ to P: (x_1, x_2, x_3); see Fig. 5.1.

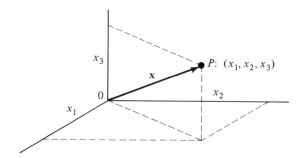

FIGURE 5.1.   A vector in three dimensions.

There are three immediate consequences of the definition of norm. For all **x** and each scalar k:

(i) $\|\mathbf{x}\| \geq 0$

(ii) $\|\mathbf{x}\| = 0$ if and only if $\mathbf{x} = \mathbf{0}$ (5.2.4)

(iii) $\|k\mathbf{x}\| = \text{abs}(k)\|\mathbf{x}\|.$

Property (iii) is proved as follows; by definition we have

$$\|k\mathbf{x}\|^2 = \sum_{i=1}^{n} (kx_i)^2$$

$$= k^2 \sum_{i=1}^{n} x_i^2 = k^2 \|\mathbf{x}\|^2. \tag{5.2.5}$$

Since $\sqrt{k^2} = \text{abs}(k)$, (iii) is proved.

A concept closely related to the norm of **x** is the *inner product* of **x** and **y**, written $\langle \mathbf{x}, \mathbf{y} \rangle$; that is,

$$\langle \mathbf{x}, \mathbf{y} \rangle = \sum_{i=1}^{n} x_i y_i. \tag{5.2.6}$$

Like the norm, $\langle \mathbf{x}, \mathbf{y} \rangle$ is a scalar. A common alternative notation for the inner product is $\mathbf{x} \cdot \mathbf{y}$, which is read "the *dot product* of $\mathbf{x}$ and $\mathbf{y}$." So

$$\langle \mathbf{x}, \mathbf{y} \rangle = \mathbf{x} \cdot \mathbf{y}. \qquad (5.2.7)$$

An extremely useful observation results from the identity

$$\mathbf{x}^T \mathbf{y} = [\mathbf{x} \cdot \mathbf{y}]_{1 \times 1}. \qquad (5.2.8)$$

We usually drop the brackets around $\mathbf{x} \cdot \mathbf{y}$ and write Eq. 5.2.8 in the technically incorrect way

$$\mathbf{x}^T \mathbf{y} = \mathbf{x} \cdot \mathbf{y} = \langle \mathbf{x}, \mathbf{y} \rangle. \qquad (5.2.9)$$

Normally, no confusion results from this notational abuse. It is easy to prove the following:

(i) $\|\mathbf{x}\|^2 = \langle \mathbf{x}, \mathbf{x} \rangle = \mathbf{x}^T \mathbf{x}$

(ii) $\langle \mathbf{x}, \mathbf{y} \rangle = \langle \mathbf{y}, \mathbf{x} \rangle$

(iii) $\langle k\mathbf{x}, \mathbf{y} \rangle = \langle \mathbf{x}, k\mathbf{y} \rangle = k\langle \mathbf{x}, \mathbf{y} \rangle$ $\qquad (5.2.10)$

(iv) $\langle \mathbf{x} + \mathbf{y}, \mathbf{z} \rangle = \langle \mathbf{x}, \mathbf{z} \rangle + \langle \mathbf{y}, \mathbf{z} \rangle.$

We illustrate the proofs by writing out the details that establish (iv). We have

$$\begin{aligned} \langle \mathbf{x} + \mathbf{y}, \mathbf{z} \rangle &= (\mathbf{x} + \mathbf{y})^T \mathbf{z} \\ &= (\mathbf{x}^T + \mathbf{y}^T)\mathbf{z} \\ &= \mathbf{x}^T \mathbf{z} + \mathbf{y}^T \mathbf{z} = \langle \mathbf{x}, \mathbf{z} \rangle + \langle \mathbf{y}, \mathbf{z} \rangle. \end{aligned} \qquad (5.2.11)$$

Properties (i) and (ii) lead to the deeper result which we now state as a theorem.

Theorem 5.1:   *If $\mathbf{A}$ has full rank, where $\mathbf{A}$ is $m \times n$, then $\mathbf{A}^T\mathbf{A}$ is invertible.*

PROOF: Although $\mathbf{A}$ need not be square, $\mathbf{A}^T\mathbf{A}$ is $n \times n$ and it is sensible to ask whether $\mathbf{A}^T\mathbf{A}$ is singular or not. Now $\mathbf{A}^T\mathbf{A}$ is invertible if and only if $\mathbf{A}^T\mathbf{A}\mathbf{x} = \mathbf{0}$ has only the trivial solution $\mathbf{x} = \mathbf{0}$. So, by way of contradiction, suppose that

$$\mathbf{A}^T\mathbf{A}\mathbf{x}_0 = \mathbf{0}, \qquad \mathbf{x}_0 \neq \mathbf{0}. \qquad (5.2.12)$$

Then, multiplying on the left by $\mathbf{x}_0$, we obtain

$$\mathbf{x}_0^T \mathbf{A}^T\mathbf{A}\mathbf{x}_0 = \mathbf{x}_0^T \mathbf{0} = 0. \qquad (5.2.13)$$

However, $\mathbf{x}_0^T \mathbf{A}^T\mathbf{A}\mathbf{x}_0 = (\mathbf{A}\mathbf{x}_0)^T \mathbf{A}\mathbf{x}_0 = \|\mathbf{A}\mathbf{x}_0\|^2$ by (i) of Eqs. 5.2.10. Thus Eq. 5.2.13 asserts

$$\|\mathbf{A}\mathbf{x}_0\|^2 = 0 \qquad (5.2.14)$$

which, by property (ii) of Eqs. 5.2.4, asserts that $\mathbf{A}\mathbf{x}_0 = \mathbf{0}$. But $\mathbf{A}$ has full rank. Hence, $\mathbf{Ax_0} = \mathbf{0}$ implies that $\mathbf{x_0} = \mathbf{0}$, contradicting Eq. 5.2.12. Therefore, $A^T A$ is invertible.

Another interesting consequence of Eq. 5.2.9 is

$$\langle \mathbf{Ax}, \mathbf{y} \rangle = \langle \mathbf{x}, A^T\mathbf{y} \rangle. \tag{5.2.15}$$

For, by Eq. 5.2.7,

$$\langle \mathbf{Ax}, \mathbf{y} \rangle = (\mathbf{Ax})^T\mathbf{y}$$
$$= \mathbf{x}^T A^T\mathbf{y} = \langle \mathbf{x}, A^T\mathbf{y} \rangle. \tag{5.2.16}$$

Just as norm generalizes length, inner product generalizes direction. For, let $\mathbf{x}$ and $\mathbf{y}$ be the vectors

$$\mathbf{x} = \begin{bmatrix} x_1 \\ x_2 \end{bmatrix}, \qquad \mathbf{y} = \begin{bmatrix} y_1 \\ y_2 \end{bmatrix} \tag{5.2.17}$$

and let $P: (x_1, x_2)$, $Q: (y_1, y_2)$ denote the points at the ends of $\mathbf{x}$ and $\mathbf{y}$, respectively (see Fig. 5.2). The side $\mathbf{L}$ in the triangle QOP of Fig. 5.2 has length equal to the norm of $\mathbf{y} - \mathbf{x}$. So, by the law of cosines,

$$\| \mathbf{y} - \mathbf{x} \|^2 = \| \mathbf{x} \|^2 + \| \mathbf{y} \|^2 - 2 \| \mathbf{x} \| \| \mathbf{y} \| \cos \theta. \tag{5.2.18}$$

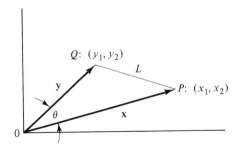

FIGURE 5.2.   A triangle with sides **x**, **y**, and *L*.

However,

$$\| \mathbf{y} - \mathbf{x} \|^2 = \langle \mathbf{y} - \mathbf{x}, \mathbf{y} - \mathbf{x} \rangle$$
$$= \langle \mathbf{y}, \mathbf{y} \rangle + \langle -\mathbf{x}, \mathbf{y} \rangle + \langle \mathbf{y}, -\mathbf{x} \rangle + \langle -\mathbf{x}, -\mathbf{x} \rangle$$
$$= \| \mathbf{y} \|^2 - 2\langle \mathbf{x}, \mathbf{y} \rangle + \| \mathbf{x} \|^2. \tag{5.2.19}$$

Comparing this expression for $\| \mathbf{y} - \mathbf{x} \|^2$ to the one given in Eq. 5.2.18, we deduce that

$$\langle \mathbf{x}, \mathbf{y} \rangle = \| \mathbf{x} \| \| \mathbf{y} \| \cos \theta. \tag{5.2.20}$$

Although this equation was derived assuming $\mathbf{x}$ and $\mathbf{y}$ to be two-dimensional vectors, it is trivially true if $\mathbf{x}$ and $\mathbf{y}$ are one-dimensional, and easy to prove if they are three-dimensional. For $n > 3$, we use Eq. 5.2.20 to define the cosine of the angle between **x** and **y**. Moreover, if **x** is perpendicular to **y**, $\cos \theta = 0$ and hence $\langle \mathbf{x}, \mathbf{y} \rangle = 0$. This motivates the definition of "orthogonality" in n-dimensions. We say that **x** is *orthogonal* to **y** and write $\mathbf{x} \perp \mathbf{y}$ if $\langle \mathbf{x}, \mathbf{y} \rangle = 0$ and hence, the zero vector is orthogonal to all vectors. It is the only such vector, for if **x** is orthogonal to every vector it is orthogonal to itself and hence $\langle \mathbf{x}, \mathbf{x} \rangle = 0$; but $\langle \mathbf{x}, \mathbf{x} \rangle = \|\mathbf{x}\|^2$ and therefore $\mathbf{x} = \mathbf{0}$.

Since abs $(\cos \theta) < 1$, Eq. 5.2.20 suggests* the inequality

$$\text{abs } \langle \mathbf{x}, \mathbf{y} \rangle \leq \|\mathbf{x}\| \|\mathbf{y}\|, \tag{5.2.21}$$

called the *Cauchy–Schwarz inequality*.

A theorem that is familiar to us all is the *Pythagorean theorem*. It states that

$$\|\mathbf{x} + \mathbf{y}\|^2 = \|\mathbf{x}\|^2 + \|\mathbf{y}\|^2 \tag{5.2.22}$$

if and only if $\mathbf{x} \perp \mathbf{y}$. An example will contain its proof.

Example 5.2.1:   Prove the Pythagorean theorem.

SOLUTION: Since

$$\begin{aligned}
\|\mathbf{x} + \mathbf{y}\|^2 &= \langle \mathbf{x} + \mathbf{y}, \mathbf{x} + \mathbf{y} \rangle \\
&= \langle \mathbf{x}, \mathbf{x} \rangle + 2\langle \mathbf{x}, \mathbf{y} \rangle + \langle \mathbf{y}, \mathbf{y} \rangle \\
&= \|\mathbf{x}\|^2 + 2\langle \mathbf{x}, \mathbf{y} \rangle + \|\mathbf{y}\|^2,
\end{aligned}$$

Eq. 5.2.22 follows if and only if $\langle \mathbf{x}, \mathbf{y} \rangle = 0$, and hence $\mathbf{x} \perp \mathbf{y}$.   ■

Example 5.2.2:   Compute the norms of

$$\mathbf{x} = \begin{bmatrix} 1 \\ 1 \\ -1 \\ 2 \end{bmatrix}, \qquad \mathbf{y} = \begin{bmatrix} 3 \\ -1 \\ 0 \\ -1 \end{bmatrix}.$$

Verify that $\mathbf{x} \perp \mathbf{y}$ and that Eq. 5.2.22 holds.

SOLUTION: We compute

$$\|\mathbf{x}\|^2 = 1 + 1 + 1 + 4 = 7$$
$$\|\mathbf{y}\|^2 = 9 + 1 + 0 + 1 = 11$$
$$\|\mathbf{x} + \mathbf{y}\|^2 = 16 + 0 + 1 + 1 = 18.$$

Thus, $\|\mathbf{x} + \mathbf{y}\|^2 = \|\mathbf{x}\|^2 + \|\mathbf{y}\|^2$. Also, $\langle \mathbf{x}, \mathbf{y} \rangle = 3 - 1 + 0 - 2 = 0$.   ■

Example 5.2.3:   For every pair of nonzero scalars α and β, $\alpha\mathbf{x} \perp \beta\mathbf{y}$ if and only if $\mathbf{x} \perp \mathbf{y}$.

---

*A simple proof, valid for n-dimensions, is outlined in Problem 29 of this section.

SOLUTION: We have

$$\langle \alpha \mathbf{x}, \beta \mathbf{y} \rangle = \alpha \beta \langle \mathbf{x}, \mathbf{y} \rangle,$$

from which the conclusion follows.   ■

Example 5.2.4   (The triangle equality): Show that the third side of a triangle has shorter length than the sum of the lengths of the other two sides, as shown.

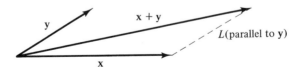

SOLUTION: If $\mathbf{x}$ and $\mathbf{y}$ are two sides of a triangle, $\mathbf{x} + \mathbf{y}$ is the third side. But

$$\begin{aligned}
\|\mathbf{x} + \mathbf{y}\|^2 &= \|\mathbf{x}\|^2 + \|\mathbf{y}\|^2 + 2\langle \mathbf{x}, \mathbf{y} \rangle \\
&\leq \|\mathbf{x}\|^2 + \|\mathbf{y}\|^2 + 2\|\mathbf{x}\|\,\|\mathbf{y}\| \\
&= (\|\mathbf{x}\| + \|\mathbf{y}\|)^2
\end{aligned}$$

by the Cauchy–Schwarz inequality. Hence,

$$\|\mathbf{x} + \mathbf{y}\| \leq \|\mathbf{x}\| + \|\mathbf{y}\|,$$

and this is the required inequality.   ■

PROBLEMS

1. Verify (i) and (ii) in properties 5.2.4.

2. Verify (i)–(iii) in properties 5.2.10.

3. Show that $\|\mathbf{y} - \mathbf{x}\|^2 = \|\mathbf{y}\|^2 + \|\mathbf{x}\| - 2\langle \mathbf{x}, \mathbf{y} \rangle$ by using definitions 5.2.2 and 5.2.6.

4. Prove the Cauchy–Schwarz inequality 5.2.21 from the definitions 5.2.2 and 5.2.6 when $\mathbf{x}$ and $\mathbf{y}$ have only two components.

Find the norm.

5. $\begin{bmatrix} 1 \\ -1 \\ 1 \end{bmatrix}$

6. $\mathbf{0}$

7. $\begin{bmatrix} 1 \\ 1 \\ \vdots \\ 1 \end{bmatrix}$

8. $\mathbf{e}_k$

9. $\begin{bmatrix} 1/\sqrt{3} \\ -1/\sqrt{3} \\ \sqrt{3}/3 \end{bmatrix}$

Find the inner product.

10. $\left\langle \begin{bmatrix} 1 \\ -1 \\ 1 \end{bmatrix}, \begin{bmatrix} 0 \\ 1 \\ 1 \end{bmatrix} \right\rangle$

11. $\left\langle \begin{bmatrix} \cos\theta \\ \sin\theta \end{bmatrix}, \begin{bmatrix} -\sin\theta \\ \cos\theta \end{bmatrix} \right\rangle$

12. $\langle \mathbf{e}_i, \mathbf{e}_j \rangle \; i \ne j$

13. $\langle \mathbf{e}_i, \mathbf{e}_i \rangle$

14. $\langle \mathbf{x}, \mathbf{x} \rangle$

15. $\left\langle \begin{bmatrix} \cos\theta \\ \sin\theta \end{bmatrix}, \begin{bmatrix} \cos\theta \\ \sin\theta \end{bmatrix} \right\rangle$

16. $\langle \mathbf{0}, \mathbf{x} \rangle$

17. $\langle \mathbf{e}_i, \mathbf{x} \rangle$

18. $\left\langle \begin{bmatrix} \sqrt{y} \\ \sqrt{x} \end{bmatrix}, \begin{bmatrix} \sqrt{x} \\ \sqrt{y} \end{bmatrix} \right\rangle$

19. Apply the Cauchy–Schwarz inequality to $\left\langle \begin{bmatrix} \sqrt{y} \\ \sqrt{x} \end{bmatrix}, \begin{bmatrix} \sqrt{x} \\ \sqrt{y} \end{bmatrix} \right\rangle$ and thereby deduce that

$$\sqrt{xy} \le \frac{x+y}{2}, \qquad 0 \le x, \quad 0 \le y.$$

20. If $\mathbf{x} \perp \mathbf{y}$ and $\mathbf{y} \perp \mathbf{z}$, does it follow that $\mathbf{x} \perp \mathbf{z}$? Explain.

21. If $\mathbf{x} \perp \mathbf{y}$ and $\mathbf{y} \perp \mathbf{z}$, is $\mathbf{y} \perp \mathbf{z}$? Explain.

22. If $\mathbf{x} \perp \mathbf{y}$ and $\mathbf{x} \perp \mathbf{z}$, is $\mathbf{x} \perp (\mathbf{y} + \mathbf{z})$? Explain.

23. Find α so that $(\mathbf{b} - \alpha\mathbf{u}) \perp \mathbf{u}$.

24. Suppose that $\|\mathbf{u}\| = 1$. Show that $(\mathbf{u}\mathbf{u}^T)^2 = \mathbf{u}\mathbf{u}^T$.

25. Show that every solution of $\mathbf{A}\mathbf{x} = \mathbf{b}$ is orthogonal to every solution of $\mathbf{y}^T\mathbf{A} = \mathbf{0}^T$.

26. Suppose that $\mathbf{b} = \Sigma\, a_i\mathbf{x}_i$ and $\mathbf{u} \perp \mathbf{x}_i$ for each i. Show that $\mathbf{u} \perp \mathbf{b}$.

Suppose that $\mathbf{x}$ is not a multiple of $\mathbf{y}$.

27. Show that for each real λ,

$$\|\mathbf{x} + \lambda\mathbf{y}\|^2 = \|\mathbf{x}\|^2 + 2\lambda\langle\mathbf{x}, \mathbf{y}\rangle + \lambda^2\|\mathbf{y}\|^2.$$

28. Set $a = \|\mathbf{y}\|^2$, $b = \langle\mathbf{x}, \mathbf{y}\rangle$, $c = \|\mathbf{x}\|^2$ and using Problem 27 explain why $a\lambda^2 + 2b\lambda + c$ never vanishes (as a function of λ), or is identically zero.

29. Conclude from No. 28 that $b^2 - ac < 0$ and thus prove the Cauchy–Schwarz inequality (Eq. 5.2.21), if $\mathbf{x}$ is not a multiple of $\mathbf{y}$.

30. Verify that Eq. 5.2.21 is an equality if $\mathbf{x} = k\mathbf{y}$.

5.3   Orthogonal Sets and Matrices

The set $\{\mathbf{x}_1, \mathbf{x}_2, \ldots, \mathbf{x}_k\}$ is *orthogonal* if the vectors are mutually orthogonal; that is, for each i and j,

$$\mathbf{x}_i \perp \mathbf{x}_j, \qquad i \ne j. \tag{5.3.1}$$

An *orthonormal* set, $\{q_1, q_2, \ldots, q_k\}$ is an orthogonal set in which each vector has norm one. So $\{q_1, q_2, \ldots q_k\}$ is orthonormal if

$$\langle q_i, q_j \rangle = q_i \cdot q_j = \delta_{ij} \tag{5.3.2}$$

where δ_{ij} is the *Kronecker delta*: $\delta_{ij} = 0$ if $i \neq j$ and $\delta_{ij} = 1$ if $i = j$. The unit vectors $e_i, e_2, \ldots, e_k$ in the directions of $x_1, x_2, \ldots, x_k$, respectively, are the most natural orthonormal set.

The pair

$$x = \begin{bmatrix} 1 \\ 1 \\ 1 \end{bmatrix}, \qquad y = \begin{bmatrix} 1 \\ 0 \\ -1 \end{bmatrix} \tag{5.3.3}$$

form an orthogonal, but not orthonormal, set. If each of these vectors is divided by length, the resulting pair does form an orthonormal set. In that case,

$$\frac{x}{\|x\|} = \frac{1}{\sqrt{3}} \begin{bmatrix} 1 \\ 1 \\ 1 \end{bmatrix}, \qquad \frac{y}{\|y\|} = \frac{1}{\sqrt{2}} \begin{bmatrix} 1 \\ 0 \\ -1 \end{bmatrix} \tag{5.3.4}$$

since $\|x\|^2 = 3$ *and* $\|y\|^2 = 2$. This modest example illustrates a general principle: *If* $\{v_1, v_2, \ldots, v_k\}$ *is an orthogonal set of nonzero vectors, then the set* $\{q_1, q_2, \ldots, q_k\}$, *where* $q_i = v_i/\|v_i\|$, *is an orthonormal set.*

Let Q be a square matrix whose columns form an orthonormal set. By the definitions of orthogonality and matrix multiplication,

$$QQ^T = Q^TQ = I \tag{5.3.5}$$

and hence Q is nonsingular and $Q^T = Q^{-1}$. Such matrices are called *orthogonal*.* Orthogonal matrices have many interesting properties. Some are illustrated in the examples, and others are included in the Problems.

Example 5.3.1:   Show that

$$Q = \begin{bmatrix} \cos \theta & -\sin \theta \\ \sin \theta & \cos \theta \end{bmatrix}$$

is an orthogonal matrix for all θ.

SOLUTION: Clearly, we can write

$$QQ^T = \begin{bmatrix} \cos \theta & -\sin \theta \\ \sin \theta & \cos \theta \end{bmatrix} \begin{bmatrix} \cos \theta & \sin \theta \\ -\sin \theta & \cos \theta \end{bmatrix} = I.$$

---

*A more appropriate name would have been orthonormal.

Similarly,

$$Q^T Q = I.$$

Thus, Q is orthogonal.   ∎

Example 5.3.2:   If Q is orthogonal, show that $|Q| = \pm 1$.

SOLUTION: From Eq. 5.3.5 we have

$$|QQ^T| = |I| = 1.$$

But

$$|QQ^T| = |Q||Q^T| = |Q|^2$$

since

$$|Q^T| = |Q|.$$

Therefore,

$$|Q|^2 = 1.   ∎$$

Example 5.3.3:   For each x and y show that

$$\langle Qx, Qy \rangle = \langle x, y \rangle.$$

SOLUTION: We have

$$\langle Qx, Qy \rangle = \langle x, Q^T Qy \rangle$$
$$= \langle x, y \rangle.   ∎$$

Example 5.3.4:   For each x, show that

$$\| Qx \| = \| x \|.$$

SOLUTION: Substitute $y = x$ in (see Example 5.3.3)

$$\langle Qx, Qy \rangle = \langle x, y \rangle$$

and recall the formula

$$\| x \|^2 = \langle x, x \rangle.$$

Thus,

$$\langle Qx, Qx \rangle = \langle x, x \rangle$$

or

$$\| Qx \| = \| x \|.   ∎$$

Example 5.3.5:   Show that

$$\mathbf{b} = \begin{bmatrix} 2 \\ 0 \\ 1 \\ 1 \end{bmatrix}$$

is a linear combination of the orthonormal vectors

$$\mathbf{q}_1 = \frac{1}{2} \begin{bmatrix} 1 \\ 1 \\ 1 \\ 1 \end{bmatrix}, \qquad \mathbf{q}_2 = \frac{1}{2} \begin{bmatrix} 1 \\ -1 \\ -1 \\ 1 \end{bmatrix}, \qquad \mathbf{q}_3 = \frac{1}{2} \begin{bmatrix} -1 \\ 1 \\ -1 \\ 1 \end{bmatrix}.$$

SOLUTION: We need to find scalars x_1, x_2, x_3 such that

$$x_1\mathbf{q}_1 + x_2\mathbf{q}_2 + x_3\mathbf{q}_3 = \mathbf{b} \tag{1}$$

There are two apparently different but essentially equivalent methods. We illustrate both.

Method 1. Set

$$\mathbf{Q} = [\mathbf{q}_1, \mathbf{q}_2, \mathbf{q}_3], \qquad \mathbf{x} = \begin{bmatrix} x_1 \\ x_2 \\ x_3 \end{bmatrix}.$$

Then (1) may be written

$$\mathbf{Qx} = \mathbf{b}.$$

Since $\mathbf{Q}$ is not a square matrix, $\mathbf{Q}$ is not orthogonal. However, $\mathbf{Q}^T\mathbf{Q} = \mathbf{I}_3$ and hence, we deduce that

$$\mathbf{x} = \mathbf{Q}^T\mathbf{Qx} = \mathbf{Q}^T\mathbf{b}$$

$$= \frac{1}{2} \begin{bmatrix} 1 & 1 & 1 & 1 \\ 1 & -1 & -1 & 1 \\ -1 & 1 & -1 & 1 \end{bmatrix} \begin{bmatrix} 2 \\ 0 \\ 1 \\ 1 \end{bmatrix} = \begin{bmatrix} 2 \\ 1 \\ -1 \end{bmatrix}$$

Thus, $x_1 = 2$, $x_2 = 1$, and $x_3 = -1$; hence, $2\mathbf{q}_1 + \mathbf{q}_2 - \mathbf{q}_3 = \mathbf{b}$.
Method 2. Multiply (1) by $\mathbf{q}_i^T$ ($i = 1, 2, 3$) and, because $\mathbf{q}_i^T\mathbf{q}_{ij} = \delta_{ij}$,

$$x_1\mathbf{q}_1^T\mathbf{q}_1 = x_1 = \mathbf{q}_1^T\mathbf{b} = 2$$

$$x_2\mathbf{q}_2^T\mathbf{q}_2 = x_2 = \mathbf{q}_2^T\mathbf{b} = 1$$

$$x_3\mathbf{q}_3^T\mathbf{q}_3 = x_3 = \mathbf{q}_3^T\mathbf{b} = -1. \quad \blacksquare$$

5.3.1   THE GRAM–SCHMIDT PROCESS AND THE *Q-R* FACTORIZATION THEOREM

From the linearly independent set $\{\mathbf{a}_1, \mathbf{a}_2, \ldots, \mathbf{a}_n\}$ it is always possible to construct an orthonormal set $\{\mathbf{q}_1, \mathbf{q}_2, \ldots, \mathbf{q}_n\}$ so that every vector which is a linear combination of the vectors in one of these sets is also a linear combination of the vectors in the other. The method we choose to construct $\{\mathbf{q}_1, \mathbf{q}_2, \dots, \mathbf{q}_n\}$ uses an algorithm known as the *Gram–Schmidt process*. The steps are these:

Step 1. Define

$$\mathbf{v}_1 = \mathbf{a}_1. \tag{5.3.6}$$

The norm of $\mathbf{v}_1$ is

$$r_1 = \|\mathbf{v}_1\|.$$

Then

$$\mathbf{q}_1 = \frac{\mathbf{v}_1}{r_1} \tag{5.3.7}$$

is of unit norm.

Step 2. Define

$$\mathbf{v}_2 = \mathbf{a}_2 - (\mathbf{q}_1 \cdot \mathbf{a}_2)\mathbf{q}_1. \tag{5.3.8}$$

The norm of $\mathbf{v}_2$ is

$$r_2 = \|\mathbf{v}_2\|.$$

Then

$$\mathbf{q}_2 = \frac{\mathbf{v}_2}{r_2} \tag{5.3.9}$$

is of unit norm.

Step 3. Define

$$\mathbf{v}_3 = \mathbf{a}_3 - (\mathbf{q}_1 \cdot \mathbf{a}_3)\mathbf{q}_1 - (\mathbf{q}_2 \cdot \mathbf{a}_3)\mathbf{q}_2. \tag{5.3.10}$$

The norm is $\mathbf{v}_3$ is

$$r_3 = \|\mathbf{v}_3\|.$$

Then

$$\mathbf{q}_3 = \frac{\mathbf{v}_3}{r_3}, \tag{5.3.11}$$

so $\mathbf{q}_3$ is of unit norm.

ith step. Define*

$$\mathbf{v}_i = \mathbf{a}_i - \sum_{k=1}^{i-1} (\mathbf{q}_k \cdot \mathbf{a}_i)\mathbf{q}_k, \qquad i > 1. \tag{5.3.12}$$

---

*It is standard practice to define the empty sum $\Sigma_{k=1}^{0}$ as zero. Then Eq. 5.3.12 holds even for $i = 1$.

The norm of $\mathbf{v}_i$ is

$$r_i = \| \mathbf{v}_i \|.$$

Then

$$\mathbf{q}_i = \frac{\mathbf{v}_i}{r_i} \qquad (5.3.13)$$

so $\mathbf{q}_i$ is of unit norm.

As long as $r_i \neq 0$, these steps produce $\mathbf{q}_1, \mathbf{q}_2, \ldots, \mathbf{q}_n$, in that order, and $\| \mathbf{q}_i \| = 1$. Therefore, if $r_i \neq 0$ for all i, $\mathbf{q}_i \perp \mathbf{q}_j$, $i \neq j$. We shall return to these items later.

Example 5.3.6:   Use the Gram–Schmidt process to construct an orthonormal set from

$$\mathbf{a}_1 = \begin{bmatrix} 1 \\ 1 \\ 1 \\ 1 \end{bmatrix}, \qquad \mathbf{a}_2 = \begin{bmatrix} 1 \\ 0 \\ 0 \\ 1 \end{bmatrix}, \qquad \mathbf{a}_3 = \begin{bmatrix} -1 \\ 0 \\ -2 \\ 1 \end{bmatrix}.$$

SOLUTION: We simply follow the steps in the process outlined above:
Step 1

$$\mathbf{v}_1 = \begin{bmatrix} 1 \\ 1 \\ 1 \\ 1 \end{bmatrix} = \mathbf{a}_1$$

$$r_1 = \| \mathbf{v}_1 \| = \sqrt{4} = 2$$

$$\mathbf{q}_1 = \frac{1}{2} \begin{bmatrix} 1 \\ 1 \\ 1 \\ 1 \end{bmatrix}.$$

Step 2

$$\mathbf{v}_2 = \mathbf{a}_2 - (\mathbf{q}_1 \cdot \mathbf{a}_2)\mathbf{q}_1$$

$$= \begin{bmatrix} 1 \\ 0 \\ 0 \\ 1 \end{bmatrix} - \begin{bmatrix} \frac{1}{2} \\ \frac{1}{2} \\ \frac{1}{2} \\ \frac{1}{2} \end{bmatrix} \cdot \begin{bmatrix} 1 \\ 0 \\ 0 \\ 1 \end{bmatrix} \quad \mathbf{q}_1 = \frac{1}{2} \begin{bmatrix} 1 \\ -1 \\ -1 \\ 1 \end{bmatrix}$$

$$r_2 = \| \mathbf{v}_2 \| = 1$$

$$\mathbf{q}_2 = \frac{1}{2} \begin{bmatrix} 1 \\ -1 \\ -1 \\ 1 \end{bmatrix}.$$

Step 3

$$\mathbf{v}_3 = \mathbf{a}_3 - (\mathbf{q}_1 \cdot \mathbf{a}_3)\mathbf{q}_1 - (\mathbf{q}_2 \cdot \mathbf{a}_3)\mathbf{q}_2$$

$$= \begin{bmatrix} -1 \\ 0 \\ -2 \\ 1 \end{bmatrix} - \begin{bmatrix} \frac{1}{2} \\ \frac{1}{2} \\ \frac{1}{2} \\ \frac{1}{2} \end{bmatrix} \cdot \begin{bmatrix} -1 \\ 0 \\ -2 \\ 1 \end{bmatrix} \mathbf{q}_1 - \begin{bmatrix} \frac{1}{2} \\ -\frac{1}{2} \\ -\frac{1}{2} \\ \frac{1}{2} \end{bmatrix} \cdot \begin{bmatrix} -1 \\ 0 \\ -2 \\ 1 \end{bmatrix} \mathbf{q}_2 = \begin{bmatrix} -1 \\ 1 \\ -1 \\ 1 \end{bmatrix}$$

$$r_3 = \| \mathbf{v}_3 \| = 2$$

$$\mathbf{q}_3 = \frac{1}{2} \begin{bmatrix} -1 \\ 1 \\ -1 \\ 1 \end{bmatrix}.$$

It is easy to check that $\mathbf{q}_i \cdot \mathbf{q}_j = \delta_{ij}$. ∎

Example 5.3.7:   Find the orthonormal vectors corresponding to

$$\mathbf{a}_1 = \begin{bmatrix} 1 \\ 1 \end{bmatrix}, \qquad \mathbf{a}_2 = \begin{bmatrix} 1 \\ 0 \end{bmatrix}.$$

SOLUTION: We have

$$\mathbf{v}_1 = \mathbf{a}_1 = \begin{bmatrix} 1 \\ 1 \end{bmatrix}$$

$$r_1 = \| \mathbf{v}_1 \| = \sqrt{2}$$

$$\mathbf{q}_1 = \frac{\sqrt{2}}{2} \begin{bmatrix} 1 \\ 1 \end{bmatrix}.$$

Next

$$\mathbf{v}_2 = \begin{bmatrix} 1 \\ 0 \end{bmatrix} - \frac{\sqrt{2}}{2} \begin{bmatrix} 1 \\ 1 \end{bmatrix} \cdot \begin{bmatrix} 1 \\ 0 \end{bmatrix} \mathbf{q}_1 = \begin{bmatrix} \frac{1}{2} \\ -\frac{1}{2} \end{bmatrix}$$

$$r_2 = \| \mathbf{v}_2 \| = \frac{\sqrt{2}}{2}$$

$$\mathbf{q}_2 = \frac{\sqrt{2}}{2} \begin{bmatrix} 1 \\ -1 \end{bmatrix}. \quad ∎$$

Example 5.3.8:   Find the orthonormal vectors corresponding to

$$\mathbf{a}_1 = \begin{bmatrix} 1 \\ 0 \end{bmatrix}, \qquad \mathbf{a}_2 = \begin{bmatrix} 1 \\ 1 \end{bmatrix}.$$

SOLUTION: Now

$$\mathbf{v}_1 = \begin{bmatrix} 1 \\ 0 \end{bmatrix} = \mathbf{q}_1$$

and

$$\mathbf{v}_2 = \begin{bmatrix} 1 \\ 1 \end{bmatrix} - \begin{bmatrix} 1 \\ 0 \end{bmatrix} \cdot \begin{bmatrix} 1 \\ 1 \end{bmatrix} \mathbf{q}_1 = \begin{bmatrix} 0 \\ 1 \end{bmatrix} = \mathbf{q}_2. \quad ∎$$

These last two examples illustrate that the order of the vectors $\mathbf{a}_1$, $\mathbf{a}_2, \ldots, \mathbf{a}_n$ plays an important role in determining the $\mathbf{q}_i$, but that once the order of the vectors $\mathbf{a}_1, \mathbf{a}_2, \ldots, \mathbf{a}_n$ is fixed, the $\mathbf{q}_i$ and their order is uniquely determined by the algorithm.

Although it is not obvious, Eq. 5.3.12 can be written in matrix form. The easiest way to see this is, first replace $\mathbf{v}_i$ by $r_i\mathbf{q}_i$ and then solve for $\mathbf{a}_i$; thus,

$$\mathbf{a}_i = r_i\mathbf{q}_i + \sum_{k=1}^{i-1} (\mathbf{q}_k \cdot \mathbf{a}_i)\mathbf{q}_k. \tag{5.3.14}$$

Now let $\mathbf{A} = [\mathbf{a}_1, \mathbf{a}_2, \ldots, \mathbf{a}_n]$ and $\mathbf{Q} = [\mathbf{q}_1, \mathbf{q}_2, \ldots, \mathbf{q}_n]$. Let

$$\mathbf{R} = \begin{bmatrix} r_1 & r_{12} & \cdots & r_{1n} \\ 0 & r_2 & & r_{2n} \\ \vdots & & & \\ 0 & 0 & \cdots & r_n \end{bmatrix} \tag{5.3.15}$$

$$r_{ki} = \mathbf{q}_k \cdot \mathbf{a}_i, \qquad k \leq i - 1. \tag{5.3.16}$$

Then

$$\mathbf{A} = \mathbf{QR}. \tag{5.3.17}$$

Equation 5.3.17 is the *Q-R factorization* of $\mathbf{A}$.

Example 5.3.9: Find the *Q-R* factorization of

(a)
$$\mathbf{A} = \begin{bmatrix} 1 & 1 & -1 \\ 1 & 0 & 0 \\ 1 & 0 & -2 \\ 1 & 1 & 1 \end{bmatrix}$$

(b)
$$\mathbf{A} = \begin{bmatrix} 1 & 1 \\ 1 & 0 \end{bmatrix}$$

(c)
$$\mathbf{A} = \begin{bmatrix} 1 & 1 \\ 0 & 1 \end{bmatrix}$$

SOLUTION: These matrices are constructed from the vectors given in Examples 5.3.6 to 5.3.8, respectively.

(a)
$$\mathbf{A} = \begin{bmatrix} \frac{1}{2} & \frac{1}{2} & -\frac{1}{2} \\ \frac{1}{2} & -\frac{1}{2} & \frac{1}{2} \\ \frac{1}{2} & -\frac{1}{2} & -\frac{1}{2} \\ \frac{1}{2} & \frac{1}{2} & \frac{1}{2} \end{bmatrix} \begin{bmatrix} 2 & 1 & -1 \\ 0 & 1 & 1 \\ 0 & 0 & 2 \end{bmatrix}$$

Note that the columns of $\mathbf{Q}$ are just $\mathbf{q}_1$, $\mathbf{q}_2$, $\mathbf{q}_3$ of Example 5.3.6. The diagonal entries in $\mathbf{R}$ are r_1, r_2, r_3. The remaining entries in $\mathbf{R}$ were computed using Eq. 5.3.16 as follows:

$$r_{12} = \mathbf{q}_1 \cdot \mathbf{a}_2 = \frac{1}{2}\begin{bmatrix} 1 \\ 1 \\ 1 \\ 1 \end{bmatrix} \cdot \begin{bmatrix} 1 \\ 0 \\ 0 \\ 1 \end{bmatrix} = 1$$

$$r_{13} = \mathbf{q}_1 \cdot \mathbf{a}_3 = \frac{1}{2} \begin{bmatrix} 1 \\ 1 \\ 1 \\ 1 \end{bmatrix} \cdot \begin{bmatrix} -1 \\ 0 \\ -2 \\ 1 \end{bmatrix} = -1$$

$$r_{23} = \mathbf{q}_2 \cdot \mathbf{a}_3 = \frac{1}{2} \begin{bmatrix} 1 \\ -1 \\ -1 \\ 1 \end{bmatrix} \cdot \begin{bmatrix} -1 \\ 0 \\ -2 \\ 1 \end{bmatrix} = 1$$

(b) and (c)

$$\begin{bmatrix} 1 & 1 \\ 1 & 0 \end{bmatrix} = \begin{bmatrix} \sqrt{2}/2 & \sqrt{2}/2 \\ \sqrt{2}/2 & -\sqrt{2}/2 \end{bmatrix} \begin{bmatrix} \sqrt{2} & \sqrt{2}/2 \\ 0 & \sqrt{2}/2 \end{bmatrix}$$

$$= \frac{1}{2} \begin{bmatrix} 1 & 1 \\ 1 & -1 \end{bmatrix} \begin{bmatrix} 2 & 1 \\ 0 & 1 \end{bmatrix}$$

$$\begin{bmatrix} 1 & 1 \\ 0 & 1 \end{bmatrix} = \begin{bmatrix} 1 & 0 \\ 0 & 1 \end{bmatrix} \begin{bmatrix} 1 & 1 \\ 0 & 1 \end{bmatrix} \quad \blacksquare$$

Thus, the factorization of **A** may be interpreted as the matrix counterpart of the Gram–Schmidt process. It is time then to prove:

1. $\mathbf{v}_i \neq \mathbf{0}$ for all i and therefore each $\mathbf{q}_i$ may be defined.
2. $\mathbf{q}_i \perp \mathbf{q}_j$ if $i \neq j$.

Both proofs are inductive and, naturally, use the defining formula (Eq. 5.3.12).

Proof of (1). Since $\{\mathbf{a}_1, \mathbf{a}_2, \ldots, \mathbf{a}_n\}$ is linearly independent, $\mathbf{a}_1 \neq \mathbf{0}$. Then $\mathbf{v}_1 = \mathbf{a}_1 \neq \mathbf{0}$. Now,

$$\mathbf{v}_2 = \mathbf{a}_1 - (\mathbf{q}_1 \cdot \mathbf{a}_2)\mathbf{q}_1$$

$$= \mathbf{a}_2 - \frac{\mathbf{q}_1 \cdot \mathbf{a}_2}{r_1}\mathbf{v}_1 = \mathbf{a}_2 - \frac{\mathbf{q}_1 \cdot \mathbf{a}_2}{r_1}\mathbf{a}_1.$$

Therefore, if $\mathbf{v}_2 = \mathbf{0}$, we would have $\{\mathbf{a}_1, \mathbf{a}_2\}$ linearly dependent. This is not the case, so $\mathbf{v}_2 \neq \mathbf{0}$. The reader may supply the proof that $\mathbf{v}_3$ is also nonzero, and so on.

Proof of (2). To show that $\mathbf{q}_1 \perp \mathbf{q}_2$ we show that $\mathbf{q}_1 \perp \mathbf{v}_2$, for

$$\mathbf{v}_2 = \mathbf{a}_2 - (\mathbf{q}_1 \cdot \mathbf{a}_2)\mathbf{q}_1$$

implies that

$$\mathbf{q}_1 \cdot \mathbf{v}_2 = \mathbf{q}_1 \cdot \mathbf{a}_2 - (\mathbf{q}_1 \cdot \mathbf{a}_2)(\mathbf{q}_1 \cdot \mathbf{q}_1).$$

But $\mathbf{q}_1 \cdot \mathbf{q}_1 = \|\mathbf{q}_1\|^2 = 1$. Hence, $\mathbf{q}_1 \cdot \mathbf{v}_2 = 0$. Again, we invite the reader to use the fact that $\mathbf{q}_1 \perp \mathbf{q}_2$ and Eq. 5.3.12 to show that $\mathbf{v}_3 \perp \mathbf{q}_1$ and $\mathbf{v}_3 \perp \mathbf{q}_2$. From this step, the complete induction is reasonably straightforward.

We thus have the following theorem.

Theorem 5.2: *If* $\mathbf{A}$ *is* $m \times n$ *and has full rank,** *then*

$$\mathbf{A} = \mathbf{QR} \qquad (5.3.18)$$

where $\mathbf{Q}$ *is* $m \times n$ *with orthonormal columns and* $\mathbf{R}$ *is an* $n \times n$, *non-singular, upper triangular matrix.*

Since $\mathbf{R}$ is nonsingular and upper triangular, it has an inverse $\mathbf{S} = \mathbf{R}^{-1}$ which is also upper triangular. Therefore, Eq. 5.3.18 implies that

$$\mathbf{AS} = \mathbf{Q}. \qquad (5.3.19)$$

The reader may verify that this equation proves that $\mathbf{q}_i$ is a linear combination of $\mathbf{a}_1, \mathbf{a}_2, \ldots, \mathbf{a}_i$ just as $\mathbf{A} = \mathbf{QR}$ shows that $\mathbf{a}_i$ is a linear combination of $\mathbf{q}_1, \mathbf{q}_2, \ldots, \mathbf{q}_i$. Note also that although $\mathbf{Q}$ is not generally a square matrix, $\mathbf{Q}^T\mathbf{Q} = \mathbf{I}_n$, and $\mathbf{QQ}^T$ is singular unless $\mathbf{Q}$ is $n \times n$, in which case $\mathbf{Q}^T = \mathbf{Q}^{-1}$. It is interesting to observe that if the Q-R factorization of $\mathbf{A}$ is known, then

$$\mathbf{Ax} = \mathbf{b} \text{ implies } \mathbf{QRx} = \mathbf{b} \qquad (5.3.20)$$

and therefore

$$\mathbf{Q}^T\mathbf{QRx} = \mathbf{Q}^T\mathbf{b}. \qquad (5.3.21)$$

Since $\mathbf{Q}^T\mathbf{Q} = \mathbf{I}$,

$$\mathbf{Ax} = \mathbf{b} \text{ implies } \mathbf{Rx} = \mathbf{Q}^T\mathbf{b}. \qquad (5.3.22)$$

The latter system is easily solved by back-substitution since $\mathbf{R}$ is upper triangular.

5.3.2   PROJECTION MATRICES

Any square matrix $\mathbf{P}$ satisfying the two conditions

(i) $$\mathbf{P}^T = \mathbf{P}$$

(ii) $$\mathbf{P}^2 = \mathbf{P} \qquad (5.3.23)$$

is a *projection matrix* (a *projection*, for short). It is a trivial observation that $\mathbf{I}_n$ and $\mathbf{O}_{n \times n}$ are projections.

Example 5.3.10:   Verify that the following symmetric matrix is a projection:

$$\frac{1}{n}\mathbf{J}_n = \frac{1}{n}\begin{bmatrix} 1 & 1 & \cdots & 1 \\ 1 & 1 & \cdots & 1 \\ \vdots & & & \\ 1 & 1 & \cdots & 1 \end{bmatrix}_{n \times n}.$$

---

*The columns of $\mathbf{A}$ form a linearly independent set.

SOLUTION: That the given matrix is a projection follows from the fact that

$$\mathbf{J}_n^2 = n\mathbf{J}_n. \quad \blacksquare$$

Example 5.3.11:   Show that all projections except **I** are singular.

SOLUTION: Suppose that **P** is a projection and $\mathbf{P}^{-1}$ exists. Then, from $\mathbf{P}^2 = \mathbf{P}$ we obtain by multiplication

$$\mathbf{P}^{-1}(\mathbf{P}^2) = \mathbf{P}^{-1}\mathbf{P} = \mathbf{I}.$$

But, $\mathbf{P}^{-1}\mathbf{P}^2 = \mathbf{P}$. So $\mathbf{P} = \mathbf{I}$ follows from the existence of $\mathbf{P}^{-1}$. $\blacksquare$

Example 5.3.12:   Suppose that $\|\mathbf{u}\| = 1$. Show that $\mathbf{u}\mathbf{u}^T$ is a projection.

SOLUTION: We easily verify that $\mathbf{u}\mathbf{u}^T$ is symmetric. Also,

$$(\mathbf{u}\mathbf{u}^T)^2 = (\mathbf{u}\mathbf{u}^T)(\mathbf{u}\mathbf{u}^T)$$
$$= \mathbf{u}(\mathbf{u}^T\mathbf{u})\mathbf{u}^T.$$

Since

$$\mathbf{u}^T\mathbf{u} = \mathbf{u} \cdot \mathbf{u} = \|\mathbf{u}\|^2 = 1,$$

it follows that

$$(\mathbf{u}^T\mathbf{u})^2 = \mathbf{u}\mathbf{u}^T. \quad \blacksquare$$

The next example provides a motivation for using the word "projection" for matrices satisfying Eq. 5.3.23, at least in the special case where $\mathbf{P} = \mathbf{u}\mathbf{u}^T$. Consider Fig. 5.3, wherein **u** is a unit vector making a 30° angle with the negative x axis, and **b** is a vector terminating at $(\sqrt{3}, 1)$. The vector **b** makes a 30° angle with the positive x axis. The "projection of **b** onto the line defined by **u**" is the vector denoted by **p**, which in this case points opposite to **u**. By elementary geometry, the coordinates of the terminus of **p** are $(\sqrt{3}/2, -\frac{1}{2})$

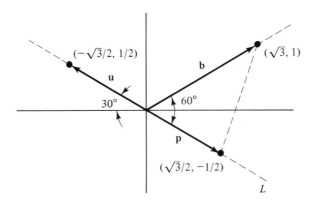

FIGURE 5.3.   **p** is the projection of **b** on the line L.

Now consider

$$(\mathbf{uu}^T)\mathbf{b} = \begin{bmatrix} -\sqrt{3}/2 \\ 1/2 \end{bmatrix} [-\sqrt{3}/2, \ 1/2] \begin{bmatrix} \sqrt{3} \\ 1 \end{bmatrix}$$

$$= \begin{bmatrix} 3/4 & -\sqrt{3}/4 \\ -\sqrt{3}/4 & 1/4 \end{bmatrix} \begin{bmatrix} \sqrt{3} \\ 1 \end{bmatrix} = \begin{bmatrix} \sqrt{3}/2 \\ -1/2 \end{bmatrix}.$$

Hence, $(\mathbf{uu}^T)\mathbf{b} = \mathbf{p}$. That is, $\mathbf{uu}^T$ projects an arbitrary vector $\mathbf{b}$ onto the line determined by $\mathbf{u}$.

We conclude this section with an example and a theorem, both of which are essential to the material in Section 5.4.

Example 5.3.13: Verify that for every $\mathbf{A}_{m \times n}$ with full rank,

$$\mathbf{P} = \mathbf{A}(\mathbf{A}^T\mathbf{A})^{-1}\mathbf{A}^T$$

is a projection.

SOLUTION: If $\mathbf{A}$ has full rank, then $\mathbf{A}^T\mathbf{A}$ is invertible (see Theorem 5.1). [Of course, $(\mathbf{A}^T\mathbf{A})^{-1}$ cannot be written as $\mathbf{A}^{-1}(\mathbf{A}^T)^{-1}$ since neither $\mathbf{A}$ nor $\mathbf{A}^T$ are necessarily square matrices.] Now

$$\mathbf{P}^T = \{\mathbf{A}(\mathbf{A}^T\mathbf{A})^{-1}\mathbf{A}^T\}^T$$

$$= \mathbf{A}\{(\mathbf{A}^T\mathbf{A})^{-1}\}^T\mathbf{A}^T = \mathbf{P}$$

since $(\mathbf{A}^T\mathbf{A})^{-1}$ is symmetric. More interesting is

$$\mathbf{P}^2 = \{\mathbf{A}(\mathbf{A}^T\mathbf{A})^{-1}\mathbf{A}^T\}\{\mathbf{A}(\mathbf{A}^T\mathbf{A})^{-1}\mathbf{A}^T\}$$

$$= \mathbf{A}(\mathbf{A}^T\mathbf{A})^{-1}(\mathbf{A}^T\mathbf{A})(\mathbf{A}^T\mathbf{A})^{-1}\mathbf{A}^T$$

$$= \mathbf{A}(\mathbf{A}^T\mathbf{A})^{-1}\mathbf{A}^T = \mathbf{P}. \quad \blacksquare$$

Theorem 5.3: *If* $\mathbf{P}$ *is a projection, then for every* $\mathbf{b}$ *and* $\mathbf{y}$,

$$\mathbf{Py} \perp (\mathbf{Pb} - \mathbf{b}). \tag{5.3.24}$$

PROOF: We show the theorem to be true by showing that $(\mathbf{Py})^T(\mathbf{Pb} - \mathbf{b}) = 0$. But

$$(\mathbf{Py})^T(\mathbf{Pb} - \mathbf{b}) = \mathbf{y}^T\mathbf{P}^T(\mathbf{Pb} - \mathbf{b})$$

$$= \mathbf{y}^T\mathbf{P}(\mathbf{Pb} - \mathbf{b})$$

since $\mathbf{P}\mathbf{P}^T = \mathbf{P}$. Hence,

$$(\mathbf{Py})^T(\mathbf{Pb} - \mathbf{b}) = \mathbf{y}^T(\mathbf{P}^2\mathbf{b} - \mathbf{Pb})$$

$$= \mathbf{y}^T(\mathbf{Pb} - \mathbf{Pb})$$

since $\mathbf{P}^2 = \mathbf{P}$. Thus, we have

$$(\mathbf{Py})^T(\mathbf{Pb} - \mathbf{b}) = \mathbf{y}^T\mathbf{0}$$

$$= 0.$$

PROBLEMS

1. Show that

$$P = \begin{bmatrix} \sin\theta\cos\theta & \sin\theta\sin\theta & \cos\theta \\ -\sin\theta & \cos\theta & 0 \\ \cos\theta\cos\theta & \cos\theta\sin\theta & -\sin\theta \end{bmatrix}$$

is orthogonal.

Use the Gram–Schmidt process to orthogonalize each set.

2.
$$\mathbf{a}_1 = \begin{bmatrix} 1 \\ 1 \\ 1 \end{bmatrix}, \mathbf{a}_2 = \begin{bmatrix} 1 \\ 1 \\ 0 \end{bmatrix}$$

3.
$$\mathbf{a}_1 = \begin{bmatrix} 1 \\ 1 \\ 0 \end{bmatrix}, \mathbf{a}_2 = \begin{bmatrix} 1 \\ 1 \\ 1 \end{bmatrix}$$

4.
$$\mathbf{a}_1 = \begin{bmatrix} 1 \\ 0 \\ 0 \end{bmatrix}, \mathbf{a}_2 = \begin{bmatrix} 1 \\ 1 \\ 0 \end{bmatrix}, \mathbf{a}_3 = \begin{bmatrix} 0 \\ 1 \\ 1 \end{bmatrix}$$

5.
$$\mathbf{a}_1 = \begin{bmatrix} 1 \\ 0 \\ 1 \end{bmatrix}, \mathbf{a}_2 = \begin{bmatrix} 1 \\ 0 \\ -1 \end{bmatrix}, \mathbf{a}_3 = \begin{bmatrix} 1 \\ 1 \\ 1 \end{bmatrix}$$

6.
$$\mathbf{a}_1 = \begin{bmatrix} 1 \\ 0 \\ 0 \end{bmatrix}, \mathbf{a}_2 = \begin{bmatrix} 1 \\ 1 \\ 1 \end{bmatrix}, \mathbf{a}_3 = \begin{bmatrix} \alpha_1 \\ \alpha_2 \\ \alpha_3 \end{bmatrix}$$

Write the equivalent Q-R factorization of $\mathbf{A}$, the matrix whose columns are the vectors $\mathbf{a}_1, \mathbf{a}_2, \ldots$, for each set of vectors.

7. The set of Problem 2.

8. The set of Problem 3.

9. The set of Problem 4.

10. The set of Problem 5.

11. The set of Problem 6.

12. Show that every orthogonal set is linearly independent. *Hint:* Use method 2 of Example 5.3.5 or

$$c_1\mathbf{q}_1 + c_2\mathbf{q}_2 + \cdots + c_k\mathbf{q}_k = \mathbf{0}.$$

13. Show that every solution of $\mathbf{Ax} = \mathbf{0}$ is orthogonal to the rows of $\mathbf{A}$. Use this result to find $\mathbf{z}$ orthogonal to $\mathbf{x}$ and $\mathbf{y}$ in Eq. 5.3.3.

14. Show that $\mathbf{v}_3 \neq \mathbf{0}$. Knowing that $\mathbf{q}_1 \perp \mathbf{q}_2$, show that Eq. 5.3.10 implies that $\mathbf{q}_1 \perp \mathbf{q}_3$ and $\mathbf{q}_2 \perp \mathbf{q}_3$.

15. Explain why $\mathbf{A} = \mathbf{QR}$ shows that each column $\mathbf{a}_i$ of $\mathbf{A}$ is a linear combination of the first i columns of $\mathbf{Q}$, namely, $\mathbf{q}_1, \mathbf{q}_2, \ldots, \mathbf{q}_i$.

16. Suppose that $\mathbf{A}$ is $k \times n$, $k > n$. Show that $\mathbf{AA}^T$ must be singular. *Hint:* Rank $\mathbf{AA}^T \leq$ number of columns of $\mathbf{AA}^T$ and rank $\mathbf{A} \leq n$.

17. If $\mathbf{P}$ is a projection, show that $\mathbf{P}^k$ is also a projection for each k.

18. If $\mathbf{P}_1$ and $\mathbf{P}_2$ are projections, is $\mathbf{P}_1\mathbf{P}_2$ a projection? If $\mathbf{P}_1\mathbf{P}_2 = \mathbf{P}_2\mathbf{P}_1$, is $\mathbf{P}_1\mathbf{P}_2$ a projection?

19. If $\mathbf{P}_1$ and $\mathbf{P}_2$ are projections, is $\mathbf{P}_1 + \mathbf{P}_2$ a projection? Explain.

20. Find $\mathbf{u}$ so that $\|\mathbf{u}\| = 1$ and $\mathbf{uu}^T = (1/n)\mathbf{J}_n$ (see Example 5.3.10).

21. Let $\mathbf{A} = \begin{bmatrix} 1 \\ 1 \\ 1 \end{bmatrix}$ in Example 5.3.13. What is $\mathbf{P}$? Show that $\mathbf{P}$ cannot be defined if $\mathbf{A} = [1, 1]$. (*Note:* [1, 1] does not have full rank.)

22. Let $\mathbf{A} = \begin{bmatrix} 1 & 1 \\ 0 & 1 \end{bmatrix}$ in Example 5.3.13. Compute $\mathbf{P}$ using this $\mathbf{A}$.

23. If $\mathbf{P}$ is a projection, then so is $(\mathbf{I} - \mathbf{P})$. Prove this theorem.

24. Solve the system $\mathbf{Ax} = \mathbf{b}$ where

$$\mathbf{A} = \begin{bmatrix} 1 & 1 & -1 \\ 1 & 0 & 0 \\ 1 & 0 & -2 \\ 1 & 1 & 1 \end{bmatrix}, \qquad \mathbf{b} = \begin{bmatrix} 2 \\ -1 \\ 1 \\ 0 \end{bmatrix}$$

by using Eq. 5.3.22 and the Q-R factorization of $\mathbf{A}$ given in Example 5.3.9.

5.4   Least Squares Fit of Data

A problem common to most experimental scientists is the fitting of curves to data. Often we have theoretical reasons for believing that the output of some experiment is related to the input by a specific functional dependence containing suitable parameters. For instance, suppose that theory predicts that

$$y = ax + b \tag{5.4.1}$$

relates the input x to the output y in an experiment that generates the data of Table 5.1. The parameters a and b in Eq. 5.4.1 are related to the physical properties of the material used in the experiment. If these data

TABLE 5.1   Experimental Data

x = input	−1	0	1	2
y = output	−0.5	0.5	1.1	2.1

are plotted, as in Fig. 5.4, the linear hypothesis expressed in Eq. 5.4.1 is reasonable. The question is; What are the values of a and b that best fit the data? Indeed, what does "best" mean?

The data overdetermines a and b, for we have the following:

$$\begin{array}{lll} x = -1: & -a + b = -0.5 \\ x = 0: & b = 0.5 \\ x = 1: & a + b = 1.1 \\ x = 2: & 2a + b = 2.1. \end{array} \tag{5.4.2}$$

These four equations for a and b are inconsistent. Statisticians have deter-

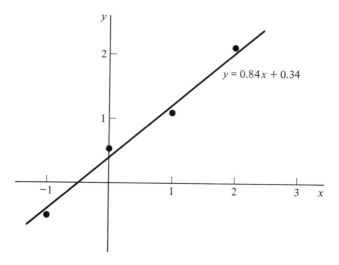

FIGURE 5.4.   The data of Table 5.1 and the line of best fit.

mined that under certain reasonably broad assumptions one can do no better than choosing a and b so as to minimize the sum of the squares of the "deviations." In the present illustration this sum of squares of the deviations is

$$(-a + b + 0.5)^2 + (b - 0.5)^2 + (a + b - 1.1)^2 + (2a + b - 2.1)^2.$$

Figure 5.5 illustrates the meaning of deviation in a slightly more general setting.

To present the ideas above most generally, suppose that **A** has full rank and we are interested in the system

$$\mathbf{Ax} = \mathbf{b}, \tag{5.4.3}$$

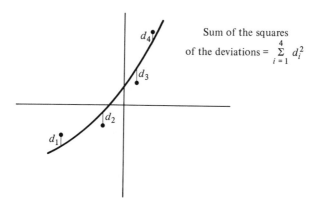

FIGURE 5.5.   The deviations.

which may be inconsistent. The vector $\mathbf{r}$ is the *residual* and is defined as

$$\mathbf{r} = \mathbf{b} - \mathbf{Ax}. \qquad (5.4.4)$$

We are interested in finding $\mathbf{x}$ so that $\|\mathbf{r}\|$ is as small as possible. This may be phrased in two equivalent ways:

1. Find $\mathbf{x} = \hat{\mathbf{x}}$ so that $\|\mathbf{Ax} - \mathbf{b}\|$ is minimized by $\hat{\mathbf{x}}$.
2. Find $\mathbf{x} = \hat{\mathbf{x}}$ such that for all $\mathbf{z}$,

$$\|\mathbf{A}\hat{\mathbf{x}} - \mathbf{b}\| \leq \|\mathbf{Az} - \mathbf{b}\|. \qquad (5.4.5)$$

We shall prove in Section 5.4.1 that

$$\hat{\mathbf{x}} = (\mathbf{A}^T\mathbf{A})^{-1}\mathbf{A}^T\mathbf{b} \qquad (5.4.6)$$

is the solution of inequality 5.4.5. Note that $\mathbf{A}$ need not be square so that $(\mathbf{A}^T\mathbf{A})^{-1}$ cannot be simplified without further restrictions on $\mathbf{A}$. For instance, if $\mathbf{A}^{-1}$ exists, then

$$(\mathbf{A}^T\mathbf{A})^{-1} = \mathbf{A}^{-1}(\mathbf{A}^T)^{-1} \qquad (5.4.7)$$

and thus

$$\hat{\mathbf{x}} = \mathbf{A}^{-1}\mathbf{b}. \qquad (5.4.8)$$

In this case $\mathbf{A}\hat{\mathbf{x}} = \mathbf{b}$ and thus inequality 5.4.3 reduces to the triviality, $0 \leq \|\mathbf{Az} - \mathbf{b}\|$ for all $\mathbf{z}$. The system

$$\mathbf{A}^T\mathbf{Ax} = \mathbf{A}^T\mathbf{b}, \qquad (5.4.9)$$

obtained formally by multiplying Eq. 5.4.3 by $\mathbf{A}^T$, are the *normal equations*. We are asuming that $\mathbf{A}$ has full rank. Therefore, $(\mathbf{A}^T\mathbf{A})^{-1}$ exists (see Theorem 5.1 and Example 5.3.13) and thus the normal equations have the unique solution $\mathbf{x}$ as given by Eq. 5.4.6.

One of the following examples generalizes the process and solves the problem of fitting a straight line to n data points. This technique is often called *linear regression*.

Example 5.4.1:   Find the line of best fit for the data of Table 5.1.

SOLUTION: The system 5.4.2 takes the form

$$\begin{bmatrix} 1 & -1 \\ 1 & 0 \\ 1 & 1 \\ 1 & 2 \end{bmatrix} \begin{bmatrix} b \\ a \end{bmatrix} = \begin{bmatrix} -0.5 \\ 0.5 \\ 1.1 \\ 2.1 \end{bmatrix},$$

so the normal equations are

$$\begin{bmatrix} 4 & 2 \\ 2 & 6 \end{bmatrix} \begin{bmatrix} b \\ a \end{bmatrix} = \begin{bmatrix} 3.2 \\ 5.8 \end{bmatrix}.$$

Hence,

$$\begin{bmatrix} a \\ b \end{bmatrix} = \begin{bmatrix} 0.84 \\ 0.38 \end{bmatrix}.$$

Thus, the line of best fit is

$$y = 0.84x + 0.38. \quad \blacksquare$$

Example 5.4.2:   Given n points $\{(x_i, y_i)\}$, find the line

$$y = ax + b$$

which is the best fit of data in the sense of inequality 5.4.5.

SOLUTION: The system is of full rank and is overdetermined:

$$\begin{bmatrix} 1 & x_1 \\ 1 & x_2 \\ \vdots & \vdots \\ 1 & x_n \end{bmatrix} \begin{bmatrix} b \\ a \end{bmatrix} = \begin{bmatrix} y_1 \\ y_2 \\ \vdots \\ y_n \end{bmatrix}.$$

The corresponding normal equations are

$$\begin{bmatrix} n & \Sigma x_i \\ \Sigma x_i & \Sigma x_i^2 \end{bmatrix} \begin{bmatrix} b \\ a \end{bmatrix} = \begin{bmatrix} \Sigma y_i \\ \Sigma x_i y_i \end{bmatrix}.$$

Let $\bar{x}$ be the mean of $x_1, x_2, \ldots, x_n$ and $\bar{y}$ the mean of $y_1, y_2, \ldots, y_n$. That is,

$$\bar{x} = \frac{\Sigma x_i}{n}, \qquad \bar{y} = \frac{\Sigma y_i}{n}.$$

Then applying elementary row operations to the augmented matrix results in

$$\begin{bmatrix} n & \Sigma x_i & \Sigma y_i \\ \Sigma x_i & \Sigma x_i^2 & \Sigma x_i y_i \end{bmatrix} \longrightarrow \begin{bmatrix} 1 & \bar{x} & \bar{y} \\ \bar{x} & \Sigma x_i^2/n & \Sigma x_i y_i/n \end{bmatrix}.$$

Hence,

$$a = \frac{\dfrac{\Sigma x_i y_i}{n} - \bar{x}\bar{y}}{\dfrac{\Sigma x_i^2}{n} - \bar{x}^2}.$$

We then have

$$\begin{bmatrix} 1 & \bar{x} & \bar{y} \\ 0 & 1 & a \end{bmatrix} \longrightarrow \begin{bmatrix} 1 & 0 & \bar{y} - a\bar{x} \\ 0 & 1 & a \end{bmatrix},$$

so that

$$b = \bar{y} - a\bar{x}.$$

As we show in the problems, a can be rewritten as

$$a = \frac{\Sigma (x_i - \bar{x})(y_i - \bar{y})}{\Sigma (x_i - \bar{x})^2}. \quad \blacksquare$$

The above computation of a and b is extremely useful and commonplace in data handling. The next example illustrates how the technique can be applied to more general curve fitting.

Example 5.4.3: Plot the data

x	0	0.5	1	1.5	2	2.5
y	3	2.34	1.82	1.42	1.1	0.86

and find the best estimate of the parameters M and k assuming that

$$y = Me^{-kx}.$$

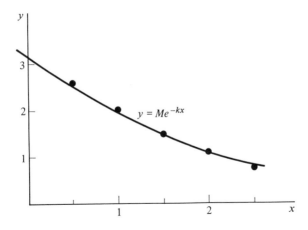

SOLUTION: The data are plotted as shown. The expected curve is sketched. We can cast this problem into a linear regression by taking logarithms. Then the given equation is equivalent to

$$z = \ln y = -kx + \ln M,$$

where we let $a = -k$ and $b = \ln M$. The data can now be presented as follows:

x	0	0.5	1	1.5	2	2.5
$z = \ln y$	1.1	0.85	0.60	0.35	0.10	−0.15

We compute

$$\bar{x} = \frac{\Sigma x_i}{6} = 1.25, \qquad \bar{z} = \frac{\Sigma z_i}{6} = 0.47$$

$$\frac{\Sigma x_i^2}{6} = 2.29 \qquad \frac{\Sigma x_i z_i}{6} = 0.23.$$

Therefore,

$$a = -0.49, \qquad b = 1.06.$$

Hence,

$$M = e^b = 2.94$$

and

$$y = 2.94e^{-0.49x}. \quad \blacksquare$$

5.4.1   MINIMIZING $\|\mathbf{Ax} - \mathbf{b}\|$

Recall Theorem 5.3 which asserts that if $\mathbf{P}$ is a projection and $\mathbf{b}$ and $\mathbf{y}$ are arbitrary vectors,

$$\mathbf{Py} \perp (\mathbf{Pb} - \mathbf{b}). \tag{5.4.10}$$

Therefore, by the Pythagorean theorem (Example 5.2.1),

$$\|\mathbf{Pb} - \mathbf{b} + \mathbf{Py}\|^2 = \|\mathbf{Pb} - \mathbf{b}\|^2 + \|\mathbf{Py}\|^2. \tag{5.4.11}$$

Suppose that $\mathbf{z}$ is a given vector. Define $\mathbf{y} = \mathbf{Az}$. Then $\mathbf{A}^T\mathbf{y} = \mathbf{A}^T\mathbf{Az}$ and hence

$$\mathbf{z} = (\mathbf{A}^T\mathbf{A})^{-1}\mathbf{A}^T\mathbf{y} \tag{5.4.12}$$

and it follows that

$$\mathbf{Az} = \mathbf{A}(\mathbf{A}^T\mathbf{A})^{-1}\mathbf{A}^T\mathbf{y}. \tag{5.4.13}$$

Let $\mathbf{P} = \mathbf{A}(\mathbf{A}^T\mathbf{A})^{-1}\mathbf{A}^T$. Then $\mathbf{P}$ is a projection (see Example 5.3.13). So, using Eq. 5.4.13,

$$\mathbf{Az} = \mathbf{Py}. \tag{5.4.14}$$

Therefore,

$$\begin{aligned}
\|\mathbf{Az} - \mathbf{b}\|^2 &= \|\mathbf{Az} - \mathbf{b} + \mathbf{Pb} - \mathbf{Pb}\|^2 \\
&= \|\mathbf{Py} - \mathbf{b} + \mathbf{Pb} - \mathbf{Pb}\|^2 \tag{5.4.15}
\end{aligned}$$

using Eq. 5.4.14. It then follows that

$$\begin{aligned}
\|\mathbf{Az} - \mathbf{b}\|^2 &= \|\mathbf{Pb} - \mathbf{b} + \mathbf{Py} - \mathbf{Pb}\|^2 \\
&= \|\mathbf{Pb} - \mathbf{b} + \mathbf{P}(\mathbf{y} - \mathbf{b})\|^2 \\
&= \|\mathbf{Pb} - \mathbf{b}\|^2 + \|\mathbf{P}(\mathbf{y} - \mathbf{b})\|^2. \tag{5.4.16}
\end{aligned}$$

The last equality is Eq. 5.4.11, in which $\mathbf{y} - \mathbf{b}$ plays the role of $\mathbf{y}$. Thus, for every $\mathbf{z}$,

$$\|\mathbf{Az} - \mathbf{b}\|^2 \geq \|\mathbf{Pb} - \mathbf{b}\|^2. \tag{5.4.17}$$

But $\mathbf{Pb} = \mathbf{A}(\mathbf{A}^T\mathbf{A})\mathbf{A}^T\mathbf{b}$ by definition of $\mathbf{P}$, and by definition of $\hat{\mathbf{x}}$, Eq. 5.4.6, $\mathbf{Pb} = \mathbf{A}\hat{\mathbf{x}}$. Hence, inequality 5.4.17 is

$$\| \mathbf{Az} - \mathbf{b} \|^2 \geq \| \mathbf{A}\hat{\mathbf{x}} - \mathbf{b} \|^2 \qquad (5.4.18)$$

and we have proved that $\hat{\mathbf{x}}$ is the vector minimizing $\| \mathbf{Ax} - \mathbf{b} \|$.

PROBLEMS

1. Show that $\dfrac{\Sigma (x_i - \bar{x})^2}{n} = \dfrac{\Sigma x_i^2}{n} - \bar{x}^2$.

2. Show that $\dfrac{\Sigma (x_i - \bar{x})(y_i - \bar{y})}{n} = \dfrac{\Sigma x_i y_i}{n} - \bar{x}\bar{y}$

3. Use Problems 1 and 2 above to show that Eq. 5.4.9 can be written as

$$a = \frac{\Sigma (x_i - \bar{x})(y_i - \bar{y})}{\Sigma (x_i - \bar{x})^2}.$$

4. Given the data $(x_1, y_1), (x_2, y_2), \ldots, (x_n, y_n)$, find the normal equations for the system

$$\alpha + \beta x_1 + \gamma x_1^2 = y_1$$
$$\alpha + \beta x_2 + \gamma x_2^2 = y_2$$
$$\vdots$$
$$\alpha + \beta x_n + \gamma x_n^2 = y_n.$$

5. Course grades are based on the following conversion from letter grades to numerical grades; $A = 4$, $A- = 3.7$, $B+ = 3.3$, $B = 3$, $B- = 2.7$, and so on. The final grades for a class of students in ENG 106 and ENG 107 are tabulated by using the conversion above.

ENG 106	ENG 107	ENG 106	ENG 107
2.0	1.3	2.7	3.0
3.3	3.3	4.0	4.0
3.7	3.3	3.7	3.0
2.0	2.0	3.0	2.7
2.3	1.7	2.3	3.0

Calculate the least squares regression line for these data. Predict the grade in ENG 107 for a student who received a B+ in ENG 106.

6. In Table A4 the Bessel function $J_0(x)$ has been tabulated for various values of x. Data from this table follow.

x	J_0
6.0	0.15065
6.2	0.20175
6.4	0.24331
6.6	0.27404
6.8	0.29310

Find the least squares regression line and use this line to predict $J_0(6.3)$ and $J_0(7.0)$. Compare with the actual values in the table.

7. Same as Problem 6 but assume a quadratic fit $\alpha + \beta x + \gamma x^2 = y$. Estimate $J_0(6.3)$ and $J_0(7.0)$ and compare with the answers in Problem 6.

5.5   Eigenvalues and Eigenvectors

Suppose that $x(t)$ and $y(t)$ are unknown, differentiable functions of t which satisfy the differential equations

$$x' = x + y$$
$$y' = 4x + y. \tag{5.5.1}$$

As is usual in the case of constant-coefficient, linear differential equations, we assume exponential solutions. So set

$$x(t) = u_1 e^{\lambda t}$$
$$y(t) = u_2 e^{\lambda t} \tag{5.5.2}$$

where u_1, u_2, and λ are constants. After substitution of these functions into Eqs. 5.5.1, we obtain

$$\lambda u_1 e^{\lambda t} = u_1 e^{\lambda t} + u_2 e^{\lambda t}$$
$$\lambda u_2 e^{\lambda t} = 4u_1 e^{\lambda t} + u_2 e^{\lambda t}. \tag{5.5.3}$$

Since $e^{\lambda t} > 0$ for all λ and t, Eqs. 5.5.3 can be simplified to

$$\lambda u_1 = u_1 + u_2$$
$$\lambda u_2 = 4u_1 + u_2. \tag{5.5.4}$$

.ese are the equations determining λ, u_1, u_2 and therefore $x(t)$ and $y(t)$. In matrix-vector form these equations can be written as

$$\begin{bmatrix} 1 & 1 \\ 4 & 1 \end{bmatrix} \begin{bmatrix} u_1 \\ u_2 \end{bmatrix} = \lambda \begin{bmatrix} u_1 \\ u_2 \end{bmatrix}. \tag{5.5.5}$$

This system is an example of the *algebraic eigenvalue–engenvector problem:* For the square matrix **A**, find scalars λ and nonzero vectors **x** such that

$$\mathbf{A}\mathbf{x} = \lambda\mathbf{x}. \tag{5.5.6}$$

The scalar λ represents the *eigenvalues* of **A**, and the corresponding nonzero vectors **x** are the *eigenvectors*. From $\mathbf{A}\mathbf{x} = \lambda\mathbf{x}$ we deduce that $\mathbf{A}\mathbf{x} - \lambda\mathbf{x} = \mathbf{0}$, or, using $\lambda\mathbf{x} = \lambda\mathbf{I}\mathbf{x}$,

$$(\mathbf{A} - \lambda\mathbf{I})\mathbf{x} = \mathbf{0}. \tag{5.5.7}$$

This homogeneous system is equivalent to the original formulation $\mathbf{A}\mathbf{x} = \lambda\mathbf{x}$ but has the advantage of separating the computation of λ from that of **x**, for Eq. 5.5.7 has a nontrivial solution if and only if

$$|\mathbf{A} - \lambda\mathbf{I}| = 0. \tag{5.5.8}$$

Example 5.5.1:   Find the eigenvalues of the matrix

$$\mathbf{A} = \begin{bmatrix} 1 & 1 \\ 4 & 1 \end{bmatrix}$$

appearing in Eq. 5.5.5.

SOLUTION: We have

$$|\mathbf{A} - \lambda\mathbf{I}| = \begin{vmatrix} 1 - \lambda & 1 \\ 4 & 1 - \lambda \end{vmatrix}$$

$$= (1 - \lambda)^2 - 4 = (\lambda - 3)(\lambda + 1) = 0.$$

Hence, the eigenvalues are

$$\lambda_1 = 3, \qquad \lambda_2 = -1. \quad \blacksquare$$

Example 5.5.2:   Find the eigenvalues of

$$\mathbf{A} = \begin{bmatrix} 1 & 1 \\ 0 & 1 \end{bmatrix}.$$

SOLUTION: In this case

$$|\mathbf{A} - \lambda\mathbf{I}| = \begin{vmatrix} 1 - \lambda & 1 \\ 0 & 1 - \lambda \end{vmatrix}$$

$$= (1 - \lambda)^2 = 0.$$

Hence, the only eigenvalue* is

$$\lambda = 1. \quad \blacksquare$$

---

*It is usual to say that $\lambda = 1$ is a double root of $(1 - \lambda)^2 = 0$. We would then write $\lambda_1 = \lambda_2 = 1$.

Example 5.5.3:   Find the eigenvalues of

$$\mathbf{A} = \begin{bmatrix} -1 & 2 & 2 \\ 2 & 2 & 2 \\ -3 & -6 & -6 \end{bmatrix}.$$

SOLUTION: Here

$$|\mathbf{A} - \lambda\mathbf{I}| = \begin{vmatrix} -1 - \lambda & 2 & 2 \\ 2 & 2 - \lambda & 2 \\ -3 & -6 & -6 - \lambda \end{vmatrix}.$$

After some simplification

$$|\mathbf{A} - \lambda\mathbf{I}| = \begin{vmatrix} -1 - \lambda & 2 & 0 \\ 2 & 2 - \lambda & \lambda \\ -1 & -4 - \lambda & 0 \end{vmatrix}$$

$$= -\lambda \begin{vmatrix} -1 - \lambda & 2 \\ -1 & -4 - \lambda \end{vmatrix}$$

$$= -\lambda(\lambda + 2)(\lambda + 3) = 0.$$

So the eigenvalues are

$$\lambda_1 = 0, \qquad \lambda_2 = -2, \qquad \lambda_3 = -3. \quad \blacksquare$$

These examples suggest that $|\mathbf{A} - \lambda\mathbf{I}|$ is a polynomial of degree n in λ when $\mathbf{A}$ is $n \times n$. Indeed, for constants $c_0, c_1, \ldots, c_{n-1}$ which are functions of the entries in $\mathbf{A}$, we can show that

$$|\mathbf{A} - \lambda\mathbf{I}| = \begin{vmatrix} a_{11} - \lambda & a_{12} & \cdots & a_{1n} \\ a_{21} & a_{22} - \lambda & \cdots & a_{2n} \\ \vdots & & & \\ a_{n1} & a_{n2} & & a_{nn} - \lambda \end{vmatrix}$$

$$= (-\lambda)^n + c_{n-1}(-\lambda)^{n-1} + \cdots + c_1(-\lambda) + c_0.$$

$$(5.5.9)$$

We write

$$C(\lambda) = |\mathbf{A} - \lambda\mathbf{I}| \qquad\qquad (5.5.10)$$

and call $C(\lambda)$ the *characteristic polynomial* of $\mathbf{A}$; it is a polynomial of degree n and it has n roots, some or all of which may be repeated, some or all of which may be complex.

Once an eigenvalue has been determined, say $\lambda = \lambda_1$, then $\mathbf{A} - \lambda_1\mathbf{I}$ is a specific, singular matrix and the homogeneous system 5.5.7 may be solved. Here are some illustrations.

Example 5.5.4:   Find the eigenvectors for the matrix in Example 5.5.2,

$$\mathbf{A} = \begin{bmatrix} 1 & 1 \\ 0 & 1 \end{bmatrix}.$$

SOLUTION: In that example, $C(\lambda) = (1 - \lambda)^2$, so there is only one distinct eigenvalue, $\lambda = 1$. Therefore, the eigenvectors of $\mathbf{A}$ satisfy

$$\begin{bmatrix} 0 & 1 \\ 0 & 0 \end{bmatrix} \mathbf{x} = \mathbf{0}.$$

Therefore, for all nonzero choices of the scalar k, the eigenvectors are given by

$$\mathbf{x} = k \begin{bmatrix} 1 \\ 0 \end{bmatrix}. \quad \blacksquare$$

Example 5.5.5:   Find the eigenvalues and eigenvectors of the identity matrix,

$$\mathbf{I}_2 = \begin{bmatrix} 1 & 0 \\ 0 & 1 \end{bmatrix}.$$

SOLUTION: In this problem, as in the one just preceding, $C(\lambda) = (1 - \lambda)^2$. Hence, the eigenvectors are the nontrivial solutions of

$$\begin{bmatrix} 0 & 0 \\ 0 & 0 \end{bmatrix} \mathbf{x} = \mathbf{0}.$$

From the two basic solutions

$$\mathbf{x}_1 = \begin{bmatrix} 1 \\ 0 \end{bmatrix}, \qquad \mathbf{x}_2 = \begin{bmatrix} 0 \\ 1 \end{bmatrix}$$

we form the family of solutions

$$c_1 \mathbf{x}_1 + c_2 \mathbf{x}_2 = \begin{bmatrix} c_1 \\ c_2 \end{bmatrix}.$$

for all c_1 and c_2 except $c_1 = c_2 = 0$. $\blacksquare$

Example 5.5.6:   Find the eigenvectors of the matrix of Example 5.5.3,

$$\mathbf{A} = \begin{bmatrix} -1 & 2 & 2 \\ 2 & 2 & 2 \\ -3 & -6 & -6 \end{bmatrix}.$$

SOLUTION: We have already found that

$$C(\lambda) = -\lambda (\lambda + 2)(\lambda + 3).$$

We have, therefore, three cases

(i) Set $\lambda_1 = 0$. Then

$$\begin{bmatrix} -1 & 2 & 2 \\ 2 & 2 & 2 \\ -3 & -6 & -6 \end{bmatrix} \mathbf{x} = \mathbf{0}.$$

which, after some elementary row operations, is equivalent to

$$\begin{bmatrix} 1 & 0 & 0 \\ 0 & 1 & 1 \\ 0 & 0 & 0 \end{bmatrix} \mathbf{x} = \mathbf{0}.$$

So the basic solution, corresponding to $\lambda_1 = 0$, is

$$\mathbf{x}_1 = \begin{bmatrix} 0 \\ -1 \\ 1 \end{bmatrix}.$$

(ii) For $\lambda_2 = -2$, we have

$$\mathbf{A} + 2\mathbf{I} = \begin{bmatrix} 1 & 2 & 2 \\ 2 & 4 & 2 \\ -3 & -6 & -4 \end{bmatrix} \longrightarrow \cdots \longrightarrow \begin{bmatrix} 1 & 2 & 0 \\ 0 & 0 & 1 \\ 0 & 0 & 0 \end{bmatrix}.$$

Hence, the only basic solution, corresponding to $\lambda_2 = -2$, is

$$\mathbf{x}_2 = \begin{bmatrix} -2 \\ 1 \\ 0 \end{bmatrix}.$$

(iii) For $\lambda_3 = -3$, we have

$$\mathbf{A} + 3\mathbf{I} = \begin{bmatrix} 2 & 2 & 2 \\ 2 & 5 & 2 \\ -3 & -6 & -3 \end{bmatrix} \longrightarrow \cdots \longrightarrow \begin{bmatrix} 1 & 0 & 1 \\ 0 & 1 & 0 \\ 0 & 0 & 0 \end{bmatrix}.$$

Here the only basic solution, corresponding to $\lambda_3 = -3$, is

$$\mathbf{x}_3 = \begin{bmatrix} -1 \\ 0 \\ 1 \end{bmatrix}. \quad \blacksquare$$

Example 5.5.7:   Find the eigenvectors for the matrix

$$\mathbf{A} = \begin{bmatrix} 1 & 1 \\ 4 & 1 \end{bmatrix}$$

of Example 5.5.1 and thereby solve the system of differential equations 5.5.1.

SOLUTION: We have already found that $\lambda_1 = -1$, and $\lambda_2 = 3$.

(i) For $\lambda_1 = -1$ we have

$$\begin{bmatrix} 2 & 1 \\ 4 & 2 \end{bmatrix} \longrightarrow \cdots \longrightarrow \begin{bmatrix} 1 & \frac{1}{2} \\ 0 & 0 \end{bmatrix},$$

so

$$\mathbf{u}_1 = \begin{bmatrix} -\frac{1}{2} \\ 1 \end{bmatrix}.$$

(ii) For $\lambda_2 = 3$ we have

$$\begin{bmatrix} -2 & 1 \\ 4 & -2 \end{bmatrix} \longrightarrow \cdots \longrightarrow \begin{bmatrix} 1 & -\frac{1}{2} \\ 0 & 0 \end{bmatrix}.$$

Thus,

$$\mathbf{u}_2 = \begin{bmatrix} \frac{1}{2} \\ 1 \end{bmatrix}.$$

For the eigenvalue $\lambda_1 = -1$ we have

$$x_1(t) = -\tfrac{1}{2} e^{-t}, \qquad y_1(t) = e^{-t}$$

and for $\lambda_2 = 3$

$$x_2(t) = \tfrac{1}{2} e^{3t}, \qquad y_2(t) = e^{3t}.$$

We shall show later that the general solution of Eqs. 5.5.1 is a superposition of these solutions as follows:

$$x(t) = -\tfrac{1}{2} c_1 e^{-t} + \tfrac{1}{2} c_2 e^{3t}$$
$$y(t) = c_1 e^{-t} + c_2 e^{3t}. \quad \blacksquare$$

5.5.1   SOME THEORETICAL CONSIDERATIONS

We know that

$$C(\lambda) = |\mathbf{A} - \lambda \mathbf{I}|$$
$$= (-\lambda)^n + c_{n-1}(-\lambda)^{n-1} + \cdots + c_1(-\lambda) + c_0$$

$$(5.5.11)$$

is an identity in λ; by setting $\lambda = 0$ we learn that

$$C(0) = |\mathbf{A}| = c_0. \tag{5.5.12}$$

On the other hand, if we denote the roots of $C(\lambda) = 0$ as $\lambda_1, \lambda_2, \ldots, \lambda_n$, by elementary algebra $C(\lambda)$ has the factored form

$$C(\lambda) = (\lambda_1 - \lambda)(\lambda_2 - \lambda) \cdots (\lambda_n - \lambda). \tag{5.5.13}$$

Hence,

$$C(0) = \lambda_1 \cdot \lambda_2 \cdots \lambda_n. \tag{5.5.14}$$

This leads to the surprising result

$$|\mathbf{A}| = \lambda_1 \cdot \lambda_2 \cdots \lambda_n. \tag{5.5.15}$$

Theorem 5.5:   *The determinant of* $\mathbf{A}$ *is the product of its eigenvalues.*

Since $|\mathbf{A}| = 0$ if and only if $\mathbf{A}$ is singular, Theorem 5.5 leads to the following corollary.

Corollary 5.6: **A** *is singular if and only if it has a zero eigenvalue.*

Equation 5.5.13 reveals another connection between the eigenvalues and the entries of **A**. It follows from Eq. 5.5.13 that

$$C(\lambda) = (-\lambda)^n + (\lambda_1 + \lambda_2 + \lambda_3 + \cdots + \lambda_n)(-\lambda)^{n-1} + \cdots$$

(5.5.16)

by multiplying the n factors together. Hence,

$$c_{n-1} = \lambda_1 + \lambda_2 + \cdots + \lambda_n.$$

(5.5.17)

In problems 51 and 52 we show that

$$C(\lambda) = (-\lambda)^n + (a_{11} + a_{22} + \cdots + a_{nn})(-\lambda)^{n-1} + \cdots.$$

(5.5.18)

Since the coefficients of $C(\lambda)$ are the same regardless of how they are represented, Eqs. 5.5.16 and 5.5.17 show that

$$\lambda_1 + \lambda_2 + \cdots + \lambda_n = a_{11} + a_{22} + \cdots + a_{nn}.$$

(5.5.19)

The latter sum, $a_{11} + a_{22} + \cdots + a_{nn}$, the sum of the diagonal elements of **A**, is known as the *trace* of **A** and is written tr **A**. Thus, analogous to Theorem 5.5, we have the following theorem.

Theorem 5.7: *The trace of* **A** *is the sum of its eigenvalues.*

This theorem can sometimes be used to find eigenvalues in certain special cases. It is most effective when used in conjunction with the next theorem.

Theorem 5.8: *If* λ_1 *is an eigenvalue of* **A** *and*

$$\text{rank } (\mathbf{A} - \lambda_1 \mathbf{I}) = n - k,$$

then λ_1 *is repeated* k *times as a root of* $C(\lambda) = 0$.

PROOF: In fact, λ_1 may be repeated more than k times, but in any case, λ_1 is a factor of $C(\lambda)$. The proof is not difficult and is left to the reader (see Problem 32 of Section 5.6).

Example 5.5.8:   Find the eigenvalues of

$$\mathbf{J}_n = \begin{bmatrix} 1 & 1 & \cdots & 1 \\ 1 & 1 & \cdots & 1 \\ \vdots & & & \\ 1 & 1 & \cdots & 1 \end{bmatrix}_{n \times n}.$$

SOLUTION: Since $\mathbf{J}_n$ is singular, $\lambda = 0$ is an eigenvalue from Corollary 5.6. Also, rank $(\mathbf{J}_n - 0\mathbf{I}) = $ rank $\mathbf{J}_n = 1$, the root $\lambda = 0$ of $C(\lambda)$ is repeated $n - 1$ times. Therefore, by Theorem 5.7,

$$\text{tr } \mathbf{J}_n = n = 0 + 0 + \cdots + \lambda_n$$

and the sole remaining eigenvalue is

$$\lambda_n = n. \quad \blacksquare$$

It is often possible to find the eigenvalues of **B** from the eigenvalues of **A** if **A** and **B** are related. We state without proof (see Problem 64 for a partial proof) the most remarkable theorem of this type.

Theorem 5.9:   *Let*

$$p(x) = a_0 x^n + a_1 x^{n-1} + \cdots + a_n \qquad (5.5.20)$$

be a polynomial of degree n. Define

$$p(\mathbf{A}) = a_0 \mathbf{A}^n + a_1 \mathbf{A}^{n-1} + \cdots + a_n \mathbf{I}. \qquad (5.5.21)$$

*Then if $\lambda_1, \lambda_2, \ldots, \lambda_n$ are the eigenvalues of **A**, $p(\lambda_1), p(\lambda_2), \ldots, p(\lambda_n)$ are the eigenvalues of $p(\mathbf{A})$. Moreover, the eigenvectors of **A** are the eigenvectors of $p(\mathbf{A})$.*

An illustration of this follows.

Example 5.5.9:   Find the eigenvalues of

$$\mathbf{K}_n = \begin{bmatrix} 0 & 1 & 1 & \cdots & 1 \\ 1 & 0 & 1 & \cdots & 1 \\ \vdots & \vdots & & & \vdots \\ 1 & 1 & 1 & \cdots & 0 \end{bmatrix}_{n \times n}.$$

SOLUTION: We recognize that $\mathbf{K}_n = \mathbf{J}_n - \mathbf{I}$, where $\mathbf{J}_n$ is defined in Example 5.5.8. So if $p(x) = x - 1, p(\mathbf{J}_n) = \mathbf{J}_n - \mathbf{I}$ and Theorem 5.9 is relevent. The eigenvalues of $\mathbf{J}_n$ were shown to be

$$0 = \lambda_1 = \lambda_2 = \cdots = \lambda_{n-1}, \qquad n = \lambda_n$$

so the eigenvalues of $\mathbf{K}_n$ are

$$-1 = \lambda_1 = \lambda_2 = \cdots = \lambda_{n-1}, \qquad n - 1 = \lambda_n.$$

The characteristic polynomial of $\mathbf{K}_n$ is, therefore,

$$C(\lambda) = (-1 - \lambda)^{n-1}(-n + 1 - \lambda) = (-1)^n (\lambda + 1)^{n-1}(\lambda + n - 1). \quad \blacksquare$$

PROBLEMS

Find the characteristic polynomial and the eigenvalues for each matrix.

1. $\begin{bmatrix} 1 & 4 \\ 2 & 3 \end{bmatrix}$ **2.** $\begin{bmatrix} 0 & 4 \\ 1 & 0 \end{bmatrix}$ **3.** $\begin{bmatrix} 2 & 0 \\ 0 & -1 \end{bmatrix}$

4. $\begin{bmatrix} 0 & 3 \\ 3 & 8 \end{bmatrix}$ **5.** $\begin{bmatrix} 2 & 2 \\ -1 & -1 \end{bmatrix}$ **6.** $\begin{bmatrix} 5 & 4 \\ 4 & -1 \end{bmatrix}$

7. $\begin{bmatrix} 2 & -2 \\ 2 & 2 \end{bmatrix}$ **8.** $[1]$ **9.** $\mathbf{O}_{n \times n}$

10. $\mathbf{I}_n$ **11.** $\begin{bmatrix} 3 & 1 \\ 5 & -1 \end{bmatrix}$ **12.** $\begin{bmatrix} -1 & 3 & 0 \\ 3 & 7 & 0 \\ 0 & 0 & 6 \end{bmatrix}$

For the matrix in Problem 12:

13. Find the eigenvalues of $\mathbf{A}^T$.

14. Find the eigenvalues of $\mathbf{A}^{-1}$.

15. Find the eigenvalues of $\mathbf{A}^2$.

16. What are the connections between the eigenvalues of $\mathbf{A}$ and those of $\mathbf{A}^T$? $\mathbf{A}^{-1}$? $\mathbf{A}^2$?

Find the characteristic polynomial and the eigenvalues for each matrix.

17. $\begin{bmatrix} 2 & 2 & 0 \\ 1 & 2 & 1 \\ 1 & 2 & 1 \end{bmatrix}$ **18.** $\begin{bmatrix} 1 & 2 & 4 \\ 0 & 1 & 0 \\ 0 & 2 & 1 \end{bmatrix}$ **19.** $\begin{bmatrix} 1 & 1 & 0 \\ 0 & 1 & 1 \\ 0 & 0 & 1 \end{bmatrix}$

20. $\begin{bmatrix} 3 & 0 & 1 \\ 0 & 2 & 0 \\ 5 & 0 & -1 \end{bmatrix}$ **21.** $\begin{bmatrix} 10 & 8 & 0 \\ 8 & -2 & 0 \\ 0 & 0 & 4 \end{bmatrix}$ **22.** $\begin{bmatrix} 1 & 1 & -1 & 2 \\ 0 & 2 & 0 & 1 \\ 0 & 0 & -1 & 1 \\ 0 & 0 & 0 & 0 \end{bmatrix}$

23. $\begin{bmatrix} \cos \theta & \sin \theta \\ -\sin \theta & \cos \theta \end{bmatrix}$ **24.** $\begin{bmatrix} 0 & 1 \\ -b & -a \end{bmatrix}$

25. $\begin{bmatrix} 0 & 1 & 0 \\ 0 & 0 & 1 \\ -c & -b & -a \end{bmatrix}$ **26.** $\begin{bmatrix} \cos \theta & \sin \theta \\ \sin \theta & \cos \theta \end{bmatrix}$

Find the eigenvectors for the matrix in each problem. State if each has n linearly independent eigenvectors.

27. Problem 1   **28.** Problem 3   **29.** Problem 5

30. Problem 7   **31.** Problem 9   **32.** Problem 11

33. Problem 12   **34.** Problem 17   **35.** Problem 19

36. Problem 20   **37.** Problem 21   **38.** Problem 22

Verify Theorems 5.5 and 5.7 for the matrix in each problem.

39. Example 5.5.1   **40.** Example 5.5.2

41. Example 5.5.3   **42.** Example 5.5.4

Show without use of Theorem 5.9 that

43. If $\mathbf{B} = \alpha \mathbf{A}$, the eigenvalues of $\mathbf{B}$ are $\alpha \lambda_1, \alpha \lambda_2, \ldots, \alpha \lambda_n$.

44. If $\mathbf{B} = \mathbf{A} - k\mathbf{I}$, the eigenvalues of $\mathbf{B}$ are $\lambda_1 - k, \lambda_2 - k, \ldots, \lambda_n - k$.

45. How do the eigenvectors of $\mathbf{B}$ relate to those of $\mathbf{A}$ in Problem 43?

46. Show that the characteristic polynomials of $\mathbf{A}$ and $\mathbf{S}^{-1}\mathbf{A}\mathbf{S}$ are identical.

47. Show that the eigenvalues of $\mathbf{A}^T$ and $\mathbf{A}$ are identical. Are the eigenvectors the same?

48. If $\mathbf{A}$ is invertible and $\mathbf{B} = \mathbf{A}^{-1}$, show that the eigenvalues of $\mathbf{B}$ are

$$\frac{1}{\lambda_1}, \frac{1}{\lambda_2}, \cdots, \frac{1}{\lambda_n}.$$

49. Suppose that $\mathbf{u} \neq \mathbf{0}$ and $\mathbf{v} \neq \mathbf{0}$. If $\mathbf{u}^T\mathbf{v} \neq 0$, show that the characteristic polynomial of $\mathbf{u}\mathbf{v}^T$ is $C(\lambda) = (-\lambda)^{n-1}(\alpha - \lambda)$, where $\alpha = \mathbf{u}^T\mathbf{v}$. *Hint:* $\mathbf{u}\mathbf{v}^T$ is of rank 1 and $\mathbf{u}$ is an eigenvector of $\mathbf{u}\mathbf{v}^T$.

50. (a) Use definition 5.5.11 to show that

$$(-\lambda)^n C\left(\frac{1}{\lambda}\right) = c_0(-\lambda)^n + c_1(-\lambda)^{n-1} + \cdots + 1.$$

(b) Suppose that $\mathbf{A}^{-1}$ exists and $C^{-1}(\lambda)$ is its characteristic polynomial. Use

$$C^{-1}(\lambda) = |\mathbf{A}^{-1} - \lambda\mathbf{I}| = |\mathbf{A}^{-1}||\mathbf{I} - \lambda\mathbf{A}|$$

and part (a) to show that

$$C^{-1}(\lambda) = (-\lambda)^n + c_1|\mathbf{A}^{-1}|(-\lambda)^{n-1} + \cdots$$

(c) Prove that

$$c_1 = |\mathbf{A}| \operatorname{tr} \mathbf{A}^{-1} = |\mathbf{A}|\left(\frac{1}{\lambda_1} + \frac{1}{\lambda_2} + \cdots + \frac{1}{\lambda_n}\right).$$

51. Refer to the definition of a determinant to show that

$$|\mathbf{A} - \lambda\mathbf{I}| = (a_{11} - \lambda)(a_{22} - \lambda) \cdots (a_{nn} - \lambda) + Q_{n-2}(\lambda),$$

where $Q_{n-2}(\lambda)$ is a polynomial of degree $n - 2$ in λ.

52. Use the result in Problem 51 to prove Eq. 5.5.18.

53. Let $C(\lambda)$ be the characteristic polynomial of $\mathbf{A}$ and set $\mathbf{B} = C(\mathbf{A})$. Show that the eigenvalues of $\mathbf{B}$ are all zero.

54. If $\mathbf{P}$ is a projection matrix and $\lambda \neq 0$ is an eigenvalue of $\mathbf{P}$, show that $\lambda = 1$. *Hint:* Multiply $\mathbf{P}\mathbf{x} = \lambda\mathbf{x}$ by $\mathbf{P}$ and simplify.

55. If $\mathbf{M}^2 = \mathbf{I}$, show that the eigenvalues of $\mathbf{M}$ are either 1 or -1. (Use the hint in Problem 54.) If $\|\mathbf{u}\| = 1$, show that $\mathbf{I} - 2\mathbf{u}\mathbf{u}^T$ is just such a matrix.

56. Show that the characteristic polynomial of

$$\mathbf{A} = \begin{bmatrix} 0 & 1 & \cdots & 0 \\ 0 & 0 & \cdots & 0 \\ \vdots & \vdots & & \\ & & & \\ 0 & 0 & \cdots & 1 \\ -a_n & -a_{n-1} & \cdots & a_1 \end{bmatrix}$$

is

$$C(\lambda) = (-1)^n(\lambda^n + a_1\lambda^{n-1} + a_2\lambda^{n-2} + \cdots + a_n).$$

Use the result of Problem 56 to construct a matrix whose characteristic polynomial is given by each of the following.

57. $\lambda^2 - 1$ 58. $\lambda^2 + 1$

59. $\lambda^3 + \lambda^2 + \lambda + 1$ 60. $\lambda^4 - 1$

61. $(\lambda - \lambda_1)(\lambda - \lambda_2)$ 62. $(\lambda_1 - \lambda)(\lambda_2 - \lambda)(\lambda_3 - \lambda)$

63. Suppose that $A x_0 = \lambda_0 x_0$, $x_0 \neq 0$. Show that

$$(\alpha A^k)x_0 = (\alpha\lambda_0^k)x_0$$

for each integer $k = 1, 2, \ldots$. Hence, show that $p(A)x_0 = p(\lambda_0)x_0$. [This proves that every $p(\lambda_0)$ is an eigenvalue of $p(A)$. It does not prove that every eigenvalue of $p(A)$ is obtained in this manner. It also shows that x_0 is always an eigenvector of $p(A)$ but does not prove that every eigenvector of $p(A)$ is an eigenvector of A.]

64. Show that Problem 63 establishes Theorem 5.9 if A has n distinct eigenvalues.

65. Suppose that A is $n \times n$ and rank $(A - \lambda_1 I) = n - k$, $k > 0$. Show that $\lambda = \lambda_1$ is an eigenvalue and that there are k linearly independent eigenvectors associated with this eigenvalue.

5.6 Symmetric and Simple Matrices

A matrix can have no more than n linearly independent eigenvectors* but may have as few as one. The example

$$A = \begin{bmatrix} 2 & 1 & 0 & \cdots & 0 \\ 0 & 2 & 1 & \cdots & 0 \\ \vdots & \vdots & & & \\ 0 & 0 & & \cdots & 1 \\ 0 & 0 & & \cdots & 2 \end{bmatrix}_{n \times n} \qquad (5.6.1)$$

illustrates this point. Its characteristic polynomial is $(2 - \lambda)^n$ and, hence, $\lambda = 2$ is the only distinct eigenvalue. The system

$$(A - 2I)x = \begin{bmatrix} 0 & 1 & 0 & \cdots & 0 \\ 0 & 0 & 1 & \cdots & 0 \\ \vdots & \vdots & & & \\ 1 & 0 & & \cdots & 1 \\ 0 & 0 & & \cdots & 0 \end{bmatrix} x = 0 \qquad (5.6.2)$$

---

*Since each eigenvector is a vector with n entries, no set of $n + 1$ or more such vectors can be linearly independent.

has only one linearly independent solution, the eigenvector

$$\mathbf{x}_1 = \mathbf{e}_1 = \begin{bmatrix} 1 \\ 0 \\ \vdots \\ 0 \end{bmatrix}. \tag{5.6.3}$$

A matrix with n linearly independent eigenvectors is a *simple* matrix. The matrices $\mathbf{I}_n$ and those in Example 5.5.1 and 5.5.3 are simple. A matrix that is not simple is *defective*. The matrices in Example 5.5.2 and Eq. 5.6.1 are defective. In fact, all matrices of the form

$$\mathbf{A} = \begin{bmatrix} \alpha & 1 & 0 & 0 & \cdots & 0 \\ 0 & \alpha & * & * & \cdots & * \\ 0 & 0 & \alpha & * & & * \\ \vdots & & & & & \\ 0 & 0 & 0 & 0 & \cdots & \alpha \end{bmatrix} \tag{5.6.4}$$

are defective because $\lambda = \alpha$ is the only eigenvalue and, hence, the only eigenvectors are the nontrivial solutions of $(\mathbf{A} - \alpha\mathbf{I})\mathbf{x} = \mathbf{0}$. The reader may easily prove that this homogeneous system can have at most $n - 1$ linearly independent solutions. The problems exhibit various families of defective matrices.

One of the most remarkable theorems in matrix theory is that real, symmetric matrices are simple.*

Example 5.6.1:   Show that the following symmetric matrices are simple by finding n linearly independent eigenvectors for each.

$$\text{(a) } \mathbf{J}_n = \begin{bmatrix} 1 & 1 & \cdots & 1 \\ 1 & 1 & \cdots & 1 \\ \vdots & & & \\ 1 & 1 & \cdots & 1 \end{bmatrix}$$

$$\text{(b) } \mathbf{K}_n = \mathbf{J}_n - \mathbf{I}$$

SOLUTION: (a) The eigenvalue $\lambda = 0$ of $\mathbf{J}_n$ has $n - 1$ eigenvectors corresponding to it; namely,

$$\mathbf{x}_1 = \begin{bmatrix} -1 \\ 1 \\ 0 \\ \vdots \\ 0 \end{bmatrix}, \quad \mathbf{x}_2 = \begin{bmatrix} -1 \\ 0 \\ 1 \\ \vdots \\ 0 \end{bmatrix}, \quad \cdots, \quad \mathbf{x}_{n-1} = \begin{bmatrix} -1 \\ 0 \\ 0 \\ \vdots \\ 1 \end{bmatrix},$$

---

*We forgo the proof in this text.

the $n - 1$ basic solutions of $\mathbf{J}_n\mathbf{x} = \mathbf{0}$. The eigenvector corresponding to the remaining eigenvalue $\lambda = n$ can be found by inspection, for

$$\mathbf{J}_n \begin{bmatrix} 1 \\ 1 \\ \vdots \\ 1 \end{bmatrix} = \mathbf{n} \begin{bmatrix} 1 \\ 1 \\ \vdots \\ 1 \end{bmatrix}.$$

Thus, $\mathbf{J}_n$ is simple.

(b) For the matrix $\mathbf{K}_n$, we note that if $\mathbf{x}$ is an eigenvector of $\mathbf{J}_n$, $\mathbf{J}_n\mathbf{x} = \lambda\mathbf{x}$. Then,

$$\mathbf{K}_n\mathbf{x} = (\mathbf{J}_n - \mathbf{I})\mathbf{x}$$
$$= \mathbf{J}_n\mathbf{x} - \mathbf{x}$$
$$= \lambda\mathbf{x} - \mathbf{x} = (\lambda - 1)\mathbf{x}$$

and $\mathbf{x}$ is an eigenvector of $\mathbf{K}_n$ as well. So the n linearly independent eigenvectors of $\mathbf{J}_n$ are eigenvectors of $\mathbf{K}_n$ and hence $\mathbf{K}_n$ is simple. ∎

Example 5.6.2: Show that the following matrix is simple:

$$\mathbf{A} = \begin{bmatrix} 1 & 0 & 1 & 0 \\ 0 & 1 & 0 & 1 \\ 1 & 0 & 1 & 0 \\ 0 & 1 & 0 & 1 \end{bmatrix}.$$

SOLUTION: Since $\mathbf{A}$ is singular, $\lambda = 0$ is an eigenvalue and we compute two independent eigenvectors

$$\mathbf{x}_1 = \begin{bmatrix} -1 \\ 0 \\ 1 \\ 0 \end{bmatrix}, \qquad \mathbf{x}_2 = \begin{bmatrix} 0 \\ -1 \\ 0 \\ 1 \end{bmatrix}.$$

By inspection,

$$\mathbf{A} \begin{bmatrix} 1 \\ 0 \\ 1 \\ 0 \end{bmatrix} = 2 \begin{bmatrix} 1 \\ 0 \\ 1 \\ 0 \end{bmatrix}, \qquad \mathbf{A} \begin{bmatrix} 0 \\ 1 \\ 0 \\ 1 \end{bmatrix} = 2 \begin{bmatrix} 0 \\ 1 \\ 0 \\ 1 \end{bmatrix},$$

so that

$$\mathbf{x}_3 = \begin{bmatrix} 1 \\ 0 \\ 1 \\ 0 \end{bmatrix}, \qquad \mathbf{x}_4 = \begin{bmatrix} 0 \\ 1 \\ 0 \\ 1 \end{bmatrix}$$

are eigenvectors. The set $(\mathbf{x}_1, \mathbf{x}_2, \mathbf{x}_3, \mathbf{x}_4)$ is linearly independent. ∎

Here are three useful facts about real symmetric matrices.

1. They are simple.
2. Their eigenvalues are real.
3. Eigenvectors corresponding to different eigenvalues are orthogonal.

We prove (2) and (3) in Section 5.6.1. Note in Example 5.6.1 that $\mathbf{x}_n = [1, 1, \ldots, 1]^T$ is orthogonal to $\mathbf{x}_1, \mathbf{x}_2, \ldots, \mathbf{x}_{n-1}$ because $\mathbf{x}_n$ corresponds to the eigenvalue $\lambda = n$, while the others correspond to $\lambda = 0$. In Example 5.6.2, $\mathbf{x}_3$ and $\mathbf{x}_4$ are orthogonal to $\mathbf{x}_1$ and $\mathbf{x}_2$ and, by accident, are also mutually orthogonal. In regard to this last point, we can always use the Gram–Schmidt process to orthogonalize eigenvectors corresponding to the same eigenvalues since linear combinations of eigenvectors corresponding to the same eigenvalue are eigenvectors.

Example 5.6.3:   Orthogonalize the eigenvectors $\mathbf{x}_1, \mathbf{x}_2, \ldots, \mathbf{x}_{n-1}$ corresponding to $\lambda = 0$ for the matrix $\mathbf{J}_n$ of Example 5.6.1.

SOLUTION: We are interested in the orthogonal set of eigenvectors corresponding to $\lambda = 0$, not necessarily an orthonormal set. The reader is invited to verify that the Gram–Schmidt process yields orthonormal vectors proportional to

$$
\mathbf{y}_1 = \begin{bmatrix} -1 \\ 1 \\ 0 \\ \vdots \\ 0 \end{bmatrix}, \quad \mathbf{y}_2 = \begin{bmatrix} -1 \\ -1 \\ 2 \\ \vdots \\ 0 \end{bmatrix}, \quad \ldots, \quad \mathbf{y}_{n-1} = \begin{bmatrix} -1 \\ -1 \\ \vdots \\ -1 \\ n-1 \end{bmatrix}.
$$

Note that $\mathbf{J}_n \mathbf{y}_k = \mathbf{0}$ for $k = 1, 2, \ldots, n - 1$, and that $\{\mathbf{y}_1, \mathbf{y}_2, \ldots, \mathbf{y}_{n-1}, \mathbf{x}_n\}$ with

$$
\mathbf{x}_n = \begin{bmatrix} 1 \\ 1 \\ \vdots \\ 1 \end{bmatrix}
$$

form an orthogonal set of eigenvectors of $\mathbf{A}$.   ∎

5.6.1   COMPLEX VECTOR ALGEBRA

It is convenient at this point to review complex numbers with an eye towards extending our matrix theory into the complex domain. We briefly remind the reader that the complex number α is defined as

$$\alpha = a + ib \tag{5.6.5}$$

where a and b are real and $i = \sqrt{-1}$. The *complex conjugate* of α is $\bar{\alpha}$, defined

$$\bar{\alpha} = a - ib. \tag{5.6.6}$$

Hence, α is a real number if and only if $\alpha = \overline{\alpha}$. The extension of these ideas to matrices and vectors is straightforward. Suppose that

$$\mathbf{A} = (a_{ij})_{m \times n} = \begin{bmatrix} a_{11} & a_{12} & \cdots & a_{1n} \\ a_{21} & a_{22} & \cdots & a_{2n} \\ \vdots & \vdots & & \\ a_{m1} & a_{m2} & \cdots & a_{mn} \end{bmatrix}. \tag{5.6.7}$$

Then, by definition,

$$\overline{\mathbf{A}} = (\overline{a}_{ij})_{mxn}. \tag{5.6.8}$$

A special case of the definition above is

$$\mathbf{x} = \begin{bmatrix} x_1 \\ x_2 \\ \vdots \\ x_n \end{bmatrix} \text{ implies } \overline{\mathbf{x}} = \begin{bmatrix} \overline{x}_1 \\ \overline{x}_2 \\ \vdots \\ \overline{x}_n \end{bmatrix}. \tag{5.6.9}$$

From the definition of $\overline{\alpha}$, we find that

$$\alpha \overline{\alpha} = \overline{\alpha} \alpha = a^2 + b^2. \tag{5.6.10}$$

Since a and b are real numbers, $\alpha \overline{\alpha}$ is nonnegative; indeed, $\alpha \overline{\alpha}$ is zero if and only if $\alpha = 0$. There is a vector analog of Eq. 5.6.10. Consider

$$\mathbf{x}^T \overline{\mathbf{x}} = x_1 \overline{x}_1 + x_2 \overline{x}_2 + \cdots + x_n \overline{x}_n$$
$$= |x_1|^2 + |x_2|^2 + \cdots + |x_n|^2. \tag{5.6.11}$$

Hence, $\mathbf{x}^T \overline{\mathbf{x}} \geq 0$ and $\mathbf{x}^T \overline{\mathbf{x}} = 0$ if and only if $\mathbf{x} = \mathbf{0}$.

The complex conjugation operator applied to matrix algebra yields the following results:

(1) $\overline{\mathbf{AB}} = \overline{\mathbf{A}}\,\overline{\mathbf{B}}$

(2) $\overline{\mathbf{Ax}} = \overline{\mathbf{A}}\overline{\mathbf{x}}$

(3) $\overline{\alpha \mathbf{A}} = \overline{\alpha}\overline{\mathbf{A}}$ (5.6.12)

(4) $\overline{\mathbf{A} + \mathbf{B}} = \overline{\mathbf{A}} + \overline{\mathbf{B}}$

(5) $|\overline{\mathbf{A}}| = \overline{|\mathbf{A}|}$

All five results follow from the definition of $\overline{\mathbf{A}}$ and $\overline{\alpha}$ (see the Problems at the end of this section).

5.6.2 SOME THEORETICAL CONSIDERATIONS

We are now in a position to prove the following theorem.

Theorem 5.10:   *Real symmetric matrices have real eigenvalues.*

PROOF: Suppose that $\mathbf{x}$ is one eigenvector of $\mathbf{A}$ corresponding to the eigenvalue λ, so that

$$\mathbf{A}\mathbf{x} = \lambda \mathbf{x}. \tag{5.6.13}$$

Take the complex conjugate of both sides and find

$$\mathbf{A}\bar{\mathbf{x}} = \bar{\lambda}\bar{\mathbf{x}} \tag{5.6.14}$$

because $\overline{\mathbf{A}\mathbf{x}} = \bar{\mathbf{A}}\bar{\mathbf{x}} = \mathbf{A}\bar{\mathbf{x}}$ since $\mathbf{A}$ has only real entries. Now, consider

$$(\mathbf{A}\mathbf{x})^T\bar{\mathbf{x}} = (\lambda\mathbf{x})^T\bar{\mathbf{x}} = \lambda\mathbf{x}^T\bar{\mathbf{x}} \tag{5.6.15}$$

and

$$(\mathbf{A}\mathbf{x})^T\bar{\mathbf{x}} = \mathbf{x}^T(\mathbf{A}^T\bar{\mathbf{x}}) = \mathbf{x}^T\mathbf{A}\bar{\mathbf{x}} \tag{5.6.16}$$

since $\mathbf{A}^T = \mathbf{A}$. Thus, using Eq. 5.6.14, we have

$$(\mathbf{A}\mathbf{x})^T\bar{\mathbf{x}} = \mathbf{x}^T(\bar{\lambda}\bar{\mathbf{x}}) = \bar{\lambda}\mathbf{x}^T\bar{\mathbf{x}}. \tag{5.6.17}$$

From Eqs. 5.6.15 and 5.6.17, we can write

$$\lambda\mathbf{x}^T\bar{\mathbf{x}} = \bar{\lambda}\mathbf{x}^T\bar{\mathbf{x}}. \tag{5.6.18}$$

However, $\mathbf{x}^T\bar{\mathbf{x}} > 0$ since $\mathbf{x} \neq \mathbf{0}$, so that Eq. 5.6.18 implies that $\lambda = \bar{\lambda}$, which in turn means that λ is real.

Theorem 5.11:   *If $\mathbf{A}$ is real symmetric and $\lambda \neq \mu$ are two eigenvalues of $\mathbf{A}$ with corresponding eigenvectors $\mathbf{x}$ and $\mathbf{y}$, respectively, then $\mathbf{x} \perp \mathbf{y}$.*

Before we present the proof, we should note that $\mathbf{x}$ and $\mathbf{y}$ may always be taken as real vectors. The reason for this is that

$$\mathbf{A}\mathbf{x} = \lambda\mathbf{x} \tag{5.6.19}$$

and

$$\mathbf{A}\mathbf{y} = \mu\mathbf{y} \tag{5.6.20}$$

with λ and μ real, always have real solutions. If $\mathbf{x}$ is a real eigenvector of $\mathbf{A}$, then $i\mathbf{x}$ is an eigenvector of $\mathbf{A}$ which is not real. We explicitly exclude this possibility by convention—we use only the real solution to the homogeneous equations.

PROOF: Consider $(\mathbf{A}\mathbf{x})^T\mathbf{y}$. On the one hand, using Eq. 5.6.19,

$$\begin{aligned}(\mathbf{A}\mathbf{x})^T\mathbf{y} &= (\lambda\mathbf{x})^T\mathbf{y} \\ &= \lambda\mathbf{x}^T\mathbf{y},\end{aligned} \tag{5.6.21}$$

and on the other hand,

$$\begin{aligned}(\mathbf{A}\mathbf{x})^T\mathbf{y} &= (\mathbf{x}^T\mathbf{A}^T)\mathbf{y} \\ &= \mathbf{x}^T\mathbf{A}\mathbf{y}\end{aligned} \tag{5.6.22}$$

since $\mathbf{A}^T = \mathbf{A}$. Thus, from Eq. 5.6.20,

$$(\mathbf{Ax})^T\mathbf{y} = \mathbf{x}^T\mu\mathbf{y} \tag{5.6.23}$$

Therefore, $\lambda\mathbf{x}^T\mathbf{y} = \mu\mathbf{x}^T\mathbf{y}$ and $\mu \neq \lambda$. Hence, $\mathbf{x}^T\mathbf{y} = 0$, which implies that $\mathbf{x} \perp \mathbf{y}$.

5.6.3   SIMPLE MATRICES

Suppose that $\mathbf{A}$ is simple and $\mathbf{x}_1, \mathbf{x}_2, \ldots, \mathbf{x}_n$ are n linearly independent eigenvectors of $\mathbf{A}$ corresponding to the eigenvalues $\lambda_1, \lambda_2, \ldots, \lambda_n$, respectively. That is

$$\mathbf{Ax}_i = \lambda_i\mathbf{x}_i, \qquad i = 1, 2, \ldots, n. \tag{5.6.24}$$

Define $\mathbf{S}$ as

$$\mathbf{S} = [\mathbf{x}_1, \mathbf{x}_2, \ldots, \mathbf{x}_n] \tag{5.6.25}$$

so that $\mathbf{S}$ is the matrix whose columns are the eigenvectors of $\mathbf{A}$. Now, using Eq. 5.6.24, we see that

$$
\begin{aligned}
\mathbf{AS} &= \mathbf{A}[\mathbf{x}_1, \mathbf{x}_2, \ldots, \mathbf{x}_n] \\
&= [\mathbf{Ax}_1, \mathbf{Ax}_2, \ldots, \mathbf{Ax}_n] \\
&= [\lambda_1\mathbf{x}_1, \lambda_2\mathbf{x}_2, \ldots, \lambda_n\mathbf{x}_n] \\
&= \mathbf{S}\begin{bmatrix} \lambda_1 & 0 & \cdots & 0 \\ 0 & \lambda_2 & \cdots & 0 \\ \vdots & \vdots & & \\ 0 & 0 & \cdots & \lambda_n \end{bmatrix}.
\end{aligned}
\tag{5.6.26}
$$

Since the columns of $\mathbf{S}$ are linearly independent, $\mathbf{S}^{-1}$ exists and we have

$$\mathbf{S}^{-1}\mathbf{AS} = \begin{bmatrix} \lambda_1 & 0 & \cdots & 0 \\ 0 & \lambda_2 & \cdots & 0 \\ \vdots & \vdots & & \\ 0 & 0 & \cdots & \lambda_n \end{bmatrix}. \tag{5.6.27}$$

It is common to write

$$\mathbf{\Lambda} = \begin{bmatrix} \lambda_1 & 0 & \cdots & 0 \\ 0 & \lambda_2 & \cdots & 0 \\ \vdots & \vdots & & \\ 0 & 0 & \cdots & \lambda_n \end{bmatrix}. \tag{5.6.28}$$

Theorem 5.12:   *If* $\mathbf{A}$ *is simple, there exists a nonsingular matrix* $\mathbf{S}$ *such that*

$$\mathbf{S}^{-1}\mathbf{AS} = \mathbf{\Lambda}. \tag{5.6.29}$$

The columns of **S** *are eigenvectors of* **A** *and the diagonal entries of* **Λ** *are their corresponding eigenvalues.*

Example 5.6.4:   Verify Theorem 5.12 for

$$\mathbf{A} = \mathbf{J}_3 = \begin{bmatrix} 1 & 1 & 1 \\ 1 & 1 & 1 \\ 1 & 1 & 1 \end{bmatrix}.$$

SOLUTION: The eigenvectors of $\mathbf{J}_3$ are

$$\begin{bmatrix} -1 \\ 1 \\ 0 \end{bmatrix} \qquad \begin{bmatrix} -1 \\ 0 \\ 1 \end{bmatrix} \qquad \begin{bmatrix} 1 \\ 1 \\ 1 \end{bmatrix}$$

and hence,

$$\mathbf{S} = \begin{bmatrix} -1 & -1 & 1 \\ 1 & 0 & 1 \\ 0 & 1 & 1 \end{bmatrix}, \qquad \mathbf{S}^{-1} = \frac{1}{3} \begin{bmatrix} -1 & 2 & -1 \\ -1 & -1 & 2 \\ 1 & 1 & 1 \end{bmatrix},$$

so

$$\mathbf{S}^{-1} \mathbf{J}_3 \mathbf{S} = \mathbf{S}^{-1} \begin{bmatrix} 0 & 0 & 3 \\ 0 & 0 & 3 \\ 0 & 0 & 3 \end{bmatrix} = \begin{bmatrix} 0 & 0 & 0 \\ 0 & 0 & 0 \\ 0 & 0 & 3 \end{bmatrix},$$

as required.   ∎

Which matrices are simple? We have asserted without proof that symmetric matrices are simple (and incidentally, **Λ** is a real matrix in that case). We now show that matrices with distinct eigenvalues are simple. The whole proof hinges on showing that the set $\{\mathbf{x}_1, \mathbf{x}_2, \cdots, \mathbf{x}_n\}$ of eigenvectors corresponding to $\lambda_1, \lambda_2, \cdots, \lambda_n$ is linearly independent if $\lambda_i \neq \lambda_j$. The heart of the proof is seen by considering the special case of $\mathbf{A}_{3\times3}$. So, suppose that $\mathbf{x}_1$, $\mathbf{x}_2$, $\mathbf{x}_3$, are eigenvectors of **A** and

$$c_1\mathbf{x}_1 + c_2\mathbf{x}_2 + c_3\mathbf{x}_3 = \mathbf{0}. \tag{5.6.30}$$

We will multiply this equation by $\mathbf{A} - \lambda_1 \mathbf{I}$. Note that

$$(\mathbf{A} - \lambda_1 \mathbf{I})\mathbf{x}_2 = \mathbf{A}\mathbf{x}_2 - \lambda_1\mathbf{x}_2$$
$$= \lambda_2\mathbf{x}_2 - \lambda_1\mathbf{x}_2 = (\lambda_2 - \lambda_1)\mathbf{x}_2 \tag{5.6.31}$$

and, similarly,

$$(\mathbf{A} - \lambda_1 \mathbf{I})\mathbf{x}_3 = (\lambda_3 - \lambda_1)\mathbf{x}_3. \tag{5.6.32}$$

Therefore,

$$c_1(\mathbf{A} - \lambda_1 \mathbf{I})\mathbf{x}_1 + c_2(\mathbf{A} - \lambda_1 \mathbf{I})\mathbf{x}_2 + c_3(\mathbf{A} - \lambda_1 \mathbf{I})\mathbf{x}_3 = \mathbf{0} \tag{5.6.33}$$

leads to

$$c_2(\lambda_2 - \lambda_1)\mathbf{x}_2 + c_3(\lambda_3 - \lambda_1)\mathbf{x}_3 = \mathbf{0}. \tag{5.6.34}$$

Now multiply this equation by $(\mathbf{A} - \lambda_2\mathbf{I})$. So

$$c_2(\lambda_2 - \lambda_1)(\mathbf{A} - \lambda_2\mathbf{I})\mathbf{x}_2 + c_3(\lambda_3 - \lambda_1)(\mathbf{A} - \lambda_2\mathbf{I})\mathbf{x}_3 = \mathbf{0}, \qquad (5.6.35)$$

which leads to

$$c_3(\lambda_3 - \lambda_1)(\lambda_3 - \lambda_2)\mathbf{x}_3 = \mathbf{0}. \qquad (5.6.36)$$

Since $\lambda_3 \neq \lambda_1$ and $\lambda_3 \neq \lambda_2$, Eq. 5.6.36 implies that $c_3 = 0$. This in turn implies that $c_2 = 0$, from Eq. 5.6.34. But then $c_1 = 0$ from Eq. 5.6.30, so $\{\mathbf{x}_1, \mathbf{x}_2, \mathbf{x}_3\}$ is linearly independent. This proves:

Theorem 5.13:   *If **A** has n distinct eigenvalues, then **A** is simple.*

PROBLEMS

1. Compute $\mathbf{x}^T\overline{\mathbf{x}}$ for $\mathbf{x} = \begin{bmatrix} 1 \\ i \\ -1 \end{bmatrix}$. Compute $\mathbf{x}^T\mathbf{x}$ for this $\mathbf{x}$.

2. Compute $\mathbf{x}^T\overline{\mathbf{x}}$ and $\mathbf{x}^T\mathbf{x}$ for $\mathbf{x} = \begin{bmatrix} 1 \\ i \end{bmatrix}$.

3. Prove item (1) in Eq. 5.6.12.

4. Prove item (2) in Eq. 5.6.12.

5. Prove item (3) in Eq. 5.6.12.

6. Prove item (4) in Eq. 5.6.12.

7. Show that if $\mathbf{A}$ has eigenvalues $\lambda_1, \lambda_2, \ldots, \lambda_n$, then $\overline{\mathbf{A}}$ has eigenvalues $\overline{\lambda}_1, \overline{\lambda}_2, \ldots, \overline{\lambda}_n$.

8. Use Problem 7 to show that $|\overline{\mathbf{A}}| = \overline{|\mathbf{A}|}$ [see item (5) in Eq. 5.6.12].

9. Show that the diagonal entries of $\overline{\mathbf{A}}\mathbf{A}$, where $\mathbf{A}$ is $n \times n$, are nonnegative, real numbers. Are the off-diagonal entries of $\overline{\mathbf{A}}\mathbf{A}$ necessarily real?

10. Show that

$$\mathbf{x}_1 = \begin{bmatrix} -1 \\ 1 \\ 0 \end{bmatrix}, \qquad \mathbf{x}_2 = \begin{bmatrix} -1 \\ -1 \\ 2 \end{bmatrix}, \qquad \mathbf{x}_3 = \begin{bmatrix} 1 \\ 1 \\ 1 \end{bmatrix}$$

is a set of linearly independent eigenvectors of $\mathbf{J}_3$. Verify Theorem 5.12 for this set.

Verify that each matrix is defective.

11. $\begin{bmatrix} \alpha & 1 \\ 0 & \alpha \end{bmatrix}$ 12. $\begin{bmatrix} \alpha & 1 & 0 \\ 0 & \alpha & * \\ 0 & 0 & \beta \end{bmatrix}$ 13. $\begin{bmatrix} \alpha & 1 & 0 & 0 \\ 0 & \alpha & * & * \\ 0 & 0 & \gamma & * \\ 0 & 0 & 0 & \delta \end{bmatrix}$

Verify, by computing the eigenvectors, that each matrix is simple.

14. $\begin{bmatrix} \alpha & * \\ 0 & \beta \end{bmatrix}$, $\alpha \neq \beta$

15. $\begin{bmatrix} \alpha & 1 & 0 \\ 0 & \beta & 1 \\ 0 & 0 & \gamma \end{bmatrix}$, $\alpha \neq \beta \neq \gamma$

16. If $\mathbf{u}^T\mathbf{v} \neq 0$, show that $\mathbf{uv}^T$ is simple. Verify that $\mathbf{uv}^T$ is defective if $\mathbf{u} = \begin{bmatrix} 1 \\ -1 \end{bmatrix}$ and $\mathbf{v}^T = [1, 1]$.

17. If $\mathbf{Q}$ is orthogonal, show that $|\lambda| = 1$. *Hint*: Consider $(\mathbf{Qx})^T(\mathbf{Q\bar{x}})$.

18. Let $\mathbf{P}$ be a projection matrix. Show that rank $\mathbf{P} = \text{tr } \mathbf{A}$. *Hint:* Use Problem 54 of Section 5.5.

19. Let $C(\lambda)$ be the characteristic polynomial of $\mathbf{A}$. For the matrices of Problems 12 and 15, show that $C(\mathbf{A}) = \mathbf{O}$.

20. Suppose that $\mathbf{A}$ is real and skew-symmetric, $\mathbf{A}^T = -\mathbf{A}$. Following the lines of the proof of Theorem 5.10, show that $-\bar{\lambda}\mathbf{x}^T\bar{\mathbf{x}} = \lambda\mathbf{x}^T\bar{\mathbf{x}}$. Why does this prove that the eigenvalues of $\mathbf{A}$ are pure imaginary? Explain why $\mathbf{A}^2$ has nonpositive eigenvalues.

21. Find an example of a symmetric matrix with at least one nonreal entry which does not have real eigenvalues.

In the proof of Theorem 5.12, explain

22. $\mathbf{A}[\mathbf{x}_1, \mathbf{x}_2, \ldots, \mathbf{x}_n] = [\mathbf{Ax}_1, \mathbf{Ax}_2, \ldots, \mathbf{Ax}_n]$

23. $[\lambda_1, \mathbf{x}_1, \lambda_2\mathbf{x}_2, \ldots, \lambda_n\mathbf{x}_n] = [\mathbf{x}_1, \mathbf{x}_2, \ldots, \mathbf{x}_n]\begin{bmatrix} \lambda_1 & 0 & 0 \\ 0 & \lambda_2 & 0 \\ \vdots & \vdots & \\ 0 & 0 & \lambda_n \end{bmatrix}$

24. Suppose that $\mathbf{Ax} = \lambda\mathbf{x}$ and $\mathbf{Ay} = \mu\mathbf{y}$, $\lambda \neq \mu$. Show that $(\mathbf{A} - \lambda\mathbf{I})\mathbf{y} = (\mu - \lambda)\mathbf{y}$.

25. What is the converse of Theorem 5.13? If it is true, prove it. If it is false, give an example illustrating its falsity.

Suppose that $\mathbf{A}$ is simple and $\mathbf{S}^{-1}\mathbf{AS} = \Lambda$. Show that:

26. $\mathbf{A}^n = \mathbf{S}\Lambda^n\mathbf{S}^{-1}$

27. $p(\mathbf{A}) = \mathbf{S}p(\Lambda)\mathbf{S}^{-1}$ where $p(x) = a_0x^n + a_1x^{n-1} + \cdots + a_n$

28. $\mathbf{A}^{-1} = \mathbf{S}\Lambda^{-1}\mathbf{S}^{-1}$ if $\mathbf{A}^{-1}$ exists.

29. $C(\mathbf{A}) = \mathbf{O}$ (use Problem 53 of Section 5.5 and Problem 27, above).

30. Let $\mathbf{x}_1, \mathbf{x}_2, \ldots, \mathbf{x}_k$ be eigenvectors of $\mathbf{A}$ corresponding to $\lambda = \lambda_0$ and $[\mathbf{x}_1, \mathbf{x}_2, \ldots, \mathbf{x}_k, \mathbf{y}_{k+1}, \ldots, \mathbf{y}_n] = \mathbf{T}$ have linearly independent columns. Show that

$$T^{-1}AT = k \begin{bmatrix} \lambda_0 & 0 & \cdots & 0 & \vdots \\ 0 & \lambda_0 & \cdots & 0 & \vdots \\ \vdots & & & & \vdots & B \\ 0 & 0 & \cdots & \lambda_0 & \vdots \\ \hline & & O & & \vdots & C \end{bmatrix}.$$

Hint: Argue by analogy with the text preceding Eq. 5.6.7.

31. Under the assumptions in Problem 30, show that

$$T^{-1}AT - \lambda I = \left[\begin{array}{c|c} (\lambda_0 - \lambda)I & B \\ \hline O & C - \lambda I \end{array} \right]$$

Hence, show that

$$|T^{-1}AT - \lambda I| = |A - \lambda I|$$

implies that

$$|A - \lambda I| = (\lambda_0 - \lambda)^k P(\lambda).$$

32. Use the results of Problem 65 in Section 5.5 and Problem 31 to prove Theorem 5.8.

5.7 Systems of Linear Differential Equations: the Homogeneous Case

The system of linear, first-order differential equations

$$\begin{aligned} x_1' &= x_1 + x_2 \\ x_2' &= 4x_1 + x_2 \end{aligned} \tag{5.7.1}$$

of Section 5.5 is a special case of

$$x' = Ax \tag{5.7.2}$$

where

$$x = \begin{bmatrix} x_1(t) \\ x_2(t) \\ \vdots \\ x_n(t) \end{bmatrix}, \qquad x' = \begin{bmatrix} x_1'(t) \\ x_2'(t) \\ \vdots \\ x_n'(t) \end{bmatrix} \tag{5.7.3}$$

and **A** is a constant matrix. This system is *homogeneous*. The system

$$x' = Ax + f, \tag{5.7.4}$$

where

$$\mathbf{f} = \begin{bmatrix} f(t) \\ f_2(t) \\ \vdots \\ f_n(t) \end{bmatrix} \neq \mathbf{0} \tag{5.7.5}$$

is a vector of known functions, is *nonhomogeneous*. A knowledge of the "general" solution of the homogeneous problem will be shown to lead to a complete solution of the nonhomogeneous system.

The *initial-value problem* is the system

$$\mathbf{x}' = \mathbf{Ax} + \mathbf{f}, \qquad \mathbf{x}(t_0) = \mathbf{x}_0. \tag{5.7.6}$$

Here $\mathbf{x}_0$ is a given, fixed vector of constants. By the simple substitution $\tau = t - t_0$, we can convert the system 5.7.6 into a standard form

$$\mathbf{x}' = \mathbf{Ax} + \mathbf{f}, \qquad \mathbf{x}(0) = \mathbf{x}_0. \tag{5.7.7}$$

Unless otherwise specified, we shall assume that $t_0 = 0$, as in Eq. 5.7.7. In this section we wish to study a homogeneous system, so we use $\mathbf{f} = \mathbf{0}$.

The vector function, with $\mathbf{u}$ constant,

$$\mathbf{x}(t) = \mathbf{u}e^{\lambda t} \tag{5.7.8}$$

is a solution of $\mathbf{x}' = \mathbf{Ax}$ if and only if

$$\lambda \mathbf{u}e^{\lambda t} = \mathbf{Au}e^{\lambda t}. \tag{5.7.9}$$

Since $e^{\lambda t} > 0$, this system is equivalent to

$$\mathbf{Au} = \lambda \mathbf{u}. \tag{5.7.10}$$

We assume that $\mathbf{u} \neq \mathbf{0}$, for otherwise $\mathbf{x}(t) = \mathbf{0}$, and this is a trivial solution of Eq. 5.7.2. Thus, we can find a solution to the homogeneous problem if we can solve the corresponding algebraic eigenvalue-eigenvector problem, Eq. 5.7.10. We shall assume that $\mathbf{A}$ is simple and hence that $\mathbf{A}$ has n linearly independent eigenvectors $\mathbf{u}_1, \mathbf{u}_2, \ldots, \mathbf{u}_n$ corresponding to the eigenvalues $\lambda_1, \lambda_2, \ldots, \lambda_n$. Hence, system 5.7.2 has n solutions

$$\mathbf{x}_i(t) = \mathbf{u}_i e^{\lambda_i t}, \qquad i = 1, 2, \ldots, n. \tag{5.7.11}$$

By taking linear combinations of the solutions (Eq. 5.7.11), we generate an infinite family of solutions,

$$\mathbf{x}(t) = c_1 \mathbf{x}_1(t) + c_2 \mathbf{x}_2(t) + \cdots + c_n \mathbf{x}_n(t)$$

$$= \sum_{k=1}^{n} c_k \mathbf{u}_k e^{\lambda_k t}. \tag{5.7.12}$$

If we wish to choose a function from the family 5.7.12 which assumes the value x_0 at $t = 0$, we must solve

$$x(0) = x_0 = \sum_{k=1}^{n} c_k u_k \qquad (5.7.13)$$

which is always possible since there are n linearly independent u_k. We may rewrite Eq. 5.7.13 as

$$Uc = x_0 \qquad (5.7.14)$$

where

$$U = [u_1, u_2, \ldots, u_n], \qquad c = \begin{bmatrix} c_1 \\ c_2 \\ \vdots \\ c_n \end{bmatrix}. \qquad (5.7.15)$$

Because we can use Eq. 5.7.12 to solve any initial-value problem, we call $x(t)$ (see Eq. 5.7.12) the *general solution* of $x' = Ax$.

Example 5.7.1:   Find the general solution of the system 5.7.1 and then solve

$$x' = \begin{bmatrix} 1 & 1 \\ 4 & 1 \end{bmatrix} x, \qquad x(0) = \begin{bmatrix} 1 \\ 1 \end{bmatrix}.$$

SOLUTION: In Example 5.5.7 we have found eigenvectors for A; corresponding to $\lambda_1 = -1$ and $\lambda_2 = 3$,

$$u_1 = \begin{bmatrix} -\frac{1}{2} \\ 1 \end{bmatrix} \quad \text{and} \quad u_2 = \begin{bmatrix} \frac{1}{2} \\ 1 \end{bmatrix}$$

The general solution is then

$$x(t) = c_1 \begin{bmatrix} -\frac{1}{2} \\ 1 \end{bmatrix} e^{-t} + c_2 \begin{bmatrix} \frac{1}{2} \\ 1 \end{bmatrix} e^{3t}.$$

We find c_1 and c_2 by solving

$$x(0) = \begin{bmatrix} 1 \\ 1 \end{bmatrix} = \begin{bmatrix} -\frac{1}{2} & \frac{1}{2} \\ 1 & 1 \end{bmatrix} \begin{bmatrix} c_1 \\ c_2 \end{bmatrix}.$$

Hence,

$$c = \begin{bmatrix} -\frac{1}{2} \\ \frac{3}{2} \end{bmatrix}.$$

Therefore,

$$x(t) = -\frac{1}{2} \begin{bmatrix} -\frac{1}{2} \\ 1 \end{bmatrix} e^{-t} + \frac{3}{2} \begin{bmatrix} \frac{1}{2} \\ 1 \end{bmatrix} e^{3t}$$

solves the initial-value problem.   ■

Example 5.7.2:   Solve the initial-value problem

$$\mathbf{x}' = \begin{bmatrix} 0 & 1 & 0 \\ 0 & 0 & 1 \\ 2 & 1 & -2 \end{bmatrix} \mathbf{x}, \qquad \mathbf{x}(0) = \begin{bmatrix} 1 \\ 0 \\ 1 \end{bmatrix}.$$

SOLUTION: After some labor, we find

$$C(\lambda) = -(\lambda - 1)(\lambda + 1)(\lambda + 2)$$

and therefore,

$$\lambda_1 = 1, \qquad \lambda_2 = -1, \qquad \lambda_3 = -2.$$

We compute

$$\mathbf{u}_1 = \begin{bmatrix} 1 \\ -1 \\ 1 \end{bmatrix}, \qquad \mathbf{u}_2 = \begin{bmatrix} 1 \\ 1 \\ 1 \end{bmatrix}, \qquad \mathbf{u}_3 = \begin{bmatrix} 1 \\ -2 \\ 4 \end{bmatrix}.$$

The general solution is then

$$\mathbf{x}(t) = c_1 \begin{bmatrix} 1 \\ -1 \\ 1 \end{bmatrix} e^{-t} + c_2 \begin{bmatrix} 1 \\ 1 \\ 1 \end{bmatrix} e^t + c_3 \begin{bmatrix} 1 \\ -2 \\ 4 \end{bmatrix} e^{-2t}.$$

Hence,

$$\mathbf{x}(0) = \begin{bmatrix} 1 \\ 0 \\ 1 \end{bmatrix} = \begin{bmatrix} 1 & 1 & 1 \\ -1 & 1 & -2 \\ 1 & 1 & 4 \end{bmatrix} \mathbf{c}$$

yields

$$\mathbf{c} = \begin{bmatrix} \frac{1}{2} \\ \frac{1}{2} \\ 0 \end{bmatrix}.$$

Finally,

$$\mathbf{x}(t) = \frac{1}{2} \begin{bmatrix} 1 \\ -1 \\ 1 \end{bmatrix} e^{-t} + \frac{1}{2} \begin{bmatrix} 1 \\ 1 \\ 1 \end{bmatrix} e^t$$

$$= \begin{bmatrix} \frac{1}{2}(e^t + e^{-t}) \\ \frac{1}{2}(e^t - e^{-t}) \\ \frac{1}{2}(e^t + e^{-t}) \end{bmatrix} = \begin{bmatrix} \cosh t \\ \sinh t \\ \cosh t \end{bmatrix}$$

is the required solution.   ■

Example 5.7.3:   Show that an eigenvalue-eigenvector problem results when solving for the displacements of the two masses shown.

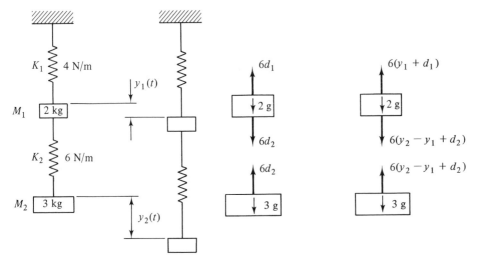

SOLUTION: We isolate the two masses and show all forces acting on each. The distances d_1 and d_2 are the amounts the springs are stretched while in static equilibrium. Using Newton's second law we may write:

Static equilibrium   *In motion*

$$0 = 6d_2 - 4d_1 + 2g \qquad 2y_1'' = 6(y_2 - y_1 + d_2) + 2g - 4(y_1 + d_1)$$
$$0 = 3g - 6d_2 \qquad 3y_2'' = 3g - 6(y_2 - y_1 + d_2)$$

These equations may be simplified to

$$y_1'' = -5y_1 + 3y_2$$
$$y_2'' = 2y_1 - 2y_2.$$

These two equations can be written as the single matrix equation

$$\mathbf{y}'' = \mathbf{Ay},$$

where

$$\mathbf{y} = \begin{bmatrix} y_1 \\ y_2 \end{bmatrix}, \qquad \mathbf{A} = \begin{bmatrix} -5 & 3 \\ 2 & -2 \end{bmatrix}.$$

We note that the coefficients of the two independent variables are all constants; hence, as is usual, we assume a solution in the form

$$\mathbf{y} = \mathbf{u}e^{mt},$$

where $\mathbf{u}$ is a constant vector to be determined, and m is a scalar to be determined. Our differential equation is then

$$\mathbf{u}m^2 e^{mt} = \mathbf{Au}e^{mt}$$

or

$$\mathbf{Au} = \lambda \mathbf{u}$$

where the parameter $\lambda = m^2$. This is an eigenvalue problem. The problem is solved by finding the eigenvalues λ_i and corresponding eigenvectors $\mathbf{x}_i$. The solutions $y_1(t)$ and $y_2(t)$ are then determined. ∎

Example 5.7.4:   Solve the eigenvalue–eigenvector problem obtained in Example 5.7.3 and find the solutions for $y_1(t)$ and $y_2(t)$.

SOLUTION: The eigenvalues are

$$\lambda_1 = \frac{-7 + \sqrt{33}}{2} = -0.6277, \qquad \lambda_2 = \frac{-7 - \sqrt{33}}{2} = -6.372.$$

The two eigenvectors are then

$$\mathbf{u}_1 = \begin{bmatrix} 1 \\ 1.46 \end{bmatrix}, \qquad \mathbf{u}_2 = \begin{bmatrix} 1 \\ -0.457 \end{bmatrix},$$

where the first component was arbitrarily chosen to be unity. The solutions $y_1(t)$ and $y_2(t)$ will now be determined. The constant m is related to the eigenvalues by $m^2 = \lambda$. Thus,

$$m_1^2 = -0.6277 \text{ and } m_2^2 = -6.372.$$

These give

$$m_1 = \pm 0.7923i, \qquad m_2 = \pm 2.524i.$$

We use both positive and negative roots and write the solution as

$$\mathbf{y}(t) = \mathbf{u}_1(c_1 e^{0.7923it} + d_1 e^{-0.7923it}) + \mathbf{u}_2(c_2 e^{2.524it} + d_2 e^{-2.524it}),$$

where we have superimposed all possible solutions introducing the arbitrary constants $c_1, d_1, c_2,$ and d_2, to obtain the most general solution. The arbitrary constants are then calculated from initial conditions.

The components of the solution vector can be written as

$$y_1(t) = a_1 \cos 0.7923t + b_1 \sin 0.7923t + a_2 \cos 2.524t + b_2 \sin 2.524t$$

$$y_2(t) = 1.46[a_1 \cos 0.7923t + b_1 \sin 0.7923t]$$

$$- 0.457[a_2 \cos 2.524t + b_2 \sin 2.524t]$$

Note that if we had made the eigenvectors of unit length, the arbitrary constants would simply change accordingly for a particular set of initial conditions. ∎

PROBLEMS

Find the general solution of the differential system $\mathbf{x}' = \mathbf{Ax}$ for each matrix $\mathbf{A}$.

1. $\begin{bmatrix} 1 & 3 \\ 1 & -3 \end{bmatrix}$

2. $\begin{bmatrix} 1 & 1 \\ 3 & -1 \end{bmatrix}$

3. $\begin{bmatrix} 1 & \alpha^2 - 1 \\ 1 & -1 \end{bmatrix} \alpha \neq 0$

4. $\begin{bmatrix} 2 & 1 \\ 2 & 3 \end{bmatrix}$

5. $\begin{bmatrix} 4 & -3 & -2 \\ 2 & -1 & -2 \\ 3 & -3 & -1 \end{bmatrix}$

6. $\begin{bmatrix} 1 & 1 & 1 \\ 0 & 2 & 1 \\ 0 & 0 & 3 \end{bmatrix}$

7. $\begin{bmatrix} 1 & 1 & 1 \\ 1 & -1 & 1 \\ 0 & 0 & 0 \end{bmatrix}$

8. $\begin{bmatrix} 1 & 1 & -1 \\ 0 & 0 & 1 \\ 0 & -2 & -3 \end{bmatrix}$

Solve the initial-value problem $\mathbf{x}' = \mathbf{Ax}$, $\mathbf{x}(0) = \mathbf{x_0}$ for each system.

9. Problem 1, $\mathbf{x_0} = \begin{bmatrix} 1 \\ 0 \end{bmatrix}$

10. Problem 1, $\mathbf{x_0} = \begin{bmatrix} 0 \\ 1 \end{bmatrix}$

11. Problem 1, $\mathbf{x_0} = \begin{bmatrix} 1 \\ 1 \end{bmatrix}$

12. Problem 8, $\mathbf{x_0} = \begin{bmatrix} 0 \\ 1 \\ 0 \end{bmatrix}$

13. Problem 8, $\mathbf{x_0} = \begin{bmatrix} 1 \\ 0 \\ 1 \end{bmatrix}$

14. Problem 8, $\mathbf{x_0} = \begin{bmatrix} 0 \\ 0 \\ 1 \end{bmatrix}$

15. Problem 8, $\mathbf{x_0} = \begin{bmatrix} -1 \\ 2 \\ -2 \end{bmatrix}$

Consider the electrical circuit shown.

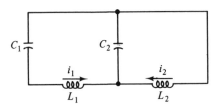

16. Derive the differential equations that describe the currents $i_1(t)$ and $i_2(t)$.

17. Write the differential equations in the matrix from $\mathbf{i}'' = \mathbf{Ai}$ and identify the elements in the coefficient matrix $\mathbf{A}$.

18. Let $\mathbf{i} = \mathbf{x}e^{mt}$ and show that an eigenvalue problem results.

19. Find the eigenvalues and unit eigenvectors if $L_1 = 1$, $L_2 = 2$, $C_1 = 0.02$, and $C_2 = 0.01$.

20. Determine the general form of the solutions $i_1(t)$ and $i_2(t)$.

21. Find the specific solutions for $i_1(t)$ and $i_2(t)$ if $i_1(0) = 1$, $i_2(0) = 0$, $i_1'(0) = 0$, and $i_2'(0) = 0$.

In Example 5.7.3, without assigning the spring constants and masses any specific values, determine the elements in the coefficient matrix **A**.

Use $M_1 = 2$, $M_2 = 2$, $K_1 = 12$, and $K_2 = 8$.

22. What are the eigenvalues and unit eigenvectors?

23. What is the most general solution?

24. Find the specific solutions for $y_1(t)$ and $y_2(t)$ if $y_1(0) = 0$, $y_2(0) = 0$, $y_1'(0) = 0$, and $\dot{y}_2(0) = 10$.

The following apply to either Problem 21 or 24. Find the specific solutions for each.

25. $L_1 = \frac{10}{3}$, $L_2 = 5$, $C_1 = \frac{1}{200}$, $C_2 = \frac{1}{300}$, $i_1(0) = 0$, $i_2(0) = 0$, $i_1'(0) = 50$, $i_2'(0) = 0$.

26. $M_1 = 1$, $M_2 = 4$, $K_1 = 20$, $K_2 = 40$, $y_1(0) = 2$, $y_2(0) = 0$, $y_1'(0) = 0$, $y_2'(0) = 0$.

5.8 Systems of Linear Equations: The Nonhomogeneous Case

In this section we find the general solution of the nonhomogeneous system

$$\mathbf{x}' = \mathbf{A}\mathbf{x} + \mathbf{f}, \qquad \mathbf{f} \neq \mathbf{0}. \tag{5.8.1}$$

We call

$$\mathbf{x}' = \mathbf{A}\mathbf{x} \tag{5.8.2}$$

the *associated homogeneous system* and assume that its general solution is

$$\mathbf{x}(t) = \sum_{k=1}^{n} c_k \mathbf{u}_k e^{\lambda_k t} \tag{5.8.3}$$

or equivalently,

$$\mathbf{x}(t) = [\mathbf{u}_1 e^{\lambda_1 t}, \mathbf{u}_2 e^{\lambda_2 t}, \cdots, \mathbf{u}_n e^{\lambda_n t}] \begin{bmatrix} c_1 \\ c_2 \\ \vdots \\ c_n \end{bmatrix}. \tag{5.8.4}$$

We are assuming that **A** is simple and that $\{\mathbf{u}_1, \mathbf{u}_2, \ldots, \mathbf{u}_n\}$ is a linearly independent set of eigenvectors of **A**. Write

$$\mathbf{\Phi}(t) = [\mathbf{u}_1 e^{\lambda_1 t}, \mathbf{u}_2 e^{\lambda_2 t}, \ldots, \mathbf{u}_n e^{\lambda_n t}]. \tag{5.8.5}$$

Then the general solution of $\mathbf{x}' = \mathbf{A}\mathbf{x}$ may be written as

$$\mathbf{x}(t) = \mathbf{\Phi}(t)\mathbf{c}, \qquad \mathbf{c} = \begin{bmatrix} c_1 \\ c_2 \\ \vdots \\ c_n \end{bmatrix}. \tag{5.8.6}$$

The matrix $\Phi(t)$, whose columns are solutions of $x' = Ax$, is a *fundamental matrix* of $x' = Ax$ if $\Phi(0)$ is nonsingular. This is the case considered here since $\Phi(0) = [u_1, u_2, \ldots, u_n]$.

We use $\Phi(t)$ to find a solution of the nonhomogeneous system 5.8.1 by the method of *variation of parameters*. Assume that there exists a function $u(t)$ such that

$$x_p(t) = \Phi(t)u(t) \tag{5.8.7}$$

is a solution of $x' = Ax + f$. Then

$$x_p'(t) = \Phi'(t)u(t) + \Phi(t)u'(t). \tag{5.8.8}$$

However,

$$\begin{aligned}
\Phi'(t) &= [\lambda_1 u_1 e^{\lambda_1 t}, \ldots, \lambda_n u_n e^{\lambda_n t}] \\
&= [Au_1 e^{\lambda_1 t}, \ldots, Au_n e^{\lambda_n t}] \\
&= A[u_1 e^{\lambda_1 t}, \ldots, u_n e^{\lambda_n t}] = A\Phi(t).
\end{aligned} \tag{5.8.9}$$

Therefore, from Eqs. 5.8.8 and 5.8.9,

$$\begin{aligned}
x_p'(t) &= A\Phi(t)u(t) + \Phi(t)u'(t) \\
&= A\Phi(t)u(t) + f(t)
\end{aligned} \tag{5.8.10}$$

since we are assuming that $x_p'(t) = Ax_p(t) + f(t)$. Equation 5.8.10 simplifies to

$$\Phi(t)u'(t) = f(t), \tag{5.8.11}$$

a necessary and sufficient condition that $\Phi(t)u(t)$ is a solution of Eq. 5.8.1.

Example 5.8.1:   Find a particular solution of

$$x' = \begin{bmatrix} 1 & 0 \\ -1 & 3 \end{bmatrix} x + \begin{bmatrix} e^t \\ 1 \end{bmatrix}$$

using the variation-of-parameters formula 5.8.11.

SOLUTION: After some labor we determine

$$\Phi(t) = \begin{bmatrix} 2e^t & 0 \\ e^t & e^{3t} \end{bmatrix}.$$

Note that

$$\begin{aligned}
\Phi'(t) &= \begin{bmatrix} 2e^t & 0 \\ e^t & 3e^{3t} \end{bmatrix} \\
&= \begin{bmatrix} 1 & 0 \\ -1 & 3 \end{bmatrix}\begin{bmatrix} 2e^t & 0 \\ e^t & e^{3t} \end{bmatrix},
\end{aligned}$$

as required by Eq. 5.8.9, a good check on our arithmetic. Equation 5.8.11 requires that

$$\begin{bmatrix} 2e^t & 0 \\ e^t & 3e^{3t} \end{bmatrix} u' = \begin{bmatrix} e^t \\ 1 \end{bmatrix},$$

which, by elementary row operations, is equivalent to

$$\begin{bmatrix} 2e^t & 0 \\ 0 & e^{3t} \end{bmatrix} \mathbf{u}' = \begin{bmatrix} e^t \\ 1 - e^t/2 \end{bmatrix}.$$

Hence,

$$\mathbf{u}'(t) = \begin{bmatrix} \frac{1}{2} \\ e^{-3t} - e^{-2t}/2 \end{bmatrix}$$

and therefore,

$$\mathbf{u}(t) = \begin{bmatrix} t/2 \\ -e^{-3t}/3 + e^{-2t}/4 \end{bmatrix}.$$

We compute

$$\mathbf{x}_p(t) = \Phi(t)\mathbf{u}(t)$$

$$= \begin{bmatrix} te^t \\ te^t/2 - \frac{1}{3} + e^t/4 \end{bmatrix}.$$

The reader is invited to verify that

$$\mathbf{x}_p' = A\mathbf{x}_p + \mathbf{f}. \quad \blacksquare$$

Example 5.8.2:   Find a particular solution of

$$\mathbf{x}' = \begin{bmatrix} 0 & -1 \\ -1 & 0 \end{bmatrix}\mathbf{x} + \begin{bmatrix} e^{2t} \\ e^{2t} \end{bmatrix}.$$

SOLUTION: For this system, the fundamental matrix is

$$\Phi(t) = \begin{bmatrix} e^t & e^{-t} \\ -e^t & e^{-t} \end{bmatrix}$$

and therefore,

$$\Phi(t)\mathbf{u}'(t) = e^{2t}\begin{bmatrix} 1 \\ 1 \end{bmatrix}$$

determines $\mathbf{u}'(t)$. Elementary row operations lead to the equivalent system

$$\begin{bmatrix} e^t & 0 \\ 0 & e^{-t} \end{bmatrix}\mathbf{u}'(t) = \begin{bmatrix} 0 \\ e^{2t} \end{bmatrix},$$

from which

$$\mathbf{u}'(t) = \begin{bmatrix} 0 \\ e^{3t} \end{bmatrix}.$$

Therefore,

$$\mathbf{u}(t) = \begin{bmatrix} 0 \\ e^{3t}/3 \end{bmatrix}$$

and

$$\mathbf{x}_p(t) = \begin{bmatrix} e^t & e^{-t} \\ -e^t & e^{-t} \end{bmatrix} \begin{bmatrix} 0 \\ e^{3t}/3 \end{bmatrix} = \frac{1}{3} \begin{bmatrix} e^{2t} \\ e^{2t} \end{bmatrix} = \frac{1}{3} e^{2t} \begin{bmatrix} 1 \\ 1 \end{bmatrix}. \quad \blacksquare$$

It is proved in advanced texts on differential equations (see also Problem 9). that $\mathbf{\Phi}(t)$ is invertible for each t. Hence, Eq. 5.8.11 can be written

$$\mathbf{u}'(t) = \mathbf{\Phi}^{-1}(t)\mathbf{f}(t) \tag{5.8.12}$$

and therefore,

$$\mathbf{u}(t) = \int_0^t \mathbf{\Phi}^{-1}(s)\mathbf{f}(s) \, ds + \mathbf{c}. \tag{5.8.13}$$

Theorem 5.14: *The function*

$$\mathbf{x}_p(t) = \mathbf{\Phi}(t)\mathbf{u}(t), \tag{5.8.14}$$

where

$$\mathbf{u}(t) = \int_0^t \mathbf{\Phi}^{-1}(s)\mathbf{f}(s) \, ds \tag{5.8.15}$$

is a solution of $\mathbf{x}' = \mathbf{Ax} + \mathbf{f}$, *which vanishes at* $t = 0$.

Example 5.8.3:   Find a particular solution of the system in Example 5.8.1 which vanishes at $t = 0$.

SOLUTION: In Example 5.8.1 we computed

$$\mathbf{u}'(t) = \begin{bmatrix} \frac{1}{2} \\ e^{-3t} - e^{-2t}/2 \end{bmatrix}.$$

Hence,

$$\mathbf{u}(t) = \int_0^t \begin{bmatrix} \frac{1}{2} \\ e^{-3s} - e^{-2s}/2 \end{bmatrix} ds$$

$$= \begin{bmatrix} \frac{t}{2} \\ -e^{-3t}/3 + e^{-2t}/4 + \frac{1}{12} \end{bmatrix}.$$

From this we find that

$$\mathbf{x}_p(t) = \mathbf{\Phi}(t)\mathbf{u}(t) = \begin{bmatrix} te^t \\ te^t/2 - \frac{1}{3} + e^t/4 + e^{3t}/12 \end{bmatrix}. \quad \blacksquare$$

We define the *general solution* of $\mathbf{x}' = \mathbf{Ax} + \mathbf{f}$ by the equation

$$\mathbf{x}(t) = \mathbf{\Phi}(t)\mathbf{c} + \mathbf{x}_p(t), \tag{5.8.16}$$

in which $\mathbf{x}_p(t)$ is any particular solution of $\mathbf{x}' = \mathbf{Ax} + \mathbf{f}$ and $\mathbf{\Phi}(t)\mathbf{c}$ is the general solution of the associated homogeneous system $\mathbf{x}' = \mathbf{Ax}$. Thus, the general solution of the system in Example 5.8.1 is

$$\mathbf{x}(t) = \begin{bmatrix} 2e^t & 0 \\ e^t & e^{3t} \end{bmatrix} \mathbf{c} + \begin{bmatrix} te^t \\ te^t/2 - \frac{1}{3} + e^t/4 \end{bmatrix}.$$

5.8.1 SPECIAL METHODS

In solving $\mathbf{x}' = \mathbf{Ax} + \mathbf{f}$, we look for any particular solution of $\mathbf{x}' = \mathbf{Ax} + \mathbf{f}$, to which we then add the general solution of $\mathbf{x}' = \mathbf{Ax}$. The method of the variation of parameters always yields a particular solution once a knowledge of the general solution of $\mathbf{x}' = \mathbf{Ax}$ is available. It sometimes happens that we may entirely bypass this general solution and discover a particular solution by inspection. This happens frequently enough to warrant the inclusion of this technique in its own section. The method is known as the *method of undetermined coefficients*. It takes full advantage of the "forcing" function $\mathbf{f}(t)$. For instance, in the system

$$\mathbf{x}' = \mathbf{Ax} - \mathbf{b} \tag{5.8.17}$$

where $\mathbf{b}$ is a constant, we might reasonably expect a constant solution, say

$$\mathbf{x}(t) = \mathbf{k}. \tag{5.8.18}$$

Since $\mathbf{x}'(t) = \mathbf{0}$ when $\mathbf{x}(t) = \mathbf{k}$, $\mathbf{k}$ must satisfy $\mathbf{0} = \mathbf{Ax} - \mathbf{b}$. If $\mathbf{A}^{-1}$ exists, then $\mathbf{k} = \mathbf{A}^{-1}\mathbf{b}$ and

$$\mathbf{x}(t) = \mathbf{A}^{-1}\mathbf{b} \tag{5.8.19}$$

solves Eq. 5.8.17. Consider the next theorem, which provides a general context for the example above.

Theorem 5.15: *The system*

$$\mathbf{x}' = \mathbf{Ax} - \mathbf{b}e^{\omega t} \tag{5.8.20}$$

has a particular solution

$$\mathbf{x}_p(t) = \mathbf{k}e^{\omega t} \tag{5.8.21}$$

if ω is not an eigenvalue of $\mathbf{A}$.

PROOF: The proof consists of simply substituting $\mathbf{x}_p$ as given by Eq. 5.8.21 into Eq. 5.8.20. This leads to

$$(\mathbf{A} - \omega\mathbf{I})\mathbf{k} = \mathbf{b}, \tag{5.8.22}$$

which has a solution if $(\mathbf{A} - \omega\mathbf{I})$ is nonsingular, that is, if ω is not an eigenvalue of $\mathbf{A}$. If $\omega = 0$ in Eq. 5.8.20, then we are in the case given by Eq. 5.8.17. If $\omega = 0$ is not an eigenvalue of $\mathbf{A}$, then $\mathbf{A}$ is nonsingular and Eqs. 5.8.19 and 5.8.22 are equivalent.

Example 5.8.4:   Find a particular solution of

$$\mathbf{x}' = \begin{bmatrix} 1 & -1 \\ -1 & 0 \end{bmatrix} \mathbf{x} + e^{2t} \begin{bmatrix} 1 \\ 1 \end{bmatrix}.$$

SOLUTION: We try

$$\mathbf{x}_p(t) = \mathbf{k}e^{2t}.$$

Then

$$2\,\mathbf{k}e^{2t} = \begin{bmatrix} 0 & -1 \\ -1 & 0 \end{bmatrix} \mathbf{k}e^{2t} + e^{2t} \begin{bmatrix} 1 \\ 1 \end{bmatrix}$$

determines $\mathbf{k}$. We divide by e^{2t} and collect terms to obtain

$$\begin{bmatrix} -2 & -1 \\ -1 & -2 \end{bmatrix} \mathbf{k} = \begin{bmatrix} -1 \\ -1 \end{bmatrix}.$$

That is,

$$\mathbf{k} = \tfrac{1}{3}\begin{bmatrix} 1 \\ 1 \end{bmatrix}, \qquad \mathbf{x}_p(t) = \tfrac{1}{3}e^{2t}\begin{bmatrix} 1 \\ 1 \end{bmatrix}.$$

Note that this equation is precisely Eq. 5.8.22, since

$$-\mathbf{b}e^{\omega t} = \begin{bmatrix} 1 \\ 1 \end{bmatrix} e^{2t}. \quad \blacksquare$$

The uses of Theorem 5.15 can be extended in a number of ways. Consider the systems

(a) $\qquad\qquad\qquad\qquad \mathbf{x}' = \mathbf{A}\mathbf{x} - \mathbf{b}e^{i\omega t}$

(b) $\qquad\qquad\qquad\qquad \mathbf{x}' = \mathbf{A}\mathbf{x} - \mathbf{b}\cos\omega t \qquad\qquad$ (5.8.23)

(c) $\qquad\qquad\qquad\qquad \mathbf{x}' = \mathbf{A}\mathbf{x} - \mathbf{b}\sin\omega t$

where $\mathbf{A}$ is a real matrix, $\mathbf{b}$ a real constant vector, and ω is a real scalar. If we assume that $\mathbf{x}_p(t)$ solves Eq. 5.8.23a, then

$$\mathbf{x}_p'(t) = \mathbf{A}\mathbf{x}_p(t) - \mathbf{b}e^{i\omega t} \qquad\qquad (5.8.24)$$

and by taking real and imaginary parts*

$$[\operatorname{Re} \mathbf{x}_p(t)]' = \mathbf{A}[\operatorname{Re} \mathbf{x}_p(t)] - \mathbf{b}\cos\omega t$$
$$[\operatorname{Im} \mathbf{x}_p(t)]' = \mathbf{A}[\operatorname{Im} \mathbf{x}_p(t)] - \mathbf{b}\sin\omega t. \qquad (5.8.25)$$

The expressions in Eq. 5.8.25 are interpreted as asserting that the real part of a solution of Eq. 5.8.23a solves Eq. 5.8.23b, the imaginary part solves Eq. 5.8.23c.

---

*We write $\mathbf{x}_p(t) = \operatorname{Re} \mathbf{x}_p(t) + i \operatorname{Im} \mathbf{x}_p(t)$; then $\operatorname{Re} e^{i\omega t} = \cos\omega t$, $\operatorname{Im} e^{i\omega t} = \sin\omega t$, since $e^{i\omega t} = \cos\omega t + i \sin\omega t$.

Example 5.8.5:   Using undetermined coefficients, find a particular solution of

$$\mathbf{x'} = \begin{bmatrix} 0 & -2 \\ 1 & -1 \end{bmatrix} \mathbf{x} + \sin t \begin{bmatrix} 0 \\ 1 \end{bmatrix}.$$

SOLUTION: We solve

$$\mathbf{x'} = \begin{bmatrix} 0 & -2 \\ 1 & -1 \end{bmatrix} \mathbf{x} + e^{it} \begin{bmatrix} 0 \\ 1 \end{bmatrix}$$

by assuming as solution of the form

$$\mathbf{x}_p(t) = \mathbf{k} e^{it}$$

and then using the imaginary part of the solution. For this function

$$ie^{it}\mathbf{k} = e^{it} \begin{bmatrix} 0 & -2 \\ 1 & -1 \end{bmatrix} \mathbf{k} + e^{it} \begin{bmatrix} 0 \\ 1 \end{bmatrix}$$

and hence,

$$\begin{bmatrix} -i & -2 \\ 1 & -1-i \end{bmatrix} \mathbf{k} = \begin{bmatrix} 0 \\ -1 \end{bmatrix}.$$

Therefore,

$$\mathbf{k} = \begin{bmatrix} -1+i \\ \frac{1}{2}(1+i) \end{bmatrix}$$

and thus

$$\mathbf{x}_p(t) = e^{it} \begin{bmatrix} -1+i \\ (1+i)/2 \end{bmatrix}$$

$$= (\cos t + i \sin t) \begin{bmatrix} -1+i \\ (1+i)/2 \end{bmatrix}$$

$$= \begin{bmatrix} -\cos t - \sin t \\ (\cos t - \sin t)/2 \end{bmatrix} + i \begin{bmatrix} -\sin t + \cos t \\ (\cos t + \sin t)/2 \end{bmatrix},$$

so

$$\text{Im } \mathbf{x}_p(t) = \begin{bmatrix} \cos t - \sin t \\ (\cos t + \sin t)/2 \end{bmatrix} \quad\blacksquare$$

The usefulness of Theorem 5.15 is further enhanced by the *principle of superposition*, which is presented in the following theorem.

Theorem 5.16:   *If* $\mathbf{x}_1(t)$ *and* $\mathbf{x}_2(t)$, *solve*

$$\mathbf{x'} = \mathbf{Ax} + \mathbf{f}_1 \tag{5.8.26}$$

and

$$\mathbf{x'} = \mathbf{Ax} + \mathbf{f}_2, \tag{5.8.27}$$

respectively, then

$$\mathbf{x}_p(t) = \alpha_1 \mathbf{x}_1(t) + \alpha_2 \mathbf{x}_2(t) \qquad (5.8.28)$$

solves

$$\mathbf{x}' = \mathbf{A}\mathbf{x} + \alpha_1 \mathbf{f}_1 + \alpha_2 \mathbf{f}_2. \qquad (5.8.29)$$

PROOF: We have

$$\begin{aligned}
\mathbf{x}_p'(t) &= \alpha_1 \mathbf{x}_1'(t) + \alpha_2 \mathbf{x}_2'(t) \\
&= \alpha_1[\mathbf{A}\mathbf{x}_1(t) + \mathbf{f}_1] + \alpha_2[\mathbf{A}\mathbf{x}_2(t) + \mathbf{f}_2] \\
&= \mathbf{A}[\alpha_1 \mathbf{x}_1(t) + \alpha_2 \mathbf{x}_2(t)] + \alpha_1 \mathbf{f}_1 + \alpha_2 \mathbf{f}_2. \qquad (5.8.30)
\end{aligned}$$

Example 5.8.6:   Find a particular solution of

$$\mathbf{x}' = \begin{bmatrix} 0 & -2 \\ 1 & -1 \end{bmatrix} \mathbf{x} + (1 + \cos t) \begin{bmatrix} 0 \\ 1 \end{bmatrix}.$$

SOLUTION: We set

$$\mathbf{f}_1 = \begin{bmatrix} 0 \\ 1 \end{bmatrix}, \qquad \mathbf{f}_2 = \cos t \begin{bmatrix} 0 \\ 1 \end{bmatrix}.$$

For the system

$$\mathbf{x}' = \mathbf{A}\mathbf{x} + \begin{bmatrix} 0 \\ 1 \end{bmatrix}$$

we try $\mathbf{x}_1(t) = \mathbf{k}$, which leads to

$$\mathbf{x}_1(t) = \begin{bmatrix} 1 \\ 0 \end{bmatrix}.$$

For the system

$$\mathbf{x}' = \mathbf{A}\mathbf{x} + \cos t \begin{bmatrix} 0 \\ 1 \end{bmatrix}$$

we use the real part of $\mathbf{x}_p(t)$ as computed in Example 5.8.5. So

$$\mathbf{x}_2(t) = \begin{bmatrix} -\cos t - \sin t \\ (\cos t - \sin t)/2 \end{bmatrix}$$

and therefore

$$\mathbf{x}_p(t) = \begin{bmatrix} 1 \\ 0 \end{bmatrix} + \begin{bmatrix} -\cos t - \sin t \\ (\cos t - \sin t)/2 \end{bmatrix} = \begin{bmatrix} 1 - \cos t - \sin t \\ (\cos t - \sin t)/2 \end{bmatrix}$$

is the required solution.   ■

5.8.2   INITIAL-VALUE PROBLEMS

The general solution provides sufficiently many solutions to solve any problem

$$\mathbf{x}' = \mathbf{A}\mathbf{x} + \mathbf{f}, \qquad \mathbf{x}(t_0) = \mathbf{x}_0. \qquad (5.8.31)$$

For, as we have seen earlier,

$$\mathbf{x}(t) = \mathbf{\Phi}(t)\mathbf{c} + \mathbf{x}_p(t) \tag{5.8.32}$$

is a solution of $\mathbf{x}' = \mathbf{A}\mathbf{x} + \mathbf{f}$ if $\mathbf{\Phi}(t)$ is a fundamental matrix of $\mathbf{x}' = \mathbf{A}\mathbf{x}$ and $\mathbf{x}_p(t)$ is a particular solution of $\mathbf{x}' = \mathbf{A}\mathbf{x} + \mathbf{f}$. Then

$$\mathbf{x}(t_0) = \mathbf{x}_0 = \mathbf{\Phi}(t_0)\mathbf{c} + \mathbf{x}_p(t_0) \tag{5.8.33}$$

has the solution for **c**,

$$\mathbf{c} = \mathbf{\Phi}^{-1}(t_0)\{\mathbf{x}_0 - \mathbf{x}_p(t_0)\}. \tag{5.8.34}$$

Example 5.8.7:   Find a solution of the initial-value problem

$$\mathbf{x}' = \begin{bmatrix} 1 & 0 \\ -1 & 3 \end{bmatrix}\mathbf{x} + \begin{bmatrix} e^t \\ 1 \end{bmatrix}, \qquad \mathbf{x}(0) = \begin{bmatrix} 2 \\ 1 \end{bmatrix}.$$

SOLUTION: The system is solved in Example 5.8.3, in which we found the general solution

$$\mathbf{x}(t) = \begin{bmatrix} 2e^t & 0 \\ e^t & e^{3t} \end{bmatrix}\mathbf{c} + \begin{bmatrix} te^t \\ te^t/2 - \frac{1}{3} + e^t/4 \end{bmatrix}.$$

We find **c** from

$$\mathbf{x}(0) = \begin{bmatrix} 2 \\ 1 \end{bmatrix} = \begin{bmatrix} 2 & 1 \\ 1 & 1 \end{bmatrix}\mathbf{c} + \begin{bmatrix} 0 \\ -\frac{1}{12} \end{bmatrix}.$$

Hence,

$$\mathbf{c} = \begin{bmatrix} 1 \\ \frac{1}{12} \end{bmatrix}$$

and

$$\mathbf{x}(t) = \begin{bmatrix} 2e^t & 0 \\ e^t & e^{3t} \end{bmatrix}\begin{bmatrix} 1 \\ \frac{1}{12} \end{bmatrix} + \begin{bmatrix} te^t \\ te^t/2 - \frac{1}{3} + e^t/4 \end{bmatrix}$$

$$= \begin{bmatrix} 2e^t + te^t \\ 5e^t/4 + e^{3t}/12 + te^t/2 - \frac{1}{3} \end{bmatrix}$$

is the required solution.   ∎

From a mathematical viewpoint there remains the issue of whether initial-value problems can have more than one solution. Suppose that $\mathbf{x}(t)$ and $\mathbf{y}(t)$ are both solutions of $\mathbf{x}' = \mathbf{A}\mathbf{x} + \mathbf{f}$, $\mathbf{x}(t_0) = \mathbf{x}_0$. Then

$$\mathbf{x}' = \mathbf{A}\mathbf{x} + \mathbf{f}, \qquad \mathbf{y}' = \mathbf{A}\mathbf{y} + \mathbf{f}, \tag{5.8.35}$$

so that by subtraction

$$(\mathbf{x} - \mathbf{y})' = \mathbf{A}(\mathbf{x} - \mathbf{y}). \tag{5.8.36}$$

Also $\mathbf{x}(t_0) - \mathbf{y}(t_0) = \mathbf{x}_0 - \mathbf{x}_0 = \mathbf{0}$. So setting $\mathbf{z}(t) = \mathbf{x}(t) - \mathbf{y}(t)$,

$$\mathbf{z}'(t) = \mathbf{A}\mathbf{z}(t)$$

$$\mathbf{z}(t_0) = \mathbf{0}. \tag{5.8.37}$$

Equation 5.8.37 is the homogeneous, initial-value problem and one can prove that it has only the trivial solution, $\mathbf{z}(t) = \mathbf{0}$. Hence, there is one and only one solution to any initial-value problem.

PROBLEMS

Find the general solution of $\mathbf{x}' = \begin{bmatrix} 0 & 1 \\ 1 & 0 \end{bmatrix} \mathbf{x} + \mathbf{f}$, where $\mathbf{f}$ is given by

1. $\begin{bmatrix} 1 \\ -1 \end{bmatrix}$ **2.** $\begin{bmatrix} 0 \\ e^t \end{bmatrix}$ **3.** $\begin{bmatrix} 1 \\ 1 \end{bmatrix}$

4. $\begin{bmatrix} 1 \\ 1 - e^t \end{bmatrix}$ **5.** $\begin{bmatrix} t \\ t \end{bmatrix}$

Given the fundamental matrix

$$\mathbf{\Phi}(t) = e^t \begin{bmatrix} 1 & -1 + t \\ 1 & t \end{bmatrix}$$

of the system $\mathbf{x}' = \mathbf{A}\mathbf{x}$, where $\mathbf{A} = \begin{bmatrix} 0 & 1 \\ -1 & 2 \end{bmatrix}$, find the solutions of $\mathbf{x}' = \mathbf{A}\mathbf{x} + \mathbf{f}$, $\mathbf{x}(0) = \mathbf{x}_0$, where

6. $\mathbf{f}(t) = \begin{bmatrix} 0 \\ 1 \end{bmatrix}$, $\mathbf{x}_0 = \begin{bmatrix} 0 \\ 1 \end{bmatrix}$ **7.** $\mathbf{f}(t) = \begin{bmatrix} 1 \\ 1 \end{bmatrix}$, $\mathbf{x}_0 = \begin{bmatrix} 0 \\ 1 \end{bmatrix}$

8. $\mathbf{f}(t) = \begin{bmatrix} 1 \\ -1 \end{bmatrix} e^{-t}$, $\mathbf{x}_0 = \begin{bmatrix} 0 \\ 0 \end{bmatrix}$

9. Show that $\mathbf{\Psi} = \mathbf{\Phi}\mathbf{B}$ satisfies $\mathbf{\Phi}' = \mathbf{A}\mathbf{\Phi}$, where $\mathbf{B}$ is an invertible constant matrix.

10. Verify by direct substitution that

$$\mathbf{x}_p(t) = \begin{bmatrix} te^t \\ te^t/2 - \frac{1}{3} + e^t/4 \end{bmatrix}$$

is a solution of the system in Example 5.8.1.

11. Verify by direct substitution that

$$\mathbf{x}_p(t) = \begin{bmatrix} te^t \\ te^t/2 - \frac{1}{3} + e^t/4 - e^{3t}/12 \end{bmatrix}$$

is a solution of the system in Example 5.8.1.

Use the method of variation of parameters to find a particular solution of

12. The system in Example 5.8.4.

13. The system in Example 5.8.5.

14. The system in Example 5.8.6.

15. Solve $\mathbf{x}' = \mathbf{Ax} + \mathbf{b}$ by the method of variation of parameters—assume that $\Phi(t)$ is a fundamental matrix for $\mathbf{x}' = \mathbf{Ax}$.

Use the method of undetermined coefficients to find a particular solution of

$$\mathbf{x}' = \begin{bmatrix} 1 & 0 \\ -1 & 3 \end{bmatrix} \mathbf{x} + \mathbf{f}, \text{ where } \mathbf{f} \text{ is given by}$$

16. $\begin{bmatrix} 1 \\ 1 \end{bmatrix}$ **17.** $\begin{bmatrix} 1 \\ 1 - e^{-t} \end{bmatrix}$ **18.** $\begin{bmatrix} \sin t \\ \cos t \end{bmatrix}$

Use the method of undetermined coefficients to find a particular solution of

$$\mathbf{x}' = \begin{bmatrix} 0 & -2 \\ 1 & -1 \end{bmatrix} \mathbf{x} + \mathbf{f}, \text{ where}$$

19. $\mathbf{f}(t) = \cos t \begin{bmatrix} 1 \\ 0 \end{bmatrix}$ **20.** $\mathbf{f}(t) = \begin{bmatrix} \cos t \\ \sin t \end{bmatrix}$

21. Using the method of undetermined coefficients, find a particular solution of

$$\mathbf{x}' = \begin{bmatrix} 2 & 1 & 0 \\ 0 & 2 & 0 \\ 0 & 0 & -1 \end{bmatrix} \mathbf{x} + \begin{bmatrix} e^t \\ 1 \\ 0 \end{bmatrix}.$$

22. Suppose that $\Phi(t)$ is a fundamental matrix of $\mathbf{x}' = \mathbf{Ax}$ and that $\Phi(0) = \mathbf{I}$. Differentiate

$$\mathbf{x}_p(t) = \int_0^t \Phi(t - s) \mathbf{f}(s) \, ds$$

to show that $\mathbf{x}_p(t)$ is a solution of

$$\mathbf{x}' = \mathbf{Ax} + \mathbf{f}, \qquad \mathbf{x}(0) = \mathbf{0}.$$

6

Vector Analysis

6.1 Introduction

One of the major changes in undergraduate science curricula, and in engineering curricula in particular, brought about by the modern space age has been the introduction of vector analysis into several courses. The use of vector analysis comes rather naturally, since many of the quantities encountered in the modeling of physical phenomena are vector quantities; examples of such quantities are velocity, acceleration, force, electric and magnetic fields, and heat flux. It is not absolutely necessary that we use vector analysis when working with these quantities; we could use the components of the vectors and continue to manipulate scalars. Vector analysis, though, simplifies many of the operations demanded in the solution of problems or in the derivation of mathematical models; thus, it has been introduced in most undergraduate science curricula.

When vectors are used in the physical sciences they usually refer to descriptions in two or three dimensions. Under such restrictions, special notations and operations, different from those in Chapters 4 and 5, seem to be useful and convenient. The notation for vectors is one such example and the operation of the "cross product" is another. The purpose of this chapter is part a review of vectors in the "classical" notation, part an introduction to vector analysis as it is used in dynamics, and part a presentation of the vector calculus, including vector operators and the vector integral theorems of Green and Stokes.

6.2   Vector Algebra

6.2.1   DEFINITIONS

A quantity that is completely defined by both magnitude and direction is a *vector*. Force is such a quantity. We must be careful, though, since not all quantities that have both magnitude and direction are vectors. Stress is such a quantity; it is not completely characterized by a magnitude and a direction and thus it is not a vector. (It is a second-order tensor, a quantity that we will not study in this text.) A *scalar* is a quantity that is completely characterized by only a magnitude. Temperature is one such quantity.

A vector will be denoted in bold face, e.g., **A**. The magnitude of the vector will be denoted by the italic letter, e.g., A or with bars, e.g., $|\mathbf{A}|$. A vector **A** is graphically represented by an arrow that points in the direction of **A** and whose length is proportional to the magnitude of **A**, as in Fig. 6.1. Two vectors are equal if they have the same magnitude and direction; they need not act at the same location. A vector with the same magnitude as **A** but acting in the opposite direction will be denoted $-\mathbf{A}$. Although we allow vectors to originate from an arbitrary point in space, as in Fig. 6.1, we will, later on, assume that vectors emanate from the origin of coordinates, in general.

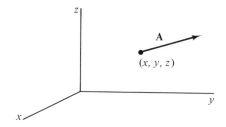

FIGURE 6.1.   Graphical representation of a vector.

A *unit vector* is a vector having a magnitude of 1. If we divide a vector **A** by the magnitude A we obtain a unit vector in the direction of **A**. Such a unit vector will be denoted $\mathbf{i}_A$, where the boldface lowercase signifies a unit vector. It is given by

$$\mathbf{i}_A = \frac{\mathbf{A}}{A}.\tag{6.2.1}$$

Any vector can be represented by its unit vector times its magnitude.

Three unit vectors which are used extensively are the unit vectors in the coordinate directions of the rectangular, Cartesian reference frame (this reference frame will be used primarily in this chapter). No subscripts will be used to denote these unit vectors. They are **i**, **j**, and **k** acting in the x, y, and z directions, respectively, and are shown in Fig. 6.2. In terms of the vectors of

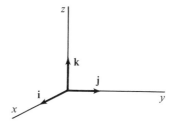

FIGURE 6.2.   The three unit vectors, **i**, **j**, and **k**.

Chapter 5, we see that

$$\mathbf{i} = \begin{bmatrix} 1 \\ 0 \\ 0 \end{bmatrix} = \mathbf{e}_1, \quad \mathbf{j} = \begin{bmatrix} 0 \\ 1 \\ 0 \end{bmatrix} = \mathbf{e}_2, \quad \mathbf{k} = \begin{bmatrix} 0 \\ 0 \\ 1 \end{bmatrix} = \mathbf{e}_3. \quad (6.2.2)$$

6.2.2   ADDITION AND SUBTRACTION

Two vectors **A** and **B** are added by placing the beginning of one at the tip of the other, as shown in Fig. 6.3. The sum **A** + **B** is the vector obtained by connecting the point of beginning of the first vector with the tip of the second vector, as shown. If the two vectors to be added are the sides of a parallelogram, their sum is given by the diagonal as shown in the figure. It is clear from the geometry that vector addition is commutative,

$$\mathbf{A} + \mathbf{B} = \mathbf{B} + \mathbf{A}. \quad (6.2.3)$$

Subtraction of vectors may be taken as a special case of addition; that is,

$$\mathbf{A} - \mathbf{B} = \mathbf{A} + (-\mathbf{B}). \quad (6.2.4)$$

Subtraction is illustrated in Fig. 6.4. Note that **A** − **B** is the other diagonal (see Fig. 6.3) of the parallelogram whose sides are **A** and **B**.

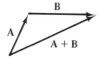
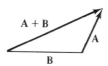
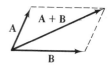

FIGURE 6.3.   Vector addition.

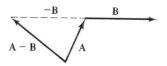
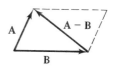

FIGURE 6.4.   Vector subtraction.

Finally, we show in Fig. 6.5 a graphical demonstration that vector addition is associative; that is,

$$A + (B + C) = (A + B) + C. \qquad (6.2.5)$$

The resultant vector is written $A + B + C$ since this sum is independent of the grouping.

FIGURE 6.5.   The associative property of vector addition.

Example 6.2.1:   Prove that the diagonals of a parallelogram bisect each other,

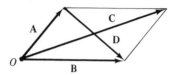

SOLUTION: From the parallelogram shown, we observe that

$$A + B = C, \qquad B - A = D.$$

The vector from point O to the intersection of the two diagonals is some fraction of C, say mC; and the vector representing part of the shorter diagonal is assumed to be nD. If m and n are both $\frac{1}{2}$, then the diagonals bisect each other.

Now, using the triangle formed with the vectors shown, we can write

$$A + nD = mC.$$

Substituting for the vectors C and D, the equation above becomes

$$A + n(B - A) = m(A + B).$$

Rearranging, we have

$$(1 - n - m)A = (m - n)B.$$

The quantity on the left is a vector in the direction of A, and the quantity on the right is a vector in the direction of B. Since the direction of A is different from the direction of B, we must demand that the coefficients be zero. Thus,

$$1 - n - m = 0$$

$$m - n = 0.$$

The solution to this set of equations is

$$n = m = \tfrac{1}{2}.$$

Hence, the diagonals bisect each other. ∎

6.2.3 COMPONENTS OF A VECTOR

So far we have not actually written a vector with magnitude and direction. To do so, we must choose a coordinate system and express the vector in terms of its *components*, which are the projections of the vector along the three coordinate axes. We illustrate this using a rectangular, Cartesian coordinate system. Consider the beginning of the vector to be at the origin, as shown in Fig. 6.6. The projections on the x, y, and z axes are A_x, A_y, and A_z, respectively. Using the unit vectors defined earlier, we can then write the vector **A** as

$$\mathbf{A} = A_x\mathbf{i} + A_y\mathbf{j} + A_z\mathbf{k}. \tag{6.2.6}$$

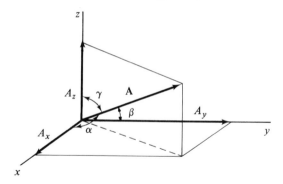

FIGURE 6.6. The components of a vector.

The vector **A** makes angles a, β, and γ with the x, y, and z axes, respectively. Thus, we have

$$A_x = A \cos \alpha, \qquad A_y = A \cos \beta, \qquad A_z = A \cos \gamma, \tag{6.2.7}$$

where A, the magnitude of **A**, is related geometrically to the components by

$$A = \sqrt{A_x^2 + A_y^2 + A_z^2}. \tag{6.2.8}$$

In the notation of Chapters 4 and 5,

$$\mathbf{A} = \begin{bmatrix} A_x \\ A_y \\ A_z \end{bmatrix} = \begin{bmatrix} A \cos \alpha \\ A \cos \beta \\ A \cos \gamma \end{bmatrix}, \tag{6.2.9}$$

where $A = \|\mathbf{A}\|$.

Substituting Eqs. 6.2.7 into Eq. 6.2.8, we have the familiar result,

$$\cos^2 \alpha + \cos^2 \beta + \cos^2 \gamma = 1. \tag{6.2.10}$$

Also, if we use Eqs. 6.2.7 in Eq. 6.2.6, we see that

$$\mathbf{A} = A(\cos \alpha \, \mathbf{i} + \cos \beta \, \mathbf{j} + \cos \gamma \, \mathbf{k}). \qquad (6.2.11)$$

Comparing this with Eq. 6.2.1, can express the unit vector $\mathbf{i}_A$ as

$$\mathbf{i}_A = \cos \alpha \, \mathbf{i} + \cos \beta \, \mathbf{j} + \cos \gamma \, \mathbf{k}. \qquad (6.2.12)$$

The quantities $\cos \alpha$, $\cos \beta$, and $\cos \gamma$ are often denoted ℓ, m, and n, respectively, and called the *direction cosines* of $\mathbf{A}$.

Two other coordinate systems are naturally encountered in physical situations, a cylindrical coordinate system and a spherical coordinate system. These coordinate systems will be presented in Section 6.5.

Example 6.2.2: Find a unit vector in the direction of $\mathbf{A} = 2\mathbf{i} + 3\mathbf{j} + 6\mathbf{k}$.

SOLUTION: The magnitude of the vector $\mathbf{A}$ is

$$A = \sqrt{A_x^2 + A_y^2 + A_z^2}$$
$$= \sqrt{2^2 + 3^2 + 6^2} = 7.$$

The unit vector is then

$$\mathbf{i}_A = \frac{\mathbf{A}}{A} = \frac{2\mathbf{i} + 3\mathbf{j} + 6\mathbf{k}}{7} = \frac{2}{7}\mathbf{i} + \frac{3}{7}\mathbf{j} + \frac{6}{7}\mathbf{k}.$$

6.2.4 MULTIPLICATION

There are three distinct multiplications which can be defined for vectors. First, a vector may be multiplied by a scalar. Consider the vector $\mathbf{A}$ multiplied by the scalar ϕ. The scalar ϕ simply multiplies each component of the vector and there results

$$\phi\mathbf{A} = \phi A_x \mathbf{i} + \phi A_y \mathbf{j} + \phi A_z \mathbf{k} \qquad (6.2.13)$$

in rectangular coordinates. The resultant vector acts in the same direction as the vector $\mathbf{A}$, unless ϕ is negative, in which case $\phi\mathbf{A}$ acts in the opposite direction of $\mathbf{A}$.

The second multiplication is the *scalar product*, also known as the *dot product*. It involves the multiplication of two vectors so that a scalar quantity results. The scalar product is defined to be the product of the magnitudes of the two vectors and the cosine of the angle between the two vectors. This is written as

$$\mathbf{A} \cdot \mathbf{B} = AB \cos \theta, \qquad (6.2.14)$$

where θ is the angle shown in Fig. 6.7. Note that the scalar product $\mathbf{A} \cdot \mathbf{B}$ is equal to the length of $\mathbf{B}$ multiplied by the projection of $\mathbf{A}$ on $\mathbf{B}$, or the length of $\mathbf{A}$ multiplied by the projection of $\mathbf{B}$ on $\mathbf{A}$. We recall that the scalar quantity

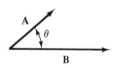 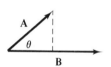 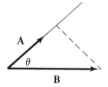

FIGURE 6.7. The dot product.

work was defined in much the same way, that is, force multiplied by the distance the force moved in the direction of the force, or

$$W = \mathbf{F} \cdot \mathbf{d}. \tag{6.2.15}$$

If the two vectors **A** and **B** are perpendicular, so that $\theta = 90°$, then $\mathbf{A} \cdot \mathbf{B} = 0$. This particular property of the dot product will be used quite often in applications. From the definition of the dot product we note that the dot product is commutative, so that

$$\mathbf{A} \cdot \mathbf{B} = \mathbf{B} \cdot \mathbf{A}, \tag{6.2.16}$$

The dot product is also distributive, that is,

$$\mathbf{A} \cdot (\mathbf{B} + \mathbf{C}) = \mathbf{A} \cdot \mathbf{B} + \mathbf{A} \cdot \mathbf{C}. \tag{6.2.17}$$

Using the unit vectors **i**, **j**, and **k**, the definition of the dot product yields

$$\mathbf{i} \cdot \mathbf{i} = 1, \qquad \mathbf{j} \cdot \mathbf{j} = 1, \qquad \mathbf{k} \cdot \mathbf{k} = 1 \tag{6.2.18}$$

and

$$\mathbf{i} \cdot \mathbf{j} = 0, \qquad \mathbf{i} \cdot \mathbf{k} = 0, \qquad \mathbf{j} \cdot \mathbf{k} = 0. \tag{6.2.19}$$

These allow us to express the dot product in rectangular coordinates as

$$\mathbf{A} \cdot \mathbf{B} = (A_x\mathbf{i} + A_y\mathbf{j} + A_z\mathbf{k}) \cdot (B_x\mathbf{i} + B_y\mathbf{j} + B_z\mathbf{k})$$
$$= A_xB_x + A_yB_y + A_zB_z. \tag{6.2.20}$$

The dot product of a vector **A** with itself can be written as

$$\mathbf{A} \cdot \mathbf{A} = A^2 = A_x^2 + A_y^2 + A_z^2. \tag{6.2.21}$$

Finally, in our discussion of the dot product we note that the component of **A** in the *x* direction is found by taking the dot product of **A** with **i**; that is,

$$\mathbf{A} \cdot \mathbf{i} = (A_x\mathbf{i} + A_y\mathbf{j} + A_z\mathbf{k}) \cdot \mathbf{i} = A_x. \tag{6.2.22}$$

Similarly,

$$\mathbf{A} \cdot \mathbf{j} = (A_x\mathbf{i} + A_y\mathbf{j} + A_z\mathbf{k}) \cdot \mathbf{j} = A_y. \tag{6.2.23}$$

In general, the component of a vector in any direction is given by the dot product of the vector with a unit vector in the desired direction.

Example 6.2.3:   Using the definition of the dot product of two vectors, show that $\cos(\alpha - \beta) = \cos\alpha\cos\beta + \sin\alpha\sin\beta$.

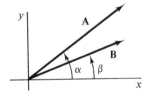

SOLUTION: The dot product of the two vectors **A** and **B** is

$$\mathbf{A} \cdot \mathbf{B} = AB\cos\theta,$$

where θ is the angle between the two vectors, that is

$$\theta = \alpha - \beta.$$

We know that

$$A = \sqrt{A_x^2 + A_y^2}, \qquad B = \sqrt{B_x^2 + B_y^2}$$

and

$$\mathbf{A} \cdot \mathbf{B} = A_x B_x + A_y B_y.$$

Thus,

$$\cos\theta = \frac{\mathbf{A}\cdot\mathbf{B}}{AB} = \frac{A_x B_x + A_y B_y}{\sqrt{A_x^2 + A_y^2}\,\sqrt{B_x^2 + B_y^2}}.$$

This can be written as

$$\cos(\alpha - \beta) = \frac{A_x}{\sqrt{A_x^2 + A_y^2}}\frac{B_x}{\sqrt{B_x^2 + B_y^2}} + \frac{A_y}{\sqrt{A_x^2 + A_y^2}}\frac{B_y}{\sqrt{B_x^2 + B_y^2}}$$

$$= \cos\alpha\cos\beta + \sin\alpha\sin\beta$$

and the trigonometric identity is verified.   ∎

Example 6.2.4:   Find the projection of **A** on **B**, if $\mathbf{A} = 12\mathbf{i} - 3\mathbf{j} + 6\mathbf{k}$ and $\mathbf{B} = 2\mathbf{i} + 4\mathbf{j} + 4\mathbf{k}$.

SOLUTION: Let us first find a unit vector $\mathbf{i}_B$ in the direction of **B**. Then the projection of **A** on **B** will be $\mathbf{A} \cdot \mathbf{i}_B$. We have

$$\mathbf{i}_B = \frac{\mathbf{B}}{B} = \frac{2\mathbf{i} + 4\mathbf{j} + 4\mathbf{k}}{\sqrt{2^2 + 4^2 + 4^2}} = \frac{2\mathbf{i} + 4\mathbf{j} + 4\mathbf{k}}{6}.$$

The projection of **A** on **B** is then

$$\mathbf{A} \cdot \mathbf{i}_B = (12\mathbf{i} - 3\mathbf{j} + 6\mathbf{k}) \cdot (\tfrac{1}{3}\mathbf{i} + \tfrac{2}{3}\mathbf{j} + \tfrac{2}{3}\mathbf{k}) = 4 - 2 + 4 = 6.$$

To relate this to our study of matrices consider the following. The projection of **A** on **B** is the magnitude of the vector obtained by "projecting **A** on **B**." In Chapter 5 we found that $\mathbf{uu}^T\mathbf{b}$ is the projection vector in the direction of $\mathbf{u}$. We set $\mathbf{u} = \mathbf{i}_B$ and $\mathbf{b} = \mathbf{A}$. Then

$$\mathbf{u} = \begin{bmatrix} \frac{2}{6} \\ \frac{4}{6} \\ \frac{4}{6} \end{bmatrix}, \qquad \mathbf{b} = \begin{bmatrix} 12 \\ -3 \\ 6 \end{bmatrix}.$$

Hence,

$$\mathbf{uu}^T\mathbf{b} = \frac{1}{9}\begin{bmatrix} 1 & 2 & 2 \\ 2 & 4 & 4 \\ 2 & 4 & 4 \end{bmatrix}\begin{bmatrix} 12 \\ -3 \\ 6 \end{bmatrix} = \frac{1}{9}\begin{bmatrix} 18 \\ 36 \\ 36 \end{bmatrix} = \begin{bmatrix} 2 \\ 4 \\ 4 \end{bmatrix}.$$

The magnitude of this vector is

$$\|\mathbf{uu}^T\mathbf{b}\| = \sqrt{4 + 16 + 16} = 6. \quad \blacksquare$$

The third multiplication operation is the *vector* product, also called the *cross product*. It is a vector, the magnitude of which is defined to be the product of the magnitudes of the two vectors comprising the product and the sine of the angle between them. The product acts in a direction perpendicular to the plane of its two factors so that the three vectors form a right-handed set of vectors. We write the cross product as

$$\mathbf{C} = \mathbf{A} \times \mathbf{B}. \tag{6.2.24}$$

The magnitude of $\mathbf{C}$ is given by

$$C = AB \sin \theta. \tag{6.2.25}$$

The vectors are shown in Fig. 6.8. We see that the magnitude of $\mathbf{C}$ is equal to the area of the parallelogram with sides $\mathbf{A}$ and $\mathbf{B}$. The cross product is an operation that has no obvious extension to dimensions higher than three.

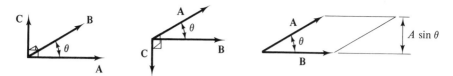

FIGURE 6.8. The cross product.

There are two other common techniques for determining the sense of the vector $\mathbf{C}$. First, $\mathbf{C}$ acts in the direction of the advance of a right-handed screw as it is turned from $\mathbf{A}$ to $\mathbf{B}$. Second, if the fingers curl $\mathbf{A}$ into $\mathbf{B}$, the thumb will point in the direction of $\mathbf{C}$.

From the definition we see that the cross product is not commutative, since

$$\mathbf{A} \times \mathbf{B} = -\mathbf{B} \times \mathbf{A}. \tag{6.2.26}$$

However, it is true that

$$\mathbf{A} \times (\mathbf{B} + \mathbf{C}) = \mathbf{A} \times \mathbf{B} + \mathbf{A} \times \mathbf{C}. \tag{6.2.27}$$

If two vectors act in the same direction, the angle θ is zero degrees and the cross product vanishes. It follows that

$$\mathbf{A} \times \mathbf{A} = \mathbf{0}. \tag{6.2.28}$$

The unit vectors $\mathbf{i}$, $\mathbf{j}$, and $\mathbf{k}$ form the cross products

$$\mathbf{i} \times \mathbf{i} = 0, \quad \mathbf{j} \times \mathbf{j} = 0, \quad \mathbf{k} \times \mathbf{k} = 0 \tag{6.2.29}$$

and

$$\begin{aligned} \mathbf{i} \times \mathbf{j} &= \mathbf{k}, & \mathbf{j} \times \mathbf{k} &= \mathbf{i}, & \mathbf{k} \times \mathbf{i} &= \mathbf{j}, \\ \mathbf{j} \times \mathbf{i} &= -\mathbf{k}, & \mathbf{k} \times \mathbf{j} &= -\mathbf{i}, & \mathbf{i} \times \mathbf{k} &= -\mathbf{j}. \end{aligned} \tag{6.2.30}$$

These relationships are easily remembered by visualizing a display of unit vectors. The cross product of a unit vector into its neighbor is the following vector when going clockwise, and is the negative of the following vector when going counterclockwise.

Using the relationship above we can express the cross product of $\mathbf{A}$ and $\mathbf{B}$ in rectangular coordinates as

$$\begin{aligned} \mathbf{A} \times \mathbf{B} &= (A_x\mathbf{i} + A_y\mathbf{j} + A_z\mathbf{k}) \times (B_x\mathbf{i} + B_y\mathbf{j} + B_z\mathbf{k}) \\ &= (A_yB_z - A_zB_y)\mathbf{i} + (A_zB_x - A_xB_z)\mathbf{j} + (A_xB_y - A_yB_x)\mathbf{k}. \end{aligned} \tag{6.2.31}$$

A convenient way to recall this expansion of the cross product is to utilize a determinant formed by the unit vectors, the components of $\mathbf{A}$, and the components of $\mathbf{B}$. The cross product of $\mathbf{A} \times \mathbf{B}$ is then related to the determinant by

$$\mathbf{A} \times \mathbf{B} = \begin{vmatrix} \mathbf{i} & \mathbf{j} & \mathbf{k} \\ A_x & A_y & A_z \\ B_x & B_y & B_z \end{vmatrix}. \tag{6.2.32}$$

in which we may expand the determinant by cofactors of the first row.

Two applications of the cross product are the torque $\mathbf{T}$ produced about a point by a force $\mathbf{F}$ acting at the distance $\mathbf{r}$ from the point, and the velocity $\mathbf{V}$ induced by an angular velocity $\boldsymbol{\omega}$ at a point $\mathbf{r}$ from the axis of rotation. The magnitude of the torque is given by the magnitude of $\mathbf{F}$ multiplied by the perpendicular distance from the point to the line of action of the force, that is,

$$T = Fd, \qquad (6.2.33)$$

where d is $r \sin \theta$ (see Fig. 6.9a). It can be represented by a vector normal to the plane of $\mathbf{F}$ and $\mathbf{r}$, given by

$$\mathbf{T} = \mathbf{r} \times \mathbf{F}. \qquad (6.2.34)$$

We are often interested in the torque produced about an axis, for example the z axis. It is the vector $\mathbf{T}$ dotted with the unit vector in the direction of the axis. About the z axis it is

$$T_z = \mathbf{T} \cdot \mathbf{k}. \qquad (6.2.35)$$

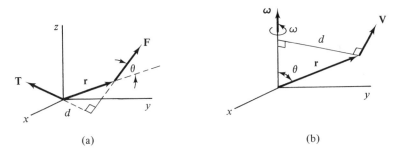

(a)   (b)

FIGURE 6.9. Examples of the cross product.

The magnitude of the velocity induced by an angular velocity $\boldsymbol{\omega}$ is the magnitude ω of the angular velocity multiplied by the perpendicular distance d from the axis to the point where the velocity is desired, as shown in Fig. 6.9b. If $\mathbf{r}$ is the position vector from the origin of a coordinate system to the point where the velocity is desired, then $r \sin \theta$ is the distance d, where θ is the angle between $\boldsymbol{\omega}$ and $\mathbf{r}$. The velocity $\mathbf{V}$ is then given by

$$\mathbf{V} = \boldsymbol{\omega} \times \mathbf{r}, \qquad (6.2.36)$$

where in the figure we have let the axis of rotation be the z axis. Note that the vector $\mathbf{V}$ is perpendicular to the plane of $\boldsymbol{\omega}$ and $\mathbf{r}$.

Example 6.2.5: Find a unit vector $\mathbf{i}_C$ perpendicular to the plane of $\mathbf{A}$ and $\mathbf{B}$, if $\mathbf{A} = 2\mathbf{i} + 3\mathbf{j}$ and $\mathbf{B} = \mathbf{i} - \mathbf{j} + 2\mathbf{k}$.

SOLUTION: Let $\mathbf{C} = \mathbf{A} \times \mathbf{B}$. Then $\mathbf{C}$ is perpendicular to the plane of $\mathbf{A}$ and $\mathbf{B}$. It is given by

$$\mathbf{C} = \mathbf{A} \times \mathbf{B} = \begin{vmatrix} \mathbf{i} & \mathbf{j} & \mathbf{k} \\ 2 & 3 & 0 \\ 1 & -1 & 2 \end{vmatrix} = 6\mathbf{i} - 4\mathbf{j} - 5\mathbf{k}.$$

The unit vector is then

$$i_C = \frac{C}{C} = \frac{6i - 4j - 5k}{\sqrt{6^2 + 4^2 + 5^2}} = 0.684i - 0.456j - 0.570k. \quad \blacksquare$$

We conclude this section by presenting the *scalar triple product*. Consider three vectors **A**, **B**, and **C**, shown in Fig. 6.10. The scalar triple product is the dot product of one of the vectors with the cross product of the remaining two. For example, the product $(\mathbf{A} \times \mathbf{B}) \cdot \mathbf{C}$ is a scalar quantity given by

$$(\mathbf{A} \times \mathbf{B}) \cdot \mathbf{C} = ABC \sin \alpha \sin \beta, \qquad (6.2.37)$$

where α is the angle between **A** and **B**, and β is the angle between $\mathbf{A} \times \mathbf{B}$ and **C**. The quantity $AB \sin \alpha$ is the area of the parallelogram with sides **A** and **B**. The quantity $C \cos \beta$ is the component of **C** in a direction perpendicular to the parallelogram with sides **A** and **B**. Thus the scalar triple product represents the volume of the parallelepiped with sides **A**, **B**, and **C**. Since the volume is the same regardless of how we form the product, we see that

$$(\mathbf{A} \times \mathbf{B}) \cdot \mathbf{C} = \mathbf{A} \cdot (\mathbf{B} \times \mathbf{C}) = (\mathbf{C} \times \mathbf{A}) \cdot \mathbf{B}. \qquad (6.2.38)$$

Also, the parentheses in the equation above are usually omitted since the cross product must be performed first. If the dot product were performed first, the quantity would be meaningless since the cross product requires two vectors.

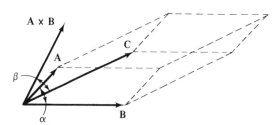

FIGURE 6.10.   The scalar triple product.

Using rectangular coordinates the scalar triple product is

$$\mathbf{A} \times \mathbf{B} \cdot \mathbf{C} = C_x(A_y B_z - A_z B_y) + C_y(A_z B_x - A_x B_z) + C_z(A_x B_y - A_y B_x)$$

$$= \begin{vmatrix} A_x & A_y & A_z \\ B_x & B_y & B_z \\ C_x & C_y & C_z \end{vmatrix}. \qquad (6.2.39)$$

Example 6.2.6:   For the vectors $\mathbf{A} = 3i - 2j + k$ and $\mathbf{B} = 2i - k$, determine (a) $\mathbf{A} \cdot \mathbf{B}$, and (b) $\mathbf{A} \times \mathbf{B} \cdot \mathbf{A}$.

SOLUTION:

(a) The dot product is given by

$$\mathbf{A} \cdot \mathbf{B} = (3i - 2j + k) \cdot (2i - k)$$

$$= 6\mathbf{i} \cdot \mathbf{i} - 3\mathbf{i} \cdot \mathbf{k} - 4\mathbf{j} \cdot \mathbf{i} + 2\mathbf{j} \cdot \mathbf{k} + 2\mathbf{k} \cdot \mathbf{i} - \mathbf{k} \cdot \mathbf{k}$$
$$= 6 - 1 = 5.$$

(b) To perform the indicated product we must first find $\mathbf{A} \times \mathbf{B}$. It is

$$\mathbf{A} \times \mathbf{B} = (A_y B_z - A_z B_y)\mathbf{i} + (A_z B_x - A_x B_z)\mathbf{j} + (A_x B_y - A_y B_x)\mathbf{k}$$
$$= [(-2)(-1) - 1 \cdot 0]\mathbf{i} + [1 \cdot 2 - 3(-1)]\mathbf{j} + [3.0 - (-2)(2)]\mathbf{k}$$
$$= 2\mathbf{i} + 5\mathbf{j} + 4\mathbf{k}.$$

We then dot this vector with $\mathbf{A}$ and obtain

$$\mathbf{A} \times \mathbf{B} \cdot \mathbf{A} = (2\mathbf{i} + 5\mathbf{j} + 4\mathbf{k}) \cdot (3\mathbf{i} - 2\mathbf{j} + \mathbf{k}) = 6 - 10 + 4 = 0.$$

We are not surprised that we get zero, since the vector $\mathbf{A} \times \mathbf{B}$ is perpendicular to $\mathbf{A}$, and the dot product of two perpendicular vectors is always zero, since the cosine of the angle between the two vectors is zero. ■

Example 6.2.7:   Find an equivalent vector expression for the vector triple product $(\mathbf{A} \times \mathbf{B}) \times \mathbf{C}$.

SOLUTION: We expand the triple product in rectangular coordinates. First, the cross product $\mathbf{A} \times \mathbf{B}$ is

$$\mathbf{A} \times \mathbf{B} = (A_y B_z - A_z B_y)\mathbf{i} + (A_z B_x - A_x B_z)\mathbf{j} + (A_x B_y - A_y B_x)\mathbf{k}.$$

Now, write the cross product of the vector above with $\mathbf{C}$. It is

$$
\begin{aligned}
(\mathbf{A} \times \mathbf{B}) \times \mathbf{C} &= [(A_z B_x - A_x B_z)C_z - (A_x B_y - A_y B_x)C_y]\mathbf{i} \\
&\quad + [(A_x B_y - A_y B_x)C_x - (A_y B_z - A_z B_y)C_z]\mathbf{j} \\
&\quad + [(A_y B_z - A_z B_y)C_y - (A_z B_x - A_x B_z)C_x]\mathbf{k}.
\end{aligned}
$$

The above can be rearranged in the form

$$
\begin{aligned}
(\mathbf{A} \times \mathbf{B}) \times \mathbf{C} &= (A_z C_z + A_y C_y + A_x C_x)B_x\mathbf{i} - (B_z C_z + B_y C_y + B_x C_x)A_x\mathbf{i} \\
&\quad + (A_x C_x + A_z C_z + A_y C_y)B_y\mathbf{j} - (B_x C_x + B_z C_z + B_y C_y)A_y\mathbf{j} \\
&\quad + (A_y C_y + A_x C_x + A_z C_z)B_z\mathbf{k} - (B_y C_y + B_x C_x + B_z C_z)A_z\mathbf{k},
\end{aligned}
$$

where the last terms in the parentheses have been inserted so that they cancel each other but help to form a dot product. Now we recognize that the equation above can be written as

$$
\begin{aligned}
(\mathbf{A} \times \mathbf{B}) \times \mathbf{C} &= (\mathbf{A} \cdot \mathbf{C})B_x\mathbf{i} + (\mathbf{A} \cdot \mathbf{C})B_y\mathbf{j} + (\mathbf{A} \cdot \mathbf{C})B_z\mathbf{k} \\
&\quad - (\mathbf{B} \cdot \mathbf{C})A_x\mathbf{i} - (\mathbf{B} \cdot \mathbf{C})A_y\mathbf{j} - (\mathbf{B} \cdot \mathbf{C})A_z\mathbf{k}
\end{aligned}
$$

or, finally,

$$(\mathbf{A} \times \mathbf{B}) \times \mathbf{C} = (\mathbf{A} \cdot \mathbf{C})\mathbf{B} - (\mathbf{B} \cdot \mathbf{C})\mathbf{A}.$$

Similarly, we can show that

$$\mathbf{A} \times (\mathbf{B} \times \mathbf{C}) = (\mathbf{A} \cdot \mathbf{C})\mathbf{B} - (\mathbf{B} \cdot \mathbf{A})\mathbf{C}.$$

Note that

$$(\mathbf{A} \times \mathbf{B}) \times \mathbf{C} \neq \mathbf{A} \times (\mathbf{B} \times \mathbf{C}).$$

unless **A**, **B**, and **C** are rather special vectors. For instance, $(\mathbf{A} \times \mathbf{B}) \times \mathbf{C} = \mathbf{A} \times (\mathbf{B} \times \mathbf{C})$ if any of the vectors is zero. ■

PROBLEMS

1. State which of the following quantities are vectors.

   (a) Volume   (b) Position of a particle

   (c) Force   (d) Energy

   (e) Momentum   (f) Color

   (g) Pressure   (h) Frequency

   (i) Magnetic field intensity   (j) Centrifugal acceleration

   (k) Voltage

Two vectors **A** and **B** act as shown. Find $\mathbf{A} + \mathbf{B}$ and $\mathbf{A} - \mathbf{B}$ for each by determining both magnitude and direction.

2. 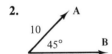   **3.** 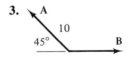   **4.**

Express each vector in component form, and write an expression for the unit vector that acts in the direction of each.

5. 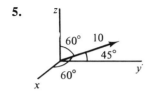   **6.** 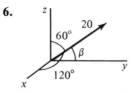   **7.**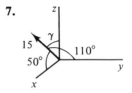

Given the vectors $\mathbf{A} = 2\mathbf{i} - 4\mathbf{j} - 4\mathbf{k}$, $\mathbf{B} = 4\mathbf{i} + 7\mathbf{j} - 4\mathbf{k}$, and $\mathbf{C} = 3\mathbf{i} - 4\mathbf{k}$. Find

8. $\mathbf{A} + \mathbf{B}$ **9.** $\mathbf{A} - \mathbf{C}$ **10.** $\mathbf{A} + \mathbf{B} - \mathbf{C}$

11. $|\mathbf{A} - \mathbf{B}|$ **12.** $\mathbf{A} \cdot \mathbf{B}$ **13.** $\mathbf{A} \times \mathbf{C}$

14. $|\mathbf{A} \times \mathbf{B}|$ **15.** $\mathbf{A} \times \mathbf{B} \cdot \mathbf{C}$ **16.** $\mathbf{A} \cdot \mathbf{A} \times \mathbf{B}$

17. $(\mathbf{A} \times \mathbf{B}) \times \mathbf{C}$ **18.** $\mathbf{A} \times (\mathbf{B} \times \mathbf{C})$ **19.** $|\mathbf{A} \times (\mathbf{B} \times \mathbf{C})|$

Two vector fields are given by $\mathbf{A} = x\mathbf{i} - y\mathbf{j} + 2t\mathbf{k}$ and $\mathbf{B} = (x^2 - z^2)\mathbf{i} - y^2\mathbf{j}$. Find each quantity at the point $(0, 2, 2)$ at $t = 2$.

20. $\mathbf{A} \cdot \mathbf{B}$ **21.** $\mathbf{A} \times \mathbf{B}$

22. $|(\mathbf{A} \times \mathbf{B}) \times \mathbf{A}|$ **23.** $\mathbf{A} \cdot \mathbf{A} \times \mathbf{B}$

24. Show that the diagonals of a rhombus (a parallelogram with equal sides) are perpendicular.

Verify each trigonometric identity.

25. $\cos(\alpha + \beta) = \cos \alpha \cos \beta - \sin \alpha \sin \beta$

26. $\sin (\alpha - \beta) = \sin \alpha \cos \beta - \sin \beta \cos \alpha$

27. $\sin (\alpha + \beta) = \sin \alpha \cos \beta + \sin \beta \cos \alpha$

Find the projection of $\mathbf{A}$ on $\mathbf{B}$ if

28. $\mathbf{A} = 3\mathbf{i} - 6\mathbf{j} + 2\mathbf{k}$, $\mathbf{B} = 7\mathbf{i} - 4\mathbf{j} + 4\mathbf{k}$

29. $\mathbf{A} = 3\mathbf{i} - 6\mathbf{j} + 9\mathbf{k}$, $\mathbf{B} = 4\mathbf{i} - 4\mathbf{j} + 2\mathbf{k}$

30. $\mathbf{A} = 4\mathbf{i} - 3\mathbf{j} + 7\mathbf{k}$, $\mathbf{B} = 2\mathbf{i} - 5\mathbf{j} - 7\mathbf{k}$

31. Determine a unit vector $\mathbf{i}_c$ perpendicular to the plane of $\mathbf{A} = 3\mathbf{i} + 6\mathbf{j} - \mathbf{k}$ and $\mathbf{B} = 2\mathbf{i} - 3\mathbf{j} + 4\mathbf{k}$.

32. Find a unit vector $\mathbf{i}_c$ perpendicular to both $\mathbf{A} = 3\mathbf{i} - 2\mathbf{j}$ and $\mathbf{B} = \mathbf{i} - 2\mathbf{j} + \mathbf{k}$.

33. Determine m such that $\mathbf{A} = 2\mathbf{i} - m\mathbf{j} + \mathbf{k}$ is perpendicular to $\mathbf{B} = 3\mathbf{i} - 2\mathbf{j}$.

34. The direction cosines of a vector $\mathbf{A}$ of length 15 are $\frac{1}{3}, \frac{2}{3}, -\frac{2}{3}$. Find the component of $\mathbf{A}$ along the line passing through the points $(1, 3, 2)$ and $(3, -2, 6)$.

35. Find the equation of the plane perpendicular to $\mathbf{B} = 3\mathbf{i} + 2\mathbf{j} - 4\mathbf{k}$. (Let $\mathbf{r} = x\mathbf{i} + y\mathbf{j} + z\mathbf{k}$ be a point on the plane.)

36. An object is moved from the point $(3, 2, -4)$ to the point $(5, 0, 6)$, where the distance is measured in meters. If the force acting on the object is $\mathbf{F} = 3\mathbf{i} - 10\mathbf{j}$ newtons, determine the work done.

37. An object weighs 10 newtons and falls 10 m while a force of $3\mathbf{i} - 5\mathbf{j}$ newtons acts on the object. Find the work done if the z axis is positive upward. Include the work done by the weight.

38. A rigid device is rotating with a speed of 45 rad/s about an axis oriented by the direction cosines $\frac{7}{9}$, $-\frac{4}{9}$, and $\frac{4}{9}$. Determine the velocity of a point on the device located by the position vector $\mathbf{r} = 2\mathbf{i} + 3\mathbf{j} - \mathbf{k}$ meters.

39. The velocity at the point $(-4, 2, -3)$, distances measured in meters, due to an angular velocity $\boldsymbol{\omega}$ is measured to be $\mathbf{V} = 10\mathbf{i} + 20\mathbf{j}$ m/s. What is $\boldsymbol{\omega}$ if $\omega_x = 2$ rad/s?

40. A force of 50 N acts at a point located by the position vector $\mathbf{r} = 4\mathbf{i} - 2\mathbf{j} + 4\mathbf{k}$ meters. The line of action of the force is oriented by the unit vector $\mathbf{i}_F = \frac{2}{3}\mathbf{i} -$

$\frac{2}{3}\mathbf{j} + \frac{1}{3}\mathbf{k}$. Determine the moment of the force about the a) x axis, and b) a line oriented by the unit vector $\mathbf{i}_L = -\frac{2}{3}\mathbf{i} + \frac{2}{3}\mathbf{j} - \frac{1}{3}\mathbf{k}$.

6.3   *Vector Differentiation*

6.3.1   ORDINARY DIFFERENTIATION

We study vector functions of one or more scalar variables. In this section we examine differentiation of vector functions of one variable. The derivative of the vector $\mathbf{u}(t)$ with respect to t is defined, as usual, by

$$\frac{d\mathbf{u}}{dt} = \lim_{\Delta t \to 0} \frac{\mathbf{u}(t + \Delta t) - \mathbf{u}(t)}{\Delta t}, \qquad (6.3.1)$$

where

$$\mathbf{u}(t + \Delta t) - \mathbf{u}(t) = \Delta \mathbf{u}. \qquad (6.3.2)$$

This is illustrated in Fig. 6.11. Note that the direction of $\Delta\mathbf{u}$ is, in general, unrelated to the direction of $\mathbf{u}(t)$.

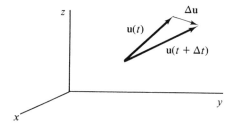

FIGURE 6.11.   Vectors used in the definition of the derivative $d\mathbf{u}/dt$.

From this definition it follows that the sums and products involving vector quantities can be differentiated as in ordinary calculus; that is,

$$\frac{d}{dt}(\phi\mathbf{u}) = \phi\frac{d\mathbf{u}}{dt} + \mathbf{u}\frac{d\phi}{dt}$$

$$\frac{d}{dt}(\mathbf{u}\cdot\mathbf{v}) = \mathbf{u}\cdot\frac{d\mathbf{v}}{dt} + \mathbf{v}\cdot\frac{d\mathbf{u}}{dt} \qquad (6.3.3)$$

$$\frac{d}{dt}(\mathbf{u}\times\mathbf{v}) = \mathbf{u}\times\frac{d\mathbf{v}}{dt} + \frac{d\mathbf{u}}{dt}\times v.$$

If we express the vector $\mathbf{u}(t)$ in rectangular coordinates, as

$$\mathbf{u}(t) = u_x\mathbf{i} + u_y\mathbf{j} + u_z\mathbf{k}, \qquad (6.3.4)$$

if can be differentiated term by term to yield

$$\frac{d\mathbf{u}}{dt} = \frac{du_x}{dt}\mathbf{i} + \frac{du_y}{dt}\mathbf{j} + \frac{du_z}{dt}\mathbf{k} \tag{6.3.5}$$

provided that the unit vectors $\mathbf{i}$, $\mathbf{j}$, and $\mathbf{k}$ are independent of t. If t represents time, such a reference frame is referred to as an *inertial reference frame*.

We shall illustrate differentiation by considering the motion of a particle in a noninertial reference frame. Let us calculate the velocity and acceleration of such a particle. The particle occupies the position (x, y, z) measured in the noninertial xyz reference frame which is rotating with an angular velocity $\boldsymbol{\omega}$, as shown in Fig. 6.12. The xyz reference frame is located by the vector $\mathbf{s}$ relative to the inertial XYZ reference frame.* The velocity $\mathbf{V}$ referred to the XYZ frame is

$$\mathbf{V} = \frac{d}{dt}(\mathbf{s} + \mathbf{r}) = \frac{d\mathbf{s}}{dt} + \frac{d\mathbf{r}}{dt}. \tag{6.3.6}$$

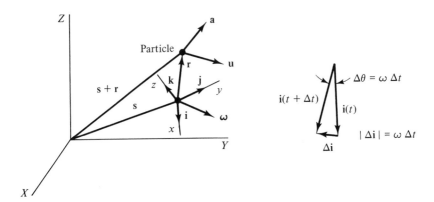

FIGURE 6.12.   Motion referred to a noninertial reference frame.

The quantity $d\mathbf{s}/dt$ is the velocity of the xyz reference frame and is denoted $\mathbf{V}_{\text{ref}}$. The vector $d\mathbf{r}/dt$ is, using $\mathbf{r} = x\mathbf{i} + y\mathbf{j} + z\mathbf{k}$,

$$\frac{d\mathbf{r}}{dt} = \frac{dx}{dt}\mathbf{i} + \frac{dy}{dt}\mathbf{j} + \frac{dz}{dt}\mathbf{k} + x\frac{d\mathbf{i}}{dt} + y\frac{d\mathbf{j}}{dt} + z\frac{d\mathbf{k}}{dt}. \tag{6.3.7}$$

To determine an expression for the time derivatives of the unit vectors, which are due to the angular velocity $\boldsymbol{\omega}$ of the xyz frame, consider the unit vector $\mathbf{i}$ to rotate through a small angle during the time Δt, illustrated in Fig. 6.12.

---

*This reference frame is attached to the ground in the case of a projectile or a rotating device; it is attached to the sun when describing the motion of satellites.

Using the definition of a derivative, there results

$$\frac{d\mathbf{i}}{dt} = \lim_{\Delta t \to 0} \frac{\mathbf{i}(t + \Delta t) - \mathbf{i}(t)}{\Delta t}$$

$$= \lim_{\Delta t \to 0} \frac{\Delta \mathbf{i}}{\Delta t} = \lim_{\Delta t \to 0} \frac{\omega \Delta t \left(\dfrac{\boldsymbol{\omega} \times \mathbf{i}}{\omega} \right)}{\Delta t} = \boldsymbol{\omega} \times \mathbf{i}, \qquad (6.3.8)$$

where the quantity $\boldsymbol{\omega} \times \mathbf{i}/\omega$ is a unit vector perpendicular to $\mathbf{i}$ in the direction of $\Delta \mathbf{i}$. Similarly,

$$\frac{d\mathbf{j}}{dt} = \boldsymbol{\omega} \times \mathbf{j}, \qquad \frac{d\mathbf{k}}{dt} = \boldsymbol{\omega} \times \mathbf{k}. \qquad (6.3.9)$$

Substituting these and Eq. 6.3.7 into Eq. 6.3.6, we have

$$\mathbf{V} = \mathbf{V}_{\text{ref}} + \frac{dx}{dt}\mathbf{i} + \frac{dy}{dt}\mathbf{j} + \frac{dz}{dt}\mathbf{k} + x\boldsymbol{\omega} \times \mathbf{i} + y\boldsymbol{\omega} \times \mathbf{j} + z\boldsymbol{\omega} \times \mathbf{k}. \tag{6.3.10}$$

The velocity $\mathbf{v}$ of the particle relative to the xyz frame is

$$\mathbf{v} = \frac{dx}{dt}\mathbf{i} + \frac{dy}{dt}\mathbf{j} + \frac{dz}{dt}\mathbf{k}. \qquad (6.3.11)$$

Hence, we can write the expression for the absolute velocity as

$$\mathbf{V} = \mathbf{V}_{\text{ref}} + \mathbf{v} + \boldsymbol{\omega} \times \mathbf{r}. \qquad (6.3.12)$$

The absolute acceleration $\mathbf{A}$ is obtained by differentiating $\mathbf{V}$ with respect to time to obtain

$$\mathbf{A} = \frac{d\mathbf{V}}{dt} = \frac{d\mathbf{V}_{\text{ref}}}{dt} + \frac{d\mathbf{v}}{dt} + \frac{d\boldsymbol{\omega}}{dt} \times \mathbf{r} + \boldsymbol{\omega} \times \frac{d\mathbf{r}}{dt}. \qquad (6.3.13)$$

In this equation

$$\frac{d\mathbf{V}_{\text{ref}}}{dt} = \mathbf{A}_{\text{ref}} \qquad (6.3.14)$$

$$\frac{d\mathbf{v}}{dt} = \frac{d}{dt}(v_x\mathbf{i} + v_y\mathbf{j} + v_z\mathbf{k})$$

$$= \frac{dv_x}{dt}\mathbf{i} + \frac{dv_y}{dt}\mathbf{j} + \frac{dv_z}{dt}\mathbf{k} + v_x\frac{d\mathbf{i}}{dt} + v_y\frac{d\mathbf{j}}{dt} + v_z\frac{d\mathbf{k}}{dt}$$

$$= \mathbf{a} + \boldsymbol{\omega} \times \mathbf{v} \qquad (6.3.15)$$

$$\frac{d\mathbf{r}}{dt} = \mathbf{v} + \boldsymbol{\omega} \times \mathbf{r}, \qquad (6.3.16)$$

where $\mathbf{a}$ is the acceleration of the particle observed in the xyz frame. The absolute acceleration is then

$$\mathbf{A} = \mathbf{A}_{\text{ref}} + \mathbf{a} + \boldsymbol{\omega} \times \mathbf{v} + \frac{d\boldsymbol{\omega}}{dt} \times \mathbf{r} + \boldsymbol{\omega} \times (\mathbf{v} + \boldsymbol{\omega} \times \mathbf{r}).$$

(6.3.17)

This is reorganized in the form

$$\mathbf{A} = \mathbf{A}_{\text{ref}} + \mathbf{a} + 2\boldsymbol{\omega} \times \mathbf{v} + \boldsymbol{\omega} \times (\boldsymbol{\omega} \times \mathbf{r}) + \frac{d\boldsymbol{\omega}}{dt} \times \mathbf{r}.$$

(6.3.18)

The quantity $2\boldsymbol{\omega} \times \mathbf{v}$ is often referred to as the *Coriolis acceleration,* and $d\boldsymbol{\omega}/dt$ is the angular acceleration of the xyz frame. For a rigid body $\mathbf{a}$ and $\mathbf{v}$ are zero.

Example 6.3.1:   Using the definition of a derivative, show that

$$\frac{d}{dt}(\mathbf{u} \cdot \mathbf{v}) = \mathbf{u} \cdot \frac{d\mathbf{v}}{dt} + \mathbf{v} \cdot \frac{d\mathbf{u}}{dt}.$$

SOLUTION: The definition of a derivative allows us to write

$$\frac{d}{dt}(\mathbf{u} \cdot \mathbf{v}) = \lim_{\Delta t \to 0} \frac{\mathbf{u}(t + \Delta t) \cdot \mathbf{v}(t + \Delta t) - \mathbf{u}(t) \cdot \mathbf{v}(t)}{\Delta t}.$$

But we know that (see Fig. 6.11)

$$\mathbf{u}(t + \Delta t) - \mathbf{u}(t) = \Delta \mathbf{u}$$

$$\mathbf{v}(t + \Delta t) - \mathbf{v}(t) = \Delta \mathbf{v}.$$

Substituting for $\mathbf{u}(t + \Delta t)$ and $\mathbf{v}(t + \Delta t)$, there results

$$\frac{d}{dt}(\mathbf{u} \cdot \mathbf{v}) = \lim_{\Delta t \to 0} \frac{[\Delta \mathbf{u} + \mathbf{u}(t)] \cdot [\Delta \mathbf{v} + \mathbf{v}(t)] - \mathbf{u}(t) \cdot \mathbf{v}(t)}{\Delta t}.$$

This product is expanded to yield

$$\frac{d}{dt}(\mathbf{u} \cdot \mathbf{v}) = \lim_{\Delta t \to 0} \frac{\Delta \mathbf{u} \cdot \Delta \mathbf{v} + \mathbf{u} \cdot \Delta \mathbf{v} + \mathbf{v} \cdot \Delta \mathbf{u} + \mathbf{u}\!\!\!/\mathbf{v} - \mathbf{u}\!\!\!/\mathbf{v}}{\Delta t}.$$

In the limit as $\Delta t \to 0$, both $\Delta \mathbf{u} \to \mathbf{0}$ and $\Delta \mathbf{v} \to \mathbf{0}$. Hence,

$$\lim_{\Delta t \to 0} \frac{\Delta \mathbf{u} \cdot \Delta \mathbf{v}}{\Delta t} \to 0.$$

We are left with

$$\frac{d}{dt}(\mathbf{u} \cdot \mathbf{v}) = \lim_{\Delta t \to 0} \left(\mathbf{u} \cdot \frac{\Delta \mathbf{v}}{\Delta t} + \mathbf{v} \cdot \frac{\Delta \mathbf{u}}{\Delta t} \right)$$

$$= \mathbf{u} \cdot \frac{d\mathbf{v}}{dt} + \mathbf{v} \cdot \frac{d\mathbf{u}}{dt}$$

and the given relationship is proved.   ■

Example 6.3.2:   The position of a particle is given by $\mathbf{r} = t^2\mathbf{i} + 2\mathbf{j} + 5(t - 1)\mathbf{k}$ meters, measured in a reference frame that has no translational velocity but that has an angular velocity of 20 rad/s about the z axis. Determine the absolute velocity at $t = 2$ s.

SOLUTION: Given that $\mathbf{V}_{\text{ref}} = \mathbf{0}$ the absolute velocity is

$$\mathbf{V} = \mathbf{v} + \boldsymbol{\omega} \times \mathbf{r}.$$

The velocity, as viewed from the rotating reference frame, is

$$\mathbf{v} = \frac{d\mathbf{r}}{dt} = \frac{d}{dt}[t^2\mathbf{i} + 2\mathbf{j} + 5(t - 1)\mathbf{k}]$$

$$= 2t\mathbf{i} + 5\mathbf{k}.$$

The contribution due to the angular velocity is

$$\boldsymbol{\omega} \times \mathbf{r} = 20\mathbf{k} \times [t^2\mathbf{i} + 2\mathbf{j} + 5(t - 1)\mathbf{k}]$$

$$= 20t^2\mathbf{j} - 40\mathbf{i}.$$

Thus, the absolute velocity is

$$\mathbf{V} = 2t\mathbf{i} + 5\mathbf{k} + 20t^2\mathbf{j} - 40\mathbf{i}$$

$$= (2t - 40)\mathbf{i} + 20t^2\mathbf{j} + 5\mathbf{k}.$$

At $t = 2$ s this becomes

$$\mathbf{V} = -36\mathbf{i} + 80\mathbf{j} + 5\mathbf{k} \qquad \text{m/s.} \quad \blacksquare$$

Example 6.3.3:   A person is walking toward the center of a merry-go-round along a radial line at a constant rate of 6 m/s. The angular velocity of the merry-go-round is 1.2 rad/s. Calculate the absolute acceleration when the person reaches a position 3 m from the axis of rotation.

SOLUTION: The acceleration $\mathbf{A}_{\text{ref}}$ is assumed to be zero, as is the angular acceleration $d\boldsymbol{\omega}/dt$ of the merry-go-round. Also, the acceleration $\mathbf{a}$ of the person relative to the merry-go-round is zero. Thus, the absolute acceleration is

$$\mathbf{A} = 2\boldsymbol{\omega} \times \mathbf{v} + \boldsymbol{\omega} \times (\boldsymbol{\omega} \times \mathbf{r}).$$

Attach the xyz reference frame to the merry-go-round with the z axis vertical and the person walking along the x axis toward the origin. Then

$$\boldsymbol{\omega} = 1.2\mathbf{k}, \qquad \mathbf{r} = 3\mathbf{i}, \qquad \mathbf{v} = -6\mathbf{i}.$$

The absolute acceleration is then

$$\mathbf{A} = 2[1.2\mathbf{k} \times (-6\mathbf{i})] + 1.2\mathbf{k} \times (1.2\mathbf{k} \times 3\mathbf{i})$$

$$= -4.32\mathbf{i} - 14.4\mathbf{j} \qquad \text{m/s}^2.$$

Note the y component of acceleration that is normal to the direction of motion, which makes the person sense a tugging in that direction. $\blacksquare$

6.3.2   PARTIAL DIFFERENTIATION

Many phenomena require that a quantity be defined at all points of some region of interest. The quantity may also vary with time. Such quantities are often referred to as *field quantities:* electric fields, magnetic fields, velocity fields, and pressure fields are examples. Partial derivatives are necessary when describing fields. Consider a vector function $\mathbf{u}(x, y, z, t)$.

The partial derivative of $\mathbf{u}$ with respect to x is defined to be

$$\frac{\partial \mathbf{u}}{\partial x} = \lim_{\Delta x \to 0} \frac{\mathbf{u}(x + \Delta x, y, z, t) - \mathbf{u}(x, y, z, t)}{\Delta x}. \qquad (6.3.19)$$

In terms of the components we have

$$\frac{\partial \mathbf{u}}{\partial x} = \frac{\partial u_x}{\partial x}\mathbf{i} + \frac{\partial u_y}{\partial x}\mathbf{j} + \frac{\partial u_z}{\partial x}\mathbf{k} \qquad (6.3.20)$$

where each component could be a function of x, y, z, and t.

The incremental quantity $\Delta \mathbf{u}$ between the two points (x, y, z) and $(x + \Delta x, y + \Delta y, z + \Delta z)$ at the same instant in time is

$$\Delta \mathbf{u} = \frac{\partial \mathbf{u}}{\partial x}\Delta x + \frac{\partial \mathbf{u}}{\partial y}\Delta y + \frac{\partial \mathbf{u}}{\partial z}\Delta z. \qquad (6.3.21)$$

At a fixed point in space $\Delta \mathbf{u}$ is given by

$$\Delta \mathbf{u} = \frac{\partial \mathbf{u}}{\partial t}\Delta t. \qquad (6.3.22)$$

If we are interested in the acceleration of a particular particle in a region fully occupied by particles, a *continuum,* we write the incremental velocity $\Delta \mathbf{v}$ between two points, shown in Fig. 6.13, as

$$\Delta \mathbf{v} = \frac{\partial \mathbf{v}}{\partial x}\Delta x + \frac{\partial \mathbf{v}}{\partial y}\Delta y + \frac{\partial \mathbf{v}}{\partial z}\Delta z + \frac{\partial \mathbf{v}}{\partial t}\Delta t, \qquad (6.3.23)$$

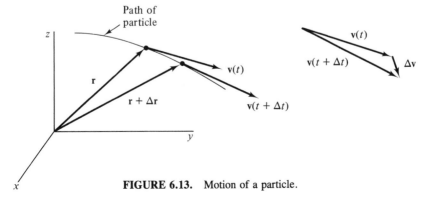

FIGURE 6.13.   Motion of a particle.

where we recognize that not only is the position of the particle changing but so is time increasing. Acceleration is defined by

$$\mathbf{a} = \frac{d\mathbf{v}}{dt} = \lim_{\Delta t \to 0} \frac{\mathbf{v}(t + \Delta t) - \mathbf{v}(t)}{\Delta t} = \lim_{\Delta t \to 0} \frac{\Delta \mathbf{v}}{\Delta t}. \qquad (6.3.24)$$

Using the expression from Eq. (6.3.23), we have

$$\frac{d\mathbf{v}}{dt} = \lim_{\Delta t \to 0} \left[\frac{\partial \mathbf{v}}{\partial x} \frac{\Delta x}{\Delta t} + \frac{\partial \mathbf{v}}{\partial y} \frac{\Delta y}{\Delta t} + \frac{\partial \mathbf{v}}{\partial z} \frac{\Delta z}{\Delta t} \right] + \frac{\partial \mathbf{v}}{\partial t}. \qquad (6.3.25)$$

Realizing that we are following a particular particle,

$$\lim_{\Delta t \to 0} \frac{\Delta x}{\Delta t} = v_x, \qquad \lim_{\Delta t \to 0} \frac{\Delta y}{\Delta t} = v_y, \qquad \lim_{\Delta t \to 0} \frac{\Delta z}{\Delta t} = v_z. \qquad (6.3.26)$$

Then there follows

$$\mathbf{a} = \frac{D\mathbf{v}}{Dt} = v_x \frac{\partial \mathbf{v}}{\partial x} + v_y \frac{\partial \mathbf{v}}{\partial y} + v_z \frac{\partial \mathbf{v}}{\partial z} + \frac{\partial \mathbf{v}}{\partial t}, \qquad (6.3.27)$$

where we have adopted the popular convention to use D/Dt to emphasize that we have followed a material particle. It is called the *material* or *substantial derivative,* and from Eq. 6.3.27 is observed to be

$$\frac{D}{Dt} = v_x \frac{\partial}{\partial x} + v_y \frac{\partial}{\partial y} + v_z \frac{\partial}{\partial z} + \frac{\partial}{\partial t}. \qquad (6.3.28)$$

We may form derivatives in a similar manner for any quantity of interest. For example, the rate of change of temperature of a particle is given by

$$\frac{DT}{Dt} = v_x \frac{\partial T}{\partial x} + v_y \frac{\partial T}{\partial y} + v_z \frac{\partial T}{\partial z} + \frac{\partial T}{\partial t}. \qquad (6.3.29)$$

Example 6.3.4: A velocity field is given by $\mathbf{v} = x^2 \mathbf{i} + xy \mathbf{j} + 2t^2 \mathbf{k}$ m/s. Determine the acceleration at the point (2, 1, 0) meters and $t = 2$ s.

SOLUTION: The acceleration is given by

$$\mathbf{a} = \frac{D\mathbf{v}}{Dt} = \left[v_x \frac{\partial}{\partial x} + v_y \frac{\partial}{\partial y} + v_z \frac{\partial}{\partial z} + \frac{\partial}{\partial t} \right] \mathbf{v}$$

$$= \left[x^2 \frac{\partial}{\partial x} + xy \frac{\partial}{\partial y} + 2t^2 \frac{\partial}{\partial z} + \frac{\partial}{\partial t} \right] (x^2 \mathbf{i} + xy \mathbf{j} + 2t^2 \mathbf{k})$$

$$= x^2 (2x \mathbf{i} + y \mathbf{j}) + xy(x \mathbf{j}) + 2t^2 \cdot 0 + 4t \mathbf{k}$$

$$= 2x^3 \mathbf{i} + 2x^2 y \mathbf{j} + 4t \mathbf{k}.$$

At the point (2, 1, 0) and at $t = 2$ s, there results

$$\mathbf{a} = 16\mathbf{i} + 8\mathbf{j} + 8\mathbf{k} \qquad \text{m/s}^2. \quad \blacksquare$$

PROBLEMS

1. By using the definition of the derivative show that

$$\frac{d}{dt}(\phi\mathbf{u}) = \phi\frac{d\mathbf{u}}{dt} + \mathbf{u}\frac{d\phi}{dt}.$$

Given the two vectors $\mathbf{u} = 2t\mathbf{i} + t^2\mathbf{k}$ and $\mathbf{v} = \cos 5t\mathbf{i} + \sin 5t\mathbf{j} - 10\mathbf{k}$. At $t = 2$, evaluate the following

2. $\dfrac{d\mathbf{u}}{dt}$

3. $\dfrac{d\mathbf{v}}{dt}$

4. $\dfrac{d}{dt}(\mathbf{u}\cdot\mathbf{v})$

5. $\dfrac{d}{dt}(\mathbf{u}\times\mathbf{v})$

6. $\dfrac{d^2}{dt^2}(\mathbf{u}\cdot\mathbf{v})$

7. Find a unit vector in the direction of $d\mathbf{u}/dt$ if $\mathbf{u} = 2t^2\mathbf{i} - 3t\mathbf{j}$, at $t = 1$.

8. The velocity of a particle of water moving down a dishwasher arm at a distance of 0.2 m is 10 m/s. It is decellerating at a rate of 30 m/s². The arm is rotating at 30 rad/s. Determine the absolute acceleration of the particle.

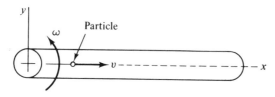

9. Show why it is usually acceptable to consider a reference frame attached to the earth an an inertial reference frame. The radius of the earth is 6400 km.

10. The wind is blowing straight south at 90 km/hr. At a latitude of 45°, calculate the magnitudes of the Coriolis acceleration and the $\boldsymbol{\omega} \times (\boldsymbol{\omega} \times \mathbf{r})$ component of the acceleration of an air particle.

11. A velocity field is given by $\mathbf{v} = x^2\mathbf{i} - 2xy\mathbf{j} + 4t\mathbf{k}$ m/s. Determine the acceleration at the point $(2, 1, -4)$ meters.

12. A temperature field is calculated to be $T(x, y, z, t) = e^{-0.1t}\sin 5x$. Determine the rate at which the temperature of a particle is changing if $\mathbf{v} = 10\mathbf{i} - 5\mathbf{j}$ m/s. Evaluate DT/Dt at $x = 2$ m and $t = 10$ s.

6.4 *The Gradient*

When studying phenomena that occur in a region of interest certain variables often change from point to point, and this change must usually be accounted for. Consider a scalar variable represented at the point (x, y, z) by the function $\phi(x, y, z)$.* This could be the temperature, for example. The incre-

---

*We use rectangular coordinates in this section. Cylindrical and spherical coordinates will be presented in Section 6.5.

mental change in ϕ, as we move to a neighboring point $(x + \Delta x, y + \Delta y, z + \Delta z)$, is given by

$$\Delta\phi = \frac{\partial\phi}{\partial x}\Delta x + \frac{\partial\phi}{\partial y}\Delta y + \frac{\partial\phi}{\partial z}\Delta z, \qquad (6.4.1)$$

where $\partial\phi/\partial x$, $\partial\phi/\partial y$, and $\partial\phi/\partial z$ represent the rate of change of ϕ in the x, y, and z directions, respectively. If we divide by the incremental distance between the two points, shown in Fig. 6.14, we have, using $|\Delta\mathbf{r}| = \Delta r$,

$$\frac{\Delta\phi}{\Delta r} = \frac{\partial\phi}{\partial x}\frac{\Delta x}{\Delta r} + \frac{\partial\phi}{\partial y}\frac{\Delta y}{\Delta r} + \frac{\partial\phi}{\partial z}\frac{\Delta z}{\Delta r}. \qquad (6.4.2)$$

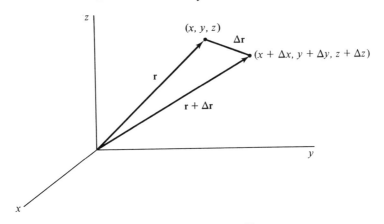

FIGURE 6.14. Change in the position vector.

Now we can let Δx, Δy, and Δz approach zero and we arrive at the derivative of ϕ in the direction of $\Delta\mathbf{r}$,

$$\frac{d\phi}{dr} = \frac{\partial\phi}{\partial x}\frac{dx}{dr} + \frac{\partial\phi}{\partial y}\frac{dy}{dr} + \frac{\partial\phi}{\partial z}\frac{dz}{dr}. \qquad (6.4.3)$$

This is the chain rule applied to $\phi[x(r), y(r), z(r)]$. The form of this result suggests that it may be written as the dot product of two vectors; that is,

$$\frac{d\phi}{dr} = \left(\frac{\partial\phi}{\partial x}\mathbf{i} + \frac{\partial\phi}{\partial y}\mathbf{j} + \frac{\partial\phi}{\partial z}\mathbf{k}\right) \cdot \left(\frac{dx}{dr}\mathbf{i} + \frac{dy}{dr}\mathbf{j} + \frac{dz}{dr}\mathbf{k}\right). \qquad (6.4.4)$$

Recognizing that

$$d\mathbf{r} = dx\mathbf{i} + dy\mathbf{j} + dz\mathbf{k}, \qquad (6.4.5)$$

we can write Eq. (6.4.4) as

$$\frac{d\phi}{dr} = \left(\frac{\partial\phi}{\partial x}\mathbf{i} + \frac{\partial\phi}{\partial y}\mathbf{j} + \frac{\partial\phi}{\partial z}\mathbf{k}\right) \cdot \frac{d\mathbf{r}}{dr}. \qquad (6.4.6)$$

The vector in parentheses is called the *gradient* of ϕ and is usually written

$$\nabla\phi = \text{grad } \phi = \frac{\partial\phi}{\partial x}\mathbf{i} + \frac{\partial\phi}{\partial y}\mathbf{j} + \frac{\partial\phi}{\partial z}\mathbf{k}. \tag{6.4.7}$$

The symbol ∇ is called *del* and is the vector differential operator

$$\nabla = \frac{\partial}{\partial x}\mathbf{i} + \frac{\partial}{\partial y}\mathbf{j} + \frac{\partial}{\partial z}\mathbf{k}. \tag{6.4.8}$$

The quantity $d\mathbf{r}/dr$ is obviously a unit vector in the direction of $d\mathbf{r}$. Thus, returning to Eq. 6.4.6 we observe that the rate of change of ϕ in a particular direction is given by $\nabla\phi$ dotted with a unit vector in that direction; that is,

$$\frac{d\phi}{dn} = \nabla\phi \cdot \mathbf{i}_n, \tag{6.4.9}$$

where $\mathbf{i}_n$ is a unit vector in the n direction.

Example 6.4.1:   Find the derivative of the function $\phi = x^2 - 2xy + z^2$ at the point $(2, -1, 1)$ in the direction of the vector $\mathbf{A} = 2\mathbf{i} - 4\mathbf{j} + 4\mathbf{k}$.

SOLUTION: To find the derivative of a function in a particular direction we use Eq. 6.4.9. We must first find the gradient of the function. It is

$$\nabla\phi = \left(\frac{\partial}{\partial x}\mathbf{i} + \frac{\partial}{\partial y}\mathbf{j} + \frac{\partial}{\partial z}\mathbf{k}\right)(x^2 - 2xy + z^2)$$

$$= (2x - 2y)\mathbf{i} - 2x\mathbf{j} + 2z\mathbf{k}.$$

At the point $(2, -1, 1)$, it is

$$\nabla\phi = 6\mathbf{i} - 4\mathbf{j} + 2\mathbf{k}.$$

The unit vector in the desired direction is

$$\mathbf{i}_n = \frac{\mathbf{A}}{A} = \frac{2\mathbf{i} - 4\mathbf{j} + 4\mathbf{k}}{6} = \frac{1}{3}\mathbf{i} - \frac{2}{3}\mathbf{j} + \frac{2}{3}\mathbf{k}.$$

Finally, the derivative in the direction of $\mathbf{A}$ is

$$\frac{d\phi}{dn} = \nabla\phi \cdot \mathbf{i}_n$$

$$= (6\mathbf{i} - 4\mathbf{j} + 2\mathbf{k}) \cdot (\tfrac{1}{3}\mathbf{i} - \tfrac{2}{3}\mathbf{j} + \tfrac{2}{3}\mathbf{k})$$

$$= 2 + \tfrac{8}{3} + \tfrac{4}{3} = 6. \quad \blacksquare$$

Another important property of $\nabla\phi$ is that $\nabla\phi$ is normal to a constant ϕ surface. To show this, consider the constant ϕ surface and the differential displacement vector $d\mathbf{r}$, illustrated in Fig. 6.15. If $\nabla\phi$ is normal to a constant ϕ surface, then $\nabla\phi \cdot d\mathbf{r}$ will be zero since $d\mathbf{r}$ is a vector that lies in the

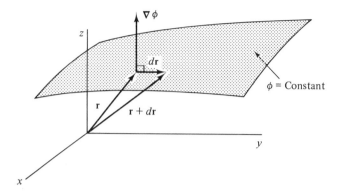

FIGURE 6.15. Constant ϕ surface.

surface. The quantity $\nabla\phi \cdot d\mathbf{r}$ is given by (see Eqs. 6.4.5 and 6.4.7)

$$\nabla\phi \cdot d\mathbf{r} = \frac{\partial\phi}{\partial x} dx + \frac{\partial\phi}{\partial y} dy + \frac{\partial\phi}{\partial z} dz. \qquad (6.4.10)$$

We recognize that this expression is simply $d\phi$. But $d\phi = 0$ along a constant ϕ surface; thus, $\nabla\phi \cdot d\mathbf{r} = 0$ and $\nabla\phi$ is normal to a constant ϕ surface.

We also note that $\nabla\phi$ points in a direction in which the derivative of ϕ is numerically the greatest since Eq. 6.4.9 shows that $d\phi/dn$ is maximum when $\mathbf{i}_n$ is in the direction of $\nabla\phi$. Because of this, $\nabla\phi$ may be referred to as the *maximum directional derivative*.

Example 6.4.2: Find a unit vector $\mathbf{i}_n$ normal to the surface represented by the equation $x^2 - 8y^2 + z^2 = 0$ at the point $(8, 1, 4)$.

SOLUTION: We know that the gradient $\nabla\phi$ is normal to a constant ϕ surface. So, with

$$\phi = x^2 - 8y^2 + z^2,$$

we form the gradient, to get

$$\nabla\phi = \frac{\partial\phi}{\partial x}\mathbf{i} + \frac{\partial\phi}{\partial y}\mathbf{j} + \frac{\partial\phi}{\partial z}\mathbf{k} = 2x\mathbf{i} - 16y\mathbf{j} + 2z\mathbf{k}.$$

At the point $(8, 1, 4)$ we have

$$\nabla\phi = 16\mathbf{i} - 16\mathbf{j} + 8\mathbf{k}.$$

This vector is normal to the surface at $(8, 1, 4)$. To find the unit vector, we simply divide the vector by its magnitude, obtaining

$$\mathbf{i}_n = \frac{\nabla\phi}{|\nabla\phi|} = \frac{16\mathbf{i} - 16\mathbf{j} + 8\mathbf{k}}{\sqrt{256 + 256 + 64}} = \frac{2}{3}\mathbf{i} - \frac{2}{3}\mathbf{j} + \frac{1}{3}\mathbf{k}. \quad \blacksquare$$

Example 6.4.3:   Find the equation of the plane which is tangent to the surface $x^2 + y^2 - z^2 = 4$ at the point $(1, 2, -1)$.

SOLUTION: The gradient ϕ is normal to a constant ϕ surface. Hence, with $\phi = x^2 + y^2 - z^2$, the vector

$$\nabla\phi = \frac{\partial\phi}{\partial x}\mathbf{i} + \frac{\partial\phi}{\partial y}\mathbf{j} + \frac{\partial\phi}{\partial z}\mathbf{k} = 2x\mathbf{i} + 2y\mathbf{j} - 2z\mathbf{k}$$

is normal to the given surface. At the point $(1, 2, -1)$ the normal vector is

$$\nabla\phi = 2\mathbf{i} + 4\mathbf{j} + 2\mathbf{k}.$$

Consider the sketch shown. The vector to the given point $\mathbf{r}_0 = \mathbf{i} + 2\mathbf{j} - \mathbf{k}$ subtracted from the vector to the general point $\mathbf{r} = x\mathbf{i} + y\mathbf{j} + z\mathbf{k}$ is a vector in the desired plane. It is

$$\mathbf{r} - \mathbf{r}_0 = (x\mathbf{i} + y\mathbf{j} + z\mathbf{k}) - (\mathbf{i} + 2\mathbf{j} - \mathbf{k})$$
$$= (x - 1)\mathbf{i} + (y - 2)\mathbf{j} + (z + 1)\mathbf{k}.$$

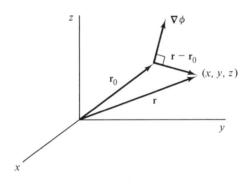

This vector, when dotted with a vector normal to it, namely $\nabla\phi$, must yield zero; that is,

$$\nabla\phi \cdot (\mathbf{r} - \mathbf{r}_0) = (2\mathbf{i} + 4\mathbf{j} + 2\mathbf{k}) \cdot [(x - 1)\mathbf{i} + (y - 2)\mathbf{j} + (z + 1)\mathbf{k}]$$
$$= 2(x - 1) + 4(y - 2) + 2(z + 1) = 0.$$

Thus, the tangent plane is given by

$$x + 2y + z = 4. \quad \blacksquare$$

The vector character of the del operator suggests that we form the dot and cross products with ∇ and a vector function. Consider a general vector function $\mathbf{u}(x, y, z)$ in which each component is a function of x, y, and z. The dot product of the ∇ operator with $\mathbf{u}(x, y, z)$ is written in rectangular coordinates* as

$$\nabla \cdot \mathbf{u} = \left(\frac{\partial}{\partial x} \mathbf{i} + \frac{\partial}{\partial y} \mathbf{j} + \frac{\partial}{\partial z} \mathbf{k} \right) \cdot (u_x \mathbf{i} + u_y \mathbf{j} + u_z \mathbf{k})$$

$$= \frac{\partial u_x}{\partial x} + \frac{\partial u_y}{\partial y} + \frac{\partial u_z}{\partial z}. \tag{6.4.11}$$

It is known as the *divergence* of the vector field **u**.

The cross product in rectangular coordinates is

$$\nabla \times \mathbf{u} = \left(\frac{\partial}{\partial x} \mathbf{i} + \frac{\partial}{\partial y} \mathbf{j} + \frac{\partial}{\partial z} \mathbf{k} \right) \times (u_x \mathbf{i} + u_y \mathbf{j} + u_z \mathbf{k})$$

$$= \left(\frac{\partial u_z}{\partial y} - \frac{\partial u_y}{\partial z} \right) \mathbf{i} + \left(\frac{\partial u_x}{\partial z} - \frac{\partial u_z}{\partial x} \right) \mathbf{j} + \left(\frac{\partial u_y}{\partial x} - \frac{\partial u_x}{\partial y} \right) \mathbf{k}$$

$$\tag{6.4.12}$$

and is known as the *curl* of the vector field **u**. The curl may be expressed as a determinant,

$$\nabla \times \mathbf{u} = \begin{vmatrix} \mathbf{i} & \mathbf{j} & \mathbf{k} \\ \dfrac{\partial}{\partial x} & \dfrac{\partial}{\partial y} & \dfrac{\partial}{\partial z} \\ u_x & u_y & u_z \end{vmatrix}. \tag{6.4.13}$$

The divergence and the curl of a vector function appear quite often in the derivation of the mathematical models for various physical phenomena. For example, let us determine the rate at which material is leaving the incremental volume shown in Fig. 6.16. The volume of material crossing a face in a time period Δt is indicated as the component of velocity normal to a face multiplied by the area of the face and the time Δt. If we account for all the material leaving the element, we have

$$\text{net loss} = \left(v_x + \frac{\partial v_x}{\partial x} \Delta x \right) \Delta y \, \Delta z \, \Delta t - v_x \Delta y \, \Delta z \, \Delta t + \left(v_y + \frac{\partial v_y}{\partial y} \Delta y \right) \Delta x \, \Delta z \, \Delta t$$

$$- v_y \Delta x \, \Delta z \, \Delta t + \left(v_z + \frac{\partial v_z}{\partial z} \Delta z \right) \Delta x \, \Delta y \, \Delta t - v_z \Delta x \, \Delta y \, \Delta t$$

$$= \left(\frac{\partial v_x}{\partial x} + \frac{\partial v_y}{\partial y} + \frac{\partial v_z}{\partial z} \right) \Delta x \, \Delta y \, \Delta z \, \Delta t. \tag{6.4.14}$$

If we divide by the elemental volume $\Delta x \, \Delta y \, \Delta z$ and the time increment Δt,

---

*Expressions in cylindrical and spherical coordinates will be given in Section 6.5.

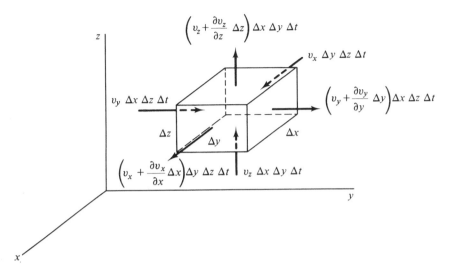

FIGURE 6.16.   Flow from an incremental volume.

there results

$$\text{rate of loss per unit volume} = \frac{\partial v_x}{\partial x} + \frac{\partial v_y}{\partial y} + \frac{\partial v_z}{\partial z} = \boldsymbol{\nabla} \cdot \mathbf{v}.$$

$$(6.4.15)$$

For an incompressible material, the amount of material in a volume remains constant; thus, the rate of loss must be zero; that is,

$$\boldsymbol{\nabla} \cdot \mathbf{v} = 0 \qquad\qquad (6.4.16)$$

for an incompressible material. It is the *continuity equation*. The same equation applies to a static electric field, in which case **v** represents the current density.

As we let Δx, Δy, and Δz shrink to zero, we note that the volume element approaches a point. If we consider material or electric current to occupy all points in a region of interest, then the divergence is valid at a point, and it represents the flux (quantity per second) emanating per unit volume.

For a physical interpretation of the curl, let us consider a rectangle undergoing motion while a material is deforming, displayed in Fig. 6.17. The velocity components at P are v_x and v_y, at Q they are $[v_x + (\partial v_x/\partial x)\Delta x]$ and $[v_y + (\partial v_y/\partial x)\Delta x]$, and at R they are $[v_x + (\partial v_x/\partial y)\Delta y]$ and $[v_y + (\partial v_y/\partial y)\Delta y]$. Point P will move to P' a distance $v_y \Delta t$ above P and a distance $v_x \Delta t$ to the right of P; Q will move to Q' a distance $[v_y + (\partial v_y/\partial x)\Delta x]\Delta t$ above Q; and R' will move a distance $[v_x + (\partial v_x/\partial y)\Delta y]\Delta t$ to the right of R. The quantity $(d/dt)[(\alpha + \beta)/2)]$, approximated by $(\Delta\alpha + \Delta\beta)/(2\Delta t)$ (the angles $\Delta\alpha$ and $\Delta\beta$ are shown),

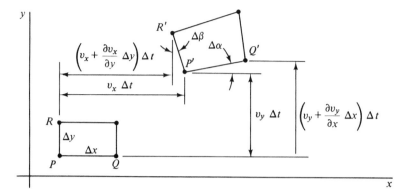

FIGURE 6.17. Displacement of a material element due to velocity components v_x and v_y.

would represent the rate at which the element is rotating. In terms of the velocity components, referring to the figure, we have

$$\frac{d}{dt}\left(\frac{\alpha + \beta}{2}\right)$$

$$\cong \frac{\Delta\alpha + \Delta\beta}{2\Delta t}$$

$$= \frac{\left[\left(v_y + \frac{\partial v_y}{\partial x}\Delta x\right)\Delta t - v_y\Delta t\right]\Big/\Delta x + \left[v_x\Delta t - \left(v_x + \frac{\partial v_x}{\partial y}\Delta y\right)\Delta t\right]\Big/\Delta y}{2\Delta t}$$

$$= \frac{1}{2}\left(\frac{\partial v_y}{\partial x} - \frac{\partial v_x}{\partial y}\right), \tag{6.4.17}$$

where we have used $\Delta\alpha \cong \tan \Delta\alpha$ since $\Delta\alpha$ is small. Thus, we see that the z component of $\nabla \times \mathbf{v}$, which is $[(\partial v_y/\partial x) - (\partial v_x/\partial y)]$, represents twice the rate of rotation of a material element about the z axis. Likewise, the x and y components of $\nabla \times \mathbf{v}$ represent twice the rate of rotation about the x and y axes, respectively. If we let $\mathbf{\Omega}$ be the angular velocity (rate of rotation), then

$$\mathbf{\Omega} = \tfrac{1}{2}\nabla \times \mathbf{v}. \tag{6.4.18}$$

As we let Δx and Δy again approach zero, we note that the element again approaches a point. Thus, the curl of a vector function is valid at a point and represents twice the rate at which a material element occupying the point is rotating. In electric and magnetic fields, the curl does not possess this physical meaning; it does, however, appear quite often and for a static field the electric current density **J** is given by the curl of the magnetic field intensity **H**; that is,

$$\mathbf{J} = \nabla \times \mathbf{H}. \tag{6.4.19}$$

There are several combinations of vector operations involving the ∇ operator which are encountered in applications. A very common one is the divergence of the gradient of a scalar function, written as $\nabla \cdot \nabla\phi$. In rectangular coordinates it is

$$\nabla \cdot \nabla\phi = \left(\frac{\partial}{\partial x}\mathbf{i} + \frac{\partial}{\partial y}\mathbf{j} + \frac{\partial}{\partial z}\mathbf{k} \right) \cdot \left(\frac{\partial\phi}{\partial x}\mathbf{i} + \frac{\partial\phi}{\partial y}\mathbf{j} + \frac{\partial\phi}{\partial z}\mathbf{k} \right)$$

$$= \frac{\partial^2\phi}{\partial x^2} + \frac{\partial^2\phi}{\partial y^2} + \frac{\partial^2\phi}{\partial z^2}. \tag{6.4.20}$$

It is usually written $\nabla^2\phi$ and is called the *Laplacian* of ϕ. If it is zero, that is

$$\nabla^2\phi = 0, \tag{6.4.21}$$

the equation is referred to as *Laplace's equation*.

The divergence of the curl of a vector function, and the curl of the gradient of a scalar function are also quantities of interest, but they can be shown to be zero by expanding in rectangular coordinates. Written out, they are

$$\nabla \cdot \nabla \times \mathbf{u} = 0$$

$$\nabla \times \nabla\phi = \mathbf{0}. \tag{6.4.22}$$

Two special kinds of vector fields exist. One is a *solenoidal vector field*, in which the divergence is zero, that is,

$$\nabla \cdot \mathbf{u} = 0, \tag{6.4.23}$$

and the other is an *irrotational* (or *conservative*) *vector field*, in which the curl is zero, that is

$$\nabla \times \mathbf{u} = \mathbf{0}. \tag{6.4.24}$$

If the vector field **u** is given by the gradient of a scalar function ϕ, that is,

$$\mathbf{u} = \nabla\phi, \tag{6.4.25}$$

then, according to Eq. 6.4.22, the curl of **u** is zero and **u** is irrotational. The function ϕ is referred to as the *scalar potential function* of the vector field **u**.

Several vector identities are often useful, and these are presented in Table 6.1. They can be verified by expanding in a particular coordinate system.

TABLE 6.1 Some Vector Identities

$$\nabla \times \nabla\phi = 0$$

$$\nabla \cdot \nabla \times \mathbf{u} = 0$$

$$\nabla \cdot (\phi\mathbf{u}) = \nabla\phi \cdot \mathbf{u} + \phi\nabla \cdot \mathbf{u}$$

$$\nabla \times (\phi\mathbf{u}) = \nabla\phi \times \mathbf{u} + \phi\nabla \times \mathbf{u}$$

$$\nabla \times (\nabla \times \mathbf{u}) = \nabla(\nabla \cdot \mathbf{u}) - \nabla^2\mathbf{u}$$

$$\mathbf{u} \times (\nabla \times \mathbf{u}) = \tfrac{1}{2}\nabla u^2 - (\mathbf{u} \cdot \nabla)\mathbf{u}$$

$$\nabla \cdot (\mathbf{u} \times \mathbf{v}) = (\nabla \times \mathbf{u}) \cdot \mathbf{v} - \mathbf{u} \cdot (\nabla \times \mathbf{v})$$

$$\nabla \times (\mathbf{u} \times \mathbf{v}) = \mathbf{u}(\nabla \cdot \mathbf{v}) - \mathbf{v}(\nabla \cdot \mathbf{u}) + (\mathbf{v} \cdot \nabla)\mathbf{u} - (\mathbf{u} \cdot \nabla)\mathbf{v}$$

$$\nabla(\mathbf{u} \cdot \mathbf{v}) = (\mathbf{u} \cdot \nabla)\mathbf{v} + (\mathbf{v} \cdot \nabla)\mathbf{u} + \mathbf{u} \times (\nabla \times \mathbf{v}) + \mathbf{v} \times (\nabla \times \mathbf{u})$$

Example 6.4.4:   A vector field is given by $\mathbf{u} = y^2\mathbf{i} + 2xy\mathbf{j} - z^2\mathbf{k}$. Determine the divergence of $\mathbf{u}$ and curl of $\mathbf{u}$ at the point $(1, 2, 1)$. Also, determine if the vector field is solenoidal or irrotational.

SOLUTION: The divergence of $\mathbf{u}$ is given by Eq. 6.4.11. It is

$$\nabla \cdot \mathbf{u} = \frac{\partial u_x}{\partial x} + \frac{\partial u_y}{\partial y} + \frac{\partial u_z}{\partial z}$$

$$= 0 + 2x - 2z.$$

At the point $(1, 2, 1)$ this scalar function has the value

$$\nabla \cdot \mathbf{u} = 2 - 2 = 0.$$

The curl of $\mathbf{u}$ is given by Eq. 6.4.12. It is

$$\nabla \times \mathbf{u} = \left(\frac{\partial u_z}{\partial y} - \frac{\partial u_y}{\partial z}\right)\mathbf{i} + \left(\frac{\partial u_x}{\partial z} - \frac{\partial u_z}{\partial x}\right)\mathbf{j} + \left(\frac{\partial u_y}{\partial x} - \frac{\partial u_x}{\partial y}\right)\mathbf{k}$$

$$= 0\mathbf{i} + 0\mathbf{j} + (2y - 2y)\mathbf{k}$$

$$= \mathbf{0}.$$

The curl of $\mathbf{u}$ is zero at all points in the field; hence, it is an irrotational vector field. However, $\nabla \cdot \mathbf{u}$ is not zero at *all* points in the field; thus, $\mathbf{u}$ is not solenoidal. ■

Example 6.4.5:   For the vector field $\mathbf{u} = y^2\mathbf{i} + 2xy\mathbf{j} - z^2\mathbf{k}$, find the associated scalar potential function $\phi(x, y, z)$, providing that one exists.

SOLUTION: The scalar potential function $\phi(x, y, z)$ is related to the vector field by

$$\nabla\phi = \mathbf{u}$$

providing that the curl of $\mathbf{u}$ is zero. The curl of $\mathbf{u}$ was shown to be zero in Example 6.4.4; hence, a potential function ϕ does exist. Writing the above using rectangular components, we have

$$\frac{\partial\phi}{\partial x}\mathbf{i} + \frac{\partial\phi}{\partial y}\mathbf{j} + \frac{\partial\phi}{\partial z}\mathbf{k} = y^2\mathbf{i} + 2xy\mathbf{j} - z^2\mathbf{k}.$$

This vector equation contains three scalar equations which result from equating the x component, the y component, and the z component, respectively from each side of the equation. This gives

$$\frac{\partial\phi}{\partial x} = y^2, \qquad \frac{\partial\phi}{\partial y} = 2xy, \qquad \frac{\partial\phi}{\partial z} = -z^2.$$

The first of these is integrated to give the solution

$$\phi(x, y, z) = xy^2 + f(y, z).$$

Note that in solving partial differential equations the "constant of integration" is a function. In the first equation we are differentiating with respect to x, holding y and z fixed; thus, this "constant of integration" may be a function of y and z, namely $f(y, z)$. Now, substitute the solution above into the second equation and obtain

$$2xy + \frac{\partial f}{\partial y} = 2xy.$$

This results in $\partial f/\partial y = 0$, which means that f does not depend on y. Thus, f must be at most a function of z. So substitute the solution into the third equation, and there results

$$\frac{df}{dz} = -z^2,$$

where we have used an ordinary derivative since $f = f(z)$. This equation is integrated to give

$$f(z) = -\frac{z^3}{3} + C,$$

where C is a constant of integration. Finally, the scalar potential function is

$$\phi(x, y, z) = xy^2 - \frac{z^3}{3} + C.$$

To prove that $\nabla\phi = \mathbf{u}$, let us find the gradient of ϕ. It is

$$\nabla\phi = \frac{\partial\phi}{\partial x}\mathbf{i} + \frac{\partial\phi}{\partial y}\mathbf{j} + \frac{\partial\phi}{\partial z}\mathbf{k} = y^2\mathbf{i} + 2xy\mathbf{j} - z^2\mathbf{k}.$$

This is equal to the given vector function $\mathbf{u}$, completing the proof. ■

PROBLEMS

Find the gradient of each scalar function. Use $\mathbf{r} = x\mathbf{i} + y\mathbf{j} + z\mathbf{k}$ if required.

1. $\phi = x^2 + y^2$ **2.** $\phi = 2xy$ **3.** $\phi = r^2$

4. $\phi = e^x \sin 2y$ **5.** $\phi = x^2 + 2xy - z^2$ **6.** $\phi = \ln r$

7. $\phi = 1/r$ **8.** $\phi = \tan^{-1} y/x$ **9.** $\phi = r^n$

Find a unit vector $\mathbf{i}_n$ normal to each surface at the point indicated.

10. $x^2 + y^2 = 5$, $(2, 1, 0)$ **11.** $r = 5$, $(4, 0, 3)$

12. $2x^2 - y^2 = 7$, $(2, 1, -1)$ **13.** $x^2 + yz = 3$, $(2, -1, 1)$

14. $x + y^2 - 2z^2 = 6$, $(4, 2, 1)$ **15.** $x^2y + yz = 6$, $(2, 3, -2)$

Determine the equation of the plane tangent to the given surface at the point indicated.

16. $x^2 + y^2 + z^2 = 25$, $(3, 4, 0)$ **17.** $r = 6$, $(2, 4, 4)$

18. $x^2 - 2xy = 0$, $(2, 2, 1)$ **19.** $xy^2 - zx + y^2 = 0$, $(1, -1, 2)$

The temperature in a region of interest is determined to be given by the function $T = x^2 + xy + yz$. At the point $(2, 1, 4)$, answer the following questions. What is the unit vector that points in the direction of maximum change of temperature? What is the value of the derivative of the temperature

20. In the x direction?

21. In the direction of the vector $\mathbf{i} - 2\mathbf{j} + 2\mathbf{k}$?

22. In the direction of $\mathbf{i} + \mathbf{j} + \mathbf{k}$?

Find the divergence of each vector field at the point $(2, 1, -1)$.

23. $\mathbf{u} = x^2\mathbf{i} + yz\mathbf{j} + y^2\mathbf{k}$ **24.** $\mathbf{u} = yi + xz\mathbf{j} + xy\mathbf{k}$

25. $\mathbf{u} = xi + yj + zk$ **26.** $\mathbf{u} = xyi + y^2\mathbf{j} + z^2\mathbf{k}$

27. $\mathbf{u} = \mathbf{r}/r$ **28.** $\mathbf{u} = \mathbf{r}/r^3$

29. Show that $\nabla \cdot (\phi\mathbf{u}) = \phi\nabla \cdot \mathbf{u} + \mathbf{u} \cdot \nabla\phi$ by expanding in rectangular coordinates.

30. One person claims that the velocity field in a certain water flow is $\mathbf{v} = x^2\mathbf{i} - y^2\mathbf{j} + 2\mathbf{k}$, and another claims that it is $\mathbf{v} = y^2\mathbf{i} - x^2\mathbf{j} + 2\mathbf{k}$. Which one is obviously wrong and why?

31. It is known that the x component of velocity in a certain plane water flow (no z component of velocity) is given by x^2. Determine the velocity vector if $v_y = 0$ along the x axis.

Find the curl of each vector field at the point $(-2, 4, 1)$.

32. $\mathbf{u} = x^2\mathbf{i} + y^2\mathbf{j} + z^2\mathbf{k}$ **33.** $\mathbf{u} = y^2\mathbf{i} + 2xy\mathbf{j} + z^2\mathbf{k}$

34. $\mathbf{u} = xyi + y^2\mathbf{j} + xz\mathbf{k}$ **35.** $\mathbf{u} = \sin y \, \mathbf{i} + x \cos y \, \mathbf{j}$

36. $\mathbf{u} = e^x \sin y \, \mathbf{i} + e^x \cos y \, \mathbf{j} + e^x\mathbf{k}$ **37.** $\mathbf{u} = \mathbf{r}/r^3$

Using the vector functions $\mathbf{u} = xyi + y^2\mathbf{j} + zk$ and $\mathbf{v} = x^2\mathbf{i} + xyj + yzk$, evaluate each function at the point $(-1, 2, 2)$.

38. $\nabla \cdot \mathbf{u}$ **39.** $\nabla \cdot \mathbf{v}$ **40.** $\nabla \times \mathbf{u}$

41. $\nabla \times \mathbf{v}$ **42.** $\nabla \cdot \mathbf{u} \times \mathbf{v}$ **43.** $(\nabla \times \mathbf{u}) \times \mathbf{v}$

44. $\nabla \times (\mathbf{u} \times \mathbf{v})$ **45.** $\mathbf{u} \times (\nabla \times \mathbf{v})$ **46.** $(\mathbf{u} \times \nabla) \times \mathbf{v}$

47. $(\mathbf{u} \cdot \nabla)\mathbf{v}$ **48.** $\nabla(\mathbf{u} \cdot \mathbf{v})$ **49.** $(\mathbf{v} \cdot \nabla)\mathbf{v}$

Determine if each vector field is solenoidal and/or irrotational.

50. $x\mathbf{i} + y\mathbf{j} + z\mathbf{k}$ **51.** $x\mathbf{i} - 2y\mathbf{j} + z\mathbf{k}$

52. $y\mathbf{i} + x\mathbf{j}$ **53.** $x^2\mathbf{i} + y^2\mathbf{j} + z^2\mathbf{k}$

54. $y^2\mathbf{i} + 2xy\mathbf{j} + z^2\mathbf{k}$ **55.** $yz\mathbf{i} + xz\mathbf{j} + xy\mathbf{k}$

56. $\sin y\,\mathbf{i} + \sin x\,\mathbf{j} + e^3\mathbf{k}$

57. $x^2y\mathbf{i} + y^2x\mathbf{j} + z^2\mathbf{k}$ **58.** $\mathbf{r}/r^3$

Verify each vector identity by expanding in rectangular coordinates.

59. $\nabla \times \nabla\phi = 0$ **60.** $\nabla \cdot \nabla \times \mathbf{u} = 0$

61. $\nabla \cdot (\phi\mathbf{u}) = \nabla\phi \cdot \mathbf{u} + \phi\nabla \cdot \mathbf{u}$ **62.** $\nabla \times (\phi\mathbf{u}) = \nabla\phi \times \mathbf{u} + \phi\nabla \times \mathbf{u}$

63. $\nabla \times (\mathbf{u} \times \mathbf{v}) = \mathbf{u}(\nabla \cdot \mathbf{v}) - \mathbf{v}(\nabla \cdot \mathbf{u}) + (\mathbf{v} \cdot \nabla)\mathbf{u} - (\mathbf{u} \cdot \nabla)\mathbf{v}$

64. $\nabla \times (\nabla \times \mathbf{u}) = (\nabla(\nabla \cdot \mathbf{u}) - \nabla^2\mathbf{u}$

65. $\nabla \cdot (\mathbf{u} \times \mathbf{v}) = \nabla \times \mathbf{u} \cdot \mathbf{v} - \mathbf{u} \cdot \nabla \times \mathbf{v}$

66. $\mathbf{u} \times (\nabla \times \mathbf{u}) = \frac{1}{2}\nabla\mathbf{u}^2 - (\mathbf{u} \cdot \nabla)\mathbf{u}$

Determine the scalar potential function ϕ, provided that one exists, associated with each vector field.

67. $\mathbf{u} = x\mathbf{i} + y\mathbf{j} + z\mathbf{k}$ **68.** $\mathbf{u} = x^2\mathbf{i} + y^2\mathbf{j} + z^2\mathbf{k}$

69. $\mathbf{u} = y^2\mathbf{i} + 2xy\mathbf{j} + z\mathbf{k}$ **70.** $\mathbf{u} = e^x \sin y\,\mathbf{i} + e^x \cos y\,\mathbf{j}$

71. $\mathbf{u} = 2x \sin y\,\mathbf{i} + x^2 \cos y\,\mathbf{j} + z^2\mathbf{k}$ **72.** $\mathbf{u} = 2xz\mathbf{i} + y^2\mathbf{j} + x^2\mathbf{k}$

6.5 Cylindrical and Spherical Coordinates

There are several coordinate systems that are convenient to use with the various geometries encountered in physical applications. The most common is the rectangular, Cartesian coordinate system (often referred to as simply the Cartesian coordinate system or the rectangular coordinate system), used primarily in this text. There are situations, however, when solutions become much simpler if a coordinate system is chosen which is more natural to the problem at hand. Two other coordinate systems that attract much attention are the cylindrical coordinate system and the spherical coordinate system. We shall relate the rectangular coordinates to both the cylindrical and spherical coordinates, and express the various vector quantities of previous sections in cylindrical and spherical coordinates.

The cylindrical coordinates* (r, θ, z), with respective orthogonal unit vectors $\mathbf{i}_r$, $\mathbf{i}_\theta$, and $\mathbf{i}_z$, and the spherical coordinates (r, θ, ϕ), with respective orthogonal unit vectors $\mathbf{i}_r$, $\mathbf{i}_\theta$, and $\mathbf{i}_\phi$, are shown in Fig. 6.18. A vector is expressed in cylindrical coordinates as

$$\mathbf{A} = A_r\mathbf{i}_r + A_\theta\mathbf{i}_\theta + A_z\mathbf{i}_z, \tag{6.5.1}$$

where the components A_r, A_θ, and A_z are functions of r, θ, and z. In spherical coordinates a vector is expressed as

$$\mathbf{A} = A_r\mathbf{i}_r + A_\theta\mathbf{i}_\theta + A_\phi\mathbf{i}_\phi, \tag{6.5.2}$$

where A_r, A_θ, and A_ϕ are functions of r, θ, and ϕ.

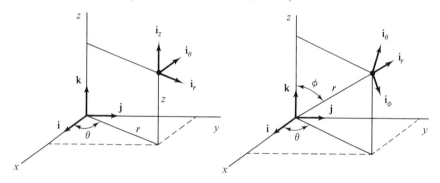

FIGURE 6.18.   The cylindrical and spherical coordinate systems.

We have, in previous sections, expressed all vector quantities in rectangular coordinates. Let us transform some of the more important quantities to cylindrical and spherical coordinates. We will do this first for cylindrical coordinates.

The cylindrical coordinates are related to rectangular coordinates by (refer to Fig. 6.18)

$$x = r \cos \theta, \qquad y = r \sin \theta, \qquad z = z, \tag{6.5.3}$$

where we are careful[†] to note that $r \neq |\mathbf{r}|$, $\mathbf{r}$ being the position vector. From the geometry of Fig. 6.18 we can write

$$\mathbf{i}_r = \cos \theta \, \mathbf{i} + \sin \theta \, \mathbf{j}$$
$$\mathbf{i}_\theta = -\sin \theta \, \mathbf{i} + \cos \theta \, \mathbf{j} \tag{6.5.4}$$
$$\mathbf{i}_z = \mathbf{k}.$$

---

*Note that in cylindrical coordinates it is conventional to use r as the distance from the z axis to the point of interest. Do not confuse it with the distance from the origin $|\mathbf{r}|$.

[†]This is a rather unfortunate choice, but it is the most conventional. Occasionally, ρ is used in place of r, which helps avoid confusion.

These three equations can be solved simultaneously to give

$$\mathbf{i} = \cos \theta \, \mathbf{i}_r - \sin \theta \, \mathbf{i}_\theta$$

$$\mathbf{j} = \sin \theta \, \mathbf{i}_r + \cos \theta \, \mathbf{i}_\theta \qquad (6.5.5)$$

$$\mathbf{k} = \mathbf{i}_z.$$

We have thus related the unit vectors in the cylindrical and rectangular coordinate systems. They are collected in Table 6.2.

TABLE 6.2 *Relationship of Cylindrical Coordinates to Rectangular Coordinates*

Cylindrical
$x = r \cos \theta \qquad r = \sqrt{x^2 + y^2}$ $y = r \sin \theta \qquad \theta = \tan^{-1} y/x$ $z = z \qquad\qquad z = z$
$\mathbf{i}_r = \cos \theta \, \mathbf{i} + \sin \theta \, \mathbf{j}$ $\mathbf{i}_\theta = -\sin \theta \, \mathbf{i} + \sin \theta \, \mathbf{j}$ $\mathbf{i}_z = \mathbf{k}$
$\mathbf{i} = \cos \theta \, \mathbf{i}_r - \sin \theta \, \mathbf{i}_\theta$ $\mathbf{j} = \sin \theta \, \mathbf{i}_r + \cos \theta \, \mathbf{i}_\theta$ $\mathbf{k} = \mathbf{i}_z$

To express the gradient of the scalar function Φ in cylindrical coordinates, we observe from Fig. 6.19 that

$$d\mathbf{r} = dr \, \mathbf{i}_r + r \, d\theta \, \mathbf{i}_\theta + dz \, \mathbf{i}_z. \qquad (6.5.6)$$

The quantity $d\Phi$ is, by the chain rule,

$$d\Phi = \frac{\partial \Phi}{\partial r} \, dr + \frac{\partial \Phi}{\partial \theta} \, d\theta + \frac{\partial \Phi}{\partial z} \, dz. \qquad (6.5.7)$$

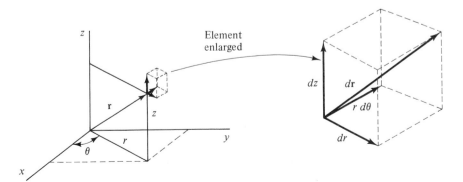

FIGURE 6.19.   Differential changes in cylindrical coordinates.

The gradient of Φ is the vector

$$\nabla\Phi = \lambda_r \mathbf{i}_r + \lambda_\theta \mathbf{i}_\theta + \lambda_z \mathbf{i}_z, \tag{6.5.8}$$

where λ_r, λ_θ, and λ_z are the components of $\nabla\theta$ that we wish to determine. We refer to Eq. 6.4.10 and recognize that

$$d\Phi = \nabla\Phi \cdot d\mathbf{r}. \tag{6.5.9}$$

Substituting the preceding expressions for $d\Phi$, $\nabla\Phi$, and $d\mathbf{r}$ into this equation results in

$$\frac{\partial\Phi}{\partial r}\,dr + \frac{\partial\Phi}{\partial\theta}\,d\theta + \frac{\partial\Phi}{\partial z}\,dz = (\lambda_r \mathbf{i}_r + \lambda_\theta \mathbf{i}_\theta + \lambda_z \mathbf{i}_z)\cdot(dr\mathbf{i}_r + r\,d\theta\,\mathbf{i}_\theta + dz\mathbf{i}_z)$$

$$= \lambda_r\,dr + \lambda_\theta r\,d\theta + \lambda_z\,dz. \tag{6.5.10}$$

Since r, θ, and z are independent quantities, the coefficients of the differential quantities allow us to write.

$$\lambda_r = \frac{\partial\Phi}{\partial r}, \qquad r\lambda_\theta = \frac{\partial\Phi}{\partial\theta}, \qquad \lambda_z = \frac{\partial\Phi}{\partial z}. \tag{6.5.11}$$

Hence, the gradient of Φ, in cylindrical coordinates, becomes

$$\nabla\Phi = \frac{\partial\Phi}{\partial r}\mathbf{i}_r + \frac{1}{r}\frac{\partial\Phi}{\partial\theta}\mathbf{i}_\theta + \frac{\partial\Phi}{\partial z}\mathbf{i}_z. \tag{6.5.12}$$

The gradient operator ∇ is, from the equation above,

$$\nabla = \frac{\partial}{\partial r}\mathbf{i}_r + \frac{1}{r}\frac{\partial}{\partial\theta}\mathbf{i}_\theta + \frac{\partial}{\partial z}\mathbf{i}_z. \tag{6.5.13}$$

Now we wish to find an expression for the divergence $\nabla \cdot \mathbf{u}$. In cylindrical coordinates, it is

$$\nabla \cdot \mathbf{u} = \left(\frac{\partial}{\partial r}\mathbf{i}_r + \frac{1}{r}\frac{\partial}{\partial\theta}\mathbf{i}_\theta + \frac{\partial}{\partial\theta}\mathbf{i}_z\right)\cdot(u_r\mathbf{i}_r + u_\theta\mathbf{i}_\theta + u_z\mathbf{i}_z). \tag{6.5.14}$$

When we perform the dot products above, we must be sure to account for the changes in $\mathbf{i}_r$ and $\mathbf{i}_\theta$ as the angle θ changes; that is, the quantities $\partial\mathbf{i}_r/\partial\theta$ and $\partial\mathbf{i}_\theta/\partial\theta$ are not zero. For example, consider the term $[(1/r)(\partial/\partial\theta)\mathbf{i}_\theta]\cdot(u_r\mathbf{i}_r)$. It yields

$$\left(\frac{1}{r}\frac{\partial}{\partial\theta}\mathbf{i}_\theta\right)\cdot(u_r\mathbf{i}_r) = \frac{1}{r}\frac{\partial u}{\partial\theta}\underset{\displaystyle 0}{\cancel{\mathbf{i}_\theta\cdot\mathbf{i}_r}} + \frac{u_r}{r}\mathbf{i}_\theta\cdot\frac{\partial\mathbf{i}_r}{\partial\theta}. \tag{6.5.15}$$

The product $\mathbf{i}_\theta \cdot \mathbf{i}_r = 0$ since $\mathbf{i}_\theta$ is normal to $\mathbf{i}_r$. The other term, however, is not zero. By referring to Fig. 6.20, we see that

$$\frac{\partial\mathbf{i}_r}{\partial\theta} = \lim_{\Delta\theta\to 0}\frac{\Delta\mathbf{i}_r}{\Delta\theta} = \lim_{\Delta\theta\to 0}\frac{\Delta\theta\mathbf{i}_\theta}{\Delta\theta} = \mathbf{i}_\theta$$

$$\frac{\partial\mathbf{i}_\theta}{\partial\theta} = \lim_{\Delta\theta\to 0}\frac{\Delta\mathbf{i}_\theta}{\Delta\theta} = \lim_{\Delta\theta\to 0}\frac{-\Delta\theta\mathbf{i}_r}{\Delta\theta} = -\mathbf{i}_r. \tag{6.5.16}$$

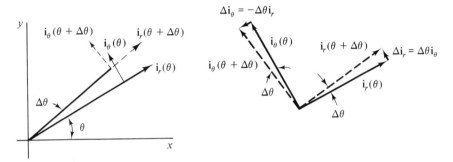

FIGURE 6.20.   Change in unit vectors with the angle θ.

Since $\mathbf{i}_z$ never changes direction, $\partial \mathbf{i}_z / \partial \theta = \mathbf{0}$. Recalling that

$$\mathbf{i}_r \cdot \mathbf{i}_r = \mathbf{i}_\theta \cdot \mathbf{i}_\theta = \mathbf{i}_z \cdot \mathbf{i}_z = 1, \qquad \mathbf{i}_r \cdot \mathbf{i}_\theta = \mathbf{i}_r \cdot \mathbf{i}_z = \mathbf{i}_\theta \cdot \mathbf{i}_z = 0,$$

(6.5.17)

the divergence is then, referring to Eqs. 6.5.14 and 6.5.16,

$$\boldsymbol{\nabla} \cdot \mathbf{u} = \frac{\partial u_r}{\partial r} + \frac{1}{r}\frac{\partial u_\theta}{\partial \theta} + \frac{\partial u_z}{\partial z} + \frac{u_r}{r}\mathbf{i}_r \cdot \frac{\partial \mathbf{i}_r}{\partial \theta} + \frac{u_\theta}{r}\mathbf{i}_\theta \cdot \overset{0}{\cancel{\frac{\partial \mathbf{i}_\theta}{\partial \theta}}}$$

$$= \frac{\partial u_r}{\partial r} + \frac{1}{r}\frac{\partial u_\theta}{\partial \theta} + \frac{\partial u_z}{\partial z} + \frac{u_r}{r}.$$

(6.5.18)

This can be rewritten in the more conventional form

$$\boldsymbol{\nabla} \cdot \mathbf{u} = \frac{1}{r}\frac{\partial}{\partial r}(r u_r) + \frac{1}{r}\frac{\partial u_\theta}{\partial \theta} + \frac{\partial u_z}{\partial z}.$$

(6.5.19)

We now express the curl $\boldsymbol{\nabla} \times \mathbf{u}$ in cylindrical coordinates. It is

$$\boldsymbol{\nabla} \times \mathbf{u} = \left(\frac{\partial}{\partial r}\mathbf{i}_r + \frac{1}{r}\frac{\partial}{\partial \theta}\mathbf{i}_\theta + \frac{\partial}{\partial z}\mathbf{i}_z \right) \times (u_r\mathbf{i}_r + u_\theta\mathbf{i}_\theta + u_z\mathbf{i}_z).$$

(6.5.20)

Carrying out the cross products term by term, we have

$$\boldsymbol{\nabla} \times \mathbf{u} = \left(\frac{1}{r}\frac{\partial u_z}{\partial \theta} - \frac{\partial u_\theta}{\partial z} \right)\mathbf{i}_r + \left(\frac{\partial u_r}{\partial z} - \frac{\partial u_z}{\partial r} \right)\mathbf{i}_\theta + \left(\frac{\partial u_\theta}{\partial r} - \frac{1}{r}\frac{\partial u_r}{\partial \theta} \right)\mathbf{i}_z$$

$$+ \frac{u_r}{r}\mathbf{i}_r \times \overset{0}{\cancel{\frac{\partial \mathbf{i}_r}{\partial \theta}}} + \frac{u_\theta}{r}\mathbf{i}_\theta \times \frac{\partial \mathbf{i}_\theta}{\partial \theta},$$

(6.5.21)

where we have used Eq. 6.5.16 and

$$\mathbf{i}_r \times \mathbf{i}_r = \mathbf{i}_\theta \times \mathbf{i}_\theta = \mathbf{i}_z \times \mathbf{i}_z = 0,$$

$$\mathbf{i}_r \times \mathbf{i}_\theta = \mathbf{i}_z, \ \mathbf{i}_\theta \times \mathbf{i}_z = \mathbf{i}_r, \ \mathbf{i}_z \times \mathbf{i}_r = \mathbf{i}_\theta.$$

(6.5.22)

Using Eqs. 6.5.16, writing $(\partial u_\theta / \partial r) + (u_\theta / r) = (1/r)(\partial / \partial r)(ru_\theta)$, we get

$$\boldsymbol{\nabla} \times \mathbf{u} = \left[\frac{1}{r} \frac{\partial u_z}{\partial \theta} - \frac{\partial u_\theta}{\partial z} \right] \mathbf{i}_r + \left[\frac{\partial u_r}{\partial z} - \frac{\partial u_z}{\partial r} \right] \mathbf{i}_\theta + \left[\frac{1}{r} \frac{\partial}{\partial r} (ru_\theta) - \frac{1}{r} \frac{\partial u_r}{\partial \theta} \right] \mathbf{i}_z.$$

$$(6.5.23)$$

Finally, the Laplacian of a scalar function Φ, in cylindrical coordinates, is

$$\boldsymbol{\nabla} \cdot \boldsymbol{\nabla}\Phi = \nabla^2 \Phi = \left(\frac{\partial}{\partial r} \mathbf{i}_r + \frac{1}{r} \frac{\partial}{\partial \theta} \mathbf{i}_\theta + \frac{\partial}{\partial z} \mathbf{i}_z \right) \cdot \left(\frac{\partial \Phi}{\partial r} \mathbf{i}_r + \frac{1}{r} \frac{\partial \Phi}{\partial \theta} \mathbf{i}_\theta + \frac{\partial \Phi}{\partial z} \mathbf{i}_z \right)$$

$$= \frac{\partial^2 \Phi}{\partial r^2} + \frac{1}{r^2} \frac{\partial^2 \Phi}{\partial \theta^2} + \frac{\partial^2 \Phi}{\partial z^2} + \frac{1}{r} \frac{\partial \Phi}{\partial r} \mathbf{i}_\theta \cdot \frac{\partial \mathbf{i}_r}{\partial \theta}$$

$$= \frac{1}{r} \frac{\partial}{\partial r} \left(r \frac{\partial \Phi}{\partial r} \right) + \frac{1}{r^2} \frac{\partial^2 \Phi}{\partial \theta^2} + \frac{\partial^2 \Phi}{\partial z^2}. \qquad (6.5.24)$$

Example 6.5.1: A particle is positioned by

$$\mathbf{r} = x\mathbf{i} + y\mathbf{j} + z\mathbf{k}$$

in rectangular coordinates. Express $\mathbf{r}$ in cylindrical coordinates.

SOLUTION: We use Eq. 6.5.3 to write

$$\mathbf{r} = r \cos \theta \, \mathbf{i} + r \sin \theta \, \mathbf{j} + z\mathbf{k}$$

and then Eq. 6.5.4 to obtain

$$\mathbf{r} = r\mathbf{i}_r + z\mathbf{i}_z. \quad \blacksquare$$

Example 6.5.2: Express $\mathbf{u} = 2x\mathbf{i} - z\mathbf{j} + y\mathbf{k}$ in cylindrical coordinates.

SOLUTION: Using Eqs. 6.5.3 and 6.5.4, we obtain

$$\mathbf{u} = 2r \cos \theta (\cos \theta \, \mathbf{i}_r - \sin \theta \, \mathbf{i}_\theta) - z(\sin \theta \, \mathbf{i}_r + \cos \theta \, \mathbf{i}_\theta) + r \sin \theta \, \mathbf{i}_z.$$

This is rearranged in the conventional form

$$\mathbf{u} = (2r \cos^2 \theta - z \sin \theta)\mathbf{i}_r - (2r \cos \theta \sin \theta + z \cos \theta)\mathbf{i}_\theta + r \sin \theta \, \mathbf{i}_z. \quad \blacksquare$$

Example 6.5.3: A particle moves in three-dimensional space. Determine an expression for its acceleration in cylindrical coordinates.

SOLUTION: The particle is positioned by the vector

$$\mathbf{r} = r\mathbf{i}_r + z\mathbf{i}_z.$$

The velocity is found by differentiating with respect to time; that is,

$$\mathbf{v} = \frac{d\mathbf{r}}{dt} = \frac{dr}{dt}\mathbf{i}_r + r\frac{d\mathbf{i}}{dt}r + \frac{dz}{dt}\mathbf{i}_z.$$

We find an expression for $d\mathbf{i}_r / dt$ by using Eq. 6.5.4 to get

$$\frac{d\mathbf{i}_r}{dt} = -\sin \theta \frac{d\theta}{dt}\mathbf{i} + \cos \theta \frac{d\theta}{dt}\mathbf{j} = \frac{d\theta}{dt}(-\sin \theta \, \mathbf{i} + \cos \theta \, \mathbf{j}) = \dot\theta\mathbf{i}_\theta.$$

Thus, using a dot to denote time differentiation,
$$\mathbf{v} = \dot{r}\mathbf{i}_r + r\dot{\theta}\mathbf{i}_\theta + \dot{z}\mathbf{i}_z.$$

Differentiate again with respect to time. We have
$$\mathbf{a} = \ddot{r}\mathbf{i}_r + \dot{r}\frac{d\mathbf{i}_r}{dt} + \dot{r}\dot{\theta}\mathbf{i}_\theta + r\ddot{\theta}\mathbf{i}_\theta + r\dot{\theta}\frac{d\mathbf{i}_\theta}{dt} + \ddot{z}\mathbf{i}_z.$$

The quantity $d\mathbf{i}_\theta/dt$ is (see Eq. 6.5.4)
$$\frac{d\mathbf{i}_\theta}{dt} = (-\cos\theta\,\mathbf{i} - \sin\theta\,\mathbf{j})\frac{d\theta}{dt} = -\mathbf{i}_r\dot{\theta}$$

The acceleration is then
$$\mathbf{a} = \ddot{r}\mathbf{i}_r + \dot{r}\dot{\theta}\mathbf{i}_\theta + \dot{r}\dot{\theta}\mathbf{i}_\theta + r\ddot{\theta}\mathbf{i}_\theta - r\dot{\theta}^2\mathbf{i}_r + \ddot{z}\mathbf{i}_z$$
$$= (\ddot{r} - r\dot{\theta}^2)\mathbf{i}_r + (2\dot{r}\dot{\theta} + r\ddot{\theta})\mathbf{i}_\theta + \ddot{z}\mathbf{i}_z. \quad \blacksquare$$

If we follow the same procedure using spherical coordinates we find that the coordinates are related by
$$x = r\sin\phi\cos\theta, \quad y = r\sin\phi\sin\theta, \quad z = r\cos\phi. \qquad (6.5.25)$$

The unit vectors are related by the following equations:
$$\mathbf{i}_r = \sin\phi\cos\theta\,\mathbf{i} + \sin\phi\sin\theta\,\mathbf{j} + \cos\phi\,\mathbf{k}$$
$$\mathbf{i}_\theta = -\sin\theta\,\mathbf{i} + \cos\theta\,\mathbf{j} \qquad (6.5.26)$$
$$\mathbf{i}_\phi = \cos\phi\cos\theta\,\mathbf{i} + \cos\phi\sin\theta\,\mathbf{j} - \sin\phi\,\mathbf{k}$$

and
$$\mathbf{i} = \sin\phi\cos\theta\,\mathbf{i}_r - \sin\theta\,\mathbf{i}_\theta + \cos\phi\cos\theta\,\mathbf{i}_\phi$$
$$\mathbf{j} = \sin\phi\sin\theta\,\mathbf{i}_r + \cos\theta\,\mathbf{i}_\theta + \cos\phi\sin\theta\,\mathbf{i}_\phi \qquad (6.5.27)$$
$$\mathbf{k} = \cos\phi\,\mathbf{i}_r - \sin\phi\,\mathbf{i}_\phi.$$

Table 6.3 relates the unit vectors in spherical and rectangular coordinates.

TABLE 6.3 *Relationship of Spherical Coordinates to Rectangular Coordinates*

Spherical
$x = r\sin\phi\cos\theta \qquad r = \sqrt{x^2 + y^2 + z^2}$
$y = r\sin\phi\sin\theta \qquad \theta = \tan^{-1}y/x$
$z = r\sin\phi \qquad \phi = \tan^{-1}\sqrt{x^2 + y^2}/z$
$\mathbf{i}_r = \sin\phi\cos\theta\,\mathbf{i} + \sin\phi\sin\theta\,\mathbf{j} + \cos\phi\,\mathbf{k}$
$\mathbf{i}_\theta = -\sin\theta\,\mathbf{i} + \cos\theta\,\mathbf{j}$
$\mathbf{i}_\phi = \cos\phi\cos\theta\,\mathbf{i} + \cos\phi\sin\theta\,\mathbf{j} - \sin\phi\,\mathbf{k}$
$\mathbf{i} = \sin\phi\cos\theta\,\mathbf{i}_r - \sin\theta\,\mathbf{i}_\theta + \cos\phi\cos\phi\,\mathbf{i}_\phi$
$\mathbf{j} = \sin\phi\sin\theta\,\mathbf{i}_r + \cos\theta\,\mathbf{i}_\theta + \cos\phi\sin\theta\,\mathbf{i}_\phi$
$\mathbf{k} = \cos\,\mathbf{i}_r - \sin\,\mathbf{i}_\phi$

Using Fig. 6.21, the gradient of the scalar function Φ is found to be

$$\nabla \Phi = \frac{\partial \Phi}{\partial r}\mathbf{i}_r + \frac{1}{r \sin \phi}\frac{\partial \Phi}{\partial \theta}\mathbf{i}_\theta + \frac{1}{r}\frac{\partial \Phi}{\partial \phi}\mathbf{i}_\phi. \tag{6.5.28}$$

allowing us to write

$$\nabla = \frac{\partial}{\partial r}\mathbf{i}_r + \frac{1}{r \sin \phi}\frac{\partial}{\partial \theta}\mathbf{i}_\theta + \frac{1}{r}\frac{\partial}{\partial \phi}\mathbf{i}_\phi. \tag{6.5.29}$$

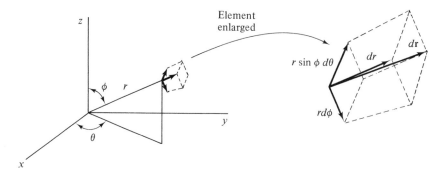

FIGURE 6.21.   Differential changes in spherical coordinates.

The divergence of a vector field is

$$\nabla \cdot \mathbf{u} = \frac{1}{r^2}\frac{\partial}{\partial r}(r^2 u_r) + \frac{1}{r \sin \phi}\frac{\partial u_\theta}{\partial \theta} + \frac{1}{r \sin \phi}\frac{\partial}{\partial \phi}(u_\phi \sin \phi)$$

$$\tag{6.5.30}$$

and the curl is

$$\nabla \times \mathbf{u} = \frac{1}{r \sin \phi}\left[\frac{\partial}{\partial \phi}(u_\theta \sin \phi) - \frac{\partial u_\phi}{\partial \theta}\right]\mathbf{i}_r + \frac{1}{r}\left[\frac{\partial}{\partial r}(r u_\phi) - \frac{\partial u_r}{\partial \phi}\right]\mathbf{i}_\theta$$

$$+ \frac{1}{r}\left[\frac{1}{\sin \phi}\frac{\partial u_r}{\partial \theta} - \frac{\partial}{\partial r}(r u_\theta)\right]\mathbf{i}_\phi. \tag{6.5.31}$$

The Laplacian of a scalar function Φ is

$$\nabla^2 \Phi = \frac{1}{r^2}\frac{\partial}{\partial r}\left(r^2 \frac{\partial \Phi}{\partial r}\right) + \frac{1}{r^2 \sin^2 \phi}\frac{\partial^2 \Phi}{\partial \theta^2}$$

$$+ \frac{1}{r^2 \sin \phi}\frac{\partial}{\partial \phi}\left(\sin \phi \frac{\partial \Phi}{\partial \phi}\right). \tag{6.5.32}$$

Example 6.5.4:   Express the position vector

$$\mathbf{r} = x\mathbf{i} + y\mathbf{j} + z\mathbf{k}$$

in spherical coordinates.

TABLE 6.4 Relationships Involving ∇ in Rectangular, Cylindrical, and Spherical Coordinates

Rectangular

$$\nabla\Phi = \frac{\partial\Phi}{\partial x}\mathbf{i} + \frac{\partial\Phi}{\partial y}\mathbf{j} + \frac{\partial\Phi}{\partial z}\mathbf{k}$$

$$\nabla\cdot\mathbf{u} = \frac{\partial u_x}{\partial x} + \frac{\partial u_y}{\partial y} + \frac{\partial u_z}{\partial z}$$

$$\nabla\times\mathbf{u} = \left(\frac{\partial u_z}{\partial y} - \frac{\partial u_y}{\partial z}\right)\mathbf{i} + \left(\frac{\partial u_x}{\partial z} - \frac{\partial u_z}{\partial x}\right)\mathbf{j} + \left(\frac{\partial u_y}{\partial x} - \frac{\partial u_x}{\partial y}\right)\mathbf{k}$$

$$\nabla^2\Phi = \frac{\partial^2\Phi}{\partial x^2} + \frac{\partial^2\Phi}{\partial y^2} + \frac{\partial^2\Phi}{\partial z^2}$$

$$\nabla^2\mathbf{u} = \nabla^2 u_x\mathbf{i} + \nabla^2 u_y\mathbf{j} + \nabla^2 u_z\mathbf{k}$$

Cylindrical

$$\nabla\Phi = \frac{\partial\Phi}{\partial r}\mathbf{i}_r + \frac{1}{r}\frac{\partial\Phi}{\partial\theta}\mathbf{i}_\theta + \frac{\partial\Phi}{\partial z}\mathbf{i}_z$$

$$\nabla\cdot\mathbf{u} = \frac{1}{r}\frac{\partial}{\partial r}(ru_r) + \frac{1}{r}\frac{\partial u_\theta}{\partial\theta} + \frac{\partial u_z}{\partial z}$$

$$\nabla\times\mathbf{u} = \left[\frac{1}{r}\frac{\partial u_z}{\partial\theta} - \frac{\partial u_\theta}{\partial z}\right]\mathbf{i}_r + \left[\frac{\partial u_r}{\partial z} - \frac{\partial u_z}{\partial r}\right]\mathbf{i}_\theta + \left[\frac{1}{r}\frac{\partial}{\partial r}(ru_\theta) - \frac{1}{r}\frac{\partial u_r}{\partial\theta}\right]\mathbf{i}_z$$

$$\nabla^2\Phi = \frac{1}{r}\frac{\partial}{\partial r}\left(r\frac{\partial\Phi}{\partial r}\right) + \frac{1}{r^2}\frac{\partial^2\Phi}{\partial\theta^2} + \frac{\partial^2\Phi}{\partial z^2}$$

$$\nabla^2\mathbf{u} = \left(\nabla^2 u_r - \frac{u_r}{r^2} - \frac{2}{r^2}\frac{\partial u_\theta}{\partial\theta}\right)\mathbf{i}_r + \left(\nabla^2 u_\theta - \frac{u_\theta}{r^2} + \frac{2}{r^2}\frac{\partial u_r}{\partial\theta}\right)\mathbf{i}_\theta + \nabla^2 u_z\mathbf{i}_z$$

Spherical

$$\nabla\Phi = \frac{\partial\Phi}{\partial r}\mathbf{i}_r + \frac{1}{r\sin\phi}\frac{\partial\Phi}{\partial\theta}\mathbf{i}_\theta + \frac{1}{r}\frac{\partial\Phi}{\partial\phi}\mathbf{i}_\phi$$

$$\nabla\cdot\mathbf{u} = \frac{1}{r^2}\frac{\partial}{\partial r}(r^2 u_r) + \frac{1}{r\sin\phi}\frac{\partial u_\theta}{\partial\theta} + \frac{1}{r\sin\phi}\frac{\partial}{\partial\phi}(u_\phi\sin\phi)$$

$$\nabla\times\mathbf{u} = \frac{1}{r\sin\phi}\left[\frac{\partial}{\partial\phi}(u_\theta\sin\phi) - \frac{\partial u_\phi}{\partial\theta}\right]\mathbf{i}_r + \frac{1}{r}\left[\frac{\partial}{\partial r}(ru_\phi) - \frac{\partial u_r}{\partial\phi}\right]\mathbf{i}_\theta + \frac{1}{r}\left[\frac{1}{\sin\phi}\frac{\partial u_r}{\partial\theta} - \frac{\partial}{\partial r}(ru_\theta)\right]\mathbf{i}_\phi$$

$$\nabla^2\Phi = \frac{1}{r^2}\frac{\partial}{\partial r}\left(r^2\frac{\partial\Phi}{\partial r}\right) + \frac{1}{r^2\sin^2\phi}\frac{\partial^2\Phi}{\partial\theta^2} + \frac{1}{r^2\sin\phi}\frac{\partial}{\partial\phi}\left(\sin\phi\frac{\partial\Phi}{\partial\phi}\right)$$

$$\nabla^2\mathbf{u} = \left[\nabla^2 u_r - \frac{2u_r}{r^2} - \frac{2}{r^2\sin\phi}\frac{\partial u_\theta}{\partial\theta} - \frac{2}{r^2\sin\phi}\frac{\partial}{\partial\phi}(u_\phi\sin\phi)\right]\mathbf{i}_r$$

$$+ \left[\nabla^2 u_\theta - \frac{u_\theta}{r^2\sin\phi} + \frac{2\cos\phi}{r^2\sin^2\phi}\frac{\partial u_\theta}{\partial\theta} + \frac{2}{r^2\sin\theta}\frac{\partial u_r}{\partial\theta}\right]\mathbf{i}_\theta$$

$$+ \left[\nabla^2 u_\theta - \frac{2\cos\phi}{r^2\sin\phi}\frac{\partial u_\theta}{\partial\theta} - \frac{u_\phi}{r^2\sin^2\phi} + \frac{2}{r^2}\frac{\partial u_r}{\partial\phi}\right]\mathbf{i}_\phi$$

SOLUTION: Note that
$$\mathbf{r} = r \sin \phi \cos \theta \, \mathbf{i} + r \sin \phi \sin \theta \, \mathbf{j} + r \cos \phi \, \mathbf{k}$$
$$= r \mathbf{i}_r$$

from Eqs. 6.5.25 and 6.5.26.   ∎

Example 6.5.5:   Express the vector $\mathbf{u} = 2x\mathbf{i} - z\mathbf{j} + y\mathbf{k}$ in spherical coordinates.

SOLUTION: For spherical coordinates use Eqs. 6.5.25 and 6.5.27. There results

$$\mathbf{u} = 2r \sin \phi \cos \theta (\sin \phi \cos \theta \, \mathbf{i}_r - \sin \theta \, \mathbf{i}_\theta + \cos \phi \cos \theta \, \mathbf{i}_\phi)$$
$$- r \cos \phi (\sin \phi \sin \theta \, \mathbf{i}_r + \cos \theta \, \mathbf{i}_\theta + \cos \phi \sin \theta \, \mathbf{i}_\phi)$$
$$+ r \sin \phi \sin \theta (\cos \phi \, \mathbf{i}_r - \sin \phi \, \mathbf{i}_\phi)$$
$$= 2r \sin^2 \phi \cos^2 \theta \, \mathbf{i}_r + r \cos \theta (2 \sin \phi \sin \theta - \cos \phi) \mathbf{i}_\theta$$
$$+ r(2 \sin \phi \cos \phi \cos^2 \theta - \sin \theta) \mathbf{i}_\phi. \quad ∎$$

Note the relatively complex forms that the vector takes when expressed in cylindrical and spherical coordinates. This, however, is not always the case; the vector $\mathbf{u} = x\mathbf{i} + x\mathbf{j} + z\mathbf{k}$ becomes simply $\mathbf{u} = r\mathbf{i}_r$ in spherical coordinates. We shall obviously choose the particular coordinate system that simplifies the analysis.

The relationships above involving the gradient operator ∇ are collected in Table 6.4.

PROBLEMS

1. By using Eqs. 6.5.4, show that $d\mathbf{i}_r/dt = \dot{\theta}\mathbf{i}_\theta$ and $d\mathbf{i}_\theta/dt = -\dot{\theta}\mathbf{i}_r$. Sketch the unit vectors at two neighboring points and graphically display $\Delta\mathbf{i}_r$ and $\Delta\mathbf{i}_\theta$.

Find an expression for each of the following at the same point. The subscript c identifies cylindrical coordinates and subscript s spherical coordinates.

2. $\mathbf{i} \cdot \mathbf{i}_{rc}$

3. $\mathbf{i} \cdot \mathbf{i}_{rs}$

4. $\mathbf{j} \cdot \mathbf{i}_{rs}$

5. $\mathbf{i}_{rc} \cdot \mathbf{i}_{rs}$

6. $\mathbf{i}_{\theta c} \cdot \mathbf{i}_\phi$

7. $\mathbf{i}_{\theta c} \cdot \mathbf{i}_{\theta s}$

8. $\mathbf{i}_z \cdot \mathbf{i}_{rs}$

9. $\mathbf{i}_{rs} \cdot \mathbf{i}_{\theta s}$

10. $\mathbf{i}_\phi \cdot \mathbf{i}_{rc}$

Show that the unit vectors are orthogonal in

11. The cylindrical coordinate system.

12. The spherical coordinate system.

13. Relate the cylindrical coordinates at a point to the spherical coordinates at the same point.

14. A point is established in three-dimensional space by the intersection of three surfaces; for eample, in rectangular coordinates they are three planes. What are the surfaces in (a) cylindrical coordinates, and (b) spherical coordinates? Sketch the intersecting surfaces for all three coordinate systems.

Express each vector as indicated.

15. $\mathbf{u} = 2r\mathbf{i}_r + r \sin \phi \, \mathbf{i}_\theta + r^2 \sin \phi \, \mathbf{i}_\phi$ in rectangular coordinates.

16. $\mathbf{u} = r_s\mathbf{i}_\theta$ in rectangular coordinates.

17. $\mathbf{u} = 2z\mathbf{i} + x\mathbf{j} + y\mathbf{k}$ in spherical coordinates.

18. $\mathbf{u} = 2z\mathbf{i} + x\mathbf{j} + y\mathbf{k}$ in cylindrical coordinates.

19. Express the square of the differential arc length, ds^2, in all three coordinate systems.

20. Following the procedure of Section 6.5, derive the expression for the gradient of scalar function Φ in spherical coordinates, Eq. 6.5.28.

Determine the scalar potential function provided that one exists, associated with each equation.

21. $\mathbf{u} = r\mathbf{i}_r + \mathbf{i}_z$

22. $\mathbf{u} = \left(A - \dfrac{B}{r^2}\right) \cos \theta \, \mathbf{i}_r - \left(A + \dfrac{B}{r^2}\right)\mathbf{i}_\theta$ (cylindrical coordinates)

23. $\mathbf{u} = \left(A - \dfrac{B}{r^3}\right) \cos \phi \, \mathbf{i}_r - \left(A + \dfrac{B}{2r^3}\right) \sin \phi \, \mathbf{i}_\phi$

6.6   Integral Theorems

Many of the derivations of the mathematical models used to describe physical phenomena make use of integral theorems, theorems that enable us to transform surface integrals to volume integrals or line integrals to surface integrals. In this section, we present the more commonly used integral theorems, with emphasis on the divergence theorem and Stokes' theorem, the two most important ones.

6.6.1   THE DIVERGENCE THEOREM

The *divergence theorem* (also referred to as *Gauss' theorem*) states that if a volume V is completely enclosed by the surface S, then for the vector function $\mathbf{u}(x, y, z)$, which is continuous with continuous derivatives,

$$\iiint_V \nabla \cdot \mathbf{u} \, dV = \oiint_S \mathbf{u} \cdot \mathbf{n} \, dS, \tag{6.6.1}$$

where n is an outward pointing unit vector normal to the elemental area dS.

In rectangular component form the divergence theorem is

$$\iiint_V \left(\frac{\partial u_x}{\partial x} + \frac{\partial u_y}{\partial y} + \frac{\partial u_z}{dz} \right) dx\, dy\, dz = \oiint_S (u_x \mathbf{i} + u_y \mathbf{j} + u_z \mathbf{k}) \cdot \mathbf{n}\, dS.$$

$$(6.6.2)$$

To prove the validity of this equation, consider the volume V of Fig. 6.22. Let S be a special surface which has the property that any line drawn parallel to a coordinate axis intersects S in at most two points. Let the equation of the lower surface S_1 be given by $f(x, y)$ and of the upper surface S_2 by $g(x, y)$, and the projection of the surface on the xy plane be denoted R. Then the third term of the volume integral of Eq. 6.6.2 can be written as

$$\iiint_V \frac{\partial u_z}{\partial z} dx\, dy\, dz = \iint_R \left[\int_{f(x, y)}^{g(x, y)} \frac{\partial u_z}{\partial z} dz \right] dx\, dy. \qquad (6.6.3)$$

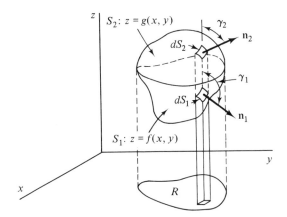

FIGURE 6.22.   Volume used in proof of the divergence theorem.

The integral in the brackets is integrated to give (we hold x and y fixed in this integration)

$$\int_{f(x, y)}^{g(x, y)} \frac{\partial u_z}{\partial z} dz = \int_{f(x, y)}^{g(x, y)} du_z = u_z(x, y, g) - u_z(x, y, f). \qquad (6.6.4)$$

Our integral then becomes

$$\iiint_V \frac{\partial u_z}{\partial z} dx\, dy\, dz = \iint_R [u_z(x, y, g) - u_z(x, y, f)]\, dx\, dy. \qquad (6.6.5)$$

The unit vector $\mathbf{n}$ is related to the direction cosines by $\mathbf{n} = \cos \alpha\, \mathbf{i} + \cos \beta\, \mathbf{j} + \cos \gamma\, \mathbf{k}$. Hence, for the upper surface S_2, we have

$$\cos \gamma_2\, dS_2 = \mathbf{n}_2 \cdot \mathbf{k}\, dS_2 = dx\, dy. \qquad (6.6.6)$$

For the lower surface S_1, realizing that γ_1 is an obtuse angle so that cos γ_1 is negative, there results

$$\cos \gamma_1 \, dS_1 = \mathbf{n}_1 \cdot \mathbf{k} \, dS_1 = -dx \, dy. \tag{6.6.7}$$

Now, with the results above substituted for $dx \, dy$, we can write Eq. 6.6.5 as

$$\iiint_V \frac{\partial u_z}{\partial z} \, dx \, dy \, dz = \iint_{S_2} u_z(x, y, g) \, \mathbf{n}_2 \cdot \mathbf{k} \, dS_2 + \iint_{S_1} u_z(x, y, f) \, \mathbf{n}_1 \cdot \mathbf{k} \, dS_1$$

$$= \iint_{S_2} u_z \mathbf{n} \cdot \mathbf{k} \, dS_2 + \iint_{S_1} u_z \mathbf{n} \cdot \mathbf{k} \, dS_1$$

$$= \oiint_S u_z \mathbf{n} \cdot \mathbf{k} \, dS, \tag{6.6.8}$$

where the complete surface S is equal to $S_1 + S_2$.

Similarly, by taking volume strips parallel to the x axis and the y axis, we can show that

$$\iiint_V \frac{\partial u_x}{\partial x} \, dx \, dy \, dz = \oiint_S u_x \mathbf{n} \cdot \mathbf{i} \, dS$$

$$\iiint_V \frac{\partial u_y}{\partial y} \, dx \, dy \, dz = \oiint_S u_y \mathbf{n} \cdot \mathbf{j} \, dS. \tag{6.6.9}$$

Summing Eqs. 6.6.8 and 6.6.9, we have

$$\iiint_V \left[\frac{\partial u_x}{\partial x} + \frac{\partial u_y}{\partial y} + \frac{\partial u_z}{\partial z} \right] dx \, dy \, dz = \oiint_S [u_x \mathbf{n} \cdot \mathbf{i} + u_y \mathbf{n} \cdot \mathbf{j} + u_z \mathbf{n} \cdot \mathbf{k}] \, dS$$

$$= \oiint_S [u_x \mathbf{i} + u_y \mathbf{j} + u_z \mathbf{k}] \cdot \mathbf{n} \, dS. \tag{6.6.10}$$

This is identical to Eq. 6.6.2, and the divergence theorem is shown to be valid. If the surface is not the special surface of Fig. 6.22, divide the volume into subvolumes, each of which satisfies the special condition. Then argue that the divergence theorem is valid for the original region.

The divergence theorem is often used to define the divergence, rather than Eq. 6.4.11, which utilizes rectangular coordinates. For an incremental volume ΔV, the divergence theorem takes the form

$$\overline{\nabla \cdot \mathbf{u}} \, \Delta V = \oiint_{\Delta S} \mathbf{u} \cdot \mathbf{n} \, dS, \tag{6.6.11}$$

where ΔS is the incremental surface area surrounding ΔV, and $\overline{\nabla \cdot \mathbf{u}}$ is the average value of $\nabla \cdot \mathbf{u}$ in ΔV. If we then allow ΔV to shrink to a point, $\overline{\nabla \cdot \mathbf{u}}$

becomes $\mathbf{\nabla} \cdot \mathbf{u}$ at the point, and there results

$$\mathbf{\nabla} \cdot \mathbf{u} = \lim_{\Delta V \to 0} \frac{\displaystyle\oiint_{\Delta S} \mathbf{u} \cdot \mathbf{n} \, dS}{\Delta V}. \qquad (6.6.12)$$

This definition of the divergence is obviously independent of any particular coordinate system.

By letting the vector function $\mathbf{u}$ of the divergence theorem take on various forms, such as $\mathbf{\nabla}\phi$, $\phi\mathbf{i}$, and $\psi\mathbf{\nabla}\phi$, we can derive other useful integral formulas. These will be included in the examples and problems. Also, Table 6.5 tabulates these formulas.

TABLE 6.5 Integral Formulas

$$\iiint_V \mathbf{\nabla} \cdot \mathbf{u} \, dV = \oiint_S \mathbf{u} \cdot \mathbf{n} \, dS$$

$$\iiint_V \nabla^2\phi \, dV = \oiint_S \frac{\partial \phi}{\partial n} \, dS$$

$$\iiint_V \mathbf{\nabla}\phi \, dV = \oiint_S \phi \mathbf{n} \, dS$$

$$\iiint_V \mathbf{\nabla} \times \mathbf{u} \, dV = \oiint_S \mathbf{n} \times \mathbf{u} \, dS$$

$$\iiint_V (\psi\nabla^2\phi - \phi\nabla^2\psi) \, dV = \oiint_S \left(\psi\frac{\partial \phi}{\partial n} - \phi\frac{\partial \psi}{\partial n}\right) dS$$

$$\iiint_V (\psi\nabla^2\phi + \mathbf{\nabla}\phi \cdot \mathbf{\nabla}\psi) \, dV = \oiint_S \psi\frac{\partial \phi}{\partial n} \, dS$$

$$\iiint_S (\mathbf{\nabla} \times \mathbf{u}) \cdot \mathbf{n} \, dS = \oint_C \mathbf{u} \cdot d\mathbf{l}$$

$$\iint_S (\mathbf{n} \times \mathbf{\nabla}) \times \mathbf{u} \, dS = -\oint_C \mathbf{u} \times d\mathbf{l}$$

6.6.2 STOKES' THEOREM

Let a surface S be surrounded by a simple curve C as shown in Fig. 6.23. For the vector function $\mathbf{u}(x, y, z)$. *Stokes' theorem* states that

$$\oint_C \mathbf{u} \cdot d\mathbf{l} = \iint_S (\mathbf{\nabla} \times \mathbf{u}) \cdot \mathbf{n} \, dS, \qquad (6.6.13)$$

where $d\mathbf{l}$ is a directed line element of C and $\mathbf{n}$ is a unit vector normal to dS.

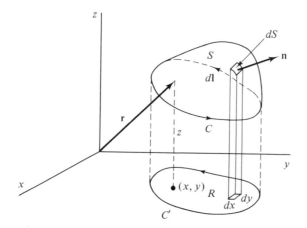

FIGURE 6.23.   Surface used in the proof of Stokes' theorem.

Using rectangular coordinates, this can be written as

$$\oint_C u_x \, dx + u_y dy + u_z \, dz = \iint_S \left[\left(\frac{\partial u_z}{\partial y} - \frac{\partial u_y}{\partial z} \right) \mathbf{i} \cdot \mathbf{n} + \left(\frac{\partial u_x}{\partial z} - \frac{\partial u_z}{\partial x} \right) \mathbf{j} \cdot \mathbf{n} \right.$$
$$\left. + \left(\frac{\partial u_y}{\partial x} - \frac{\partial u_x}{\partial y} \right) \mathbf{k} \cdot \mathbf{n} \right] dS. \qquad (6.6.14)$$

We will show that the terms involving u_x are equal; that is,

$$\oint_C u_x \, dx = \iint_S \left[\frac{\partial u_x}{\partial z} \mathbf{j} \cdot \mathbf{n} - \frac{\partial u_x}{\partial y} \mathbf{k} \cdot \mathbf{n} \right] dS. \qquad (6.6.15)$$

To show this, assume that the projection of S on the xy plane forms a simple curve C', which is intersected by lines parallel to the y axis only twice. Let the surface S be located by the function $z = f(x, y)$; then a position vector to a point on S is

$$\mathbf{r} = x\mathbf{i} + y\mathbf{j} + z\mathbf{k}$$
$$= x\mathbf{i} + y\mathbf{j} + f(x, y)\mathbf{k} \qquad (6.6.16)$$

If we increment y an amount Δy the position vector locating a neighboring point on S becomes $\mathbf{r} + \Delta \mathbf{r} = x\mathbf{i} + (y + \Delta y)\mathbf{j} + (f + \Delta f)\mathbf{k}$, so that $\Delta \mathbf{r}$ is a vector approximately tangent to the surface S. In the limit as $\Delta y \to 0$, it is tangent. Hence, the vector

$$\frac{\partial \mathbf{r}}{\partial y} = \lim_{\Delta y \to 0} \frac{\Delta \mathbf{r}}{\Delta y} = \mathbf{j} + \frac{\partial f}{\partial y} \mathbf{k} \qquad (6.6.17)$$

is tangent to S and thus normal to $\mathbf{n}$. We can then write

$$\mathbf{n} \cdot \frac{\partial \mathbf{r}}{\partial y} = \mathbf{n} \cdot \mathbf{j} + \frac{\partial f}{\partial y} \mathbf{n} \cdot \mathbf{k} = 0, \qquad (6.6.18)$$

so

$$\mathbf{n} \cdot \mathbf{j} = -\frac{\partial f}{\partial y}\mathbf{n} \cdot \mathbf{k}. \qquad (6.6.19)$$

Substitute this back into Eq. 6.6.15 and obtain

$$\oint_C u_x \, dx = -\iint_S \left[\frac{\partial u_x}{\partial z}\frac{\partial f}{\partial y} + \frac{\partial u_x}{\partial y}\right]\mathbf{n} \cdot \mathbf{k} \, dS. \qquad (6.6.20)$$

Now, on the surface S,

$$u_x = u_x[x, y, f(x, y)] = g(x, y). \qquad (6.6.21)$$

Using the chain rule from calculus we have, using $z = f(x, y)$,

$$\frac{\partial g}{\partial y} = \frac{\partial u_x}{\partial y} + \frac{\partial u_x}{\partial z}\frac{\partial f}{\partial y}. \qquad (6.6.22)$$

Equation 6.6.20 can then be written in the form (see Eq. 6.6.6)

$$\oint_C u_x \, dx = -\iint_R \frac{\partial g}{\partial y} \, dx \, dy. \qquad (6.6.23)$$

The area integral above can be written as* (see Fig. 6.24)

$$\iint_R \frac{\partial g}{\partial y} \, dx \, dy = \int_{x_1}^{x_2}\left[\int_{h_1(x)}^{h_2(x)} \frac{\partial g}{\partial y} \, dy\right] dx = \int_{x_1}^{x_2} [g(x, h_2) - g(x, h_1)] \, dx$$

$$= -\int_{C_{2'}} g \, dx - \int_{C_{1'}} g \, dx, \qquad (6.6.24)$$

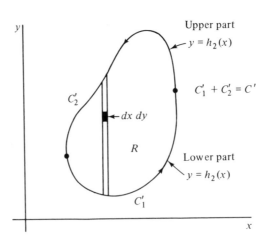

FIGURE 6.24.   Plane surface R from Fig. 6.23.

---

*This can also be accomplished by using Green's theorem in the plane, derived in Example 6.6.3, by letting $\phi = 0$ in that theorem.

where the negative sign on the C_2' integral is necessary to account for changing the direction of integration. Since $C_1' + C_2' = C'$, we see that

$$\oint_C u_x \, dx = \int_{C'} g \, dx. \tag{6.6.25}$$

From Eq. 6.6.21 we see that g on C' is the same as u_x on C. Thus, our proof of Eq. 6.6.15 is complete.

Similarly, by projections on the other coordinate planes, we can verify that

$$\oint_C u_y \, dy = \iint_S \left[\frac{\partial u_y}{\partial x} \mathbf{k} \cdot \mathbf{n} - \frac{\partial u_y}{\partial z} \mathbf{i} \cdot \mathbf{n} \right] dS$$

$$\oint_C u_z \, dz = \iint_S \left[\frac{\partial u_z}{\partial y} \mathbf{i} \cdot \mathbf{n} - \frac{\partial u_z}{\partial x} \mathbf{j} \cdot \mathbf{n} \right] dS. \tag{6.6.26}$$

If we add Eq. 6.6.15 to the two equations above, Eq. 6.6.14 results and our proof of Stokes' theorem is accomplished, a rather difficult task!

The scalar quantity resulting from the integration in Stokes' theorem is called the *circulation* of the vector $\mathbf{u}$ around the curve C. It is usually designated Γ and is

$$\Gamma = \oint_C \mathbf{u} \cdot d\mathbf{l}. \tag{6.6.27}$$

It is of particular interest in aerodynamics since the quantity $\rho \Gamma U$ (ρ is the density of air and U is the speed) gives the magnitude of the lift on an airfoil. Note that for an irrotational vector field the circulation is zero.

Example 6.6.1:   In the divergence theorem let the vector function $\mathbf{u} = \phi \mathbf{i}$, where ϕ is a scalar function. Derive the resulting integral theorem.

SOLUTION: The divergence theorem is given by Eq. 6.6.1. Let $\mathbf{u} = \phi \mathbf{i}$ and there results

$$\iiint_V \boldsymbol{\nabla} \cdot (\phi \mathbf{i}) \, dV = \oiint_S \phi \mathbf{i} \cdot \mathbf{n} \, dS.$$

The unit vector $\mathbf{i}$ is constant and thus $\boldsymbol{\nabla} \cdot (\phi \mathbf{i}) = \mathbf{i} \cdot \boldsymbol{\nabla} \phi$ (see Table 6.1). Removing the constant $\mathbf{i}$ from the integrals yields

$$\mathbf{i} \cdot \iiint_V \boldsymbol{\nabla} \phi \, dV = \mathbf{i} \cdot \oiint_S \phi \mathbf{n} \, dS.$$

This can be rewritten as

$$\mathbf{i} \cdot \left[\iiint_V \boldsymbol{\nabla} \phi \, dV - \oiint_S \phi \mathbf{n} \, dS \right] = 0.$$

Since **i** is never zero and the quantity in brackets is not, in general, perpendicular to **i**, we must demand that the quantity in brackets be zero. Consequently,

$$\iiint_V \nabla\phi \, dV = \oiint_S \phi\mathbf{n} \, dS.$$

This is another useful form Gauss' theorem. ∎

Example 6.6.2:   Let $\mathbf{u} = \psi\nabla\phi$ in the divergence theorem, and then let $\mathbf{u} = \phi\nabla\psi$. Subtract the resulting equations, thereby deriving Green's theorem.

SOLUTION: Substituting $\mathbf{u} = \psi\nabla\phi$ into the divergence theorem given by Eq. 6.6.1, we have

$$\iiint_V \nabla \cdot (\psi\nabla\phi) \, dV = \oiint_S \psi\nabla\phi \cdot \mathbf{n} \, dS.$$

Using Table 6.1, we can write

$$\nabla \cdot (\psi\nabla\phi) = \nabla\psi \cdot \nabla\phi + \psi\nabla^2\phi = \nabla\psi \cdot \nabla\phi + \psi\nabla^2\phi.$$

The divergence theorem takes the form, using Eq. 6.4.9,

$$\iiint_V [\psi\nabla^2\phi + \nabla\psi \cdot \nabla\phi] \, dV = \oiint_S \psi\frac{\partial\phi}{\partial n} \, dS.$$

Now, with $\mathbf{u} = \phi\nabla\psi$, we find that

$$\iiint_V [\phi\nabla^2\psi + \nabla\psi \cdot \nabla\phi] \, dV = \oiint_S \phi\frac{\partial\psi}{\partial n} \, dS.$$

Subtract the two equations above and obtain

$$\iiint_V [\psi\nabla^2\phi - \phi\nabla^2\psi] \, dV = \oiint_S \left[\psi\frac{\partial\phi}{\partial n} - \phi\frac{\partial\psi}{\partial n}\right] \, dS.$$

This is known as *Green's theorem*, or alternately, the *second form of Green's theorem*. ∎

Example 6.6.3:   Let the volume V be the simply connected region shown at the top of page 363. Derive the resulting form of the divergence theorem if

$$\mathbf{u} = \mathbf{u}(x, y) = \phi\mathbf{i} + \psi\mathbf{j}.$$

SOLUTION: The divergence theorem, Eq. 6.6.1, takes the form

$$\iint_R \left[\int_0^H \nabla \cdot \mathbf{u} \, dz\right] dx \, dy = \oiint_S \mathbf{u} \cdot \mathbf{n} \, dS.$$

The quantity $\nabla \cdot \mathbf{u}$ is independent of z, since $\mathbf{u} = \mathbf{u}(x, y)$. Thus,

$$\int_0^H \nabla \cdot \mathbf{u} \, dz = H\nabla \cdot \mathbf{u}.$$

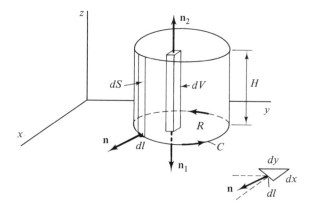

Also, on the top surface $\mathbf{u} \cdot \mathbf{n}_2 = \mathbf{u} \cdot \mathbf{k} = 0$, since $\mathbf{u}$ has no z component. Likewise, on the bottom surface $\mathbf{u} \cdot \mathbf{n}_1 = 0$. Consequently, only the side surface contributes to the surface integral. For this side surface we can write $dS = H \, dl$ and perform the integration around the closed curve C. The divergence theorem then takes the form

$$H \iint_R \nabla \cdot \mathbf{u} \, dx \, dy = \oint_C \mathbf{u} \cdot \mathbf{n}H \, dl.$$

Since

$$\mathbf{u} = \phi\mathbf{i} + \psi\mathbf{j} \quad \text{and} \quad \mathbf{n} = \frac{dy}{dl}\mathbf{i} - \frac{dx}{dl}\mathbf{j}$$

(see the small sketch in the figure of this example), there results

$$\nabla \cdot \mathbf{u} = \frac{\partial \phi}{\partial x} + \frac{\partial \psi}{\partial y}$$

$$\mathbf{u} \cdot \mathbf{n} = \phi\frac{dy}{dl} - \psi\frac{dx}{dl}.$$

Finally, we have the useful result

$$\iint_R \left(\frac{\partial \phi}{\partial x} + \frac{\partial \psi}{\partial y}\right) dy \, dx = \oint_C (\phi \, dy - \psi \, dx).$$

It is known as *Green's theorem in the plane*. ∎

Example 6.6.4:   Determine the circulation of the vector function $\mathbf{u} = 2y\mathbf{i} + x\mathbf{j}$ around the curve shown, by (a) direct integration, and (b) Stokes' theorem.

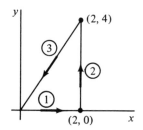

SOLUTION: (a) The circulation is given by

$$\Gamma = \oint_C \mathbf{u} \cdot d\mathbf{l} = \oint_C u_x \, dx + u_y \, dy + u_z \, dz.$$

For the three parts of the curve, we have

$$\Gamma = \int_① u_x \, dx + u_y \, \cancel{dy}^{\,0} + u_z \, \cancel{dz}^{\,0} + \int_② u_x \, \cancel{dx}^{\,0} + u_y \, dy + u_z \, \cancel{dz}^{\,0}$$

$$+ \int_③ u_x \, dx + u_y \, dy + u_z \, \cancel{dz}^{\,0}.$$

Along part 1, $u_x = 2y = 0$; along part 2, $u_y = x = 2$; and along part 3, $2x = y$, so $2dx = dy$. Thus, we have

$$\Gamma = \int_0^4 2dy + \int_2^0 (2 \cdot 2x \, dx + x \, 2dx)$$

$$= 2 \cdot 4 + 3x^2 \Big|_2^0 = -4.$$

(b) Using Stokes' theorem, we have

$$\Gamma = \iint_S \nabla \times \mathbf{u} \cdot \mathbf{n} \, dS$$

$$= \iint_S \left[\left(\frac{\partial u_z}{\partial y} - \cancel{\frac{\partial u_y}{\partial z}}^{\,0} \right) \mathbf{i} + \left(\frac{\partial u_x}{\partial z} - \cancel{\frac{\partial u_z}{\partial x}}^{\,0} \right) \mathbf{j} + \left(\frac{\partial u_y}{\partial x} - \frac{\partial u_x}{\partial y} \right) \mathbf{k} \right] \cdot \mathbf{k} \, dS$$

$$= \iint_S (1 - 2)\mathbf{k} \cdot \mathbf{k} \, dS = -\iint dS = -4. \quad ■$$

PROBLEMS

By using the divergence theorem, evaluate $\oiint_S \mathbf{u} \cdot \mathbf{n} \, dS$, where

1. $\mathbf{u} = x\mathbf{i} + y\mathbf{j} + z\mathbf{k}$ and S is the sphere $x^2 + y^2 + z^2 = 9$.

2. $\mathbf{u} = xy\mathbf{i} + xz\mathbf{j} + (1 - z)\mathbf{k}$ and S is the unit cube bounded by $z = 0$, $y = 0$, $z = 0$, $x = 1$, $y = 1$, and $z = 1$.

3. $\mathbf{u} = x\mathbf{i} + x\mathbf{j} + z^2\mathbf{k}$ and S is the cylinder $x^2 + y^2 = 4$ bounded by $z = 0$ and $z = 8$.

Recognizing that $\mathbf{i} \cdot \mathbf{n} \, dS = dy \, dz$, $\mathbf{j} \cdot \mathbf{n} \, dS = dx \, dz$, and $\mathbf{k} \cdot \mathbf{n} \, dS = dx \, dy$, evaluate the following using the divergence theorem.

4. $\iint_S (x\ dy\ dz + 2y\ dx\ dz + y^2\ dx\ dy)$, where S is the sphere $x^2 + y^2 + z^2 = 4$.

5. $\iint_S (x^2\ dy\ dz + 2xy\ dx\ dz + xy\ dx\ dy)$, where S is the cube of Problem 2.

6. $\iint_S z^2\ dx\ dy$, where S is the cylinder of Problem 3.

7. Let $\mathbf{u} = \nabla\phi$, and derive one of the forms of the divergence theorem given in Table 6.5.

8. With $\mathbf{u} = \mathbf{v} \times \mathbf{i}$, derive one of the forms of the divergence theorem given in Table 6.5.

Assume that ϕ is a harmonic function, that is, ϕ satisfies Laplace's equation $\nabla^2\phi = 0$. Show that

9. $\oint\!\!\!\oint_S \dfrac{\partial\phi}{\partial n}\ dS = 0$ $\qquad\qquad$ **10.** $\iiint_V \nabla\phi \cdot \nabla\phi\ dV = \oint\!\!\!\oint_S \phi\dfrac{\partial\phi}{\partial n}\ dS.$

11. If no fluid is being introduced into a volume V, that is, there are no sources or sinks, the conservation of mass is written in integral form as

$$-\iiint_V \frac{\partial\rho}{\partial t}\ dV = \oint\!\!\!\oint_S \rho\mathbf{v}\cdot\mathbf{n}\ dS,$$

where the surface S surrounds V, $\rho(x, y, z, t)$ is the density (mass per unit volume), and $\mathbf{v}(x, y, z, t)$ is the velocity. Convert the area integral to a volume integral and combine the two volume integrals. Then, since the equation is valid for any arbitrary volume, extract the differential form of the conservation of mass.

12. The integral form of the energy equation of a stationary material equates the rate of change of energy contained by the material to the rate at which heat enters the material by conduction; that is,

$$\iiint_V \rho\frac{\partial e}{\partial t}dV = -\oint\!\!\!\oint_S \mathbf{q}\cdot\mathbf{n}\ dS,$$

where ρ is the density, e is the internal energy, and $\mathbf{q}$ is the heat flux. Empirical evidence allows us to write

$$\Delta e = C\ \Delta T \quad \text{and} \quad \mathbf{q} = -k\ \nabla T,$$

where C is the specific heat and k is the conductivity. If this is true for any arbitrary volume, derive the differential heat equation if the coefficients are assumed constant.

13. Derive Green's theorem in the plane by letting $\mathbf{u} = \phi\mathbf{i} + \psi\mathbf{j}$ and S be the xy plane in Stokes' theorem.

14. Calculate the circulation of the vector $\mathbf{u} = y^2\mathbf{i} + xy\mathbf{j} + z^2\mathbf{k}$ around a triangle with vertices at the origin, $(2, 2, 0)$ and $(0, 2, 0)$, by (a) direct integration and (b) using Stokes' theorem.

15. Calculate the circulation of $\mathbf{u} = y\mathbf{i} - x\mathbf{j} + z\mathbf{k}$ around a unit circle in the xy plane with center at the origin by (a) direct integration, and (b) using Stokes' theorem.

Evaluate the circulation of each vector function around the curve specified. Use either direct integration or Stokes' theorem.

16. $\mathbf{u} = 2z\mathbf{i} + y\mathbf{j} + x\mathbf{k}$; the triangle with vertices at the origin, $(1, 0, 0)$ and $(0, 0, 4)$.

17. $\mathbf{u} = 2xy\mathbf{i} + y^2z\mathbf{j} + xy\mathbf{k}$; the rectangle with corners at $(0, 0, 0)$, $(0, 4, 0)$, $(6, 4, 0)$, $(6, 0, 0)$

18. $\mathbf{u} = x^2\mathbf{i} + y^2\mathbf{j} + z^2\mathbf{k}$; the unit circle in the xy plane with center at the origin.

7

Fourier Series

7.1 *Introduction*

We have seen in Chapter 1 that nonhomogeneous differential equations with constant coefficients containing sinusoidal input functions (e.g., $A \sin \omega t$) can be solved quite easily for any input frequency ω. There are many examples, however, of periodic input functions that are not sinusoidal. Figure 7.1 illustrates four common ones. The voltage input to a circuit or the force on a spring–mass system may be periodic but possess discontinuities such as those illustrated. The object of this chapter is to present a technique for solving such problems and others connected to the solution of certain boundary-value problems in the theory of partial differential equations.

The technique of this chapter employs series of the form

$$\frac{a_0}{2} + \sum_{n=1}^{\infty} \left(a_n \cos \frac{n\pi t}{T} + b_n \sin \frac{n\pi t}{T} \right), \qquad (7.1.1)$$

the so-called *trigonometric series*. Unlike power series, such series present

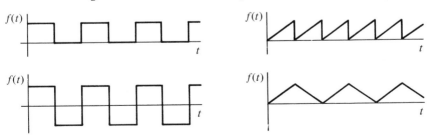

FIGURE 7.1 Some periodic input functions.

many pitfalls and subtleties. A complete theory of trigonometric series is beyond the scope of this text and most works on applications of mathematics to the physical sciences. We make our task tractable by narrowing our scope to those principles that bear directly on our interests.

Let $f(t)$ be sectionally continuous in the interval $-T < t < T$ so that in this interval $f(t)$ has at most a finite number of discontinuities. At each point of discontinuity the right- and left-hand limits exist; that is, at the end points $-T$ and T of the interval $-T < t < T$ we define $f(-T^+)$ and $f(T^-)$ as limits from the right and left, respectively, according to the following expressions:

$$f(-T^+) = \lim_{\substack{t \to -T \\ t > -T}} f(t), \qquad f(T^-) = \lim_{\substack{t \to T \\ t < T}} f(t) \qquad (7.1.2)$$

and insist that $f(-T^+)$ and $f(T^-)$ exist also. Then the following sets of *Fourier coefficients* of $f(t)$ in $-T < t < T$ exist:

$$a_0 = \frac{1}{T} \int_{-T}^{T} f(t) \, dt$$

$$a_n = \frac{1}{T} \int_{-T}^{T} f(t) \cos \frac{n \pi t}{T} \, dt \qquad (7.1.3)$$

$$b_n = \frac{1}{T} \int_{-T}^{T} f(t) \sin \frac{n \pi t}{T} \, dt, \qquad n = 1, 2, 3, \ldots$$

The trigonometric series 7.1.1, defined by using these coefficients, is the *Fourier series* expansion of $f(t)$ in $-T < t < T$. In this case we write

$$f(t) \sim \frac{a_0}{2} + \sum_{n=1}^{\infty} \left(a_n \cos \frac{n \pi t}{T} + b_n \sin \frac{n \pi t}{T} \right). \qquad (7.1.4)$$

This representation means only that the coefficients in the series are the Fourier coefficients of $f(t)$ as computed in Eq. 7.1.3. We shall concern ourselves in the next section with the question of when "$\sim$" may be replaced with "$=$"; conditions on $f(t)$ which are sufficient to permit this replacement are known as *Fourier theorems*.

We conclude this introduction with an example that illustrates one of the difficulties under which we labor. In the next section we shall show that $f(t) = t$, $-\pi < t < \pi$ has the Fourier series representation

$$t = 2 \sum_{n=1}^{\infty} \frac{(-1)^{n+1}}{n} \sin nt, \qquad (7.1.5)$$

where the series converges for all t, $-\pi < t < \pi$. Now $f'(t) = 1$. But if we differentiate the series 7.1.5 term by term, we obtain

$$2 \sum_{n=1}^{\infty} (-1)^{n+1} \cos nt, \qquad (7.1.6)$$

which diverges in $-\pi < t < \pi$ since the nth term, $(-i)^{n+1} \cos nt$, does not tend to zero as n tends to infinity. Moreover, it is not even the Fourier series representation of $f'(t) = 1$. This is in sharp contrast to the "nice" results we are accustomed to in working with power and Frobenius series.

PROBLEMS

1. What is the Fourier representation of $f(t) = 1$, $-\pi < t < \pi$?

2. Verify the representation, Eq. 7.1.5, by using Eq. 7.1.3.

3. Does the series, (Eq. 7.1.5) converge if t is exterior to $-\pi < t < \pi$? At $t = \pi$? At $t = -\pi$? To what values?

4. Show that the Fourier series representation given as Eq. 7.1.4 may be written

$$f(t) \sim \frac{1}{2T} \int_{-T}^{T} f(t)\, dt + \frac{1}{T} \sum_{n=1}^{\infty} \int_{-T}^{T} f(s) \cos \frac{n\pi t}{T}(s - t)\, dt.$$

5. Explain how

$$\frac{\pi}{4} = 1 - \frac{1}{3} + \frac{1}{5} - \frac{1}{7} + \cdots.$$

follows from Eq. 7.1.5. *Hint*: Pick $t = \pi/2$. Note that this result also follows from

$$\tan^{-1} x = x - \frac{x^3}{3} + \frac{x^5}{5} - \frac{x^7}{7} + \cdots, \qquad -1 < x \le 1.$$

6. What is the Fourier series expansion of $f(t) = -1$, $-T < t < T$?

7.2   A Fourier Theorem

As we have remarked in the introduction, we shall assume throughout this chapter that $f(t)$ is sectionally continuous in $-T < t < T$. Whether $f(t)$ is defined at the end points $-T$ or T or defined exterior* to $(-T, T)$ is a matter of indifference. For if the Fourier series of $f(t)$ converges to $f(t)$ in $(-T, T)$ it converges almost everywhere since it is periodic with period $2T$. Hence, unless $f(t)$ is also periodic, the series will converge, not to $f(t)$, but to its "periodic extension." Let us make this idea more precise. First, we make the following stipulation:

(1) If t_0 is a point of discontinuity of $f(t)$, $-T < t_0 < T$, then redefine $f(t_0)$, if necessary, so that

$$f(t_0) = \tfrac{1}{2}[f(t_0^-) + f(t_0^+)]. \tag{7.2.1}$$

---

*The notation $(-T, T)$ means the set of t, $-T < t < T$. Thus, the exterior of $(-T, T)$ means those t, $t \ge T$ or $t \le -T$.

In other words, we shall assume that in $(-T, T)$ the function $f(t)$ is always the average of the right- and left-hand limits at t. Of course, if t is a point of continuity of $f(t)$, then $f(t^+) = f(t^-)$ and hence Eq. 7.2.1 is also true at points of continuity. The *periodic extension* $\widetilde{f}(t)$ of $f(t)$ is defined

$$(2) \qquad\qquad \widetilde{f}(t) = f(t), \qquad\qquad -T < t < T \qquad\qquad (7.2.2)$$

$$(3) \qquad\qquad \widetilde{f}(t + 2T) = \widetilde{f}(t) \qquad \text{for all } t \qquad\qquad (7.2.3)$$

$$(4) \qquad\qquad \widetilde{f}(T) = \widetilde{f}(-T) = \tfrac{1}{2}[f(-T^+) + f(T^-)]. \qquad\qquad (7.2.4)$$

Condition (2) requires $\widetilde{f}(t)$ and $f(t)$ to agree on the *fundamental interval* $(-T, T)$. Condition (3) extends the definition of $f(t)$ so that $\widetilde{f}(t)$ is defined everywhere and is periodic with period $2T$. Condition (4) is somewhat more subtle. Essentially, it forces stipulation (1) (see Eq. 7.2.1) on $\widetilde{f}(t)$ at the points $\pm nT$ (see Examples 7.2.1 and 7.2.2).

Example 7.2.1: Sketch the periodic extension of $f(t) = t/\pi$, $-\pi < t < \pi$.

SOLUTION: In this example, $f(\pi^-) = 1$ and $f(-\pi^+) = -1$, so that $\widetilde{f}(\pi) = \widetilde{f}(-\pi) = 0$. The graph of $\widetilde{f}(t)$ follows.

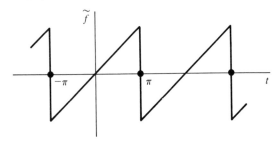

Note that the effect of condition (4) (See Eq. 7.2.4) is to force $\widetilde{f}(t)$ to have the average of its values at all t; in particular, $\widetilde{f}(n\pi) = \widetilde{f}(-n\pi) = 0$ for all n. ∎

Example 7.2.2: Sketch the periodic extension of $f(t) = 0$ for $t < 0$, $f(t) = 1$ for $t > 0$, if the fundamental interval is $(-1, 1)$.

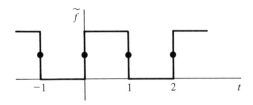

SOLUTION: There are two preliminary steps. First, we redefine $f(t)$ at $t = 0$; to wit,

$$f(0) = \frac{1 + 0}{2} = \frac{1}{2}.$$

Second, since $f(1) = 1$ and $f(-1) = 0$, we set

$$\widetilde{f}(-1) = \widetilde{f}(1) = \frac{1 + 0}{2} = \frac{1}{2}$$

The graph of $f(t)$ is as shown.   ■

A Fourier theorem is a set of conditions sufficient to imply the convergence of the Fourier series of $f(t)$ to some function closely "related" to $f(t)$. The following is one such theorem.

Theorem 7.1:   *Suppose that $f(t)$ and $f'(t)$ are sectionally continuous in $-T < t < T$. Then the Fourier series of $f(t)$ converges to the periodic extension of $f(t)$, that is, $\widetilde{f}(t)$, for all t.*

We offer no proof for this theorem.* Note, however, that the Fourier series for the functions given in Examples 7.2.1 and 7.2.2 converge to the functions portrayed in the respective figures of those examples. Thus, Eq. 7.1.4 with an equal sign is a consequence of this theorem.

There is another observation relevant to Theorem 7.1; in the interval $-T < t < T$, $\widetilde{f}(t) = f(t)$. Thus, the convergence of the Fourier series of $f(t)$ is to $f(t)$ in $(-T, T)$.

PROBLEMS

The following sketches define a function in some interval $-T < t < T$. Complete the sketch for the periodic extension of this function and indicate the value of the function at points of discontinuity.

Sketch the periodic extension of each function.

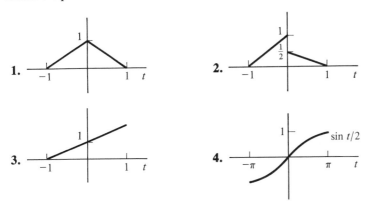

1.

2.

3.

4.

---

*A proof is given in many textbooks on Fourier series.

5.

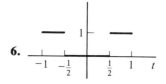

6.

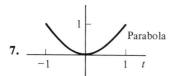

7.

Parabola

8. $f(t) = \begin{cases} -1, & -\pi < t < 0 \\ 1, & 0 < t < \pi \end{cases}$

9. $f(t) = t + 1, \quad -\pi < t < \pi$

10. $f(t) = \begin{cases} t + \pi, & -\pi < t < 0 \\ -t + \pi, & 0 < t < \pi \end{cases}$

11. $f(t) = |\sin t|, \quad -\pi < t < \pi$

12. $f(t) = \begin{cases} 0, & -2 < t < 0 \\ \sin \pi t/2, & 0 < t < 2 \end{cases}$

13. $f(t) = t^2, \quad -\pi < t < \pi$

14. $f(t) = \begin{cases} -1, & -1 < t < -\frac{1}{2} \\ 0, & -\frac{1}{2} < t < \frac{1}{2} \\ 1, & \frac{1}{2} < t < 1 \end{cases}$

15. $f(t) = |t|, \quad -1 < t < 1$

16. $f(t) = \begin{cases} 0, & -\pi < t < 0 \\ \sin t, & 0 < t < \pi \end{cases}$

17. $f(t) = \begin{cases} -1, & -1 < t < 0 \\ 1, & 0 < t < 1 \end{cases}$

18. $f(t) = \cos t, \quad -\pi < t < \pi$

19. $f(t) = \sin 2t, \quad -\pi < t < \pi$

20. $f(t) = \tan t, \quad -\frac{\pi}{2} < t < \frac{\pi}{2}$

21. $f(t) = |t|, \quad -1 < t < 1$

22. Explain why $f(t) = \sqrt{|t|}$ is continuous in $-1 < t < 1$ but $f'(t)$ is not sectionally continuous in this interval.

23. Explain why $f(t) = |t|^{3/2}$ is continuous and $f'(t)$ is also continuous in $-1 < t < 1$. Contrast this with Problem 22.

24. Is $\ln |\tan t/2|$ sectionally continuous in $0 < t < \pi/4$? Explain.

25. Is

$$f(t) = \begin{cases} \ln |\tan t/2|, & 0 < \epsilon \le |t| < \pi/4 \\ 0, & |t| < \epsilon \end{cases}$$

sectionally continuous in $0 < t < \pi/4$? Explain.

7.3 The Computation of the Fourier Coefficients

7.3.1 KRONECKER'S METHOD

We shall be faced with integrations of the type

$$\int x^k \cos \frac{n\pi x}{L} \, dx \qquad (7.3.1)$$

for various small positive integer values of k. This type of integration is accomplished by repeated integration by parts. We wish to diminish the tedious details inherent in such computations. So consider the integration-by-parts formula

$$\int g(x)f(x) \, dx = g(x) \int f(x) \, dx - \int \left[g'(x) \int f(x) \, dx \right] dx. \qquad (7.3.2)$$

Let

$$F_1(x) = \int f(x) \, dx$$

$$F_2(x) = \int F_1(x) \, dx$$

$$\vdots$$

$$F_n(x) = \int F_{n-1}(x) \, dx. \qquad (7.3.3)$$

Then Eq. 7.3.2 is

$$\int g(x)f(x) \, dx = g(x)F_1(x) - \int g'(x)F_1(x) \, dx, \qquad (7.3.4)$$

from which

$$\int g(x)f(x) \, dx = g(x)F_1(x) - g'(x)F_2(x) + \int g''(x)F_2(x) \, dx \qquad (7.3.5)$$

follows by another integration by parts. This may be repeated indefinitely, leading to

$$\int g(x)f(x) \, dx = g(x)F_1(x) - g'(x)F_2(x) + g''(x)F_3(x) + \cdots. \qquad (7.3.6)$$

Note that each term on the right-hand side of Eq. 7.3.6 comes from the preceding term by differentiation of the g function and an indefinite integration of the f function as well as an alternation of sign.

Example 7.3.1:   Compute $\int_{-\pi}^{\pi} x \cos nx \, dx$.

SOLUTION: We integrate by parts as follows:

$$\int_{-\pi}^{\pi} x \cos nx \, dx = \frac{x}{n} \sin nx \, \bigg|_{-\pi}^{\pi} - 1\left(-\frac{1}{n^2} \cos nx\right)\bigg|_{-\pi}^{\pi}$$

$$= 0 + \frac{1}{n^2}(\cos n\pi - \cos n\pi) = 0. \quad \blacksquare$$

Example 7.3.2:   Compute $\int_{-\pi}^{\pi} x^2 \cos nx \, dx$.

SOLUTION: For this example, integration by parts yields

$$\int_{-\pi}^{\pi} x^2 \cos nx \, dx = \left[\frac{x^2}{n} \sin nx - 2x\left(-\frac{1}{n^2} \cos nx\right) + 2\left(-\frac{1}{n^3} \sin nx\right)\right]_{-\pi}^{\pi}$$

$$= \frac{2}{n^2}(\pi \cos n\pi + \pi \cos n\pi) = \frac{4\pi}{n^2}(-1)^n. \quad \blacksquare$$

Example 7.3.3:   Use Kronecker's method and integrate $\int e^x \cos ax \, dx$.

SOLUTION: Let $g(x) = e^x$. Then

$$\int e^x \cos ax \, dx = e^x \frac{1}{a} \sin ax - e^x\left(-\frac{1}{a^2} \cos ax\right) + e^x\left(\frac{-1}{a^3} \sin ax\right) + \cdots$$

$$= e^x\left(\frac{1}{a} \sin ax + \frac{1}{a^2} \cos ax - \frac{1}{a^3} \sin ax + \cdots\right)$$

$$= e^x \sin ax\left(\frac{1}{a} - \frac{1}{a^3} + \cdots\right) + e^x \cos ax\left(\frac{1}{a^2} - \frac{1}{a^4} + \cdots\right)$$

$$= e^x \frac{1}{a} \frac{1}{1 + 1/a^2} \sin ax + e^x \frac{1}{a^2} \frac{1}{1 + 1/a^2} \cos ax$$

$$= \frac{e^x}{a^2 + 1}(a \sin ax + \cos ax). \quad \blacksquare$$

PROBLEMS

Find a general formula for each integral as a function of the positive integer n.

1. $\int x^n \cos ax \, dx$

2. $\int x^n \sin ax \, dx$

3. $\int x^n e^{bx} \, dx$

4. $\int x^n \sinh bx \, dx$

5. $\int x^n \cosh bx \, dx$

6. $\int x^n(ax + b)^\alpha \, dx$

Find each integral using as a model the work in Example 7.3.3.

7. $\int e^{bx} \cos ax \, dx$ **8.** $\int e^{bx} \sin ax \, dx$

7.3.2 SOME EXPANSIONS

In this section we will find some Fourier series expansions of several of the more common functions, applying the theory of the previous sections.

Example 7.3.4: Write the Fourier series representation of the periodic function $f(t)$ if in one period

$$f(t) = t, \qquad -\pi < t < \pi.$$

SOLUTION: For this example, $T = \pi$. For a_n we have

$$a_0 = \frac{1}{\pi} \int_{-\pi}^{\pi} f(t) \, dt = \frac{1}{\pi} \int_{-\pi}^{\pi} t \, dt = \frac{t^2}{2\pi} \bigg|_{-\pi}^{\pi} = 0$$

$$a_n = \frac{1}{\pi} \int_{-\pi}^{\pi} f(t) \cos nt \, dt, \qquad n = 1, 2, 3, \ldots$$

$$= \frac{1}{\pi} \int_{-\pi}^{\pi} t \cos nt \, dt = \frac{1}{\pi} \left[\frac{t}{n} \sin nt + \frac{1}{n^2} \cos nt \right]_{-\pi}^{\pi} = 0,$$

recognizing that $\cos n\pi = \cos(-n\pi)$ and $\sin n\pi = -\sin(-n\pi) = 0$. For b_n we have

$$b_n = \frac{1}{\pi} \int_{-\pi}^{\pi} f(t) \sin nt \, dt, \qquad n = 1, 2, 3, \ldots$$

$$= \frac{1}{\pi} \int_{-\pi}^{\pi} t \sin nt \, dt = \frac{1}{\pi} \left[-\frac{t}{n} \cos nt + \frac{1}{n^2} \sin nt \right]_{-\pi}^{\pi} = -\frac{2}{n} \cos n\pi.$$

The Fourier series representation has only sine terms. It is given by

$$f(t) = -2 \sum_{n=1}^{\infty} \frac{(-1)^n}{n} \sin nt,$$

where we have used $\cos n\pi = (-1)^n$. Writing out several terms, we have

$$f(t) = -2[-\sin t + \tfrac{1}{2} \sin 2t - \tfrac{1}{3} \sin 3t + \cdots]$$

$$= 2 \sin t - \sin 2t + \tfrac{2}{3} \sin 3t - \cdots.$$

Note the sketch, showing the increasing accuracy with which the terms approximate the $f(t)$. Notice also the close approximation using three terms. Obviously, using a computer and keeping, say 50 terms, a remarkably good approximation can result using Fourier series.

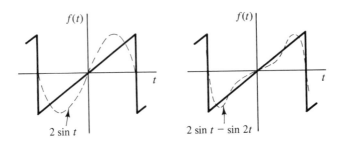

2 sin t

2 sin t − sin 2t

2 sin t − sin 2t + $\frac{2}{3}$ sin 3t ∎

Example 7.3.5: Find the Fourier series expansion for the periodic function $f(t)$ if in one period

$$f(t) = \begin{cases} 0, & -\pi < t < 0 \\ t, & 0 < t < \pi. \end{cases}$$

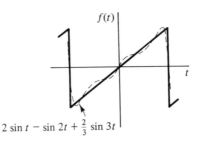

SOLUTION: The period is again 2π; thus, $T = \pi$. The Fourier coefficients are given by

$$a_0 = \frac{1}{\pi} \int_{-\pi}^{\pi} f(t)\, dt = \frac{1}{\pi} \int_{-0}^{\pi} t\, dt = \frac{\pi}{2}$$

$$a_n = \frac{1}{\pi} \int_{-\pi}^{\pi} f(t) \cos nt\, dt = \frac{1}{\pi} \int_{-\pi}^{0} 0 \cos nt\, dt + \frac{1}{\pi} \int_{0}^{\pi} t \cos nt\, dt$$

$$= \frac{1}{\pi}\left[\frac{t}{n} \sin nt + \frac{1}{n^2} \cos nt \right]_0^{\pi} = \frac{1}{\pi n^2}(\cos n\pi - 1), \qquad n = 1, 2, 3, \dots$$

$$b_n = \frac{1}{\pi} \int_{-\pi}^{\pi} f(t) \sin nt\, dt = \frac{1}{\pi} \int_{-\pi}^{0} 0 \sin nt\, dt + \frac{1}{\pi} \int_{0}^{\pi} t \sin nt\, dt$$

$$= \frac{1}{\pi}\left[-\frac{t}{n} \cos nt + \frac{1}{n^2} \sin nt \right]_0^{\pi} = -\frac{1}{n} \cos n\pi, \qquad n = 1, 2, 3, \dots$$

The Fourier series representation is, then, using $\cos n\pi = (-1)^n$,

$$f(t) = \frac{\pi}{4} + \sum_{n=1}^{\infty} \left[\frac{(-1)^n - 1}{\pi n^2} \cos nt - \frac{(-1)^n}{n} \sin nt \right]$$

$$= \frac{\pi}{4} - \frac{2}{\pi} \cos t - \frac{2}{9\pi} \cos 3t + \cdots + \sin t - \frac{1}{2} \sin 2t + \frac{1}{3} \sin 3t + \cdots.$$

A computer program has been written with n equal to 5, 10, and 20, respectively, in the equation above. The results are plotted on the following graph.

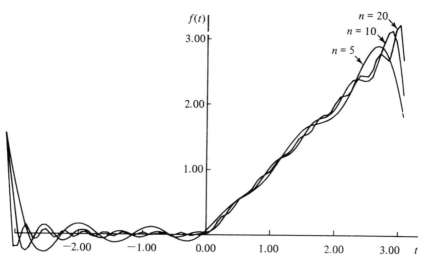

```
           PROGRAM FOUIER (INPUT,OUTPUT)
           DIMENSION IBUF (257)
CC         INITIALIZE THE PLOT ROUTINE
           CALL PLOTS(IBUF,257,5)
           CALL PLOT(0.0,1.5,-3)
           PI = 3.14159265
           J = 5
CC         THIS PROGRAM DRAWS F(T) FOR THREE DIFFERENT J
           DO 10 K=1,3
           T = - PI
           F = 0.0
           CALL PLOT (0.0,0.0,3)
CC         PLOT 100 POINTS
           DO 20 L=1,100
CC         SUM THE FOURIER SERIES FROM 1 TO J
           DO 30 I=1,J
           A = I
           F = F +((-1.)**I-1.0)/(PI*A*A)*COS(A*T)-(((-1.)**I)/A)*SIN(A*T)
30         CONTINUE
           F = F + PI/4.0
           PRINT 40,T,J,F
40         FORMAT (F15.4,I10,F15.4)
```

```
CC        PLOT F(T) AND INCREMENT T
          Y = F
          X = T + PI
          CALL PLOT (X,Y,2)
          T = PI/50. + T
          F = 0.0
20        CONTINUE
CC        RAISE THE PEN AND DO ANOTHER PLOT WITH A NEW J
          CALL PLOT (0.0,0.0,3)
          IF (J.EQ. 10) J = 20
          IF (J.EQ.5) J = 10
10        CONTINUE
CC        DRAW THE AXIS
          CALL AXIS (PI ,0.0,4HF(T),4,3.,90.,0.0,1.0)
          AAA = PI - 3.0
          CALL AXIS(AAA,0.0,1HT,-1,6.,0.0,-3.,1.0)
          CALL PLOT (2.0,0.0,999)
          END
```

Example 7.3.6:   Find the Fourier series for the periodic extension of

$$f(t) = \begin{cases} \sin t, & 0 \le t \le \pi \\ 0, & \pi \le t \le 2\pi. \end{cases}$$

SOLUTION: The period is 2π and the Fourier coefficients are computed as usual except for the fact that a_1 and b_1 must be computed separately—as we shall see. We have

$$a_0 = \frac{1}{\pi} \int_0^\pi \sin t \, dt = \frac{1}{\pi}(-\cos t)\Big|_0^\pi = \frac{2}{\pi}.$$

For $n \ne 1$:

$$a_n = \frac{1}{\pi} \int_0^\pi \sin t \cos nt \, dt$$

$$= \frac{1}{2\pi} \int_0^\pi [\sin (t + nt) + \sin (t - nt)] \, dt$$

$$= -\frac{1}{2\pi} \left[\frac{\cos (n + 1)t}{n + 1} - \frac{\cos (n - 1)t}{n - 1} \right]_0^\pi$$

$$= -\frac{1}{2\pi} \left[\frac{(-1)^{n+1}}{n + 1} - \frac{(-1)^{n-1}}{n - 1} \right] + \frac{1}{2\pi} \left[\frac{1}{n + 1} - \frac{1}{n - 1} \right]$$

$$= \frac{1}{\pi(n^2 - 1)} [(-1)^{n+1} - 1].$$

$$b_n = \frac{1}{\pi} \int_0^\pi \sin t \sin nt \, dt$$

$$= \frac{1}{2\pi} \int_0^\pi [-\cos(n+1)t + \cos(n-1)t]\, dt$$

$$= \frac{1}{2\pi} \left[\frac{-\sin(n+1)t}{n+1} + \frac{\sin(n-1)t}{n-1} \right]_0^\pi = 0.$$

For $n = 1$ the expressions above are not defined; hence, the integration is performed specifically for $n = 1$:

$$a_1 = \frac{1}{\pi} \int_0^\pi \sin t \cos t\, dt$$

$$= \frac{1}{\pi} \frac{\sin^2 t}{2} \bigg|_0^\pi = 0.$$

$$b_1 = \frac{1}{\pi} \int_0^\pi \sin t \sin t\, dt = \frac{1}{\pi} \int_0^\pi \left(\frac{1}{2} - \frac{1}{2} \cos 2t \right) dt$$

$$= \frac{1}{\pi} \left(\frac{1}{2} t - \frac{1}{4} \sin 2t \right) \bigg|_0^\pi = \frac{1}{2}.$$

Therefore, when all this information is incorporated in the Fourier series, we obtain the expansion

$$\widetilde{f}(t) = \frac{1}{\pi} + \frac{1}{2} \sin t + \frac{1}{\pi} \sum_{n=2}^\infty \frac{(-1)^{n+1}}{n^2 - 1} \cos nt$$

$$= \frac{1}{\pi} + \frac{1}{2} \sin t - \frac{2}{\pi} \sum_{n=1}^\infty \frac{\cos 2nt}{4n^2 - 1}.$$

The two series representations for $\widetilde{f}(t)$ are equal because $(-1)^{2k+1} - 1 = -2$ and $(-1)^{2k} - 1 = 0$. This series converges everywhere to the periodic function sketched in the example. For $t = \pi/2$, we have

$$\sin \frac{\pi}{2} = \frac{1}{\pi} + \frac{1}{2} \sin \frac{\pi}{2} - \frac{2}{\pi} \sum_{n=1}^\infty \frac{(-1)^n}{4n^2 - 1},$$

which leads to

$$\frac{\pi}{4} = \frac{1}{2} - \sum_{n=1}^\infty \frac{(-1)^n}{4n^2 - 1} = \frac{1}{2} + \frac{1}{3} - \frac{1}{15} + \frac{1}{35} - \frac{1}{63} + \cdots .$$

The function $\widetilde{f}(t)$ of this example is useful in the theory of diodes. ∎

PROBLEMS

Write the Fourier series representation for each periodic function. One period is defined for each. Express the answer as a series using the summation symbol.

1. $f(t) = \begin{cases} -t, & -\pi < t < 0 \\ t, & 0 < t < \pi \end{cases}$

2. $f(t) = t^2, \quad -\pi < t < \pi$

3. $f(t) = \cos \dfrac{t}{2}, \quad -\pi < t < \pi$

4. $f(t) = t + 2\pi, \quad -2 < t < 2\pi$

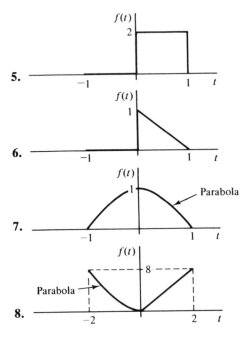

5.

6.

7.

8.

9. Problem 7 of Section 7.1

11. Problem 9 of Section 7.1

13. Problem 13 of Section 7.1

10. Problem 8 of Section 7.1

12. Problem 11 of Section 7.1

7.3.3 EVEN AND ODD FUNCTIONS

The Fourier series expansions of even and odd functions can be accomplished with significantly less effort than needed for functions without either of these symmetries. Recall that an even function is one that satisfies the condition

$$f(-t) = f(t) \qquad (7.3.7)$$

and hence exhibits a graph symmetric with respect to the vertical axis. An odd function satisfies

$$f(-t) = -f(t) \qquad (7.3.8)$$

and is therefore symmetric with respect to the origin. The functions $\cos t$, $t^2 - 1$, $\tan^2 t$, k, $|t|$ are even; the functions $\sin t$, $\tan t$, t, $t|t|$ are odd. Some even and odd functions are displayed in Fig. 7.2. It should be obvious from the definitions that sums of even (odd) functions are even (odd). The product of two even or two odd functions is even. However, the product of an even and an odd function is odd; for suppose that $f(t)$ is even and $g(t)$ is odd and $h = fg$. Then

$$h(-t) = g(-t)f(-t) = -g(t)f(t) = -h(t). \qquad (7.3.9)$$

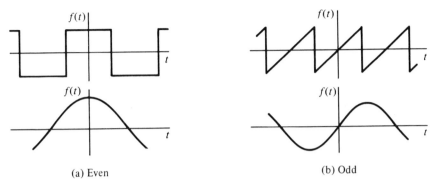

(a) Even (b) Odd

FIGURE 7.2.   Some even and odd functions.

The relationship of Eqs. 7.3.7 and 7.3.8 to the computations of the Fourier coefficients arises from the next formulas. Again, $f(t)$ is even and $g(t)$ is odd. Then

$$\int_{-T}^{T} f(t) \, dt = 2 \int_{0}^{T} f(t) \, dt \qquad (7.3.10)$$

and

$$\int_{-T}^{T} g(t) \, dt = 0. \qquad (7.3.11)$$

The proof is easy. We have

$$\int_{-T}^{T} f(t) \, dt = \int_{-T}^{0} f(t) \, dt + \int_{0}^{T} f(t) \, dt$$

$$= -\int_{T}^{0} f(-s) \, ds + \int_{0}^{T} f(t) \, dt \qquad (7.3.12)$$

by the change of variables $-s = t$, $-ds = dt$. Hence,

$$\int_{-T}^{T} f(t) \, dt = \int_{0}^{T} f(-s) \, ds + \int_{0}^{T} f(t) \, dt$$

$$= \int_{0}^{T} f(s) \, ds + \int_{0}^{T} f(t) \, dt \qquad (7.3.13)$$

since $f(t)$ is even. These last two integrals are the same because s and t are dummy variables. Similarly,

$$\int_{-T}^{T} g(t) \, dt = \int_{0}^{T} g(-s) \, ds + \int_{0}^{T} g(t) \, dt$$

$$= -\int_{0}^{T} g(s) \, ds + \int_{0}^{T} g(t) \, dt = 0 \qquad (7.3.14)$$

because $g(-s) = -g(s)$.

We leave it to the reader to verify:

1. An even function is continuous at $t = 0$, redefining $f(0)$ by Eq. 7.2.1, if necessary.
2. The value (average value, if necessary) at the origin of an odd function is zero.
3. The derivative of an even (odd) function is odd (even).

In view of the above, particularly Eqs. 7.3.10 and 7.3.11, it is easy to see that if $f(t)$ is an even function,

$$f(t) \sim \frac{a_0}{2} + \sum_{n=1}^{\infty} a_n \cos \frac{n\pi t}{T} \qquad (7.3.15)$$

where

$$a_0 = \frac{2}{T} \int_0^T f(t)\, dt, \qquad a_n = \frac{2}{T} \int_0^T f(t) \cos \frac{n\pi t}{T}\, dt. \qquad (7.3.16)$$

If $f(t)$ is an odd function,

$$f(t) \sim \sum_{n=1}^{\infty} b_n \sin \frac{n\pi t}{T} \qquad (7.3.17)$$

where

$$b_n = \frac{2}{T} \int_0^T f(t) \sin \frac{n\pi t}{T}\, dt. \qquad (7.3.18)$$

From the point of view of a physical system, the periodic input function sketched in Fig. 7.3 is neither even or odd. A function may be even or odd depending on where the vertical axis, $t = 0$, is drawn. In Fig. 7.4 we can clearly see the impact of the placement of $t = 0$; it generates an even function $f_1(t)$ in (a), an odd function $f_2(t)$ in (b), and $f_3(t)$ in (c) which is neither even nor odd. The next example illustrates how this observation may be exploited.

FIGURE 7.3.   A periodic input.

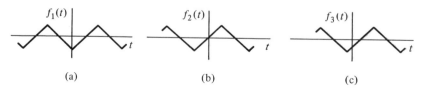

(a)   (b)   (c)

FIGURE 7.4.   An input expressed as various functions.

Example 7.3.7:   A periodic forcing function acts on a spring–mass system as shown. Find a sine-series representation by considering the function to be odd, and a cosine-series representation by considering the function to be even.

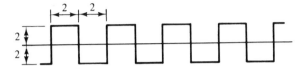

SOLUTION: If the $t = 0$ location is selected as shown, the resulting odd function can be written, for one period, as

$$f_1(t) = \begin{cases} -2, & -2 < t < 0 \\ 2, & 0 < t < 2. \end{cases}$$

For an odd function we know that

$$a_n = 0.$$

Hence, we are left with the task of finding b_n. We have, using $T = 2$,

$$b_n = \frac{2}{T} \int_0^T f_1(t) \sin \frac{n\pi t}{T} \, dt, \quad n = 1, 2, 3, \ldots$$

$$= \frac{2}{2} \int_0^2 2 \sin \frac{n\pi t}{2} \, dt = -\frac{4}{n\pi} \cos \frac{n\pi t}{2} \Big|_0^2 = -\frac{4}{n\pi} (\cos n\pi - 1).$$

The Fourier sine series is, then, again substituting $\cos n\pi = (-1)^n$,

$$f_1(t) = \sum_{n=1}^{\infty} \frac{4[1 - (-1)^n]}{n\pi} \sin \frac{n\pi t}{2}$$

$$= \frac{8}{\pi} \sin \frac{\pi t}{2} - \frac{8}{3\pi} \sin \frac{3\pi t}{2} + \frac{8}{5\pi} \sin \frac{5\pi t}{2} - \cdots.$$

If we select the $t = 0$ location as displayed, an even function results. Over one period it is

$$f_2(t) = \begin{cases} -2, & -2 < t < -1 \\ 2, & -1 < t < 1 \\ -2, & 1 < t < 2. \end{cases}$$

For an even function we know that

$$b_n = 0.$$

The coefficients a_n are found from

$$a_n = \frac{2}{T} \int_0^T f_2(t) \cos \frac{n\pi t}{T} dt; \qquad n = 1, 2, 3, \ldots.$$

$$= \frac{2}{2} \left[\int_0^1 2 \cos \frac{n\pi t}{2} dt + \int_1^2 (-2) \cos \frac{n\pi t}{2} dt \right]$$

$$= \frac{4}{n\pi} \sin \frac{n\pi t}{2} \Big|_0^1 - \frac{4}{n\pi} \sin \frac{n\pi t}{2} \Big|_1^2 = \frac{8}{n\pi} \sin \frac{n\pi}{2}.$$

The result for $n = 0$ is found from

$$a_0 = \frac{2}{T} \int_0^T f_2(t) dt$$

$$= \frac{2}{2} \left[\int_0^1 2 dt + \int_1^2 (-2) dt \right] = 2 - 2 = 0.$$

Finally, the Fourier cosine series is

$$f_2(t) = \sum_{n=1}^{\infty} \frac{8}{n\pi} \sin \frac{n\pi}{2} \cos \frac{n\pi t}{2}$$

$$= \frac{8}{\pi} \cos \frac{\pi t}{2} - \frac{8}{3\pi} \cos \frac{3\pi t}{2} + \frac{8}{5\pi} \cos \frac{5\pi t}{2} + \cdots . \quad \blacksquare$$

We can take a somewhat different view of the problem in the preceding example. The relationship between $f_1(t)$ and $f_2(t)$ is

$$f_1(t + 1) = f_2(t). \tag{7.3.19}$$

Hence, the odd expansion in Example 7.3.7 is just a "shifted" version of the even expansion. Indeed,

$$f_1(t + 1) = f_2(t) = \sum_{n=1}^{\infty} 4 \frac{[1 - (-1)^n]}{n\pi} \sin \frac{n\pi(t + 1)}{2}$$

$$= \sum_{n=1}^{\infty} 4 \frac{[1 - (-1)^n]}{n\pi} \left(\sin \frac{n\pi}{2} \cos \frac{n\pi t}{2} + \cos \frac{n\pi}{2} \sin \frac{n\pi t}{2} \right)$$

$$= \frac{8}{\pi} \sum_{n=1}^{\infty} \frac{(-1)^{n-1}}{2n - 1} \cos \frac{2n - 1}{2} \pi t,$$

which is an even expansion, equivalent to that above.

PROBLEMS

1. In Problems 1 to 8 of Section 7.3.2, (a) which of the functions are even, (b) which of the functions are odd, (c) which of the functions could be made even by shifting the vertical axis, and (d) which of the functions could be made odd by shifting the vertical axis?

Expand each periodic function in a Fourier sine series and a Fourier cosine series.

2. $f(t) = 4t, \quad 0 < t < \pi$

3. $f(t) = \begin{cases} 10, & 0 < t < \pi \\ 0, & \pi < t < 2\pi \end{cases}$

4. $f(t) = \sin t, \quad 0 < t < \pi$

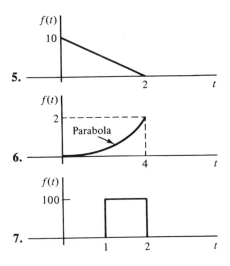

5.

6.

7.

8. Show that the periodic extension of an even function must be continuous at $t = 0$.

9. Show that the period extension of an odd function is zero at $t = 0$.

10. Use the definition of derivative to explain why the derivative of an odd (even) function is even (odd).

7.3.4   HALF-RANGE EXPANSIONS

In modeling some physical phenomena it is necessary that we consider the values of a function only in the interval 0 to T. This is especially true when considering partial differential equations, as we shall do in Chapter 8. There is no condition of periodicity on the function, since there is no interest in the function outside the interval 0 to T. Consequently, we can extend the function arbitrarily to include the interval $-T$ to 0. Consider the function $f(t)$ shown in Fig. 7.5. If we extend it as in part (b), an even function results; an extension as in part (c) results in an odd function. Since these functions are defined differently in $(-T, 0)$ we denote them with different subscripts: f_e for an even extension, f_o for an odd extension. Note that the Fourier series for $f_e(t)$ contains only cosine terms and contains only sine terms for $f_o(t)$. Both series converge to $f(t)$ in $0 < t < T$. Such series expansions are known as *half-range expansions*. An example will illustrate such expansions.

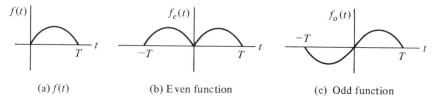

(a) $f(t)$ (b) Even function   (c) Odd function

FIGURE 7.5   Extension of a function.

Example 7.3.8:   A function $f(t)$ is defined only over the range $0 < t < 4$ as

$$f(t) = \begin{cases} t, & 0 < t < 2 \\ 4 - t, & 2 < t < 4. \end{cases}$$

Find the half-range cosine and sine expansions of $f(t)$.

SOLUTION: A half-range cosine expansion is found by forming a symmetric extension $f(t)$. The b_n of the Fourier series is zero. The coefficients a_n are

$$a_n = \frac{2}{T} \int_0^T f(t) \cos \frac{n\pi t}{T} \, dt, \qquad n = 1, 2, 3, \ldots$$

$$= \frac{2}{4} \int_0^2 t \cos \frac{n\pi t}{4} \, dt + \frac{2}{4} \int_2^4 (4 - t) \cos \frac{n\pi t}{4} \, dt$$

$$= \frac{1}{2} \left[\frac{4t}{n\pi} \sin \frac{n\pi t}{4} + \frac{16}{\pi^2 n^2} \cos \frac{n\pi t}{4} \right]_0^2 + \frac{1}{2} \left[\frac{16}{n\pi} \sin \frac{n\pi t}{4} \right]_2^4$$

$$\qquad\qquad\qquad\qquad - \frac{1}{2} \left[\frac{4t}{n\pi} \sin \frac{n\pi t}{4} + \frac{16}{n^2 \pi^2} \cos \frac{n\pi t}{4} \right]_2^4$$

$$= -\frac{8}{n^2 \pi^2} \left[1 + \cos n\pi - 2 \cos \frac{n\pi}{2} \right].$$

For $n = 0$ the coefficient a_0 is

$$a_0 = \tfrac{1}{2} \int_0^2 t \, dt + \tfrac{1}{2} \int_2^4 (4 - t) \, dt = 2.$$

The half-range cosine expansion is then

$$f(t) = 1 + \sum_{n=1}^{\infty} \frac{8}{n^2 \pi^2} \left(2 \cos \frac{n\pi}{2} - \cos n\pi - 1 \right) \cos \frac{n\pi t}{4}$$

$$= 1 - \frac{8}{\pi^2} \left[\cos \frac{\pi t}{2} + \frac{1}{9} \cos \frac{3\pi t}{2} + \cdots \right], \qquad 0 < t < 4.$$

It is an even periodic extension that graphs as follows:

SOLUTION: Clearly, $f_1(t) + f_2(t) = \tilde{f}_e(t)$ as displayed below, where, as usual, $\tilde{f}_e(t)$ represents the even extension of $f(t) = \sin t$, $0 < t < \pi$. But

$$f_1(t + \pi) = f_2(t)$$

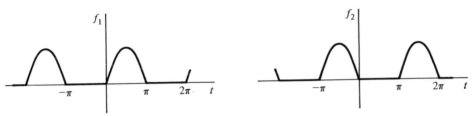

and, from Example 7.3.6,

$$f_1(t) = \frac{1}{\pi} + \frac{1}{2} \sin t - \frac{2}{\pi} \sum_{n=1}^{\infty} \frac{\cos 2nt}{4n^2 - 1}.$$

Therefore,

$$f_2(t) = f_1(t + \pi) = \frac{1}{\pi} + \frac{1}{2} \sin (t + \pi) - \frac{2}{\pi} \sum_{n=1}^{\infty} \frac{\cos 2n(t + \pi)}{4n^2 - 1}.$$

Since $\sin (t + \pi) = -\sin t$ and $\cos [2n(t + \pi)] = \cos 2nt$, we have

$$f_2(t) = \frac{1}{\pi} - \frac{1}{2} \sin t - \frac{2}{\pi} \sum_{n=1}^{\infty} \frac{\cos 2nt}{4n^2 - 1}.$$

Finally, without a single integration, there results

$$\tilde{f}_e(t) = f_1(t) + f_2(t)$$

$$= \frac{2}{\pi} - \frac{4}{\pi} \sum_{n=1}^{\infty} \frac{\cos 2nt}{4n^2 - 1}. \quad \blacksquare$$

It is also useful to derive the effects of a change of scale in t. For instance, if

$$f(t) \sim \frac{a_0}{2} + \sum_{n=1}^{\infty} \left(a_n \cos \frac{n\pi t}{T} + b_n \sin \frac{n\pi t}{T} \right), \qquad (7.3.24)$$

then the period of the series is $2T$. Let

$$t = \frac{T}{\tau} \hat{t}. \qquad (7.3.35)$$

Then

$$\hat{f}(\hat{t}) = f\left(\frac{T}{\tau} \hat{t}\right) \sim \frac{a_0}{2} + \sum_{n=1}^{\infty} \left(a_n \cos \frac{n\pi \hat{t}}{\tau} + b_n \sin \frac{n\pi \hat{t}}{\tau} \right) \qquad (7.3.26)$$

is the series representing $\hat{f}(\hat{t})$ with period 2τ. The changes $\tau = 1$ and $\tau = \pi$ are most common and lead to expansions with period 2 and 2π, respectively.

Example 7.3.10:   Find the Fourier series expansion of the even periodic extension
of

$$g(t) = \begin{cases} t, & 0 \le t < 1 \\ 2 - t, & 1 \le t < 2. \end{cases}$$

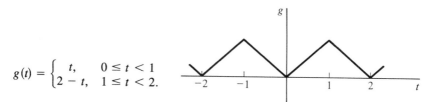

SOLUTION: This periodic input resembles the input in Example 7.3.8. Here the
period is 4; in Example 7.3.8 it is 8. This suggests the scale change $2\hat{t} = t$. So if

$$f(t) = \begin{cases} t, & 0 \le t < 2 \\ 4 - t, & 2 \le t < 4 \end{cases}$$

$$\hat{f}(\hat{t}) = f(2\hat{t}) = \begin{cases} 2\hat{t}, & 0 \le 2\hat{t} < 2 \\ 4 - 2\hat{t}, & 2 \le 2\hat{t} < 4. \end{cases}$$

Note that $g(\hat{t}) = \hat{f}(\hat{t})/2$. So

$$g(\hat{t}) = \begin{cases} \hat{t}, & 0 \le \hat{t} < 1 \\ 2 - \hat{t}, & 1 \le \hat{t} < 2. \end{cases}$$

But from $g(\hat{t}) = \hat{f}(\hat{t})/2$ we have (see Example 7.3.8)

$$g(\hat{t}) = \frac{1}{2}\left[1 - \frac{8}{\pi^2}\left(\cos \pi\hat{t} + \frac{1}{9}\cos 3\pi\hat{t} + \cdots\right)\right].$$

Replacing $\hat{t}$ by t yields

$$g(t) = \frac{1}{2} - \frac{4}{\pi^2}\left(\cos \pi t + \frac{1}{9}\cos 3\pi t + \cdots\right), \qquad 0 \le t < 2. \quad \blacksquare$$

PROBLEMS

1. Let

$$f(t) = \begin{cases} 0, & -\pi < t < 0 \\ f_1(t), & 0 < t < \pi \end{cases}$$

have the expansion

$$f(t) = \frac{a_0}{2} + \sum_{n=1}^{\infty} a_n \cos nt + b_n \sin nt.$$

(a) Prove that

$$f(-t) = \begin{cases} f_1(-t), & -\pi < t < 0 \\ 0, & 0 < t < \pi \end{cases}$$

and, by use of formulas for the Fourier coefficients, that

$$f(-t) = \frac{a_0}{2} + \sum_{n=1}^{\infty} a_n \cos nt - b_n \sin nt, \qquad \pi < t < \pi.$$

(b) Verify that

$$f_e(t) = a_o + 2 \sum_{n=1}^{\infty} a_n \cos nt, \qquad -\pi < t < \pi,$$

where $f_e(t)$ is the even extension of $f_1(t)$, $0 < t < \pi$.

2. Use the results of Problem 1 and the expansion of

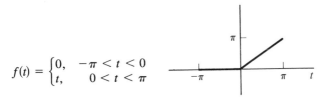

$$f(t) = \begin{cases} 0, & -\pi < t < 0 \\ t, & 0 < t < \pi \end{cases}$$

which is

$$\frac{\pi}{4} + \sum_{n=1}^{\infty} \frac{(-1)^n - 1}{\pi n^2} \cos nt - \frac{(-1)^2}{n} \sin nt$$

to obtain the expansion of

$$f(t) = |t|, \qquad -\pi < t < \pi.$$

3. Use the result in Problems 1 and 2 and the methods of this section to find the Fourier expansion of

$$f(t) = \begin{cases} t + 1, & -1 < t < 0 \\ -t + 1, & 0 < t < 1. \end{cases}$$

4. The Fourier expansion of

$$\hat{f}(t) = \begin{cases} -1, & -\pi < t < 0 \\ 1, & 0 < t < \pi \end{cases}$$

is

$$\frac{4}{\pi} \sum_{n=1}^{\infty} \frac{\sin (zn - 1)t}{2n - 1}.$$

Use this result to obtain the following expansion:

$$f(t) = \begin{cases} 0, & -\pi < t < 0 \\ 1, & 0 < t < \pi \end{cases}$$

by observing that $f(t) = [1 + \hat{f}(t)]/2$.

5. Use the information given in Problem 4 and find the expansion of

$$f(t) = \begin{cases} -1, & -\pi < t < 0 \\ 0, & 0 < t < \pi. \end{cases}$$

6. If $f(t)$ is constructed as in Problem 1, describe the function $f(t) - f(-t)$.

7. Use Problems 2 and 6 to derive

$$t = 2 \sum_{n=1}^{\infty} \frac{(-1)^{n-1}}{n} \sin nt, \qquad -\pi < t < \pi.$$

7.4   Forced Oscillations

We shall now consider an important application involving an external force acting on a spring-mass system. The differential equation describing this motion is

$$M\frac{d^2 y}{dt^2} + C\frac{dy}{dt} + Ky = F(t). \tag{7.4.1}$$

If the input function $F(t)$ is a sine or cosine function, the steady-state solution is a harmonic motion having the frequency of the input function. We will now see that if $F(t)$ is periodic with frequency ω but is not a sine or cosine function, then the steady-state solution to Eq. 7.4.1 will contain the input frequency ω and multiples of this frequency contained in the terms of a Fourier series expansion of $F(t)$. If one of these higher frequencies is close to the natural frequency of an underdamped system, then the particular term containing that frequency may play the dominant role in the system response. This is somewhat surprising, since the input frequency may be considerably lower than the natural frequency of the system; yet that input could lead to serious problems if it is not purely sinusoidal. This will be illustrated with an example.

Example 7.4.1:   Consider the force $F(t)$ acting on the spring-mass system shown at the top of page 393. Determine the steady-state response to this forcing function.

SOLUTION: The coefficients in the Fourier series expansion of an odd forcing function $F(t)$ are (see Example 7.3.7)

$$a_n = 0$$

$$b_n = \frac{2}{1} \int_0^1 100 \sin \frac{n\pi t}{1} \, dt = -\frac{200}{n\pi} \cos n\pi t \Big|_0^1 = -\frac{200}{n\pi}(\cos n\pi - 1),$$

$$n = 1, 2, , \ldots .$$

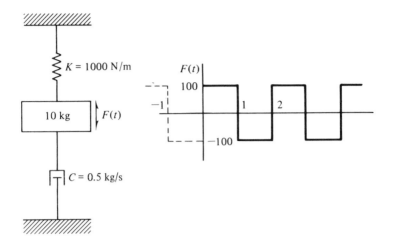

The Fourier series representation of $F(t)$ is then

$$F(t) = \sum_{n=1}^{\infty} \frac{200}{n\pi}(1 - \cos n\pi) \sin n\pi t$$

$$= \frac{400}{\pi} \sin \pi t - \frac{400}{3\pi} \sin 3\pi t + \frac{80}{\pi} \sin 5\pi t - \cdots.$$

The differential equation can then be written

$$10\frac{d^2 y}{dt^2} + 0.5\frac{dy}{dt} + 1000y = \frac{400}{\pi} \sin \pi t - \frac{400}{3\pi} \sin 3\pi t + \frac{80}{\pi} \sin 5\pi t - \cdots.$$

Because the differential equation is linear, we can first find the particular solution $(y_p)_1$ corresponding to the first term on the right, then $(y_p)_2$ corresponding to the second term, and so on. Finally, the steady-state solution is

$$y_p(t) = (y_p)_1 + (y_p)_2 + \cdots.$$

Doing this for the three terms shown, using the methods developed earlier, we have

$$(y_p)_1 = 0.141 \sin \pi t - 2.5 \times 10^{-4} \cos \pi t$$

$$(y_p)_2 = -0.376 \sin 3\pi t + 1.56 \times 10^{-3} \cos 3\pi t$$

$$(y_p)_3 = -0.0174 \sin 5\pi t - 9.35 \times 10^{-5} \cos 5\pi t.$$

Actually, rather than solving the problem each time for each term, we could have found a $(y_p)_n$ corresponding to the term $[-(200/n\pi)(\cos n\pi - 1) \sin n\pi t]$ as a general function of n. Note the amplitude of the sine term in $(y_p)_2$. It obviously dominates the solution, as displayed in a sketch of $y_p(t)$:

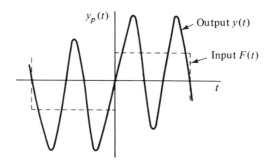

Yet $(y_p)_2$ has an annular frequency of 3π rad/s, whereas the frequency of the input function was π rad/s. This happened because the natural frequency of the undamped system was 10 rad/s, very close to the frequency of the second sine term in the Fourier series expansion. Hence, it is this overtone that resonates with the system, and not the fundamental. Overtones may dominate the steady-state response for any under-damped system that is forced with a periodic function having a frequency smaller than the natural frequency of the system. ■

PROBLEMS

Find the steady-state solution to Eq. 7.4.1 for each of the following.

1. $M = 2,$ $C = 0,$ $K = 8,$ $F(t) = \sin 4t$

2. $M = 2,$ $C = 0,,$ $K = 2,$ $F(t) = \cos 2t$

3. $M = 1,$ $C = 0,$ $K = 16,$ $F(t) = \sin t + \cos 2t$

4. $M = 1,$ $C = 0,$ $K = 25,$ $F(t) = \cos 2t + \frac{1}{10} \sin 4t$

5. $M = 4,$ $C = 0,$ $K = 36,$ $F(t) = \sum\limits_{n=1}^{N} a_n \cos nt$

6. $M = 4,$ $C = 4,$ $K = 36,$ $F(t) = \sin 2t$

7. $M = 1,$ $C = 2,$ $K = 4,$ $F(t) = \cos t$

8. $M = 1,$ $C = 12,$ $K = 16,$ $F(t) = \sum\limits_{n=1}^{N} b_n \sin nt$

9. $M = 2,$ $C = 2,$ $K = 8,$ $F(t) = \sin t + \frac{1}{10} \cos 2t$

10. $M = 2,$ $C = 16,$ $K = 32,$ $F(t) = \begin{cases} t & -\pi/2 < t < \pi/2 \\ \pi - t & \pi/2 < t < 3\pi/2 \end{cases}$ and

$$F(t + 2\pi) = F(t).$$

11. What is the steady-state response of the mass to the forcing function shown?

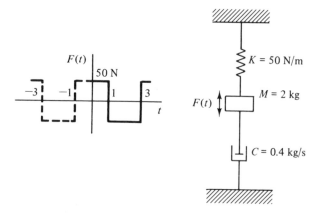

12. Determine the steady-state current in the circuit shown.

7.5   *Miscellaneous Expansion Techniques*

7.5.1   INTEGRATION

Term-by-term integration of a Fourier series is a valuable method for generating new expansions. This technique is valid under surprisingly weak conditions, due in part to the "smoothing" effect of integration.

Theorem 7.2:   *Suppose that $f(t)$ is sectionally continuous in $-\pi < t < \pi$ and is periodic with period 2π. Let $f(t)$ have the expansion*

$$f(t) \sim \sum_{n=1}^{\infty} (a_n \cos nt + b_n \sin nt). \qquad (7.5.1)$$

Then

$$\int_0^t f(s)\, ds = \sum_{n=1}^{\infty} \frac{b_n}{n} + \sum_{n=1}^{\infty} \left(-\frac{b_n}{n} \cos nt + \frac{a_n}{n} \sin nt \right). \qquad (7.5.2)$$

PROOF: Set

$$F(t) = \int_0^t f(s)\, ds \qquad (7.5.3)$$

and verify $F(t + 2\pi) = F(t)$ as follows:

$$F(t + 2\pi) = \int_0^{t+2\pi} f(s)\ ds$$

$$= \int_0^t f(s)\ ds + \int_t^{t+2\pi} f(s)\ ds. \qquad (7.5.4)$$

But $f(t)$ is periodic with period 2π, so that

$$\int_t^{t+2\pi} f(s)\ ds = \int_{-\pi}^{\pi} f(s)\ ds = 0 \qquad (7.5.5)$$

since $1/\pi \int_{-\pi}^{\pi} f(s)\ ds = a_0$, which is zero from Eq. 7.5.1. Therefore, Eq. 7.5.4 becomes $F(t + 2\pi) = F(t)$. The integral of a sectionally continuous function is continuous from Eq. 7.5.3 and $F'(t) = f(t)$ from this same equation. Hence, $F'(t)$ is sectionally continuous. By the Fourier theorem (Theorem 7.1) we have

$$F(t) = \frac{A_0}{2} + \sum_{n=1}^{\infty} (A_n \cos nt + B_n \sin nt) \qquad (7.5.6)$$

valid for all t. Here

$$A_n = \frac{1}{\pi} \int_{-\pi}^{\pi} F(t) \cos nt\ dt, \qquad B_n = \frac{1}{\pi} \int_{-\pi}^{\pi} F(t) \sin nt\ dt. \quad (7.5.7)$$

The formulas 7.5.7 are amenable to an integration by parts. There results

$$A_n = \frac{1}{\pi} \int_{-\pi}^{\pi} F(t) \cos nt\ dt$$

$$= \frac{1}{\pi} F(t) \frac{\sin nt}{n} \Big|_{-\pi}^{\pi} - \frac{1}{\pi} \int_{-\pi}^{\pi} f(t) \frac{\sin nt}{n}\ dt$$

$$= -\frac{b_n}{n}, \qquad n = 1, 2, \ldots \qquad (7.5.8)$$

Similarly,

$$B_n = \frac{1}{\pi} \int_{-\pi}^{\pi} F(t) \sin nt\ dt$$

$$= \frac{1}{\pi} F(t) \left(-\frac{\cos nt}{n} \right) \Big|_{-\pi}^{\pi} + \frac{1}{\pi} \int_{-\pi}^{\pi} f(t) \frac{\cos nt}{n}\ dt$$

$$= \frac{a_n}{n}, \qquad n = 1, 2, \ldots \qquad (7.5.9)$$

because $F(\pi) = F(-\pi + 2\pi) = F(-\pi)$ and $\cos ns = \cos(-ns)$ so that the integrated term is zero. When these values are substituted in Eq. 7.5.6, we obtain

$$F(t) = \frac{A_0}{2} + \sum_{n=1}^{\infty}\left(-\frac{b_n}{n}\cos nt + \frac{a_n}{n}\sin nt\right). \qquad (7.5.10)$$

Now set $t = 0$ to obtain an expression for A_0:

$$F(0) = \int_0^0 f(t)\,dt = 0 = \frac{A_0}{2} - \sum_{n=1}^{\infty}\frac{b_n}{n}, \qquad (7.5.11)$$

so that

$$\frac{A_0}{2} = \sum_{n=1}^{\infty}\frac{b_n}{n}. \qquad (7.5.12)$$

Hence, Eq. 7.5.2 is established.

It is very important to note that Eq. 7.5.2 is just the term-by-term integration of relation 7.5.1; one need not memorize Fourier coefficient formulas in Eq. 7.5.2.

Example 7.5.1: Find the Fourier series expansion of the even periodic extension of $f(t) = t^2$, $-\pi < t < \pi$. Assume the expansion

$$t = 2\sum_{n=1}^{\infty}\frac{(1)^{n-1}}{n}\sin nt.$$

SOLUTION: We obtain the result by integration:

$$\int_0^t s\,ds = 2\sum_{n=1}^{\infty}\frac{(-1)^{n-1}}{n}\int_0^t \sin ns\,ds$$

$$= 2\sum_{n=1}^{\infty}\frac{(-1)^{n-1}}{n^2}(-\cos ns)\Big|_0^t$$

$$= 2\sum_{n=1}^{\infty}\frac{(-1)^{n-1}}{n^2} - 2\sum_{n=1}^{\infty}\frac{(-1)^{n-1}}{n^2}\cos nt.$$

Of course, $\int_0^t s\,ds = t^2/2$, so that

$$\frac{t^2}{2} = 2\sum_{n=1}^{\infty}\frac{(-1)^{n-1}}{n^2} - 2\sum_{n=1}^{\infty}\frac{(-1)^{n-1}}{n^2}\cos nt.$$

The sum $2\sum_{n=1}^{\infty}[(-1)^{n-1}/n^2]$ may be evaluated by recalling that it is $a_0/2$ for the Fourier expansion of $t^2/2$. That is,

$$a_0 = \frac{1}{\pi}\int_{-\pi}^{\pi}\frac{s^2}{2}\,ds = \frac{1}{\pi}\frac{s^3}{6}\Big|_{-\pi}^{\pi}$$

$$= \frac{1}{6\pi}[\pi^3 - (-\pi)^3] = \frac{\pi^2}{3}.$$

Hence,

$$\frac{a_0}{2} = 2 \sum_{n=1}^{\infty} \frac{(-1)^{n-1}}{n^2} = \frac{\pi^2}{6},$$

so

$$\frac{t^2}{2} = \frac{\pi^2}{6} - 2 \sum_{n=1}^{\infty} \frac{(-1)^{n-1}}{n^2} \cos nt. \quad \blacksquare$$

Example 7.5.2:   Find the Fourier expansion of the odd periodic extension of t^3, $-\pi < t < \pi$.

SOLUTION: From the result of Example 7.5.1 we have

$$\frac{t^2}{2} - \frac{\pi^2}{6} = \sum_{n=1}^{\infty} \frac{-2(-1)^{n-1}}{n^2} \cos nt.$$

This is in the form for which Theorem 7.2 is applicable, so

$$\int_0^t \left(\frac{s^2}{2} - \frac{\pi^2}{6} \right) ds = \frac{t^3}{6} - \frac{\pi^2 t}{6}$$

$$= -2 \sum_{n=1}^{\infty} \frac{(-1)^{n-1}}{n^3} \sin nt.$$

Therefore,

$$t^3 = \pi^2 t - 12 \sum_{n=1}^{\infty} \frac{(-1)^{n-1}}{n^3} \sin nt,$$

which is not yet a pure Fourier series because of the $\pi^2 t$ term. We remedy this defect by using the Fourier expansion of t given in Example 7.5.1. We have

$$t^3 = \pi^2 2 \sum_{n=1}^{\infty} \frac{(-1)^{n-1}}{n} \sin nt - 12 \sum_{n=1}^{\infty} \frac{(-1)^{n-1}}{n^3} \sin nt$$

$$= \sum_{n=1}^{\infty} \left(\frac{2\pi^2}{n} - \frac{12}{n^3} \right) (-1)^{n-1} \sin nt. \quad \blacksquare$$

In summary, note these facts:

1. $\sum_{n=1}^{\infty} b_n/n$ converges and is the value $A_0/2$; that is

$$\frac{1}{2\pi} \int_{-\pi}^{\pi} F(s) \, ds = \sum_{n=1}^{\infty} \frac{b_n}{n}. \tag{7.5.13}$$

2. The Fourier series representing $f(t)$ need not converge to $f(t)$, yet the Fourier series representing $F(t)$ converges to $F(t)$ for all t.

3. If

$$f(t) \sim \frac{a_0}{2} + \sum_{n=1}^{\infty} (a_n \cos nt + b_n \sin nt), \tag{7.5.14}$$

we apply the integration to the function $f(t) - a_0/2$ because

$$f(t) - \frac{a_0}{2} \sim \sum_{n=1}^{\infty} (a_n \cos nt + b_n \sin nt). \qquad (7.5.15)$$

PROBLEMS

Use the techniques of this section to obtain the Fourier expansions of the integrals of the following functions.

1. Section 7.2, Problem 1.

2. Section 7.2, Problem 3.

3. Section 7.2, Problem 5.

4. Section 7.2, Problem 6.

5. Section 7.2, Problem 9.

6. Section 7.2, Problem 13.

7. Section 7.2, Problem 14.

8. Example 7.3.5.

9. Example 7.3.6.

10. Section 7.3, Problem 4.

11. Section 7.3, Problem 7.

12. Show that we may derive

$$\frac{\pi^2 x - x^3}{12} = \sum_{n=1}^{\infty} (-1)^{n+1} \frac{\sin nx}{n^3}$$

by integration of

$$\frac{\pi^2 - 3x^2}{12} = \sum_{n=1}^{\infty} (-1)^{n+1} \frac{\cos nx}{n^2}$$

7.5.2   DIFFERENTIATION

Term-by-term differentiation of a Fourier series does not lead to the Fourier series of the differentiated function even when that derivative has a Fourier series, unless suitable restrictive hypotheses are placed on the given function and its derivatives. This is in marked contrast to term-by-term integration and is illustrated quite convincingly by Eqs. 7.1.4 and 7.1.5. The following theorem incorporates sufficient conditions to permit term-by-term differentiation.

Theorem 7.3:   *Suppose that in* $-\pi < t < \pi, f(t)$ *is continuous,* $f'(t)$ *and* $f''(t)$ *are sectionally continuous, and* $f(-\pi) = f(\pi)$. *Then*

$$f(t) = \frac{a_0}{2} + \sum_{n=1}^{\infty} a_n \cos nt + b_n \sin nt \qquad (7.5.16)$$

implies that

$$f'(t) = \frac{d}{dt}\left(\frac{a_0}{2}\right) + \sum_{n=1}^{\infty} \frac{d}{dt}(a_n \cos nt + b_n \sin nt)$$

$$= \sum_{n=1}^{\infty} nb_n \cos nt - na_n \sin nt. \tag{7.5.17}$$

PROOF: We know that df/dt has a convergent Fourier series by Theorem 7.1, in which theorem we use f' for f and f'' for f'. (This is the reason we require f'' to be sectionally continuous.) We express the Fourier coefficients of $f'(t)$ by α_n and β_n so that

$$f'(t) = \frac{\alpha_0}{2} + \sum_{n=1}^{\infty} \alpha_n \cos nt + \beta_n \sin nt, \tag{7.5.18}$$

where, among other things,

$$\alpha_0 = \frac{1}{\pi} \int_{-\pi}^{\pi} f'(s)\, ds$$

$$= \frac{1}{\pi}[f(\pi) - f(-\pi)] = 0, \tag{7.5.19}$$

by hypothesis. By Theorem 7.2, we may integrate Eq. 7.5.18 term by term to obtain

$$\int_0^t f'(s)\, ds = f(t) - f(0)$$

$$= \sum_{n=1}^{\infty} \frac{\beta_n}{n} + \sum_{n=1}^{\infty} -\frac{\beta_n}{n} \cos nt + \frac{\alpha_n}{n} \sin nt. \tag{7.5.20}$$

But Eq. 7.5.16 is the Fourier expansion of $f(t)$ in $-\pi < t < \pi$. Therefore, comparing the coefficients in Eqs. 7.5.16 and 7.5.20, we find

$$a_n = -\frac{\beta_n}{n}, \qquad b_n = \frac{\alpha_n}{n}, \qquad n = 1, 2, \ldots \tag{7.5.21}$$

We obtain the conclusion (Eq. 7.5.17) by substitution of the coefficient relations (Eq. 7.5.21) into Eq. 7.5.18.

Example 7.5.3:   Find the Fourier series of the periodic extension of

$$g(t) = \begin{cases} 0, & -\pi < t < 0 \\ \cos t, & 0 < t < \pi. \end{cases}$$

SOLUTION: The structure of $g(t)$ suggests examining the function

$$f(t) = \begin{cases} 0, & -\pi < t < 0 \\ \sin t, & 0 < t < \pi. \end{cases}$$

In Example 7.3.6 we have shown that

$$\widetilde{f}(t) = \frac{1}{\pi} + \frac{1}{2} \sin t - \frac{2}{\pi} \sum_{n=1}^{\infty} \frac{\cos 2nt}{4n^2 - 1}.$$

Moreover, $f(\pi) = f(-\pi) = 0$ and $f(t)$ is continuous. Also, all the derivatives of $f(t)$ are sectionally continuous. Hence, we may apply Theorem 7.3 to obtain

$$\widetilde{g}(t) = \frac{1}{2} \cos t + \frac{4}{\pi} \sum_{n=1}^{\infty} \frac{n \sin 2nt}{4n^2 - 1},$$

where $\widetilde{g}(t)$ is the periodic extension of $g(t)$. Note, incidentally, that

$$\widetilde{g}(0) = \frac{g(0^+) + g(0^-)}{2} = \frac{1}{2}$$

and this is precisely the value of the Fourier series at $t = 0$.   ∎

PROBLEMS

1. Let $g(t)$ be the function defined in Example 7.5.3. Find $g'(t)$. To what extent does $g'(t)$ resemble

$$f(t) = \begin{cases} \sin t, & 0 \le t < \pi \\ 0, & -\pi \le t < 0? \end{cases}$$

Differentiate the Fourier series expansion for $g(t)$ and explain why it does not resemble the Fourier series for $-f(t)$.

2. Show that in $-\pi < t < \pi$, $t \ne 0$,

$$\frac{d}{dt} |\sin t| = \begin{cases} -\cos t, & -\pi < t < 0 \\ \cos t, & 0 < t < \pi. \end{cases}$$

Sketch $d/dt |\sin t|$ and find its Fourier series. Is Theorem 7.3 applicable?

3. What hypotheses are sufficient to guarantee k-fold term-by-term differentiation of

$$f(t) = \frac{a_0}{2} + \sum_{n=1}^{\infty} a_n \cos nt + b_n \sin nt?$$

7.5.3   FOURIER SERIES FROM POWER SERIES*

Consider the function $\ln (1 + z)$. We know that

---

*The material in this section requires some knowledge of the theory of the functions of a complex variable, a topic we explore in Chapter 10.

$$\ln (1 + z) = z - \frac{z^2}{2} + \frac{z^3}{3} - \cdots \tag{7.5.22}$$

is valid for all z, $|z| \leq 1$ except $z = -1$. On the unit circle $|z| = 1$ we may write $z = e^{i\theta}$ and hence,

$$\ln (1 + e^{i\theta}) = e^{i\theta} - \tfrac{1}{2}e^{2i\theta} + \tfrac{1}{3}e^{3i\theta} - \cdots \tag{7.5.23}$$

except for $z = -1$, which corresponds to $\theta = \pi$. Now

$$e^{i\theta} = \cos \theta + i \sin \theta, \tag{7.5.24}$$

so that $e^{in\theta} = \cos n\theta + i \sin n\theta$ and

$$1 + e^{i\theta} = 1 + \cos \theta + i \sin \theta$$

$$= 2\left(\cos^2 \frac{\theta}{2} + i \sin \frac{\theta}{2} \cos \frac{\theta}{2}\right)$$

$$= 2 \cos \frac{\theta}{2}\left(\cos \frac{\theta}{2} + i \sin \frac{\theta}{2}\right) = 2 \cos \frac{\theta}{2} e^{i\theta/2}.$$

$$\tag{7.5.25}$$

Now

$$\ln u = \ln |u| + i \arg u, \tag{7.5.26}$$

so that

$$\ln (1 + e^{i\theta}) = \ln \left| 2 \cos \frac{\theta}{2} \right| + i\frac{\theta}{2}, \tag{7.5.27}$$

which follows by taking logarithms of Eq. 7.5.25. Thus, from Eqs. 7.5.23, 7.5.24, and 7.5.27, we have

$$\ln \left| 2 \cos \frac{\theta}{2} \right| + i\frac{\theta}{2} = \cos \theta - \frac{1}{2} \cos 2\theta + \cdots$$

$$+ i\left(\sin \theta - \frac{1}{2} \sin 2\theta + \cdots\right)$$

$$\tag{7.5.28}$$

and therefore, changing θ to t,

$$\ln \left| 2 \cos \frac{t}{2} \right| = \cos t - \frac{1}{2} \cos 2t + \frac{1}{3} \cos 3t + \cdots \tag{7.5.29}$$

$$\frac{t}{2} = \sin t - \frac{1}{2} \sin 2t + \frac{1}{3} \sin 3t + \cdots \tag{7.5.30}$$

Both expansions are convergent in $-\pi < t < \pi$ to their respective functions. In this interval $|2 \cos t/2| = 2 \cos t/2$ but $\ln (2 \cos t/2)$ is not sectionally continuous. Recall that our Fourier theorem is a sufficient condition for convergence. Equation 7.5.29 shows that it is certainly not a necessary one.

An interesting variation on Eq. 7.5.29 arises from the substitution $t = x - \pi$. Then

$$\ln \left(2 \cos \frac{x - \pi}{2}\right) = \ln \left(2 \sin \frac{x}{2}\right)$$

$$= \sum_{n=1}^{\infty} \frac{(-1)^{n-1}}{n} \cos n(x - \pi)$$

$$= \sum_{n=1}^{\infty} \frac{(-1)^{n-1}(-1)^n}{n} \cos nx. \qquad (7.5.31)$$

Therefore, replacing x with t,

$$-\ln \left(2 \sin \frac{t}{2}\right) = \sum_{n=1}^{\infty} \frac{1}{n} \cos nt, \qquad (7.5.32)$$

which is valid* in $0 < t < 2\pi$. Adding the functions and their representations in Eqs. 7.5.29 and 7.5.32 yields

$$-\ln \tan \frac{t}{2} = 2 \sum_{n=1}^{\infty} \frac{1}{2n - 1} \cos (2n - 1)t. \qquad (7.5.33)$$

Another example arises from consideration of

$$\frac{a}{a - z} = \frac{1}{1 - z/a}$$

$$= 1 + \frac{z}{a} + \frac{z^2}{a^2} + \cdots$$

$$= 1 + \frac{\cos \theta}{a} + \frac{\cos 2\theta}{a^2} + \cdots + i\left(\frac{\sin \theta}{a} + \frac{\sin 2\theta}{a^2} + \cdots\right).$$

$$(7.5.34)$$

But

$$\frac{a}{a - e^{i\theta}} = \frac{a}{a - \cos \theta - i \sin \theta}$$

$$= a \frac{(a - \cos \theta) + i \sin \theta}{(a - \cos \theta)^2 + \sin^2 \theta}$$

$$= a \frac{a - \cos \theta + i \sin \theta}{a^2 - 2a \cos \theta + 1}. \qquad (7.5.35)$$

---

*Since $-\pi < t < \pi$ becomes $-\pi < x - \pi < \pi$, we have $0 < x < 2\pi$.

Separating real and imaginary parts and using Eq. 7.5.34 results in the two expansions

$$a\frac{a - \cos t}{a^2 - 2a \cos t + 1} = \sum_{n=0}^{\infty} a^{-n} \cos nt \qquad (7.5.36)$$

$$\frac{a \sin t}{a^2 - 2a \cos t + 1} = \sum_{n=1}^{\infty} a^{-n} \sin nt. \qquad (7.5.37)$$

The expansion are valid for all t, assuming that $a > 1$.

PROBLEMS

1. Explain why $\ln |2 \cos t/2|$ and $\ln (\tan t/2)$ in $-\pi < t < \pi$ or in $0 < t < \pi$ are not sectionally continuous.

In each problem use ideas of this section to construct $f(t)$ for the given series.

2. $1 + \sum_{n=1}^{\infty} \dfrac{\cos nt}{n!}$

3. $\sum_{n=1}^{\infty} (-1)^{n+1} \dfrac{\sin 2nt}{(2n)!}$

4. $1 + \sum_{n=1}^{\infty} (-1)^n \dfrac{\cos (2n + 1)t}{(2n + 1)!}$

5. $1 + \sum_{n=1}^{\infty} \dfrac{\cos 2nt}{(2n)!}$

6. Use Eq. 7.5.36 to find the Fourier series expansion of

$$f(t) = \frac{1}{a^2 - 2a \cos t + 1}.$$

Hint: Subtract $\frac{1}{2}$ from both sides of Eq. 7.5.36.

7. Equations 7.5.36 and 7.5.37 are valid for $a > 1$. Find $f(t)$ given

(a) $\sum_{n=1}^{\infty} b^n \cos nt, \quad b < 1$

(b) $\sum_{n=1}^{\infty} b^n \sin nt, \quad b < 1.$

What Fourier series expansions arise from considerations of the power series of each function?

8. $\dfrac{a}{(a - z)^2}, \quad a < 1$

9. $\dfrac{a^2}{a^2 - z^2}, \quad a < 1$

10. e^{-z}

11. $\sin z$

12. $\cosh z$

13. $\tan^{-1} z$

8

Partial Differential Equations

8.1   Introduction

The physical systems studied thus far have been described primarily by ordinary differential equations. We are now interested in studying phenomena that require partial derivatives in the describing equations. Partial differential equations arise when the dependent variable is a function of two or more independent variables. The assumption of lumped parameters in a physical problem usually leads to ordinary differential equations, whereas the assumption of a continuously distributed quantity, a field, generally leads to a partial differential equation. A field approach is quite common now in such undergraduate courses as deformable solids, electromagnetics, and fluid mechanics; hence, the study of partial differential equations is often included in undergraduate programs. Many applications (fluid flow, heat transfer, wave motion) involve second-order equations; for this reason we place great emphasis on such equations.

The order of the highest derivative is again the order of the equation. The questions of linearity and homogeneity are answered as before in ordinary differential equations. Solutions are superposable as long as the equation is linear and homogeneous. In general, the number of solutions of a partial differential equation is very large. The unique solution corresponding to a particular physical problem is obtained by use of additional information arising from the physical situation. If this information is given on the boundary as *boundary conditions,* a *boundary-value problem* results. If the information is given at one instant as *initial conditions,* an *initial-value problem* results. A *well-posed problem* has just the right number of these conditions specified to determine a solution. We shall not delve into the mathematical theory of formulating a well-posed problem. We shall, instead, rely on our physical understanding to determine problems that are well posed. We caution the reader that:

1. A problem that has too many boundary and/or initial conditions specified is not well posed and is an overspecified problem.
2. A problem that has too few boundary and/or initial conditions does not possess a unique solution.

In general, a partial differential equation with independent variables x and t which is second order in each of the variables requires two conditions (this could be dependent on time t) at some x location (or x locations) and two conditions at some time t, usually $t = 0$.

We present a mathematical tool by way of physical motivation. We shall derive the describing equations of some common phenomena to illustrate the modeling process; other phenomena could have been chosen such as those encountered in magnetic fields, elasticity, fluid flows, aerodynamics, diffusion of pollutants, and so on. An analytical solution technique will be reviewed in this chapter. In the next chapter numerical methods will be reviewed so that approximate solutions may be obtained to problems that cannot be solved analytically.

We shall be particularly concerned with second-order partial differential equations involving two independent variables, because of the frequency with which they appear. The general form is written as

$$A\frac{\partial^2 u}{\partial x^2} + B\frac{\partial^2 u}{\partial x \, \partial y} + C\frac{\partial^2 u}{\partial y^2} + D\frac{\partial u}{\partial x} + E\frac{\partial u}{\partial y} + Fu = G, \quad (8.1.1)$$

where the coefficients may depend on x and y. The equations are classified according to the coefficients A, B, and C. They are said to be

$$
\begin{array}{llll}
(1) & \text{Elliptic} & \text{if } B^2 - 4AC < 0. & \\
(2) & \text{Parabolic} & \text{if } B^2 - 4AC = 0. & (8.1.2) \\
(3) & \text{Hyperbolic} & \text{if } B^2 - 4AC > 0. &
\end{array}
$$

We shall derive equations of each class and illustrate the different types of solutions for each. The type of boundary conditions that is specified depends on the class of the partial differential equation. That is, for an elliptic equation the function (or its derivative) will be specified around the entire boundary enclosing a region of interest, whereas for the hyperbolic and parabolic equations the function cannot be specified around an entire boundary. It is also possible to have an elliptic equation in part of a region of interest and a hyperbolic equation in the remaining part. A discontinuity separates the two parts of such regions; a shock wave is an example of such a discontinuity.

In the following three sections we shall derive the mathematical equations that describe several phenomena of general interest. The remaining sections will be devoted to the solutions of these equations.

PROBLEMS

Classify each equation.

1. The wave equation: $\dfrac{\partial^2 u}{\partial t^2} = a^2 \dfrac{\partial^2 u}{\partial x^2}$.

2. The heat equation: $\dfrac{\partial u}{\partial t} = C \dfrac{\partial^2 u}{\partial x^2}$.

3. Laplace's equation: $\dfrac{\partial^2 u}{\partial x^2} + \dfrac{\partial^2 u}{\partial y^2} = 0$.

4. Poisson's equation: $\dfrac{\partial^2 u}{\partial x^2} + \dfrac{\partial^2 u}{\partial y^2} = f(x, y)$.

5. $\dfrac{\partial^2 u}{\partial x^2} - \dfrac{\partial^2 u}{\partial x\, \partial y} - \dfrac{\partial^2 u}{\partial y^2} = 0$

6. $(1 - x)\dfrac{\partial^2 u}{\partial x^2} + 2y\dfrac{\partial^2 u}{\partial x\, \partial y} + (1 + x)\dfrac{\partial^2 u}{\partial y^2} = 0$

7. $\dfrac{\partial^2 u}{\partial x^2} + \sqrt{1 + \left(\dfrac{\partial u}{\partial x}\right)^2}\,\dfrac{\partial^2 u}{\partial y^2} + k\dfrac{\partial u}{\partial y} = G(x, y)$

8. $\left(\dfrac{\partial u}{\partial x}\right)^2 = u(x, y)$

9. $\dfrac{du}{dx} = u(x)$

Verify each statement.

10. $u(x, y) = e^x \sin y$ is a solution of Laplace's equation, $\nabla^2 u = 0$.

11. $T(x, t) = e^{-kt} \sin x$ is a solution of the parabolic heat equation, $\partial T/\partial t = k\partial^2 T/\partial x^2$.

12. $u(x, t) = \sin \omega x \sin \omega at$ is a solution of the wave equation, $\partial^2 u/\partial t^2 = a^2 \partial^2 u/\partial x^2$.

8.2   Wave Motion

One of the first phenomena to be modeled with a partial differential equation was wave motion. Wave motion occurs in a variety of physical situations; these include vibrating strings, vibrating membranes (drum heads), waves traveling through a solid bar, waves traveling through a solid media (earthquakes), acoustic waves, water waves, compression waves (shock waves), electromagnetic radiation, vibrating beams, and oscillating shafts, to mention a few. We shall illustrate wave motion with several examples.

8.2.1   VIBRATION OF A STRETCHED, FLEXIBLE STRING

The motion of a tightly stretched, flexible string was modeled with a partial differential equation approximately 250 years ago. It still serves as an excellent introductory example. We shall derive the equation that describes the motion and then in later sections present methods of solution.

Suppose that we wish to describe the position for all time of the string shown in Fig. 8.1. In fact, we shall seek a describing equation for the deflection u of the string for any position x and for any time t. The initial and boundary conditions will be considered in detail when the solution is presented.

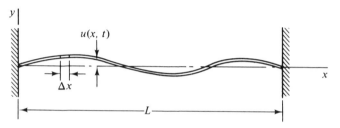

FIGURE 8.1.   Deformed, flexible string at an instant t.

Consider an element of the string at a particular instant enlarged in Fig. 8.2. We shall make the following assumptions:

1. The string offers no resistance to bending so that no shearing force exists on a surface normal to the string.
2. The tension P is so large that the weight of the string is negligible.
3. Every element of the string moves normal to the x axis.
4. The slope of the deflection curve is small.
5. The mass m per unit length of the string is constant.
6. The effects of friction are negligible.

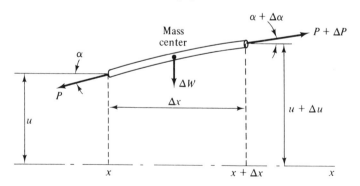

FIGURE 8.2.   Small element of the vibrating string.

Newton's second law states that the net force acting on a body of constant mass equals the mass M of the body multiplied by the acceleration **a** of the center of mass of the body. This is expressed as

$$\sum \mathbf{F} = M\mathbf{a}. \tag{8.2.1}$$

Consider the forces acting in the x direction on the element of the string. By assumption 3 there is no acceleration of the element in the x direction; hence,

$$\sum F_x = 0 \tag{8.2.2}$$

or, referring to Fig. 8.2,

$$(P + \Delta P) \cos (\alpha + \Delta\alpha) - P \cos \alpha = 0. \tag{8.2.3}$$

By assumption 4 we have

$$\cos \alpha \cong \cos (\alpha + \Delta\alpha) \cong 1. \tag{8.2.4}$$

Equation 8.2.3 then gives us

$$\Delta P = 0, \tag{8.2.5}$$

showing us that the tension is constant along the string.

For the y direction we have, neglecting friction and the weight of the string,

$$P \sin (\alpha + \Delta\alpha) - P \sin \alpha = m \,\Delta x \frac{\partial^2}{\partial t^2}\left(u + \frac{\Delta u}{2}\right), \tag{8.2.6}$$

where $m \,\Delta x$ is the mass of the element and $\partial^2/\partial t^2(u + \Delta u/2)$ is the acceleration of the mass center. Again, by assumption 4 we have

$$\sin (\alpha + \Delta\alpha) \cong \tan (\alpha + \Delta\alpha) = \frac{\partial u}{\partial x}(x + \Delta x, t)$$

$$\sin \alpha \cong \tan \alpha = \frac{\partial u}{\partial x}(x, t). \tag{8.2.7}$$

Equation 8.2.6 can then be written as

$$P\left[\frac{\partial u}{\partial x}(x + \Delta x, t) - \frac{\partial u}{\partial x}(x, t)\right] = m \,\Delta x \frac{\partial^2}{\partial t^2}\left(u + \frac{\Delta u}{2}\right) \tag{8.2.8}$$

or, equivalently.

$$P\frac{\dfrac{\partial u}{\partial x}(x + \Delta x, t) - \dfrac{\partial u}{\partial x}(x, t)}{\Delta x} = m \frac{\partial^2}{\partial t^2}\left(u + \frac{\Delta u}{2}\right) \tag{8.2.9}$$

Now, we let $\Delta x \rightarrow 0$, which also implies that $\Delta u \rightarrow 0$. Then, by definition,

$$\lim_{\Delta x \to 0} \frac{\frac{\partial u}{\partial x}(x + \Delta x, t) - \frac{\partial u}{\partial x}(x, t)}{\Delta x} = \frac{\partial^2 u}{\partial x^2}, \tag{8.2.10}$$

and our describing equation becomes

$$P \frac{\partial^2 u}{\partial x^2} = m \frac{\partial^2 u}{\partial t^2}. \tag{8.2.11}$$

This is usually written in the form

$$\frac{\partial^2 u}{\partial t^2} = a^2 \frac{\partial^2 u}{\partial x^2}, \tag{8.2.12}$$

where we have set

$$a = \sqrt{\frac{P}{m}}. \tag{8.2.13}$$

Equation 8.2.12 is the *one-dimensional wave equation* and a is the *wave speed*. It is a transverse wave; that is, it moves normal to the string. This hyperbolic equation will be solved in a subsequent section.

8.2.2   THE VIBRATING MEMBRANE

A stretched vibrating membrane, such as a drumhead, is simply an extension into a second space dimension of the vibrating-string problem. We shall derive a partial differential equation that describes the deflection u of the membrane for any position (x, y) and for any time t. The simplest equation results if the following assumptions are made:

1. The membrane offers no resistance to bending, so shearing stresses are absent.
2. The tension τ per unit length is so large that the weight of the membrane is negligible.
3. Every element of the membrane moves normal to the xy plane.
4. The slope of the deflection surface is small.
5. The mass m of the membrane per unit area is constant.
6. Frictional effects are neglected.

With these assumptions we can now apply Newton's second law to a typical element of the membrane as shown in Fig. 8.3. Assumption 3 leads to the conclusion that τ is constant throughout the membrane, since there are

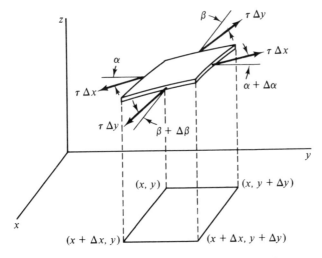

FIGURE 8.3. Element from a stretched, flexible membrane.

no accelerations of the element in the x and y directions. This is shown on the element. In the z direction we have

$$\sum F_z = Ma_z. \tag{8.2.14}$$

For each element this becomes

$$\tau\,\Delta x \sin(\alpha + \Delta\alpha) - \tau\,\Delta x \sin \alpha$$

$$+ \tau\,\Delta y \sin(\beta + \Delta\beta) - \tau\,\Delta y \sin \beta = m\,\Delta x\,\Delta y \frac{\partial^2 u}{\partial t^2}, \tag{8.2.15}$$

where the mass of the element is $m\,\Delta x\,\Delta y$ and the acceleration a_z is $\partial^2 u/\partial t^2$. For small angles

$$\sin(\alpha + \Delta\alpha) \cong \tan(\alpha + \Delta\alpha) = \frac{\partial u}{\partial y}\left(x + \frac{\Delta x}{2}, y + \Delta y, t\right)$$

$$\sin \alpha \cong \tan \alpha = \frac{\partial u}{\partial y}\left(x + \frac{\Delta x}{2}, y, t\right)$$

$$\sin(\beta + \Delta\beta) \cong \tan(\beta + \Delta\beta) = \frac{\partial u}{\partial x}\left(x + \Delta x, y + \frac{\Delta y}{2}, t\right) \tag{8.2.16}$$

$$\sin \beta \cong \tan \beta = \frac{\partial u}{\partial x}\left(x, y + \frac{\Delta y}{2}, t\right).$$

We can then write Eq. 8.2.15 as

$$\tau \, \Delta x \left[\frac{\partial u}{\partial y}\left(x + \frac{\Delta x}{2}, y + \Delta y, t\right) - \frac{\partial u}{\partial y}\left(x + \frac{\Delta x}{2}, y, t\right) \right]$$

$$+ \tau \, \Delta y \left[\frac{\partial u}{\partial x}\left(x + \Delta x, y + \frac{\Delta y}{2}, t\right) - \frac{\partial u}{\partial x}\left(x, y + \frac{\Delta y}{2}, t\right) \right] = m \, \Delta x \, \Delta y \, \frac{\partial^2 u}{\partial t^2}$$

$$(8.2.17)$$

or, by dividing by $\Delta x \, \Delta y$,

$$\tau \left[\frac{\dfrac{\partial u}{\partial y}\left(x + \dfrac{\Delta x}{2}, y + \Delta y, t\right) - \dfrac{\partial u}{\partial y}\left(x + \dfrac{\Delta x}{2}, y, t\right)}{\Delta y} \right.$$

$$\left. + \frac{\dfrac{\partial u}{\partial x}\left(x + \Delta x, y + \dfrac{\Delta y}{2}, t\right) - \dfrac{\partial u}{\partial x}\left(x, y + \dfrac{\Delta y}{2}, t\right)}{\Delta x} \right] = m \frac{\partial^2 u}{\partial t^2}.$$

$$(8.2.18)$$

Taking the limit as $\Delta x \to 0$ and $\Delta y \to 0$, we arrive at

$$\frac{\partial^2 u}{\partial t^2} = a^2 \left(\frac{\partial^2 u}{\partial x^2} + \frac{\partial^2 u}{\partial y^2} \right), \tag{8.2.19}$$

where

$$a = \sqrt{\frac{\tau}{m}}. \tag{8.2.20}$$

Equation 8.2.19 is the *two-dimensional wave equation* and a is the wave speed.

8.2.3   LONGITUDINAL VIBRATIONS OF AN ELASTIC BAR

As another example of wave motion, let us determine the equation describing the motion of an elastic bar (steel, for example) that is subjected to an initial displacement or velocity, such as striking the bar on the end with a hammer, Fig. 8.4. We make the following assumptions:

1.   The bar has a constant cross-sectional area A in the unstrained state.
2.   All cross-sectional planes remain plane.
3.   Hooke's law may be used to relate stress and strain.

We let $u(x, t)$ denote the displacement of the plane of particles that were at x at $t = 0$. Consider the element of the bar between x_1 and x_2, shown in Fig. 8.5. We assume that the bar has mass per unit volume (ρ). The force exerted on the element at x_1 is, by Hooke's law,

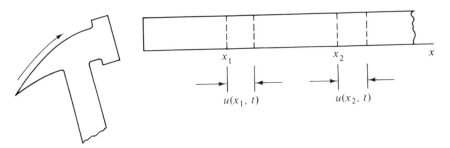

FIGURE 8.4. Wave motion in an elastic bar.

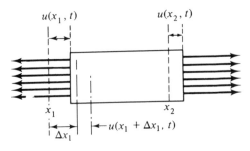

FIGURE 8.5. Element of an elastic bar.

$$F_x = \text{area} \times \text{stress} = \text{area} \times E \times \text{strain}, \qquad (8.2.21)$$

where E is the modulus of elasticity. The strain ϵ at x_1 is given by

$$\epsilon = \frac{\text{elongation}}{\text{unstrained length}}. \qquad (8.2.22)$$

Thus, for Δx_1 small, we have the strain at x_1 as

$$\epsilon = \frac{u(x_1 + \Delta x_1, t) - u(x_1, t)}{\Delta x} \qquad (8.2.23)$$

Letting $\Delta x_1 \to 0$, we find that

$$\epsilon = \frac{\partial u}{\partial x}. \qquad (8.2.24)$$

Returning to the element, the force acting in the x direction is

$$F_x = AE\left[\frac{\partial u}{\partial x}(x_2, t) - \frac{\partial u}{\partial x}(x_1, t) \right]. \qquad (8.2.25)$$

Newton's second law states that

$$F_x = ma = \rho A (x_2 - x_1) \frac{\partial^2 u}{\partial t^2}. \qquad (8.2.26)$$

Hence, Eqs. 8.2.25 and 8.2.26 give

$$\rho A (x_2 - x_1) \frac{\partial^2 u}{\partial t^2} = AE \left[\frac{\partial u}{\partial x}(x_2, t) - \frac{\partial u}{\partial x}(x_1, t) \right]. \qquad (8.2.27)$$

We divide Eq. 8.2.27 by $(x_2 - x_1)$ and let $x_1 \to x_2$, to give

$$\frac{\partial^2 u}{\partial t^2} = a^2 \frac{\partial^2 u}{\partial x^2}, \qquad (8.2.28)$$

where the longitudinal wave speed a is given by

$$a = \sqrt{\frac{E}{\rho}}. \qquad (8.2.29)$$

Therefore, longitudinal displacements in an elastic bar may be described by the one-dimensional wave equation with wave speed $\sqrt{E/\rho}$.

8.2.4 TRANSMISSION-LINE EQUATIONS

As a final example of wave motion, we derive the transmission-line equations. Electricity flows in the transmission line shown in Fig. 8.6, resulting in a current flow between conductors due to the capacitance and conductance between the conductors. The cable also possesses both resistance and inductance resulting in voltage drops along the line. We shall choose the following symbols in our analysis:

$v(x, t)$ = voltage at any point along the line

$i(x, t)$ = current at any point along the line

R = resistance per meter

L = self-inductance per meter

C = capacitance per meter

G = conductance per meter.

The voltage drop over the incremental length Δx at a particular instant (see Eqs. 1.4.3) is

$$\Delta v = v(x + \Delta x, t) - v(x, t) = -iR \, \Delta x - L \, \Delta x \frac{\partial i}{\partial t}. \qquad (8.2.30)$$

Dividing by Δx and taking the limit as $\Delta x \to 0$ yields the partial differential equation relating $v(x, t)$ and $i(x, t)$,

$$\frac{\partial v}{\partial x} + iR + L \frac{\partial i}{\partial t} = 0. \qquad (8.2.31)$$

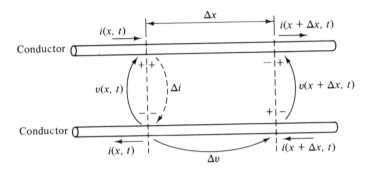

(a) Actual element

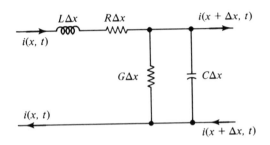

(b) Equivalent circuit

FIGURE 8.6.   Element from a transmission line.

Now, let us find an expression for the change in the current over the length Δx. The current change is

$$\Delta i = i(x + \Delta x, t) - i(x, t) = -G\,\Delta x\,v - C\,\Delta x\frac{\partial v}{\partial t}. \quad (8.2.32)$$

Again, dividing by Δx and taking the limit as $\Delta x \to 0$ gives a second equation,

$$\frac{\partial i}{\partial x} + v\,G + C\frac{\partial v}{\partial t} = 0. \quad (8.2.33)$$

Take the partial derivative of Eq. 8.2.31 with respect to x and of Eq. 8.2.33 with respect to t. Then, multiplying the second equation by L and subtracting the resulting two equations, using $\partial^2 i/\partial x\,\partial t = \partial^2 i/\partial t\,\partial x$, presents us with

$$\frac{\partial^2 v}{\partial x^2} + R\frac{\partial i}{\partial x} = LG\frac{\partial v}{\partial t} + LC\frac{\partial^2 v}{\partial t^2}. \quad (8.2.34)$$

Then, substituting for $\partial i/\partial x$ from Eq. 8.2.33 results in an equation for $v(x, t)$ only. It is

$$\frac{\partial^2 v}{\partial x^2} = LC\frac{\partial^2 v}{\partial t^2} + (LG + RC)\frac{\partial v}{\partial t} + RGv. \quad (8.2.35)$$

Take the partial derivative of Eq. 8.2.31 with respect to t and multiply by C; take the partial derivative of Eq. 8.2.33 with respect to x, subtract the resulting two equations and substitute for $\partial v/\partial x$ from Eq. 8.2.31; there results

$$\frac{\partial^2 i}{\partial x^2} = LC\frac{\partial^2 i}{\partial t^2} + (LG + RC)\frac{\partial i}{\partial t} + RGi. \qquad (8.2.36)$$

The two equations above are difficult to solve in the general form presented; two special cases are of interest. First, there are conditions under which the self-inductance and leakage due to the conductance between conductors are negligible; that is, $L \cong 0$, and $G \cong 0$. Then our equations become

$$\frac{\partial^2 v}{\partial x^2} = RC\frac{\partial v}{\partial t}, \qquad \frac{\partial^2 i}{\partial x^2} = RC\frac{\partial i}{\partial t}. \qquad (8.2.37)$$

Second, under conditions of high frequency, a time derivative increases* the magnitude of a term; that is, $\partial^2 i/\partial t^2 \gg \partial i/\partial t \gg i$. Thus, our general equations can be approximated by

$$\frac{\partial^2 v}{\partial t^2} = \frac{1}{LC}\frac{\partial^2 v}{\partial x^2}, \qquad \frac{\partial^2 i}{\partial t^2} = \frac{1}{LC}\frac{\partial^2 i}{\partial x^2}. \qquad (8.2.38)$$

These latter two equations are wave equations with $\sqrt{1/LC}$ in units of meters/second.

Although we shall not discuss any other wave phenomenon, it is well for the reader to be aware that sound waves, light waves, water waves, quantum-mechanical systems, and many other physical systems are described, at least in part, by a wave equation.

PROBLEMS

1. In arriving at the equation describing the motion of a vibrating string, the weight was assumed to be negligible. Include the weight of the string in the derivation and determine the describing equation. Classify the equation.

---

*As an example, consider the term $\sin(\omega t + x/L)$ where $\omega \gg 1$. Then

$$\frac{\partial}{\partial t}\left[\sin\left(\omega t + \frac{x}{L}\right)\right] = \omega\cos\left(\omega t + \frac{x}{L}\right).$$

We see that

$$\left|\omega\cos\left(\omega t + \frac{x}{L}\right)\right| \gg \left|\sin\left(\omega t + \frac{x}{L}\right)\right|.$$

2. Derive the describing equation for a stretched string subject to gravity loading and viscous drag. Viscous drag per unit length of string may be expressed by $c(\partial u/\partial t)$; the drag force is proportional to the velocity. Classify the resulting equation.

3. A tightly stretched string, with its ends fixed at the points $(0, 0)$ and $(2L, 0)$, hangs at rest under its own weight. The y axis points vertically upward. Find the describing equation for the position $u(x)$ of the string. Is the following expression a solution?

$$u(x) = \frac{g}{2a^2}(x - L)^2 - \frac{gL^2}{2a^2},$$

where $a^2 = P/m$. If so, show that the depth of the vertex of the parabola (i.e., the lowest point) varies directly with m (mass per unit length) and L^2, and inversely with P, the tension.

4. Derive the torsional vibration equation for a circular shaft by applying the basic law which states that $I\alpha = \Sigma T$, where α is the angular acceleration, T is the torque $(T = GJ\theta/L$, where θ is the angle of twist of the shaft of length L and J and G are constants), and I is the mass moment of inertia $(I = k^2 m$, where the radius of gyration $k = \sqrt{J/A}$ and m is the mass of the shaft). Choose an infinitesimal element of the shaft of length Δx, sum the torques acting on it, and using ρ as the mass density, show that this wave equation results,

$$\frac{\partial^2 \theta}{\partial t^2} = \frac{G}{\rho} \frac{\partial^2 \theta}{\partial x^2}.$$

5. An unloaded beam will undergo vibrations when subjected to an initial disturbance. Derive the appropriate partial differential equation which describes the motion using Newton's second law applied to an infinitesimal section of the beam. Assume the inertial force to be a distributed load acting on the beam. A uniformly distributed load w is related to the vertical deflection $y(x, t)$ of the beam by $w = -EI\partial^4 y/\partial x^4$, where E and I are constants.

6. For the special situation in which $LG = RC$, show that the transmission-line equation 8.2.36 reduces to the wave equation

$$\frac{\partial^2 u}{\partial t^2} = a^2 \frac{\partial^2 u}{\partial x^2} \quad \text{if we let} \quad i(x, t) = e^{-abt} u(x, t),$$

where $a^2 = 1/LC$ and $b^2 = RG$.

7. For low frequency and negligibly small G, show that the *telegraph equations* result:

$$\frac{\partial v}{\partial t} = \frac{1}{RC} \frac{\partial^2 v}{\partial x^2}, \qquad \frac{\partial i}{\partial t} = \frac{1}{RC} \frac{\partial^2 i}{\partial x^2}.$$

8.3   Diffusion

Another class of physical problems can be characterized by diffusion equations. Diffusion may be likened to a spreading, smearing, or mixing. A physical system that has a high concentration of some substance in volume A and a low concentration in volume B may be subject to the *diffusion* of the substance so that the concentrations in A and B approach equality. This phenomenon is exhibited by the tendency of a body toward a uniform temperature. One of the most common diffusion processes that is encountered is the transfer of energy in the form of heat.

From thermodynamics we learn that heat is thermal energy in transit. It may be transmitted by conduction (when two bodies are in contact), by convection (when a body is in contact with a liquid or a gas), and by radiation (when energy is transmitted by energy waves). We shall consider the first of these mechanisms in some detail. Experimental observations have shown that we may make the following two statements:

1.   Heat flows in the direction of decreasing temperature.
2.   The rate at which energy in the form of heat is transferred through an area is proportional to the area and to the temperature gradient normal to the area.

These statements may be expressed analytically. The heat flux through an area A oriented normal to the x axis is

$$Q = -KA\frac{\partial T}{\partial x},\tag{8.3.1}$$

where Q (watts, W) is the heat flux, $\partial T/\partial x$ is the temperature gradient normal to A, and K (W/m · °C) is a constant of proportionality called the *thermal conductivity*. The minus sign is present since heat is transferred in the direction opposite the temperature gradient.

The energy (usually called an internal energy) gained or lost by a body of mass m that undergoes a uniform temperature change ΔT may be expressed as

$$\Delta E = Cm\,\Delta T,\tag{8.3.2}$$

where ΔE (J) is the energy change of the body and C (J/kg · °C) is a constant of proportionality called the *specific heat*.

Conservation of energy is a fundamental law of nature. We use this law to make an energy balance on the element in Fig. 8.7. The density ρ of the element is used to determine its mass, namely,

$$m = \rho\,\Delta x\,\Delta y\,\Delta z.\tag{8.3.3}$$

By energy balance we mean that the net energy flowing into the element in time Δt must equal the increase in energy in the element in Δt. For

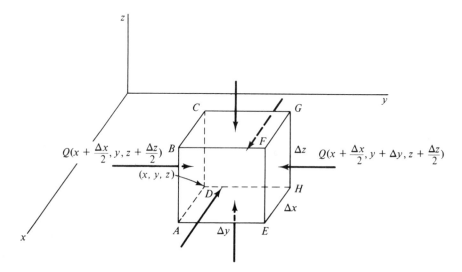

FIGURE 8.7.   Element of mass.

simplicity, we assume that there are no sources inside the element. Equation 8.3.2 gives the change in energy in the element as

$$\Delta E = Cm\,\Delta T = C\rho\,\Delta x\,\Delta y\,\Delta z\,\Delta T. \tag{8.3.4}$$

The energy that flows into the element through face *ABCD* in Δt is, by Eq. 8.3.1,

$$\Delta E_{ABCD} = Q_{ABCD}\,\Delta t = -K\,\Delta x\,\Delta z\,\Delta t\,\frac{\partial T}{\partial y}\bigg|_{\substack{x+\Delta x/2\\ y\\ z+\Delta z/2}}, \tag{8.3.5}$$

where we have approximated the temperature derivative by the value at the center of the face. The flow into the element through face *EFGH* is

$$\Delta E_{EFGH} = K\,\Delta x\,\Delta z\,\Delta t\,\frac{\partial T}{\partial y}\bigg|_{\substack{x+\Delta x/2\\ y+\Delta y\\ z+\Delta z/2}}. \tag{8.3.6}$$

Similar expressions are found for the other four faces. The energy balance then provides us with

$$\Delta E = \Delta E_{ABCD} + \Delta E_{EFGH} + \Delta E_{ADHE} + \Delta E_{BCGF} + \Delta E_{DHGC} + \Delta E_{BFEA} \tag{8.3.7}$$

or, using Eqs. 8.3.5, 8.3.6, and their counterparts for the *x* and *z* directions,

$$C\rho\,\Delta x\,\Delta y\,\Delta z\,\Delta T = K\,\Delta x\,\Delta z\,\Delta t\left(\frac{\partial T}{\partial y}\bigg|_{\substack{x+\Delta x/2\\ y+\Delta y\\ z+\Delta z/2}} - \frac{\partial T}{\partial y}\bigg|_{\substack{x+\Delta x/2\\ y\\ z+\Delta z/2}}\right)$$

$$+ K\,\Delta y\,\Delta z\,\Delta t\left(\left.\frac{\partial T}{\partial x}\right|_{\substack{x+\Delta x \\ y+\Delta y/2 \\ z\,+\,\Delta z/2}} - \left.\frac{\partial T}{\partial x}\right|_{\substack{x \\ y+\Delta y/2 \\ z\,+\,\Delta z/2}}\right)$$

$$+ K\,\Delta x\,\Delta y\,\Delta t\left(\left.\frac{\partial T}{\partial z}\right|_{\substack{x+\Delta x/2 \\ y+\Delta y/2 \\ z\,+\,\Delta z}} - \left.\frac{\partial T}{\partial z}\right|_{\substack{x+\Delta x/2 \\ y+\Delta y/2 \\ z}}\right). \qquad (8.3.8)$$

Both sides of the equation are divided by $C\rho\,\Delta x\,\Delta y\,\Delta z\,\Delta t$; then, let $\Delta x \to 0$, $\Delta y \to 0$, $\Delta z \to 0$, $\Delta t \to 0$. There results

$$\frac{\partial T}{\partial t} = k\left[\frac{\partial^2 T}{\partial x^2} + \frac{\partial^2 T}{\partial y^2} + \frac{\partial^2 T}{\partial z^2}\right], \qquad (8.3.9)$$

where $k = K/C\rho$ is called the *thermal diffusivity* and is assumed constant. It has dimensions of square meters per second (m^2/s). Equation 8.3.9 is a *diffusion equation*.

Two special cases of the diffusion equation are of particular interest. For instance, a number of situations involve time and only one coordinate, say x, as in a long, slender rod with insulated sides. The *one-dimensional heat equation* then results. It is given by

$$\frac{\partial T}{\partial t} = k\frac{\partial^2 T}{\partial x^2}, \qquad (8.3.10)$$

which is a parabolic equation.

In some situations $\partial T/\partial t$ is zero and we have a steady-state condition; then we no longer have a diffusion equation, but the equation

$$\frac{\partial^2 T}{\partial x^2} + \frac{\partial^2 T}{\partial y^2} + \frac{\partial^2 T}{\partial z^2} = 0. \qquad (8.3.11)$$

This equation is known as *Laplace's equation*. It is sometimes written in the shorthand form

$$\nabla^2 T = 0. \qquad (8.3.12)$$

If the temperature depends only on two coordinates x and y, as in a thin rectangular plate, an elliptic equation is encountered:

$$\frac{\partial^2 T}{\partial x^2} + \frac{\partial^2 T}{\partial y^2} = 0. \qquad (8.3.13)$$

Cylindrical or spherical coordinates (see Fig. 8.8) should be used in certain geometries. It is then convenient to express $\nabla^2 T$ in cylindrical coordinates as

$$\nabla^2 T = \frac{1}{r}\frac{\partial}{\partial r}\left(r\frac{\partial T}{\partial r}\right) + \frac{1}{r^2}\frac{\partial^2 T}{\partial \theta^2} + \frac{\partial^2 T}{\partial z^2}, \qquad (8.3.14)$$

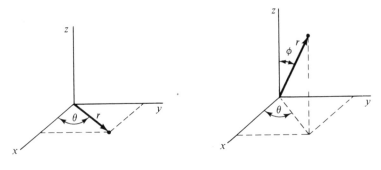

| (a) Cylindrical coordinates | (b) Spherical coordinates |

FIGURE 8.8.   Cylindrical and spherical coordinates.

and in spherical coordinates as

$$\nabla^2 T = \frac{1}{r^2}\frac{\partial}{\partial r}\left(r^2\frac{\partial T}{\partial r}\right) + \frac{1}{r^2\sin^2\phi}\frac{\partial^2 T}{\partial\theta^2} + \frac{1}{r\sin\phi}\frac{\partial}{\partial\phi}\left(\sin\phi\frac{\partial T}{\partial\phi}\right). \qquad (8.3.15)$$

See Table 6.5.

Our discussion of heat transfer has included heat conduction only. Radiative and convective forms of heat transfer would necessarily lead to other partial differential equations. We have also assumed no heat sources in the volume of interest, and have assumed the conductivity K to be constant. Finally, the specification of boundary and initial conditions makes our problem statement complete. These will be reserved for a later section in which a solution to the diffusion equation is presented.

PROBLEMS

1. Use an elemental slice of length Δx of a long, slender, laterally insulated rod and derive the one-dimensional heat equation

$$\frac{\partial T}{\partial t} = k\frac{\partial^2 T}{\partial x^2}$$

   using Eqs. 8.3.1 and 8.3.2.

2. Modify Eq. 8.3.10 to account for internal heat generation within the rod. The rate of heat generation is denoted ϕ (W/m³).

3. Allow the sides of a long, slender circular rod to transfer heat by convection. The convective rate of heat loss is given by $Q = hA(T - T_f)$, where h (W/m² · °C) is the convection coefficient, A is the surface area, and T_f is the temperature of the surrounding fluid. Derive the describing partial differential equation. (*Hint:* Apply an energy balance to an elemental slide of the rod).

4. The tip of a 2-m-long slender rod with lateral surface insulated is dipped into a hot liquid at 200°C. What differential equation describes the temperature? After a long time, what is the temperature distribution in the rod if the other end is held at 0°C? The lateral surfaces of the rod are insulated.

5. The conductivity K in the derivation of Eq. 8.3.10 was assumed constant. Let K be a function of x and let C and ρ be constants. Write the appropriate describing equation.

6. Write the one-dimensional heat equation that could be used to determine the temperature in (a) a flat circular disk with the flat surfaces insulated, and (b) in a sphere with initial temperature a function of r only.

7. Determine the steady-state temperature distribution in (a) a flat circular disk with sides held at 100°C with the flat surfaces insulated, and (b) a sphere with the outer surface held at 100°C.

8. Use a hollow cylinder of thickness Δr and derive the one-dimensional heat equation for a solid cylinder assuming that $T = T(r, t)$.

9. Use a hollow sphere of thickness Δr and derive the one-dimensional heat equation for a solid sphere assuming that $T = T(r, t)$.

8.4   Gravitational Potential

There are a number of physical situations that are modeled by Laplace's equation. We choose the force of attraction of particles to demonstrate its derivation. The law of gravitation states that a lumped mass m located at the point (X, Y, Z) attracts a unit mass located at the point (x, y, z) (see Fig. 8.9), with a force directed along the line connecting the two points with magnitude given by

$$F = -\frac{km}{r^2}, \tag{8.4.1}$$

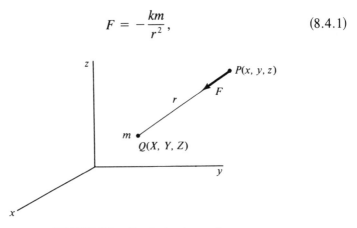

FIGURE 8.9. Gravitational attraction.

where k is a positive constant and the negative sign indicates that the force acts toward the mass m. The distance between the two points is provided by the expression

$$r = \sqrt{(x - X)^2 + (y - Y)^2 + (z - Z)^2}, \qquad (8.4.2)$$

positive being from Q to P.

A gravitational potential ϕ is defined by

$$\phi = \frac{km}{r}. \qquad (8.4.3)$$

This allows the force F acting on a unit mass at P due to a mass at Q to be related to ϕ by the equation

$$F = \frac{\partial \phi}{\partial r} = -\frac{km}{r^2}. \qquad (8.4.4)$$

Now, let the mass m be fixed in space and let the unit mass move to various locations $P(x, y, z)$. The potential function ϕ is then a function of x, y, and z. If we let P move along a direction parallel to the x axis, then

$$\frac{\partial \phi}{\partial x} = \frac{\partial \phi}{\partial r} \frac{\partial r}{\partial x}$$

$$= -\frac{km}{r^2} \frac{1}{2} (2)(x - X)[(x - X)^2 + (y - Y)^2 + (z - Z)^2]^{-1/2}$$

$$= -\frac{km}{r^2} \frac{x - X}{r}$$

$$= F \cos \alpha = F_x, \qquad (8.4.5)$$

where α is the angle between r and the x axis, and F_x is the projection of F in the x direction. Similarly, for the other two directions,

$$F_y = \frac{\partial \phi}{\partial y}, \qquad F_z = \frac{\partial \phi}{\partial z}. \qquad (8.4.6)$$

The discussion above is now extended to include a distributed mass throughout a volume V. The potential $d\phi$ due to an incremental mass dm is written, following Eq. 8.4.3, as

$$d\phi = \frac{k\rho \, dV}{r}, \qquad (8.4.7)$$

where ρ is the mass per unit volume. Letting $dV = dx \, dy \, dz$, we have

$$\phi = k \iiint_V \frac{\rho \, dx \, dy \, dz}{[(x - X)^2 + (y - Y)^2 + (z - Z)^2]^{1/2}}. \qquad (8.4.8)$$

This is differentiated to give the force components. For example, F_x is given by

$$F_x = \frac{\partial \phi}{\partial x} = -k \iiint\limits_V \frac{x - X}{r} \frac{\rho}{r^2} \, dx \, dy \, dz. \qquad (8.4.9)$$

This represents the x component of the total force exerted on a unit mass located outside the volume V at $P(x, y, z)$ due to the distributed mass in volume V.

If we now differentiate Eq. 8.4.9 again with respect to x, we find that

$$\frac{\partial^2 \phi}{\partial x^2} = -k \iiint\limits_V \left[\frac{1}{r^3} - \frac{3(x - X)^2}{r^5} \right] \rho \, dx \, dy \, dz. \qquad (8.4.10)$$

We can also show that

$$\frac{\partial^2 \phi}{\partial y^2} = -k \iiint\limits_V \left[\frac{1}{r^3} - \frac{3(y - Y)^2}{r^5} \right] \rho \, dx \, dy \, dz$$

$$\frac{\partial^2 \phi}{\partial z^2} = -k \iiint\limits_V \left[\frac{1}{r^3} - \frac{3(z - Z)^2}{r^5} \right] \rho \, dx \, dy \, dz. \qquad (8.4.11)$$

The sum of the bracketed terms inside the three integrals above is identically zero, using Eq. 8.4.2. Hence, Laplace's equation

$$\frac{\partial^2 \phi}{\partial x^2} + \frac{\partial^2 \phi}{\partial y^2} + \frac{\partial^2 \phi}{\partial z^2} = 0, \qquad (8.4.12)$$

results, or, in our shorthand notation,

$$\nabla^2 \phi = 0. \qquad (8.4.13)$$

Laplace's equation is also satisfied by a magnetic potential function and an electric potential function at points not occupied by magnetic poles or electric charges. We have already observed in Section 8.3 that the steady-state, heat-conduction problem leads to Laplace's equation. Finally, the flow of an incompressible fluid with negligible viscous effects also leads to Laplace's equation.

We have now derived several partial differential equations that describe a variety of physical phenomena. This modeling process is quite difficult to perform in a situation that is new and different. The confidence gained in deriving the equations of this chapter and in finding solutions, as we shall presently do, will hopefully allow the reader to derive and solve other partial differential equations arising in applications.

PROBLEMS

1. Differentiate Eq. 8.4.8 and show that Eq. 8.4.9 results. Also verify Eq. 8.4.10.

2. Express Laplace's equation using spherical coordinates. Assume that $\phi = \phi(r, \theta)$ (see Table 6.4).

8.5   The D'Alembert Solution of the Wave Equation

It is possible to solve all the partial differential equations that we have derived in this chapter by a general method, the separation of variables. The wave equation can, however, be solved by a special technique that will be presented in this section. It gives a quick look at the motion of a wave. We obtain a general solution to the wave equation

$$\frac{\partial^2 u}{\partial t^2} = a^2 \frac{\partial^2 u}{\partial x^2} \tag{8.5.1}$$

by an appropriate transformation of variables. Introduce the new independent variables

$$\xi = x - at, \qquad \eta = x + at. \tag{8.5.2}$$

Then, using the chain rule we find that

$$\frac{\partial u}{\partial x} = \frac{\partial u}{\partial \xi} \frac{\partial \xi}{\partial x} + \frac{\partial u}{\partial \eta} \frac{\partial \eta}{\partial x} = \frac{\partial u}{\partial \xi} + \frac{\partial u}{\partial \eta}$$

$$\frac{\partial u}{\partial t} = \frac{\partial u}{\partial \xi} \frac{\partial \xi}{\partial t} + \frac{\partial u}{\partial \eta} \frac{\partial \eta}{\partial t} = -a \frac{\partial u}{\partial \xi} + a \frac{\partial u}{\partial \eta} \tag{8.5.3}$$

and

$$\frac{\partial^2 u}{\partial x^2} = \frac{\partial \left(\frac{\partial u}{\partial x} \right)}{\partial \xi} \frac{\partial \xi}{\partial x} + \frac{\partial \left(\frac{\partial u}{\partial x} \right)}{\partial \eta} \frac{\partial \eta}{\partial x} = \frac{\partial^2 u}{\partial \xi^2} + 2 \frac{\partial^2 u}{\partial \xi \partial \eta} + \frac{\partial^2 u}{\partial \eta^2}$$

$$\frac{\partial^2 u}{\partial t^2} = \frac{\partial \left(\frac{\partial u}{\partial t} \right)}{\partial \xi} \frac{\partial \xi}{\partial t} + \frac{\partial \left(\frac{\partial u}{\partial t} \right)}{\partial \eta} \frac{\partial \eta}{\partial t} = a^2 \frac{\partial^2 u}{\partial \xi^2} - 2a^2 \frac{\partial^2 u}{\partial \xi \partial \eta} + a^2 \frac{\partial^2 u}{\partial \eta^2}. \tag{8.5.4}$$

Substitute the expressions above into the wave equation to obtain

$$a^2 \left[\frac{\partial^2 u}{\partial \xi^2} - 2 \frac{\partial^2 u}{\partial \xi \partial \eta} + \frac{\partial^2 u}{\partial \eta^2} \right] = a^2 \left[\frac{\partial^2 u}{\partial \xi^2} + 2 \frac{\partial^2 u}{\partial \xi \partial \eta} + \frac{\partial^2 u}{\partial \eta^2} \right],$$

$$\tag{8.5.5}$$

and there results

$$\frac{\partial^2 u}{\partial \xi \, \partial \eta} = 0. \tag{8.5.6}$$

Integration with respect to ξ gives

$$\frac{\partial u}{\partial \eta} = h(\eta), \tag{8.5.7}$$

where $h(\eta)$ is an arbitrary function of η (for an ordinary differential equation, this would be a constant). A second integration yields

$$u(\xi, \eta) = \int h(\eta) \, d\eta + g(\xi). \tag{8.5.8}$$

The integral is a function of η only and is replaced by $f(\eta)$, so the solution is

$$u(\xi, \eta) = g(\xi) + f(\eta) \tag{8.5.9}$$

or, equivalently.

$$u(x, t) = g(x - at) + f(x + at). \tag{8.5.10}$$

This is the *D'Alembert solution* of the wave equation.

Inspection of the equation above shows the wave nature of the solution. Consider an infinite string, stretched from $-\infty$ to $+\infty$, with an initial displacement $u(x, 0) = g(x) + f(x)$, as shown in Fig. 8.10. At some later time $t = t_1$ the curves $g(x)$ and $f(x)$ will simply be displaced to the right and left, respectively, a distance at_1. The original deflection curves move without distortion at the speed of propagation a.

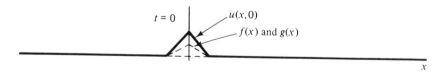

(a) Initial displacement.

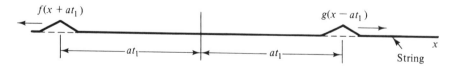

(b) Displacement after a time t_1.

FIGURE 8.10.   Traveling wave in a string.

To determine the form of the functions $g(x)$ and $f(x)$ when $u(x, 0)$ is given, we use the initial conditions. The term $\partial^2 u / \partial t^2$ in the wave equation demands that two conditions be given at $t = 0$. Let us assume, for example, that the initial velocity is zero and that the initial displacement is given by

$$u(x, 0) = f(x) + g(x) = \phi(x). \qquad (8.5.11)$$

The velocity is

$$\frac{\partial u}{\partial t} = \frac{dg}{d\xi}\frac{\partial \xi}{\partial t} + \frac{df}{d\eta}\frac{\partial \eta}{\partial t} \qquad (8.5.12)$$

At $t = 0$ this becomes (see Eqs. 8.5.2 and 8.5.10)

$$\frac{\partial u}{\partial t} = \frac{dg}{dx}(-a) + \frac{df}{dx}(a) = 0. \qquad (8.5.13)$$

Hence, we have the requirement that

$$\frac{dg}{dx} = \frac{df}{dx}, \qquad (8.5.14)$$

which is integrated to provide us with

$$g = f + C. \qquad (8.5.15)$$

Inserting this in Eq. 8.5.11 gives

$$f(x) = \frac{\phi(x)}{2} - \frac{C}{2}, \qquad (8.5.16)$$

so that

$$g(x) = \frac{\phi(x)}{2} + \frac{C}{2}. \qquad (8.5.17)$$

Finally, replacing x in $f(x)$ with $x + at$ and x in $g(x)$ with $x - at$, there results the specific solution for the prescribed initial conditions,

$$u(x, t) = \tfrac{1}{2}\phi(x - at) + \tfrac{1}{2}\phi(x + at). \qquad (8.5.18)$$

Our result shows that, for the infinite string, two initial conditions are sufficient to determine a solution. A finite string will be discussed in the following section.

Example 8.5.1:   Consider that the string in this article is given an initial velocity $\theta(x)$ and zero initial displacement. Determine the form of the solution.

SOLUTION: The velocity is given by Eq. 8.5.12:

$$\frac{\partial u}{\partial t} = \frac{dg}{d\xi}\frac{\partial \xi}{\partial t} + \frac{df}{d\eta}\frac{\partial \eta}{\partial t}.$$

At $t = 0$ this takes the form

$$\theta(x) = a\frac{df}{dx} - a\frac{dg}{dx}.$$

This is integrated to yield

$$f - g = \frac{1}{a}\int_0^x \theta(s)\, ds + C,$$

where s is a dummy variable of integration. The initial displacement is zero, giving

$$u(x, 0) = f(x) + g(x) = 0$$

or,

$$f(x) = -g(x).$$

The constant of integration C is thus evaluated as

$$C = 2f(0) = -2g(0).$$

Combining this with the relation above results in

$$f(x) = \frac{1}{2a}\int_0^x \theta(s)\, ds + f(0)$$

$$g(x) = -\frac{1}{2a}\int_0^x \theta(s)\, ds + g(0).$$

Returning to Eq. 8.5.10, we can obtain the solution $u(x, t)$ using the forms above for $f(x)$ and $g(x)$ simply by replacing x by the appropriate quantity. We then have the solution

$$u(x, t) = \frac{1}{2a}\left[\int_0^{x+at} \theta(s)\, ds - \int_0^{x-at} \theta(s)\, ds\right]$$

$$= \frac{1}{2a}\left[\int_0^{x+at} \theta(s)\, ds + \int_{x-at}^0 \theta(s)\, ds\right]$$

$$= \frac{1}{2a}\int_{x-at}^{x+at} \theta(s)\, ds.$$

For a given $\theta(x)$ this expression provides us with a solution. ∎

Example 8.5.2:   An infinite string is subjected to the initial displacement

$$\phi(x) = \frac{0.02}{1 + 9x^2}.$$

Find an expression for the subsequent motion of the string if it is released from rest. The tension is 20 N and the mass per unit length is 5×10^{-4} kg/m. Also, sketch the solution for $t = 0$, $t = 0.002$ s, and $t = 0.01$ s.

SOLUTION: The motion is given by the solution of this section. Equation 8.5.18 gives it as

$$u(x, t) = \frac{1}{2} \frac{0.02}{1 + 9(x - at)^2} + \frac{1}{2} \frac{0.02}{1 + 9(x + at)^2}.$$

The wave speed a is given by

$$a = \sqrt{\frac{P}{m}} = \sqrt{\frac{20}{5 \times 10^{-4}}} = 200 \text{ m/s}.$$

The solution is then

$$u(x, t) = \frac{0.01}{1 + 9(x - 200t)^2} + \frac{0.01}{1 + 9(x + 200t)^2}.$$

The sketches are presented on the following figure.

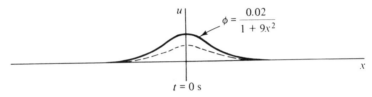

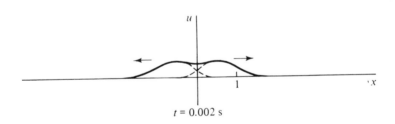

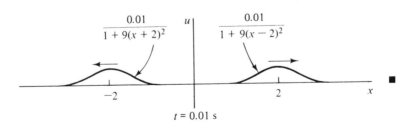

PROBLEMS

1. A very long string is given an initial displacement $\phi(x)$ and an initial velocity $\theta(x)$. Determine the general form of the solution for $u(x, t)$. Compare with the solution 8.5.18 and that of Example 8.5.1.

2. An infinite string with a mass of 0.03 kg/m is stretched with a force of 300 N. It is subjected to an initial displacement of $\cos x$ for $-\pi/2 < x < \pi/2$ and zero for all other x and released from rest. Determine the subsequent displacement of the string and sketch the solution for $t = 0.1$ s and 0.1 s.

3. An infinite string is subject to the initial displacement

$$\phi(x) = \begin{cases} 0, & x < -1 \\ 0.2(x + 1), & -1 \le x \le 0 \\ 0.2(1 - x), & 0 < x \le 1 \\ 0, & 1 < x. \end{cases}$$

If the string is released from rest, find $\phi(x, t)$. At what time t_0 is the displacement zero at $x = 0$? Sketch $\phi(x)$ at $t = t_0/2$ and at $t = 2t_0$.

4. An infinite string is given the initial velocity

$$\theta(x) = \begin{cases} 0, & x < -1 \\ 10(x + 1), & -1 \le x \le 0 \\ 10(1 - x), & 0 < x \le 1 \\ 0, & 1 < x. \end{cases}$$

If the string has zero initial displacement and $a = 40$, find $\phi(x, t)$. Sketch the displacement at $t = 0.025$ s and $t = 0.25$ s.

8.6   *Separation of Variables*

We shall now present a powerful technique used to solve many of the partial differential equations encountered in physical applications in which the domains of interest are finite. It is the method of *separation of variables*. Even though it has limitations, it is widely used. It involves the idea of reducing a more difficult problem to several simpler problems; here, we shall reduce a partial differential equation to several ordinary differential equations for which we already have a method of solution. Then, hopefully, by satisfying the initial and boundary conditions, a solution to the partial differential equation can be found.

To illustrate the details of the method, let us use the mathematical description of a finite string of length L that is fixed at both ends and is released from rest with an initial displacement. The motion of the string is described by the wave equation

$$\frac{\partial^2 u}{\partial t^2} = a^2 \frac{\partial^2 u}{\partial x^2}. \tag{8.6.1}$$

We shall, as usual, consider the wave speed a to be a constant. The boundary conditions of fixed ends may be written as

$$u(0, t) = 0 \tag{8.6.2}$$

and

$$u(L, t) = 0. \qquad (8.6.3)$$

Since the string is released from rest, the initial velocity is zero; hence,

$$\frac{\partial u}{\partial t}(x, 0) = 0. \qquad (8.6.4)$$

The initial displacement will be denoted by $f(x)$. We then have

$$u(x, 0) = f(x). \qquad (8.6.5)$$

We assume that the solution of our problem can be written in the separated form

$$u(x, t) = X(x)T(t); \qquad (8.6.6)$$

that is, the x variable separates from the t variable. Substitution of this relationship into Eq. 8.6.1 yields

$$X(x)T''(t) = a^2 X''(x)T(t), \qquad (8.6.7)$$

where the primes denote differentiation with respect to the associated independent variable. Rewriting Eq. 8.6.7 results in

$$\frac{T''}{a^2 T} = \frac{X''}{X}. \qquad (8.6.8)$$

The left side of this equation is a function of t only and the right side is a function of x only. Thus, as we vary t holding x fixed, the right side cannot change; this means that $T''(t)/a^2 T(t)$ must be the same for all t. As we vary x holding t fixed the left side must not change. Thus, the quantity $X''(x)/X(x)$ must be the same for all x. Therefore, both sides must equal the same *constant* value μ, sometimes called the *separation constant*. Equation 8.6.8 may then be written as two ordinary differential equations:

$$T'' - \mu a^2 T = 0 \qquad (8.6.9)$$

$$X'' - \mu X = 0. \qquad (8.6.10)$$

We note at this point that we have *separated* the *variables* and reduced a *partial differential* equation to two *ordinary differential equations*. If the boundary conditions can be satisfied, then we have succeeded with our separation of variables. We shall assume that we need to consider μ only as a real number. Thus, we are left with the three cases:

$$\mu > 0, \qquad \mu = 0, \qquad \mu < 0. \qquad (8.6.11)$$

For any nonzero value of μ, we know that the solutions of these second-order ordinary differential equations are of the form e^{mt} and e^{nx}, respectively (see

Section 1.6). The characteristic equations are

$$m^2 - \mu a^2 = 0 \tag{8.6.12}$$

$$n^2 - \mu = 0. \tag{8.6.13}$$

The roots are

$$m_1 = a\sqrt{\mu}, \qquad m_2 = -a\sqrt{\mu} \tag{8.6.14}$$

$$n_1 = \sqrt{\mu}, \qquad n_2 = -\sqrt{\mu}. \tag{8.6.15}$$

The resulting solutions are

$$T(t) = c_1 e^{\sqrt{\mu}at} + c_2 e^{-\sqrt{\mu}at}. \tag{8.6.16}$$

and

$$X(x) = c_3 e^{\sqrt{\mu}x} + c_4 e^{-\sqrt{\mu}x}. \tag{8.6.17}$$

Now let us consider the three cases, $\mu > 0$, $\mu = 0$, and $\mu < 0$. For $\mu > 0$, we have the result that $\sqrt{\mu}$ is a real number and the general solution is

$$u(x,t) = T(t)X(x) = (c_1 e^{\sqrt{\mu}at} + c_2 e^{-\sqrt{\mu}at})(c_3 e^{\sqrt{\mu}x} + c_4 e^{-\sqrt{\mu}x}), \tag{8.6.18}$$

which is a decaying or growing exponential. The derivative of Eq. 8.6.18 with respect to time yields the velocity and it, too, is growing or decaying with respect to time. This, of course, means that the kinetic energy of an element of the string is increasing or decreasing in time, as is the total kinetic energy. However, energy remains constant; therefore, this solution violates the basic law of physical conservation of energy. The solution also does not give the desired wave motion and the boundary and initial conditions cannot be satisfied; thus, we cannot have $\mu > 0$. Similar arguments prohibit the use of $\mu = 0$. Hence, we are left with $\mu < 0$. For simplicity, let

$$\sqrt{\mu} = i\beta, \tag{8.6.19}$$

where β is a real number and i is $\sqrt{-1}$. In this case, Eq. 8.6.16 becomes

$$T(t) = c_1 e^{i\beta at} + c_2 e^{-i\beta at}, \tag{8.6.20}$$

and Eq. 8.6.17 becomes

$$X(x) = c_3 e^{i\beta x} + c_4 e^{-i\beta x}. \tag{8.6.21}$$

Using the relation

$$e^{i\theta} = \cos\theta + i\sin\theta, \tag{8.6.22}$$

Eqs. 8.6.20 and 8.6.21 may be rewritten as

$$T(t) = A\sin\beta at + B\cos\beta at \tag{8.6.23}$$

and

$$X(x) = C \sin \beta x + D \cos \beta x, \qquad (8.6.24)$$

where A, B, C, and D are new constants. The relation of the new constants in terms of the constants c_1, c_2, c_3, and c_4 is left as an exercise.

Now that we have solutions to Eqs. 8.6.9 and 8.6.10 that are periodic in time and space, let us attempt to satisfy the boundary conditions and initial conditions given in Eqs. 8.6.2 through 8.6.5. Our solution thus far is

$$u(x, t) = (A \sin \beta a t + B \cos \beta a t)(C \sin \beta x + D \cos \beta x).$$

$$(8.6.25)$$

The boundary condition $u(0, t) = 0$ states that u is zero for all t at $x = 0$; that is,

$$u(0,t) = (A \sin \beta a t + \beta \cos \beta a t)D = 0. \qquad (8.6.26)$$

The only way this is possible is if $D = 0$. Hence, we are left with

$$u(x, t) = (A \sin \beta a t + \beta \cos \beta a t)C \sin \beta x, \qquad (8.6.27)$$

The boundary condition $u(L, t) = 0$ states that u is zero for all t at $x = L$; this is expressed as

$$u(L, t) = (A \sin \beta a t + B \cos \beta a t)C \sin \beta L = 0, \qquad (8.6.28)$$

which is possible if and only if

$$\sin \beta L = 0. \qquad (8.6.29)$$

For this to be true, we must have

$$\beta L = n\pi, \qquad n = 1, 2, 3, \ldots \qquad (8.6.30)$$

or $\beta = n\pi/L$; the quantity β is called an *eigenvalue*. When the β is substituted back into $\sin \beta x$, the function $\sin n\pi x/L$ is called an *eigenfunction*. Each eigenvalue corresponding to a particular value of n produces a unique eigenfunction. Note that the $n = 0$ eigenvalue ($\mu = 0$) has already been eliminated as a possible solution, so it is not included here. The solution given in Eq. 8.6.27 may now be written as

$$u(x, t) = \left(A \sin \frac{n\pi a t}{L} + B \cos \frac{n\pi a t}{L}\right)C \sin \frac{n\pi x}{L}. \qquad (8.6.31)$$

For simplicity, let us make the substitutions

$$AC = a_n, \qquad BC = b_n \qquad (8.6.32)$$

since each value of n may require different constants. Equation 8.6.31 is then

$$u_n(x, t) = \left(a_n \sin \frac{n\pi a t}{L} + b_n \cos \frac{n\pi a t}{L}\right) \sin \frac{n\pi x}{L}, \qquad (8.6.33)$$

where the subscript n has been added to $u(x, t)$ to allow for a different function for each value of n.

For the vibrating string, each value of n results in harmonic motion of the string with frequency $na/2L$ cycles per second (hertz). For $n = 1$ the *fundamental mode* results, and for $n > 1$ *overtones* result (see Fig. 8.11). *Nodes* are those points of the string which do not move. The velocity $\partial u_n / \partial t$ is then

$$\frac{\partial u_n}{\partial t} = \frac{n\pi a}{L}\left(a_n \cos \frac{n\pi at}{L} - b_n \sin \frac{n\pi at}{L}\right) \sin \frac{n\pi x}{L}. \qquad (8.6.34)$$

Thus, to satisfy the boundary conditions 8.6.4,

$$\frac{\partial u_n}{\partial t}(x, 0) = \frac{n\pi a}{L} a_n \sin \frac{n\pi x}{L} = 0 \qquad (8.6.35)$$

for all x, we must have $a_n = 0$. We are now left with

$$u_n(x, t) = b_n \cos \frac{n\pi at}{L} \sin \frac{n\pi x}{L}. \qquad (8.6.36)$$

Finally, we must satisfy boundary condition (8.6.5),

$$u_n(x, 0) = f(x). \qquad (8.6.37)$$

But, unless $f(x)$ is a multiple of $\sin (n\pi x/L)$, no one value of n will satisfy Eq. 8.6.37. How do we then satisfy the boundary condition $u(x, 0) = f(x)$ if $f(x)$ is not a sine function?

Equation 8.6.36 is a solution of Eq. 8.6.1 and satisfies Eqs. 8.6.2 through 8.6.4 for all n, $n = 1, 2, 3, \ldots$. Hence, any linear combination of any of

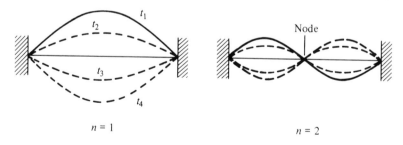

$n = 1$ $\qquad\qquad\qquad\qquad$ $n = 2$

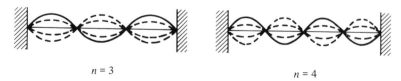

$n = 3$ $\qquad\qquad\qquad\qquad$ $n = 4$

FIGURE 8.11.   Harmonic motion. The solution of various values of time t is as shown.

the solutions

$$u_n(x, t) = b_n \cos \frac{n\pi a t}{L} \sin \frac{n\pi x}{L}, \qquad n = 1, 2, 3, \ldots, \qquad (8.6.38)$$

is also a solution, since the describing equation is linear and therefore super-position is possible. If we assume that for the most general function $f(x)$ we need to consider all values of n, then we should try

$$u(x, t) = \sum_{n=1}^{\infty} u_n(x, t) = \sum_{n=1}^{\infty} b_n \cos \frac{n\pi a t}{L} \sin \frac{n\pi x}{L}. \qquad (8.6.39)$$

To match the initial conditions (8.6.5) we have

$$u(x, 0) = \sum_{n=1}^{\infty} b_n \sin \frac{n\pi x}{L} = f(x). \qquad (8.6.40)$$

If constants b_n can be determined to satisfy Eq. 8.6.40, then Eq. 8.6.39 represents a solution for those domains in which Eq. 8.6.39 converges. The series in Eq. 8.6.40 is a *Fourier sine series*. It was presented in Section 7.3.3, but the essential features will be repeated here.

To find the b_n's, multiply the right side of Eq. 8.6.40 by $\sin (m\pi x/L)$ to give

$$\sin \frac{m\pi x}{L} \sum_{n=1}^{\infty} b_n \sin \frac{n\pi x}{L} = f(x) \sin \frac{m\pi x}{L}. \qquad (8.6.41)$$

Now integrate both sides of Eq. 8.6.41 from $x = 0$ to $x = L$. We may take $\sin m\pi x/L$ inside the sum, since it is a constant as far as the summation is concerned. The integral and the summation may be switched if the series converges properly. This may be done for most functions of interest in physical applications. Thus, we have

$$\sum_{n=1}^{\infty} b_n \int_0^L \sin \frac{n\pi x}{L} \sin \frac{m\pi x}{L} \, dx = \int_0^L f(x) \sin \frac{m\pi x}{L} \, dx. \qquad (8.6.42)$$

With the use of trigonometric identities* we can verify that

$$\int_0^L \sin \frac{n\pi x}{L} \sin \frac{m\pi x}{L} \, dx = \begin{cases} 0, & \text{if } m \neq n \\ \dfrac{L}{2}, & \text{if } m = n. \end{cases} \qquad (8.6.43)$$

Hence, Eq. 8.6.42 gives us

$$b_n = \frac{2}{L} \int_0^L f(x) \sin \frac{n\pi x}{L} \, dx \qquad (8.6.44)$$

---

*Use the trigonometric identities $\sin \alpha \sin \beta = \frac{1}{2} [\cos(\alpha - \beta) - \cos (\alpha + \beta)]$ and $\sin^2 \alpha = \frac{1}{2} - \frac{1}{2} \cos 2\alpha$.

if $f(x)$ may be expressed by

$$f(x) = \sum_{n=1}^{\infty} b_n \sin \frac{n\pi x}{L}. \qquad (8.6.45)$$

Equation 8.6.45 gives the Fourier sine series representation of $f(x)$ with the coefficients given by Eq. 8.6.44. Examples will illustrate the use of the above equations for particular functions $f(x)$.

Example 8.6.1:   A tight string 2 m long with $a = 30$ m/s is initially at rest but is given an initial velocity of $300 \sin 4\pi x$ from its equilibrium position. Determine the maximum displacement at the $x = \frac{1}{8}$ m location of the string.

SOLUTION: We assume that the solution to the describing differential equation

$$\frac{\partial^2 u}{\partial t^2} = 900 \frac{\partial^2 u}{\partial x^2}$$

can be separated as

$$u(x, t) = T(t)X(x).$$

Following the procedure of the previous section, we substitute into the describing equation to obtain

$$\frac{1}{900} \frac{T''}{T} = \frac{X''}{X} = -\beta^2,$$

where we have chosen the separation constant to be $-\beta^2$ so that an oscillatory motion will result. The two ordinary differential equations that result are

$$T'' + 900\beta^2 T = 0, \qquad X'' + \beta^2 X = 0.$$

The general solutions to the equations above are

$$T(t) = A \sin 30\,\beta t + B \cos 30\beta t$$

$$X(x) = C \sin \beta x + D \cos \beta x.$$

The solution for $u(x, t)$ is

$$u(x, t) = (A \sin 30\beta t + B \cos 30\beta t)(C \sin \beta x + D \cos \beta x).$$

The end at $x = 0$ remains motionless; that is, $u(0, t) = 0$. Hence,

$$u(0, t) = (A \sin 30\,\beta t + B \cos 30\beta t)(0 + D) = 0.$$

Thus, $D = 0$. The initial displacement $u(x, 0) = 0$. Hence,

$$u(x, 0) = (0 + B)C \sin \beta x = 0.$$

Thus, $B = 0$. The solution reduces to

$$u(x, t) = AC \sin 30\beta t \sin \beta x.$$

The initial velocity $\partial u/\partial t$ is given as $300 \sin 4\pi x$. We then have, at $t = 0$,

$$\frac{\partial u}{\partial t} = 30\beta AC \sin \beta x = 300 \sin 4\pi x.$$

This gives

$$\beta = 4\pi, \qquad AC = \frac{300}{30(4\pi)} = \frac{2.5}{\pi}.$$

The solution for the displacement is finally

$$u(x, t) = \frac{2.5}{\pi} \sin 120\pi t \sin 4\pi x.$$

We have not imposed the condition that the end at $x = 2$ m is motionless. Put $x = 2$ in the expression above and it is obvious that this boundary condition is satisfied; thus we have found an accceptable solution.

The maximum displacement at $x = 1/8$ m occurs when $\sin 120\pi t = 1$. thus, the maximum displacement is

$$u_{max} = \frac{2.5}{\pi} \text{ m.}$$

Note that we did not find it necessary to use the general expression given by Eq. 8.6.39. We could have, but it would have required more work to obtain a solution. This happened because the initial condition was given as a sine function. Other functions may require the more general form given by Eq. 8.6.39. ∎

Example 8.6.2:   Determine several coefficients in the series solution for $u(x, t)$ if

$$f(x) = \begin{cases} 0.1x, & 0 \le x \le 1 \\ 0.2 - 0.1x, & 1 < x \le 2. \end{cases}$$

The string is 2 m long. Use the boundary and initial conditions of Section 8.6.

SOLUTION: The solution for the displacement of the string is given by Eq. 8.6.39. It is

$$u(x, t) = \sum_{n=1}^{\infty} b_n \cos \frac{n\pi at}{2} \sin \frac{n\pi x}{2},$$

where we have used $L = 2$ m. The coefficients b_n are related to the initial displacement $f(x)$ by Eq. 8.6.44,

$$b_n = \frac{2}{2} \int_0^2 f(x) \sin \frac{n\pi x}{2} dx.$$

Substituting for $f(x)$ results in

$$b_n = 0.1 \int_0^1 x \sin \frac{n\pi x}{2} dx + 0.1 \int_1^2 (2 - x) \sin \frac{n\pi x}{2} dx.$$

Performing the integrations (integration by parts* is required) gives

$$b_n = 0.1 \left[-\frac{2x}{n\pi} \cos \frac{n\pi x}{2} + \frac{4}{n^2\pi^2} \sin \frac{n\pi x}{2} \right]_0^1$$

$$+ 0.1 \left[-\frac{4}{n\pi} \cos \frac{n\pi x}{2} + \frac{2x}{n\pi} \cos \frac{n\pi x}{2} - \frac{4}{n^2\pi^2} \sin \frac{n\pi x}{2} \right]_1^2.$$

By being careful in reducing this result, we have

$$b_n = \frac{0.8}{\pi^2 n^2} \sin \frac{n\pi}{2}.$$

This gives several b_n's as

$$b_1 = \frac{0.8}{\pi^2}, \quad b_2 = 0, \quad b_3 = -\frac{0.8}{9\pi^2}, \quad b_4 = 0, \quad b_5 = \frac{0.8}{25\pi^2}.$$

The solution is, finally,

$$u(x, t) = \frac{0.8}{\pi^2} \left[\cos \frac{\pi a t}{2} \sin \frac{\pi x}{2} - \frac{1}{9} \cos \frac{3\pi a t}{2} \sin \frac{3\pi x}{2} \right.$$

$$\left. + \frac{1}{25} \cos \frac{5\pi a t}{2} \sin \frac{5\pi x}{2} + \cdots \right].$$

We see that the amplitude of each term is getting smaller and smaller. A good approximation results if we keep several terms (say five) and simply ignore the rest. This, in fact, was done before the advent of the computer. With the computer many more terms can be retained, with accurate numbers resulting from the calculations. A computer plot of the solution above is shown for $a = 100$ m/s. One hundred terms were retained.

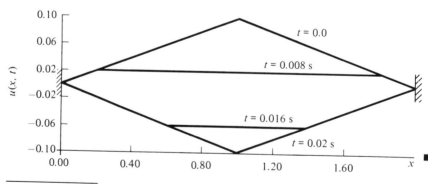

*We shall integrate $\int_0^\pi x \sin x \, dx$ by parts. Let $u = x$ and $dv = \sin x \, dx$. Then $du = dx$ and $v = -\cos x$. The integral is then $\int_0^\pi x \sin x \, dx = -x \cos x \Big|_0^\pi + \int_0^\pi \cos x \, dx = \pi$.

Example 8.6.3:   A tight string, π m long and fixed at both ends, is given an initial displacement $f(x)$ and an initial velocity $g(x)$. Find an expression for $u(x, t)$.

SOLUTION: We follow the steps of Section 8.6 and find the general solution to be

$$u(x, t) = (A \sin \beta at + B \cos \beta at)(C \sin \beta x + D \cos \beta x).$$

Using the boundary condition in which the left end is fixed, that is, $u(0, t) = 0$, we have $D = 0$. We also have the boundary condition $u(\pi, t) = 0$, giving

$$0 = (A \sin \beta at + B \cos \beta at)C \sin \beta \pi.$$

If we let $C = 0$, a trivial solution results, $u(x, t) = 0$. Thus, we must let

$$\beta \pi = n\pi$$

or $\beta = n$, an integer. The general solution is then

$$u_n(x, t) = (a_n \sin nat + b_n \cos nat)\sin nx$$

where the subscript n on $u_n(x, t)$ allows for a different $u(x, t)$ for each value of n. The most general $u(x, t)$ is then found by superposing all of the $u_n(x, t)$; that is,

$$u(x, t) = \sum_{n=1}^{\infty} u_n(x, t) = \sum_{n=1}^{\infty} (a_n \sin nat + b_n \cos nat)\sin nx, \qquad (1)$$

Now, to satisfy the initial displacement, we require that

$$u(x, 0) = \sum_{n=1}^{\infty} b_n \sin nx = f(x).$$

Multiply by $\sin mx$ and integrate from 0 to π. Using the results indicated in Eq. 8.6.43, we have

$$b_n = \frac{2}{\pi} \int_0^{\pi} f(x) \sin nx \, dx. \qquad (2)$$

Next, to satisfy the initial velocity, we must have

$$\frac{\partial u}{\partial t}(x, 0) = \sum_{n=1}^{\infty} a_n an \sin nx = g(x).$$

Again, multiply by $\sin mx$ and integrate from 0 to π. Then

$$a_n = \frac{2}{an\pi} \int_0^{\pi} g(x) \sin nx \, dx. \qquad (3)$$

Our solution is now complete. It is given by Eq. 1 with the b_n provided by Eq. 2 and the a_n by Eq. 3. If $f(x)$ and $g(x)$ are specified, numerical values for each b_n and a_n results.   ■

Example 8.6.4:   A tight string, π m long, is fixed at the left end but the right end moves, with displacement $0.2 \sin 15t$. Find $u(x, t)$ if the wave speed is 30 m/s and state the initial conditions if a solution using separation of variables is to be possible.

SOLUTION: Separation of variables leads to the general solution as

$$u(x, t) = (A \sin 30\beta t + B \cos 30\beta t)(C \sin \beta x + D \cos \beta x).$$

The left end is fixed, requiring that $u(0, t) = 0$. Hence, $D = 0$. The right end moves with the displacement $0.2 \sin 15t$; that is,

$$u(x, t) = 0.2 \sin 15t = (A \sin 30\beta t + B \cos 30\beta t)C \sin \beta \pi.$$

This can be satisfied if we let

$$B = 0, \quad \beta = \frac{1}{2} \quad AC = 0.2.$$

The resulting solution for $u(x, t)$ is

$$u(x, t) = 0.2 \sin 15t \sin \frac{x}{2}.$$

The initial displacement $u(x, 0)$ must be zero and the initial velocity must be

$$\frac{\partial u}{\partial t}(x, 0) = 3 \sin \frac{x}{2}.$$

Any other set of initial conditions would not allow a solution using separation of variables. ■

Example 8.6.5:   A tight string is fixed at both ends. A forcing function (such as wind blowing over a wire), applied normal to the string, is given by $\mathcal{F}(t) = Km \sin \omega t$ kilograms per meter of length. Show that resonance occurs whenever $\omega = an\pi/L$.

SOLUTION: The forcing function $\mathcal{F}(t)$ multiplied by the distance Δx can be added to the right-hand side of Eq. 8.2.8. Dividing by $m\Delta x$ results in

$$a^2 \frac{\partial^2 u}{\partial x^2} = \frac{\partial^2 u}{\partial t^2} + K \sin \omega t,$$

where $a^2 = P/m$. This is a nonhomogeneous partial differential equation, since the last term does not contain the dependent variable $u(x, t)$. As with ordinary differential equations that are linear, we can find a particular solution and add it to a family of solutions of the associated homogeneous equation to obtain a set of solutions of the nonhomogeneous equation.

We assume that the forcing function produces a displacement having the same frequency as the forcing function, as is the case with lumped systems. This suggests that the particular solution has the form

$$u_p(x, t) = X(x) \sin \omega t.$$

Substituting this into the partial differential equation gives

$$a^2 X'' \sin \omega t = -X\omega^2 \sin \omega t + K \sin \omega t.$$

The $\sin \omega t$ divides out and we are left with the ordinary differential equation

$$X'' + \frac{\omega^2}{a^2} X = K.$$

The general solution of this nonhomogeneous differential equation is (see Chapter 1)

$$X(x) = c_1 \sin \frac{\omega}{a}x + c_2 \cos \frac{\omega}{a}x + \frac{Ka^2}{\omega^2}.$$

We will force this solution to satisfy the end conditions that apply to the string. Hence,

$$X(0) = 0 = c_2 + \frac{Ka^2}{\omega^2}$$

$$X(L) = 0 = c_1 \sin \frac{\omega L}{a} + c_2 \cos \frac{\omega L}{a} + \frac{Ka^2}{\omega^2}.$$

The equations above give

$$c_2 = -\frac{Ka^2}{\omega^2}, \qquad c_1 = \frac{\dfrac{Ka^2}{\omega^2}\left(\cos \dfrac{\omega L}{a} - 1\right)}{\sin (\omega L/a)}.$$

The particular solution is then

$$u_p(x, t) = \frac{Ka^2}{\omega^2}\left[\frac{\cos \dfrac{\omega L}{a} - 1}{\sin (\omega L/a)} \sin \frac{\omega x}{a} - \cos \frac{\omega x}{a} + 1\right] \sin \omega t$$

The amplitude of the above solution becomes infinite whenever $\sin \omega L/a = 0$ and $\cos \omega L/a \neq 1$. This occurs if and only if

$$\frac{\omega L}{a} = (2n - 1)\pi.$$

Hence, if the input frequency is such that

$$\omega = \frac{(2n - 1)\pi a}{L}, \qquad n = 1, 2, 3, \ldots ,$$

the amplitude of the resulting motion becomes infinitely large. This input frequency is the natural frequency corresponding to the fundamental mode or one of the significant overtones of the string, depending on the value of n. Thus, we see that a number of input frequencies can lead to resonance in the string. This is true of all phenomena modeled by the wave equation. Although we have neglected any type of damping, the ideas presented in this example carry over to the realistic cases of problems involving small damping. ∎

PROBLEMS

1. Express the solution (8.6.36) in terms of the solution (8.5.10). What are f and g?

2. Determine the general solution for the wave equation using separation of variables assuming that the separation constant is zero. Show that this solution cannot satisfy the boundary and/or initial conditions.

3. Verify that

$$u(x, t) = b_n \cos \frac{n\pi a t}{L} \sin \frac{n\pi x}{L}$$

is a solution to Eq. 8.6.1, and the conditions (8.6.2) through (8.6.4).

4. Find the constants A, B, C, and D in Eqs. 8.6.23 and 8.6.24 in terms of the constants c_1, c_2, c_3, and c_4 in Eqs. 8.6.20 and 8.6.21.

5. Determine the relationship of the fundamental frequency of a vibrating string to the mass per unit length, the length of the string, and the tension in the string.

6. If, for a vibrating wire, the original displacement of the 2-m-long stationary wire is given by a) $0.1 \sin x\pi/2$, b) $0.1 \sin 3\pi x/2$, and c) $0.1(\sin \pi x/2 - \sin 3\pi x/2)$, find the displacement function $u(x, t)$. Both ends are fixed, $P = 50$ N, and the mass per unit length is 0.01 kg/m. With what frequency does the wire oscillate? Write the eigenvalue and eigenfunction for part (a).

7. The initial displacement in a 2-m-long string is given by $0.2 \sin \pi x$ and released from rest. Calculate the maximum velocity in the string and state its location.

8. A string π m long is stretched until the wave speed is 40 m/s. It is given an initial velocity of $4 \sin x$ from its equilibrium position. Determine the maximum displacement and state its location and when it occurs.

8. A string 4 m long is stretched until the wave speed is 40 m/s. It is given an initial velocity of $4 \sin x$ from its equilibrium position. Determine the maximum displacement and state its location and when it occurs.

9. A string 4 m long is stretched, resulting in a wave speed of 60 m/s. It is given an initial displacement of $0.2 \sin \pi x/4$ and an initial velocity of $20 \sin \pi x/4$. Find the solution representing the displacement of the string.

10. A 4-m-long stretched string, with $a = 20$ m/s, is fixed at each end. The string is started off by an initial displacement $u(x, 0) = 0.2 \sin \pi x/4$. The initial velocity is zero. Determine the solution for $u(x, t)$.

11. Suppose that we wish to generate the same string vibration as in Problem 10 (a standing half-sine wave with the same amplitude), but we want to start with a zero-displacement, non-zero velocity condition, that is, $u(x, 0) = 0$, $\partial u/\partial t\,(x, 0) = g(x)$. What should $g(x)$ be?

12. For $u(x, 0) = 0.1 \sin \pi x/4$ and $\partial u/\partial t\,(x, 0) = 10 \sin \pi x/4$, what are the arbitrary constants? What is the maximum displacement value $u_{max}(x, t)$, and where does it occur? Let $a = 40$ m/s and $L = 4$ m in the tight string.

Suppose that a tight string is subjected to the following conditions: $u(0, t) = 0$, $u(L, t) = 0$, $\partial u/\partial t\,(x, 0) = 0$. Calculate the first three nonzero terms of the solution $u(x, t)$ if

13. $u(x, 0) = k$

14. $u(x, 0) = \begin{cases} k, & 0 < x < L/2 \\ 0, & L/2 < x < L \end{cases}$

15. $u(x, 0) = \begin{cases} kx, & 0 < x < L/2 \\ k(L - x), & L/2 < x < L \end{cases}$

16. A string π m long is started into motion by giving the middle one-half an initial velocity of 20 m/s. The string is stretched until the wave speed is 60 m/s. Determine the resulting displacement of the string as a function of x and t.

17. The right end of a 6-m-long wire, which is stretched until the wave speed is 60 m/s, is continually moved with the displacement $0.5 \cos 4\pi t$. What is the maximum amplitude of the resulting displacement?

18. The wind is blowing over some suspension cables on a bridge, causing a force that is approximated by the function $0.02 \sin 21\pi t$. Is resonance possible if the force in the cable is 40,000 N, the cable has a mass of 10 kg/m, and it is 15 m long?

19. A circular shaft π m long is fixed at both ends. The middle of the shaft is twisted through an angle α, the remainder of the shaft through an angle proportional to the distance from the nearest end, and then the shaft is released from rest. Determine the subsequent motion expressed as $\theta(x, t)$. Problem 4 of Section 8.2 gives the appropriate wave equation.

8.7 Solution of the Diffusion Equation

This section is devoted to a solution of the diffusion equation developed in Section 8.3. Recall that the diffusion equation is

$$\frac{\partial T}{\partial t} = k\left(\frac{\partial^2 T}{\partial x^2} + \frac{\partial^2 T}{\partial y^2} + \frac{\partial^2 T}{\partial z^2}\right). \tag{8.7.1}$$

This very important phenomenon will be illustrated by heat transfer. The procedure developed for the wave equation is used, but the solution is quite different, owing to the presence of the first derivative with respect to time rather than the second derivative. This requires only one initial condition instead of the two required by the wave equation. We illustrate the solution technique in three separate situations.

8.7.1 A LONG, INSULATED ROD WITH ENDS AT FIXED TEMPERATURES

A long rod, shown in Fig. 8.12, is subjected to an initial temperature distribution along its axis; the rod is insulated on the lateral surface, and the

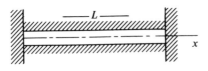

FIGURE 8.12. Heated rod.

ends of the rod are kept at the same constant temperature.* The insulation prevents heat flux in the radial direction; hence, the temperature will depend on the x coordinate only. The describing equation is then the one-dimensional heat equation, given by Eq. 8.3.10, as

$$\frac{\partial T}{\partial t} = k \frac{\partial^2 T}{\partial x^2}. \tag{8.7.2}$$

We hold the ends at $T = 0°$. These boundary conditions are expressed as

$$T(0, t) = 0, \qquad T(L, t) = 0. \tag{8.7.3}$$

Let the initial temperature distribution be represented by

$$T(x, 0) = f(x). \tag{8.7.4}$$

Following the procedure developed for the solution of the wave equation, we assume that the variables separate; that is,

$$T(x, t) = \theta(t)X(x). \tag{8.7.5}$$

Substitution of Eq. 8.7.5 into Eq. 8.7.2 yields

$$\theta' X = k\theta X'', \tag{8.7.6}$$

where $\theta' = d\theta/dt$ and $X'' = d^2X/dx^2$. This is rearranged as

$$\frac{\theta'}{k\theta} = \frac{X''}{X}. \tag{8.7.7}$$

Since the left side is a function of t only and the right side is a function of x only, we set Eq. 8.7.7 equal to a constant λ. This gives

$$\theta' - \lambda k\theta = 0 \tag{8.7.8}$$

and

$$X'' - \lambda X = 0. \tag{8.7.9}$$

The solution of Eq. 8.7.8 is of the form

$$\theta(t) = c_1 e^{\lambda k t}. \tag{8.7.10}$$

Equation 8.7.9 yields the solution

$$X(x) = c_2 e^{\sqrt{\lambda}x} + c_3 e^{-\sqrt{\lambda}x}. \tag{8.7.11}$$

---

*We choose the temperature of the ends in the illustration to be 0°C. Note, however, that they could be held at any temperature T_0. Since it is necessary to have the ends maintained at zero, we simply define a new variable $\theta = T - T_0$, so that $\theta = 0$ at both ends. We would then find a solution for $\theta(x, t)$ with the desired temperature given by $T(x, t) = \theta(x, t) + T_0$.

Again, we must decide whether

$$\lambda > 0, \qquad \lambda = 0, \qquad \lambda < 0. \qquad (8.7.12)$$

For $\lambda > 0$, Eq. 8.7.10 shows that the solution has unbounded temperature for large t due to exponential growth; of course, this is not physically possible. For $\lambda = 0$, the solution is independent of time. Again our physical intuition tells us this is not possible. Therefore, we are left with $\lambda < 0$. Let

$$\beta^2 = -\lambda \qquad (8.7.13)$$

so that

$$\beta^2 > 0 \qquad (8.7.14)$$

The solutions, Eqs. 8.7.10 and 8.7.11 may then be written as

$$\theta(t) = Ae^{-\beta^2 kt} \qquad (8.7.15)$$

and

$$X(x) = B \sin \beta x + C \cos \beta x, \qquad (8.7.16)$$

where A, B, and C are constants to be determined. Therefore, the solution is

$$T(x, t) = Ae^{-\beta^2 kt}[B \sin \beta x + C \cos \beta x]. \qquad (8.7.17)$$

The first condition of Eq. 8.7.3 implies that

$$C = 0. \qquad (8.7.18)$$

Therefore, the solution reduces to

$$T(x, t) = De^{-\beta^2 kt} \sin \beta x, \qquad (8.7.19)$$

where $D = AB$. The second boundary condition of Eq. 8.7.3 requires that

$$\sin \beta L = 0. \qquad (8.7.20)$$

This is satisfied if

$$\beta L = n\pi, \qquad \text{or } \beta = n\pi x/L. \qquad (8.7.21)$$

The constant β is the eigenvalue, and the function $\sin n\pi x/L$ is the eigenfunction. The solution is now

$$T(x, t) = \sum_{n=1}^{\infty} T_n(x, t) = \sum_{n=1}^{\infty} D_n e^{-kn^2\pi^2 t/L^2} \sin \frac{n\pi x}{L}. \qquad (8.7.22)$$

The initial condition, 8.7.4, will be satisfied at $t = 0$ if

$$T(x, 0) = f(x) = \sum_{n=1}^{\infty} D_n \sin \frac{n\pi x}{L}, \qquad (8.7.23)$$

that is, if $f(x)$ can be expanded in a convergent Fourier sine series. If such is the case, the coefficients will be given by (see Eq. 8.6.44)

$$D_n = \frac{2}{L} \int_0^L f(x) \sin \frac{n\pi x}{L} dx \qquad (8.7.24)$$

and the separation-of-variables technique is successful.

It should be noted again that all solutions of partial differential equations cannot be found by separation of variables; in fact, it is only a very special set of boundary conditions that allows us to separate the variables. For example, Eq. 8.7.20 would obviously not be useful in satisfying the boundary condition $T(L, t) = 20t$. Separation of variables would then be futile. Numerical methods could be used to find a solution, or other analytical techniques not covered in this book would be necessary.

Example 8.7.1:   A long copper rod with insulated lateral surface has its left end maintained at a temperature of 0°C and its right end, at $x = 2$ m, maintained at 100°C. Determine the temperature as a function of x and t if the initial condition is given by

$$T(x, 0) = f(x) = \begin{cases} 100x, & 0 < x < 1 \\ 100, & 1 < x < 2. \end{cases}$$

The thermal diffusivity for copper is $k = 1.14 \times 10^{-4}$ m^2/s.

SOLUTION: We again assume the variables separate as

$$T(x, t) = \theta(t)X(x),$$

with the resulting equation,

$$\frac{1}{k}\frac{\theta'}{\theta} = \frac{X''}{X} = \lambda.$$

In this problem the eigenvalue $\lambda = 0$ will play an important role. The solution for $\lambda = 0$ is

$$\theta(t) = C_1, \qquad X(x) = A_1 x + B_1$$

resulting in

$$T(x, t) = C_1(A_1 x + B_1).$$

To satisfy the two end conditions $T(0, t) = 0$ and $T(2, t) = 100$, it is necessary to require $B_1 = 0$ and $A_1 C_1 = 50$. Then

$$T(x, t) = 50x. \qquad (1)$$

This solution is, of course, independent of time, but we will find it quite useful.

Now, we return to the case that allows for exponential decay of temperature, namely $\lambda = -\beta^2$. For this eigenvalue (see Eq. 8.7.17) the solution is

$$T(x, t) = Ae^{-\beta^2 kt}[B \sin \beta x + C \cos \beta x]. \qquad (2)$$

We can superimpose the above two solutions and obtain the more general solution

$$T(x, t) = 50x + Ae^{-\beta^2 kt}[B \sin \beta x + C \cos \beta x].$$

Now let us satisfy the boundary conditions. The left-end condition $T(0, t) = 0$ demands that $C = 0$. The right-end condition demands that

$$100 = 100 + ABe^{-\beta^2 kt} \sin \beta L.$$

This requires $\sin \beta L = 0$, which occurs whenever

$$\beta L = n\pi \quad \text{or} \quad \beta = n\pi/L, \qquad n = 1, 2, 3, \ldots.$$

The solution is then

$$T(x, t) = 50x + \sum_{n=1}^{\infty} D_n e^{-n^2 \pi^2 kt/4} \sin \frac{n\pi x}{2},$$

using $L = 2$. Note that this satisfies the describing equation (8.7.2) and the two boundary conditions. Finally, it must satisfy the initial condition

$$f(x) = 50x + \sum_{n=1}^{\infty} D_n \sin \frac{n\pi x}{2}.$$

We see that if the function $[f(x) - 50x]$ can be expanded in a convergent Fourier sine series, then the solution will be complete. The Fourier coefficients are

$$D_n = \frac{2}{L} \int_0^L [f(x) - 50x] \sin \frac{n\pi x}{L} \, dx$$

$$= \frac{2}{2} \int_0^1 (100x - 50x) \sin \frac{n\pi x}{2} \, dx + \frac{2}{2} \int_1^2 (100 - 50x) \sin \frac{n\pi x}{2} \, dx$$

$$= 50 \left[-\frac{2x}{n\pi} \cos \frac{n\pi x}{2} + \frac{4}{n^2 \pi^2} \sin \frac{n\pi x}{2} \right]_0^1 - \frac{200}{n\pi} \cos \frac{n\pi x}{2} \Big|_1^2$$

$$- 50 \left[-\frac{2x}{n\pi} \cos \frac{n\pi x}{2} + \frac{4}{n^2 \pi^2} \sin \frac{n\pi x}{2} \right]_1^2$$

$$= \frac{400}{n^2 \pi^2} \sin \frac{n\pi}{2}.$$

Using $k = 1.14 \times 10^{-4}$ m/s for copper, we have

$$T(x, t) = 50x + \sum_{n=1}^{\infty} \frac{40.5}{n^2} \sin \frac{n\pi}{2} e^{-2.81 \times 10^{-4} n^2 t} \sin \frac{n\pi x}{2}.$$

which converges for all $t \geq 0$ and all k, $0 \leq x < 2$. Note that the time t is measured in seconds. ■

8.7.2   A LONG, TOTALLY INSULATED ROD

The lateral sides of the long rod are again insulated so that heat transfer occurs only in the x direction along the rod. The temperature in the rod is described by the one-dimensional heat equation

$$\frac{\partial T}{\partial t} = k \frac{\partial^2 T}{\partial x^2}. \tag{8.7.25}$$

For this problem, we have an initial temperature distribution given by

$$T(x, 0) = f(x). \qquad (8.7.26)$$

Since the rod is totally insulated, the heat flux across the end faces is zero. This condition gives, with the use of Eq. 8.3.1,

$$\frac{\partial T}{\partial x}(0, t) = 0, \qquad \frac{\partial T}{\partial x}(L, t) = 0. \qquad (8.7.27)$$

We assume that the variables separate,

$$T(x, t) = \theta(t)X(x). \qquad (8.7.28)$$

Substitute into Eq. 8.7.25, to obtain

$$\frac{\theta'}{k\theta} = \frac{X''}{X} = -\beta^2, \qquad (8.7.29)$$

where $-\beta^2$ is a negative real number. Equation 8.7.29 gives

$$\theta' = -\beta^2 k\theta \qquad (8.7.30)$$

and

$$X'' + \beta^2 X = 0. \qquad (8.7.31)$$

The equations have solutions in the form

$$\theta(t) = Ae^{-\beta^2 kt} \qquad (8.7.32)$$

and

$$X(x) = B \sin \beta x + C \cos \beta x. \qquad (8.7.33)$$

The first boundary condition of 8.7.27 implies that $B = 0$, and the second requires

$$\frac{\partial X}{\partial x}(L) = -C\beta \sin \beta L = 0. \qquad (8.7.34)$$

This can be satisfied by setting

$$\sin \beta L = 0; \qquad (8.7.35)$$

hence, the eigenvalues are

$$\beta = \frac{n\pi}{L}, \qquad n = 0, 1, 2, \ldots . \qquad (8.7.36)$$

Thus, the independent solutions are of the form

$$T_n(x, t) = a_n e^{-n^2\pi^2 kt/L^2} \cos \frac{n\pi x}{L}, \qquad (8.7.37)$$

where the constant a_n replaces AC. A family of solutions is then

$$T(x,\ t) = \sum_{n=0}^{\infty} a_n e^{-(n^2 \pi^2 k/L^2)t} \cos \frac{n\pi x}{L}. \qquad (8.7.38)$$

Note that we retain the $\beta = 0$ eigenvalue in the series.

The initial condition is given by Eq. 8.7.26. It demands that

$$f(x) = \sum_{n=0}^{\infty} a_n \cos \frac{n\pi x}{L}. \qquad (8.7.39)$$

Using trigonometric identities (see Eq. 8.6.43) we can show that

$$\int_0^L \cos \frac{n\pi x}{L} \cos \frac{m\pi x}{L}\, dx = \begin{cases} 0, & m \neq n \\ L/2, & m = n \neq 0 \\ L, & m = n = 0. \end{cases} \qquad (8.7.40)$$

Multiply both sides of Eq. 8.7.39 by cos $m\pi x/L$ and integrate from 0 to L. We then have*

$$a_0 = \frac{1}{L} \int_0^L f(x)\, dx, \qquad a_n = \frac{2}{L} \int_0^L f(x) \cos \frac{n\pi x}{L}\, dx. \qquad (8.7.41)$$

The solution is finally

$$T(x,\ t) = \sum_{n=0}^{\infty} a_n e^{-(n^2 \pi^2 k/L^2)t} \cos \frac{n\pi x}{L}. \qquad (8.7.42)$$

Thus, the temperature distribution can be determined provided that $f(x)$ has a convergent Fourier cosine series.

Example 8.7.2:   A long, laterally insulated stainless steel rod has heat generation occurring within the rod at the constant rate of 4140 W/m^3. The right end is insulated and the left end is maintained at 0°C. Find an expression for $T(x,\ t)$ if the initial temperature distribution is

$$T(x,\ 0) = f(x) = \begin{cases} 100x, & 0 < x < 1 \\ 200 - 100x, & 1 < x < 2 \end{cases}$$

for the 2-m-long rod. Use the specific heat $C = 460$ J/kg $\cdot$ °C, $\rho = 7820$ kg/m^3, and $k = 3.86 \times 10^{-6}$ m^2/s.

SOLUTION: To find the appropriate describing equation, we must account for the heat generated in the infinitesimal element of Fig. 8.7. To Eq. 8.3.8 we add a heat-generation term,

$$\phi(x,\ y,\ z,\ t)\, \Delta x\, \Delta y\, \Delta z\, \Delta t$$

---

*Note that it is often the practice to define a_0 as $a_0 = (2/L) \int_0^L f(x)\, dx$ and then to write the solution as $T(x,\ t) = a_0/2 + \sum_{n=1}^{\infty} a_n e^{-n^2 \pi^2 kt/L^2} \cos (n\pi x/L)$. This was done in Chapter 7. Both methods are, of course, equivalent.

where $\phi(x, y, z, t)$ is the amount of heat generated per volume per unit time. The one-dimensional heat equation then takes the form

$$\frac{\partial T}{\partial t} = k\frac{\partial^2 T}{\partial x^2} + \frac{\phi}{\rho C}.$$

For the present example the describing equation is

$$\frac{\partial T}{\partial t} = k\frac{\partial^2 T}{\partial x^2} + \frac{4140}{7820 \cdot 460}.$$

This nonhomogeneous, partial differential equation is solved by finding a particular solution and adding it to the solution of the homogeneous equation

$$\frac{\partial T}{\partial t} = k\frac{\partial^2 T}{\partial x^2}.$$

The solution of the homogeneous equation is

$$T(x, t) = Ae^{-\beta^2 kt}[B \sin \beta x + C \cos \beta x].$$

The left-end boundary condition is $T(0, t) = 0$, resulting in $C = 0$. The insulated right end requires that $\partial T/\partial x\,(L, t) = 0$. This results in $\cos \beta L = 0$. Thus, the quantity βL must equal $\pi/2, 3\pi/2, 5\pi/2, \ldots$. This is accomplished by using

$$\beta = \frac{(2n - 1)\pi}{2L}. \qquad n = 1, 2, 3, \ldots.$$

The homogeneous solution is, using $k = 3.86 \times 10^{-6}$ and $L = 2$,

$$T(x, t) = \sum_{n=1}^{\infty} D_n e^{-2.38 \times 10^{-6}(2n-1)^2 t} \sin\left(\frac{2n - 1}{4}\,\pi x\right).$$

To find the particular solution, we note that the generation of heat is independent of time. Since the homogeneous solution decays to zero with time, we anticipate that the heat-generation term will lead to a steady-state temperature distribution. Thus, we assume the particular solution is independent of time, that is

$$T_p(x, t) = g(x).$$

Substituting this into the describing equation leads to

$$0 = 3.86 \times 10^{-6}g'' + 1.15 \times 10^{-3}.$$

The solution of this ordinary differential equation is

$$g(x) = -149x^2 + c_1 x + c_2.$$

This solution must also satisfy the boundary condition at the left end, yielding $c_2 = 0$ and the boundary condition at the right end ($g' = 0$), giving $c_1 = 596$. The complete solution, which must satisfy the initial condition, is

$$T(x, t) = -149x^2 + 596x + \sum_{n=1}^{\infty} D_n e^{-2.38 \times 10^{-6}(2n-1)^2 t} \sin\left(\frac{2n - 1}{4}\,\pi x\right).$$

To find the unknown coefficients D_n we use the initial condition, which states that

$$f(x) = -149x^2 + 596x + \sum_{n=1}^{\infty} D_n \sin\left(\frac{2n-1}{4}\pi x\right).$$

The coefficients are then

$$D_n = \frac{2}{2} \int_0^2 [f(x) + 149x^2 - 596x] \sin\left(\frac{2n-1}{4}\pi x\right) dx$$

$$= \int_0^1 (149x^2 - 496x) \sin\left(\frac{2n-1}{4}\pi x\right) dx$$

$$+ \int_1^2 (149x^2 - 696x + 200) \sin\left(\frac{2n-1}{4}\pi x\right) dx.$$

The integrals can be integrated by parts and the solution is thereby completed.   ∎

8.7.3   TWO-DIMENSIONAL HEAT CONDUCTION IN A LONG, RECTANGULAR BAR

A long, rectangular bar is bounded by the planes $x = 0$, $x = a$, $y = 0$, and $y = b$. These faces are kept at $T = 0°C$, as shown by the cross section in Fig. 8.13. The bar is heated so that the variation in the z direction may be neglected. Thus, the variation of temperature in the bar is described by

$$\frac{\partial T}{\partial t} = k\left(\frac{\partial^2 T}{\partial x^2} + \frac{\partial^2 T}{\partial y^2}\right). \tag{8.7.43}$$

The initial temperature distribution in the bar is given by

$$T(x, y, 0) = f(x, y). \tag{8.7.44}$$

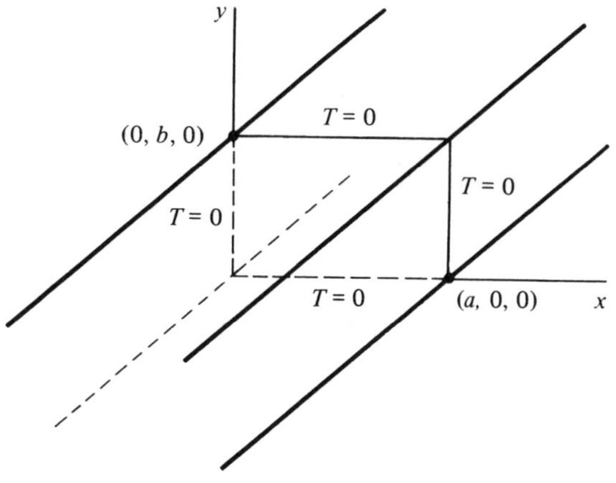

FIGURE 8.13.   Cross section of a rectangular bar.

We want to find an expression for $T(x, y, t)$. Hence, we assume that

$$T(x, y, t) = X(x)Y(y)\theta(t). \tag{8.7.45}$$

After Eq. 8.7.45 is substituted into Eq. 8.7.43, we find that

$$XY\theta' = k(X''Y\theta + XY''\theta). \tag{8.7.46}$$

Equation 8.7.46 may be rewritten as

$$\frac{X''}{X} = \frac{\theta'}{k\theta} - \frac{Y''}{Y}. \tag{8.7.47}$$

Since the left-hand side of Eq. 8.7.47 is a function of x only and the right side is a function of t and y, we may deduce both sides equal the constant value $-\lambda$. (With experience we now anticipate the minus sign.) Therefore, we have

$$X'' + \lambda X = 0 \tag{8.7.48}$$

and

$$\frac{Y''}{Y} = \frac{\theta'}{k\theta} + \lambda. \tag{8.7.49}$$

We use the same argument on Eq. 8.7.49 and set it equal to a constant $-\mu$. That is,

$$\frac{Y''}{Y} = \frac{\theta'}{k\theta} + \lambda = -\mu. \tag{8.7.50}$$

This yields the two ordinary differential equations

$$Y'' + \mu Y = 0 \tag{8.7.51}$$

and

$$\theta' + (\lambda + \mu)k\theta = 0. \tag{8.7.52}$$

The boundary conditions on $X(x)$ are

$$X(0) = 0, \qquad X(a) = 0, \tag{8.7.53}$$

since the temperature is zero at $x = 0$ and $x = a$. Consequently, the solution of Eq. 8.7.48,

$$X(x) = A \sin \sqrt{\lambda}x + B \cos \sqrt{\lambda}x, \tag{8.7.54}$$

reduces to

$$X(x) = A \sin \frac{n\pi x}{a}, \tag{8.7.55}$$

where we have used

$$\lambda = \frac{n^2\pi^2}{a^2}, \qquad n = 1, 2, 3, \ldots. \tag{8.7.56}$$

Similarly, the solution to Eq. 8.7.51 reduces to

$$Y(y) = C \sin \frac{m\pi y}{b}, \tag{8.7.57}$$

where we have employed,

$$\mu = \frac{m^2 \pi^2}{b^2}, \qquad m = 1, 2, 3, \dots \tag{8.7.58}$$

With the use of Eqs. 8.7.56 and 8.7.58 we find the solution of Eq. 8.7.52 is

$$\theta(t) = D e^{-\pi^2 k(n^2/a^2 + m^2/b^2)t}. \tag{8.7.59}$$

Equations 8.7.55, 8.7.57, and 8.7.59 may be combined to give

$$T_{mn}(x, y, t) = a_{mn} e^{-\pi^2 k(n^2/a^2 + m^2/b^2)t} \sin \frac{n\pi x}{a} \sin \frac{m\pi y}{b}, \tag{8.7.60}$$

where the constant a_{mn} replaces ACD. The most general solution is then obtained by superposition, namely,

$$T(x, y, t) = \sum_{m=1}^{\infty} \sum_{n=1}^{\infty} T_{mn}, \tag{8.7.61}$$

and we have

$$T(x, y, t) = \sum_{m=1}^{\infty} \sum_{n=1}^{\infty} a_{mn} e^{-\pi^2 k(n^2/a^2 + m^2/b^2)t} \sin \frac{n\pi x}{a} \sin \frac{m\pi y}{b}. \tag{8.7.62}$$

This is a solution if coefficients a_{mn} can be determined so that

$$T(x, y, 0) = f(x, y) = \sum_{m=1}^{\infty} \left[\sum_{n=1}^{\infty} a_{mn} \sin \frac{n\pi x}{a} \right] \sin \frac{m\pi y}{b}. \tag{8.7.63}$$

We make the grouping indicated by the brackets in Eq. 8.7.63. Thus, for a given x in the range $(0, a)$, we have a Fourier series in y. [For a given x, $f(x, y)$ is a function of y only.] Therefore, the term in the brackets is the constant b_n in the Fourier sine series. Hence,

$$\sum_{n=1}^{\infty} a_{mn} \sin \frac{n\pi x}{a} = \frac{2}{b} \int_0^b f(x, y) \sin \frac{m\pi y}{b} \, dy = F_m(x). \tag{8.7.64}$$

The right-hand side of Eq. 8.7.64 is a series of functions of x, one for each $m = 1, 2, 3, \dots$ Thus, Eq. 8.7.64 is a Fourier sine series for $F_m(x)$. Therefore, we have

$$a_{mn} = \frac{2}{a} \int_0^a F_m(x) \sin \frac{n\pi x}{a} \, dx. \tag{8.7.65}$$

Substitution of Eq. 8.7.64 into Eq. 8.7.65 yields

$$a_{mn} = \frac{4}{ab} \int_0^a \int_0^b f(x, y) \sin \frac{m\pi y}{b} \sin \frac{n\pi x}{a} \, dy \, dx. \qquad (8.7.66)$$

The solution of our problem is Eq. 8.7.62 with a_{mn} given by Eq. 8.7.66, assuming, as usual, that the various series converge.

The latter problem is an extension of the ideas presented earlier. It deals with functions of three independent variables and expansions utilizing two-dimensional Fourier series.

The applications studied so far were presented in rectangular coordinates. One consequence of our choice of problem and coordinate system is the need to expand the function representing the initial condition as a series of sine and cosine terms, the so-called Fourier series. For problems better suited to cylindrical coordinates, a series of Bessel functions is the natural tool; in spherical coordinates expansions in a series of Legendre polynomials is more natural. The following two sections will present the solutions to Laplace's equation in spherical coordinates and cylindrical coordinates, respectively.

Example 8.7.3: The edges of a thin plate are held at the temperatures shown in the sketch. Determine the steady-state temperature distribution in the plate. Assume the large flat surfaces to be insulated.

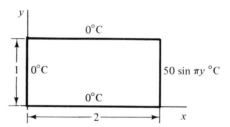

SOLUTION: The describing equation is the heat equation

$$\frac{\partial T}{\partial t} = k\left(\frac{\partial^2 T}{\partial x^2} + \frac{\partial^2 T}{\partial y^2} + \frac{\partial^2 T}{\partial z^2}\right).$$

For the steady-state situation there is no variation of temperature with time; that is, $\partial T/\partial t = 0$. For a thin plate with insulated surfaces we have $\partial^2 T/\partial z^2 = 0$. Thus,

$$\frac{\partial^2 T}{\partial x^2} + \frac{\partial^2 T}{\partial y^2} = 0.$$

This is Laplace's equation. Let us assume that the variables separate; that is,

$$T(x, y) = X(x)Y(y).$$

Then substitute into the describing equation to obtain

$$\frac{X''}{X} = -\frac{Y''}{Y} = \beta^2,$$

where we have chosen the separation constant to be positive to allow for a sinusoidal variation* with y. The ordinary differential equations that result are

$$X'' - \beta^2 X = 0, \qquad Y'' + \beta^2 Y = 0.$$

The solutions are

$$X(x) = A e^{\beta x} + B e^{-\beta x}$$

$$Y(y) = C \sin \beta y + D \cos \beta y.$$

The solution for $T(x, y)$ is then

$$T(x, y) = (A e^{\beta x} + B e^{-\beta x})(C \sin \beta y + D \cos \beta y).$$

Using $T(0, y) = 0$, $T(x, 0) = 0$, and $T(x, 1) = 0$ gives

$$0 = A + B, \qquad 0 = D, \qquad 0 = \sin \beta.$$

The final boundary condition is

$$T(2, y) = 50 \sin \pi y = (A e^{2\beta} + B e^{-2\beta}) C \sin \beta y.$$

From this condition we have

$$\beta = \pi, \qquad 50 = C(A e^{2\beta} + B e^{-2\beta}).$$

From the equations above we can solve for the constants. We have

$$B = -A, \qquad AC = \frac{50}{e^{2\pi} - e^{2\pi}} = 0.0934.$$

Finally, the expression for $T(x, y)$ is

$$T(x, y) = 0.0934(e^{\pi x} - e^{-\pi x}) \sin \pi y.$$

Note that the expression above for the temperature is independent of the material properties; it is a steady-state solution. ■

PROBLEMS

Find $T(x, t)$ in a laterally insulated, 2-m-long rod if $k = 10^{-4}$ m^2/s and

1. $T(x, 0) = 100 \sin \pi x$, $\qquad\qquad$ $T(0, t) = 0$, $\qquad$ $T(2, t) = 0$

2. $T(x, 0) = 100 \sin \pi x/4$, $\qquad\quad$ $T(0, t) = 0$, $\qquad$ $T(2, t) = 100$

3. $T(x, 0) = 80 \cos 3\pi x/4$, $\qquad\quad$ $T(0, t) = 80$, $\qquad$ $T(2, t) = 0$

4. $T(x, 0) = 200(1 + \sin \pi x)$, $\qquad$ $T(0, t) = 200$, $\quad$ $T(2, t) = 200$

5. $T(x, 0) = 100$, $\qquad\qquad\qquad\quad$ $T(0, t) = 0$, $\qquad$ $T(2, t) = 0$

---

*If the right-hand edge were held at a constant temperature we would also choose the separation constant so that $\cos \beta y$ and $\sin \beta y$ appear. This would allow a Fourier series to satisfy the edge condition.

6. $T(x, 0) = 100,$ $\qquad\qquad\qquad\qquad$ $T(0, t) = 0,$ $\qquad$ $T(2, t) = 100$

7. $T(x, 0) = \begin{cases} 100x, & 0 < x < 1 \\ 200 - 100x, & 1 < x < 2, \end{cases}$ $T(0, t) = 0,$ $\qquad$ $T(2, t) = 0$

8. $T(x, 0) = 100(2x - x^2),$ $\qquad\qquad$ $T(0, t) = 0,$ $\qquad$ $T(2, t) = 0$

9. $T(x, 0) = 50x^2,$ $\qquad\qquad\qquad\qquad$ $T(0, t) = 0,$ $\qquad$ $T(2, t) = 200$

Find the temperature at the center of the rod for each problem.

10. Problem 1 at $t = 1000$ s. $\qquad\qquad$ **11.** Problem 1 at $t = \infty$.

12. Problem 2 at $t = \infty$. $\qquad\qquad\qquad$ **13.** Problem 3 at $t = \infty$.

14. Problem 5 at $t = 1$ hr. $\qquad\qquad\quad$ **15.** Problem 2 at $t = 1$ hr.

Calculate the time needed for the center of the rod to reach $50°$ for each problem.

16. Problem 1 $\qquad\qquad$ **17.** Problem 2 $\qquad\qquad$ **18.** Problem 3

19. Problem 5 $\qquad\qquad$ **20.** Problem 7

Find the heat transfer rate from the left and if $d = 20$ cm and $K = 800$ W/m · °C for each problem.

21. Problem 1 at $t = 0$. $\qquad\qquad\qquad$ **22.** Problem 2 at $t = 0$.

23. Problem 2 at $t = 1000$ s. $\qquad\qquad$ **24.** Problem 5 at $t = 0$.

25. Problem 5 at $t = 1000$ s. $\qquad\qquad$ **26.** Problem 3 at $t = 1000$ s.

Sketch the temperature distribution at $t = 0$, 1000 s, and 10 hr for each problem.

27. Problem 1 $\qquad\qquad$ **28.** Problem 2 $\qquad\qquad$ **29.** Problem 3

30. Problem 5 $\qquad\qquad$ **31.** Problem 7

Find $T(x, t)$ in a laterally insulated, π-m-long rod if $k = 10^{-4}$ m²/s and

32. $T(x, 0) = 100 \cos \pi x,$ $\qquad$ $\dfrac{\partial T}{\partial x}(0, t) = 0,$ $\dfrac{\partial T}{\partial x}(\pi, t) = 0$

33. $T(x, 0) = 100$ $\qquad\qquad$ $\dfrac{\partial T}{\partial x}(0, t) = 0,$ $\dfrac{\partial T}{\partial x}(\pi, t) = 0$

34. $T(x, 0) = 100 \sin x/2,$ $\qquad$ $T(0, t) = 0,$ $\quad \dfrac{\partial T}{\partial x}(\pi, t) = 0$

35. $T(x, 0) = 100 \sin x,$ $\qquad\quad$ $\dfrac{\partial T}{\partial x}(0, t) = 0,$ $\quad T(\pi, t) = 0$

36. $T(x, 0) = \begin{cases} 100, & 0 < x < \pi/2, \\ 0, & \pi/2 < x < \pi, \end{cases}$ $T(0, t) = 0,$ $\quad \dfrac{\partial T}{\partial x}(\pi, t) = 0$

For Problem 36, if $d = 10$ cm and $K = 600$ W/m · °C, find

37. The maximum heat transfer rate from the left end.

38. The temperature at the center of the rod at $t = 1000$ s.

39. The time needed for the center of the rod to reach $10°$ C.

Heat generation occurs at the rate of 2000 W/m³ in a laterally insulated, 2-m-long rod. If $C = 400$ J/kg · °C, $\rho = 9000$ kg/m³, and $k = 10^{-4}$ m²/s, find the steady-state temperature distribution if

40. $T(0, t) = 0$ and $T(2, t) = 0$

41. $T(0, t) = 0$ and $\dfrac{\partial T}{\partial x}(2, t) = 0$

42. $T(0, t) = 100$ and $\dfrac{\partial T}{\partial x}(2, t) = 0$

Find $T(x, t)$ in the rod if the initial temperature distribution is constant at 100° for

43. Problem 40   **44.** Problem 41   **45.** Problem 42

Sketch the steady-state temperature distribution in a 1 m × 1 m slab if the flat surfaces are insulated and the edge conditions are as follows:

46. $T(0, y) = 0,$ $T(x, 0) = 0,$ $T(1, y) = 0,$ $T(x, 1) = 100 \sin \pi x$

47. $T(0, y) = 0,$ $T(x, 0) = 0,$ $T(1, y) = 100 \sin \pi y,$ $T(x, 1) = 0$

48. $T(0, y) = 0,$ $T(x, 0) = 0,$ $\dfrac{\partial T}{\partial x}(1, y) = 0,$ $T(x, 1) = 100$

49. $T(0, y) = 0,$ $\dfrac{\partial T}{\partial y}(x, 0) = 0,$ $\dfrac{\partial T}{\partial x}(1, y) = 0,$ $T(x, 1) = 100$

50. $T(0, y) = 100,$ $T(x, 0) = 100,$ $T(1, y) = 200,$ $T(x, 1) = 100$

51. The initial temperature distribution in a 2 m × 2 m rectangular slab is 100°C. Find $T(x, t)$ if all sides are maintained at 0° and $k = 10^{-4}$ m²/s.

8.8   Electric Potential about a Spherical Surface

Consider a spherical surface maintained at an electrical potential V. The potential depends only on ϕ and is given by the function $f(\phi)$. The equation that describes the potential in the region on either side of the spherical surface is Laplace's equation, written in spherical coordinates (shown in Fig. 8.8) as

$$\frac{\partial}{\partial r}\left(r^2 \frac{\partial V}{\partial r}\right) + \frac{1}{\sin \phi}\frac{\partial}{\partial \phi}\left(\sin \phi \frac{\partial V}{\partial \phi}\right) = 0, \qquad (8.8.1)$$

Obviously, one boundary condition requires that

$$V(r_0, \phi) = f(\phi). \qquad (8.8.2)$$

The fact that a potential exists on the spherical surface of finite radius should not lead to a potential at infinite distances from the sphere; hence, we set

$$V(\infty, \phi) = 0. \qquad (8.8.3)$$

We follow the usual procedure of separating variables; that is, assume that

$$V(r, \phi) = R(r)\Phi(\phi). \qquad (8.8.4)$$

This leads to the equations

$$\frac{1}{R}\frac{d}{dr}\left(r^2\frac{dR}{dr}\right) = -\frac{1}{\phi\sin\theta}\frac{d}{d\phi}(\Phi\sin\phi) = \mu, \qquad (8.8.5)$$

which can be written as, letting $\cos\phi = x$, so that $\Phi = \Phi(x)$,

$$r^2R'' + 2rR' - \mu R = 0 \qquad (8.8.6)$$
$$(1 - x^2)\Phi'' - 2x\Phi + \mu\Phi = 0.$$

The first of these is recognized as the Cauchy–Euler equation (see Section 1.10) and has the solution

$$R(r) = c_1 r^{-1/2+\sqrt{\mu + 1/4}} + c_2 r^{-1/2-\sqrt{\mu+1/4}}. \qquad (8.8.7)$$

This is put in better form by letting $-\frac{1}{2} + \sqrt{\mu + \frac{1}{4}} = n$. Then

$$R(r) = c_1 r^n + \frac{c_2}{r^{n+1}}. \qquad (8.8.8)$$

The equation for Φ becomes Legendre's equation (see Section 2.3.2)

$$(1 - x^2)\Phi'' - 2x\Phi' + n(n + 1)\Phi = 0, \qquad (8.8.9)$$

where n must be a positive integer for a proper solution to exist. The general solution to this equation is

$$\Phi(x) = c_3 P_n(x) + c_4 Q_n(x). \qquad (8.8.10)$$

Since $Q_n(x) \to \infty$ as $x \to 1$ (see Eq. 2.3.35), we set $c_4 = 0$. This results in the following solution for $V(r, x)$:

$$V(r, x) = \sum_{n=0}^{\infty} V_n(r, x) = \sum_{n=0}^{\infty} [A_n r^n P_n(x) + B_n r^{-(n+1)} P_n(x)]. \qquad (8.8.11)$$

Let us first consider points inside the spherical surface. The constants $B_n = 0$ if a finite potential is to exist at $r = 0$. We are left with

$$V(r, x) = \sum_{n=0}^{\infty} A_n r^n P_n(x). \qquad (8.8.12)$$

This equation must satisfy the boundary condition

$$V(r_0, x) = f(x) = \sum_{n=0}^{\infty} A_n r_0^n P_n(x). \qquad (8.8.13)$$

The unknown coefficients A_n are found by using the property

$$\int_{-1}^{1} P_m(x)P_n(x)\, dx = \begin{cases} 0 & m \neq n \\ \dfrac{2}{2n + 1} & m = n. \end{cases} \qquad (8.8.14)$$

Multiply both sides of Eq. 8.8.12 by $P_m(x)\,dx$ and integrate from -1 to 1. This gives

$$A_n = \frac{2n+1}{2r_0^n} \int_{-1}^{1} f(x) P_n(x)\,dx. \tag{8.8.15}$$

For a prescribed $f(\phi)$, using $\cos\phi = x$, Eq. 8.8.12 provides us with the solution for interior points with the constants A_n given by Eq. 8.8.15.

For exterior points we require that $A_n = 0$ in Eq. 8.8.11, so the solution is bounded as $x \to \infty$. This leaves the solution

$$V(r,x) = \sum_{n=0}^{\infty} B_n r^{-(n+1)} P_n(x). \tag{8.8.16}$$

This equation must also satisfy the boundary condition

$$f(x) = \sum_{n=0}^{\infty} B_n r_0^{-(n+1)} P_n(x). \tag{8.8.17}$$

Using property 8.8.14, the B_n's are given by

$$B_n = \frac{2n+1}{2} r_0^{n+1} \int_{-1}^{1} f(x) P_n(x)\,dx. \tag{8.8.18}$$

If $f(x)$ is a constant we must evaluate $\int_{-1}^{1} P_n(x)\,dx$. Using Eq. 2.3.31 we can show that

$$\int_{-1}^{1} P_0(x)\,dx = 2, \quad \int_{-1}^{1} P_n(x)\,dx = 0, \quad n = 1, 2, 3, \ldots . \tag{8.8.19}$$

An example will illustrate the application of this presentation for a specific $f(x)$.

Example 8.8.1:   Find the electric potential inside a spherical surface of radius r_0 if the hemispherical surface when $\pi > \phi > \pi/2$ is maintained at a constant potential V_0 and the hemispherical surface when $\pi/2 > \phi > 0$ is maintained at zero potential.

SOLUTION: Inside the sphere of radius r_0, the solution is

$$V(r,x) = \sum_{n=0}^{\infty} A_n r^n P_n(x),$$

where $x = \cos\phi$. The coefficients A_n are given by Eq. 8.8.15,

$$A_n = \frac{2n+1}{2r_0^n} \int_{-1}^{1} f(x)\,P_n(x)\,dx$$

$$= \frac{2n+1}{2r_0^n} \left[\int_{-1}^{0} V_0 P_n(x)\,dx + \int_{0}^{1} 0 \cdot P_n(x)\,dx \right]$$

$$= \frac{2n+1}{2r_0^n} V_0 \int_{-1}^{0} P_n(x)\,dx,$$

where we have used $V = V_0$ for $\pi > \phi > \pi/2$ and $V = 0$ for $\pi/2 > \phi > 0$. Several A_n's can be evaluated, to give (see Eq. 2.3.31)

$$A_0 = \frac{V_0}{2}, \quad A_1 = -\frac{3\,V_0}{4r_0}, \quad A_2 = 0, \quad A_3 = \frac{7\,V_0}{16r_0^3}, \quad A_4 = 0, \quad A_5 = -\frac{11\,V_0}{32r_0^5}.$$

This provides us with the solution, letting $\cos \phi = x$,

$$V(r, \phi) = A_0 P_0 + A_1 r P_1 + A_2 r^2 P_2 + \cdots$$

$$= V_0 \left[\frac{1}{2} - \frac{3}{4} \frac{r}{r_0} \cos \phi + \frac{7}{16} \left(\frac{r}{r_0} \right)^3 P_3(\cos \phi) \right.$$

$$\left. - \frac{11}{32} \left(\frac{r}{r_0} \right)^5 P_5 (\cos \phi) + \cdots \right],$$

where the Legendre polynomials are given by Eq. 2.3.31. Note that the expression above could be used to give a reasonable approximation to the temperature in a solid sphere if the hemispheres are maintained at T_0 and zero degrees, respectively, since Laplace's equation also describes the temperature distribution in a solid body. ∎

A complete analysis requires an investigation of the convergence properties of series of Legendre polynomials, a subject best left for advanced texts.

PROBLEMS

1. The temperature of a spherical surface 0.2 m in diameter is maintained at a temperature of 250°C. This surface is interior to a very large mass. Find an expression for the temperature distribution inside and outside the surface.

2. The temperature on the surface of a 1-m-diameter sphere is $100 \cos \phi$°C. What is the temperature distribution inside the sphere?

3. Find the potential field between two concentric spheres if the potential of the outer sphere is maintained at $V = 100$ and the potential of the inner sphere is maintained at zero. The radii are 2 m and 1 m, respectively.

8.9   Heat Transfer in a Cylindrical Body

Boundary-value problems involving a boundary condition applied to a circular cylindrical surface are encountered quite often in physical situations. The solution of such problems invariably involve Bessel functions, which were introduced in Section 2.10. We shall use the problem of finding the steady-state temperature distribution in the cylinder shown in Fig. 8.14 as an example. Other exercises are included in the Problems.

The partial differential equation describing the phenomenon illustrated in Fig. 8.14 is

$$\frac{\partial T}{\partial t} = k \, \nabla^2 T, \tag{8.9.1}$$

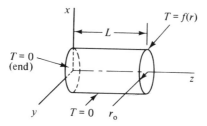

FIGURE 8.14.   Circular cylinder with boundary conditions.

where we have assumed constant material properties. For a steady-state situation using cylindrical coordinates (see Eq. 8.3.14), this becomes

$$\frac{\partial^2 T}{\partial r^2} + \frac{1}{r}\frac{\partial T}{\partial r} + \frac{\partial^2 T}{\partial z^2} = 0, \tag{8.9.2}$$

where, considering the boundary conditions shown in the figure, we have assumed the temperature to be independent of θ. We assume a separated solution of the form

$$T(r, z) = R(r)Z(z), \tag{8.9.3}$$

which leads to the equations

$$\frac{1}{R}\left(R'' + \frac{1}{r}R'\right) = -\frac{Z''}{Z} = -\mu^2, \tag{8.9.4}$$

where a negative sign is chosen on the separation constant since we anticipate an exponential variation with z. We are thus confronted with solving the two ordinary differential equations

$$R'' + \frac{1}{r}R' + \mu^2 R = 0 \tag{8.9.5}$$

$$Z'' - \mu^2 Z = 0. \tag{8.9.6}$$

The solution to Eq. 8.9.6 is simply

$$Z(z) = c_1 e^{\mu z} + c_2 e^{-\mu z} \tag{8.9.7}$$

for $\mu > 0$; for $\mu = 0$, it is

$$Z(z) = c_1' z + c_2'. \tag{8.9.8}$$

This solution may be of use. We note that Eq. 8.9.5 is close to being Bessel's equation 2.10.1 with $\lambda = 0$. By substituting $x = \mu r$, Eq. 8.9.5 becomes

$$x^2 R'' + x R' + x^2 R = 0, \tag{8.9.9}$$

which is Bessel's equation with $\lambda = 0$. It possesses the general solution

$$R(x) = c_3 J_0(x) + c_4 Y_0(x), \tag{8.9.10}$$

where $J_0(x)$ and $Y_0(x)$ are Bessel functions of the first and second kind, respectively. We know (see Fig. 2.5) that $Y_0(x)$ is singular at $x = 0$. (This corresponds to $r = 0$.) Hence, we require that $c_4 = 0$, and the solution to our problem is

$$T(r, z) = J_0(\mu r)[Ae^{\mu z} + Be^{-\mu z}].\qquad(8.9.11)$$

The temperature on the surface at $z = 0$ is maintained at zero degrees. This gives $B = -A$ from the equation above. The temperature at $r = r_0$ is also maintained at zero degrees; that is,

$$T(r_0, z) = 0 = AJ_0(\mu r_0)[e^{\mu z} - e^{-\mu z}].\qquad(8.9.12)$$

The Bessel function $J_0(\mu r_0)$ has infinitely many roots none of which are zero. These roots permit a solution of Eq. 8.9.12 analogous to the trigonometric situation. Since none of the roots is zero, the $\mu = 0$ eigenvalue is not of use. Let the nth root be designated μ_n. Four such roots are shown in Fig. 2.4 and are given numerically in the Appendix.

Returning to Eq. 8.9.11, our solution is now

$$T(r, z) = \sum_{n=1}^{\infty} T_n(r, z) = \sum_{n=1}^{\infty} J_0(\mu_n r)A_n[e^{\mu_n z} - e^{-\mu_n z}].\qquad(8.9.13)$$

This solution must allow the final end condition to be satisfied. It is

$$T(r, L) = f(r) = \sum_{n=1}^{\infty} A_n J_0(\mu_n r)[e^{\mu_n L} - e^{-\mu_n L}].\qquad(8.9.14)$$

Once more we assume that the series converges. We must now use the property that

$$\int_0^b xJ_j(\mu_n x)J_j(\mu_m x)\, dx = \begin{cases} 0 & n \neq m \\ \dfrac{b^2}{2}J_{j+1}^2(\mu_n b), & n = m, \end{cases}\qquad(8.9.15)$$

where the μ_n are the roots of the equation $J_j(\mu r_0) = 0$. This permits the coefficients A_n to be determined from

$$A_n = \frac{2(e^{\mu_n L} - e^{-\mu_n L})^{-1}}{r_0^2 J_1^2(\mu_n r_0)} \int_0^{r_0} rf(r) J_0(\mu_n r)\, dr,\qquad(8.9.16)$$

letting $j = 0$. This completes the solution. For a specified $f(r)$ for the temperature on the right end, Eq. 8.9.13 gives the temperature at any interior point if the coefficients are evaluated using Eq. 8.9.16. This process will be illustrated with an example.

Example 8.9.1: Determine the steady-state temperature distribution in a 2-unit-long, 4-unit-diameter circular cylinder with one end maintained at 0°C, and the other end at $100r$°C, and the lateral surface insulated.

SOLUTION: Following the solution procedure outlined in the previous section, the solution is

$$T(r, z) = J_0(\mu r)[Ae^{\mu z} + Be^{-\mu z}].$$

The temperature at the base where $z = 0$ is zero. Thus, $B = -A$ and

$$T(r, z) = AJ_0(\mu r)[e^{\mu z} - e^{-\mu z}].$$

On the lateral surface where $r = 2$, the heat transfer is zero, requiring that

$$\frac{\partial T}{\partial r}(2, z) = 0 = AJ_0'(2\mu)[e^{\mu z} - e^{-\mu z}]$$

or

$$J_0'(2\mu) = 0.$$

There are infinitely many values of μ that meet this condition, the first of which is $\mu = 0$. Let the nth one be μ_n, the eigenvalue. The solution corresponding to this eigenvalue is

$$T_n(r, z) = A_n J_0(\mu_n r)[e^{\mu_n z} - e^{-\mu_n z}],$$

for $\mu_n > 0$; for $\mu_1 = 0$, the solution is, using Eq. 8.9.8,

$$T_1(r, z) = A_1 z.$$

The general solution is then found by superimposing all the individual solutions, resulting in

$$T(r, z) = \sum_{n=1}^{\infty} T_n(r, z) = A_1 z + \sum_{n=2}^{\infty} A_n J_0(\mu_n r)[e^{\mu_n z} - e^{-\mu_n z}].$$

The remaining boundary condition is that the end at $z = 2$ is maintained at $100r\,°C$; that is,

$$T(r, 2) = 100r = 2A_1 + \sum_{n=2}^{\infty} A_n J_0(\mu_n r)[e^{2\mu_n} - e^{-2\mu_n}].$$

We must be careful, however, and not assume that the A_n in this series are given by Eq. 8.9.16; they are not, since the roots μ_n are not to the equation $J_0(\mu r_0) = 0$, but to $J_0'(\mu r_0) = 0$. The property analogous to Eq. 8.9.15 takes the form

$$\int_0^{r_0} x J_j(\mu_n x) J_j(\mu_m x)\, dx = \begin{cases} 0, & n \neq m \\ \dfrac{\mu_n^2 r_0^2 - j^2}{2\mu_n^2} J_j^2(\mu_n r_0), & n = m \end{cases}$$

whenever μ_n are the roots of $J_j'(\mu r_0) = 0$. The coefficients A_n are then given by

$$A_n = \frac{2(e^{2\mu_n} - e^{-2\mu_n})^{-1}}{r_0^2 J_0^2(\mu_n r_0)} \int_0^{r_0} r f(r) J_0(\mu_n r)\, dr,$$

where $j = 0$ and $f(r) = 100r$. For the first root, $\mu_1 = 0$, the coefficient is

$$A_1 = \frac{2}{r_0^2} \int_0^{r_0} r f(r)\, dr.$$

Some of the coefficients are, using $\mu_1 = 0$, $\mu_2 = 1.916$, $\mu_3 = 3.508$,

$$A_1 = \frac{2}{2^2} \int_0^2 r(100r) \, dr = \frac{400}{3}$$

$$A_2 = \frac{2(e^{3.832} - e^{-3.832})^{-1}}{2^2 \times 0.403^2} \int_0^2 r(100r) J_0(1.916r) \, dr$$

$$= 6.68 \int_0^2 r^2 J_0(1.916r) \, dr = 0.951 \int_0^{3.832} x^2 J_0(x) \, dx$$

$$A_3 = \frac{2(e^{7.016} - e^{-7.016})^{-1}}{2^2 \times 0.300^2} \int_0^2 r(100r) J_0(3.508r) \, dr$$

$$= 0.501 \int_0^2 r^2 J_0(3.508r) \, dr = 0.0117 \int_0^{7.016} x^2 J_0(x) \, dx.$$

The integrals above could be easily evaluated by use of a computer integration scheme. Such a scheme will be presented in Chapter 9. The solution is then

$$T(r, z) = \frac{400}{3}z + A_2 J_0(1.916r)[e^{1.916z} - e^{-1.916z}]$$

$$+ A_3 J_0(3.508r)[e^{3.508z} - e^{-3.508z}] + \cdots. \quad \blacksquare$$

PROBLEMS

1. A right circular cylinder is 1 m long and 2 m in diameter. Its left end and lateral surface are maintained at a temperature of 0°C and its right end at 100°C. Find an expression for its temperature at any interior point. Calculate the first three coefficients in the series expansion.

2. Determine the solution for the temperature as a function of r and t in a circular cylinder of radius r_0 with insulated (or infinitely long) ends if the initial temperature distribution is a function $f(r)$ of r only and the lateral surface is maintained at 0°C (see Eq. 8.3.14).

3. An aluminum circular cylinder 50 mm in diameter with ends insulated is initially at 100°C. Approximate the temperature at the center of the cylinder after 2 s if the lateral surface is kept at 0°C. For aluminum, $k = 8.6 \times 10^{-5}$ m²/s.

4. A circular cylinder 1 m in radius is completely insulated and has an initial temperature distribution $100r$ °C. Find an expression for the temperature as a function of r and t. Write integral expressions for at least three coefficients in the series expansion.

9

Numerical Methods

9.1 Introduction

In previous chapters we presented analytical solution techniques to both ordinary and partial differential equations. More often than not, problems are encountered for which the describing differential equations are extremely difficult, if not impossible, to solve analytically. Fortunately, since the latter part of the 1950s, the digital computer has become an increasingly useful tool for solving differential equations, whether they be ordinary or partial, linear or nonlinear, homogeneous or nonhomogeneous, or first order or fourth order. It is, of course, not always a simple matter to solve a differential equation, or a set of differential equations, using numerical methods. A numerical technique can be very intricate and difficult to understand, requiring substantial computer capability. Some techniques exist only in the literature or in advanced texts on the subject. We will, however, present several of the simplest methods for solving both ordinary and partial differential equations.

This chapter is intended to present some fundamental ideas in numerical methods and is not meant to be exhaustive. Textbooks should be consulted for more complete treatments. Some sample computer programs will be presented; however, it is assumed that the reader is capable of using the computer, so the numerical methods outlined can be applied to the solution of real problems.

The numerical solution to a problem is quite different from the analytical solution. The analytical solution provides the value of the dependent variable for any value of the independent variable; that is, for the simple spring-mass system the analytical solution is

$$y(t) = A \sin \omega t + B \cos \omega t. \tag{9.1.1}$$

We can choose any value of t and determine the displacement of the mass. Equation 9.1.1 is a solution of*

$$\ddot{y} + \omega^2 y = 0. \tag{9.1.2}$$

If Eq. 9.1.2 is solved numerically, the time interval of interest is divided into a predetermined number of increments, not necessarily of equal length. Initial conditions are necessary to "start" the solution at $t = 0$; then the solution is "generated" by solving numerically for the dependent variable y at each incremental step. This is done by using one of a host of numerical methods, all of which allow one to predict the value of the dependent variable at the $(i + 1)$ increment knowing its value at the ith increment [and possibly the $(i - 1)$ and $(i - 2)$ increments, depending on the method chosen]. The derivatives of the dependent variable may also be required in this process. After the solution is completed, the results are presented either in graphical or tabular form.

For a sufficiently small step size, the numerical solution to Eq. 9.1.2 closely approximates the analytical solution given by Eq. 9.1.1. However, difficulties are encountered which are fairly common in numerical work. After one "debugs" a computer program, which may turn one's hair gray prematurely, a numerical solution may become "unstable"; that is, as the solution progresses from one step to the next, the numerical results may begin to oscillate in an uncontrolled manner. This is referred to as a *numerical instability*. If the step size is changed, the stability characteristic changes. The objective is, for a particular numerical method, to choose an appropriate step size such that the solution is reasonably accurate and such that no instability results.

The problem of *truncation error*, which will be discussed in Section 9.4 arises when a series of computational steps is cut short prematurely with the hope that the terms omitted are negligible. Truncation error depends on the method used and is minimized by retaining additional terms in the series of computational steps. Choosing a different numerical technique, with less truncation error, is often a feasible alternative.

Another difficulty that always exists in numerical work is that of *round-off error*. Numerical computations are rounded off[†] to a particular number of digits at each step in a numerical process, whether it be a *fixed-point system*, in which numbers are expressed with a fixed number of decimal places (e.g., 0.1734, 69.3712), or a *floating-point system*, in which numbers are expressed

---

*In this chapter we shall often use the notation $\dot{y} = dy/dt$, $\dot{y} = dy/dx$, or $y' = dy/dx$.

[†] The general rule for rounding off is best reviewed by giving examples. If we round off to three digits, $62.55 \rightarrow 62.6$, $62.45 \rightarrow 62.4$, $0.01724 \rightarrow 0.0172$, $0.017251 \rightarrow 0.0173$, and $99.97 \rightarrow 100$.

with a fixed number of significant digits (e.g., 3.22×10^4, 5.00×10^{-10}). Round-off error accumulates in computations and thus increases with an increasing number of steps. Consequently, we are limited in the number of steps in solving a particular problem if we wish to keep the round-off error from destroying the accuracy of the solution.

Usually, various choices in step size are used and the numerical results compared. The best solution is then chosen. Of course, the larger the step size, the shorter the computer time required, which leads to savings in computer costs. Thus, one must choose a small-enough step size to guarantee accurate results: not so small as to give excessive round-off error, and not so small as to incur high computer costs.

Another restraint in the numerical solution of problems is the size of the computer. A computer has only a particular number of "bytes" in which information can be stored; normally, a character in a text requires one byte. In any numerical solution the total number of bytes necessary to solve the problem must not exceed the "memory" of the computer in which information is stored. In the past, this was a definite limitation; now, only in the solution of very complex problems does one experience a lack of memory space.

9.2   Finite-Difference Operators

A knowledge of finite-difference operators is helpful in understanding and deriving the vast variety of equations necessary when using numerical methods. The most common difference operator is the *forward difference operator* Δ, defined by

$$\Delta f_i = f_{i+1} - f_i, \tag{9.2.1}$$

where we use the abbreviation $f_i = f(x_i)$ (see Fig. 9.1). In this chapter all increments will be equal so that for each i,

$$x_{i+1} - x_i = \Delta x = h. \tag{9.2.2}$$

In addition to the forward difference operator, we define two operators, ∇ and δ. The *backward difference operator* ∇ is defined by

$$\nabla f_i = f_i - f_{i-1} \tag{9.2.3}$$

and the *central difference operator* δ by

$$\delta f_i = f_{i+1/2} - f_{i-1/2}. \tag{9.2.4}$$

In the latter definition, $f_{i+1/2} = f(x_i + \Delta x/2)$ and similarly, $f_{i-1/2} = f(x_i - \Delta x/2)$. (These two values of f are not generally found in the tabulation of f, nor does a computer calculate values at half steps; they are of theoretical interest in our development of the various difference formulas.)

The differences described above are referred to as the *first differences*.

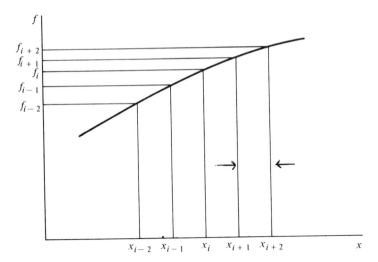

FIGURE 9.1. The function $f(x)$.

The second forward difference is

$$\Delta^2 f_i = \Delta(\Delta f_i)$$
$$= \Delta(f_{i+1} - f_i)$$
$$= \Delta f_{i+1} - \Delta f_i$$
$$= f_{i+2} - f_{i+1} - f_{i+1} + f_i$$
$$= f_{i+2} - 2f_{i+1} + f_i. \tag{9.2.5}$$

The second backward difference is

$$\nabla^2 f_i = f_i - 2f_{i-1} + f_{i-2}, \tag{9.2.6}$$

which follows as in the derivation of Eq. 9.2.5. The second central difference is

$$\delta^2 f_i = \delta(\delta f_i)$$
$$= \delta(f_{i+1/2} - f_{i-1/2}) = \delta f_{i+1/2} - \delta f_{i-1/2}. \tag{9.2.7}$$

However, since $f_{i+1/2} = f(x_i + \Delta x/2)$, we see that

$$\delta f_{i+1/2} = f\left(x_i + \frac{\Delta x}{2} + \frac{\Delta x}{2}\right) - f\left(x_i + \frac{\Delta x}{2} - \frac{\Delta x}{2}\right)$$
$$= f(x_i + \Delta x) - f(x_i) = f_{i+1} - f_i \tag{9.2.8}$$

and similarly, $\delta f_{i-1/2} = f_i - f_{i-1}$. We then have

$$\delta^2 f_i = (f_{i+1} - f_i) - (f_i - f_{i-1})$$
$$= f_{i+1} - 2f_i + f_{i-1}. \tag{9.2.9}$$

Continuing to the third differences, we find

$$\Delta^3 f_i = f_{i+3} - 3f_{i+2} + 3f_{i+1} - f_i \tag{9.2.10}$$

$$\nabla^3 f_i = f_i - 3f_{i-1} + 3f_{i-2} - f_{i-3} \tag{9.2.11}$$

$$\delta^3 f_i = f_{i+3/2} - 3f_{i+1/2} + 3f_{i-1/2} - f_{i-3/2}. \tag{9.2.12}$$

Table 9.1 exhibits these differences.

The reader is asked to construct a similar table for the backward differences. An examination of Table 9.1 shows that the entries in each position are the same in part (a) as in part (b); only the "names" for the entries are different. So, for instance, $\delta^4 f_2 = \Delta^4 f_0$. Note that in Table 9.1(a), the entries with the same subscript on f lie on sloping lines downward, while in part (b) they lie on horizontal lines. For backward differences they lie on sloping lines upward.

TABLE 9.1. Forward and Central Differences

(a) Forward differences

x	f	Δ	Δ^2	Δ^3	Δ^4
x_0	f_0				
		Δf_0			
x_1	f_1		$\Delta^2 f_0$		
		Δf_1		$\Delta^3 f_0$	
x_2	f_2		$\Delta^2 f_1$		$\Delta^4 f_0$
		Δf_2		$\Delta^3 f_1$	
x_3	f_3		$\Delta^2 f_2$		
		Δf_3			
x_4	f_4				

(b) Central differences

x	f	δ	δ^2	δ^3	δ^4
x_0	f_0				
		$\delta f_{1/2}$			
x_1	f_1		$\delta^2 f_1$		
		$\delta f_{3/2}$		$\delta^3 f_{3/2}$	
x_2	f_2		$\delta^2 f_2$		$\delta^4 f_2$
		$\delta f_{5/2}$		$\delta^3 f_{5/2}$	
x_3	f_3		$\delta^2 f_3$		
		$\delta f_{7/2}$			
x_4	f_4				

Example 9.2.1:   Compute a table of differences through Δ^3 for the function $f(x) = x^2 - 1$ using $h = 1$ and $x_0 = -1$.

SOLUTION: Using forward differences we can compute the following:

x	f	Δ	Δ^2	Δ^3
-1	0			
		-1		
0	-1		2	
		1		0
1	0		2	
		3		
2	3			

■

In the example above, it should be noted that it is much easier to compute the entries using forward differences rather than central differences. The usefulness of central differences will become apparent as we continue in our study.

Another useful operator is the *E operator*, defined by

$$Ef_i = f_{i+1}. \qquad (9.2.13)$$

Clearly, for each integer $n > 0$, the definition above implies that

$$E^n f_i = f_{i+n}. \qquad (9.2.14)$$

It is convenient to define $E^\alpha f_i$ for noninteger values of α. We set

$$E^\alpha f_i = f_{i+\alpha} = f(x_i + \alpha\,\Delta x), \qquad (9.2.15)$$

which reduces to Eq. 9.2.14 for $\alpha = n$. It also follows that

$$E^{-1} f_i = f_{i-1}, \qquad E^{1/2} f_i = f_{i+1/2}, \qquad E^{-1/2} f_i = f_{i-1/2}. \qquad (9.2.16)$$

The E operator can be related to the difference operators by observing that

$$\Delta f_i = f_{i+1} - f_i = Ef_i - f_i = (E - 1)f_i. \qquad (9.2.17)$$

We see then that the operator Δ operating on f_i is equal to $(E - 1)$ operating on f_i. We conclude that

$$\Delta = E - 1. \qquad (9.2.18)$$

Similarly, we can show that

$$\nabla = 1 - E^{-1} \qquad (9.2.19)$$

$$\delta = E^{1/2} - E^{-1/2} \qquad (9.2.20)$$

Rewritten, we have

$$E = \Delta + 1 \qquad (9.2.21)$$

$$E^{-1} = 1 - \nabla. \qquad (9.2.22)$$

We easily verify, by using the definitions, that

$$\nabla E = E\nabla = \Delta = \delta E^{1/2}. \tag{9.2.23}$$

Another operator, the *averaging operator* μ, is defined by

$$\mu = \tfrac{1}{2}(E^{1/2} + E^{-1/2}). \tag{9.2.24}$$

A variety of equations relating various operators are presented in Table 9.2. Note that after the operators have been separated from the function they operate on, we can treat them as algebraic quantities. We can manipulate them into various expressions to give any desired form. This will be illustrated with Examples and Problems. A word of caution is also in order; namely, the operators operate on a function, such as ∇f_i. This order must be retained since $\nabla f_i \neq f_i \nabla$.

TABLE 9.2. The Operators

First-order operators:	$\Delta f_i = f_{i+1} - f_i$ $\nabla f_i = f_i - f_{i-1}$ $\delta f_i = f_{i+1/2} - f_{i-1/2}$ $E f_i = f_{i+1}$ $\mu f_i = \tfrac{1}{2}(f_{i+1/2} + f_{i-1/2})$
Second-order operators:	$\Delta^2 f_i = f_{i+2} - 2f_{i+1} + f_i$ $\nabla^2 f_i = f_i - 2f_{i-1} + f_{i-2}$ $\delta^2 f_i = f_{i+1} - 2f_i + f_{i-1}$ $E^2 f_i = f_{i+2}$
Third-order operators:	$\Delta^3 f_i = f_{i+3} - 3f_{i+2} + 3f_{i+1} - f_i$ $\nabla^3 f_i = f_i - 3f_{i-1} + 3f_{i-2} - f_{i-3}$ $\delta^3 f_i = f_{i+3/2} - 3f_{i+1/2} + 3f_{i-1/2} - f_{i-3/2}$ $E^3 f_i = f_{i+3}$

Example 9.2.2: Derive the relationships $\Delta = \delta^2/2 + \delta\sqrt{1 + \delta^2/4}$, $\nabla = -\delta^2/2 + \delta\sqrt{1 + \delta^2/4}$, and $\mu = \sqrt{1 + \delta^2/4}$.

SOLUTION: The definition of the central difference operator is

$$\delta f_i = f_{i+1/2} - f_{i-1/2} = (E^{1/2} - E^{-1/2})f_i.$$

Hence,

$$\delta = E^{1/2} - E^{-1/2}.$$

Using $E = 1 + \Delta$, we have

$$\delta = \sqrt{1 + \Delta} - \frac{1}{\sqrt{1 + \Delta}}.$$

Squaring both sides gives

$$\delta^2 = 1 + \Delta + \frac{1}{1 + \Delta} - 2$$

or

$$\delta^2 + 2 = \frac{(1 + \Delta)^2 + 1}{1 + \Delta}.$$

Put in standard quadratic form,

$$\Delta^2 - \delta^2\Delta - \delta^2 = 0.$$

The quadratic formula for the positive root gives

$$\Delta = \frac{\delta^2}{2} + \frac{1}{2}\sqrt{\delta^4 + 4\delta^2} = \frac{\delta^2}{2} + \delta\sqrt{1 + \frac{\delta^2}{4}}.$$

Similarly, using $E^{-1} = 1 - \nabla$, we find that

$$\delta = \frac{1}{\sqrt{1 - \nabla}} - \sqrt{1 - \nabla}.$$

After writing this in the standard quadratic form, the positive root is

$$\nabla = -\frac{\delta^2}{2} + \delta\sqrt{1 + \frac{\delta^2}{4}}.$$

Now, μ can be written as

$$\mu = \tfrac{1}{2}(E^{1/2} + E^{-1/2})$$

or, squaring both sides,

$$4\mu^2 = E + 2 + E^{-1}.$$

Also, if we square the initial expression in this example for δ, we have

$$\delta^2 = E - 2 + E^{-1} = E + 2 + E^{-1} - 4 = 4\mu^2 - 4.$$

Thus,

$$\mu^2 = 1 + \frac{\delta^2}{4}$$

or

$$\mu = \sqrt{1 + \frac{\delta^2}{4}}.$$

We can then write

$$\Delta = \frac{\delta^2}{2} + \delta\mu, \qquad \nabla = -\frac{\delta^2}{2} + \delta\mu. \quad \blacksquare$$

PROBLEMS

1. Derive expressions for $\Delta^4 f_i$, $\nabla^4 f_i$, and $\delta^4 f_i$.

2. Show that $\nabla\Delta f_i = \delta^2 f_i$.

3. Show that all of the difference operators commute with one another, e.g., $\Delta E = E\Delta$ and $\delta\nabla = \nabla\delta$.

4. Verify that $\nabla E = \Delta = \delta E^{1/2}$.

5. Prove that $E^{-1/2} = \mu - \delta/2$ and that $\mu\delta f_i = \frac{1}{2}(f_{i+1} - f_{i-1})$. Also, find an expression for $\mu\delta^3 f_i$.

6. Use the binomial theorem $(a + x)^n = a^n + na^{n-1}x + n(n - 1)a^{n-2}x^2/2! + n(n - 1)(n - 2)a^{n-3}x^3/3! + \cdots$ to find a series expression for Δ in terms of δ.

7. Show that $2\mu\delta = \nabla + \Delta$. Also express $(E^2 - E^{-2})$ in terms of δ and μ.

8. Derive Eq. 9.2.12.

9. Construct the difference table, similar to Table 9.1, for the backward difference operator ∇. Describe the location of all entries with the same subscript.

10. Rework Example 9.2.1 using the table for backward differences developed in Problem 9.

Set up difference tables up to ∇^3, using $h = 1$ and $x_0 = 1$, for each function.

11. x

12. kx

13. $ax^2 + bx + c$

14. Use Table 9.1 to conjecture a relationship between Δ^n and δ^n. Consider n even and n odd separately.

9.3   The Differential Operator Related to the Difference Operator

We shall now relate the various operators to the *differential operator* $D = d/dx$. In this process the Taylor series is used. Recall that

$$f(x + h) = f(x) + h\frac{df}{dx} + \frac{h^2}{2!}\frac{d^2f}{dx^2} + \frac{h^3}{3!}\frac{d^3f}{dx^3} + \cdots, \qquad (9.3.1)$$

where the derivatives are evaluated at x and the step size $\Delta x = h$. This is written, using the difference notation, as

$$f_{i+1} = f_i + hf_i' + \frac{h^2}{2!}f_i'' + \frac{h^3}{3!}f_i''' + \cdots, \qquad (9.3.2)$$

where the primes denote differentiation with respect to the independent variable. The higher-order derivatives are written as

$$D^2 = \frac{d^2}{dx^2}, \qquad D^3 = \frac{d^3}{dx^3}, \cdots. \qquad (9.3.3)$$

Then Eq. 9.3.2 can be written as

$$Ef_i = \left[1 + hD + \frac{h^2 D^2}{2!} + \frac{h^3 D^3}{3!} + \cdots\right] f_i. \tag{9.3.4}$$

We recognize that the quantity in brackets is (see Table 2.1)

$$e^{hD} = 1 + hD + \frac{h^2 D^2}{2!} + \frac{h^3 D^3}{3!} + \cdots, \tag{9.3.5}$$

which leads to

$$Ef_i = e^{hD} f_i. \tag{9.3.6}$$

This relates the operator E to the operator D,

$$E = e^{hD}. \tag{9.3.7}$$

Making the substitution

$$E = \Delta + 1, \tag{9.3.8}$$

we have

$$\Delta = e^{hD} - 1 = hD + \frac{h^2 D^2}{2!} + \frac{h^3 D^3}{3!} + \cdots. \tag{9.3.9}$$

The second forward difference is found by squaring the equation above, to obtain

$$\Delta^2 = \left(hD + \frac{h^2 D^2}{2} + \frac{h^3 D^3}{6} + \cdots\right)^2$$

$$= h^2 D^2 + h^3 D^3 + \tfrac{7}{12} h^4 D^4 + \cdots. \tag{9.3.10}$$

To find D in terms of Δ we take the natural logarithm of both sides of Eq. 9.3.7 and obtain

$$D = \frac{1}{h} \ln E = \frac{1}{h} \ln (1 + \Delta). \tag{9.3.11}$$

In series form, we have

$$\ln (1 + \Delta) = \Delta - \frac{\Delta^2}{2} + \frac{\Delta^3}{3} - \cdots. \tag{9.3.12}$$

We may now relate the differential operator to the forward difference operator; there results

$$D = \frac{1}{h}\left(\Delta - \frac{\Delta^2}{2} + \frac{\Delta^3}{3} - \cdots\right). \tag{9.3.13}$$

Squaring both sides yields

$$D^2 = \frac{1}{h^2}\left(\Delta^2 - \Delta^3 + \frac{11}{12}\Delta^4 - \frac{5}{6}\Delta^5 - \cdots\right). \qquad (9.3.14)$$

The central and backward difference operators can be related to the differential operator by using Eq. 9.3.7 to write

$$E^{1/2} = e^{hD/2}, \qquad E^{-1/2} = e^{-hD/2}, \qquad E^{-1} = e^{-hD}. \qquad (9.3.15)$$

The resulting expressions will be included in the Examples and Problems. They are summarized in Table 9.3.

TABLE 9.3. *Relationship between the Operators*

$\Delta = E - 1$
$\qquad\qquad \delta = E^{1/2} - E^{-1/2}$
$\qquad\qquad \nabla = -\dfrac{\delta^2}{2} + \delta\sqrt{1 + \dfrac{\delta^2}{4}}$

$\nabla = 1 - E^{-1}$
$\qquad\qquad 2\mu\delta = E - E^{-1}$
$\qquad\qquad \mu = \sqrt{1 + \dfrac{\delta^2}{4}}$

$2\mu = E^{1/2} + E^{-1/2}$
$\qquad\qquad \Delta = \dfrac{\delta^2}{2} + \delta\sqrt{1 + \dfrac{\delta^2}{4}}$

$D = \dfrac{1}{h}\left[\Delta - \dfrac{\Delta^2}{2} + \dfrac{\Delta^3}{3} - \cdots\right] = \dfrac{1}{h}\left[\nabla + \dfrac{\nabla^2}{2} + \dfrac{\nabla^3}{3} + \cdots\right] = \dfrac{\mu}{h}\left[\delta - \dfrac{\delta^3}{6} + \dfrac{\delta^5}{30} - \cdots\right]$

$D^2 = \dfrac{1}{h^2}\left[\Delta^2 - \Delta^3 + \dfrac{11}{12}\Delta^4 - \cdots\right] = \dfrac{1}{h^2}\left[\nabla^2 + \nabla^3 + \dfrac{11}{12}\nabla^4 + \cdots\right]$

$\qquad = \dfrac{1}{h^2}\left[\delta^2 - \dfrac{\delta^4}{12} + \dfrac{\delta^6}{90} - \cdots\right]$

$D^3 = \dfrac{1}{h^3}\left[\Delta^3 - \dfrac{3}{2}\Delta^4 + \dfrac{7}{4}\Delta^5 - \cdots\right] = \dfrac{1}{h^3}\left[\nabla^3 + \dfrac{3}{2}\nabla^4 + \dfrac{7}{4}\nabla^5 + \cdots\right]$

$\qquad = \dfrac{\mu}{h^3}\left[\delta^3 - \dfrac{\delta^5}{4} + \dfrac{7}{120}\delta^7 - \cdots\right]$

$D^4 = \dfrac{1}{h^4}\left[\Delta^4 - 2\Delta^5 + \dfrac{17}{6}\Delta^6 - \cdots\right] = \dfrac{1}{h^4}\left[\nabla^4 + 2\nabla^5 + \dfrac{17}{6}\nabla^6 + \cdots\right]$

$\qquad = \dfrac{1}{h^4}\left[\delta^4 - \dfrac{\delta^6}{6} + \dfrac{7}{240}\delta^8 - \cdots\right]$

$\Delta = hD + \dfrac{h^2}{2}D^2 + \dfrac{h^3}{6}D^3 + \cdots$
$\qquad\qquad\qquad \Delta^2 = h^2D^2 + h^3D^3 + \dfrac{7}{12}h^4D^4 + \cdots$

$\Delta^3 = h^3D^3 + \dfrac{3}{2}h^4D^4 + \dfrac{5}{4}h^5D^5 + \cdots$

$\nabla = hD - \dfrac{h^2}{2}D^2 + \dfrac{h^3}{6}D^3 + \cdots$
$\qquad\qquad\qquad \nabla^2 = h^2D^2 - h^3D^3 + \dfrac{7}{12}h^4D^4 + \cdots$

$\nabla^3 = h^3D^3 - \dfrac{3}{2}h^4D^4 + \dfrac{5}{4}h^5D^5 + \cdots$

$\mu\delta = hD + \dfrac{h^3D^3}{6} + \dfrac{h^5D^5}{120} + \cdots$
$\qquad\qquad\qquad \delta^2 = h^2D^2 + \dfrac{h^4D^4}{12} + \dfrac{h^6D^6}{360} + \cdots$

$\mu\delta^3 = h^3D^3 + \dfrac{h^5D^5}{4} + \dfrac{h^7D^7}{40} + \cdots$

The results above are used to express the first derivative of the function $f(x)$ at x_i as

$$Df_i = \frac{df_i}{dx} = \frac{1}{h}\left(\Delta f_i - \frac{\Delta^2}{2}f_i + \frac{\Delta^3}{3}f_i - \cdots\right). \qquad (9.3.16)$$

The second derivative is

$$D^2f_i = \frac{d^2f_i}{dx^2} = \frac{1}{h^2}\left(\Delta^2 f_i - \Delta^3 f_i + \frac{11}{12}\Delta^4 f_i - \cdots\right). \qquad (9.3.17)$$

Higher-order derivatives can be generated similarly.

Example 9.3.1:   Relate the differential operator D to the central difference operator δ by using Eq. 9.3.13 and the result of Example 9.2.2.

SOLUTION: We use the relationship $\Delta = \delta^2/2 + \delta\sqrt{1 + \delta^2/4}$ (see Example 9.2.2). Expand $(1 + \delta^2/4)^{1/2}$ in a series using the binomial theorem* to give

$$\Delta = \frac{\delta^2}{2} + \delta\left(1 + \frac{\delta^2}{8} - \frac{\delta^4}{128} + \cdots\right)$$

$$= \delta + \frac{\delta^2}{2} + \frac{\delta^3}{8} - \frac{\delta^5}{128} + \cdots.$$

Substitute this into Eq. 9.3.13 to find

$$D = \frac{1}{h}\left[\delta + \frac{\delta^2}{2} + \frac{\delta^3}{8} - \frac{\delta^5}{128} + \cdots - \frac{1}{2}\left(\delta + \frac{\delta^2}{2} + \frac{\delta^3}{8} - \frac{\delta^5}{128} + \cdots\right)^2\right.$$

$$\left. + \frac{1}{3}\left(\delta + \frac{\delta^2}{2} + \frac{\delta^3}{8} - \frac{\delta^5}{128} + \cdots\right)^3 + \cdots\right]$$

$$= \frac{1}{h}\left(\delta - \frac{\delta^3}{24} + \frac{3\delta^5}{640} - \cdots\right)$$

This expression allows us to relate Df_i to quantities such as $f_{i+1/2}, f_{i+1/2}, f_{i+3/2}, f_{i-5/2}$, and so on. It is more useful to introduce the averaging operator μ so that quantities with integer subscripts result. From Example 9.2.2 we have $\mu = \sqrt{1 + \delta^2/4}$. Expressing this as a series, we have

$$\mu = 1 + \frac{\delta^2}{8} - \frac{\delta^4}{128} + \cdots.$$

Now, we can write D as

$$D = \frac{\mu}{h}\left(\delta - \frac{\delta^3}{24} + \frac{3\delta^5}{640} - \cdots\right)\frac{1}{\mu} = \frac{\mu}{h}\frac{\delta - \dfrac{\delta^3}{24} + \dfrac{3\delta^5}{640} - \cdots}{1 + \dfrac{\delta^2}{8} - \dfrac{\delta^4}{128} + \cdots}.$$

---

*The binomial theorem is

$$(a + b)^n = a^n + na^{n-1}b + n(n - 1)a^{n-2}b^2/2! + \cdots,$$

which terminates if and only if n is a nonnegative integer.

Dividing one series by the other, we finally have

$$D = \frac{\mu}{h}\left(\delta - \frac{\delta^3}{6} + \frac{\delta^5}{30} - \cdots\right).$$

Using this expression, Df_i contains only integer subscripts; this permits use of this formula on a computer. ∎

Example 9.3.2: Relate the central difference operator to the differential operator by starting with a Taylor series.

SOLUTION: An alternative form of the Taylor series is

$$f\left(x + \frac{h}{2}\right) = f(x) + \frac{h}{2}f'(x) + \left(\frac{h}{2}\right)^2\frac{1}{2!}f''(x) + \left(\frac{h}{2}\right)^3\frac{1}{3!}f'''(x) + \cdots$$

where the primes denote differentiation with respect to x. In difference notation, we have

$$f_{i+1/2} = f_i + \frac{h}{2}f_i' + \frac{h^2}{8}f_i'' + \frac{h^3}{48}f_i''' + \cdots$$

$$= \left(1 + \frac{hD}{2} + \frac{h^2D^2}{8} + \frac{h^3D^3}{48} + \cdots\right)f_i.$$

Similarly,

$$f_{i-1/2} = f_i - \frac{h}{2}f_i' + \frac{h^2}{8}f_i'' - \frac{h^3}{48}f_i''' + \cdots$$

$$= \left(1 - \frac{hD}{2} + \frac{h^2D^2}{8} - \frac{h^3D^3}{48} + \cdots\right)f_i.$$

Subtracting gives

$$\delta f_i = f_{i+1/2} - f_{i-1/2} = \left(hD + \frac{h^3D^3}{24} + \frac{h^5D^5}{1920} + \cdots\right)f_i.$$

Finally,

$$\delta = hD + \frac{h^3D^3}{24} + \frac{h^5D^5}{1920} + \cdots.$$

This could also have been obtained by using Eqs. 9.2.20 and 9.3.7. We have

$$\delta = E^{1/2} - E^{-1/2} = e^{hD/2} - e^{-hD/2}.$$

Expanding the exponentials as in Eq. 9.3.5 results in the series

$$\delta = \left(1 + \frac{hD}{2} + \frac{h^2D^2}{8} + \frac{h^3D^3}{48} + \cdots\right) - \left(1 - \frac{hD}{2} + \frac{h^2D^2}{8} - \frac{h^3D^3}{48} + \cdots\right)$$

$$= hD + \frac{h^3D^3}{24} + \frac{h^5D^5}{1920} + \cdots. ∎$$

PROBLEMS

1. Verify the following expressions by squaring the appropriate series.

$$\Delta^2 = h^2 D^2 + h^3 D^3 + \tfrac{7}{12} h^4 D^4 + \cdots$$

$$D^2 = \frac{1}{h^2}\left(\Delta^2 - \Delta^3 + \frac{11}{12}\Delta^4 - \frac{5}{6}\Delta^5 + \cdots\right).$$

2. Relate the backward difference operator ∇ to the differential operator D using h as the step size. Also find ∇^2 in terms of D, and D^2 in terms of ∇. Check Table 9.3 for the correct expressions.

3. Find an expression for $\mu\delta^3$ in terms of D. Use the results of Example 9.3.2.

4. Find the relationship for D^2 in terms of δ. Check with Table 9.3.

5. Start with Taylor's series and show that $E^{-1} = e^{-hD}$.

Using any results from the examples, verify each expression given in Table 9.3.

6. D^2 in terms of δ.

7. D^3 in terms of δ and μ.

8. D^3 in terms of Δ.

9. D^4 in terms of ∇.

10. D^4 in terms of δ.

9.4   Truncation Error

We obviously cannot use all the terms in the infinite series when representing a derivative in finite-difference form, as in Eqs. 9.3.16 and 9.3.17. The series is truncated and the sum of the omitted terms is the *truncation error*. It is quite difficult to determine the sum of the omitted terms; instead, we estimate the magnitude of the first term omitted in the series. Since each term is smaller than the preceding term, we call the magnitude of the first truncated term the *order of magnitude* of the error. Its primary function is to allow a comparison of formulas. If the magnitude of the first truncated term of one formula is smaller than that of another formula, we assume the first formula to be more accurate.

If the first term truncated is $\Delta^2 f_i / 2$ in Eq. 9.3.16, the order of magnitude of the error is of order h, written symbolically as $e = o(h)$, since from Eq. 9.3.10 Δ^2 is of order h^2 and in Eq. 9.3.16 we divide $\Delta^2 f_i / 2$ by h. Hence, we can express the first derivative of a function, with $e = o(h)$, as

$$Df_i = \frac{1}{h}\Delta f_i$$

$$= \frac{1}{h}(f_{i+1} - f_i), \qquad e = o(h). \qquad (9.4.1)$$

If a smaller error is desired, an additional term is maintained and

$$Df_i = \frac{1}{h}(\Delta f_i - \Delta^2 f_i/2)$$

$$= \frac{1}{h}\left[f_{i+1} - f_i - \frac{1}{2}(f_{i+2} - 2f_{i+1} + f_i)\right]$$

$$= \frac{1}{2h}(-f_{i+2} + 4f_{i+1} - 3f_i), \qquad e = o(h^2) \qquad (9.4.2)$$

This, of course, requires additional information, the value of $f(x)$ at x_{i+2}.
The second derivative can be approximated, with $e = o(h)$, by

$$D^2 f_i = \frac{1}{h^2}\Delta^2 f_i$$

$$= \frac{1}{h^2}(f_{i+2} - 2f_{i+1} + f_i), \qquad e = o(h). \qquad (9.4.3)$$

Note that the Δ^3-term was omitted. It is of order h^3; but it is divided by h^2, hence $e = o(h)$. Maintaining an additional term in Eq. 9.3.17, the second derivative, with $e = o(h^2)$, is

$$D^2 f_i = \frac{1}{h^2}(\Delta^2 f_i - \Delta^3 f_i)$$

$$= \frac{1}{h^2}(f_{i+2} - 2f_{i+1} + f_i - f_{i+3} + 3f_{i+2} - 3f_{i+1} + f_i)$$

$$= \frac{1}{h^2}(-f_{i+3} + 4f_{i+2} - 5f_{i+1} + 2f_i), \qquad e = o(h^2).$$

$$(9.4.4)$$

Results in tabular form are presented in Table 9.4.

The error analysis above is meaningful only when the phenomenon of interest occurs over a time duration of order unity or over a length of order unity. If we are studying a phenomenon that occurs over a long time T, the time increment Δt could be quite large even for a reasonable number of steps. Or if the phenomenon occurs over a large length L, the length increment Δx could be quite large. For example, the deflection of a 300-m-high smokestack on a power plant could be reasonably calculated with increments of 3 m. We would not then say that the error gets larger with each term truncated, i.e., $o(h^3) > o(h^2)$. The same reasoning is applied to phenomena of very short duration or lengths. Then h is extremely small and the truncation error would appear to be much smaller than it actually is. The quantity that determines the error is actually the step size involved when the time duration or the length scale is of order unity; hence, to determine the error when large or small scales are encountered, we first "normalize" on the independent variable so

TABLE 9.4. The Derivatives in Finite-Difference Form

Forward	Backward	Central
$e = o(h)$		
$Df_i = \dfrac{1}{h}(f_{i+1} - f_i)$	$\dfrac{1}{h}(f_i - f_{i-1})$	
$D^2 f_i = \dfrac{1}{h^2}(f_{i+2} - 2f_{i+1} + f_i)$	$\dfrac{1}{h^2}(f_i - 2f_{i-1} + f_{i-2})$	
$D^3 f_i = \dfrac{1}{h^3}(f_{i+3} - 3f_{i+2} + 3f_{i+1} - f_i)$	$\dfrac{1}{h^3}(f_i - 3f_{i-1} + 3f_{i-2} - f_{i-3})$	
$D^4 f_i = \dfrac{1}{h^4}(f_{i+4} - 4f_{i+3} + 6f_{i+2} - 4f_{i+1} + f_i)$	$\dfrac{1}{h^4}(f_i - 4f_{i-1} + 6f_{i-2} - 4f_{i-3} + f_{i-4})$	
$e = o(h^2)$		
$Df_i = \dfrac{1}{2h}(-f_{i+2} + 4f_{i+1} - 3f_i)$	$\dfrac{1}{2h}(3f_i - 4f_{i-1} + f_{i-2})$	$\dfrac{1}{2h}(f_{i+1} - f_{i-1})$
$D^2 f_i = \dfrac{1}{h^2}(-f_{i+3} + 4f_{i+2} - 5f_{i+1} + 2f_i)$	$\dfrac{1}{h^2}(2f_i - 5f_{i-1} + 4f_{i-2} - f_{i-3})$	$\dfrac{1}{h^2}(f_{i+1} - 2f_i + f_{i-1})$
$D^3 f_i = \dfrac{1}{2h^3}(-3f_{i+4} + 14f_{i+3} - 24f_{i+2} + 18f_{i+1} - 5f_i)$	$\dfrac{1}{2h^3}(5f_i - 18f_{i-1} + 24f_{i-2} - 14f_{i-3} + 3y_{i-4})$	$\dfrac{1}{2h^3}(f_{i+2} - 2f_{i+1} + 2f_{i-1} - f_{i-2})$
$D^4 f_i = \dfrac{1}{h^4}(-2f_{i+5} + 11f_{i+4} - 24f_{i+3} + 26f_{i+2} - 14f_{i+1} + 3f_i)$	$\dfrac{1}{h^4}(3f_i - 14f_{i-1} + 26f_{i-2} - 24f_{i-3} + 11f_{i-4} - 2f_{i-5})$	$\dfrac{1}{h^4}(f_{i+2} - 4f_{i+1} + 6f_i - 4f_{i-1} + f_{i-2})$

that the phenomenon occurs over a duration or length of order unity. That is, we consider the quantity h/T or h/L to determine the order of the error. The expressions for error in Eqs. 9.4.1, 9.4.2, and 9.4.3 are based on the assumption that the time duration or length scale is of order 1.

Example 9.4.1: Find an expression for the second derivative using central differences with $e = o(h^4)$.

SOLUTION: In terms of central differences D^2 is found by squaring the expression given in Example 9.3.1 for D in terms of δ. It is

$$D^2 = \left[\frac{1}{h}\left(\delta - \frac{\delta^3}{24} + \frac{3\delta^5}{640} - \cdots\right)\right]^2$$

$$= \frac{1}{h^2}\left(\delta^2 - \frac{\delta^4}{12}\right), \qquad e = o(h^4).$$

This expression can also be found in Table 9.3. Now, we have

$$D^2 f_i = \frac{1}{h^2}\left(\delta^2 f_i - \frac{1}{12}\delta^4 f_i\right)$$

$$= \frac{1}{h^2}\left[f_{i+1} - 2f_i + f_{i-1} - \frac{1}{12}(f_{i+2} - 4f_{i+1} + 6f_i - 4f_{i-1} + f_{i-2})\right]$$

$$= \frac{1}{12h^2}(-f_{i+2} + 16f_{i+1} - 30f_i + 16f_{i-1} - f_{i-2}), \qquad e = o(h^4).$$

The relationship for $\delta^4 f_i$ is part of Problem 1 of Section 9.2.   ■

Example 9.4.2:   Write the differential equation

$$\ddot{y} - \frac{C}{M}\dot{y} + \frac{K}{M}y = A \sin \omega t$$

in difference notation using forward differences with $e = o(h^2)$.

SOLUTION: The first derivative is given by Eq. 9.4.2 as

$$\dot{y}_t = \frac{1}{2h}(-y_{i+2} + 4y_{i+1} - 3y_i).$$

The second derivative, found by maintaining the first two terms in Eq. (9.3.17), with $e = o(h^2)$, is given by Eq. 9.4.4. The differential equation is then written in difference form as

$$\frac{1}{h^2}(-y_{i+3} + 4y_{i+2} - 5y_{i+1} + 2y_i) - \frac{C}{2Mh}(-y_{i+2} + 4y_{i+1} - 3y_i) + \frac{K}{M}y_i$$

$$= A \sin \omega t_i.$$

By letting $i = 0$, y_3 is seen to be related to y_2, y_1, y_0, and t_0, and is the first value of the dependent variable $y(t)$ that could be found by the difference equation. But we do not know the values of y_2 and y_1. (The value of y_0 is known from an initial condition.) Thus, a "starting technique" is necessary to find the values y_1 and y_2. This is presented in Section 9.9. The difference equation is used to find an approximation to the solution of the differential equation. The solution would be presented in tabular form.   ■

PROBLEMS

Using forward differences, verify the expression in Table 9.4 for

1. $D^3 f_i$ with $e = o(h)$.

2. $D^4 f_i$ with $e = o(h)$.

3. Df_i with $e = o(h^2)$.

4. $D^3 f_i$ with $e = o(h^2)$.

Using backward differences, verify the expression in Table 9.4 for

5. $D^2 f_i$ with $e = o(h)$.

6. $D^3 f_i$ with $e = o(h)$.

7. $D^2 f_i$ with $e = o(h^2)$.

8. $D^4 f_i$ with $e = o(h^2)$.

Using central differences, verify the expression in Table 9.4 for

9. Df_i with $e = o(h^2)$. **10.** D^2f_i with $e = o(h^2)$.

11. D^3f_i with $e = o(h^2)$.

Derive an expression, using difference notation with $e = o(h^3)$, for

12. Df_i using forward differences.

13. D^2f_i using central differences.

14. D^3f_i using backward differences.

Estimate a value for d/dx (erf x) at $x = 1.6$ using Table A3. Check with the exact value obtained analytically. Use five significant figures. Employ

15. Central differences with $e = o(h^2)$. **16.** Central differences with $e = o(h^2)$.

17. Forward differences with $e = o(h^2)$.

18. Backward differences with $e = o(h^2)$.

Estimate a value for $d^2/dx^2 J_1(x)$ at $x = 2.0$ using Table A4 with $e = o(h^2)$. Use

19. Forward differences. **20.** Backward differences.

21. Central differences.

22. Estimate a value for $d^2/dx^2 J_0(x)$ at $x = 0$ using Table A4. Use $e = o(h^2)$.

23. Estimate a value for $d^2/dx^2 Y_1(x)$ at $x = 15.0$ with $e = o(h^2)$ using Table A4.

9.5   Numerical Integration

Since the symbol for differentiation is $D = d/dx$, it is natural to use D^{-1} to represent the operation inverse to differentiation, namely, integration; that is,

$$D^{-1}f(x) = \int f(x)\, dx \tag{9.5.1}$$

or, between the limits of x_i and x_{i+1}, this is

$$\int_{x_i}^{x_{i+1}} f(x)\, dx = D^{-1}f(x) \Big|_{x_i}^{x_{i+1}}$$

$$= D^{-1}(f_{i+1} - f_i) = \frac{E - 1}{D} f_i. \tag{9.5.2}$$

Relating this to the forward difference operator, we use Eqs. 9.2.21 and 9.3.13 to obtain

$$\int_{x_i}^{x_{i+1}} f(x)\, dx = \frac{\Delta}{\dfrac{1}{h}\left(\Delta - \dfrac{\Delta^2}{2} + \dfrac{\Delta^3}{3} - \cdots\right)} f_i$$

$$= h\left(1 + \frac{\Delta}{2} - \frac{\Delta^2}{12} + \frac{\Delta^3}{24} - \cdots\right)f_i. \qquad (9.5.3)$$

If we neglect the Δ^2-term and the higher-order terms in the parentheses above, there results

$$\int_{x_i}^{x_{i+1}} f(x)\, dx = h\left(1 + \frac{\Delta}{2}\right)f_i$$

$$= \frac{h}{2}(f_{i+1} + f_i), \qquad e = o(h^3). \qquad (9.5.4)$$

This approximation is seen to be nothing more than the average of $f(x)$ between x_{i+1} and x_i multiplied by the step size Δx (see Fig. 9.2). The smaller the step size h, the closer the approximation to the integral. The error results from the neglected $o(h^2)$ term multiplied by h; it is $e = o(h^3)$.

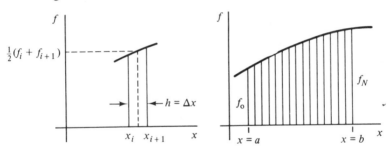

FIGURE 9.2.   The integral of $f(x)$.

To obtain the integral from $x = a$ to $x = b$, we simply add up all the areas to arrive at

$$\int_a^b f(x)\, dx \cong \frac{h}{2}[f_0 + f_1) + (f_1 + f_2) + (f_2 + f_3) + \cdots$$

$$+ (f_{N-2} + f_{N-1}) + (f_{N-1} + f_N)]$$

$$= \frac{h}{2}(f_0 + 2f_1 + 2f_2 + \cdots + 2f_{N-1} + f_N), \qquad (9.5.5)$$

where $N = (b - a)/h$. This is the *trapezoidal rule* of integration. Each element in the interval from a to b contains an error $e = o(h^3)$. Hence, assuming the interval to be of order unity, that is, $b - a = o(1)$, it follows that $N = o(1/h)$. The order of magnitude of the total error in the integration formula 9.5.5 is then $N \times o(h^3)$, or $o(h^2)$.

We can also determine an approximation to the integral between x_i and x_{i+2} as follows:

$$\int_{x_i}^{x_{i+2}} f(x)\, dx = D^{-1}f(x)\Big|_{x_i}^{x_{i+2}} = D^{-1}(f_{i+2} - f_i)$$

$$= \frac{E^2 - 1}{D} f_i$$

$$= \frac{2\Delta + \Delta^2}{(1/h)(\Delta - \Delta^2/2 + \Delta^3/3 - \cdots)} f_i$$

$$= h\left(2 + 2\Delta + \frac{\Delta^2}{3} - \frac{\Delta^4}{90} + \cdots\right) f_i. \qquad (9.5.6)$$

We keep terms up through Δ^2, so that we do not go outside the limits of integration* x_i to x_{i+2}; there results

$$\int_{x_i}^{x_{i+2}} f(x)\, dx = h(2 + 2\Delta + \Delta^2/3) f_i$$

$$= \frac{h}{3}(f_{i+2} + 4f_{i+1} + f_i), \qquad e = o(h^5). \qquad (9.5.7)$$

The error in this formula is, surprisingly, of order $o(h^5)$ because of the absence of the Δ^3-term. This small error makes this a popular integration formula. The integral from $x = a$ to $x = b$ is then

$$\int_a^b f(x)\, dx \cong \frac{h}{3}[(f_0 + 4f_1 + f_2) + (f_2 + 4f_3 + f_4) + \cdots$$

$$+ (f_{N-4} + 4f_{N-3} + f_{N-2}) + (f_{N-2} + 4f_{N-1} + f_N)]$$

$$= \frac{h}{3}(f_0 + 4f_1 + 2f_2 + 4f_3 + 2f_4 \cdots + 2f_{N-2} + 4f_{N-1} + f_N),$$

$$(9.5.8)$$

where $N = (b - a)/h$ and N is an even integer. This is *Simpson's one-third rule*. The integral has been approximated by $N/2$ pairs of elements, each pair having $e = o(h^5)$. Since $N = o(1/h)$ it follows that the order of the error for the formula (9.5.8) is $e = (N/2) \times o(h^5) = o(h^4)$. Note that the factor 2 does not change the order of the error.

Similarly, we have

$$\int_{x_i}^{x_{i+3}} f(x)\, dx = \frac{3h}{8}(f_{i+3} + 3f_{i+2} + 3f_{i+1} + f_i), \qquad e = o(h^5). \qquad (9.5.9)$$

The integration formula is then

$$\int_a^b f(x)\, dx = \frac{3h}{8}[(f_0 + 3f_1 + 3f_2 + f_3) + (f_3 + 3f_4 + 3f_5 + f_6) + \cdots$$

$$+ (f_{N-3} + 3f_{N-2} + 3f_{N-1} + f_N)]$$

---

*$\Delta^3 f_i = f_{i+3} - 3f_{i+2} + 3f_{i+1} - f_i$; but $f_{i+3} = f(x_{i+3})$ and this value of f does not lie in the interval $[x_i, x_{i+2}]$.

$$= \frac{3h}{8}(f_0 + 3f_1 + 3f_2 + 2f_3 + 3f_4 + 3f_5 + 2f_6 + \cdots$$

$$+ 2f_{N-3} + 3f_{N-2} + 3f_{N-1} + f_N), \qquad (9.5.10)$$

where $N = (b - a)/h$ and N is divisible by 3. This is *Simpson's three-eighths rule*. The error is found to be of order $o(h^4)$, essentially the same as Simpson's one-third rule.

If we desired the integral in backward difference form, for example, $\int_{x_{i-2}}^{x_i} f(x)\, dx$, we would have chosen to express E and D in terms of backward differences; if $\int_{x_{i-2}}^{x_{i+2}} f(x)\, dx$ were desired, central differences would be chosen. Examples of these will be included in the Problems and the Examples.

It is possible to establish error bounds on the numerical integration process, which are more exact than the order of magnitude. Let us first consider the trapezoidal rule of integration. The error e involved is (see Eq. 9.5.5)

$$e = \frac{h}{2}(f_0 + 2f_1 + 2f_2 + \cdots + f_N) - \int_a^b f(x)\, dx. \qquad (9.5.11)$$

We will find the error for only the first interval, letting the step size h be a variable, as shown in Fig 9.3. Using the relationship above, the error in this single strip is

$$e(t) = \frac{t - a}{2}[f(a) + f(t)] - \int_a^t f(x)\, dx. \qquad (9.5.12)$$

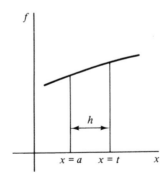

FIGURE 9.3.   Variable—with element used in the error analysis.

Differentiate this equation to obtain

$$e'(t) = \frac{1}{2}[f(a) + f(t)] + \frac{t - a}{2}f'(t) - f(t), \qquad (9.5.13)$$

where we have used the fundamental theorem of calculus to obtain

$$\frac{d}{dt} \int_a^t f(x)\, dx = f(t). \tag{9.5.14}$$

Again, we differentiate and find

$$e''(t) = \frac{t-a}{2} f''(t). \tag{9.5.15}$$

Thus, the maximum value of e'' is obtained if we replace $f''(t)$ with its maximum value in the interval, and the minimum value results when $f''(t)$ is replaced with its minimum value. This is expressed by the inequalities

$$\frac{t-a}{2} f''_{\min} \leq e''(t) \leq \frac{t-a}{2} f''_{\max}. \tag{9.5.16}$$

Now, let us integrate to find the bounds on the error $e(t)$. Integrating once gives

$$\frac{(t-a)^2}{4} f''_{\min} \leq e'(t) \leq \frac{(t-a)^2}{4} f''_{\max}. \tag{9.5.17}$$

A second integration results in

$$\frac{(t-a)^3}{12} f''_{\min} \leq e(t) \leq \frac{(t-a)^3}{12} f''_{\max}. \tag{9.5.18}$$

In terms of the step size, the error for this first step is

$$\frac{h^3}{12} f''_{\min} \leq e \leq \frac{h^3}{12} f''_{\max}. \tag{9.5.19}$$

But there are N steps in the interval of integration from $x = a$ to $x = b$. Assuming that each step has the same bounds on its error, the total accumulated error is N times that of a single step,

$$\frac{h^3}{12} N f''_{\min} \leq e \leq \frac{h^3}{12} N f''_{\max}, \tag{9.5.20}$$

where $f''_{\min}$ and $f''_{\max}$ are the smallest and largest second derivatives, respectively, in the interval of integration.

A similar analysis, using Simpson's one-third rule, leads to an error bounded by

$$\frac{h^5}{180} N f^{(iv)}_{\min} \leq e \leq \frac{h^5}{180} N f^{(iv)}_{\max}. \tag{9.5.21}$$

Example 9.5.1:   Find an approximate value for $\int_0^2 x^{1/3}\, dx$ using the trapezoidal rule of integration with eight increments.

SOLUTION: The formula for the trapezoidal rule of integration is given by Eq. 9.5.5. It is, using $h = \frac{2}{8} = \frac{1}{4}$,

$$\int_0^2 x^{1/3} \, dx \cong \tfrac{1}{8}(f_0 + 2f_1 + 2f_2 + \cdots + 2f_7 + f_8)$$

$$= \tfrac{1}{8}[0 + 2(\tfrac{1}{4})^{1/3} + 2(\tfrac{2}{4})^{1/3} + \cdots + 2(\tfrac{7}{4})^{1/3} + 2^{1/3}]$$

$$= \tfrac{1}{8}[2(.63 + .794 + .909 + 1.0 + 1.077 + 1.145 + 1.205) + 1.26]$$

$$= 1.85.$$

This compares with the exact value of

$$\int_0^2 x^{1/3} \, dx = \tfrac{3}{4}x^{4/3} \Big|_0^2 = \tfrac{3}{4}(2)^{4/3} = 1.89. \quad \blacksquare$$

Example 9.5.2: Derive the integration formula using central differences with the largest error.

SOLUTION: The integral of interest is $\int_{x_{i-1}}^{x_{i+1}} f(x) \, dx$. In difference notation it is expressed as

$$\int_{x_{i-1}}^{x_{i+1}} f(x) \, dx = D^{-1}(f_{i+1} - f_{i-1}) = \frac{(E - E^{-1})}{D} f_i$$

$$= \frac{(E^{1/2} + E^{-1/2})(E^{1/2} - E^{-1/2})}{D} f_i = \frac{\delta 2\mu}{D} f_i,$$

using the results of Example 9.2.2. With the appropriate expression from Table 9.3, we have

$$\int_{x_{i-1}}^{x_{i+1}} f(x) \, dx = \frac{2\mu\delta}{(\mu/h)(\delta - \delta^3/6 + \delta^5/30 - \cdots)} f_i.$$

Dividing, we get, neglecting terms of $o(h^4)$ in the series expansion,

$$\int_{x_{i-1}}^{x_{i+1}} f(x) \, dx = 2h\left(1 + \frac{\delta^2}{6}\right) f_i$$

$$= \frac{h}{3}(f_{i+1} + 4f_i + f_{i-1}), \qquad e = o(h^5).$$

Note that it is not correct to retain the δ^4-term in the above since it uses f_{i+2} and f_{i-2}, quantities outside the limits of integration. The integration formula is then

$$\int_a^b f(x) \, dx = \frac{h}{3}[(f_0 + 4f_1 + f_2) + (f_2 + 4f_3 + f_4) + \cdots$$

$$+ (f_{N-4} + 4f_{N-3} + f_{N-2}) + (f_{N-2} + 4f_{N-1} + f_N)]$$

$$= \frac{h}{3}(f_0 + 4f_1 + 2f_2 + 4f_3 + 2f_4 + \cdots$$

$$+ 2f_{N-2} + 4f_{N-1} + f_N).$$

This is identical to Simpson's one-third rule. $\blacksquare$

PROBLEMS

1. Express the value of the integral of $f(x)$ from x_{i-2} to x_i using backward differences.

2. Approximate the value of the integral of $f(x) = x^2$ from $x = 0$ to $x = 6$ using six steps. Use (a) the trapezoidal rule and (b) Simpson's one-third rule. Compare with the actual value found by integrating. Then, for part (a) show that the error falls within the limits established by Eq. 9.5.20.

3. Determine an approximate value for $\int_0^9 x^2 \sin{(\pi x/6)} \, dx$ using (a) the trapezoidal rule, and (b) Simpson's three-eights rules. Use nine steps.

4. Determine a value for $\int_0^2 J_0(x) \, dx$ applying (a) the trapezoidal rule (b) Simpson's one-third rule, using ten steps.

5. Find an expression for the integral of $f(x)$ from x_{i-2} to x_{i+2} using central differences. Using this expression, determine a formula for the integral $\int_a^b f(x) \, dx$. What is the order of magnitude of the error?

6. Integrate $y(t)$ from $t = 0$ to $t = 1.2$ using Simpson's one-third rule.

t	0	0.2	0.4	0.6	0.8	1.0	1.2
y		9.6	9.1	7.4	6.8	7.6	8.8	12.2

7. Integrate $y(t)$ of Problem 6 from $t = 0$ to $t = 1.2$ using Simpson's three-eights rule.

Find the value of each integral to five significant digits using a computer. Use Simpson's one-third rule.

8. $\int_0^5 (x^2 + 2) \, dx$ **9.** $\int_0^2 (x + \sin x)e^x \, dx$ **10.** $\int_0^4 xe^x \cos x \, dx$

11. $\int_0^{10} x^2 e^{-x \sin x} \, dx$ **12.** $\int_1^2 e^{x^2} \sin x \, dx$

9.6   Numerical Interpolation

We often desire information at points other than a multiple of Δx, or at points other than at the entries in a table of numbers. The value desired is f_{i+n}, where n is not an integer but some fraction such as $\frac{1}{3}$ (see Fig. 9.4). But f_{i+n} can be written in terms of E^n:

$$E^n f_i = f_{i+n}. \qquad (9.6.1)$$

In terms of the forward difference operator Δ, we have

$$(1 + \Delta)^n f_i = f_{i+n} \qquad (9.6.2)$$

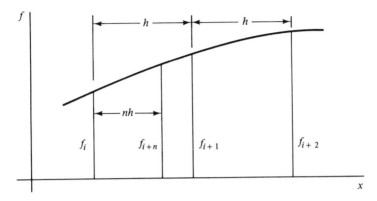

FIGURE 9.4. Numerical interpolation.

or, by using the binomial theorem,

$$(1 + \Delta)^n = 1 + n\Delta + \frac{n(n - 1)}{2}\Delta^2 + \frac{n(n - 1)(n - 2)}{6}\Delta^3 + \cdots .$$

(9.6.3)

Hence,

$$f_{i+n} = \left[1 + n\Delta + \frac{n(n - 1)}{2}\Delta^2 + \frac{n(n - 1)(n - 2)}{6}\Delta^3 + \cdots\right]f_i$$

(9.6.4)

Neglecting terms of order higher than Δ^3, this becomes

$$f_{i+n} = \left[f_i + n(f_{i+1} - f_i) + \frac{n(n - 1)}{2}(f_{i+2} - 2f_{i+1} + f_i)\right.$$
$$\left. + \frac{n(n - 1)(n - 2)}{6}(f_{i+3} - 3f_{i+2} + 3f_{i+1} - f_i)\right].$$

(9.6.5)

If we desired f_{i-n}, where n is a fraction, we can use backward differences to obtain

$$f_{i-n} = \left[f_i - n(f_i - f_{i-1}) + \frac{n(n - 1)}{2}(f_i - 2f_{i-1} + f_{i-2})\right.$$
$$\left. - \frac{n(n - 1)(n - 2)}{6}(f_i - 3f_{i-1} + 3f_{i-2} - f_{i-3})\right].$$

(9.6.6)

This formula is used to interpolate for a value near the end of a set of numbers.

Example 9.6.1: Find the value for the Bessel function $J_0(x)$ at $x = 2.06$ using numerical interpolation with (a) $e = o(h^2)$ and (b) $e = o(h^3)$. Use forward differences and four significant places.

SOLUTION: (a) Using Eq. 9.6.4 with $e = o(h^2)$, we have

$$f_{i+n} = (1 + n\Delta)f_i = f_i + n(f_{i+1} - f_i).$$

Table A4 for Bessel functions is given with $h = 0.1$. For our problem,

$$n = \frac{0.06}{0.1} = 0.6.$$

The interpolated value is then, using the ith term corresponding to $x = 2.0$,

$$J_0(2.06) = f_{i+0.6} = 0.2239 + 0.6(0.1666 - 0.2239) = 0.1895.$$

This is a *linear interpolation*, the method used most often when interpolating between tabulated values.

(b) Now, let us determine a more accurate value for $J_0(2.06)$. Equation 9.6.4 with $e = o(h^3)$ is

$$f_{i+n} = [1 + n\Delta + \tfrac{1}{2}(n)(n - 1)\Delta^2]f_i$$

$$= f_i + n(f_{i+1} - f_i) + \frac{n(n - 1)}{2}(f_{i+2} - 2f_{i+1} + f_i).$$

Again, using $n = 0.6$, we have

$$J_0(2.06) = f_{i+0.6} = 0.2239 + 0.6(0.1666 - 0.2239)$$

$$+ \frac{0.6(0.6 - 1)}{2}(0.1104 - 2 \times 0.1666 + 0.2239)$$

$$= 0.1894.$$

Note that the linear interpolation was not valid for four significant places; the next-order interpolation scheme was necessary to obtain the fourth significant place. ■

PROBLEMS

We desire the value of $J_0(x)$ at $x = 7.24$ using the information in Table A4. Approximate its value using

1. Forward differences with $e = o(h^2)$.

2. Forward differences with $e = o(h^3)$.

3. Backward differences with $e = o(h^3)$.

4. Backward differences with $e = o(h^4)$.

Determine the error in the approximation to $J_0(x)$, using the expression following Table A4 to

5. Problem 1   **6.** Problem 2   **7.** Problem 3   **8.** Problem 4

Find an approximation of erf x with $e = o(h^3)$ at

9. $x = 0.01$ **10.** $x = 2.01$ **11.** $x = 0.91$

Find an approximation, to five significant digits, at $x = 1.51$ to

12. erf x **13.** $Y_0(x)$ **14.** $J_1(x)$ **15.** $Y_1(x)$

9.7 *Roots of Equations*

It is often necessary to find roots of equations, that is, the values of x for which $f(x) = 0$. This is encountered whenever we solve the characteristic equation of ordinary differential equations with constant coefficients. It may also be necessary to find roots of equations when using numerical methods in solving differential equations. We will study one technique that is commonly used in locating roots; it is *Newton's method*, sometimes called the *Newton-Raphson* method. We make a guess at the root, say $x = x_0$. Using this value of x_0 we calculate $f(x_0)$ and $f'(x_0)$ from the given equation,

$$f(x) = 0. \tag{9.7.1}$$

Then, a Taylor series with $e = o(h^2)$ is used to predict an improved value for the root. Using two terms of the series in Eq. 9.3.1, we have, approximately,

$$f(x_0 + h) = f(x_0) + hf'(x_0). \tag{9.7.2}$$

We presume that $f(x_0)$ will not be zero, since we only guessed at the root. What we desire from Eq. (9.7.2) is that $f(x_0 + h) = 0$; then $x_1 = x_0 + h$ will be our next guess for the root. Setting $f(x_0 + h) = 0$ and solving for h, we have

$$h = -\frac{f(x_0)}{f'(x_0)}. \tag{9.7.3}$$

The next guess is then

$$x_1 = x_0 - \frac{f(x_0)}{f'(x_0)}. \tag{9.7.4}$$

Adding another iteration gives a third guess as

$$x_2 = x_1 - \frac{f(x_1)}{f'(x_1)}, \tag{9.7.5}$$

or, in general,

$$x_{n+1} = x_n - \frac{f(x_n)}{f'(x_n)}. \tag{9.7.6}$$

This process can be visualized by considering the function $f(x)$ displayed in Fig. 9.5. Let us search for the root x, shown. Assume that the first guess

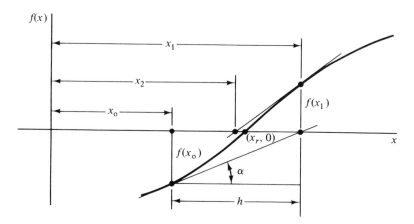

FIGURE 9.5.   Newton's method.

x_0 is too small, so that $f(x_0)$ is negative as shown and $f'(x_0)$ is positive. The first derivative $f'(x_0)$ is equal to $\tan \alpha$. Then, from Eq. 9.7.3,

$$\tan \alpha = -\frac{f(x_0)}{h}, \tag{9.7.7}$$

where h is the horizontal leg on the triangle shown. The next guess is then seen to be

$$x_1 = x_0 + h. \tag{9.7.8}$$

Repeating the steps above gives x_2 as shown. A third iteration can be added to the figure with x_3 being very close to x_r. It is obvious that this iteration process converges to the root x_r. However, there are certain functions $f(x)$ for which the initial guess must be very close to a root for convergence to that root. An example of this kind of function is shown in Fig. 9.6a. An initial guess outside the small increment Δx will lead to one of the other two roots shown and not to x_r. The root x_r would be quite difficult to find using Newton's method.

Another type of function for which Newton's method may give trouble is shown in Fig. 9.6b. By making the guess x_0, following Newton's method, the first iteration would yield x_1. The next iteration could yield a value x_2 close to the initial guess x_0. The process would just repeat itself indefinitely. To avoid an infinite loop of this nature, we should set a maximum number of iterations for our calculations.

One last word of caution is in order. Note from Eq. 9.7.3 that if we guess a point on the curve where $f'(x_0) = 0$, or approximately zero, then h is undefined or extremely large and the process may not work. Either a new guess should be attempted, or we may use Taylor series with $e = o(h^3)$, neglecting the first derivative term; in that case,

$$f(x_0 + h) = f(x_0) + \frac{h^2}{2}f''(x_0). \tag{9.7.9}$$

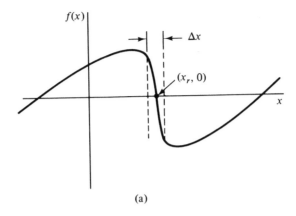

(a)

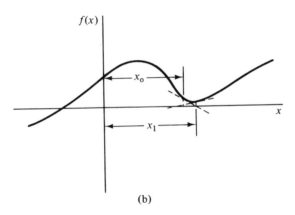

(b)

FIGURE 9.6.   Examples for which Newton's method gives trouble.

Setting $f(x_0 + h) = 0$, we have

$$h^2 = -\frac{2f(x_0)}{f''(x_0)}. \tag{9.7.10}$$

This step is then substituted into the iteration process in place of the step in which $f'(x_0) \cong 0$.

Example 9.7.1:   Find at least one root of the equation $x^5 - 10x + 100 = 0$. Carry out four iterations from an initial guess.

SOLUTION: The function $f(x)$ and its first derivative are

$$f(x) = x^5 - 10x + 100$$

$$f'(x) = 5x^4 - 10.$$

Note that the first derivative is zero at $x^4 = 2$. This gives a value $x = \pm\sqrt[4]{2}$. So, let's keep away from these points of zero slope. A positive value of $x > 1$ is no use since

$f(x)$ will always be positive, so let's try $x_0 = -2$. At $x_0 = -2$, we have

$$f(x_0) = 88$$
$$f'(x_0) = 70.$$

For the first iteration, Eq. 9.7.4 gives

$$x_1 = -2.0 - \frac{88}{70} = -3.26.$$

Using this value, we have

$$x_2 = -3.26 - \frac{-235}{555} = -2.84.$$

The third iteration gives

$$x_3 = -2.84 - \frac{-56.1}{315} = -2.66.$$

Finally, the fourth iteration results in

$$x_4 = -2.66 - \frac{-6}{240} = -2.64.$$

If a more accurate value of the root is desired, a fifth iteration is necessary. Obviously, a computer would approximate this root with extreme accuracy with multiple iterations. ∎

Note that the first derivative was required when applying Newton's method. There are situations in which the first derivative is very difficult, if not impossible, to find explicitly. For those situations we form an approximation to the first derivative using a numerical expression such as that given by Eq. 9.4.2.

PROBLEMS

Find an approximation to the root of the equation $x^3 - 5x^2 + 6x - 1 = 0$ in the neighborhood of each location. Carry out the iteration to three significant figures.

1. $x = 0$ **2.** $x = 1$ **3.** $x = 4$

Find a root to three significant figures of each equation near the point indicated.

4. $x^3 + 2x - 6 = 0,$ $x = 2$

5. $x^3 - 6x = 5,$ $x = 3$

6. $x^4 = 4x + 2,$ $x = 2$

7. $x^5 = 3x - 2,$ $x = 1$

Find a root to three significant figures of each equation.

8. $x^3 + 10x = 4$ **9.** $\cos 2x = x$

10. $x + \ln x = 10$

Find a root to five significant figures, using a computer, to the equation of

11. Problem 8   **12.** Problem 9   **13.** Problem 10

9.8 Initial-Value Problems—Ordinary Differential Equations

One of the most important and useful applications of numerical analysis is in the solution of differential equations, both ordinary and partial. There are two common problems encountered in finding the numerical solution to a differential equation. The first is: When one finds a numerical solution, is the solution acceptable; that is, is it sufficiently close to the exact solution? If one has an analytical solution, this can easily be checked; but for a problem for which an analytical solution is not known, one must be careful in concluding that a particular numerical solution is acceptable. When extending a solution from x_i to x_{i+1}, a truncation error is incurred, as discussed in Section 9.4, and as the solution is extended across the interval of interest, this error accumulates to give an accumulated truncation error. After, say, 100 steps, this error must be sufficiently small so as to give acceptable results. Obviously, all the various methods give different accumulated error. Usually, a method is chosen that requires a minimum number of steps, requiring the shortest possible computer time, yet one that does not give excessive error.

The second problem often encountered in numerical solutions to differential equations is the instability of numerical solutions. The actual solution to the problem of interest is stable (well behaved), but the errors incurred in the numerical solution are magnified in such a way that the numerical solution is obviously incompatible with the actual solution. This often results in a wildly oscillating solution in which extremely large variations occur in the dependent variable from one step to the next. When this happens, the numerical solution is unstable. By changing the step size or by changing the numerical method, a stable numerical solution can usually be found.

A numerical method that gives accurate results and is stable with the least amount of computer time often requires that it be "started" with a somewhat less accurate method and then continued with a more accurate technique. There are, of course, a host of starting techniques and methods that are used to continue a solution. We shall consider only a few methods; the first will not require starting techniques and will be the most inaccurate. However, the various methods do include the basic ideas of numerical solution of differential equations, and hence are quite important.

We shall initially focus our attention on solving first-order equations, since, every nth-order equation is equivalent to a system of n first-order equations.

Many of the examples in which ordinary differential equations describe the phenomenon of interest involve time as the independent variable. Thus, we shall use time t in place of the independent variable x of the preceding sections. Naturally, the difference operators are used as defined, with t substituted for x.

We study first-order equations which can be put in the form

$$\dot{y} = f(y, t) \tag{9.8.1}$$

where $\dot{y} = dy/dt$. If y_i and $\dot{y}_i$ at t_i are known, then Eq. 9.8.1 can be used to give y_{i+1} and $\dot{y}_{i+1}$ at t_{i+1}. We shall assume that the necessary condition is given at a particular time t_0.

9.8.1   TAYLOR'S METHOD

A simple technique for solving a first-order differential equation is to use a Taylor series, which in difference notation is

$$y_{i+1} = y_i + h\dot{y}_i + \frac{h^2}{2}\ddot{y}_i + \frac{h^3}{6}\dddot{y}_i + \cdots, \tag{9.8.2}$$

where h is the step size $(t_{i+1} - t_i.)$ This may require several derivatives at t_i depending on the order of the terms truncated. These derivatives are found by differentiating the equation

$$\dot{y} = f(y, t). \tag{9.8.3}$$

Since we consider the function f to depend on the two variables y and t, and y is a function of t, we must be careful when differentiating with respect to t. For example, consider $\dot{y} = 2y^2 t$. Then to find $\ddot{y}$ we must differentiate a product to give

$$\ddot{y} = 4y\dot{y}t + 2y^2 \tag{9.8.4}$$

and, differentiating again,

$$\dddot{y} = 4\dot{y}^2 t + 4y\ddot{y}t + 8y\dot{y}. \tag{9.8.5}$$

Higher-order derivatives follow in a like manner.

By knowing an initial condition, y_0 at $t = t_0$, the first derivative $\dot{y}_0$ is calculated from the given differential equation and $\ddot{y}_0$ and $\dddot{y}_0$ from equations similar to Eqs. 9.8.4 and 9.8.5. The value y_1 at $t = t_1$ then follows by putting $i = 0$ in Eq. 9.8.2 which is truncated appropriately. The derivatives, at $t = t_1$, are then calculated from Eqs. 9.8.3, 9.8.4, and 9.8.5. This procedure is continued to the maximum t that is of interest for as many steps as required.

This method can also be used to solve higher-order equations simply by expressing the higher-order equation as a set of first-order equations and proceeding with a simultaneous solution of the set of equations.

9.8.2 EULER'S METHOD

Euler's method results from approximating the derivative

$$\frac{dy}{dt} = \lim_{\Delta t \to 0} \frac{\Delta y}{\Delta t} \qquad (9.8.6)$$

by the difference equation

$$\Delta y \cong \dot{y} \, \Delta t \qquad (9.8.7)$$

or, in difference notation

$$y_{i+1} = y_i + h\dot{y}_i, \qquad e = o(h^2). \qquad (9.8.8)$$

This is immediately recognized as the first-order approximation of Taylor's method; thus, we would expect for the same step size that more accurate results would occur by retaining higher-order terms in Taylor's method. Euler's method is, of course, simpler to use, since we do not have to compute the higher derivatives at each point. It could also be used to solve higher-order equations, as will be illustrated later.

9.8.3 ADAMS' METHOD

Adams' method is one of the multitude of more accurate methods. It illustrates another technique for solving first-order differential equations.

The Taylor series allows us to write

$$y_{i+1} = y_i + \left(hD + \frac{h^2 D^2}{2} + \frac{h^3 D^3}{6} + \cdots \right) y_i$$

$$= y_i + \left(1 + \frac{hD}{2} + \frac{h^2 D^2}{6} + \cdots \right) hDy_i. \qquad (9.8.9)$$

Let us neglect terms of order h^5 and greater so that $e = o(h^5)$. Then, writing D in terms of ∇ (see Table 9.3), we have

$$y_{i+1} = y_i + h\left[1 + \frac{1}{2}\left(\nabla + \frac{\nabla^2}{2} + \frac{\nabla^3}{3} + \cdots \right) + \frac{1}{6}(\nabla^2 + \nabla^3 + \cdots) \right.$$

$$\left. + \frac{1}{24}(\nabla^3 + \cdots) \right] Dy_i$$

$$= y_i + h\left(1 + \frac{\nabla}{2} + \frac{5\nabla^2}{12} + \frac{3\nabla^3}{8} \right) Dy_i, \qquad e = o(h^5). \qquad (9.8.10)$$

Using the notation, $Dy_i = \dot{y}_i$, the equation above can be put in the form (using Table 9.2)

$$y_{i+1} = y_i + \frac{h}{24}(55\dot{y}_i - 59\dot{y}_{i-1} + 37\dot{y}_{i-2} - 9\dot{y}_{i-3}), \qquad e = o(h^5).$$

$$(9.8.11)$$

Adams' method uses the expression above to predict y_{i+1} in terms of previous information. This method requires several starting values, which could be obtained by Taylor's or Euler's methods, usually using smaller step sizes to maintain accuracy. Note that the first value obtained by Eq. 9.8.11 is y_4. Thus, we must use a different technique to find y_1, y_2, and y_3. If we were to use Adams' method with $h = 0.1$ we could choose Taylor's method with $e = o(h^3)$ and use $h = 0.02$ to find the starting values so that the same accuracy as that of Adams' method results. We then apply Taylor's method for 15 steps and use every fifth value for y_1, y_2, and y_3 to be used in Adams' method. Equation 9.8.11 is then used to continue the solution. The method is quite accurate, since $e = o(h^5)$.

Adams' method can be used to solve a higher-order equation by writing the higher-order equation as a set of first-order equations, or by differentiating Eq. 9.8.11 to give the higher-order derivatives. One such derivative is

$$\dot{y}_{i+1} = \dot{y}_i + \frac{h}{24}(55\ddot{y}_i - 59\ddot{y}_{i-1} + 37\ddot{y}_{i-2} - 9\ddot{y}_{i-3}). \qquad (9.8.12)$$

Others follow naturally.

9.8.4 RUNGE-KUTTA METHODS

In order to produce accurate results using Taylor's method, derivatives of higher order must be evaluated. This may be difficult, or the higher-order derivatives may be inaccurate. Adams' method requires several starting values, which may be obtained by less accurate methods, resulting in larger truncation error than desirable. Methods that require only the first-order derivative and give results with the same order of truncation error as Taylor's method maintaining the higher-order derivatives, are called the *Runge-Kutta methods*. Estimates of the first derivative must be made at points within each interval $t_i \le t \le t_{i+1}$. The prescribed first-order equation is used to provide the derivative at the interior points. The Runge-Kutta method with $e = o(h^3)$ will be developed and methods with $e = o(h^4)$ and $e = o(h^5)$ will simply be presented with no development.

Let us again consider the first-order equation $\dot{y} = f(y, t)$. All Runge-Kutta methods utilize the approximation

$$y_{i+1} = y_i + h\phi_i, \qquad (9.8.13)$$

where ϕ_i is an approximation to the slope in the interval $t_i < t \le t_{i+1}$.

Certainly, if we used $\phi_i = f_i$, the approximation for y_{i+1} would be too large for the curve in Fig. 9.7; and, if we used $\phi_i = f_{i+1}$, the approximation would be too small. Hence, the correct ϕ_i needed to give the exact y_{i+1} lies in the interval $f_i \leq \phi_i \leq f_{i+1}$. The trick is to find a technique that will give a good approximation to the correct slope ϕ_i. Let us assume that

$$\phi_i = a\xi_i + b\eta_i, \tag{9.8.14}$$

where

$$\xi_i = f(y_i, t_i) \tag{9.8.15}$$

$$\eta_i = f(y_i + qh\xi_i, t_i + ph). \tag{9.8.16}$$

The quantities a, b, p, and q are constants to be established later.

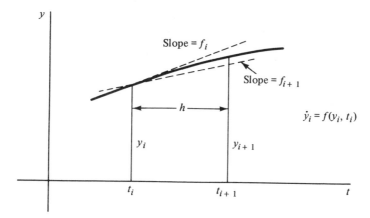

FIGURE 9.7.   Curve showing approximations to y_{i+1}, using slopes f_i and f_{i+1}.

A good approximation for η_i is found by expanding in a Taylor series, neglecting higher-order terms:

$$\eta_i = f(y_i, t_i) + \frac{\partial f}{\partial y}(y_i, t_i)\Delta y + \frac{\partial f}{\partial t}(y_i, t_i)\Delta t + o(h^2)$$

$$= f_i + qhf_i\frac{\partial f}{\partial y}(y_i, t_i) + ph\frac{\partial f}{\partial t}(y_i, t_i) + o(h^2), \tag{9.8.17}$$

where we have used $\Delta y = qhf_i$ and $\Delta t = ph$, as required by Eq. 9.8.16. Equation 9.8.13 then becomes, using $\xi_i = f_i$,

$$y_{i+1} = y_i + h\phi_i$$

$$= y_i + h(a\xi_i + b\eta_i)$$

$$= y_i + h(af_i + bf_i) + h^2\left[bqf_i\frac{\partial f}{\partial y}(y_i, t_i) + bp\frac{\partial f}{\partial t}(y_i, t_i)\right] + o(h^3),$$

$$\tag{9.8.18}$$

where we have substituted for ξ_i and η_i from Eqs. 9.8.15 and 9.8.17, respectively. Expand y_i in a Taylor series, with $e = o(h^3)$, so that

$$y_{i+1} = y_i + h\dot{y}_i + \frac{h^2}{2}\ddot{y}_i$$

$$= y_i + hf(y_i, t_i) + \frac{h^2}{2}\dot{f}(y_i, t_i). \qquad (9.8.19)$$

Now, using the chain rule,

$$\dot{f} = \frac{\partial f}{\partial y}\frac{\partial y}{\partial t} + \frac{\partial f}{\partial t}\frac{\partial t}{\partial t} = \dot{y}\frac{\partial f}{\partial y} + \frac{\partial f}{\partial t} = f\frac{\partial f}{\partial y} + \frac{\partial f}{\partial t}. \qquad (9.8.20)$$

Thus, we have

$$y_{i+1} = y_i + hf(y_i, t_i) + \frac{h^2}{2}\left[f_i\frac{\partial f}{\partial y}(y_i, t_i) + \frac{\partial f}{\partial t}(y_i, t_i)\right]. \qquad (9.8.21)$$

Comparing this with Eq. 9.8.18, we find that (equating terms in like powers of h)

$$a + b = 1, \qquad bq = \tfrac{1}{2} \qquad bp = \tfrac{1}{2}. \qquad (9.8.22)$$

These three equations contain four unknowns, hence one of them is arbitrary. It is customary to choose $b = \tfrac{1}{2}$ or $b = 1$. For $b = \tfrac{1}{2}$, we have $a = \tfrac{1}{2}, q = 1$, and $p = 1$. Then our approximation for y_{i+1} from Eq. 9.8.18 becomes

$$y_{i+1} = y_i + h(a\xi_i + b\eta_i)$$

$$= y_i + \frac{h}{2}[f(y_i, t_i) + f(y_i + hf_i, t_i + h)], \qquad e = o(h^3).$$

$$(9.8.23)$$

For $b = 1$, we have $a = 0, q = \tfrac{1}{2}$, and $p = \tfrac{1}{2}$, there results

$$y_{i+1} = y_i + h\eta_i$$

$$= y_i + hf\left(y_i + \frac{h}{2}f_i, t_i + \frac{h}{2}\right), \qquad e = o(h^3). \qquad (9.8.24)$$

Knowing y_i, t_i, and $\dot{y}_i = f_i$ we can now calculate y_{i+1} with the same accuracy obtained using Taylor's method that required us to know $\ddot{y}_i$.

The Runge-Kutta method, with $e = o(h^4)$, can be developed in a similar manner. First, the function ϕ_i is assumed to have the form

$$\phi_i = a\xi_i + b\eta_i + c\zeta_i, \qquad (9.8.25)$$

where

$$\xi_i = f(y_i, t_i)$$
$$\eta_i = f(y_i + ph\xi_i, t_i + ph) \qquad (9.8.26)$$
$$\zeta_i = f[y_i + sh\eta_i + (r - s)h\xi_i, t_i + rh].$$

Equating coefficients of the Taylor series expansions results in two arbitrary coefficients. The common choice is $a = \frac{1}{6}$, $b = \frac{2}{3}$, and $c = \frac{1}{6}$. We then have

$$y_{i+1} = y_i + \frac{h}{6}(\xi_i + 4\eta_i + \zeta_i), \qquad e = o(h^4), \qquad (9.8.27)$$

with

$$\xi_i = f(y_i, t_i)$$

$$\eta_i = f\left(y_i + \frac{h}{2}\xi_i, t_i + \frac{h}{2}\right) \qquad (9.8.28)$$

$$\zeta_i = f(y_i + 2h\eta_i - h\xi_i, t_i + h).$$

The Runge-Kutta method with $e = o(h^5)$ is perhaps the most widely used method for solving ordinary differential equations. One such method results in

$$y_{i+1} = y_i + \frac{h}{6}[\xi_i + (2 - \sqrt{2})\eta_i + (2 + \sqrt{2})\zeta_i + \omega_i], \qquad e = o(h^5),$$

$$(9.8.29)$$

where

$$\xi_i = f(y_i, t_i)$$

$$\eta_i = f\left(y_i + \frac{h}{2}\xi_i, t_i + \frac{h}{2}\right)$$

$$\zeta_i = f\left[y_i + \frac{h}{\sqrt{2}}(\xi_i - \eta_i) - \frac{h}{2}(\xi_i - 2\eta_i), t_i + \frac{h}{2}\right] \qquad (9.8.30)$$

$$\omega_i = f\left[y_i - \frac{h}{\sqrt{2}}(\eta_i - \zeta_i) + h\zeta_i, t_i + h\right].$$

Another method with $e = o(h^5)$ that is widely used gives

$$y_{i+1} = y_i + \frac{h}{6}(\xi_i + 2\eta_i + 2\zeta_i + \omega_i), \qquad e = o(h^5), \qquad (9.8.31)$$

where

$$\xi_i = f(y_i, t_i)$$

$$\eta_i = f\left(y_i + \frac{h}{2}\xi_i, t_i + \frac{h}{2}\right)$$

$$\zeta_i = f\left(y_i + \frac{h}{2}\eta_i, t_i + \frac{h}{2}\right) \qquad (9.8.32)$$

$$\omega_i = f(y_i + h\zeta_i, t_i + h).$$

In all the methods above no information is needed other than the initial condition. For example, y_i is approximated by using y_0, ξ_0, η_0, and so on. The quantities are found from the given equation with no differentiation required. These reasons, combined with the accuracy of the Runge-Kutta methods, make them extremely popular.

9.8.5   DIRECT METHOD

The final method that will be discussed is seldom used because of its inaccuracy, but it is easily understood and follows directly from the expressions for the derivatives as presented in Section 9.4. It also serves to illustrate the method used to solve partial differential equations. Let us again use as an example the first-order differential equation

$$\dot{y} = 2y^2t. \tag{9.8.33}$$

Then, from Table 9.4, with $e = o(h^2)$, and using forward differences, we have

$$\dot{y}_i = \frac{dy_i}{dt} = Dy_i = \frac{1}{2h}(-y_{i+2} + 4y_{i+1} - 3y_i). \tag{9.8.34}$$

Substitute this directly into Eq. 9.8.33, to obtain

$$\frac{1}{2h}(-y_{i+2} + 4y_{i+1} - 3y_i) = 2y_i^2 t_i. \tag{9.8.35}$$

This is rearranged to give

$$y_{i+2} = 4y_{i+1} - (3 + 4hy_it_i)y_i, \qquad e = o(h^2) \tag{9.8.36}$$

Using $i = 0$, we can determine y_2 if we know y_1 and y_0. This requires a starting technique to find y_1. We could use Euler's method, since that also has $e = o(h^2)$.

This method can easily be used to solve higher-order equations. We simply substitute from Table 9.4 for the higher-order derivatives and find an equation similar to Eq. 9.8.36 to advance the solution.

Let us now work some examples using the techniques of this section.

Example 9.8.1:   Use Euler's method to solve $\dot{y} + 2yt = 4$ if $y(0) = 0.2$. Compare with Taylor's method, $e = o(h^3)$. Use $h = 0.1$. Carry the solution out for four time steps.

SOLUTION: In Euler's method we must have the first derivative at each point; it is given by

$$\dot{y}_i = 4 - 2y_it_i.$$

The solution is approximated at each point by

$$y_{i+1} = y_i + h\dot{y}_i.$$

For the first four steps there results

$$t_0 = 0: \qquad y_0 = 0.2$$
$$t_1 = 0.1: \qquad y_1 = y_0 + h\dot{y}_0 = 0.2 + 0.1 \times 4 = 0.6$$
$$t_2 = 0.2: \qquad y_2 = y_1 + h\dot{y}_1 = 0.6 + 0.1 \times 3.88 = 0.988$$
$$t_3 = 0.3: \qquad y_3 = y_2 + h\dot{y}_2 = 0.988 + 0.1 \times 3.60 = 1.35$$
$$t_4 = 0.4: \qquad y_4 = y_3 + h\dot{y}_3 = 1.35 + 0.1 \times 3.19 = 1.67.$$

Using Taylor's method with $e = o(h^3)$, we approximate y_{i+1} using

$$y_{i+1} = y_i + h\dot{y}_i + \frac{h^2}{2}\ddot{y}_i.$$

Thus we see that we need $\ddot{y}$. It is found by differentiating the given equation, providing us with

$$\ddot{y}_i = -2\dot{y}_i t_i - 2y_i.$$

Progressing in time as in Euler's method, there results

$$t_0 = 0: \qquad y_0 = 0.2$$

$$t_1 = 0.1: \qquad y_1 = y_0 + h\dot{y}_0 + \frac{h^2}{2}\ddot{y}_0 = 0.2 + 0.1 \times 4 + 0.005 \times (-0.4)$$
$$= 0.598$$

$$t_2 = 0.2: \qquad y_2 = y_1 + h\dot{y}_1 + \frac{h^2}{2}\ddot{y}_1 = 0.598 + 0.1 \times 3.88 + 0.005 \times (-1.97)$$
$$= 0.976$$

$$t_3 = 0.3: \qquad y_3 = y_2 + h\dot{y}_2 + \frac{h^2}{2}\ddot{y}_2 = 1.32$$

$$t_4 = 0.4: \qquad y_4 = y_3 + h\dot{y}_3 + \frac{h^2}{2}\ddot{y}_3 = 1.62. \quad \blacksquare$$

Example 9.8.2:   Use a Runge-Kutta method with $e = o(h^5)$ and solve $\dot{y} + 2yt = 4$ if $y(0) = 0.2$ using $h = 0.1$. Carry out the solution for two time steps.

SOLUTION: We will choose Eq. 9.8.31 to illustrate the Runge-Kutta method. The first derivative is used at various points interior to each interval. It is found from

$$\dot{y}_i = 4 - 2y_i t_i.$$

To find y_1 we must know y_0, ξ_0, η_0, ζ_0, and ω_0. They are

$$y_0 = 0.2$$
$$\xi_0 = 4 - 2y_0 t_0 = 4 - 2 \times 0.2 \times 0 = 4$$
$$\eta_0 = 4 - 2\left(y_0 + \frac{h}{2}\xi_0\right)\left(t_0 + \frac{h}{2}\right) = 4 - 2\left(0.2 + \frac{0.1}{2} \times 4\right)\left(\frac{0.1}{2}\right) = 3.96$$

$$\zeta_0 = 4 - 2\left(y_0 + \frac{h}{2}\eta_0\right)\left(t_0 + \frac{h}{2}\right) = 4 - 2\left(0.2 + \frac{0.1}{2} \times 3.96\right)\left(\frac{0.1}{2}\right) = 3.96$$

$$\omega_0 = 4 - 2(y_0 + h\zeta_0)(t_0 + h) = 4 - 2(0.2 + 0.1 \times 3.96)(0.1) = 3.88.$$

Thus,

$$y_1 = y_0 + \frac{h}{6}(\xi_0 + 2\eta_0 + 2\zeta_0 + \omega_0)$$

$$= 0.2 + \frac{0.1}{6}(3.96 + 7.92 + 7.92 + 3.88) = 0.595.$$

To find y_2 we calculate

$$\xi_1 = 4 - 2y_1 t_1 = 4 - 2 \times 0.595 \times 0.1 = 3.88$$

$$\eta_1 = 4 - 2\left(y_1 + \frac{h}{2}\xi_1\right)\left(t_1 + \frac{h}{2}\right) = 4 - 2\left(0.595 + \frac{0.1}{2} \times 3.88\right)\left(0.1 + \frac{0.1}{2}\right)$$

$$= 3.76$$

$$\zeta_1 = 4 - 2\left(y_1 + \frac{h}{2}\eta_1\right)\left(t_1 + \frac{h}{2}\right) = 4 - 2\left(0.595 + \frac{0.1}{2} \times 3.76\right)\left(0.1 + \frac{0.1}{2}\right)$$

$$= 3.77$$

$$\omega_1 = 4 - 2(y_1 + h\zeta_1)(t_1 + h) = 4 - 2(0.595 + 0.1 \times 3.77)(0.1 + 0.1)$$

$$= 3.61.$$

Finally,

$$y_2 = y_1 + \frac{h}{6}(\xi_1 + 2\eta_1 + 2\zeta_1 + \omega_1)$$

$$= 0.595 + \frac{0.1}{6}(3.88 + 7.52 + 7.54 + 3.61) = 0.971.$$

Additional values follow. Note that the procedure above required no starting values and no higher-order derivatives, but still $e = o(h^5)$. ■

Example 9.8.3:   Use the direct method to solve the equation $\dot{y} + 2yt = 4$ if $y(0) = 0.2$ using $h = 0.1$. Use forward differences with $e = o(h^2)$. Carry out the solution for four time steps.

SOLUTION: Using the direct method we substitute for $\dot{y} = Dy_i$ from Table 9.4 with $e = o(h^2)$. We have

$$\frac{1}{2h}(-y_{i+2} + 4y_{i+1} - 3y_i) + 2y_i t_i = 4.$$

Rearranging, there results

$$y_{i+2} = 4y_{i+1} - (3 - 4ht_i)y_i - 8h.$$

The first value that we can find with the formula above is y_2. Hence, we must find y_1 by some other technique. Use Euler's method to find y_1. It is

$$y_1 = y_0 + h\dot{y}_0 = 0.2 + 0.1 \times 4 = 0.6.$$

We now can use the direct method to find

$$y_2 = 4y_1 - (3 - 4ht_0)y_0 - 8h$$
$$= 4 \times 0.6 - (3 - 0) \times 0.2 - 8 \times 0.1 = 1.0$$

$$y_3 = 4y_2 - (3 - 4ht_1)y_1 - 8h$$
$$= 4 \times 1.0 - (3 - 4 \times 0.1 \times 0.1) \times 0.6 - 8 \times 0.1 = 1.424$$

$$y_4 = 4y_3 - (3 - 4ht_2)y_2 - 8h$$
$$= 4 \times 1.424 - (3 - 4 \times 0.1 \times 0.2) \times 1.0 - 8 \times 0.1 = 1.976. \quad \blacksquare$$

These results are, of course, less accurate than those obtained using Taylor's method or the Runge–Kutta method in Examples 9.8.1 and 9.8.2.

PROBLEMS

1. Estimate $y(0.2)$ if $\dot{y} + 2yt = 4$ and $y(0) = 0.2$ using Euler's method. Use $h = 0.05$ and compare with Example 9.8.1.

2. Estimate $y(0.2)$ if $\dot{y} + 2yt = 4$ and $y(0) = 0.2$ using Taylor's method, $e = o(h^3)$. Use $h = 0.05$ and compare with Example 9.8.1.

Estimate $y(2)$ if $2\dot{y}y = t^2$ and $y(0) = 2$ using $h = 0.4$. Compare with the exact solution. Use the given method.

3. Euler's method.

4. Taylor's method, $e = o(h^3)$.

5. Adams' method.

6. Runge–Kutta method, $e = o(h^3)$. Use Eq. 9.8.24.

Find a numerical solution between $t = 0$ and $t = 1$ to $y + 4yt = t^2$ if $y(0) = 2$. Use $h = 0.2$ and the given method.

7. Euler's method.

8. Taylor's method, $e = o(h^3)$.

9. Runge–Kutta method, $e = o(h^3)$. Use Eq. 9.8.24.

10. The direct method, $e = o(h^2)$.

Find y_1 and $\dot{y}_2$ if $y^2 + 2y = 4$ and $y(0) = 0$ using $h = 0.2$ and the given method.

11. Taylor's method, $e = o(h^3)$.

12. Taylor's method, $e = o(h^4)$.

13. Runge–Kutta method, $e = o(h^3)$. Use Eq. 9.8.23.

14. Runge–Kutta method, $e = o(h^4)$.

15. Runge–Kutta method, $e = o(h^5)$.

16. Derive an "Adams' method" with order of magnitude of error $o(h^4)$.

Solve the differential equation $\dot{y}^2 + 2y = 4$ if $y(0) = 0$. Use a computer and carry out the solution to five significant figures until $t = 5$. Use the given method.

17. Euler's method.

18. Taylor's method, $e = o(h^3)$.

19. Taylor's method, $e = o(h^4)$.

20. Runge–Kutta method, $e = o(h^3)$. Use Eq. 9.8.23.

21. Runge–Kutta method, $e = o(h^4)$.

22. Runge–Kutta method, $e = o(h^5)$.

9.9   Higher-Order Equations

Taylor's method can be used to solve higher-order differential equations without representing them as a set of first-order differential equations. Consider the third-order equation

$$\dddot{y} + 4t\ddot{y} + 5y = t^2 \tag{9.9.1}$$

with three required initial conditions imposed at $t = 0$, namely, y_0, $\dot{y}_0$, and $\ddot{y}_0$. Thus, at $t = 0$ all the necessary information is known and y_1 can be found from the Taylor series, with $e = o(h^3)$,

$$y_1 = y_0 + h\dot{y}_0 + \frac{h^2}{2}\ddot{y}_0. \tag{9.9.2}$$

To find y_2 the derivatives $\dot{y}_1$ and $\ddot{y}_1$ would be needed. To find them we differentiate the Taylor series to get

$$\dot{y}_1 = \dot{y}_0 + h\ddot{y}_0 + \frac{h^2}{2}\dddot{y}_0$$

$$\ddot{y}_1 = \ddot{y}_0 + h\dddot{y}_0 + \frac{h^2}{2}\left(\frac{d^4 y}{dt^4}\right)_0. \tag{9.9.3}$$

The third derivative, $\dddot{y}_0$, is then found from Eq. 9.9.1 and $(d^4 y/dt^4)_0$ is found by differentiating Eq. 9.9.1. We can then proceed to the next step and continue through the interval of interest.

Instead of using Taylor's method directly, we could have written Eq.

9.9.1 as the following set of first-order equations:

$$\dot{y}_i = u_i$$

$$\dot{u}_i = v_i \qquad (9.9.4)$$

$$\dot{v}_i = -4t_i v_i - 5y_i + t_i^2.$$

The last of these equations results from substituting the first two into Eq. 9.9.1. The initial conditions specified at $t = 0$ are y_0, $u_0 = \dot{y}_o$, and $v_0 = \ddot{y}_0$. If Euler's method is used we have, at the first step,

$$v_1 = v_0 + \dot{v}_0 h = v_0 + (-4t_0 v_0 - 5y_0 + t_0^2)h$$

$$u_1 = u_0 + \dot{u}_o h = u_0 + v_0 h \qquad (9.9.5)$$

$$y_1 = y_0 + \dot{y}_0 h = y_0 + u_0 h, \qquad e = o(h^2).$$

With the values at $t_0 = 0$ known we can perform these calculations. This procedure is continued for all additional steps. Other methods for solving the first-order equations can also be used.

We also could have chosen the direct method by expressing Eq. 9.9.1 in finite-difference notation using the information contained in Table 9.4. For example, the forward-differencing relationships could be used to express Eq. 9.9.1 as

$$y_{i+3} - 3y_{i+2} + 3y_{i+1} - y_i + 4t_i h(y_{i+2} - 2y_{i+1} + y_i) + 5y_i h^3 = t_i^2 h^3,$$

$$e = o(h). \qquad (9.9.6)$$

This may be rewritten as

$$y_{i+3} = (3 - 4t_i h)y_{i+2} - (3 - 8t_i h)y_{i+1} + (1 - 4t_i h - 5h^3)y_i + t_i^2 h^3.$$

$$(9.9.7)$$

For $i = 0$, this becomes, using the initial condition $y = y_0$ at $t_0 = 0$,

$$y_3 = 3y_2 - 3y_1 + (1 - 5h^3)y_0. \qquad (9.9.8)$$

To find y_3 we need the starting values y_1 and y_2. They may be found by using Euler's method. Equation 9.9.7 is then used until all values of interest are determined.

A decision that must be made when solving problems numerically is how small the step size should be. The phenomenon being studied usually has a time scale T, or a length L, associated with it. The time scale T is the time necessary for a complete cycle of a periodic phenomenon, or the time required for a transient phenomenon to disappear (see Fig. 9.8). The length L may be the distance between telephone poles or the size of a capillary tube. What is necessary is that $h \ll T$ or $h \ll L$. If the numerical results using the various techniques or smaller step sizes differ considerably, this usually implies that h is not sufficiently small.

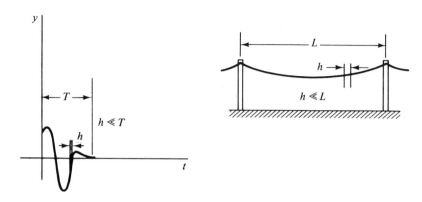

FIGURE 9.8. Examples of how the step size h should be chosen.

Example 9.9.1: Solve the differential equation $\ddot{y} - 2ty = 5$ with initial conditions $y(0) = 2$ and $\dot{y}(0) = 0$. Use Adams' method with $h = 0.2$ using Taylor's method with $e = o(h^3)$ to start the solution.

SOLUTION: Adams' method predicts the dependent variables at a forward step to be

$$y_{i+1} = y_i + \frac{h}{24}(55\dot{y}_i - 59\dot{y}_{i-1} + 37\dot{y}_{i-2} - 9\dot{y}_{i-3}).$$

Differentiating this expression results in

$$\dot{y}_{i+1} = \dot{y}_i + \frac{h}{24}(55\ddot{y}_i - 59\ddot{y}_{i-1} + 37\ddot{y}_{i-2} - 9\ddot{y}_{i-3}).$$

These two expressions can be used with $i \geq 3$; hence, we need a starting technique to give y_1, y_2, and y_3. We shall use Taylor's method with $e = o(h^3)$ to start the solution. Taylor's method uses

$$y_{i+1} = y_i + h\dot{y}_i + \frac{h^2}{2}\ddot{y}_i$$

$$\dot{y}_{i+1} = \dot{y}_i + h\ddot{y}_i + \frac{h^2}{2}\dddot{y}_i.$$

This requires the second and third derivatives; the second derivative is provided by the given differential equation,

$$\ddot{y}_i = 5 + 2t_i y_i.$$

The third derivative is found by differentiating the above and is

$$\dddot{y}_i = 2y_i + 2t_i\dot{y}_i.$$

Taylor's method provides the starting values.

$$t_1 = 0.2: \quad y_1 = y_o + h\dot{y}_o + \frac{h^2}{2}\ddot{y}_o = 2.1$$

$$\dot{y}_1 = \dot{y}_0 + h\ddot{y}_0 + \frac{h^2}{2}\dddot{y}_0 = 1.08$$

$$t_2 = 0.4: \qquad y_2 = y_1 + h\dot{y}_1 + \frac{h^2}{2}\ddot{y}_1 = 2.43$$

$$\dot{y}_2 = \dot{y}_1 + h\ddot{y}_1 + \frac{h^2}{2}\dddot{y}_1 = 2.34$$

$$t_3 = 0.6: \qquad y_3 = y_2 + h\dot{y}_2 + \frac{h^2}{2}\ddot{y}_2 = 3.04$$

$$\dot{y}_3 = \dot{y}_2 + h\ddot{y}_2 + \frac{h^2}{2}\dddot{y}_2 = 3.86.$$

Now Adams' method can be used to predict additional values. Several are as follows:

$$t_4 = 0.8: \qquad y_4 = y_3 + \frac{h}{24}(55\dot{y}_3 - 59\dot{y}_2 + 37\dot{y}_1 - 9\dot{y}_0) = 3.99$$

$$\dot{y}_4 = \dot{y}_3 + \frac{h}{24}(55\ddot{y}_3 - 59\ddot{y}_2 + 37\ddot{y}_1 - 9\ddot{y}_0) = 5.84$$

$$t_5 = 1.0: \qquad y_5 = y_4 + \frac{h}{24}(55\dot{y}_4 - 59\dot{y}_3 + 37\dot{y}_2 - 9\dot{y}_1) = 5.41$$

$$\dot{y}_5 = \dot{y}_4 + \frac{h}{24}(55\ddot{y}_4 - 59\ddot{y}_3 + 37\ddot{y}_2 - 9\ddot{y}_1) = 8.51$$

Other values can be found similarly.   ∎

Example 9.9.2:   Solve the differential equation $\ddot{y} - 2ty = 5$ with initial conditions $y(0) = 2$ and $\dot{y}(0) = 0$. Use the direct method with $e = o(h^2)$, using forward differences and $h = 0.2$. Start the solution with the values from Example 9.9.1.

SOLUTION: We write the differential equation in difference form using the relationships of Table 9.4. There results

$$\frac{1}{h^2}(-y_{i+3} + 4y_{i+2} - 5y_{i+1} + 2y_i) - 2t_i y_i = 5.$$

This is rearranged as

$$y_{i+3} = 4y_{i+2} - 5y_{i+1} + (2 - 2t_i h^2)y_i - 5h^2.$$

Letting $i = 0$, the first value that we can find from the equation above is y_3. Thus, we need to use a starting method to find y_1 and y_2. From Example 9.9.1 we have $y_1 = 2.1$ and $y_2 = 2.43$. Now we can use the equation of this example to find y_3. It is, letting $i = 0$,

$$y_3 = 4y_2 - 5y_1 + (2 - \overset{0}{\cancel{2t_0}}h^2)y_0 - 5h^2 = 3.02.$$

Two additional values are found as follows:

$$y_4 = 4y_3 - 5y_2 + (2 - 2t_1 h^2)y_1 - 5h^2 = 3.90$$

$$y_5 = 4y_4 - 5y_3 + (2 - 2t_2 h^2)y_2 - 5h^2 = 5.08.$$

This method is, of course, less accurate than the method of Example 9.9.1. It is however, easier to use, and if a smaller step size were chosen, more accurate numbers would result. ■

Example 9.9.3: Write a computer program to solve the differential equation $\ddot{y} - 2ty = 5$ using Adams' method with $h = 0.04$ if $\dot{y}(0) = 0$ and $y(0) = 2$. Use Taylor's method with $e = o(h^3)$ using $h = 0.02$ to start the solution.

SOLUTION: The language to be used is Fortran. The control statements (the first few statements necessary to put the program on a particular computer) are usually unique to each computer and are omitted here. The computer program and solution follow:

```
            PROGRAM DIFFEQ(INPUT,OUTPUT)
            DIMENSION Y(48),DY(48),D2Y(48)
            PRINT 30
30          FORMAT (1H1, 31X,*.*,9X,*..*,/,* I*5X,*T*,13X,3(*Y(T)*,  6X))
            Y(1) = 2.0
            H = 0.02
            DY(1) = 0.0
            D2Y(1) = 5.0
            T = 0.0
C           SOLVES D2Y - 2TY = 5 FOR Y HAVING THE INITIAL VALUE OF 2 AND
C           DY BEING INITIALLY EQUAL TO 0.
C           FIRST USE TAYLORS METHOD TO FIND THE STARTING VALUES.
            DO 10 I=1,6
            T = T + 0.02
            Y(I+1) = Y(I) + H*DY(I) + (H*H/2.0)*D2Y(I)
            DY(I+1) = DY(I)+H*D2Y(I) + (H*H/2.)*(2.*T*DY(I) + 2.*Y(I))
            D2Y(I+1) = 5.0 + 2.0*T*Y(I+1)
10          CONTINUE
            T = 0.0
            DO 15 I = 1,4
            Y(I) = Y(2*I-1)
            DY(I) = DY(2*I-1)
            D2Y(I) = D2Y(2*I-1)
            PRINT 40,I,T,Y(I),DY(I),D2Y(I)
            T = T + 0.04
15          CONTINUE
            T = 0.16
            H = 0.04
C           NOW USE ADAMS METHOD
            DO 20 I=4,44
            Y(I+1) = Y(I) + (H/24.)*(55.*DY(I) - 59.*DY(I-1) + 37.*DY(I-2)
         1            -9.*DY(I-3))
            DY(I+1) = DY(I) + (H/24.)*(55.*D2Y(I) - 59.*D2Y(I-1) +
         1            37.*D2Y(I-2) - 9.*D2Y(I-3))
            D2Y(I+1) = 5.0 + 2.0*T*Y(I+1)
            II = I + 1
            PRINT 40,II,T,Y(I+1),DY(I+1),D2Y(I+1)
40          FORMAT (I3,4X,F5.2,,5X,3F10.4)
            T = T + 0.04
20          CONTINUE
            END
```

i	t	$y(t)$	$\dot{y}(t)$	$\ddot{y}(t)$
1	0.00	2.0000	0.0000	5.0000
2	0.04	2.0040	0.2032	5.1603
3	0.08	2.0163	0.4129	5.3226
4	0.12	2.0371	0.6291	5.4889
5	0.16	2.0667	0.8520	5.6614
6	0.20	2.1054	1.0821	5.8422
7	0.24	2.1534	1.3196	6.0336
8	0.28	2.2110	1.5649	6.2382
9	0.32	2.2787	1.8188	6.4584
10	0.36	2.3567	2.0819	6.6968
11	0.40	2.4454	2.3548	6.9563
12	0.44	2.5452	2.6387	7.2398
13	0.48	2.6566	2.9344	7.5504
14	0.52	2.7801	3.2431	7.8913
15	0.56	2.9163	3.5661	8.2662
16	0.60	3.0656	3.9049	8.6787
17	0.64	3.2289	4.2610	9.1330
18	0.68	3.4067	4.6361	9.6332
19	0.72	3.6000	5.0323	10.1841
20	0.76	3.8096	5.4516	10.7906
21	0.80	4.0365	5.8964	11.4584
22	0.84	4.2817	6.3691	12.1933
23	0.88	4.5464	6.8728	13.0017
24	0.92	4.8320	7.4103	13.8908
25	0.96	5.1398	7.9852	14.8683
26	1.00	5.4713	8.6011	15.9427
27	1.04	5.8284	9.2620	17.1231
28	1.08	6.2129	9.9725	18.4200
29	1.12	6.6269	10.7373	19.8443
30	1.16	7.0727	11.5618	21.4087
31	1.20	7.5527	12.4520	23.1266
32	1.24	8.0698	13.4142	25.0131
33	1.28	8.6269	14.4555	27.0849
34	1.32	9.2274	15.5836	29.3603
35	1.36	9.8749	16.8072	31.8596
36	1.40	10.5733	18.1356	34.6054
37	1.44	11.3272	19.5792	37.6224
38	1.48	12.1413	21.1493	40.9384
39	1.52	13.0210	22.8586	44.5838
40	1.56	13.9720	24.7208	48.5927
41	1.60	15.0009	26.7513	53.0028
42	1.64	16.1146	28.9668	57.8557
43	1.68	17.3209	31.3861	63.1982
44	1.72	18.6284	34.0298	69.0816
45	1.76	20.0465	36.9205	75.5637

PROBLEMS

1. Compare the values for y_4 and y_5 from Adams' method in Example 9.9.1 with the values found by extending Taylor's method.

2. Write the differential equation of Example 9.9.2 in central difference form with $e = o(h^2)$ and show that

$$y_{i+1} = 5h^2 + 2(1 + t_i h^2)y_i - y_{i-1}.$$

Use this expression and find y_2, y_3, and y_4. Compare with the results of Example 9.9.2.

Estimate $y(t)$ between $t = 0$ and $t = 2$ for $\ddot{y} + y = 0$ if $y(0) = 0$ and $\dot{y}(0) = 4$. Use five steps with the given method.

3. Euler's method.   **4.** Taylor's method, $e = o(h^3)$.

5. Runge–Kutta method, $e = o(h^3)$.   **6.** The direct method, $e = o(h^2)$.

Use five steps and estimate $y(t)$ between $t = 0$ and $t = 1$ for $\ddot{y} + 2\dot{y}^2 + 10y = 0$ if $y(0) = 0$ and $\dot{y}(0) = 1.0$. Use the given method.

7. Euler's method.

8. Taylor's method, $e = o(h^3)$.   **9.** Runge–Kutta method, $e = o(h^3)$.

Solve the differential equation $\ddot{y} + 0.2\dot{y}^2 + 10y = 10 \sin 2t$ if $y(0) = 0$ and $\dot{y}(0) = 0$. With a computer, find a solution to four significant digits from $t = 0$ to $t = 8$ using the given method.

10. Euler's method.   **11.** Taylor's method, $e = o(h^3)$.

12. Adams' method.   **13.** Runge–Kutta method, $e = o(h^3)$.

14. Runge–Kutta method, $e = o(h^4)$.   **15.** Runge–Kutta method, $e = o(h^5)$.

16. Determine the maximum height a 0.2-kg ball reaches if it is thrown vertically upward at 50 m/s. Assume the drag force to be given by $0.0012\dot{y}^2$. Does the equation $\ddot{y} + 0.006\dot{y}^2 + 9.81 = 0$ describe the motion of the ball? Solve the problem with one computer program using Euler's method, Taylor's method with $e = o(h^3)$, and Adams' method. Vary the number of steps so that three significant figures are assured. (To estimate the time required, eliminate the $\dot{y}^2$ term and find t_{max}. Using this approximation an appropriate time step can be chosen.) Can you find an exact solution to the given equation?

17. Work Problem 16, but in addition calculate the speed that the ball has when it again reaches $y = 0$. Note that the differential equation must change on the downward flight.

9.10   *Boundary-Value Problems—Ordinary Differential Equations*

The initial-value problem for which all the necessary conditions are given at a particular point or instant, was considered in the previous section. Now we shall consider problems for which the conditions are given at two different positions. For example, in the hanging string problem, information for the second-order describing equation is known at $x = 0$ and $x = L$. It is a boundary-value problem. Boundary-value problems are very common in physical applications; thus, several techniques to solve them will be presented.

9.10.1 ITERATIVE METHOD

Suppose that we are solving the second-order differential equation

$$\ddot{y} = 3x\dot{y} + (x^2 - 1)y = \sin\frac{\pi x}{4}. \tag{9.10.1}$$

This requires that two conditions be given; let them be $y = 0$ at $x = 0$, and $y = 0$ at $x = 6$. Because the conditions are given at two different values of the independent variable, it is a boundary-value problem. Now, if we knew $\dot{y}$ at $x = 0$ it would be an initial-value problem and Taylor's (or any other) method could be used. So, let's assume a value for $\dot{y}_0$ and proceed as though it is an initial-value problem. Then, when $x = 6$ is reached, the boundary condition there requires that $y = 0$. Of course, in general, this condition will not be satisfied and the procedure must be repeated with another guess for $\dot{y}_0$. An interpolation (or extrapolation) scheme could be employed to zero in on the correct $\dot{y}_0$. The procedure works for both linear and nonlinear equations.

An interpolation scheme that can be employed when using a computer is derived by using a Taylor series with $e = o(h^2)$. We consider the value of y at $x = 6$, let's call it y_N, to be the dependent variable and $\dot{y}_0$ to be the independent variable. Then, using $\dot{y}_0^{(1)}$ and $\dot{y}_0^{(2)}$ to be the first two guesses leading to the values $y_N^{(1)}$ and $y_N^{(2)}$, respectively, we have

$$y_N^{(3)} = y_N^{(2)} + \frac{y_N^{(2)} - y_N^{(1)}}{\dot{y}_0^{(2)} - \dot{y}_0^{(1)}}[\dot{y}_0^{(3)} - \dot{y}_0^{(2)}]. \tag{9.10.2}$$

We set $y_N^{(3)}$ equal to zero and calculate a new guess to be

$$\dot{y}_0^{(3)} = \dot{y}_0^{(2)} - \frac{\dot{y}_0^{(2)} - \dot{y}_0^{(1)}}{y_N^{(2)} - y_N^{(1)}}y_N^{(2)}. \tag{9.10.3}$$

Using this value for $\dot{y}_0^{(3)}$, we calculate $y_N^{(3)}$. If it is not sufficiently close to zero, we go through another iteration and find a new value $y_N^{(4)}$. Each additional value should be nearer zero and the iterations are stopped when y_N is sufficiently small.

9.10.2 SUPERPOSITION

For a linear equation we can use the principle of superposition. Consider Eq. 9.10.1 with the same boundary conditions. Completely ignore the given boundary conditions and choose any arbitrary set of initial conditions, for example, $y^{(1)}(0) = 1$ and $\dot{y}^{(1)}(0) = 0$. This leads to a solution $y^{(1)}(x)$. Now, change the initial conditions to $y^{(2)}(0) = 0$ and $\dot{y}^{(2)}(0) = 1$. The solution $y^{(2)}(x)$ would follow. The solutions are now superposed, made possible because of the linear equation, to give the desired solution as

$$y(t) = c_1 y^{(1)}(x) + c_2 y^{(2)}(x). \tag{9.10.4}$$

The actual boundary conditions are then used to determine c_1 and c_2.

If a third-order equation is being solved, then three arbitrary, but different, sets of initial conditions lead to three constants to be determined by the boundary conditions. The method for solving the initial-value problems could be any of those described earlier.

A word of caution is necessary. We must be careful when we choose the two sets of initial conditions. They must be chosen so that the two solutions generated are, in fact, independent.

9.10.3   SIMULTANEOUS EQUATIONS

Let's write Eq. 9.10.1 in finite-difference form for each step in the given interval. The equations are written for each value of i and then all the equations are solved simultaneously. There are a sufficient number of equations to equal the number of unknowns y_i. In finite-difference form, using forward differences with $e = o(h)$, Eq. 9.10.1 is

$$\frac{1}{h^2}(y_{i+2} - 2y_{i+1} + y_i) + \frac{3x_i}{h}(y_{i+1} - y_i) + (x_i^2 - 1)y_i = \sin\frac{\pi}{4}x_i.$$

$$\tag{9.10.5}$$

Now write the equation for each value of i, $i = 0$ to $i = N$. Using $x_0 = 0$, $x_1 = h$, $x_2 = 2h$, and so on, and choosing $h = 1.0$ so that $N = 6$, there results

$$y_2 - 2y_1 + y_0 + 3x_0(y_1 - y_0) + (x_0^2 - 1)y_0 = \sin\frac{\pi}{4}x_0$$

$$y_3 - 2y_2 + y_1 + 3x_1(y_2 - y_1) + (x_1^2 - 1)y_1 = \sin\frac{\pi}{4}x_1$$

$$y_4 - 2y_3 + y_2 + 3x_2(y_3 - y_2) + (x_2^2 - 1)y_2 = \sin\frac{\pi}{4}x_2 \qquad (9.10.6)$$

$$y_5 - 2y_4 + y_3 + 3x_3(y_4 - y_3) + (x_3^2 - 1)y_3 = \sin\frac{\pi}{4}x_3$$

$$y_6 - 2y_5 + y_4 + 3x_4(y_5 - y_4) + (x_4^2 - 1)y_4 = \sin\frac{\pi}{4}x_4$$

Now, with $y_0 = y_6 = 0$ and $x_0 = 0$, there results

$$-2y_1 + y_2 = 0$$

$$-2y_1 + y_2 + y_3 = \sin \frac{\pi}{4}$$

$$-2y_2 + 4y_3 + y_4 = \sin \frac{\pi}{2} \qquad (9.10.7)$$

$$7y_4 + y_5 = \sin \frac{3\pi}{4}$$

$$4y_4 + 10y_5 = 0.$$

There are five equations which can be solved to give the five unknowns y_1, y_2, y_3, y_4, and y_5. If the number of steps is increased to 100, there would be 99 equations to solve simultaneously. A computer would then be used to solve the algebraic equations.

Equations 9.10.7 are often written in matrix form as

$$\begin{bmatrix} -2 & 1 & 0 & 0 & 0 \\ -2 & 1 & 1 & 0 & 0 \\ 0 & -2 & 4 & 1 & 0 \\ 0 & 0 & 0 & 7 & 1 \\ 0 & 0 & 0 & 4 & 10 \end{bmatrix} \begin{bmatrix} y_1 \\ y_2 \\ y_3 \\ y_4 \\ y_5 \end{bmatrix} = \begin{bmatrix} 0 \\ \sin \pi/4 \\ \sin \pi/2 \\ \sin 3\pi/4 \\ 0 \end{bmatrix}, \qquad (9.10.8)$$

or, using matrix notation, as

$$\mathbf{Ay} = \mathbf{B}. \qquad (9.10.9)$$

The solution is written as

$$\mathbf{y} = \mathbf{A}^{-1}\mathbf{b} \qquad (9.10.10)$$

where $\mathbf{A}^{-1}$ is the inverse of $\mathbf{A}$. The solution can be found using a variety of techniques (see, for instance, Chapter 4).

Example 9.10.1:   Solve the boundary-value problem defined by the differential equation $\ddot{y} - 10y = 0$ with boundary conditions $y(0) = 0.4$ and $y(1.2) = 0$. Choose six steps to illustrate the procedure using Taylor's method with $e = o(h^3)$.

SOLUTION: The superposition method will be used to illustrate the numerical solution of the linear equation. We can choose any arbitrary initial conditions, so choose, for the first solution $y_0^{(1)} = 1$ and $\dot{y}_0^{(1)} = 0$. The solution then proceeds as follows for the $y^{(1)}$ solution. Using $h = 0.2$ and $\ddot{y} = 10y$, we have

$$y_0 = 1.0, \qquad \ddot{y}_0 = 10y_0 = 10$$

$$\dot{y}_0 = 0.0, \qquad \dddot{y}_0 = 10\dot{y}_0 = 0$$

$$y_1 = y_0 + h\dot{y}_0 + \frac{h^2}{2}\ddot{y}_0 = 1.2$$

$$\dot{y}_1 = \dot{y}_0 + h\ddot{y}_0 + \frac{h^2}{2}\dddot{y}_0 = 2.0$$

$$y_2 = y_1 + h\dot{y}_1 + \frac{h^2}{2}\ddot{y}_1 = 1.84$$

$$\dot{y}_2 = \dot{y}_1 + h\ddot{y}_1 + \frac{h^2}{2}\dddot{y}_1 = 4.8$$

$$y_3 = 3.17, \qquad y_4 = 5.69, \qquad y_5 = 10.37, \qquad y_6 = 18.96$$

$$\dot{y}_3 = 9.44, \qquad \dot{y}_4 = 17.7, \qquad \dot{y}_5 = 32.6.$$

To find $y^{(2)}$ we choose a different set of initial values, say $y_0^{(2)} = 0.0$ and $\dot{y}_0^{(2)} = -1$. Then proceeding as before we find $y^{(2)}$ to be given by

$$y_0 = 0.0$$

$$\dot{y}_0 = -1.0$$

$$y_1 = y_0 + h\dot{y}_0 + \frac{h^2}{2}\ddot{y}_0 = -0.2$$

$$\dot{y}_1 = \dot{y}_0 + h\ddot{y}_0 + \frac{h^2}{2}\dddot{y}_0 = -1.2$$

$$y_2 = -0.48, \quad y_3 = -0.944, \quad y_4 = -1.952, \quad y_5 = -3.559, \quad y_6 = -8.04$$

$$\dot{y}_2 = -1.84, \quad \dot{y}_3 = -4.096, \quad \dot{y}_4 = -6.083, \quad \dot{y}_5 = -3.559.$$

Combine the two solutions with the usual superposition technique and obtain

$$y = Ay^{(1)} + By^{(2)}.$$

The actual boundary conditions require that

$$0.4 = A(1.0) + B(0.0)$$

$$0 = A(18.96) + B(-8.04).$$

Thus,

$$A = 0.4 \quad \text{and} \quad B = 0.943.$$

The solution is then

$$y = 0.4y^{(1)} + 1.064y^{(2)}.$$

The solution with independent solutions $y^{(1)}$ and $y^{(2)}$ are tabulated below.

x	0	0.2	0.4	0.6	0.8	1.0	1.2
$y^{(1)}$	1.0	1.2	1.84	3.17	5.69	10.37	18.96
$y^{(2)}$	0.0	−0.20	−0.48	−0.944	−1.952	−3.559	−8.04
y	0.4	0.291	0.283	0.378	0.435	0.791	0.0

■

PROBLEMS

Choose a different set of initial conditions for $y^{(2)}$ in Example 9.10.1 and show that the combined solution remains essentially unchanged.

1. Choose $y_0^{(2)} = 1$ and $\dot{y}^{(2)} = 1$.   2. Choose $y_0^{(2)} = 0$ and $\dot{y}^{(2)} = 1$.

3. Select your own initial conditions.

4. Solve the problem of Example 9.10.1 to five significant figures using the superposition method with Taylor's method, $e = o(h^3)$. A computer would be quite helpful.

A bar connects two bodies of temperatures 150°C and 0°C, respectively. Heat is lost by the surface of the bar to the 30°C surrounding air and is conducted from the hot to the colder body. The describing equation is $T'' - 0.01(T - 30) = 0$. Calculate the temperature in the 2-m-long bar. Use the superposition method with five steps. Use the given method.

5. Euler's method.   6. Taylor's method, $e = o(h^3)$.

Assume a tight telephone wire and show that the equation describing $y(x)$ is $d^2y/dx^2 - b = 0$. Express b in terms of the tension P in the wire, the mass per unit length m of the wire, and gravity g. The boundary conditions are $y = 0$ at $x = 0$ and $x = L$. Solve the problem numerically with $b = 10^{-4}$ m^{-1} and $L = 20$ m and solve for the maximum sag. Five steps are sufficient to illustrate the procedures using Euler's method. Use the given method.

7. The iterative method.   8. The superposition method.

9. Assume a loose hanging wire; then $d^2y/dx^2 - b[1 + (dy/dx)^2]^{1/2} = 0$ describes the resulting curve, a catenary. Can the superposition method be used? Using the iterative method, determine the maximum sag if $b = 10^{-3}$ m^{-1}. The boundary conditions are $y = 0$ at $x = 0$ and $y = 40$ m at $x = 100$ m. Using a computer, find an approximation to the minimum number of steps necessary for accuracy of three significant figures.

9.11 Numerical Stability

In numerical calculations the calculation of the dependent variable is dependent on all previous calculations made of this quantity, its derivatives, and the step size. Truncation and round-off errors are contained in each calculated value. If the change in the dependent variable is small for small changes in the independent variable, then the solution is stable. There are times, however, when small step changes in the independent variable lead to large changes in the dependent variable so that a condition of instability exists. It is usually possible to detect such instability, since such results will usually violate physical reasoning. It is possible to predict numerical instabilities for linear equations. It is seldom done, though, since by changing the step size or the numerical method, instabilities can usually be avoided. With the present capacity of high-speed computers, stability problems, if ever encountered, can usually be eliminated by using smaller and smaller step sizes.

When solving partial differential equations with more than one independent variable, stability may be influenced by controlling the relationship between the step sizes chosen. For example, in solving a second-order equation with t and x the independent variables, in which a numerical instability results, attempts would be made to eliminate the instability by changing the relationship between Δx and Δt. If this is not successful, a different numerical technique may eliminate the instability.

There are problems, though, for which direct attempts at a numerical solution lead to numerical instability, even though various step sizes are attempted and various methods utilized. This type of problem can often be solved by either using multiple precision* or by employing a specially devised technique.

9.12   Numerical Solution of Partial Differential Equations

In Chapter 8 an analytical technique was presented for solving partial differential equations, the separation-of-variables technique. When a solution to a partial differential equation is being sought, one should always attempt to separate the variables even though the ordinary differential equations that result may not lead to an analytical solution directly. It is always advisable to solve a set of ordinary differential equations numerically instead of the original partial differential equation. There are occasions, however, when either the equation will not separate or the boundary conditions will not admit a separated solution. For example, in the heat-conduction problem the general solution, assuming the variables separate, for the long rod was

$$T(x, t) = e^{-k\beta^2 t}[A \sin \beta x + B \cos \beta x]. \qquad (9.12.1)$$

We attempt to satisfy the end condition that $T(0, t) = 100t$, instead of $T(0, t) = 0$ as was used in Chapter 8. This requires that

$$T(0, t) = 100t = Be^{-k\beta^2 t}. \qquad (9.12.2)$$

The constant B cannot be chosen to satisfy this condition; hence, the solution 9.12.1 is not acceptable. Instead, we turn to a numerical solution of the original partial differential equation.

---

*The precision of a computer is a measure of the number of digits that can be handled simultaneously by the computer's arithmetic register. Multiple-precision calculations involve manipulating numbers whose size exceeds the precision of the arithmetic register. The price paid for keeping more significant digits is a slowdown in computation time and a loss of storage capacity.

9.12.1 THE DIFFUSION EQUATION

We can solve the diffusion problem numerically for a variety of boundary and initial conditions. The diffusion equation for a rod of length L, assuming no heat losses from the lateral surfaces, is

$$\frac{\partial T}{\partial t} = a\frac{\partial^2 T}{\partial x^2}. \tag{9.12.3}$$

Let the conditions be generalized so that $T(0, t) = f(t)$, $T(L, t) = g(t)$, $T(x, 0) = F(x)$. The function $T(x, t)$ is written in difference notation as T_{ij} and represents the temperature at $x = x_i$ and $t = t_j$. If we hold the time fixed (this is done by keeping j unchanged), the second derivative with respect to x becomes, using a central difference method with $e = o(h^2)$ and referring to Table 9.4,

$$\frac{\partial^2 T}{\partial x^2}(x_i, t_j) = \frac{1}{h^2}(T_{i+1,j} - 2T_{i,j} + T_{i-1,j}), \tag{9.12.4}$$

where h is the step size Δx.

The time step Δt is chosen as k. Then, using forward differences on the time derivative, with $e = o(k)$,

$$\frac{\partial T}{\partial t}(x_i, t_j) = \frac{1}{k}(T_{i,j+1} - T_{i,j}). \tag{9.12.5}$$

The diffusion equation 9.12.3 is then written, in difference notation,

$$\frac{1}{k}(T_{i,j+1} - T_{i,j}) = \frac{a}{h^2}(T_{i+1,j} - 2T_{i,j} + T_{i-1,j}) \tag{9.12.6}$$

or, by rearranging,

$$T_{i,j+1} = \frac{ka}{h^2}(T_{i+1,j} - 2T_{i,j} + T_{i-1,j}) + T_{i,j}. \tag{9.12.7}$$

The given boundary conditions, in difference notation, are

$$T_{0,j} = f(t_j), \qquad T_{N,j} = g(t_j), \qquad T_{i,0} = F(x_i), \tag{9.12.8}$$

where N is the total number of x steps.

A stability analysis, or some experimenting on the computer, will show that a numerical solution is unstable unless the time step and displacement step satisfy the criterion $ak/h^2 \leq \frac{1}{2}$.

The solution proceeds by determining the temperature at $t_0 = 0$ at the various x locations; that is, $T_{0,0}, T_{1,0}, T_{2,0}, T_{3,0}, \ldots, T_{N,0}$ from the given $F(x_i)$. Equation 9.12.7 then allows us to calculate, at time $t_1 = k$, the values $T_{0,1}, T_{1,1}, T_{2,1}, T_{3,1}, \ldots, T_{N,1}$. This process is continued to $t_2 = 2k$ to give $T_{0,2}, T_{1,2}, T_{2,2}, T_{3,2}, \ldots, T_{N,2}$ and repeated for all additional t_j's of interest. By choosing the number N of x steps sufficiently large, a satisfactory approximation to the actual temperature distribution should result.

The boundary conditions and the initial condition are given by the conditions 9.12.8. If an insulated boundary condition is imposed at $x = 0$, then $\partial T/\partial x(0, t) = 0$. Using a forward difference this is expressed by

$$\frac{1}{h}(T_{1,j} - T_{0,j}) = 0 \tag{9.12.9}$$

or

$$T_{1,j} = T_{0,j}. \tag{9.12.10}$$

For an insulated boundary at $x = L$, using a backward difference, we would have

$$T_{N,j} = T_{N-1,j}. \tag{9.12.11}$$

Example 9.12.1: A 1-m-long, laterally insulated rod, originally at 60°C, is subjected at one end to 500°C. Estimate the temperature in the rod as a function of time if the ends are held at 500°C and 60°C, respectively. The diffusivity is 2×10^{-6} m²/s. Use five displacement steps with a time step of 4 ks.

SOLUTION: The diffusion equation describes the heat-transfer phenomenon, hence the difference equation 9.12.7 is used to estimate the temperature at successive times. At time $t = 0$ we have

$$T_{0,0} = 500, \; T_{1,0} = 60, \; T_{2,0} = 60, \; T_{3,0} = 60, \; T_{4,0} = 60, \; T_{5,0} = 60.$$

The left end will be maintained at 500°C and the right end at 60°C. These boundary conditions are expresseed in difference form as $T_{o,j} = 500$ and $T_{5,j} = 60$. Using Eq. 9.12.7 we have, at $t = 4$ ks,

$$T_{i,1} = \frac{ak}{h^2}(T_{i+1,0} - 2T_{i,0} + T_{i-1,0}) + T_{i,0}.$$

Letting i assume the values 1, 2, 3, and 4 successively, we have with $ak/h^2 = \frac{1}{5}$,

$$T_{1,1} = \tfrac{1}{5}(T_{2,0} - 2T_{1,0} + T_{0,0}) + T_{1,0} = 148$$

$$T_{2,1} = \tfrac{1}{5}(T_{3,0} - 2T_{2,0} + T_{1,0}) + T_{2,0} = 60$$

$$T_{3,1} = T_{4,1} = T_{5,1} = 60.$$

At $t = 8$ ks, there results

$$T_{i,2} = \tfrac{1}{5}(T_{i+1,1} - 2T_{i,1} + T_{i-1,1}) + T_{i,1}.$$

For the various values of i, we have

$$T_{1,2} = \tfrac{1}{5}(T_{2,1} - 2T_{1,1} + T_{0,1}) + T_{1,1} = 201$$

$$T_{2,2} = \tfrac{1}{5}(T_{3,1} - 2T_{2,1} + T_{1,1}) + T_{2,1} = 78$$

$$T_{3,2} = T_{4,2} = T_{5,2} = 60.$$

At $t = 12$ ks, the temperature is

$$T_{i,3} = \tfrac{1}{5}(T_{i+1,2} - 2T_{i,2} + T_{i-1,2}) + T_{i,2},$$

yielding

$$T_{1,3} = 236, \quad T_{2,3} = 99, \quad T_{3,3} = 64, \quad T_{4,3} = T_{5,3} = 60.$$

At $t = 16$ ks, there results

$$T_{i,4} = \tfrac{1}{5}(T_{i+1,3} - 2T_{i,3} + T_{i-1,3}) + T_{i,3},$$

giving

$$T_{1,4} = 261, \quad T_{2,4} = 119, \quad T_{3,4} = 70, \quad T_{4,4} = 61, \quad T_{5,4} = 60.$$

At $t = 20$ ks, we find that

$$T_{1,5} = 281, \quad T_{2,5} = 138, \quad T_{3,5} = 78, \quad T_{4,5} = 63, \quad T_{5,5} = 60.$$

Temperatures at future times follow. The temperatures will eventually approach a linear distribution as predicted by the steady-state solution. ■

9.12.2   THE WAVE EQUATION

The same technique can be applied to solve the wave equation as was used in the solution of the diffusion equation. We express the wave equation

$$\frac{\partial^2 u}{\partial t^2} = a^2 \frac{\partial^2 u}{\partial x^2}, \qquad (9.12.12)$$

using central differences for both derivatives, as

$$\frac{1}{k^2}(u_{i,j+1} - 2u_{i,j} + u_{i,j-1}) = \frac{a^2}{h^2}(u_{i+1,j} - 2u_{i,j} + u_{i-1,j}). \qquad (9.12.13)$$

This is rearranged as

$$u_{i,j+1} = \frac{a^2 k^2}{h^2}(u_{i+1,j} - 2u_{i,j} + u_{i-1,j}) + 2u_{i,j} - u_{i,j-1}. \qquad (9.12.14)$$

The boundary and initial conditions are

$$u(0, t) = f(t), \qquad u(L, t) = g(t)$$
$$u(x, 0) = F(x), \qquad \frac{\partial u}{\partial t}(x, 0) = G(x), \qquad (9.12.15)$$

which, if written in difference notation, are

$$u_{0,j} = f(t_j), \qquad u_{N,j} = g(t_j)$$
$$u_{i,0} = F(x_i), \qquad u_{i,1} = u_{i,0} + kG(x_i), \qquad (9.12.16)$$

where N is the total number of x steps. The values $u_{i,0}$ and $u_{i,1}$ result from the initial conditions. The remaining values $u_{i,2}$, $u_{i,3}$, $u_{i,4}$, etc., result from Eq.

9.12.14. Hence we can find the numerical solution for $u(x, t)$. Instability is usually avoided in the numerical solution of the wave equation if $ak/h \leq 1$.

Example 9.12.2: A tight 6-m-long string is set in motion by releasing the string from rest, as shown in the figure. Find an appropriate solution for the deflection using increments of 1 m and time steps of 0.01 s. The wave speed is 100 m/s.

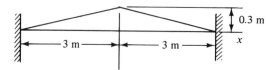

SOLUTION: The initial displacement is given by

$$u(x, 0) = \begin{cases} 0.1x & 0 < x < 3 \\ 0.1(6 - x) & 3 < x < 6 \end{cases}$$

In difference form we have

$$u_{0,0} = 0, \quad u_{1,0} = 0.1, \quad u_{2,0} = 0.2, \quad u_{3,0} = 0.3, \quad u_{4,0} = 0.2,$$
$$u_{5,0} = 0.1, \quad u_{6,0} = 0.$$

The initial velocity is zero since the string is released from rest. Using the appropriate condition listed in Eq. 9.12.16 with $G(x) = 0$, we have at $t = 0.01$,

$$u_{0,1} = 0, \quad u_{1,1} = 0.1, \quad u_{2,1} = 0.2, \quad u_{3,1} = 0.3, \quad u_{4,1} = 0.2,$$
$$u_{5,1} = 0.1, \quad u_{6,1} = 0.$$

Now, we can use Eq. 9.12.14, which marches the solution forward in time and obtain, with $a^2k^2/h^2 = 1$,

$$u_{i,j+1} = u_{i+1,j} + u_{i-1,j} - u_{i,j-1}.$$

This yields the following solution.

At $t = 0.02$:

$$u_{0,2} = 0$$
$$u_{1,2} = u_{2,1} + u_{0,1} - u_{1,0} = 0.2 + 0 - 0.1 = 0.1$$
$$u_{2,2} = u_{3,1} + u_{1,1} - u_{2,0} = 0.3 + 0.1 - 0.2 = 0.2$$
$$u_{3,2} = u_{4,1} + u_{2,1} - u_{3,0} = 0.2 + 0.2 - 0.3 = 0.1$$
$$u_{4,2} = u_{5,1} + u_{3,1} - u_{4,0} = 0.1 + 0.3 - 0.2 = 0.2$$
$$u_{5,2} = u_{6,1} + u_{4,1} - u_{5,0} = 0 + 0.2 - 0.1 = 0.1$$
$$u_{6,2} = 0.$$

At $t = 0.03$:

$$u_{0,3} = 0, \quad u_{1,3} = 0.1, \quad u_{2,3} = 0.0, \quad u_{3,3} = 0.1, \quad u_{4,3} = 0.0,$$
$$u_{5,3} = 0.1, \quad u_{6,3} = 0.$$

At $t = 0.04$:

$$u_{0,4} = 0, \quad u_{1,4} = -0.1, \quad u_{2,4} = 0.0, \quad u_{3,4} = -0.1, \quad u_{4,4} = 0.0,$$
$$u_{5,4} = -0.1, \quad u_{6,4} = 0.$$

At $t = 0.05$:

$$u_{0,5} = 0, \quad u_{1,5} = -0.1, \quad u_{2,5} = -0.2, \quad u_{3,5} = -0.1, \quad u_{4,5} = -0.2,$$
$$u_{5,5} = -0.1, \quad u_{6,5} = 0.$$

At $t = 0.06$:

$$u_{0,6} = 0, \quad u_{1,6} = -0.1, \quad u_{2,6} = -0.2, \quad u_{3,6} = -0.3, \quad u_{4,6} = -0.2,$$
$$u_{5,6} = -0.1, \quad u_{6,6} = 0.$$

Two observations are made from the results above. First, the solution remains symmetric, as it should. Second, the numerical results are significantly in error; it should be noted, however, that at $t = 0.06$ s we have completed one half a cycle. This is exactly as it should be, since the frequency is $a/2L$ cycles/second (see Fig. 8.11), and thus it takes $2L/a = 0.12$ s to complete one cycle. Substantially smaller length increments and time steps are necessary to obtain a solution that approximates the actual solution. An acceptable solution would result for this problem if we used 100 length increments and a time step size chosen so that $ak/h = 1$. ∎

Example 9.12.3: Solve the problem presented in Example 9.12.2 using the computer. Use 100 length increments. This requires that $k = 0.0006$ s if $ak/h = 1$. Print every tenth x step and every eighth time step until one complete cycle is accomplished.

SOLUTION: The Fortran program, less the control statements, and solution are as follows:

```
      PROGRAM DIFFEQ2(INPUT,OUTPUT)
      DIMENSION U(101,208),XLOC(11)
C        I=X, J=TIME
      X = 0.0
      XLOC(1) = 0.0
      DO 5 I=1,11
      XLOC(I+1) = XLOC(I) + 0.6
5     CONTINUE
      DO 10 I=1,50
      U(I,1) = 0.1*X
      X = X+0.06
10    CONTINUE
      X = 3.0
      DO 20 I=51,101
      U(I,1) = 0.1* (6.0-X)
      X = X+0.06
20    CONTINUE
C        AT X = 0, 6, U(X,T) = 0.0
      DO 15 J=1,208
      U(1,J) = 0.0
      U(101,J) = 0.
```

Location (x)

Time	0.0	0.6	1.2	1.8	2.4	3.0	3.6	4.2	4.8	5.4	6.0
0.0000	0.000	0.060	0.120	0.180	0.240	0.300	0.240	0.180	0.120	0.060	0.000
0.0048	0.000	0.060	0.120	0.180	0.240	0.252	0.240	0.180	0.120	0.060	0.000
0.0096	0.000	0.060	0.120	0.180	0.204	0.204	0.204	0.180	0.120	0.060	0.000
0.0144	0.000	0.060	0.120	0.156	0.156	0.156	0.156	0.156	0.120	0.060	0.000
0.0192	0.000	0.060	0.108	0.108	0.108	0.108	0.108	0.108	0.108	0.060	0.000
0.0240	0.000	0.060	0.060	0.060	0.060	0.060	0.060	0.060	0.060	0.060	0.000
0.0288	0.000	0.012	0.012	0.012	0.012	0.012	0.012	0.012	0.012	0.012	0.000
0.0336	0.000	-0.036	-0.036	-0.036	-0.036	-0.036	-0.036	-0.036	-0.036	-0.036	0.000
0.0384	0.000	-0.060	-0.084	-0.084	-0.084	-0.084	-0.084	-0.084	-0.084	-0.060	0.000
0.0432	0.000	-0.060	-0.120	-0.132	-0.132	-0.132	-0.132	-0.132	-0.120	-0.060	0.000
0.0480	0.000	-0.060	-0.120	-0.180	-0.180	-0.180	-0.180	-0.180	-0.120	-0.060	0.000
0.0528	0.000	-0.060	-0.120	-0.180	-0.228	-0.228	-0.228	-0.180	-0.120	-0.060	0.000
0.0576	0.000	-0.060	-0.120	-0.180	-0.240	-0.276	-0.240	-0.180	-0.120	-0.060	0.000
0.0600	0.000	-0.060	-0.120	-0.180	-0.240	-0.300	-0.240	-0.180	-0.120	-0.060	0.000
0.0624	0.000	-0.060	-0.120	-0.180	-0.240	-0.267	-0.240	-0.180	-0.120	-0.060	0.000
0.0672	0.000	-0.060	-0.120	-0.180	-0.228	-0.228	-0.228	-0.180	-0.120	-0.060	0.000
0.0720	0.000	-0.060	-0.120	-0.180	-0.180	-0.180	-0.180	-0.180	-0.120	-0.060	0.000
0.0768	0.000	-0.060	-0.120	-0.132	-0.132	-0.132	-0.132	-0.132	-0.120	-0.060	0.000
0.0816	0.000	-0.060	-0.084	-0.084	-0.084	-0.084	-0.084	-0.084	-0.084	-0.060	0.000
0.0864	0.000	-0.036	-0.036	-0.036	-0.036	-0.036	-0.036	-0.036	-0.036	-0.036	0.000
0.0912	0.000	0.012	0.012	0.012	0.012	0.012	0.012	0.012	0.012	0.012	0.000
0.0960	0.000	0.060	0.060	0.060	0.060	0.060	0.060	0.060	0.060	0.060	0.000
0.1008	0.000	0.060	0.108	0.108	0.108	0.108	0.108	0.108	0.108	0.060	0.000
0.1056	0.000	0.060	0.120	0.156	0.156	0.156	0.156	0.156	0.120	0.060	0.000
0.1104	0.000	0.060	0.120	0.180	0.204	0.204	0.204	0.180	0.120	0.060	0.000
0.1152	0.000	0.060	0.120	0.180	0.240	0.252	0.240	0.180	0.120	0.060	0.000
0.1200	0.000	0.060	0.120	0.180	0.240	0.300	0.240	0.180	0.120	0.060	0.000

```
15      CONTINUE
C       AT TIME =.0006SEC U(I,1) = U(I,2)
        DO 30 1=1,101
30      U(I,2) = U(I,1)
C       WAVE SPEED IS 100 M/S, THE TIME STEP IS 0.0006 SEC,
                AND THE DISPLACEMENT STEP IS 0.06 M.
        T = 0.0
        DO 40 J=2,206
        DO 40 I=2,100
40      U(I,J+1) = U(I+1,J) + U(I-1,J) - U(I,J-1)
        PRINT 70,XLOC
70      FORMAT (1H1,30X,*X - LOCATION*,///,4X,*TIME*,11F8.1 ,/,1X)
        DO 50 K=1,204,4
        PRINT 60,T,(U(L,K),L=1,100,10) ,U(101,K)
60      FORMAT (F8.4,11F8.3)
        T = T+ 0.0024
50      CONTINUE
        END                                                    ■
```

9.12.3 LAPLACE'S EQUATION

It is again convenient if Laplace's equation

$$\frac{\partial^2 T}{\partial x^2} + \frac{\partial^2 T}{\partial y^2} = 0 \tag{9.12.17}$$

is written in difference notation using central differences. It is then

$$\frac{1}{h^2}(T_{i+1,j} - 2T_{i,j} + T_{i-1,j}) + \frac{1}{k^2}(T_{i,j+1} - 2T_{i,j} + T_{i,j-1}) = 0. \tag{9.12.18}$$

Solving for $T_{i,j}$, and letting $h = k$ (which is not necessary but convenient), we have

$$T_{i,j} = \frac{1}{4}[T_{i,j+1} + T_{i,j-1} + T_{i+1,j} + T_{i-1,j}]. \tag{9.12.19}$$

Using central differences, we see that the temperature at a particular mesh point is the average of the four neighboring temperatures. For example (see Fig. 9.9),

$$T_{8,5} = \frac{1}{4}[T_{8,6} + T_{8,4} + T_{9,5} + T_{7,5}]. \tag{9.12.20}$$

The Laplace equation requires that the dependent variable (or its derivative) be specified at all points surrounding a given region. A typical set of boundary conditions, for a rectangular region is

$$T(0, y) = f(y), \qquad T(W, y) = g(y),$$
$$T(x, 0) = F(x), \qquad T(x, H) = G(x), \tag{9.12.21}$$

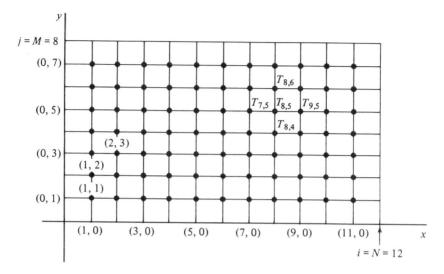

FIGURE 9.9.   Typical mesh for the solution of Laplace's equation.

where W and H are the dimensions of the rectangle. In difference notation these conditions are

$$T_{0,j} = f(y_i), \qquad T_{N,j} = g(y_i),$$
$$T_{i,0} = F(x_i), \qquad T_{i,M} = G(x_i). \tag{9.12.22}$$

The temperature is known at all the boundary mesh points, and with a 12 by 8 mesh, shown in Fig. 9.9, Eq. 9.12.19 gives 77 algebraic equations. These equations, which include the 77 unknowns $T_{i,j}$ at each of the interior points, can then be solved simultaneously. Of course, a computer would be used to solve the set of simultaneous equations; computer programs are generally available to accomplish this.

It is also possible, and often necessary, to specify that no heat transfer occurs across a boundary, so that the temperature gradient is zero. This would, of course, change the conditions 9.12.22.

It should be pointed out that there is a simple technique, especially useful before the advent of the computer, that gives a quick approximation to the solution of Laplace's equation. It is a *relaxation* method. In this method the temperatures at every interior mesh point are guessed. Then Eq. 9.12.19 is used in a systematic manner by starting, say, at the (1, 1) element, averaging for a new value and working across the first horizontal row, then going to the (1, 2) element and working across the second horizontal row, always using the most recently available values. This is continued until the values at every interior point of the complete mesh of elements are changed from the guessed values. A second iteration is then accomplished by recalculating every temperature again starting at the (1, 1) element. This iteration process is continued until the value at each point converges or, at least, does not significantly change with successive iterations. This can be done by hand for a fairly large number of mesh points and hence can provide a quick approximation to the solution of Laplace's equation.

Example 9.12.4:   A 50- by 60-mm flat plate, insulated on both flat surfaces, has its edges maintained at 0, 100, 200, and 300°C, in that order, going counterclockwise. Using the relaxation method, determine the steady-state temperature at each grid point, using a 10 × 10 mm grid.

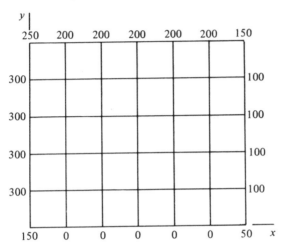

SOLUTION: The grid is set up as shown. Note that the corner temperatures are assumed to be the average of the neighboring two temperatures. The actual solution does not involve the corner temperatures. We start assuming a temperature at each grid point; the more accurate our assumption, the fewer iterations required for convergence. Let us assume the following:

$$
\begin{array}{ccccccc}
250 & 200 & 200 & 200 & 200 & 200 & 150 \\
300 & 290 & 270 & 240 & 200 & 150 & 100 \\
300 & 280 & 250 & 220 & 180 & 130 & 100 \\
300 & 200 & 150 & 100 & 100 & 100 & 100 \\
300 & 50 & 50 & 50 & 50 & 50 & 100 \\
150 & 0 & 0 & 0 & 0 & 0 & 50
\end{array}
$$

The first iteration comes by applying Eq. 9.12.19 to each of the interior grid points above. Starting at the lower left interior grid point (50°C) and continuing to the second row up (200°C), using the corrected values in the process, the following improved temperature distribution results:

$$
\begin{array}{ccccccc}
250 & 200 & 200 & 200 & 200 & 200 & 150 \\
300 & 259 & 232 & 207 & 180 & 152 & 100 \\
300 & 267 & 230 & 196 & 161 & 130 & 100 \\
300 & 217 & 163 & 135 & 117 & 110 & 100 \\
300 & 138 & 84 & 58 & 52 & 63 & 100 \\
100 & 0 & 0 & 0 & 0 & 0 & 50
\end{array}
$$

We continue the iterations until there is no significant change in additional iterations.

Three more iterations are listed below:

250	200	200	200	200	200	150
300	246	216	194	168	147	100
300	252	212	180	127	121	100
300	220	170	138	118	104	100
300	150	93	70	62	68	100
100	0	0	0	0	0	50

250	200	200	200	200	200	150
300	240	209	186	168	148	100
300	244	202	165	140	122	100
300	219	167	136	108	99	100
300	153	98	77	66	68	100
100	0	0	0	0	0	50

250	200	200	200	200	200	150
300	237	204	183	168	148	100
300	240	194	162	140	122	100
300	216	164	128	108	99	100
300	154	100	76	63	66	100
100	0	0	0	0	0	50

Note that in the last three iterations, the maximum change in temperature from one iteration to the next is 34, 15, and 8°C, respectively. Two more iterations will result in a steady-state temperature distribution, accurate to within about 1°C. ∎

PROBLEMS

1. Solve for the steady-state temperature $T_{1,\infty}$ of Example 9.12.1.

2. Predict the time necessary for the temperature of the rod in Example 9.12.1 at $x = 0.4$ to reach 145°C.

A laterally insulated fin 2.5 m long is initially at 60°C. The temperature of one end is suddenly increased to and maintained at 600°C while the temperature of the other end is maintained at 60°C. The diffusivity is $a = 10^{-5}\,\text{m}^2/\text{s}$. Use five x steps, time increments of 5 ks, and calculate the time necessary for the center of the rod to reach a temperature of

3. 90°C   **4.** 110°C   **5.** 120°C   **6.** 140°C   **7.** 160°C

If the end at $x = 2$ m of a laterally insulated steel rod is insulated and the end at $x = 0$ is suddenly subjected to a temperature of 200°C, predict the temperature at $t = 80$ s, using five displacement steps and time increments of 20 ks. Use $a = 4 \times 10^{-6}\,\text{m}^2/\text{s}$. The initial temperature is

8. 0°C   **9.** 50°C   **10.** 100°C

A 6-m-long wire, fixed at both ends, is given an initial displacement of $u_{1,0} = 0.1$, $u_{2,0} = 0.2$, $u_{3,0} = 0.3$, $u_{4,0} = 0.4$, $u_{5,0} = 0.2$ at the displacement steps. Predict the displacement at five future time steps, using $k = 0.025$ s if the wave speed is 40 m/s and the initial velocity is given by

11. 0   **12.** 4 m/s   **13.** 8 m/s

Using five x increments in a 1-m-long tight string, find an approximate solution for the displacement if the initial displacement is

$$u(x, 0) = \begin{cases} x/10, & 0 < x < 0.4 \\ 0.04, & 0.4 \leq x \leq 0.6 \\ (1 - x)/10, & 0.6 < x < 1.0. \end{cases}$$

Assume a wave speed of 50 m/s and use a time step of 0.004 s. Present the solution at five additional time steps if the initial velocity is

14. 0   **15.** 2.5 m/s   **16.** 10 m/s

A 4-m by 5-m plate is divided into 1-m squares. One long side is maintained at 100°C and the other at 200°C. The flat surfaces are insulated. Predict the steady-state temperature distribution in the plate using the relaxation method if the short sides are maintained at

17. 300°C and 0°C   **18.** 400°C and 100°C

19. Solve Problem 15 with the 100°C changed to a linear distribution varying from 0 to 300°C. The 0°C corner is adjacent to the 0°C side.

10

Complex Variables

10.1 Introduction

In the course of developing tools for the solution of the variety of problems encountered in the physical sciences, we have had many occasions to use results from the theory of functions of a complex variable. The solution of a differential equation describing the motion of a spring–mass system is an example that comes immediately to mind. Functions of a complex variable also play an important role in fluid mechanics, heat transfer, and field theory, to mention just a few areas.

In this chapter we present a survey of this important subject. We progress from complex numbers to complex variables to analytic functions to line integrals and finally to the famous residue theorem of Cauchy. This survey is not meant to be a treatise; the reader should consult any of the standard texts on the subject for a complete development.

10.2 Complex Numbers

There are many algebraic equations such as

$$z^2 - 12z + 52 = 0 \tag{10.2.1}$$

which have no solutions among the set of real numbers. We have two alternatives; either admit that there are equations with no solutions, or enlarge the set of numbers so that every algebraic equation has a solution. A great deal of experience has led us to accept the second choice. We write $i = \sqrt{-1}$ so that $i^2 = -1$ and attempt to find solutions of the form

$$z = x + iy, \tag{10.2.2}$$

where x and y are real. Equation 10.2.1 has solutions in this form:

$$z_1 = 6 + 4i$$
$$z_2 = 6 - 4i. \tag{10.2.3}$$

For each pair of real numbers x and y, z is a *complex number.* One of the outstanding mathematical achievements is the theorem of Gauss, which asserts that every equation

$$a_n z^n + a_{n-1} z^{n-1} + \cdots + a_0 = 0 \tag{10.2.4}$$

has a solution in this enlarged set.* The complex number $z = x + iy$ has x as its *real part* and y as its *imaginary part* (y is real). We write

$$\text{Re } z = x, \qquad \text{Im } z = y. \tag{10.2.5}$$

The notation $z = x + iy$ suggests a geometric interpretation for z. The point (x, y) is the plot of $z = x + iy$. Therefore, to every point in the xy-plane there is associated a complex number and to every complex number a point. The x axis is called the *real axis* and the y axis is the *imaginary axis,* as displayed in Fig. 10.1.

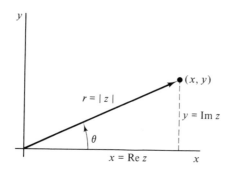

FIGURE 10.1.   The complex plane.

In terms of polar coordinates (r, θ) the variables x and y are

$$x = r \cos \theta, \qquad y = r \sin \theta. \tag{10.2.6}$$

The complex variable z is then written as

$$z = r(\cos \theta + i \sin \theta). \tag{10.2.7}$$

The quantity r is the *absolute value* of z and is denoted by $|z|$; hence,

$$r = |z| = \sqrt{x^2 + y^2}. \tag{10.2.8}$$

The angle θ, measured in radians and positive in the counterclockwise sense,

---

*From this point, it is elementary to prove that this equation actually has n solutions, counting possible duplicates.

is the *argument* of z, written arg z and given by

$$\arg z = \theta = \tan^{-1} \frac{y}{x}. \qquad (10.2.9)$$

Obviously, there are an infinite number of θ's satisfying Eq. 10.2.9 at intervals of 2π radians. We shall make the usual choice of limiting θ to the interval $0 \le \theta < 2\pi$ for its *principal value.**

A complex number is *pure imaginary* if the real part is zero. It is *real* if the imaginary part is zero. The *conjugate* of the complex number z is denoted by $\bar{z}$; it is found by changing the sign on the imaginary part of z, that is,

$$\bar{z} = x - iy. \qquad (10.2.10)$$

The conjugate is useful in manipulations involving complex numbers. An interesting observation and often useful result is that the product of a complex number and its conjugate is real. This follows from

$$z\bar{z} = (x + iy)(x - iy) = x^2 - i^2 y^2 = x^2 + y^2 \qquad (10.2.11)$$

where we have used $i^2 = -1$. Note then that

$$z\bar{z} = |z|^2. \qquad (10.2.12)$$

The addition, subtraction, multiplication, or division of two complex numbers $z_1 = x_1 + iy_1$ and $z_2 = x_2 + iy_2$ is accomplished as follows:

$$z_1 + z_2 = (x_1 + iy_1) + (x_2 + iy_2)$$
$$= (x_1 + x_2) + i(y_1 + y_2) \qquad (10.2.13)$$

$$z_1 - z_2 = (x_1 - iy_1) - (x_2 + iy_2)$$
$$= (x_1 - x_2) + i(y_1 - y_2) \qquad (10.2.14)$$

$$z_1 z_2 = (x_1 + iy_1)(x_2 + iy_2)$$
$$= (x_1 x_2 - y_1 y_2) + i(x_1 y_2 + x_2 y_1) \qquad (10.2.15)$$

$$\frac{z_1}{z_2} = \frac{x_1 + iy_1}{x_2 + iy_2} = \frac{x_1 + iy_1}{x_2 + iy_2} \frac{x_2 - iy_2}{x_2 - iy_2}$$

$$= \frac{x_1 x_2 + y_1 y_2}{x_2^2 + y_2^2} + i \frac{x_2 y_1 - x_1 y_2}{x_2^2 + y_2^2}. \qquad (10.2.16)$$

Note that the conjugate of z_2 was used to form a real number in the denominator of z_1/z_2. This last computation can also be written

$$\frac{z_1}{z_2} = \frac{z_1 \bar{z}_2}{z_2 \bar{z}_2} = \frac{z_1 \bar{z}_2}{|z_2|^2}$$

---

*Another commonly used interval is $-\pi < \theta \le \pi$.

$$= \frac{x_1x_2 + y_1y_2}{|z_2|^2} + i\frac{x_2y_1 - x_1y_2}{|z_2|^2}. \qquad (10.2.17)$$

Figure 10.1 shows clearly that

$$|x| = |\text{Re } z| \leq |z|$$
$$|y| = |\text{Im } z| \leq |z| \qquad (10.2.18)$$

Addition and subtraction is illustrated graphically in Fig. 10.2. From the parallelogram formed by the addition of the two complex numbers, we observe that

$$|z_1 + z_2| \leq |z_1| + |z_2|. \qquad (10.2.19)$$

This inequality will be quite useful in later considerations.

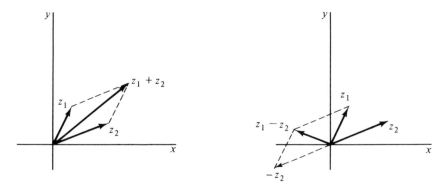

FIGURE 10.2. Addition and subtraction of two complex numbers.

When Eqs. 10.2.6 are used to write z as in Eq. 10.2.7, we say that z is in *polar* form. This form is particularly useful in computing z_1z_2, z_1/z_2, z^n, and $z^{1/n}$. Suppose that

$$z_1 = r_1(\cos \theta_1 + i \sin \theta_1), \qquad z_2 = r_2(\cos \theta_2 + i \sin \theta_2),$$
$$(10.2.20)$$

so that

$$z_1z_2 = r_1r_2[(\cos \theta_1 \cos \theta_2 - \sin \theta_1 \sin \theta_2)$$
$$+ i(\sin \theta_1 \cos \theta_2 + \sin \theta_2 \cos \theta_1)] \qquad (10.2.21)$$
$$= r_1r_2[\cos (\theta_1 + \theta_2) + i \sin (\theta_1 + \theta_2)],$$

It then follows that

$$|z_1z_2| = r_1r_2 = |z_1||z_2|, \qquad (10.2.22)$$

and*

$$\arg (z_1 z_2) = \arg z_1 + \arg z_2. \tag{10.2.23}$$

In other words, the absolute value of the product is the product of the absolute values of the factors while the argument of the product is the sum of the arguments of the factors. Note also that

$$\frac{z_1}{z_2} = \frac{r_1}{r_2}[\cos (\theta_1 - \theta_2) + i \sin (\theta_1 - \theta_2)]. \tag{10.2.24}$$

Hence,

$$\left|\frac{z_1}{z_2}\right| = \frac{|z_1|}{|z_2|}, \tag{10.2.25}$$

and*

$$\arg \left(\frac{z_1}{z_2}\right) = \arg z_1 - \arg z_2. \tag{10.2.26}$$

From repeated applications of Eq. 10.2.21 we derive the rule

$$z_1 z_2 \cdots z_n = r_1 r_2 \cdots r_n[\cos (\theta_1 + \theta_2 + \cdots + \theta_n) \\ + i \sin (\theta_1 + \theta_2 + \cdots + \theta_n)]. \tag{10.2.27}$$

An important special case occurs when $z_1 = z_2 = \cdots = z_n$. Then

$$z^n = r^n(\cos n\theta + i \sin n\theta), \qquad n = 0, 1, 2, \cdots. \tag{10.2.28}$$

The symbol $z^{1/n}$ expresses the statement that $(z^{1/n})^n = z$; that is, $z^{1/n}$ is an nth root of z. Equation 10.2.28 enables us to find the n nth roots of any complex number z. Let

$$z = r(\cos \theta + i \sin \theta). \tag{10.2.29}$$

For each nonnegative integer k, it is also true that

$$z = r[\cos (\theta + 2\pi k) + i \sin (\theta + 2\pi k)]. \tag{10.2.30}$$

So

$$z^{1/n} = z_k = r^{1/n}\left[\cos \left(\frac{\theta + 2\pi k}{n}\right) + i \sin \left(\frac{\theta + 2\pi k}{n}\right)\right] \tag{10.2.31}$$

has the property that $z_k^n = z$ according to Eq. 10.2.28. It is obvious that $z_0, z_1, \ldots, z_{n-1}$ are distinct complex numbers, unless $z = 0$. Hence, for $k = 0, 1, \ldots, n - 1$, Eq. 10.2.31 provides n distinct nth roots of z.

---

*Since arg z is multivalued, we read this equation as stating that there exist arguments of z_1 and z_2 for which the equality holds (see Problems 31 and 32).

To find n roots of $z = 1$, we define the special symbol

$$\omega_n = \cos \frac{2\pi}{n} + i \sin \frac{2\pi}{n}. \tag{10.2.32}$$

Then $\omega_n, \omega_n^2, \ldots, \omega_n^n = 1$ are the n distinct roots of 1. Note that

$$\omega_n^k = \cos \frac{2\pi k}{n} + i \sin \frac{2\pi k}{n}. \tag{10.2.33}$$

As a problem in the problem set, the reader is asked to prove that if $1 \le k < j \le n$, then $\omega_n^k \ne \omega_n^j$.

Now suppose that $z_0^n = z$, so that z_0 is an nth root of z. Then the set

$$z_0, \omega_n z_0, \ldots, \omega_n^{n-1} z_0 \tag{10.2.34}$$

is the set of nth roots of z since

$$(\omega_n^k z_0)^n = \omega_n^{kn} z_0^n = 1 \cdot z. \tag{10.2.35}$$

The examples illustrate this point.

Finally, we note the rather obvious fact that if two complex numbers are equal, the real parts and the imaginary parts are equal, respectively. Hence, an equation written in terms of complex variables includes two real equations, one found by equating the real parts from each side of the equation and the other found by equating the imaginary parts. Thus, for instance, the equation

$$a + ib = 0 \tag{10.2.36}$$

implies that $a = b = 0$.

Example 10.2.1:   Find the three cube roots of unity.

SOLUTION: These roots are $\omega_3, \omega_3^2, 1$, where

$$\omega_3 = \cos \frac{2\pi}{3} + i \sin \frac{2\pi}{3} = -\frac{1}{2} + i\frac{\sqrt{3}}{2},$$

so

$$\omega_3^2 = \left(-\frac{1}{2} + i\frac{\sqrt{3}}{2}\right)^2 = -\frac{1}{2} - i\frac{\sqrt{3}}{2}.$$

This is, of course, equal to

$$\omega_3^2 = \cos \frac{4\pi}{3} + i \sin \frac{4\pi}{3}.$$

We also have

$$\omega_3^3 = \cos \frac{2\pi \cdot 3}{3} + i \sin \frac{2\pi \cdot 3}{3} = \cos 2\pi + i \sin 2\pi = 1. \quad \blacksquare$$

Example 10.2.2:   Find the three roots of $z = -1$ using Eq. 10.2.34.

SOLUTION: Since $(-1)^3 = -1$, -1 is a cube root of -1. Hence, $-\omega_3$, $-\omega_3^2$, and -1 are the three distinct cube roots of -1; in the notation of Eq. 10.2.31, the roots are

$$z_0 = -1, \qquad z_1 = \frac{1}{2} - i\frac{\sqrt{3}}{2}, \qquad z_2 = \frac{1}{2} + i\frac{\sqrt{3}}{2}.$$

Note that

$$(-\omega_3)^3 = (-1)^3\omega_3^3 = -1. \quad \blacksquare$$

Example 10.2.3:   Find the three cube roots of $z = i$ using Eq. 10.2.34.

SOLUTION: We write i in polar form as

$$i = \cos\frac{\pi}{2} + i\sin\frac{\pi}{2}.$$

Then

$$z_0 = i^{1/3} = \cos\frac{\pi}{6} + i\sin\frac{\pi}{6} = \frac{\sqrt{3}}{2} + i\frac{1}{2}.$$

Now the remaining two roots are $z_0\omega_3$, $z_0\omega_3^2$, or

$$z_1 = \left(\frac{\sqrt{3}}{2} + i\frac{1}{2}\right)\left(-\frac{1}{2} + i\frac{\sqrt{3}}{2}\right) = -\frac{\sqrt{3}}{2} + i\frac{1}{2}$$

$$z_2 = \left(\frac{\sqrt{3}}{2} + i\frac{1}{2}\right)\left(-\frac{1}{2} - i\frac{\sqrt{3}}{2}\right) = -i. \quad \blacksquare$$

Example 10.2.4:   Determine (a) $(3 + 4i)^2$, and (b) $(3 + 4i)^{1/3}$ using Eq. 10.2.31.

SOLUTION: The number is expressed in polar form, using $r = \sqrt{3^2 + 4^2} = 5$ and $\theta = \tan^{-1}\frac{4}{3} = 53.13°$ as

$$3 + 4i = 5(\cos 53.13° + i\sin 53.13°).$$

(a)   To determine $(3 + 4i)^2$ we use Eq. 10.2.28 and find

$$(3 + 4i)^2 = 5^2(\cos 2 \times 53.13° + i\sin 2 \times 53.13°)$$

$$= 25(-0.280 + 0.960i)$$

$$= -7 + 24i.$$

We could also simply form the product

$$(3 + 4i)^2 = (3 + 4i)(3 + 4i)$$

$$= 9 - 16 + 12i + 12i = -7 + 24i.$$

(b)   There are three distinct cube roots that must be determined when evaluating $(3 + 4i)^{1/3}$. They are found by expressing $(3 + 4i)^{1/3}$ as

$$(3 + 4i)^{1/3} = 5^{1/3}\left(\cos \frac{53.13 + 360k}{3} + i \sin \frac{53.13 + 360k}{3}\right),$$

where the angles are expressed in degrees, rather than radians.
The first root is then, using $k = 0$,

$$(3 + 4i)^{1/3} = 5^{1/3}(\cos 17.71° + i \sin 17.71°)$$

$$= 1.710(0.9526 + 0.3042i)$$

$$= 1.629 + 0.5202i.$$

The second root is, using $k = 1$,

$$(3 + 4i)^{1/3} = 5^{1/3}(\cos 137.7° + i \sin 137.7°)$$

$$= 1.710(-0.7397 + 0.6729i)$$

$$= -1.265 + 1.151i.$$

The third and final root is, using $k = 2$,

$$(3 + 4i)^{1/3} = 5^{1/3}(\cos 257.7° + i \sin 257.7°)$$

$$= 1.710(-0.2129 - 0.9771i)$$

$$= -0.3641 - 1.671i.$$

It is easy to verify that if we choose $k \geq 3$, we would simply return to one of the three roots already computed. The three distinct roots are illustrated in the following diagram.

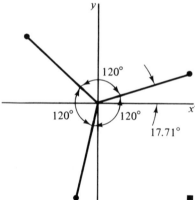

Example 10.2.5:   Express the complex number $3 + 6i$ in polar form, and also divide it by $2 - 3i$.

SOLUTION: To express $3 + 6i$ in polar form we must determine r and θ. We have

$$r = \sqrt{x^2 + y^2} = \sqrt{3^2 + 6^2} = 6.708.$$

The angle θ is found, in degrees, to be

$$\theta = \tan^{-1} \tfrac{6}{3} = 63.43°.$$

In polar form we have

$$3 + 6i = 6.708(\cos 63.43° + i \sin 63.43°).$$

The desired division is

$$\frac{3 + 6i}{2 - 3i} = \frac{3 + 6i}{2 - 3i}\frac{2 + 3i}{2 + 3i}$$

$$= \frac{6 - 18 + i(12 + 9)}{4 + 9} = \frac{1}{13}(-12 + 21i) = -0.9231 + 1.615i. \quad \blacksquare$$

Example 10.2.6:   What set of points in the complex plane (i.e., the xy plane) satisfies

$$\left|\frac{z}{z - 1}\right| = 2$$

SOLUTION: First, using Eq. 10.2.25, we can write

$$\left|\frac{z}{z - 1}\right| = \frac{|z|}{|z - 1|}.$$

Then, recognizing that the magnitude squared of a complex number is the real part squared plus the imaginary part squared, we have

$$\frac{|z|^2}{|z - 1|^2} = \frac{x^2 + y^2}{(x - 1)^2 + y^2},$$

where we have used $z - 1 = x - 1 + iy$. The desired equation is then

$$\frac{x^2 + y^2}{(x - 1)^2 + y^2} = 4$$

or

$$x^2 + y^2 = 4(x - 1)^2 + 4y^2.$$

This can be written as

$$(x - \tfrac{4}{3})^2 + y^2 = \tfrac{4}{9},$$

which is the equation of a circle of radius $\tfrac{2}{3}$ with center at $(\tfrac{4}{3}, 0)$. $\blacksquare$

PROBLEMS

Determine the angle θ, in degrees and radians, which is necessary to write each complex number in polar form.

1. $4 + 3i$ **2.** $-4 + 3i$ **3.** $4 - 3i$ **4.** $-4 - 3i$

For the complex number $z = 3 - 4i$, find each following term.

5. z^2 **6.** $z\bar{z}$ **7.** $z/\bar{z}$

8. $\left|\dfrac{z+1}{z-1}\right|$ **9.** $(z+1)(z-i)$ **10.** $|z^2|$

11. $(z-i)^2/(z-1)^2$ **12.** z^4 **13.** $z^{1/2}$

14. $z^{1/3}$ **15.** $z^{2/3}$ **16.** $\dfrac{z^2}{z^{1/2}}$

Determine the roots of each term (express in the form $a+ib$).

17. $1^{1/5}$ **18.** $-16^{1/4}$ **19.** $i^{1/3}$ **20.** $9^{1/2}$

Show that each equation represents a circle.

21. $|z|=4$ **22.** $|z-2|=2$ **23.** $|(z-1)/(z+1)|=3$

Find the equation of each curve represented by the following.

24. $|(z-1)/(z+1)|=4$ **25.** $|(z+i)/(z-i)|=2$

26. Identify the region represented by $|z-2|\le x$.

27. Show that for each n and each complex number $z\ne 1$,

$$1+z+z^2+\cdots+z^{n-1}=\frac{1-z^n}{1-z}.$$

28. Use the result in Problem 27 to find

$$1+\omega_n+\omega_n^2+\cdots+\omega_n^{n-1},$$

where ω_n is a nonreal nth root of unity.

29. Find the four solutions of $z^4+16=0$.

30. Show geometrically why $|z_1-z_2|\ge|z_1|-|z_2|$.

31. Find arguments of $z_1=-1+i$ and $z_2=1-i$ so that $\arg z_1z_2 = \arg z_1 + \arg z_2$. Explain why this equation is false if $0\le \arg z<2\pi$ is a requirement on z_1, z_2, and z_1z_2.

32. Find z_1 and z_2 so that $0\le \arg z_1$, $\arg z_2 <2\pi$, and $0\le \arg(z_1/z_2)<2\pi$ makes $\arg(z_1/z_2)=\arg z_1 - \arg z_2$ false.

10.3 Elementary Functions

Most functions of a real variable which are of interest to the natural scientist can be profitably extended to a function of a complex variable by replacing the real variable x by $z=x+iy$. This guarantees that when $y=0$, the generalized variable reduces to the original real variable. As we shall see, this simple device generates remarkable insight into our understanding of the classical functions of mathematical physics. One especially attractive example is the interconnection between the inverse tangent and the logarithm, which is presented in Eq. 10.3.25 below.

A *polynomial* is an expression

$$P_n(z) = a_n z^n + a_{n-1} z^{n-1} + \cdots + a_1 z + a_0 \qquad (10.3.1)$$

where the coefficients $a_n, a_{n-1}, \ldots, a_1, a_0$ are complex and n is a non-negative integer. These are the simplest functions. Their behavior is well understood and easy to analyze. The next class of functions comprises the *rational* functions, the quotients of polynomials:

$$Q(z) = \frac{a_n z^n + a_{n-1} z^{n-1} + \cdots + a_1 z + a_0}{b_m z^m + b_{m-1} z^{m-1} + \cdots + b_1 z + b_0}. \qquad (10.3.2)$$

The polynomial comprising the denominator of $Q(z)$ is understood to be of degree ≥ 1 so that $Q(z)$ does not formally reduce to a polynomial.

From these two classes we move to *power series*, defined as

$$f(z) = \sum_{n=0}^{\infty} a_n z^n, \qquad (10.3.3)$$

where the series is assumed convergent for all z, $|z| < R$, $R > 0$. Various tests are known that determine R. When

$$\lim_{n \to \infty} \left| \frac{a_{n+1}}{a_n} \right| = \frac{1}{R} \qquad (10.3.4)$$

exists,* the series in Eq. 10.3.3 converges for all z in $|z| < R$, and diverges for all z, $|z| > R$. No general statement, without further assumptions on either $f(z)$ or the sequence $a_0, a_1, a_2, \ldots$, can be made for those z on the *circle of convergence, $|z| = R$.*

We define e^z, $\sin z$, and $\cos z$ by the following series, each of which converges for all z:

$$e^z = 1 + z + \frac{z^2}{2!} + \frac{z^3}{3!} + \cdots \qquad (10.3.5)$$

$$\sin z = z - \frac{z^3}{3!} + \frac{z^5}{5!} - \cdots \qquad (10.3.6)$$

$$\cos z = 1 - \frac{z^2}{2!} + \frac{z^4}{4!} - \cdots . \qquad (10.3.7)$$

These definitions are chosen so that they reduce to the standard Taylor series for e^x, $\sin x$, and $\cos x$ when $y = 0$. The following is an elementary consequence of these formulas:

$$\sin z = \frac{e^{iz} - e^{-iz}}{2i}, \qquad \cos z = \frac{e^{iz} + e^{-iz}}{2}. \qquad (10.3.8)$$

---

*This quotient in Eq. 10.3.4 is either 0 or ∞ for the series in (10.3.6) and (10.3.7). Nonetheless, these series do converge for all z.

Also, we note that, letting $z = i\theta$,

$$e^{i\theta} = 1 + i\theta - \frac{\theta^2}{2!} - i\frac{\theta^3}{3!} + \frac{\theta^4}{4!} + \frac{i\theta^5}{5!} + \cdots$$

$$= \left(1 - \frac{\theta^2}{2!} + \frac{\theta^4}{4!} - \cdots\right) + i\left(\theta - \frac{\theta^3}{3!} + \frac{\theta^5}{5!} - \cdots\right)$$

$$= \cos\theta + i\sin\theta. \tag{10.3.9}$$

This leads to a very useful expression for the complex variable z. In polar form, $z = r(\cos\theta + i\sin\theta)$, so Eq. 10.3.9 allows us to write

$$z = re^{i\theta}. \tag{10.3.10}$$

This form is quite useful in obtaining powers and roots of z and in various other operations involving complex numbers.

The *hyperbolic sine* and *cosine* are defined as

$$\sinh z = \frac{e^z - e^{-z}}{2}, \qquad \cosh z = \frac{e^z + e^{-z}}{2}. \tag{10.3.11}$$

With the use of Eqs. 10.3.8 we see that

$$\sinh iz = i\sin z, \qquad \sin iz = i\sinh z,$$
$$\cosh iz = \cos z, \qquad \cos iz = \cosh z. \tag{10.3.12}$$

We can then separate the real and imaginary parts from $\sin z$ and $\cos z$, with the use of trigonometric identities, as follows:

$$\sin z = \sin(x + iy)$$
$$= \sin x \cos iy + \sin iy \cos x$$
$$= \sin x \cosh y + i\sinh y \cos x \tag{10.3.13}$$
$$\cos z = \cos(x + iy)$$
$$= \cos x \cos iy - \sin x \sin iy$$
$$= \cos x \cosh y - i\sin x \sinh y. \tag{10.3.14}$$

The *natural logarithm* of z, written $\ln z$, should be defined so that

$$e^{\ln z} = z. \tag{10.3.15}$$

Using Eq. 10.3.10, we see that

$$\ln z = \ln(re^{i\theta})$$
$$= \ln r + i\theta \tag{10.3.16}$$

is a reasonable candidate for this definition. An immediate consequence of this definition shows that

$$e^{\ln z} = e^{\ln r + i\theta}$$

$$= e^{\ln r} e^{i\theta}$$

$$= r(\cos \theta + i \sin \theta) = z \qquad (10.3.17)$$

using Eq. 10.3.9. Hence, ln z, so defined, does satisfy Eq. 10.3.15. Since θ is ∞-valued, we must restrict its value so that ln z becomes ln x when $y = 0$. The restrictions $0 \le \theta < 2\pi$ or $-\pi < \theta \le \pi$ are both used. The *principal value** of ln z results when $0 \le \theta < 2\pi$.

We are now in a position to define z^a for any complex number a. By definition

$$z^a = e^{a \ln z}. \qquad (10.3.18)$$

We leave it to the reader to verify that definition 10.3.18 agrees with the definition of z^a when the exponent is a real fraction or integer (see Problem 53).

Finally, in our discussion of elementary functions, we include the inverse trigonometric functions and inverse hyperbolic functions. Let

$$w = \sin^{-1} z. \qquad (10.3.19)$$

Then, using Eq. 10.3.8,

$$z = \sin w = \frac{e^{iw} - e^{-iw}}{2i}. \qquad (10.3.20)$$

Rearranging and multiplying by $2ie^{iw}$ gives

$$e^{2iw} - 2ize^{iw} - 1 = 0. \qquad (10.3.21)$$

This quadratic equation (let $e^{iw} = \phi$, so that $\phi^2 - 2iz\phi - 1 = 0$) has solutions

$$e^{iw} = iz + (1 - z^2)^{1/2}. \qquad (10.3.22)$$

The square root is to be understood in the same sense as Section 10.2. We solve for iw in Eq. 10.3.22 and obtain

$$w = \sin^{-1} z = -i \ln \left[iz + (1 - z^2)^{1/2} \right]. \qquad (10.3.23)$$

This expression is double-valued because of the square root and is multi-valued because of the logarithm. Two principal values result for each complex number z except for $z = 1$, in which case the square-root quantity is zero. In a similar manner we can find expressions for the other inverse functions. They are listed in the following:

---

*The principal value of ln z is discontinuous in regions containing the positive real axis because ln x is not close to ln $(x - i\epsilon)$ for small ϵ. This is true since Im $[\ln x] = 0$ but Im $[\ln (x - i\epsilon)]$ is almost 2π. For this reason, we often define ln z by selecting $-\pi < \arg \ln z < \pi$. Then ln z is continuous for all z, excluding $-\infty < z = x \le 0$.

$$\sin^{-1} z = -i \ln \left[iz + (1 - z^2)^{1/2} \right]$$

$$\cos^{-1} z = -i \ln \left[z + (z^2 - 1)^{1/2} \right] \qquad (10.3.24)$$

$$\tan^{-1} z = \frac{i}{2} \ln \frac{1 - iz}{1 + iz}$$

$$\sinh^{-1} z = \ln \left[z + (1 + z^2)^{1/2} \right]$$

$$\cosh^{-1} z = \ln \left[z + (z^2 - 1)^{1/2} \right] \qquad (10.3.25)$$

$$\tanh^{-1} z = \frac{1}{2} \ln \frac{1 + z}{1 - z}.$$

It is worthwhile to note that the exponential and logarithmic functions are sufficient to define z^a, $\sin z$, $\cos z$, $\tan z$, $\csc z$, $\sec z$, $\sin^{-1} z$, $\cos^{-1} z$, $\tan^{-1} z$, $\sinh z$, $\cosh z$, $\tanh z$, $\sinh^{-1} z$, $\cosh^{-1} z$, and $\tanh^{-1} z$; these interconnections are impossible to discover without the notion of a complex variable. Witness, for $z = x$, that the third equation in 10.3.24 is

$$\tan^{-1} x = \frac{i}{2} \ln \frac{1 - ix}{1 + ix}, \qquad (10.3.26)$$

a truly remarkable formula.

Example 10.3.1:   Find the principal value of

$$(2 + i)^{1-i}.$$

SOLUTION: Using Eq. 10.3.18, we can write

$$(2 + i)^{1-i} = e^{(1-i)\ln(2+i)}.$$

We find the principal value of $(2 + i)^{1-i}$ by using the principal value of $\ln(2 + i)$:

$$\ln(2 + i) = \ln \sqrt{5} + 0.4636i$$

since $\tan^{-1} 1/2 = 0.4636$ rad. Then

$$(2 + i)^{1-i} = e^{(1-i)(\ln \sqrt{5} + 0.4636i)} = e^{1.2683 - 0.3411i}$$

$$= e^{1.2683}(\cos 0.3411 - i \sin 0.3411)$$

$$= 3.555(0.9424 - 0.3345i)$$

$$= 3.350 - 1.189i. \quad \blacksquare$$

Example 10.3.2:   Find the principal value of i^i.

SOLUTION: We have

$$i^i = e^{i \ln i} = e^{i[\ln 1 + (\pi/2)i]} = e^{i[(\pi/2)i]} = e^{-\pi/2}$$

a result that delights the imagination, because of the appearance of e, π, and i in one simple equation. $\blacksquare$

Example 10.3.3: Using $z = 3 - 4i$, find the value or principal value of (a) e^{iz}, (b) e^{-iz}, (c) $\sin z$, and (d) $\ln z$.

SOLUTION:

(a)
$$e^{i(3-4i)} = e^{4+3i} = e^4 e^{3i}$$
$$= 54.60(\cos 3 + i \sin 3)$$
$$= 54.60(-0.990 + 0.1411i)$$
$$= -54.05 + 7.704i.$$

(b)
$$e^{-i(3-4i)} = e^{-4-3i} = e^{-4} e^{-3i}$$
$$= 0.01832[\cos(-3) + i \sin(-3)]$$
$$= 0.01832[-0.990 - 0.1411i]$$
$$= -0.01814 - 0.002585i.$$

(c)
$$\sin(3 - 4i) = \frac{e^{i(3-4i)} - e^{-i(3-4i)}}{2i}$$
$$= \frac{-54.05 + 7.704i - (-0.01814 - 0.002585i)}{2i}$$
$$= 3.853 + 27.01i.$$

(d)
$$\ln(3 - 4i) = \ln r + i\theta$$
$$= \ln 5 + i \tan^{-1} \frac{-4}{3}$$
$$= 1.609 + 5.356i,$$

where the angle θ is expressed in radians. ■

Example 10.3.4: What is the value of z so that $\sin z = 10$?

SOLUTION: From Eq. 10.3.24, we write

$$z = \sin^{-1} 10 = -i \ln [10i + (-99)^{1/2}].$$

The two roots of -99 are $3\sqrt{11}\, i$ and $-3\sqrt{11}\, i$. Hence,

$$z_1 = -i \ln [(10 + 3\sqrt{11})i], \qquad z_2 = -i \ln [(10 - 3\sqrt{11})i].$$

But if α is real, $\ln \alpha i = \ln |\alpha| + (\pi/2)i$. Hence,

$$z_1 = \frac{\pi}{2} - i \ln (10 + 3\sqrt{11}), \qquad z_2 = \frac{\pi}{2} - i \ln (10 - 3\sqrt{11}),$$

or

$$z_1 = \frac{\pi}{2} - 2.993i, \qquad z_2 = \frac{\pi}{2} + 2.993i. \quad ■$$

PROBLEMS

1. Show that $\sin z = (e^{iz} - e^{-iz})/2i$ and $\cos z = (e^{iz} + e^{-iz})/2$ using Eqs. 10.3.5 through 10.3.7.

Express each complex number in exponential form (see Eq. 10.3.10).

2. -2 **3.** $2i$ **4.** $-2i$

5. $3 + 4i$ **6.** $5 - 12i$ **7.** $-3 - 4i$

8. $-5 + 12i$ **9.** $0.213 - 2.15i$

10. Using $z = (\pi/2) - i$, show that Eq. 10.3.8 yields the same result as Eq. 10.3.13 for $\sin z$.

Find the value of e^z for each value of z.

11. $\dfrac{\pi}{2} i$ **12.** $2i$ **13.** $-\dfrac{\pi}{4} i$ **14.** $4\pi i$

15. $2 + \pi i$ **16.** $-1 - \dfrac{\pi}{4} i$

Find each quantity using Eq. 10.3.10.

17. $1^{1/5}$ **18.** $(1 - i)^{1/4}$ **19.** $(-1)^{1/3}$ **20.** $(2 + i)^3$

21. $(3 + 4i)^4$ **22.** $\sqrt{2 - i}$

For the value $z = \pi/2 - (\pi/4)i$, find each term.

23. e^{iz} **24.** $\sin z$ **25.** $\cos z$ **26.** $\sinh z$

27. $\cosh z$ **28.** $|\sin z|$ **29.** $|\tan z|$

Find the principal value of the $\ln z$ for each value for z.

30. i **31.** $3 + 4i$ **32.** $4 - 3i$ **33.** $-5 + 12i$

34. ei **35.** -4 **36.** e^i

Using the relationship that $z^a = e^{\ln z^a} = e^{a \ln z}$, find the principal value of each power.

37. i^i **38.** $(3 + 4i)^{(1-i)}$ **39.** $(4 - 3i)^{(2+i)}$ **40.** $(1 + i)^{(1+i)}$

41. $(-1 - i)^{-i/2}$

Find the values or principal values of z for each equation.

42. $\sin z = 2$ **43.** $\cos z = 4$ **44.** $e^z = -3$ **45.** $\sin z = -2i$

46. $\cos z = -2$

Show that each equation is true.

47. $\cos^{-1} z = -i \ln \left[z + (z^2 - 1)^{1/2} \right]$ **48.** $\sinh^{-1} z = \ln \left[z + (1 + z^2)^{1/2} \right]$

For $z = 2 - i$, evaluate each function.

49. $\sin^{-1}$ **50.** $\tan^{-1} z$ **51.** $\cosh^{-1} z$

52. Using the principal value of ln z, explain why ln 1 and ln $(1 - i\epsilon)$ are not close even when ϵ is very near zero.

53. In Eq. 10.3.18 set $a = n$, n a real integer. Show that $z^n = e^{n \ln z}$ is in agreement with $z^n = r^n(\cos n\theta + i \sin n\theta)$.

54. In Eq. 10.3.18 set $a = p/q$, p/q a real rational number. Show that $z^{p/q} = e^{(p/q \ln z)}$ is the same set of complex numbers as

$$z^{p/q} = r^{p/q}\left[\cos \frac{p}{q}(\theta + 2\pi k) + \sin \frac{p}{q}(\theta + 2\pi k)\right].$$

10.4   Analytic Functions

In Section 10.3 we motivated the definitions of the various "elementary" functions by requiring $f(z)$ to reduce to the standard function when $y = 0$. Although intuitively appealing, this consideration is only part of the picture. In this section we round out our presentation by showing that our newly defined functions satisfy the appropriate differential relationships:

$$\frac{de^z}{dz} = e^z, \qquad \frac{d \cos z}{dz} = -\sin z$$

and so on.

The definition of the derivative of a function $f(z)$ is

$$f'(z) = \lim_{\Delta z \to 0} \frac{f(z + \Delta z) - f(z)}{\Delta z}. \tag{10.4.1}$$

It is important to note that in the limiting process as $\Delta z \to 0$ there are an infinite number of paths that Δz can take. Some of these are sketched in Fig. 10.3. For a derivative to exist we demand that $f'(z)$ be unique as $\Delta z \to 0$,

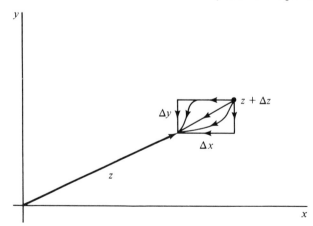

FIGURE 10.3.   Various paths for Δz to approach zero.

regardless of the path chosen. In real variables this restriction on the derivative was not necessary since only one path was used, along the x axis only. Let us illustrate the importance of this demand with the function

$$f(z) = \bar{z} = x - iy. \qquad (10.4.2)$$

The quotient in the definition of the derivative using $\Delta z = \Delta x + i \Delta y$ is

$$\frac{f(z + \Delta z) - f(z)}{\Delta z} = \frac{[(x + \Delta x) - i(y + \Delta y)] - (x - iy)}{\Delta x + i \Delta y}.$$

$$= \frac{\Delta x - i \Delta y}{\Delta x + i \Delta y} \qquad (10.4.3)$$

First, let $\Delta y = 0$ and then let $\Delta x \to 0$. Then, the quotient is $+1$. Next, let $\Delta x = 0$ and then let $\Delta y \to 0$. Now the quotient is -1. Obviously, we obtain a different value for each path. Actually, there is a different value for the quotient for each value of the slope of the line along which Δz approaches zero (see Problem 2). Since the limit is not unique, we say that the derivative does not exist. We shall now derive conditions which must hold if a function has a derivative.

Let us assume now that the derivative $f'(z)$ does exist. The real and imaginary parts of $f(z)$ are denoted by $u(x, y)$ and $v(x, y)$, respectively; that is,

$$f(z) = u(x, y) + iv(x, y). \qquad (10.4.4)$$

First, let $\Delta y = 0$ so that $\Delta z \to 0$ parallel to the x axis. From Eq. 10.4.1 with $\Delta z = \Delta x$,

$$f'(z) = \lim_{\Delta x \to 0} \frac{u(x + \Delta x, y) + iv(x + \Delta x, y) - u(x, y) - iv(x, y)}{\Delta x}$$

$$= \lim_{\Delta x \to 0} \left[\frac{u(x + \Delta x, y) - u(x, y)}{\Delta x} + i \frac{v(x + \Delta x, y) - v(x, y)}{\Delta x} \right]$$

$$= \frac{\partial u}{\partial x} + i \frac{\partial v}{\partial x}. \qquad (10.4.5)$$

Next, let $\Delta x = 0$ so that $\Delta z \to 0$ parallel to the y axis. Then, using $\Delta z = i \Delta y$,

$$f'(z) = \lim_{\Delta y \to 0} \frac{u(x, y + \Delta y) + iv(x, y + \Delta y) - u(x, y) - iv(x, y)}{i \Delta y}$$

$$= \lim_{\Delta y \to 0} \left[\frac{u(x, y + \Delta y) - u(x, y)}{i \Delta y} + \frac{v(x, y + \Delta y) - v(x, y)}{\Delta y} \right]$$

$$= -i \frac{\partial u}{\partial y} + \frac{\partial v}{\partial y}. \qquad (10.4.6)$$

For the derivative to exist, it is necessary that these two expressions for $f'(z)$ be equal. Hence,

$$\frac{\partial u}{\partial x} + i\frac{\partial v}{\partial x} = -i\frac{\partial u}{\partial y} + \frac{\partial v}{\partial y}. \tag{10.4.7}$$

Setting the real parts and the imaginary parts equal to each other, respectively, we find that

$$\frac{\partial u}{\partial x} = \frac{\partial v}{\partial y}, \qquad \frac{\partial u}{\partial y} = -\frac{\partial v}{\partial x}, \tag{10.4.8}$$

the famous *Cauchy–Riemann equations*. We have derived these equations by considering only two possible paths along which $\Delta z \to 0$. It can be shown (we shall not do so in this text) that no additional relationships are necessary to ensure the existence of the derivative. If the Cauchy–Riemann equations are satisfied at a point $z = z_0$, and the first partials of u and v are continuous at z_0, then the derivative $f'(z_0)$ exists. If $f'(z)$ exists at $z = z_0$ and at every point in a neighborhood of z_0, then the function $f(z)$ is said to be *analytic* at z_0.

Thus, the definition of analyticity puts some restrictions on the nature of the sets on which $f(z)$ is analytic.* For instance, if $f(z)$ is analytic for all z, $z < 1$ and at $z = i$ as well, then $f(z)$ is analytic at least in a domain portrayed in Fig. 10.4. If $f(z)$ is not analytic at z_0, $f(z)$ is *singular* at z_0. In most applications z_0 is an *isolated singular point,* by which we mean that in some neighborhood of z_0, $f(z)$ is analytic for $z \neq z_0$ and singular only at z_0. The most common singular points of an otherwise analytic function arise because of zeros in the denominator of a quotient. For each of the following functions, $f(z)$ has an isolated singular point at $z_0 = 0$:

$$\frac{1}{e^z - 1}, \quad \frac{1}{z(z+1)}, \quad \frac{1}{z \sin z}, \quad \tan z. \tag{10.4.9}$$

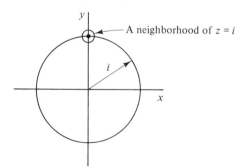

FIGURE 10.4.   A set of analyticity for some $f(z)$.

---

*We do not explore this point here. Suffice it to say that functions are analytic on open sets in the complex plane.

The rational function

$$Q(z) = \frac{a_n z^n + a_{n-1} z^{n-1} + \cdots + a_1 z + a_0}{b_m z^m + b_{m-1} z^{m-1} + \cdots + b_1 z + b_0} \qquad (10.4.10)$$

has isolated singularities at each zero of the denominator polynomial. We use "singularity" and mean "isolated singularity" unless we explicitly comment to the contrary.

Example 10.4.1:   Determine if and where the functions $z\bar{z}$ and z^2 are analytic.

SOLUTION: The function $z\bar{z}$ is written as

$$f(z) = z\bar{z} = (x + iy)(x - iy) = x^2 + y^2$$

and is a real function only; its imaginary part is zero. That is,

$$u = x^2 + y^2, \qquad v = 0.$$

The Cauchy–Riemann equations give

$$\frac{\partial u}{\partial x} = \frac{\partial v}{\partial y} \qquad \text{or} \qquad 2x = 0$$

$$\frac{\partial u}{\partial y} = -\frac{\partial v}{\partial x} \qquad \text{or} \qquad 2y = 0.$$

Hence, we see that x and y must each be zero the for Cauchy–Riemann equations to be satisfied. This is true at the origin but not in the neighborhood (however small) of the origin. Thus, the function $z\bar{z}$ is not analytic anywhere.

Now consider the function z^2. It is

$$f(z) = z^2 = (x + iy)(x + iy) = x^2 - y^2 + i2xy.$$

The real and imaginary parts are

$$u = x^2 - y^2, \qquad v = 2xy.$$

The Cauchy–Reimann equations give

$$\frac{\partial u}{\partial x} = \frac{\partial v}{\partial y} \qquad \text{or} \qquad 2x = 2x$$

$$\frac{\partial u}{\partial y} = -\frac{\partial v}{\partial x} \qquad \text{or} \qquad -2y = -2y.$$

We see that these equations are satisfied at all points in the xy plane. Hence, the function z^2 is analytic everywhere.   ∎

Example 10.4.2:   Find the regions of analyticity of the functions listed below and compute their first derivatives.

(a) e^z (b) $\ln z$ (c) $\sin z$

SOLUTION: (a) Since $e^z = e^{x+iy} = e^x e^{iy}$, we have

$$e^z = e^x \cos y + ie^x \sin y.$$

Therefore, $u = e^x \cos y$ and $v = e^x \sin y$. The verification of the Cauchy–Riemann equations is simple, so we learn that e^z is analytic for all z. Also, from Eq. 10.4.7

$$\frac{de^z}{dz} = \frac{\partial u}{\partial x} + i\frac{\partial v}{\partial x}$$

$$= e^x \cos y + ie^x \sin y = e^z.$$

(b) Here we express $\ln z = \ln r + i\theta$, $r = \sqrt{x^2 + y^2}$, and $\theta = \tan^{-1} y/x$ with $-\pi < \theta \le \pi$. Hence,

$$u = \ln \sqrt{x^2 + y^2} = \tfrac{1}{2} \ln (x^2 + y^2)$$

$$v = \tan^{-1} \frac{y}{x},$$

so

$$\frac{\partial u}{\partial x} = \frac{1}{2} \frac{2x}{x^2 + y^2} = \frac{x}{x^2 + y^2}$$

$$\frac{\partial v}{\partial y} = \frac{1/x}{1 + (y/x)^2} = \frac{x}{x^2 + y^2}.$$

Also,

$$\frac{\partial u}{\partial y} = \frac{1}{2} \frac{2y}{x^2 + y^2} = \frac{y}{x^2 + y^2}$$

$$\frac{\partial v}{\partial x} = \frac{-y/x^2}{1 + (y/x^2)} = -\frac{y}{x^2 + y^2}.$$

The Cauchy–Riemann equations are satisfied as long as $x^2 + y^2 \ne 0$ and θ is uniquely defined, say $-\pi < \theta < \pi$. Finally,

$$\frac{d}{dz} \ln z = \frac{\partial u}{\partial x} + i\frac{\partial v}{\partial y}$$

$$= \frac{x}{x^2 + y^2} - i\frac{y}{x^2 + y^2} = \frac{1}{z},$$

valid as long as $z \ne 0$ and $\ln z$ is continuous. For $-\pi < \theta \le \pi$, $\ln z$ is continuous at every z except $z = x \le 0$.
(c) Since

$$\sin z = \sin x \cosh y + i \sinh y \cos x$$

by Eq. 10.3.13, we have

$$u = \sin x \cosh y, \qquad v = \cos x \sinh y,$$

so the Cauchy–Riemann equations are easily checked and are valid for all z. Moreover,

$$\frac{d}{dz}\sin z = \frac{\partial u}{\partial x} + i\frac{\partial v}{\partial y}$$

$$= \cos x \cosh y - i \sin x \sinh y = \cos z$$

from Eq. 10.3.14. ∎

Since the definition of $f'(z)$ is formally the same as the definition of $f'(z)$, we can mirror the arguments in elementary calculus to prove that

(1)
$$\frac{d}{dz}[f(z) \pm g(z)] = f'(z) \pm g'(z) \qquad (10.4.11)$$

(2)
$$\frac{d}{dz}[kf(z)] = kf'(z) \qquad (10.4.12)$$

(3)
$$\frac{d}{dz}[f(z)g(z)] = f'(z)g(z) + g'(z)f(z) \qquad (10.4.13)$$

(4)
$$\frac{d}{dz}\left[\frac{f(z)}{g(z)}\right] = \frac{f'(z)g(z) - f(z)g'(z)}{g^2(z)} \qquad (10.4.14)$$

(5)
$$\frac{d}{dz}[f(g(z))] = \frac{df}{dg}\frac{dg}{dz} \qquad (10.4.15)$$

Also, we can show that

$$\frac{d}{dz}(a_n z^n + a_{n-1}z^{n-1} + \cdots + a_1 z + a_0)$$

$$= na_n z^{n-1} + (n-1)a_{n-1}z^{n-2} + \cdots + 2a_2 z + a_1 \quad (10.4.16)$$

by using properties (1) and (2) and the easily verified facts

$$\frac{dz}{dz} = 1, \qquad \frac{da_0}{dz} = 0. \qquad (10.4.17)$$

Example 10.4.3:   Find $d/dz \tan^{-1} z$.

SOLUTION: Since we know, by Eq. 10.3.24, that

$$\tan^{-1} z = \frac{i}{2}\ln\frac{1 - iz}{1 + iz},$$

we have

$$\frac{d}{dz}\tan^{-1} z = \frac{i}{2}\frac{1 + iz}{1 - iz}\frac{d}{dz}\left(\frac{1 - iz}{1 + iz}\right)$$

by utilizing $d/dz \ln z = 1/z$ and the chain rule, property (5) above. Since

$$\frac{d}{dz}\frac{1 - iz}{1 + iz} = \frac{(-i)(1 + iz) - i(1 - iz)}{(1 + iz)^2}$$

$$= -\frac{2i}{(1 + iz)^2},$$

we have

$$\frac{d}{dz} \tan^{-1} z = \frac{i}{2} \frac{1 + iz}{1 - iz} \frac{-zi}{(1 + iz)^2}$$

$$= \frac{1}{(1 - iz)(1 + iz)} = \frac{1}{1 + z^2}.$$

This result could be obtained somewhat more easily by assuming that

$$\ln \frac{1 - iz}{1 + iz} = \ln (1 - iz) - \ln (1 + iz).$$

But this is not a true statement without some qualifications.* ∎

10.4.1   HARMONIC FUNCTIONS

Consider, once again, the Cauchy–Riemann equations (10.4.8), from which we can deduce

$$\frac{\partial^2 u}{\partial x^2} = \frac{\partial^2 v}{\partial x \, \partial y}, \qquad \frac{\partial^2 u}{\partial y^2} = -\frac{\partial^2 v}{\partial x \, \partial y}, \qquad (10.4.18)$$

and[†]

$$\frac{\partial^2 v}{\partial y^2} = \frac{\partial^2 u}{\partial x \, \partial y}, \qquad \frac{\partial^2 v}{\partial x^2} = -\frac{\partial^2 u}{\partial x \, \partial y}. \qquad (10.4.19)$$

From Eqs. 10.4.18 we see that

$$\frac{\partial^2 u}{\partial x^2} + \frac{\partial^2 u}{\partial y^2} = 0, \qquad (10.4.20)$$

and from Eqs. 10.4.19,

$$\frac{\partial^2 v}{\partial x^2} + \frac{\partial^2 v}{\partial y^2} = 0. \qquad (10.4.21)$$

The real and imaginary parts of an analytic function satisfy Laplace's equation. Functions that satisfy Laplace's equation are called *harmonic functions*. Hence, $u(x, y)$ and $v(x, y)$ are harmonic functions. Two functions that satisfy Laplace's equation and the Cauchy-Riemann equations are known as *conjugate harmonic functions*. If one of the conjugate harmonic functions is known, the other can be found by using the Cauchy-Riemann equations. This will be illustrated by an example.

---

*See Problem 16 to see one difficulty with the "rule" $\ln z_1 z_2 = \ln z_1 + \ln z_2$.

[†]We have interchanged the order of differentiation since the second partial derivatives are assumed to be continuous.

Finally, let us show that constant u lines are normal to constant v lines if $u + iv$ is an analytic function. From the chain rule of calculus

$$du = \frac{\partial u}{\partial x} dx + \frac{\partial u}{\partial y} dy. \tag{10.4.22}$$

Along a constant u line, $du = 0$. Hence,

$$\left. \frac{dy}{dx} \right|_{u=C} = -\frac{\partial u / \partial x}{\partial u / \partial y}. \tag{10.4.23}$$

Along a constant v line, $dv = 0$, giving

$$\left. \frac{dy}{dx} \right|_{v=C} = -\frac{\partial v / \partial x}{\partial v / \partial y}. \tag{10.4.24}$$

But, using the Cauchy-Riemann equations,

$$-\frac{\partial u / \partial x}{\partial u / \partial y} = \frac{\partial v / \partial y}{\partial v / \partial x}. \tag{10.4.25}$$

The slope of the constant u line is the negative reciprocal of the slope of the constant v line. Hence, the lines are orthogonal. This property is useful in sketching constant u and v lines, as in fluid fields or electrical fields.

Example 10.4.5:   The real function $u(x, y) = Ax + By$ obviously statisfies Laplace's equation. Find its conjugate harmonic function, and write the function $f(z)$.

SOLUTION: The conjugate harmonic function, denoted $v(x, y)$, is related to $u(x, y)$ by the Cauchy-Riemann equations. Thus,

$$\frac{\partial u}{\partial x} = \frac{\partial v}{\partial y} \quad \text{or} \quad \frac{\partial v}{\partial y} = A.$$

The solution for v is

$$v = Ay + g(x),$$

where $g(x)$ is an unknown function to be determined. The relationship above must satisfy the other Cauchy-Riemann equation,

$$\frac{\partial v}{\partial x} = -\frac{\partial u}{\partial y} \quad \text{or} \quad \frac{dg}{dx} = -B.$$

Hence,

$$g(x) = -Bx + C,$$

where C is a constant of integration, to be determined by an imposed condition. Finally, the conjugate harmonic function $v(x, y)$ is

$$v(x, y) = Ay - Bx + C$$

for every choice of C. The function $f(z)$ is then

$$f(z) = Ax + By + i(Ay - Bx + C)$$
$$= A(x + iy) - Bi(x + iy) + iC$$
$$= (A - iB)z + iC$$
$$= K_1 z + K_2,$$

where K_1 and K_2 are complex constants.   ∎

10.4.2   A TECHNICAL NOTE

The definition of analyticity requires that the derivative of $f(z)$ exist in some neighborhood of z_0. It does not, aparently, place any restrictions on the behavior of this derivative. It is possible to prove* that $f'(z)$ is far from arbitrary; it is also analytic at z_0. This same proof applies to $f'(z)$ and leads to the conclusion that $f''(z)$ is also analytic at z_0. We have, therefore,

Theorem 10.1:   *If f is analytic at z_0, then so are $f'(z)$ and all the higher order derivatives.*

This theorem is supported by all of the examples we have studied. The reader should note that the analogous result for functions of a real variable is false (see Problem 17).

PROBLEMS

Compare the derivative of each function $f(z)$ using Eq. 10.4.5 with that obtained using Eq. 10.4.6.

1. z^2 **2.** $\bar{z}$ **3.** $\dfrac{1}{z + 2}$ **4.** $(z - 1)^2$

5. $\ln(z - 1)$ **6.** e^z **7.** $\bar{z}z$

8. Express a complex fraction in polar coordinates as $f(z) = u(r, \theta) + iv(r, \theta)$ and show that the Cauchy-Riemann equations can be expressed as

$$\frac{\partial u}{\partial r} = \frac{1}{r}\frac{\partial v}{\partial \theta}, \qquad \frac{\partial v}{\partial r} = -\frac{1}{r}\frac{\partial u}{\partial \theta}.$$

Hint: Sketch Δz using polar coordinates; then, note that for $\Delta\theta = 0$, $\Delta z = \Delta r(\cos\theta + i\sin\theta)$, and for $\Delta z = 0$, $\Delta z = r\Delta\theta(-\sin\theta + i\cos\theta)$.

9. Derive Laplace's equation for polar coordinates.

10. Find the conjugate harmonic function associated with $u(r, \theta) = \ln r$. Sketch some constant u and v lines.

---

*See any text on the theory of a complex variable.

Show that each function is harmonic and find the conjugate harmonic function. Also, write the analytic function $f(z)$.

11. xy **12.** $x^2 - y^2$ **13.** $e^y \sin x$ **14.** $\ln (x^2 + y^2)$

15. If v is a harmonic conjugate of u, find a harmonic conjugate of v.

16. Suppose that $\ln z = \ln r + i\theta$, $0 \le \theta < 2\pi$. Show that $\ln (-i)(-i) \neq \ln (-i) + \ln (-i)$. Hence, $\ln z_1 z_2 = \ln z_1 + \ln z_2$ may be false for the principal value of $\ln z$. Note that $\ln z_1 z_2 = \ln z_1 + \ln z_2 + 2\pi n i$, for some integer n.

17. Let $f(x) = x^2$ if $x \ge 0$ and $f(x) = -x^2$ if $x < 0$. Show that $f'(x)$ exists and is continuous, but $f''(x)$ does not exist at $x = 0$.

10.5   Complex Integration

10.5.1   ARCS AND CONTOURS

A *smooth arc* is the set of points (x, y) that satisfy

$$x = \phi(t), \qquad y = \psi(t), \qquad a \le t \le b \qquad (10.5.1)$$

where $\phi'(t)$ and $\psi'(t)$ are continuous in $[a, b]$ and do not vanish simultaneously. The circle, $x^2 + y^2 = 1$, is represented parametrically by

$$x = \cos t, \qquad y = \sin t, \qquad 0 \le t \le 2\pi. \qquad (10.5.2)$$

This is the most natural illustration of this method of representing a smooth arc in the xy plane. The representation

$$x = t, \qquad y = t^2, \qquad -\infty < t < \infty \qquad (10.5.3)$$

defines the parabola $y = x^2$. Note that a parametric representation provides an ordering to the points on the arc. A smooth arc has length given by

$$L = \int_a^b \sqrt{[\phi'(t)]^2 + [\psi'(t)]^2}\, dt. \qquad (10.5.4)$$

A *contour* is a continuous chain of smooth arcs. Figure 10.5 illustrates a variety of contours.

    A *simple closed contour,* or a *Jordan curve,* is a contour which does not intersect itself except that $\phi(a) = \phi(b)$ and $\psi(a) = \psi(b)$. A simple closed contour divides a plane into two parts, an "inside" and an "outside," and is traversed in the *positive sense* if the inside is to the left. The square portrayed in Fig. 10.5a is being traversed in the positive sense, as indicated by the direction arrows on that simple closed contour.

    Circles in the complex plane have particularly simple parametric representations, which exploit the polar and exponential forms of z. The circle $|z| = a$ is given parametrically by

$$z = a \cos \theta + ia \sin \theta, \qquad 0 \le \theta \le 2\pi \qquad (10.5.5)$$

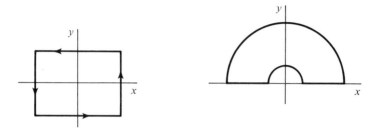

(a) Simple closed contours

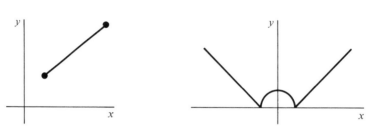

(b) Contours that are not closed

FIGURE 10.5.   Examples of contours.

or

$$z = ae^{i\theta}, \qquad\qquad 0 \le \theta \le 2\pi. \qquad (10.5.6)$$

Both formulas are to be understood in this sense: $x = a \cos \theta$, $y = a \sin \theta$, so that $z = x + iy = a \cos \theta + ia \sin \theta = ae^{i\theta}$.

A circle of radius a, centered at z_0, as shown in Fig. 10.6, is described by the equations

$$z - z_0 = ae^{i\theta} = a \cos \theta + ia \sin \theta, \qquad (10.5.7)$$

where θ is measured from an axis passing through the point $z = z_0$ parallel to the x axis.

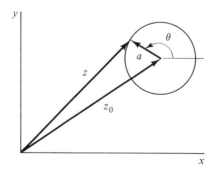

FIGURE 10.6.   The description of a circle.

10.5.2   LINE INTEGRALS

Let z_0 and z_1 be two points in the complex plane and C a contour connecting them, as shown in Fig. 10.7. We suppose that C is defined parametrically by

$$x = \phi(t), \qquad y = \psi(t), \tag{10.5.8}$$

so that

$$z_0 = \phi(a) + i\psi(a), \qquad z_1 = \phi(b) + i\psi(b). \tag{10.5.9}$$

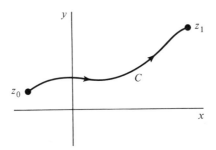

FIGURE 10.7.   The contour C joining z_0 to z_1.

The *line integral*

$$\int_C f(z) \, dz = \int_{z_0}^{z_1} f(z) \, dz \tag{10.5.10}$$

is defined by the real integrals

$$\int_C f(z) \, dz = \int_C (u + iv)(dx + i \, dy)$$

$$= \int_C (u \, dx - v \, dy) + i \int_C (v \, dx + u \, dy), \tag{10.5.11}$$

where we have written $f(z) = u + iv$. The integral relation 10.5.11 leads to several "natural conclusions": First,

$$\int_{z_0}^{z_1} f(z) \, dz = -\int_{z_1}^{z_0} f(z) \, dz, \tag{10.5.12}$$

where the path of integration for the integral on the right-hand side of Eq. 10.5.12 is the same as that on the left but traversed in the opposite direction. Also,

$$\int_{z_0}^{z} kf(z) \, dz = k \int_{z_0}^{z} f(z) \, dz. \tag{10.5.13}$$

If the contour C is a continuous chain of contours $C_1, C_2, \ldots, C_k$, such as displayed in Fig. 10.8, then

$$\int_{z_0}^{z_1} f(z)\ dz = \int_{C_1} f(z)\ dz + \int_{C_2} f(z)\ dz + \cdots + \int_{C_k} f(z)\ dz.$$

$$(10.5.14)$$

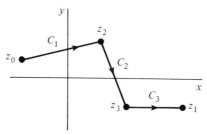

FIGURE 10.8.   A chain of contours.

Equation 10.5.11 can also be used to prove a most essential inequality:

Theorem 10.2:   *Suppose that* $|f(z)| \le M$ *along the contour C and the length of C is L; then*

$$\left| \int_C f(z)\ dz \right| \le ML.$$

$$(10.5.15)$$

A proof of this inequality can be found in a text on complex variables. Here we outline a heuristic argument based on approximating the line integral by a sum. Consider the contour shown in Fig. 10.9 and the chords joining points $z_0, z_1, \ldots, z_n$ on C. Suppose that N is very large and the points $z_1, z_2, \ldots, z_n$ are quite close. Then

$$\int_C f(z)\ dz \cong \sum_{n=1}^{N} f(z_n)\ \Delta z_n,$$

$$(10.5.16)$$

where $\Delta z_n = z_n - z_{n-1}$. From repeated use of $|z_1 + z_2| \le |z_1| + |z_2|$ and

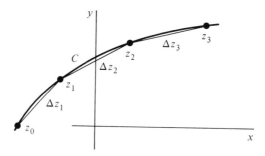

FIGURE 10.9.   A polygonal approximation to C.

Eq. 10.2.19, we have

$$\left| \int_C f(z) \, dz \right| \cong \left| \sum_{n=1}^{N} f(z_n) \, \Delta z_n \right| \le \sum_{n=1}^{N} |f(z_n)| \, |\Delta z_n|. \quad (10.5.17)$$

Now $|f(z)| \le M$ along C, so

$$\left| \int_C f(z) \, dz \right| \le M \sum_{n=1}^{N} |\Delta z_n| \le ML \quad (10.5.18)$$

since

$$\sum_{n=1}^{N} |\Delta z_n| \cong L. \quad (10.5.19)$$

When the path of integration is a simple closed contour traversed positively, we write the line integral as $\oint f(z) \, dz$; this signals an integration once around the contour in the positive sense.

Example 10.5.1:   Find the value of $\int_0^{1+i} z^2 \, dz$ along the following contours. (a) The straight line from 0 to $1 + i$. (b) The polygonal line from 0 to 1 and from 1 to $1 + i$. The contours are sketched in the figure.

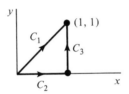

SOLUTION: Along any contour the integral can be written as

$$\int_0^{i+i} z^2 \, dz = \int_0^{1+i} [(x^2 - y^2) + 2xyi](dx + i \, dy)$$

$$= \int_0^{1+i} [(x^2 - y^2) \, dx - 2xy \, dy] + i \int_0^{1+i} [2xy \, dx + (x^2 - y^2) \, dy].$$

(a) The contour C_1 is the straight line from 0 to $1 + i$ and it has the parametric representation:

$$x = t, \qquad y = t, \qquad 0 \le t \le 1$$

with $dt = dx = dy$. We make these substitutions in the integral above to obtain

$$\int_0^{1+i} z^2 \, dz = \int_0^1 [(t^2 - t^2) \, dt - 2t^2 \, dt] + i \int_0^1 [2t^2 \, dt + (t^2 - t^2) \, dt]$$

$$= -\tfrac{2}{3} + \tfrac{2}{3} i.$$

(b) In this case, the contour is a polygonal line and this line requires two separate parameterizations. Using $z = x$ along C_2 and $z = 1 + iy$ along C_3, we can write

$$\int_0^{1+i} z^2\, dz = \int_0^1 x^2\, dx + \int_0^1 (1 + iy)^2 i\, dy.$$

This simplification follows because the contour C_2 has $y = 0$ and $dy = 0$. The contour C_3 requires $dx = 0$. Therefore,

$$\int_0^{1+i} z^2\, dz = \tfrac{1}{3} + \int_0^1 (1 - y^2 + 2yi)i\, dy$$

$$= \tfrac{1}{3} - \int_0^1 2y\, dy + i \int_0^1 (1 - y^2)\, dy$$

$$= \tfrac{1}{3} - 1 + \tfrac{2}{3}i = -\tfrac{2}{3} + \tfrac{2}{3}i. \quad \blacksquare$$

Example 10.5.2:   Evaluate $\oint dz/z$ around the unit circle with center at the origin.

SOLUTION: The simplest representation for this circle is the exponential form

$$z = e^{i\theta} \qquad \text{and} \qquad dz = ie^{i\theta}d\theta,$$

where we have noted that $r = 1$ for a unit circle with center at the origin. We then have

$$\oint \frac{dz}{z} = \int_0^{2\pi} \frac{ie^{i\theta}d\theta}{e^{i\theta}} = \int_0^{2\pi} i\, d\theta = 2\pi i.$$

This is an important integration technique and an important result which will be used quite often in the remainder of this chapter. $\blacksquare$

Example 10.5.3:   Evaluate the integral $\oint dz/z^n$ around the unit circle with center at the origin. Assume that n is a positive integer greater than unity.

SOLUTION: As in Example 10.5.2, we use $r = 1$ and the exponential form for the parametric representation of the circle:

$$z = e^{i\theta}, \qquad dz = ie^{i\theta}d\theta.$$

We then have, if $n > 1$,

$$\oint \frac{dz}{z^n} = \int_0^{2\pi} \frac{ie^{i\theta}}{e^{ni\theta}}d\theta$$

$$= i \int_0^{2\pi} e^{i\theta(1-n)}d\theta$$

$$= \frac{ie^{i\theta(1-n)}}{i(1-n)}\Big|_0^{2\pi} = \frac{1}{1-n}(1 - 1) = 0. \quad \blacksquare$$

Example 10.5.4:   Show that

$$\int_C dz = \int_{z_0}^{z_1} dz = z_1 - z_0.$$

SOLUTION: Since $f(z) = 1$, then $u = 1$, $v = 0$ and we have, using Eq. 10.5.11,

$$\int_{z_0}^{z_1} dz = \int_{z_0}^{z_1} dx + i \int_{z_0}^{z_1} dy.$$

Now suppose that C has the parametric representation

$$x = \phi(t), \qquad y = \psi(t), \qquad a \le t \le b.$$

Then $dx = \phi' \, dt$, $dy = \psi' \, dt$ and

$$\begin{aligned}
\int_{z_0}^{z_1} dz &= \int_a^b \phi'(t) \, dt + i \int_a^b \psi'(t) \, dt \\
&= [\phi(b) - \phi(a)] + i[\psi(b) - \psi(a)] \\
&= [\phi(b) + i\psi(b)] - [\phi(a) + i\psi(a)] \\
&= z_1 - z_0.
\end{aligned}$$

Note that this result is independent of the contour. ∎

10.5.3   GREEN'S THEOREM

There is an important relationship that allows us to transform a line intergral into a double integral for contours in the xy plane. It is often referred to as *Green's theorem*:

Theorem 10.3: *Suppose that C is a simple closed contour traversed in the positive direction and bounding the region R. Suppose also that u and v are continuous with continuous first partial derivatives in R. Then*

$$\oint_C u \, dx - v \, dy = -\iint_R \left(\frac{\partial v}{\partial x} + \frac{\partial u}{\partial y} \right) dx \, dy. \tag{10.5.20}$$

PROOF: Consider the curve C surrounding the region R in Fig. 10.10. Let us investigate the first part of the double integral in Green's theorem. It can be written as

$$\begin{aligned}
\iint_R \frac{\partial v}{\partial x} \, dx \, dy &= \int_{h_1}^{h_2} \int_{x_1(y)}^{x_2(y)} \frac{\partial v}{\partial x} \, dx \, dy \\
&= \int_{h_1}^{h_2} [v(x_2, y) - v(x_1, y)] \, dy \\
&= \int_{h_1}^{h_2} v(x_2, y) \, dy + \int_{h_2}^{h_1} v(x_1, y) \, dy. \tag{10.5.21}
\end{aligned}$$

The first integral on the right-hand side is the line integral of $v(x, y)$ taken along the path ABC from A to C and the second integral is the line integral of $v(x, y)$ taken along the path ADC from C to A. Note that the region R is

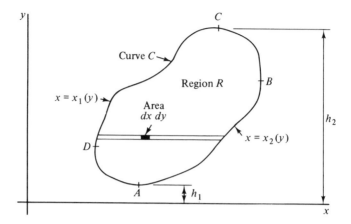

FIGURE 10.10.   Curve C surrounding region R in Green's theorem.

on the left. Hence, we can write

$$\iint\limits_R \frac{\partial v}{\partial x}\, dx\, dy = \oint_C v(x,y)\, dy. \tag{10.5.22}$$

Similarly, we can show that

$$\iint\limits_R \frac{\partial u}{\partial y}\, dx\, dy = -\oint_C u(x,y)\, dx, \tag{10.5.23}$$

and Green's theorem is proved.

It should be noted that Green's theorem may be applied to a multiply connected region by appropriately cutting the region, as shown in Fig. 10.11.

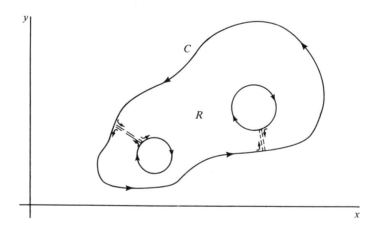

FIGURE 10.11.   Multiply connected region.

This makes a simply connected region* from the original multiply connected region. The contribution to the line integrals from the cuts is zero, since each cut is traversed twice, in opposite directions.

Example 10.5.5: Verify Green's theorem by integrating the quantities $u = x + y$ and $v = 2y$ around the unit square shown.

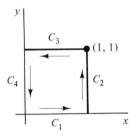

SOLUTION: Let us integrate around the closed curve C formed by the four sides of the squares. We have

$$\oint u\,dx - v\,dy = \int_{C_1} u\,dx - v\,dy + \int_{C_2} u\,dx - v\,dy$$

$$+ \int_{C_3} u\,dx - v\,dy + \int_{C_4} u\,dx - v\,dy$$

$$= \int_0^1 x\,dx + \int_0^1 -2y\,dy + \int_1^0 (x + 1)\,dx + \int_1^0 -2y\,dy,$$

where along C_1, $dy = 0$ and $y = 0$; along C_2, $dx = 0$; along C_3, $dy = 0$ and $y = 1$; and along C_4, $dx = 0$. The equation above is integrated to give

$$\oint u\,dx - v\,dy = \tfrac{1}{2} - 1 - (\tfrac{1}{2} + 1) + 1 = -1.$$

Now, using Green's theorem, let us evaluate the double integral

$$-\iint \left(\frac{\partial v}{\partial x} + \frac{\partial u}{\partial y} \right) dx\,dy.$$

Using $\partial v/\partial x = 0$ and $\partial u/\partial y = 1$, there results

$$-\iint \left(\frac{\partial v}{\partial x} + \frac{\partial u}{\partial y} \right) dx\,dy = -\iint (1)\,dx\,dy = -\text{area} = -1.$$

For the functions $u(x,y)$ and $v(x,y)$ of this example we have verified Green's theorem. ∎

---

*A simply connected region is one in which any closed curve contained in the region can be shrunk to zero without passing through points not in the region. A circular ring (like a washer) is not simply connected. A region that is not simply connected is multiply connected.

PROBLEMS

Find convenient parametric representation of each equation.

1. $\dfrac{x^2}{a^2} - \dfrac{y^2}{b^2} = 1$ **2.** $\dfrac{x^2}{a^2} + \dfrac{y^2}{b^2} = 1$

3. $y = 2x - 1$

Integrate each function around the closed curve indicated and compare with the double integral of Eq. 10.5.20.

4. $u = y$, $v = x$ around the unit square as in Example 10.5.5.

5. $u = y$, $v = -x$ around the unit circle with center at the origin.

6. $u = x^2 - y^2$, $v = -2xy$ around the triangle with vertices at $(0,0)$ $(2,0)$, $(2,2)$.

7. $u = x + 2y$, $v = x^2$ around the triangle of Problem 6.

8. $u = y^2$, $v = -x^2$ around the unit circle with center at the origin.

To show that line integrals are, in general, dependent on the limits of integration, evaluate each line integral.

9. $\displaystyle\int_{0,0}^{2,2} (x - iy)\, dz$ along a straight line connecting the two points.

$\displaystyle\int_{0,0}^{2,2} (x - iy)\, dz$ along the x axis to the point $(2,0)$ and then vertically to $(2,2)$.

10. $\displaystyle\int_{0,0}^{0,2} (x^2 + y^2)\, dz$ along the y axis.

$\displaystyle\int_{0,0}^{0,2} (x^2 + y^2)\, dz$ along the x axis to the point $(2,0)$, then along a circular arc.

To verify that the line integral of an analytic function is independent of the path, evaluate each line integral.

11. $\displaystyle\int_{0,0}^{2,2} z\, dz$ along a straight line connecting the two points.

$\displaystyle\int_{0,0}^{2,2} z\, dz$ along the z axis to the point $(2,0)$ and then vertical.

12. $\displaystyle\int_{0,0}^{0,2} z^2\, dz$ along the x axis to the point $(2,0)$ and then along a circular arc.

$\displaystyle\int_{0,0}^{0,2} z^2\, dz$ along the y axis.

Which of the following sets are simply connected?

13. The xy plane.   **14.** All z except z negative.

15. $|z| > 1$.   **16.** $0 < |z| < 1$.

17. Re $z \geq 0$.   **18.** Re $z \geq 0$ and Im $z \geq 0$.

19. All z such that $0 < \arg z < \pi/4$.

20. All z such that $0 < \arg z < \pi/4$ and $|z| > 1$.

21. Im $z > 0$ and $|z| > 1$. (Compare with Problem 14.)

10.6   Cauchy's Integral Theorem

Now let us investigate the line integral $\oint_C f(z)\, dz$, where $f(z)$ is an analytic function within a simply connected region R enclosed by the simple closed contour C. From Eq. 10.5.11, we have

$$\oint_C f(z)\, dz = \oint_C (u\, dx - v\, dy) + i\oint_C (v\, dx + u\, dy), \quad (10.6.1)$$

which we have used as the definition of $\oint f(z)\, dz$. Green's theorem allows us to transform the above into

$$\oint_C f(z)\, dz = -\iint_R \left(\frac{\partial v}{\partial x} + \frac{\partial u}{\partial y} \right) dx\, dy - i\iint_R \left(-\frac{\partial u}{\partial x} + \frac{\partial v}{\partial y} \right) dx\, dy.$$

$$(10.6.2)$$

Using the Cauchy–Reimann equations 10.4.8 we arrive at *Cauchy's integral theorem,*

$$\oint_C f(z)\, dz = 0.$$

We present it as a theorem:

Theorem 10.4:   *Let C be a simply closed contour enclosing a region R in which f(z) is analytic. Then*

$$\oint_C f(z)\, dz = 0. \quad (10.6.3)$$

If we divide the closed curve C into two parts, as shown in Fig. 10.12, Cauchy's integral theorem can be written as

$$\oint_C f(z)\, dz = \underbrace{\int_a^b f(z)\, dz}_{\text{along } C_1} + \underbrace{\int_b^a f(z)\, dz}_{\text{along } C_2}$$

$$= \underbrace{\int_a^b f(z)\, dz}_{\text{along } C_1} - \underbrace{\int_a^b f(z)\, dz}_{\text{along } C_2}, \quad (10.6.4)$$

where we have reversed the order of the integration (i.e., the direction along the contour). Thus, we have

$$\underbrace{\int_a^b f(z)\, dz}_{\text{along } C_1} = \underbrace{\int_a^b f(z)\, dz}_{\text{along } C_2}, \quad (10.6.5)$$

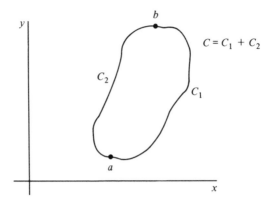

FIGURE 10.12.   Two paths from *a* to *b* enclosing a simply connected region.

showing that the value of a line integral between two points is independent of the path provided that the $f(z)$ is analytic throughout a region containing the paths. In Fig. 10.12, it is sufficient to assume that $f(z)$ is analytic in the first quadrant, for example. In Example 10.5.1 we found that

$$\int_0^{1+i} z^2 \, dz = -\tfrac{2}{3} + \tfrac{2}{3} i,$$

regardless of whether the integration is taken along the line joining 0 to $1 + i$ or the polygonal line joining 0 to 1 and then to $1 + i$. Since z^2 is analytic everywhere, this result is a consequence of Eq. 10.6.5. Indeed, we can assert that this integral is independent of the path from $(0, 0)$ to $(1, 1)$.

10.6.1   INDEFINITE INTEGRALS

The indefinite integral

$$F(z) = \int_{z_0}^{z} f(w) \, dw \tag{10.6.6}$$

defines F as a function of z as long as the contour joining z_0 to z lies entirely within a domain D which is simply connected and within which $f(z)$ is analytic. As we might reasonably expect

$$F'(z) = f(z), \qquad \text{in } D, \tag{10.6.7}$$

which means that $F(z)$ itself is analytic in D. To see how this comes about, consider the difference quotient

$$\frac{\Delta F}{\Delta z} = \frac{F(z + \Delta z) - F(z)}{\Delta z}$$

$$= \frac{1}{\Delta z} \left(\int_{z_0}^{z+\Delta z} f(w) \, dw - \int_{z_0}^{z} f(w) \, dw \right)$$

$$= \frac{1}{\Delta z} \int_{z}^{z+\Delta z} f(w) \, dw. \tag{10.6.8}$$

Also,

$$f(z) = \frac{1}{\Delta z} \int_{z}^{z+\Delta z} f(z) \, dw, \tag{10.6.9}$$

which follows from Eq. 10.5.13 and Example 10.5.4. We subtract Eq. 10.6.9 from Eq. 10.6.8 to obtain

$$\frac{\Delta F(z)}{\Delta z} - f(z) = \frac{1}{\Delta z} \int_{z}^{z+\Delta z} [f(w) - f(z)] \, dw. \tag{10.6.10}$$

Since the integral in Eq. 10.6.10 is independent of the path between z and $z + \Delta z$ we take this path as linear. As $\Delta z \to 0, f(w) \to f(z)$ and hence, for any $\epsilon > 0$, we can be assured that

$$|f(w) - f(z)| \le \epsilon \qquad \text{for } |\Delta z| \text{ small.} \tag{10.6.11}$$

Thus,

$$\left| \frac{\Delta F(z)}{\Delta z} - f(z) \right| = \frac{1}{|\Delta z|} \left| \int_{z}^{z+\Delta z} [f(w) - f(z)] \, dw \right|$$

$$\le \frac{|\Delta z|}{|\Delta z|} \epsilon = \epsilon \tag{10.6.12}$$

from Theorem 10.2. Clearly, $\epsilon \to 0$ as $\Delta z \to 0$. Hence,

$$\lim_{\Delta z \to 0} \frac{\Delta F(z)}{\Delta z} = F'(z) \tag{10.6.13}$$

by definition and $F'(z) = f(z)$ from Eq. 10.6.12.

The identity

$$\int_{a}^{b} f(z) \, dz = \int_{z_0}^{b} f(z) \, dz - \int_{z_0}^{a} f(z) \, dz$$

$$= F(b) - F(a) \tag{10.6.14}$$

is the familiar formula from elementary calculus. The importance of Eq. 10.6.14 is this: The contour integral $\int_{a}^{b} f(z) \, dz$ may be evaluated by finding an antiderivative $F(z)$ (a function satisfying $F' = f$) and computing $[F(b) - F(a)]$ instead of parameterizing the arc joining a to b and evaluating the resulting real integrals. Compare the next example with Example 10.5.1.

Example 10.6.1: Evaluate the integral

$$\int_0^{1+i} z^2 \, dz.$$

SOLUTION: Let

$$F(z) = \frac{z^3}{3}.$$

Then, since $F'(z) = z^2$, we have

$$\int_0^{1+i} z^2 \, dz = F(1+i) - F(0)$$

$$= \frac{(1+i)^3}{3} - 0 = -\frac{2}{3} + \frac{2}{3}i,$$

as expected from Example 10.5.1. Note the more general result:

$$\int_0^z w^2 \, dw = F(z) - F(0) = \frac{z^3}{3}$$

for each z. ∎

10.6.2 EQUIVALENT CONTOURS

Cauchy's integral theorem enables us to replace an integral about an arbitrary simple closed contour by an integral about a more conveniently shaped region, often a circle. Consider the integral $\oint_{C_1} f(z) \, dz$, where the contour C_1 is portrayed in Fig. 10.13. We call C_2 an equivalent contour to C_1 if

$$\oint_{C_1} f(z) \, dz = \oint_{C_2} f(z) \, dz. \tag{10.6.15}$$

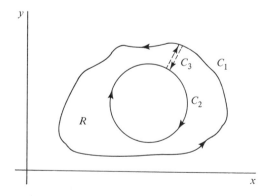

FIGURE 10.13. Equivalent contours C_1 and C_2.

This raises the question: Under what circumstances are C_1 and C_2 equivalent contours?

Suppose that $f(z)$ is analytic in the region bounded by C_1 and C_2 and on these contours, as in Fig. 10.13. Then introduce the line segment C_3 joining C_1 to C_2. Let C be the contour made up of C_1 (counterclockwise), C_3 to C_2, C_2 (clockwise), and C_3 from C_2 to C_1. By Cauchy's integral theorem,

$$\oint_C f(z)\, dz = 0. \tag{10.6.16}$$

However, by construction of C

$$\oint_C f(z)\, dz = \int_{C_1} f(z)\, dz + \int_{C_3} f(z)\, dz - \int_{C_2} f(z)\, dz - \int_{C_3} f(z)\, dz. \tag{10.6.17}$$

So, using Eq. 10.6.16 in Eq. 10.6.17, we see that C_2 is equivalent to C_1 since the two integrals on C_3 cancel.

Example 10.6.2:   Evaluate the integral $\oint f(z)\, dz$ around the circle of radius 2 with center at the origin if $f(z) = 1/(z - 1)$.

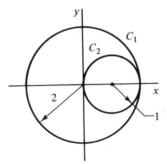

SOLUTION: The given function $f(z)$ is not analytic at $z = 1$, a point in the interior domain defined by the contour C_1. However, $f(z)$ is analytic between and on the two circles. Hence,

$$\oint_{C_1} \frac{dz}{z - 1} = \oint_{C_2} \frac{dz}{z - 1}.$$

The contour C_2 is a unit circle centered at $z = 1$. For this circle we have

$$z - 1 = e^{i\theta} \quad \text{and} \quad dz = ie^{i\theta}\, d\theta,$$

where θ is now measured with respect to a radius emanating from $z = 1$. The integral becomes

$$\oint_{C_1} \frac{dz}{z - 1} = \oint_{C_2} \frac{dz}{z - 1} = \int_0^{2\pi} \frac{ie^{i\theta}\, d\theta}{e^{i\theta}} = 2\pi i.$$

Observe that this integration is independent of the radius of the circle with center at $z = 1$; a circle of any radius would serve our purpose. Often we choose a circle of radius ϵ, a very small radius. Also, note that the integration around any curve enclosing the point $z = 1$, whether it is a circle or not, would give the value $2\pi i$.   ∎

PROBLEMS

Evaluate $\oint f(z)\, dz$ for each function, where the path of integration is the unit circle with center at the origin.

1. e^z **2.** $\sin z$ **3.** $1/z^3$ **4.** $\dfrac{1}{z - 2}$

5. $1/\bar{z}$ **6.** $\dfrac{1}{z^2 - 5z + 6}$

Evaluate $\oint f(z)\, dz$ by direct integration using each function, when the path of integration is the circle with radius 4, center at the origin.

7. $1/z$ **8.** $\dfrac{1}{z^2 - 5z + 6}$ **9.** $\dfrac{1}{z - 1}$ **10.** $\dfrac{1}{z^2 - 4}$

11. $z^2 + 1/z^2$ **12.** $\dfrac{z}{z - 1}$

10.7 Cauchy's Integral Formulas

We now apply the results of Section 10.6 to the integral

$$\oint_C \frac{f(z)}{z - z_0}\, dz. \tag{10.7.1}$$

We suppose that C is a simple closed contour defining the domain D as its interior. The point z_0 is in D and $f(z)$ is assumed analytic throughout D and on C. Figure 10.14 displays this situation. Since the integrand in Eq. 10.7.1

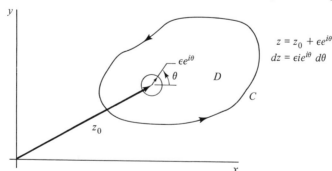

FIGURE 10.14.   Small-circle equivalent to the curve C.

has a singular point at z_0, Cauchy's integral theorem is not directly applicable. However, as we have seen in Section 10.6,

$$\oint_C \frac{f(z)}{z - z_0} \, dz = \oint_{circle} \frac{f(z)}{z - z_0} \, dz, \qquad (10.7.2)$$

where the circle is as shown in Fig. 10.14. The parameterization of the small circle with radius ϵ leads to

$$\oint_{circle} \frac{f(z)}{z - z_0} \, dz = \int_0^{2\pi} \frac{f(z_0 + \epsilon e^{i\theta})}{\epsilon e^{i\theta}} \epsilon e^{i\theta} i \, d\theta$$

$$= i \int_0^{2\pi} f(z_0 + \epsilon e^{i\theta}) \, d\theta. \qquad (10.7.3)$$

Hence, using this equation in Eq. 10.7.2, we find

$$\oint_C \frac{f(z)}{z - z_0} \, dz = i \int_0^{2\pi} f(z_0 + \epsilon e^{i\theta}) \, d\theta. \qquad (10.7.4)$$

Now, as $\epsilon \to 0$, we have $f(z_0 + \epsilon e^{i\theta}) \to f(z_0)$, which suggests that

$$i \int_0^{2\pi} f(z_0 + \epsilon e^{i\theta}) \, d\theta = i \int_0^{2\pi} f(z_0) \, d\theta = i f(z_0) \, 2\pi. \qquad (10.7.5)$$

Hence, we conjecture that

$$\oint_C \frac{f(z)}{z - z_0} \, dz = 2\pi i f(z_0). \qquad (10.7.6)$$

This is *Cauchy's integral formula,* usually written as

$$f(z_0) = \frac{1}{2\pi i} \oint_C \frac{f(z)}{z - z_0} \, dz. \qquad (10.7.7)$$

We prove Eq. 10.7.5, and thereby Eq. 10.7.7, by examining

$$\left| \int_0^{2\pi} [f(z_0 + \epsilon e^{i\theta}) - f(z_0)] \, d\theta \right| \leq M 2\pi \qquad (10.7.8)$$

by Theorem 10.2. However,

$$M = \max |f(z_0 + \epsilon e^{i\theta}) - f(z_0)| \qquad (10.7.9)$$

around the small circle $z - z_0 = \epsilon e^{i\theta}$. Since $f(z)$ is analytic, it is continuous and so $M \to 0$ as $\epsilon \to 0$. Therefore, inequality 10.7.8 actually implies the equality

$$\int_0^{2\pi} [f(z_0 + \epsilon e^{i\theta}) - f(z_0)] \, d\theta = 0 \qquad (10.7.10)$$

and Eq. 10.7.5 is proved.

We can obtain an expression for the derivative of $f(z)$ at z_0 by using Cauchy's integral formula in the definition of a derivative as follows:

$$
\begin{aligned}
f'(z_0) &= \lim_{\Delta z_0 \to 0} \frac{f(z_0 + \Delta z_0) - f(z_0)}{\Delta z_0} \\
&= \lim_{\Delta z_0 \to 0} \frac{1}{\Delta z_0} \left[\frac{1}{2\pi i} \oint_C \frac{f(z)\, dz}{z - z_0 - \Delta z_0} - \frac{1}{2\pi i} \oint_C \frac{f(z)}{z - z_0}\, dz \right] \\
&= \lim_{\Delta z_0 \to 0} \frac{1}{\Delta z_0} \left[\frac{1}{2\pi i} \oint_C f(z) \left(\frac{1}{z - z_0 - \Delta z_0} - \frac{1}{z - z_0} \right) dz \right] \\
&= \lim_{\Delta z_0 \to 0} \frac{1}{\Delta z_0} \left[\frac{\Delta z_0}{2\pi i} \oint_C \frac{f(z)\, dz}{(z - z_0 - \Delta z_0)(z - z_0)} \right] \\
&= \frac{1}{2\pi i} \oint_C \frac{f(z)}{(z - z_0)^2}\, dz.
\end{aligned}
\tag{10.7.11}
$$

(This last equality needs to be proved in a manner similar to the proof of Cauchy's integral formula, 10.7.7). In a like manner we can show that

$$
f''(z_0) = \frac{2!}{2\pi i} \oint_C \frac{f(z)}{(z - z_0)^3}\, dz
\tag{10.7.12}
$$

or, in general

$$
f^{(n)}(z_0) = \frac{n!}{2\pi i} \oint_C \frac{f(z)}{(z - z_0)^{n+1}}\, dz.
\tag{10.7.13}
$$

We often refer to the family of formulas in Eq. 10.7.13 as *Cauchy's integral formulas*.

Cauchy's integral formula 10.7.7 allows us to determine the value of an analytic function at any point z_0 interior to a simply connected region by integrating around a curve C surrounding the region. Only values of the function on the boundary are used. Thus, we note that if an analytic function is prescribed on the entire boundary of a simply connected region, the function and all its derivatives can be determined at all interior points. We can write Eq. 10.7.7 in the alternative form

$$
f(z) = \frac{1}{2\pi i} \oint_C \frac{f(w)}{w - z}\, dw,
\tag{10.7.14}
$$

where z is any interior point such as that shown in Fig. 10.15. The complex variable w is simply a dummy variable of integration that disappears in the integration process. Cauchy's integral formula is often used in this form.

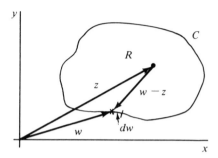

FIGURE 10.15.   Integration variables for Cauchy's integral theorem.

Example 10.7.1:   Find the value of the integral $\oint z^2/(z^2 - 1)\, dz$ around the unit circle with center at (a) $z = 1$, (b) $z = -1$, and (c) $z = \frac{1}{2}$.

SOLUTION: Using Cauchy's integral formula (Eq. 10.7.7), we must make sure that $f(z)$ is analytic in the unit circle, and that z_0 lies within the circle.
(a) With the center of the unit circle at $z = 1$, we write

$$\oint \frac{z^2}{z^2 - 1}\, dz = \oint \frac{z^2/(z + 1)}{z - 1}\, dz,$$

where we recognize that

$$f(z) = \frac{z^2}{z + 1}.$$

This function in analytic at $z = 1$ and in the unit circle. Hence, at that point

$$f(1) = \tfrac{1}{2},$$

and we have

$$\oint \frac{z^2}{z^2 - 1}\, dz = 2\pi i\left(\tfrac{1}{2}\right) = \pi i.$$

(b) with the center of the unit circle at $z = -1$, we write

$$\oint \frac{z^2}{z^2 - 1}\, dz = \oint \frac{z^2/(z - 1)}{z + 1}\, dz$$

where

$$f(z) = \frac{z^2}{z - 1} \quad \text{and} \quad f(-1) = -\tfrac{1}{2}.$$

There results

$$\oint \frac{z^2}{z^2 - 1}\, dz = 2\pi i\left(-\tfrac{1}{2}\right) = -\pi i.$$

(c) Rather than integrating around the unit circle with center at $z = \frac{1}{2}$, we can

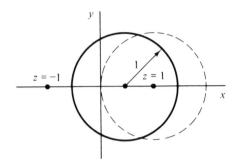

integrate around any curve enclosing the point $z = 1$ just so the curve does not enclose the other singular point at $z = -1$. Obviously, the unit circle of part (a) is an acceptable alternative curve. Hence,

$$\oint \frac{z^2}{z^2 - 1} \, dz = \pi i. \quad \blacksquare$$

Example 10.7.2:   Evaluate the integrals

$$\oint \frac{z^2 + 1}{(z - 1)^2} \, dz \quad \text{and} \quad \oint \frac{\cos z}{z^3} \, dz$$

around the circle $|z| = 2$.

SOLUTION: Using Eq. 10.7.11, we can write the first integral as

$$\oint \frac{z^2 + 1}{(z - 1)^2} \, dz = 2\pi i f'(1),$$

where

$$f(z) = z^2 + 1 \quad \text{and} \quad f'(z) = 2z.$$

Then

$$f'(1) = 2.$$

The value of the integral is then determined to be

$$\oint \frac{z^2 + 1}{(z - 1)^2} \, dz = 2\pi i (2) = 4\pi i.$$

For the second integral of the example, we have

$$\oint \frac{\cos z}{z^3} \, dz = \frac{2\pi i}{2!} f''(0),$$

where

$$f(z) = \cos z \quad \text{and} \quad f''(z) = -\cos z.$$

At the origin

$$f''(0) = -1.$$

Example 10.8.2:   Find the Taylor series representation of $\ln(1 + z)$ by noting that

$$\frac{d}{dz}\ln(1 + z) = \frac{1}{1 + z}.$$

SOLUTION: First, let us write the Taylor series expansion of $1/(1 + z)$. It is, using the results of Example 10.8.1(c)

$$\frac{1}{1 + z} = \frac{1}{1 - (-z)} = 1 - z + z^2 - z^3 + \cdots.$$

Now, we can perform the integration

$$\int d[\ln(1 + z)]dz = \int \frac{1}{1 + z}\,dz,$$

using the series expansion of $1/(1 + z)$ to obtain

$$\ln(1 + z) = \int \frac{1}{1 + z}\,dz = z - \frac{z^2}{2} + \frac{z^3}{3} - \frac{z^4}{4} + \cdots + C.$$

The constant of integration $C = 0$, since when $z = 0$, $\ln(1) = 0$. The power-series expansion is finally

$$\ln(1 + z) = z - \frac{z^2}{2} + \frac{z^3}{2} - \cdots.$$

This series is valid for $|z| < 1$ since a singularity exists at $z = -1$. ■

Example 10.8.3:   Determine the Taylor series expansion of

$$f(z) = \frac{1}{(z^2 - 3z + 2)}$$

about the origin.

SOLUTION: First, represent the function $f(z)$ as partial fractions; that is,

$$\frac{1}{z^2 - 3z + 2} = \frac{1}{(z - 2)(z - 1)} = \frac{1}{z - 2} - \frac{1}{z - 1}.$$

The series representations are then, using the results of Example 10.8.1(c),

$$\frac{1}{z - 1} = -\frac{1}{1 - z} = -(1 + z + z^2 + \cdots)$$

$$\frac{1}{z - 2} = -\frac{1}{2}\left(\frac{1}{1 - z/2}\right) = -\frac{1}{2}\left[1 + \frac{z}{2} + \left(\frac{z}{2}\right)^2 + \left(\frac{z}{2}\right)^3 + \cdots\right]$$

$$= -\frac{1}{2}\left[1 + \frac{z}{2} + \frac{z^2}{4} + \frac{z^3}{8} + \cdots\right].$$

Finally, the difference of the two series is

$$\frac{1}{z^2 - 3z - 2} = \frac{1}{2} + \frac{3}{4}z + \frac{7}{8}z^2 + \frac{15}{16}z^3 + \cdots .$$

We could also have multiplied the two series together to obtain the same result.   ∎

Example 10.8.4:   Find the Taylor series expansion of

$$f(z) = \frac{1}{z^2 - 9}$$

by expanding about the point $z = 1$.

SOLUTION: We write the function $f(z)$ in partial fractions as

$$\frac{1}{z^2 - 9} = \frac{1}{(z - 3)(z + 3)} = \frac{1}{2}\left(\frac{1}{z - 3}\right) - \frac{1}{6}\left(\frac{1}{z + 3}\right)$$

$$= -\frac{1}{6}\left[\frac{1}{2 - (z - 1)}\right] - \frac{1}{6}\left[\frac{1}{4 + (z - 1)}\right]$$

$$= -\frac{1}{12}\left[\frac{1}{1 - \dfrac{z - 1}{2}}\right] - \frac{1}{24}\left[\frac{1}{1 - \left(-\dfrac{z - 1}{4}\right)}\right].$$

Now, we can expand in a Taylor series as

$$\frac{1}{z^2 - 9} = -\frac{1}{12}\left[1 + \frac{z - 1}{2} + \left(\frac{z - 1}{2}\right)^2 + \left(\frac{z - 1}{2}\right)^3 + \cdots\right]$$

$$-\frac{1}{24}\left[1 - \frac{z - 1}{4} + \left(\frac{z - 1}{4}\right)^2 - \left(\frac{z - 1}{4}\right)^3 + \cdots\right]$$

$$= -\frac{1}{8} - \frac{1}{32}(z - 1) - \frac{3}{128}(z - 1)^2 - \frac{5}{512}(z - 1)^3 + \cdots .$$

The nearest singularity is at the point $z = 3$; hence, the radius of convergence is 2; that is $|z - 1| < 2$. This is also obtained from the first ratio since $|(z - 1)/2| < 1$ or $|z - 1| < 2$. The second ratio is convergent if $|-(z - 1)/4| < 1$ or $|z - 1| < 4$; thus, it is the first ratio that limits the radius of convergence.   ∎

PROBLEMS

Using the Taylor series, find the expansion about the origin for each function. State the radius of convergence.

1. $\cos z$ **2.** $\dfrac{1}{1 + z}$ **3.** $\ln(1 + z)$ **4.** $\dfrac{z - 1}{z + 1}$

5. $\cosh z$ **6.** $\sinh z$

For the function $1/(z - 2)$, determine the Taylor series expansion about each of the given point. Use the known series expansion for $1/(1 - z)$. State the radius of covergence for each.

7. $z = 0$ **8.** $z = 1$ **9.** $z = i$ **10.** $z = -1$

11. $z = 3$ **12.** $z = -2i$

Using known series expansions, find the Taylor series expansion about the origin of each of the following.

13. $\dfrac{1}{1 - z^2}$ **14.** $\dfrac{z - 1}{1 + z^3}$ **15.** $\dfrac{z^2 + 3}{2 - z}$ **16.** $\dfrac{1}{z^2 - 3z - 4}$

17. e^{-z^2} **18.** e^{2-z} **19.** $\sin \pi z$ **20.** $\sin z^2$

21. $\dfrac{\sin z}{1 - z}$ **22.** $e^z \cos z$ **23.** $\tan z$ **24.** $\dfrac{\sin z}{e^{-z}}$

What is the Taylor series expansion about the origin for each of the following?

25. $\displaystyle\int_0^z e^{-w^2}\, dw$ **26.** $\displaystyle\int_0^z \sin w^2\, dw$ **27.** $\displaystyle\int_0^z \dfrac{\sin w}{w}\, dw$ **28.** $\displaystyle\int_0^z \cos w^2\, dw$

29. Find the Taylor series expansion about the origin of $f(z) = \tan^{-1} z$ by recognizing that $f'(z) = 1/(1 + z^2)$.

Determine the Taylor series expansion of each function about the point $z = a$.

30. $e^z, \ a = 1$ **31.** $\dfrac{1}{1 - z}, \ a = 2$ **32.** $\sin z, \ a = \dfrac{\pi}{2}$

33. $\ln z, \ a = 1$ **34.** $\dfrac{1}{z^2 - z - 2}, \ a = 0$ **35.** $\dfrac{1}{z^2}, \ a = 1$

10.9   *Laurent Series*

There are many applications in which we wish to expand a function $f(z)$ in a series about a point $z = a$, which is a singular point. Consider the annulus shown in Fig. 10.17a. The function $f(z)$ is analytic in the annular region; however, there may be singular points inside the smaller circle or outside the larger circle. The possibility of a singular point inside the smaller circle bars us from expanding in a Taylor series, since the function $f(z)$ must be analytic at all interior points. We can apply Cauchy's integral formula to the multiply connected region by cutting the region as shown in Fig. 10.17b, thereby forming a simply connected region bounded by the curve C'. Cauchy's integral formula is then

$$f(z) = \frac{1}{2\pi i} \oint_{C'} \frac{f(w)}{w - z}\, dw$$

$$= \frac{1}{2\pi i} \oint_{C_2} \frac{f(w)}{w - z}\, dw - \frac{1}{2\pi i} \oint_{C_1} \frac{f(w)}{w - z}\, dw, \qquad (10.9.1)$$

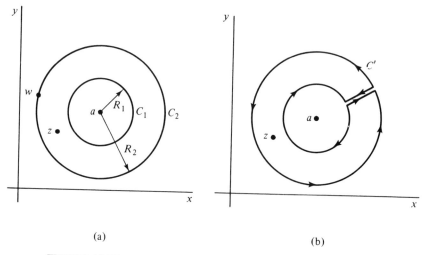

(a)   (b)

FIGURE 10.17.   Annular region inside of which a singular point exists.

where C_1 and C_2 are both traversed in the counterclockwise direction. The negative sign results because the direction of integration was reversed on C_1. Now, let us express the quantity $(w - z)^{-1}$ in the integrand of Eq. 10.9.1 in a form that results in positive powers of $(z - a)$ in the C_2 integration, and that results in negative powers in the C_1 integration. If no singular points exist inside C_1, then the coefficients of the negative powers will all be zero and a Taylor series will result. Doing this, we have

$$f(z) = \frac{1}{2\pi i} \oint_{C_2} \frac{f(w)}{w - a} \left[\frac{1}{1 - \dfrac{z - a}{w - a}} \right] dw$$

$$+ \frac{1}{2\pi i} \oint_{C_1} \frac{f(w)}{z - a} \left[\frac{1}{1 - \dfrac{w - a}{z - a}} \right] dw. \qquad (10.9.2)$$

By using arguments analogous to those used to prove the convergence of the Taylor series, we can show that Eq. 10.9.2 leads to

$$f(z) = a_0 + a_1(z - a) + a_2(z - a)^2 + \cdots$$
$$+ b_1(z - a)^{-1} + b_2(z - a)^{-2} + \cdots, \qquad (10.9.3)$$

where

$$a_n = \frac{1}{2\pi i} \oint_{C_2} \frac{f(w)}{(w - a)^{n+1}} \, dw, \qquad b_n = \frac{1}{2\pi i} \oint_{C_1} f(w)(w - a)^{n-1} \, dw.$$

$$(10.9.4)$$

The series expression 10.9.3 is a *Laurent series*. The integral expression for the coefficients a_n resembles the formulas for the derivatives of $f(z)$; but this is only superficial, for $f(z)$ may not be defined at $z = a$ and certainly $f(z)$ may not be analytic there. Note, however, that if $f(z)$ is analytic in the circle C_1, the integrand in the integral for b_n is everywhere analytic, requiring the b_n's to all be zero, a direct application of Cauchy's integral theorem. In this case the Laurent series reduces to a Taylor series.

The integral expressions 10.9.4 for the coefficients in the Laurent series are not normally used to find the coefficients. It is known that the series expansion is unique; hence, elementary techniques are usually used to find the Laurent series. This will be illustrated with an example. The region of convergence may be found, in most cases, by putting the desired $f(z)$ in the form $1/(1 - z^*)$ so that $|z^*| < 1$ establishes the region of convergence.

Example 10.9.1:   What is the Laurent series expansion of

$$f(z) = \frac{1}{z^2 - 3z + 2}$$

valid in each of the shaded regions shown?

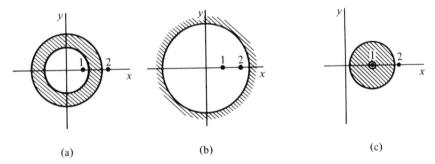

(a)   (b)   (c)

SOLUTION: (a) To obtain a Laurent series expansion in the shaded region of (a), we expand about the origin. We express the ratio in partial fractions as

$$\frac{1}{z^2 - 3z + 2} = \frac{1}{(z - 2)(z - 1)} = \frac{1}{z - 2} - \frac{1}{z - 1}$$

$$= -\frac{1}{2}\left(\frac{1}{1 - z/2}\right) - \frac{1}{z}\left(\frac{1}{1 - 1/z}\right).$$

The first fraction has a singularity at $z/2 = 1$ and can be expanded in a Taylor series that converges if $|z/2| < 1$ or $|z| < 2$. The second fraction has a singularity at $1/z = 1$ and can be expanded in a Laurent series that converges if $|1/z| < 1$ or $|z| > 1$. The two fractions are expressed in the appropriate series as

$$-\frac{1}{2}\left(\frac{1}{1 - z/2}\right) = -\frac{1}{2}\left[1 + \frac{z}{2} + \left(\frac{z}{2}\right)^2 + \left(\frac{z}{2}\right)^3 + \cdots\right]$$

$$= -\frac{1}{2} - \frac{z}{4} - \frac{z^2}{8} - \frac{z^3}{16} - \cdots$$

$$-\frac{1}{z}\left(\frac{1}{1 - 1/z}\right) = -\frac{1}{z}\left[1 + \frac{1}{z} + \left(\frac{1}{z}\right)^2 + \left(\frac{1}{z}\right)^3 + \cdots\right]$$

$$= -\frac{1}{z} - \frac{1}{z^2} - \frac{1}{z^3} - \frac{1}{z^4} - \cdots$$

where the first series is valid for $|z| < 2$ and the second series for $|z| > 1$. Adding the two expressions above yields the Laurent series

$$\frac{1}{z^2 - 3z + 2} = \cdots - \frac{1}{z^3} - \frac{1}{z^2} - \frac{1}{z} - \frac{1}{2} - \frac{z}{4} - \frac{z^2}{8} - \frac{z^3}{16} - \cdots,$$

which is valid in the region $1 < |z| < 2$.

(b) In the region exterior to the circle $|z| = 2$, we expand $1/(z - 1)$, as before,

$$\frac{1}{z - 1} = \frac{1}{z}\left(\frac{1}{1 - 1/z}\right) = \frac{1}{z} + \frac{1}{z^2} + \frac{1}{z^3} + \cdots,$$

which is valid if $|1/z| < 1$ or $|z| > 1$. Now, though, we write

$$\frac{1}{z - 2} = \frac{1}{z}\left(\frac{1}{1 - 2/z}\right) = \frac{1}{z}\left[1 + \frac{2}{z} + \left(\frac{2}{z}\right)^2 + \left(\frac{2}{z}\right)^3 + \cdots\right]$$

$$= \frac{1}{z} + \frac{2}{z^2} + \frac{4}{z^3} + \frac{8}{z^4} + \cdots,$$

which is valid if $|2/z| < 1$ or $|z| > 2$. The two series expansions above are thus valid for $|z| > 2$, and we have the Laurent series

$$\frac{1}{z^2 - 3z + 2} = \frac{1}{z^2} + \frac{3}{z^3} + \frac{7}{z^4} + \frac{15}{z^5} + \cdots,$$

valid in the region $|z| > 2$.

(c) To obtain a series expansion in the region $0 < |z - 1| < 1$, we expand about the point $z = 1$ and obtain

$$\frac{1}{z^2 - 3z + 2} = \frac{1}{z - 1}\left(-\frac{1}{2 - z}\right) = \frac{1}{z - 1}\left[\frac{-1}{1 - (z - 1)}\right]$$

$$= \frac{-1}{z - 1}[1 + (z - 1) + (z - 1)^2 + (z - 1)^3 + \cdots]$$

$$= -\frac{1}{z - 1} - 1 + (z - 1) + (z - 1)^2 + \cdots.$$

This Laurent series is valid if $0 < |z - 1| < 1$. ■

PROBLEMS

Expand each function in a Laurent series about the origin, convergent in the region $0 < |z| < R$. State the radius of convergence R.

1. $\dfrac{1}{z^2} \sin z$

2. $\dfrac{1}{z^2 - 2z}$

3. $\dfrac{1}{z(z^2 + 3z + 2)}$

4. $\dfrac{e^{z-1}}{z}$

For each function, find all Taylor series and Laurent series expansions about the point $z = a$ and state the region of convergence for each.

5. $\dfrac{1}{z}$, $\quad a = 1$

6. $e^{1/z}$, $\quad a = 0$

7. $\dfrac{1}{1 - z}$, $\quad a = 0$

8. $\dfrac{1}{1 - z}$, $\quad a = 1$

9. $\dfrac{1}{1 - z}$, $\quad a = 2$

10. $\dfrac{1}{z(z - 1)}$, $\quad a = 0$

11. $\dfrac{z}{1 - z^2}$, $\quad a = 1$

12. $\dfrac{1}{z^2 + 1}$, $\quad a = i$

13. $\dfrac{1}{(z + 1)(z - 2)}$, $\quad a = 0$

14. $\dfrac{1}{(z + 1)(z - 2)}$, $\quad a = -1$

15. $\dfrac{1}{(z + 1)(z - 2)}$, $\quad a = 2$

10.10   Residues

In this section we shall present a technique that is especially useful when evaluating certain types of real integrals. Suppose that a function $f(z)$ is singular at the point $z = a$ and is analytic at all other points within some circle with center at $z = a$. Then $f(z)$ can be expanded in the Laurent series (see Eq. 10.9.3)

$$f(z) = \cdots + \frac{b_m}{(z - a)^m} + \cdots + \frac{b_2}{(z - a)^2} + \frac{b_1}{z - a}$$
$$+ a_0 + a_1(z - a) + \cdots . \qquad (10.10.1)$$

Three cases arise. First, all the coefficients b_1, b_2, . . . are zero. Then $f(z)$ is said to have a *removable singularity*. The function $(\sin z)/z$ has a removable singularity at $z = 0$. Second, only a finite number of the b_n are nonzero. Then $f(z)$ has a *pole* at $z = a$. If $f(z)$ has a pole, then

$$f(z) = \frac{b_m}{(z - a)^m} + \cdots + \frac{b_1}{z - a} + a_0 + a_1(z - a) + \cdots \qquad (10.10.2)$$

where $b_m \neq 0$. In this case we say that the pole at $z = a$ is of *order m*. Third, if infinitely many b_n are not zero, then $f(z)$ has an *essential singularity* at $z = a$. The function $e^{1/z}$ has the Laurent expansion

$$e^{1/z} = 1 + \frac{1}{z} + \frac{1}{2! \, z^2} + \cdots + \frac{1}{n! \, z^n} + \cdots \qquad (10.10.3)$$

valid for all z, $|z| > 0$. The point $z = 0$ is an essential singularity of $e^{1/z}$. It is interesting to observe that rational fractions have poles or removable singularities as their only singularities.

From the expression 10.9.4 we see that

$$b_1 = \frac{1}{2\pi i} \oint_{C_1} f(w)\, dw. \tag{10.10.4}$$

Hence, the integral of a function $f(z)$ about some connected curve surrounding one singular point is given by

$$\oint_{C_1} f(z)\, dz = 2\pi i b_1 \tag{10.10.5}$$

where b_1 is the coefficient of the $(z - a)^{-1}$ term in the Laurent series expansion at the point $z = a$. The quantity b_1 is called the *residue* of $f(z)$ at $z = a$. Thus, to find the integral of a function about a singular point*, we simply find the Laurent series expansion and use the relationship 10.10.5. An actual integration is not necessary. If more than one singularity exists within the closed curve C, we make it simply connected by cutting it as shown in Fig. 10.18. Then an application of Cauchy's integral theorem gives

$$\oint_C f(z)\, dz + \oint_{C_1} f(z)\, dz + \oint_{C_2} f(z)\, dz + \oint_{C_3} f(z)\, dz = 0,$$

$$\tag{10.10.6}$$

since $f(z)$ is analytic at all points in the region outside the small circles and inside C. If we reverse the direction of integration on the integrals around the circles, there results

$$\oint_C f(z)\, dz = \oint_{C_1} f(z)\, dz + \oint_{C_2} f(z)\, dz + \oint_{C_3} f(z)\, dz. \tag{10.10.7}$$

In terms of the residues at the points, we have Cauchy's *residue theorem*,

$$\oint_C f(z)\, dz = 2\pi i[(b_1)_{a_1} + (b_1)_{a_2} + (b_1)_{a_3}] \tag{10.10.8}$$

where the b_1's are coefficients of the $(z - a)^{-1}$ terms of the Laurent series expansions at each of the points.

Another technique, often used to find the residue at a particular singular point, is to multiply the Laurent series (10.10.2) by $(z - a)^m$, to obtain

$(z - a)^m f(z) = b_m + b_{m-1}(z - a) + \cdots$

$$+ b_1(z - a)^{m-1} + a_0(z - a)^m + a_1(z - a)^{m+1} + \cdots. $$

$$\tag{10.10.9}$$

---

*Recall that we only consider isolated singularities.

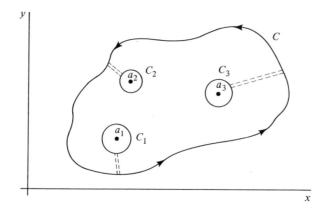

FIGURE 10.18.   Integration about a curve that surrounds singular points.

Now, if the series above is differentiated $(m - 1)$ times and we let $z = a$, the residue results; that is,

$$b_1 = \frac{1}{(m - 1)!} \left\{ \frac{d^{m-1}}{dz^{m-1}} [(z - a)^m f(z)] \right\}_{z=a}.$$

(10.10.10)

Obviously, the order of the pole must be known before this method is useful. If $m = 1$, no differentiation is required and the residue results from

$$\lim_{z \to a} (z - a) f(z).$$

The residue theorem can be used to evaluate certain real integrals. Several examples will be presented here. Consider the real integral

$$I = \int_0^{2\pi} g(\cos \theta, \sin \theta) \, d\theta$$

(10.10.11)

where $g(\cos \theta, \sin \theta)$ is a rational* function of $\cos \theta$ and $\sin \theta$ with no singularities in the interval $0 \le \theta < 2\pi$. Let us make the substitution

$$e^{i\theta} = z,$$

(10.10.12)

resulting in

$$\cos \theta = \frac{1}{2}(e^{i\theta} + e^{-i\theta}) = \frac{1}{2}\left(z + \frac{1}{z} \right)$$

$$\sin \theta = \frac{1}{2i}(e^{i\theta} - e^{-i\theta}) = \frac{1}{2i}\left(z - \frac{1}{z} \right)$$

(10.10.13)

$$d\theta = \frac{dz}{ie^{i\theta}} = \frac{dz}{iz}$$

---

*Recall that a rational function can be expressed as the ratio of two polynomials.

As θ ranges from, 0 to 2π, the complex variable z moves around the unit circle, as shown in Fig. 10.19, in the counterclockwise sense. The real integral now takes the form

$$I = \oint_C \frac{f(z)}{iz} \, dz. \tag{10.10.14}$$

The residue theorem can be applied to the integral above once $f(z)$ is given. All residues inside the unit circle must be accounted for.

FIGURE 10.19.   Paths of integration.

A second real integral that can be evaluted using the residue theorem is the integral

$$I = \int_{-\infty}^{\infty} f(x) \, dx \tag{10.10.15}$$

where $f(x)$ is the rational function

$$f(x) = \frac{p(x)}{q(x)} \tag{10.10.16}$$

and $q(x)$ has no real zeros and is of degree at least 2 greater than $p(x)$. Consider the corresponding integral

$$I_1 = \oint_C f(z) \, dz \tag{10.10.17}$$

where C is the closed path shown in Fig. 10.20. If C_1 is the semicircular part of curve C, Eq. 10.10.17 can be written as

$$I_1 = \int_{C_1} f(z) \, dz + \int_{-R}^{R} f(x) \, dx = 2\pi i \sum_{n=1}^{N} (b_1)_n, \tag{10.10.18}$$

where Cauchy's residue theorem has been used. In this equation, N represents the number of singularities in the upper half-plane contained within the semicircle. Let us now show that

$$\int_{C_1} f(z) \, dz \rightarrow 0 \tag{10.10.19}$$

as $R \rightarrow \infty$. Using Eq. 10.10.16 and the restriction that $q(z)$ is of degree at

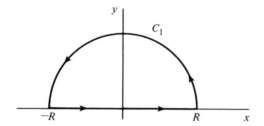

FIGURE 10.20.   Path of integration.

least 2 greater than $p(z)$, we have

$$|f(z)| = \left|\frac{p(z)}{q(z)}\right| = \frac{|p(z)|}{|q(z)|} \sim \frac{1}{R^2}. \qquad (10.10.20)$$

Then there results

$$\left|\int_{C_1} f(x)\, dz\right| \le |f_{max}|\pi R \sim \frac{1}{R}. \qquad (10.10.21)$$

from Theorem 10.2. As the radius R of the semicircle approaches ∞, we see that

$$\int_{C_1} f(z)\, dz \to 0. \qquad (10.10.22)$$

Finally,

$$\int_{-\infty}^{\infty} f(x)\, dx = 2\pi i \sum_{n=1}^{N} (b_1)_n \qquad (10.10.23)$$

where the b_1's include the residues of $f(z)$ at all singularities in the upper half-plane.

A third real integral that may be evaluated using the residue theorem is

$$I = \int_{-\infty}^{\infty} f(x) \sin mx\, dx \quad \text{or} \quad \int_{-\infty}^{\infty} f(x) \cos mx\, dx. \qquad (10.10.24)$$

Consider the complex integral

$$I_1 = \oint_C f(z) e^{imz}\, dz \qquad (10.10.25)$$

where m is positive and C is the curve of Fig. 10.20. If we limit ourselves to the upper half-plane so that $y \ge 0$,

$$|e^{imz}| = |e^{imx}||e^{-my}| = e^{-my} \le 1, \qquad (10.10.26)$$

We then have

$$|f(z)e^{imz}| = |f(z)||e^{imz}| \le |f(z)|. \qquad (10.10.27)$$

The remaining steps follow as in the previous example for $\int_{-\infty}^{\infty} f(z)\, dz$ using Fig. 10.20. This results in

$$\int_{-\infty}^{\infty} f(x)e^{imx}\, dx = 2\pi i \sum_{n=1}^{N} (b_1)_n \qquad (10.10.28)$$

where the b_1's include the residues of $[f(z)e^{imz}]$ at all singularities in the upper half-plane. Then the value of the integrals in Eq. 10.10.24 are either the real or imaginary parts of Eq. 10.10.28.

Example 10.10.1:   Find the value of the following integrals, where C is the circle $|z| = 2$.

(a) $\oint_C \dfrac{\cos z}{z^3}\, dz$ (b) $\oint_C \dfrac{dz}{z^2 + 1}$

(c) $\oint_C \dfrac{z^2 - 2}{z(z - 1)(z + 4)}\, dz$ (d) $\oint_C \dfrac{z}{(z - 1)^3(z + 3)}\, dz$

SOLUTION: (a) We expand the function $\cos z$ as

$$\cos z = 1 - \frac{z^2}{2!} + \frac{z^4}{4!} - \cdots .$$

The integrand is then

$$\frac{\cos z}{z^3} = \frac{1}{z^3} - \frac{1}{2z} + \frac{z}{4!} + \cdots .$$

The residue, the coefficient of the $1/z$ term, is

$$b_1 = -\frac{1}{2}.$$

Thus, the value of the integral is

$$\oint_C \frac{\cos z}{z}\, dz = 2\pi i \left(-\frac{1}{2} \right) = -\pi i.$$

(b) The integrand is factored as

$$\frac{1}{z^2 + 1} = \frac{1}{(z + i)(z - i)}.$$

Two singularities exist inside the circle of interest. The residue at each singularity is found to be

$$(b_1)_{z=i} = (z - i)\frac{1}{(z + i)(z - i)}\bigg|_{z=i} = \frac{1}{2i}$$

$$(b_1)_{z=-i} = (z + i)\frac{1}{(z + i)(z - i)}\bigg|_{z=-i} = -\frac{1}{2i}.$$

The value of the integral is

$$\oint_C \frac{dz}{z^2 + 1} = 2\pi i\left(\frac{1}{2i} - \frac{1}{2i}\right) = 0.$$

Moreover, this is the value of the integral around every curve that encloses the two poles.

(c) There are two poles of order 1 in the region of interest, one at $z = 0$ and the other at $z = 1$. The residue at each of these poles is

$$(b_1)_{z=0} = z\frac{z^2 - 2}{z(z - 1)(z + 4)}\bigg|_{z=0} = \frac{1}{2}$$

$$(b_1)_{z=1} = (z - 1)\frac{z^2 - 2}{z(z - 1)(z + 4)}\bigg|_{z=1} = -\frac{1}{5}.$$

The integral is

$$\oint_C \frac{z^2 - 2}{z(z - 1)(z + 4)} dz = 2\pi i\left(\frac{1}{2} - \frac{1}{5}\right) = \frac{3\pi i}{5}.$$

(d) There is one pole in the circle $|z| = 2$, a pole of order 3. The residue at that pole is (see Eq. 10.10.10)

$$b_1 = \frac{1}{2!} \frac{d^2}{dz^2}\left[(z - 1)^3 \frac{z}{(z - 1)^3(z + 3)}\right]_{z=1}$$

$$= \frac{1}{2} \frac{-6}{(z + 3)^3}\bigg|_{z=1} = -\frac{3}{64}.$$

The value of the integral is then

$$\oint_C \frac{z}{(z - 1)^3(z + 3)} dz = 2\pi i\left(-\frac{3}{64}\right) = -0.2945i. \quad \blacksquare$$

Example 10.10.2:   Evaluate the real integral

$$\int_0^{2\pi} \frac{d\theta}{(2 + \cos \theta)}.$$

SOLUTION: Using Eqs. 10.10.13, the integral is transformed as follows:

$$\int_0^{2\pi} \frac{d\theta}{2 + \cos \theta} = \oint_C \frac{dz/iz}{2 + \frac{1}{2}\left(z + \frac{1}{z}\right)} = -2i \oint \frac{dz}{z^2 + 4z + 1}$$

where C is the unit circle. The roots of the denominator are found to be

$$z = -2 \pm \sqrt{3}.$$

Hence, there is a zero at $z = -0.2679$ and at $z = -3.732$. The first of these zeros is located in the unit circle, so we must determine the residue at that zero; the second is outside the unit circle, so we ignore it. To find the residue, write the integrand as partial fractions

$$\frac{1}{z^2 + 4z + 1} = \frac{1}{(z + 0.2679)(z + 3.732)} = \frac{0.2887}{z + 0.2679} + \frac{-0.2887}{z + 3.732}.$$

The residue at the singularity in the unit circle is then the coefficient of the $(z + 0.2679)^{-1}$ term. It is 0.2887. Thus, the value of the integral is, using the residue theorem,

$$\int_0^{2\pi} \frac{d\theta}{2 + \cos\theta} = -2i(2\pi i \times 0.2887) = 3.628. \quad \blacksquare$$

Example 10.10.3:   Evaluate the real integral

$$\int_0^\infty \frac{dx}{(1 + x^2)}.$$

Note that the lower limit is zero.

SOLUTION: We consider the complex function $f(z) = 1/(1 + z^2)$. Two poles exist at the points where

$$1 + z^2 = 0.$$

They are

$$z_1 = i \quad \text{and} \quad z_2 = -i.$$

The first of these roots lies in the upper half-plane. The residue there is

$$(b_1)_{z=i} = (z - i)\frac{1}{(z - i)(z + i)}\bigg|_{z=i} = \frac{1}{2i}.$$

The value of the integral is then (refer to Eq. 10.10.23)

$$\int_{-\infty}^\infty \frac{dx}{1 + x^2} = 2\pi i\left(\frac{1}{2i}\right) = \pi.$$

Since the integrand is an even function,

$$\int_0^\infty \frac{dx}{1 + x^2} = \frac{1}{2}\int_{-\infty}^\infty \frac{dx}{1 + x^2}.$$

Hence,

$$\int_0^\infty \frac{dx}{1 + x^2} = \frac{\pi}{2}. \quad \blacksquare$$

Example 10.10.4:   Determine the value of the real integrals

$$\int_{-\infty}^\infty \frac{\cos x}{1 + x^2}\,dx \quad \text{and} \quad \int_{-\infty}^\infty \frac{\sin x}{1 + x^2}\,dx.$$

SOLUTION: To evaluate the given integrals refer to Eqs. 10.10.25 through 10.10.28. Here

$$I_1 = \oint_C \frac{e^{iz}}{1 + z^2} \, dz.$$

and C is the semicircle in Fig. 10.20. The quantity $(1 + z^2)$ has zeros as $z = \pm i$. One of these points is in the upper half-plane. The residue at $z = i$ is

$$(b_1)_{z=i} = (z - i)\frac{e^{iz}}{1 + z^2}\bigg|_{z=i} = \frac{e^{-1}}{2i} = -0.1839i.$$

The value of the integral is then

$$\int_{-\infty}^{\infty} \frac{e^{ix}}{1 + x^2} \, dx = 2\pi i(-0.1839i) = 1.188.$$

The integral can be rewritten as

$$\int_{-\infty}^{\infty} \frac{e^{ix}}{1 + x^2} \, dx = \int_{-\infty}^{\infty} \frac{\cos x}{1 + x^2} \, dx + i \int_{-\infty}^{\infty} \frac{\sin x}{1 + x^2} \, dx$$

Equating real and imaginary parts, we have

$$\int_{-\infty}^{\infty} \frac{\cos x}{1 + x^2} \, dx = 1.188 \quad \text{and} \quad \int_{-\infty}^{\infty} \frac{\sin x}{1 + x^2} \, dx = 0.$$

The result with $\sin x$ is not surprising since the integrand is an odd function, and hence

$$\int_0^{\infty} \frac{\sin x}{1 + x^2} \, dx = -\int_{-\infty}^0 \frac{\sin x}{1 + x^2} \, dx. \quad \blacksquare$$

It should be carefully noted that

$$\int_{-\infty}^{\infty} f(x) \, dx \qquad\qquad (10.10.29)$$

is an improper integral. Technically, this integral is defined as the following sum of limits, both of which must exist:

$$\int_{-\infty}^{\infty} f(x) \, dx = \lim_{R \to \infty} \int_0^R f(x) \, dx + \lim_{S \to \infty} \int_{-S}^0 f(x) \, dx. \quad (10.10.30)$$

When we use the residue theorem we are in fact computing

$$\lim_{R \to \infty} \int_{-R}^R f(x) \, dx, \qquad\qquad (10.10.31)$$

which may exist even though the limits 10.10.30 do not exist.* We call the

---

*Note $\lim_{R \to \infty} \int_{-R}^R x \, dx = 0$ but neither $\lim_{R \to \infty} \int_0^R x \, dx$ nor $\lim_{S \to \infty} \int_{-S}^0 x \, dx$ exist.

value of the limit in 10.10.31 the *Cauchy principle value* of

$$\int_{-\infty}^{\infty} f(x)\, dx.$$

Of course, if the two limits in Eq. 10.10.30 exist, then the principal value exists and is the same limit.

PROBLEMS

Find the residue of each function at each pole.

1. $\dfrac{1}{z^2 + 4}$

2. $\dfrac{z}{z^2 + 4}$

3. $\dfrac{1}{z^2} \sin 2z$

4. $\dfrac{e^z}{(z - 1)^2}$

5. $\dfrac{\cos z}{z^2 + 2z + 1}$

6. $\dfrac{z^2 + 1}{z^2 + 3z + 2}$

Evaluate each integral around the circle $|z| = 2$.

7. $\displaystyle\oint \frac{e^z}{z^4}\, dz$

8. $\displaystyle\oint \frac{\sin z}{z^3}\, dz$

9. $\displaystyle\oint \frac{z^2}{1 - z}\, dz$

10. $\displaystyle\oint \frac{z + 1}{z + i}\, dz$

11. $\displaystyle\oint \frac{z\, dz}{z^2 + 4z + 3}$

12. $\dfrac{dz}{4z^2 + 9}$

13. $\displaystyle\oint \frac{e^{1/z}}{z}\, dz$

14. $\displaystyle\oint \frac{\sin z}{z^3 - z^2}\, dz$

15. $\displaystyle\oint e^z \tan z\, dz$

16. $\displaystyle\oint \frac{z^2 + 1}{z(z + 1)^3}\, dz$

17. $\displaystyle\oint \frac{\sinh \pi z}{z^2 + 1}\, dz$

18. $\displaystyle\oint \frac{\cosh \pi z}{z^2 + z}\, dz$

Determine the value of each real integral.

19. $\displaystyle\int_0^{2\pi} \frac{\sin \theta}{1 + \cos \theta}\, d\theta$

20. $\displaystyle\int_0^{2\pi} \frac{d\theta}{(2 + \cos \theta)^2}$

21. $\displaystyle\int_0^{2\pi} \frac{d\theta}{5 - 4 \cos \theta}$

22. $\displaystyle\int_0^{2\pi} \frac{d\theta}{2 + 2 \sin \theta}$

23. $\displaystyle\int_0^{2\pi} \frac{\sin 2\theta\, d\theta}{5 + 4 \cos \theta}$

24. $\displaystyle\int_0^{2\pi} \frac{\cos 2\theta\, d\theta}{5 - 4 \cos \theta}$

Evaluate each integral.

25. $\displaystyle\int_{-\infty}^{\infty} \frac{dx}{1 + x^4}$

26. $\displaystyle\int_0^{\infty} \frac{x^2\, dx}{(1 + x^2)^2}$

27. $\displaystyle\int_{-\infty}^{\infty} \frac{1 + x}{1 + x^3}\, dx$

28. $\displaystyle\int_{-\infty}^{\infty} \frac{x^2\, dx}{x^4 + x^2 + 1}$

29. $\displaystyle\int_0^{\infty} \frac{x^2\, dx}{1 + x^6}$

30. $\displaystyle\int_{-\infty}^{\infty} \frac{dx}{x^4 + 5x^2 + 2}$

31. $\displaystyle\int_{-\infty}^{\infty} \frac{\cos 2x}{1 + x}\, dx$

32. $\displaystyle\int_{-\infty}^{\infty} \frac{\cos x}{(1 + x^2)^2}\, dx$

33. $\displaystyle\int_{-\infty}^{\infty} \frac{x \sin x}{1 + x^2}\, dx$

34. $\displaystyle\int_0^\infty \frac{\cos x}{1 + x^4}\, dx$ **35.** $\displaystyle\int_{-\infty}^\infty \frac{x \sin x}{x^2 + 3x + 2}\, dx$ **36.** $\displaystyle\int_0^\infty \frac{\cos 4x}{(1 + x^2)^2}\, dx$

37. Find the value of $\int_{-\infty}^\infty dx/(x^4 - 1)$ following the technique using the path of integration of Fig. 10.20, but integrate around the two poles on the x axis by considering the path of integration shown.

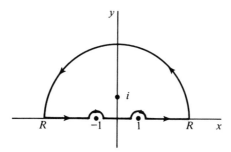

Bibliography

AYRES, F., JR., *Matrices,* Schaum Publishing Co., New York, 1962.

CHURCHILL, R. V., *Fourier Series and Boundary Value Problems,* McGraw-Hill Book Company, New York, 1941.

HILDEBRAND, F. B., *Advanced Calculus for Application,* Prentice-Hall, Inc., Englewood Cliffs, N.J., 1976.

HOVANESSIAN, S. A., and PIPES, L. A., *Digital Computer Methods in Engineering,* McGraw-Hill Book Company, New York, 1969.

ISAACSON, E., and KELLER, H. B., *Analysis of Numerical Methods,* John Wiley & Sons, Inc., New York, 1966.

KREYSZIG, E., *Advanced Engineering Mathematics,* John Wiley & Sons, Inc., New York, 1972.

POTTER, M. C., and FOSS, J. F., *Fluid Mechanics,* The Ronald Press Company, New York, 1975.

RAINVILLE, E. D., *Intermediate Differential Equations,* John Wiley & Sons, Inc., New York, 1943.

RAO, N. N., *Basic Electromagnetics with Applications,* Prentice-Hall, Inc., Englewood Cliffs, N.J., 1972.

SPIEGEL, M. R., *Laplace Transforms,* Schaum Publishing Co., New York, 1965.

SPIEGEL, M. R., *Vector Analysis,* Schaum Publishing Co., New York, 1959.

WYLIE, C. R., JR., *Advanced Engineering Mathematics,* McGraw-Hill Book Company, New York, 1951.

Appendix

TABLE A1. U.S. Engineering Units, SI Units, and Their Conversion Factors

Quantity	Engineering Units (U.S. System)	International System (SI)[a]	Conversion Factor
Length	inch	millimeter	1 in. = 25.4 mm
	foot	meter	1 ft = 0.3048 m
	mile	kilometer	1 mi = 1.609 km
Area	square inch	square centimeter	1 in.^{2} = 6.452 cm^2
	square foot	square meter	1 ft^2 = 0.09290 m^2
Volume	cubic inch	cubic centimeter	1 in.^{3} = 16.39 cm^3
	cubic foot	cubic meter	1 ft^3 = 0.02832 m^3
	gallon		1 gal = 0.004546 m^3
Mass	pound-mass	kilogram	1 lb$_m$ = 0.4536 kg
	slug		1 slug = 14.61 kg
Density	pound/cubic foot	kilogram/cubic meter	1 lb$_m$/ft^3 = 16.02 kg/m^3
Force	pound-force	newton	1 lb = 4.448 N
Work or torque	foot-pound	newton-meter	1 ft-lb = 1.356 N·m
Pressure	pound/square inch	newton/square meter	1 psi = 6895 N/m^2
	pound/square foot		1 psf = 47.88 N/m^2
Temperature	degree Fahrenheit	degree Celsius	$°F = \frac{9}{5}°C + 32$
	degree Rankine	degree Kelvin	$°R = \frac{9}{5}K$
Energy	British thermal unit	joule	1 Btu = 1055 J
	calorie		1 cal = 4.186 J
	foot-pound		1 ft-lb = 1.356 J
Power	horsepower	watt	1 hp = 745.7 W
	foot-pound/second		1 ft-lb/sec = 1.356 W
Velocity	foot/second	meter/second	1 fps = 0.3048 m/s
Acceleration	foot/second squared	meter/second squared	1 ft/sec^2 = 0.3048 m/s^2
Frequency	cycle/second	hertz	1 cps = 1.000 Hz

[a]The reversed initials in this abbreviation come from the French form of the name: *Système International.*

TABLE A2. Gamma Function, $\Gamma(\alpha) = \int_0^\infty e^{-t} t^{\alpha-1}\, dt$

α	$\Gamma(\alpha)$	α	$\Gamma(\alpha)$	α	$\Gamma(\alpha)$	α	$\Gamma(\alpha)$	α	$\Gamma(\alpha)$
1.00	1.000000	1.20	0.918169	1.40	0.887264	1.60	0.893516	1.80	0.931384
1.01	0.994326	1.21	0.915577	1.41	0.886764	1.61	0.894681	1.81	0.934076
1.02	0.988844	1.22	0.913106	1.42	0.886356	1.62	0.895924	1.82	0.936845
1.03	0.983550	1.23	0.910755	1.43	0.886036	1.63	0.897244	1.83	0.939690
1.04	0.978438	1.24	0.908521	1.44	0.885805	1.64	0.898642	1.84	0.942612
1.05	0.973504	1.25	0.906403	1.45	0.885661	1.65	0.900117	1.85	0.945611
1.06	0.968744	1.26	0.904397	1.46	0.885604	1.66	0.901668	1.86	0.948687
1.07	0.964152	1.27	0.902503	1.47	0.885633	1.67	0.903296	1.87	0.951840
1.08	0.959725	1.28	0.900719	1.48	0.885747	1.68	0.905001	1.88	0.955071
1.09	0.955459	1.29	0.899042	1.49	0.885945	1.69	0.906782	1.89	0.958380
1.10	0.951351	1.30	0.897471	1.50	0.886227	1.70	0.908639	1.90	0.961766
1.11	0.947395	1.31	0.896004	1.51	0.886592	1.71	0.910572	1.91	0.965231
1.12	0.943590	1.32	0.894640	1.52	0.887039	1.72	0.912580	1.92	0.968774
1.13	0.939931	1.33	0.893378	1.53	0.887568	1.73	0.914665	1.93	0.972397
1.14	0.936416	1.34	0.892215	1.54	0.888178	1.74	0.916826	1.94	0.976099
1.15	0.933041	1.35	0.891151	1.55	0.888869	1.75	0.919062	1.95	0.979881
1.16	0.929803	1.36	0.890184	1.56	0.889639	1.76	0.921375	1.96	0.983742
1.17	0.926700	1.37	0.889313	1.57	0.890490	1.77	0.923763	1.97	0.987685
1.18	0.923728	1.38	0.888537	1.58	0.891420	1.78	0.926227	1.98	0.991708
1.19	0.920885	1.39	0.887854	1.59	0.892428	1.79	0.928767	1.99	0.995813
1.20	0.918169	1.40	0.887264	1.60	0.893516	1.80	0.931384	2.00	1.000000

Polynomial approximation[a]:

$$\Gamma(x + 1) = 1 - 0.577191652x + 0.988205891x^2 - 0.897056937x^3 + 0.918206857x^4$$
$$-0.756704078x^5 + 0.482199394x^6 - 0.193527818x^7 + 0.035868343x^8 + \epsilon$$
$$|\epsilon| \le 3 \times 10^{-7}$$

[a]From C. Hastings, Jr., *Approximations for Digital Computers*, Princeton University Press, Princeton, N.J., 1955.

TABLE A3. *Error Function,* $\text{erf } x = \dfrac{2}{\sqrt{\pi}} \displaystyle\int_0^x e^{-t^2}\, dt$

x	Erf(x)	x	Erf(x)	x	Erf(x)
0.00	0.000000	0.68	0.663782	1.36	0.945561
0.02	0.022564	0.70	0.677801	1.38	0.949016
0.04	0.045111	0.72	0.691433	1.40	0.952285
0.06	0.067621	0.74	0.704678	1.42	0.955376
0.08	0.090078	0.76	0.717537	1.44	0.958296
0.10	0.112463	0.78	0.730010	1.46	0.961053
0.12	0.134758	0.80	0.742101	1.48	0.963654
0.14	0.156947	0.82	0.753811	1.50	0.966105
0.16	0.179012	0.84	0.765143	1.52	0.968413
0.18	0.200936	0.86	0.776100	1.54	0.970586
0.20	0.222703	0.88	0.786688	1.56	0.972628
0.22	0.244296	0.90	0.796908	1.58	0.974547
0.24	0.265700	0.92	0.806768	1.60	0.976348
0.26	0.286900	0.94	0.816271	1.62	0.978038
0.28	0.307880	0.96	0.825424	1.64	0.979622
0.30	0.328627	0.98	0.834232	1.66	0.981105
0.32	0.349126	1.00	0.842701	1.68	0.982493
0.34	0.369365	1.02	0.850838	1.70	0.983791
0.36	0.389330	1.04	0.858650	1.72	0.985003
0.38	0.409010	1.06	0.866144	1.74	0.986135
0.40	0.428392	1.08	0.873326	1.76	0.987190
0.42	0.447468	1.10	0.880205	1.78	0.988174
0.44	0.466225	1.12	0.886788	1.80	0.989091
0.46	0.484655	1.14	0.893082	1.82	0.989943
0.48	0.502750	1.16	0.899096	1.84	0.990736
0.50	0.520500	1.18	0.904837	1.86	0.991473
0.52	0.537898	1.20	0.910314	1.88	0.992156
0.54	0.554939	1.22	0.915534	1.90	0.992791
0.56	0.571616	1.24	0.920505	1.92	0.993378
0.58	0.587923	1.26	0.925236	1.94	0.993923
0.60	0.603856	1.28	0.928734	1.96	0.994427
0.62	0.619411	1.30	0.934008	1.98	0.994892
0.64	0.634586	1.32	0.938065	2.00	0.995323
0.66	0.649376	1.34	0.941913	2.02	0.995720

Rational approximation[a]:

$$\text{erf } x = 1 - [1 + 0.0705230784x + 0.0422820123x^2 + 0.0092705272x^3$$
$$+ 0.0001520143x^4 + 0.0002765672x^5 + 0.0000430638x^6]^{-16} + \epsilon$$

$$|\epsilon| \leq 3 \times 10^{-7}$$

[a]From C. Hastings, Jr., *Approximations for Digital Computers*, Princeton University Press, Princeton, N.J., 1955.

TABLE A4. *Bessel Functions*[a]

x	$J_0(x)$	$Y_0(x)$	$J_1(x)$	$Y_1(x)$
0.0	1.00000	$-\ \infty$	0.00000	$-\ \infty$
0.1	0.99750	-1.53424	0.04994	-6.45895
0.2	0.99002	-1.08114	0.09950	-3.32382
0.3	0.97763	-0.80727	0.14832	-2.29311
0.4	0.96040	-0.60602	0.19603	-1.78087
0.5	0.93847	-0.44452	0.24227	-1.47147
0.6	0.91200	-0.30851	0.28670	-1.26039
0.7	0.88120	-0.19066	0.32900	-1.10325
0.8	0.84629	-0.08680	0.36884	-0.97814
0.9	0.80752	$+0.00563$	0.40595	-0.87313
1.0	0.76520	0.08825	0.44005	-0.78121
1.1	0.71962	0.16216	0.47090	-0.79812
1.2	0.67113	0.22808	0.49829	-0.62114
1.3	0.62009	0.28654	0.52202	-0.54852
1.4	0.56686	0.33790	0.54195	-0.47915
1.5	0.51183	0.38245	0.55794	-0.41231
1.6	0.45540	0.42043	0.56990	-0.34758
1.7	0.39798	0.45203	0.57777	-0.28473
1.8	0.33999	0.47743	0.58152	-0.22366
1.9	0.28182	0.49682	0.58116	-0.16441
2.0	0.22389	0.51038	0.57672	-0.10703
2.1	0.16661	0.51829	0.56829	-0.05168
2.2	0.11036	0.52078	0.55596	$+0.00149$
2.3	0.05554	0.51808	0.53987	0.05228
2.4	$+0.00251$	0.51041	0.52019	0.10049
2.5	-0.04838	0.49807	0.49709	0.14592
2.6	-0.09680	0.48133	0.47082	0.18836
2.7	-0.14245	0.46050	0.44160	0.22763
2.8	-0.18504	0.43592	0.40971	0.26355
2.9	-0.22431	0.40791	0.37543	0.29594
3.0	-0.26005	0.37685	0.33906	0.32467
3.1	-0.29206	0.34310	0.30092	0.34963
3.2	-0.32019	0.30705	0.26134	0.37071
3.3	-0.34430	0.26909	0.22066	0.38785
3.4	-0.36430	0.22962	0.17923	0.40102
3.5	-0.38013	0.18902	0.13738	0.41019
3.6	-0.39177	0.14771	0.09547	0.41539
3.7	-0.39923	0.10607	0.05383	0.41667
3.8	-0.40256	0.06450	$+0.01282$	0.41411
3.9	-0.40183	$+0.02338$	-0.02724	0.40782
4.0	-0.39715	-0.01694	-0.06604	0.39793
4.1	-0.38867	-0.05609	-0.10327	0.38459
4.2	-0.37656	-0.09375	-0.13865	0.36801
4.3	-0.36101	-0.12960	-0.17190	0.34839
4.4	-0.34226	-0.16334	-0.20278	0.32597
4.5	-0.32054	-0.19471	-0.23106	0.30100
4.6	-0.29614	-0.22346	-0.25655	0.27375

TABLE A4. Bessel Functions[a] (Cont.)

x	$J_0(x)$	$Y_0(x)$	$J_1(x)$	$Y_1(x)$
4.7	−0.26933	−0.24939	−0.27908	0.24450
4.8	−0.24043	−0.27230	−0.29850	0.21357
4.9	−0.20974	−0.29205	−0.31469	0.18125
5.0	−0.17760	−0.30852	−0.32758	0.14786
5.1	−0.14433	−0.32160	−0.33710	0.11374
5.2	−0.11029	−0.33125	−0.34322	0.07919
5.3	−0.07580	−0.33744	−0.34596	0.04455
5.4	−0.04121	−0.34017	−0.34534	+0.01013
5.5	−0.00684	−0.33948	−0.34144	−0.02376
5.6	+0.02697	−0.33544	−0.33433	−0.05681
5.7	0.05992	−0.32816	−0.32415	−0.08872
5.8	0.09170	−0.31775	−0.31103	−0.11923
5.9	0.12203	−0.30437	−0.29514	−0.14808
6.0	0.15065	−0.28819	−0.27668	−0.17501
6.1	0.17729	−0.26943	−0.25586	−0.19981
6.2	0.20175	−0.24831	−0.23292	−0.22228
6.3	0.22381	−0.22506	−0.20809	−0.24225
6.4	0.24331	−0.19995	−0.18164	−0.25956
6.5	0.26009	−0.17324	−0.15384	−0.27409
6.6	0.27404	−0.14523	−0.12498	−0.28575
6.7	0.28506	−0.11619	−0.09534	−0.29446
6.8	0.29310	−0.08643	−0.06522	−0.30019
6.9	0.29810	−0.05625	−0.03490	−0.30292
7.0	0.30008	−0.02595	−0.00468	−0.30267
7.1	0.29905	+0.00418	+0.02515	−0.29948
7.2	0.29507	0.03385	0.05433	−0.29342
7.3	0.28822	0.06277	0.08257	−0.28459
7.4	0.27860	0.09068	0.10963	−0.27311
7.5	0.26634	0.11731	0.13525	−0.25913
7.6	0.25160	0.14243	0.15921	−0.24280
7.7	0.23456	0.16580	0.18131	−0.22432
7.8	0.21541	0.18723	0.20136	−0.20389
7.9	0.19436	0.20652	0.21918	−0.18172
8.0	0.17165	0.22352	0.23464	−0.15806
8.1	0.14752	0.23809	0.24761	−0.13315
8.2	0.12222	0.25012	0.25800	−0.10724
8.3	0.09601	0.25951	0.26574	−0.08060
8.4	0.06916	0.26622	0.27079	−0.05348
8.5	0.04194	0.27021	0.27312	−0.02617
8.6	+0.01462	0.27146	0.27275	+0.00108
8.7	−0.01252	0.27000	0.26972	0.02801
8.8	−0.03923	0.26587	0.26407	0.05436
8.9	−0.06525	0.25916	0.25590	0.07987
9.0	−0.09033	0.24994	0.24531	0.10431
9.1	−0.11424	0.23834	0.23243	0.12747
9.2	−0.13675	0.22449	0.21741	0.14911
9.3	−0.15766	0.20857	0.20041	0.16906

TABLE A4. Bessel Functions[a] **(Cont.)**

x	$J_0(x)$	$Y_0(x)$	$J_1(x)$	$Y_1(x)$
9.4	−0.17677	0.19074	0.18163	0.18714
9.5	−0.19393	0.17121	0.16126	0.20318
9.6	−0.20895	0.15018	0.13952	0.21706
9.7	−0.2̇2̇180	0.12787	0.11664	0.22866
9.8	−0.23228	0.10453	0.09284	0.23789
9.9	−0.24034	0.08038	0.06837	0.24469
10.0	−0.24594	0.05567	0.04347	0.24902
10.1	−0.24903	0.03066	+0.01840	0.25084
10.2	−0.24962	+0.00559	−0.00662	0.25019
10.3	−0.24772	−0.01930	−0.03132	0.24707
10.4	−0.24337	−0.04375	−0.05547	0.24155
10.5	−0.23665	−0.06753	−0.07885	0.23370
10.6	−0.22764	−0.09042	−0.10123	0.22363
10.7	−0.21644	−0.11219	−0.12240	0.21144
10.8	−0.20320	−0.13264	−0.14217	0.19729
10.9	−0.18806	−0.15158	−0.16035	0.18132
11.0	−0.17119	−0.16885	−0.17679	0.16371
11.1	−0.15277	−0.18428	−0.19133	0.14464
11.2	−0.13299	−0.19773	−0.20385	0.12431
11.3	−0.11201	−0.20910	−0.21426	0.10294
11.4	−0.09021	−0.21829	−0.22245	0.08074
11.5	−0.06765	−0.22523	−0.22838	0.05794
11.6	−0.04462	−0.22987	−0.23200	0.03477
11.7	−0.02133	−0.23218	−0.23330	+0.01145
11.8	+0.00197	−0.23216	−0.23228	−0.01179
11.9	0.02505	−0.22983	−0.22898	−0.03471
12.0	0.04769	−0.22524	−0.22345	−0.05710
12.1	0.06967	−0.21844	−0.21575	−0.07874
12.2	0.09077	−0.20952	−0.20598	−0.09942
12.3	0.11080	−0.19859	−0.19426	−0.11895
12.4	0.12956	−0.18578	−0.18071	−0.13714
12.5	0.14688	−0.17121	−0.16548	−0.15384
12.6	0.16261	−0.15506	−0.14874	−0.16888
12.7	0.17659	−0.13750	−0.13066	−0.18213
12.8	0.18870	−0.11870	−0.11143	−0.19347
12.9	0.19884	−0.09887	−0.09125	−0.20282
13.0	0.20693	−0.07821	−0.07032	−0.21008
13.1	0.21289	−0.05693	−0.04885	−0.21521
13.2	0.21669	−0.03524	−0.02707	−0.21817
13.3	0.21830	−0.01336	−0.00518	−0.21895
13.4	0.21773	+0.00848	+0.01660	−0.21756
13.5	0.21499	0.03008	0.03805	−0.21402
13.6	0.21013	0.05122	0.05896	−0.20839
13.7	0.20322	0.07169	0.07914	−0.20074
13.8	0.19434	0.09130	0.09839	−0.19116
13.9	0.18358	0.10986	0.11652	−0.17975
14.0	0.17107	0.12719	0.13338	−0.16664

TABLE A4. Bessel Functions[a] **(Cont.)**

x	$J_0(x)$	$Y_0(x)$	$J_1(x)$	$Y_1(x)$
14.1	0.15695	0.14314	0.14878	−0.15198
14.2	0.14137	0.15754	0.16261	−0.13592
14.3	0.12449	0.17028	0.17473	−0.11862
14.4	0.10648	0.18123	0.18503	−0.10026
14.5	0.08754	0.19030	0.19343	−0.08104
14.6	0.06786	0.19742	0.19985	−0.06115
14.7	0.04764	0.20252	0.20425	−0.04079
14.8	0.02708	0.20557	0.20660	−0.02016
14.9	+0.00639	0.20655	0.20688	+0.00053
15.0	−0.01422	0.20546	0.20510	0.02107

Polynomial Approximations[b]

$$3 \leq x \leq 3$$

$$J_0(x) = 1 - 2.2499997(x/3)^2 + 1.2656208(x/3)^4$$
$$- 0.3163866(x/3)^6 + 0.0444479(x/3)^8$$
$$- 0.0039444(x/3)^{10} + 0.0002100(x/3)^{12} + \epsilon$$

$$|\epsilon| < 5 \times 10^{-9}$$

$$0 < x \leq 3$$

$$Y_0(x) = (2/\pi) \ln (\tfrac{1}{2}x)J_0(x) + 0.36746691$$
$$+ 0.60559366(x/3)^2 - 0.74350384(x/3)^4$$
$$+ 0.25300117(x/3)^6 - 0.04261214(x/3)^8$$
$$+ 0.00427916(x/3)^{10} - 0.00024846(x/3)^{12} + \epsilon$$

$$|\epsilon| < 1.4 \times 10^{-8}$$

$$-3 \leq x \leq 3$$

$$x^{-1}J_1(x) = \tfrac{1}{2} - 0.56249985(x/3)^2 + 0.21093573(x/3)^4$$
$$- 0.03954289(x/3)^6 + 0.00443319(x/3)^8$$
$$- 0.00031761(x/3)^{10} + 0.00001109(x/3)^{12} + \epsilon$$

$$|\epsilon| < 1.3 \times 10^{-8}$$

$$0 < x \leq 3$$

$$x Y_1(x) = (2/\pi)x \ln (\tfrac{1}{2}x)J_1(x) - 0.6366198$$
$$+ 0.2212091(x/3)^2 + 2.1682709(x/3)^4$$
$$- 1.3164827(x/3)^6 + 0.3123951(x/3)^8$$
$$- 0.0400976(x/3)^{10} + 0.0027873(x/3)^{12} + \epsilon$$

$$|\epsilon| < 1.1 \times 10^{-7}$$

$$3 \leq x < \infty$$

$$J_0(x) = x^{-1/2} f_0 \cos \theta_0 \qquad Y_0(x) = x^{-1/2} f_0 \sin \theta_0$$
$$f_0 = 0.79788456 - 0.00000077(3/x) - 0.00552740(3/x)^2$$
$$- 0.00009512(3/x)^3 + 0.00137237(3/x)^4$$
$$- 0.00072805(3/x)^5 + 0.00014476(3/x)^6 + \epsilon$$

$$|\epsilon| < 1.6 \times 10^{-8}$$

TABLE A4. Bessel Functions [a] **(Cont.)**

$$\theta_0 = x - 0.78539816 - 0.04166397(3/x)$$
$$- 0.00003954(3/x)^2 + 0.00262573(3/x)^3$$
$$- 0.00054125(3/x)^4 - 0.00029333(3/x)^5$$
$$+ 0.00013558(3/x)^6 + \epsilon$$

$$|\epsilon| < 7 \times 10^{-8}$$

$$3 \le x < \infty$$

$$J_1(x) = x^{-1/2} f_1 \cos \theta_1, \qquad Y_1(x) = x^{-1/2} f_1 \sin \theta_1$$
$$f_1 = 0.79788456 + 0.00000156(3/x) + 0.01659667(3/x)^2$$
$$+ 0.00017105(3/x)^3 - 0.00249511(3/x)^4$$
$$+ 0.00113653(3/x)^5 - 0.00020033(3/x)^6 + \epsilon$$

$$|\epsilon| < 4 \times 10^{-8}$$

$$\theta_1 = x - 2.35619449 + 0.12499612(3/x)$$
$$+ 0.00005650(3/x)^2 - 0.00637879(3/x)^3$$
$$+ 0.00074348(3/x)^4 + 0.00079824(3/x)^5$$
$$- 0.00029166(3/x)^6 + \epsilon$$

$$|\epsilon| < 9 \times 10^{-8}$$

---

[a] $J_0(x) = 0$ at $x = 2.4048, 5.5201, 8.6537, 11.7915, 14.9309, 18.0711, \ldots$
$Y_0(x) = 0$ at $x = 0.8936, 3.9577, 7.0861, 10.2223, 13.3611, 16.5009, \ldots$
$J_1(x) = 0$ at $x = 3.8317, 7.0156, 10.1735, 13.3237, 16.4706, 19.6159, \ldots$
$Y_1(x) = 0$ at $x = 2.1971, 5.4297, 8.5960, 11.7492, 14.8974, 18.0434, \ldots$

[b] From *Handbook of Mathematical Functions*, National Bureau of Standards, p. 369, 1964.

Answers to Selected Problems

SECTION 1.2

2. Nonlinear, first order
4. Linear, second order, nonhomogeneous
6. Linear, first order, homogeneous
8. (After division by u) linear, first order, homogeneous
10. Linear, first order, nonhomogeneous
12. Linear, second order, nonhomogeneous
14. Linear, second order, nonhomogeneous
22. 250
24. 0.625 m

SECTION 1.3.1

2. $-(10x + C)^{-1}$ **4.** $Ke^{-\cos x}$ **6.** $\dfrac{1 + Ke^{2/x}}{1 - Ke^{2/x}}$

8. $Ke^{-x^2/10}$ **10.** $x(K - \ln x)^{-1}$

12. $\ln\left[\dfrac{u^2}{x^2} + 4\right] + \arctan\left[\dfrac{u}{2x}\right] = K - 2\ln x$

14. $u - x + C = \ln(x + 2u + 3)$ **16.** $u = \dfrac{\sin x}{\sin 2} - 1$

18. $u = x + xC/\sqrt{1 - (xC)^2}$

SECTION 1.3.2

2. Kx^{-2} **4.** $u^3 = K - x^3/3$ **6.** Ke^{-x}
10. $(1 + x^2)^{-1}$ **12.** $u = 0$

SECTION 1.3.3

4. $Ce^{-2x} + x - \frac{1}{2}$

6. $Ce^{2x} - e^x$

8. $Ce^x - \frac{1}{2}(\cos x + \sin x)$

10. $Ke^{-1/x} - \cos(1/x) + \sin(1/x)$

12. $e^{-x^2/2} \int_0^x e^{-t^2/2} \, dt + Ke^{-x^2/2}$ where $K = -\int_0^1 e^{-t^2/2} \, dt$

14. $2e^{2x} - 2$

SECTION 1.4

2. $0.01[e^{2t} - e^{-2 \times 10^4 t}]$

4. 1.386×10^{-4} sec

6. $0.03456 - 0.03056e^{-1.111 \times 10^{-4} t}$

9. 5.098 sec; 5.116 sec

11. $-83.3x + 217°C$

13. 86.3 min

SECTION 1.5.1

2. A discontinuity which is not a jump discontinuity

4. A discontinuity which is not a jump discontinuity

6. A jump discontinuity

8. No   **10.** No   **12.** Yes

SECTION 1.5.3

2. K/x **4.** K

8. Relate the Wronskian to the Wronskian of u_1 and u_2.

10. Provided $\alpha\delta + \beta\gamma \neq 0$

SECTION 1.6

2. $Ae^{-3x} + Be^{2x}$ **4.** $Ae^{ix} + Be^{-ix}$ **6.** $(Ax + B)e^{-2x}$

8. $e^{-2x}[Ae^{\alpha x} + Be^{-\alpha x}]$ where $\alpha = 2\sqrt{2}$

10. $e^{2x}[Ae^{\alpha x} + Be^{-\alpha x}]$ where $\alpha = 2\sqrt{3}$

12. $e^{-3x}[Ae^{\alpha x} + Be^{-\alpha x}]$ where $\alpha = i$

14. $Ae^{-3x} + Be^{3x}$ **16.** $A \sin(x/2) + B \cos(x/2)$

18. $(Ax + B)e^{-2x}$

20. $e^{2x}[Ae^{\alpha x} + Be^{-\alpha x}]$ where $\alpha = 2\sqrt{3}$

22. $e^{-5x/2}[Ae^{\alpha x} + Be^{-\alpha x}]$ where $\alpha = \sqrt{13}/2$

24. $A \sinh 3x + B \cosh 3x$ **26.** $A \sin(B + x/2)$

28. $e^{-2x}[A \sinh \alpha x + B \cosh \alpha x]$ where $\alpha = 2\sqrt{2}$

30. $e^{2x}[A \sinh \alpha x + B \cosh \alpha x]$ where $\alpha = 2\sqrt{3}$

32. $e^{-5x/2}[A \sinh \alpha x + B \cosh \alpha x]$ where $\alpha = \sqrt{13}/2$

34. $-4e^{-3x} + 6e^{-2x}$ **36.** $e^{2x} + e^{-2x}$ **38.** $-4e^{-3x} + 6e^{-2x}$

40. $\frac{1}{3} \sin 3x$

42. $(5e^{2x} + 3e^{-2x})/4$

44. $Ae^x + e^{-x/2}(B \cos \alpha x + C \sin \alpha x)$ where $\alpha = \sqrt{3}/2$

46. $Ax + B + Ce^x + De^{-x}$ **48.** $Ax^2 + Bx + C + De^x$

SECTION 1.7.1

2. $10(\cos t + \sin t)$ **4.** 1.54 sec

SECTION 1.7.2

2. $\frac{25}{2}(e^{-8t} - e^{-2t})$ **12.** $(\pi/4)$ sec; 70.7%
14. (a) $R < 14.14$ ohms; (b) $R > 3.54$ ohms
16. $10e^{-10^5t}(\cos 10^5t + \sin 10^5t)$

SECTION 1.8

2. $x - \frac{1}{2}$ **4.** $xe^x/2$ **6.** $(x \sin x)/6$
8. $x^2/2 - \frac{2}{81} - x/6 (\cos 3x)$
10. $(C_1 + C_2x)e^{-2x} + x^2/4 - x/4 + \frac{9}{8}$
12. $(C_1 - x/4) \cos 2x + C_2 \sin 2x$
14. $C_1e^{-3x} + C_2e^{-2x} + (3 \sin 2x - 15 \cos 2x)/52$
16. $\cos 2x - \frac{1}{3} \sin 2x + \frac{2}{3} \sin x$ **18.** $(\sin 2x - 2x \cos 2x)/4$
20. $(e^{-4x} - e^{4x} + 8xe^{4x})/32$

SECTION 1.9.2

2. 228 m   **4.** 94 sec   **6.** 80 sec   **8.** $\frac{4}{3}t \sin 3t$
10. $(2 \sin t + 37 \sin 4t)/15$ **12.** $e^{-t} - \cos t + 3 \sin t$
14. 94 sec   **16.** 0.00533 hz, 2.1 m

SECTION 1.9.3

4. $(-4 \cos 3t + 3 \sin 3t)/50$ **6.** $-0.0995 (\cos 2t + 10 \sin 2t)$
8. $(2 \cos t + 2 \sin t + \cos 2t + \sin 2t)/4$
10. $Ae^{-5t} + Be^{-2t} + 0.1385 \sin t - 0.1077 \cos t - 0.6034 \sin 2t - 0.02586 \cos 2t$
12. $e^{-0.05t}(A \cos 1.41t + B \sin 1.41t) - 0.0495(\sin 2t - 10 \cos 2t)$
14. $(1 + t)e^{-t} - \cos t$
16. $e^{-0.05t}(4.96 \sin 1.41t - 20 \cos 1.41t)$
18. $e^{-0.01t}(25 \cos 4t + 0.0625 \sin 4t) - 25 \cos 4t$
20. 2.2 m; 1.62 rad/s
22. $q(t) = 2.05 \times 10^{-6}e^{-2.618 \times 10^5t} - 14.05 \times 10^{-6}e^{-3.82 \times 10^4t} + 12 \times 10^{-6}$
$i(t) = -0.5367(e^{-2.618 \times 10^5t} - e^{-3.82 \times 10^4t})$
24. -0.591 amp   **26.** $-2.48e^{-7.5 \times 10^4t} \sin 3.23 \times 10^4t$

SECTION 1.10

2. $Ax^{-6} + Bx^{-2}$ **4.** $Ax^{-4} + Bx^3$
6. $-1 + (3x^{-4} + 4x^3)/7$ **8.** Cx^α where $\alpha = -a_1/a_0$

SECTION 1.11

4. $A \sin x + B \cos x + x \sin x + (\cos x) \ln \cos x$
6. $(Ax + B)e^{2x} + e^x + xe^{2x} \int e^{-x}/x \, dx$

8. $Ax + B/x + 2x^2/3$

10. $Cx^2 + B + x^3/3 - x^2/4 + (x^2/2) \ln x$

12. $(10t - 7)e^t/100$

SECTION 1.12.1

2. $(Ax + B)e^{-2x} + x^2 e^{-2x}/2$

4. $f(x) = x; u(x) = Ax + Bx \int F(x)/x^2 \, dx$ where $F(x) = e^{\int xp(x) \, dx}$

SECTION 1.12.2

2. $\dfrac{n(n + 1)}{1 - x^2} + \dfrac{1}{(1 - x^2)^2}$ **4.** $1 + \dfrac{1}{4x^2}(1 - 4n^2)$

6. $2n + 1 - x^2$

SECTION 1.12.3

2. $A \cos (\sin x) + B \sin (\sin x)$ **4.** $Ae^{-2/x} + Be^{-1/x}$

6. $e^{x^2}[A \cos (x^2) + B \sin (x^2)]$

SECTION 2.2

2. $1 + x + x^2/2! + \cdots + x^n/n! + \cdots$

4. $1 - x^2/2! + x^4/4! + \cdots + (-1)^n x^{2n}/(2n)! + \cdots$

6. $\ln (1 + x) = x - x^2/2 + x^3/3 \cdots + (-1)^n x^n/n + \cdots$

8. $\frac{1}{2}[1 - x/2 + x^2/4 + \cdots + (-1)^n x^n/2^n + \cdots]$

10. $-7/12 + 7x/144 + \cdots - (3^{n+1} + (-1)^n 4^{n+1})x^n/(12)^{n+1} + \cdots$

12. $1 - x^2/ + x^4/2! + \cdots + (-1)^n x^{2n}/n! + \cdots$

14. $x + x^3/3 + 2x^5/15 + \cdots$

16. $-2[(x/2)^2/2 + (x/2)^4/4 + \cdots + (x/2)^{2n}/2n + \cdots]$

18. $x - x^2 + x^3/3 - x^5/30 + \cdots$

20. $\frac{1}{4}[x + x^3/12 + x^5/80 + \cdots + 2(x/2)^{2n+1}/(2n + 1) + \cdots]$

22. $\dfrac{x^3}{3}(1 - x^2/5 + 2x^4/105 - \cdots) + C$

24. $\dfrac{-1}{4}[1 - (2x)^2/2! + (2x)^4/4! + \cdots + (-1)^n (2x)^{2n}/(2n)! + \cdots]$

26. No singular points. $R = \infty$.

28. $(0, -2)$. The singular point at the origin is removable; then $R = 2$.

30. $(-1, 0)$. $R = 1$

32. 1   **34.** 2   **36.** ∞

38. $1 - (x - 1) + (x - 1)^2 + \cdots + (-1)^n (x - 1)^n + \cdots$

40. $\dfrac{-1}{4}[(1 + \frac{1}{3}) + (1 - 1/3^2)(x - 1) + \cdots + (1 + (-1)^n/3^{n+1})(x - 1)^n + \cdots]$

SECTION 2.3

2. $R = \infty$. $b_0[1 - (kx)/1! + (kx)^2/2! + \cdots + (-kx)^n/n! + \cdots]$

4. $R = \infty$. $b_0[1 - x^2/2 + x^4/2^2 2! + \cdots + (-1)^n x^{2n}/2^n n! + \cdots]$

6. $R = 1$

8. $A \sum_0^\infty (-5x)^n/n! + B \sum_0^\infty (-x)^n/n!$

10. $x - 2 \sum \dfrac{1}{4n^2 - 1} \left(\dfrac{x}{2}\right)^{2n+1}$

12. $x + x^6/180 + x^8/6720 + \cdots$

14. $f(x) = -1/(1 - x)$. $S(x) = 1 - (x - 2) + (x - 2)^2 - (x - 2)^3 + \cdots$
$f(1.9) = 10/9 = 1.1111$, $S(1.9) = 1.1110$

16. $R = 1$. $b_0[1 - (x - 1)^2/2 + (x - 1)^3/3 + \cdots] + b_1[(x - 1) - (x - 1)^2/6 + (x - 1)^4/6 + \cdots]$

18. $R = \infty$. $u(x) = 4 - 2x - x^4/3 + x^5/10 + x^8/168 - x^9/720 + \cdots$
$u(2) \approx -1.3206$

20. $u(1) \approx 8.115$; $u(3) \approx 7.302$

SECTION 2.3.2

2. $\frac{1}{128}(35 - 1260x + 6930x^4 - 12012x^6 + 6435x^8)$

8. $AP_2(x) + B[P_2(x)Q_2(x) - 3x/2] + x/4$

10. $u(\phi) = \dfrac{1}{2}(3 \cos^2 \phi - 1)\left(A + B \ln\dfrac{1 + \cos \phi}{1 - \cos \phi}\right) + \dfrac{3}{2}B \cos \phi$

SECTION 2.3.3

6. $H_4(x) = 48(1 - 4x^2 + 4x^4/3)$

SECTION 2.5

2. 6 **4.** ∞ **6.** -9394 **8.** $3 \times 10!$

10. $r = 1$, $h = 3$, $3^n\Gamma(\frac{1}{3} + n)\Gamma(\frac{1}{3})$

12. $\Gamma(1) = 1 = 0\Gamma(0)$, a contradiction

SECTION 2.10.1

2. $J_0(x) = 1 - x^2/4 + x^4/64 + x^6/2304 + \cdots$
$J_1(x) = x/2 - x^3/16 + x^5/384 - x^7/18,432$

4. See the answer to Problem 2. **6.** $AJ_{1/4}(x) + BJ_{-1/4}(x)$

8. $AJ_{1/2}(x) + BJ_{-1/2}(x)$

SECTION 2.10.4

2. 0.2612 **4.** -0.2481 **6.** 0.0572

8. $-xJ_1(x) - 2J_0(x) + C$ **10.** $-xJ_0(x) + \int J_0(x)\, dx + C$

12. $-\left(\dfrac{5}{2x^2} + 1\right)J_1(x) + \dfrac{3}{2x}J_0(x) + \int J_0(x)\, dx + C$

18. $I_{1/2}(x) = \left(\dfrac{2}{\pi x}\right)^{1/2} \sinh x$; $I_{-1/2}(x) = \left(\dfrac{2}{\pi x}\right)^{1/2} \cosh x$

SECTION 2.11

2. $1 - x^2/2 + 5x^3/6 + x^4/24 + \cdots$

6. $2 - 2x + x^2 + A \sum_0^\infty \dfrac{(-1)^n x^n}{n!}$

8. $x - \dfrac{7x^3}{18} + \dfrac{143x^5}{1800} + \cdots + A \sum_{0}^{\infty} \dfrac{(-1)^n x^{2n}}{2^n n!}$

12. $R = \infty.\ 1 - x^2/2 + 5x^3/6 + x^4/24 + \cdots$

14. Exact solution $= f(x) = (x^2 - 5)/(1 - x).\ \ f(1.9) = 1.5444$
$S(x) = 1 - 5(x - 2) + 4(x - 2)^2 - 4(x - 2)^3 + \cdots + S(1.9) = 1.544$

SECTION 3.2

2. $1/s^2 - 3/s$ **4.** $2/(s^2 + 1)$ **6.** $0.8862s^{-3/2}$ **8.** $8/s^3 - 3/s$

10. $2/s^3 - 4/s^2 + 4/s$

12. $e^{-1}/(s - 2)$

14. $[e^{-4s}(1 - 4s^2) + 1]/2s^2$

16. $3/(s - 3)^2$

18. $(s + 2)/(s^2 + 4s + 20)$

20. $6/(s^2 + 2s + 5)$

22. $(s - 7)/(s^2 + 2s + 17)$

24. $(5s^2 + 24s + 30)/(s + 2)^3$

26. $\pi e^{-4s}/(s^2 + \pi^2)$

28. $(1 - e^{-4s})/2s^2 - e^{-4s}/s$

30. $(1 - e^{-2s})/s^2$

32. $(1 + e^{-\pi s})/(s^2 + 1) + 2e^{-\pi s}/(s^2 + 4)$

34. $(s - 3)/(s^2 - 6s + 13) - (s + 3)/(s^2 + 6s + 13)$

36. $3(s - 1)/(s^2 - 2s + 2) - 3(s + 1)/(s^2 + 2s + 2)$

38. $(s - 1)/(s^2 - 2s + 5) + (s + 1)/(s^2 + 2s + 5)$

40. $3t^2/2 + 2t$ **42.** $t(1 - t/2)e^{-t}$ **44.** $-t/2 - \frac{1}{4} + e^{2t}/4$

46. $(2e^{-2t} + e^t)/3$ **48.** $u_1(t)e^{1-t}$ **50.** $2e^{-t} \sin 2t$

52. $e^t \sinh 2t$

54. $e^{-(t-2\pi)}(4 \cos 2t - 2 \sin 2t)u_{2\pi}(t)$

56. $\left(\dfrac{2t}{3} \sin 3t - \dfrac{1}{54} \sin 3t + \dfrac{t}{18} \cos 3t\right)e^{-2t}$

SECTION 3.3

2. $s\mathcal{L}(f) - f(0) - [f(a^+) - f(a^-)]e^{-as} - [f(b^+) - f(b^-)]e^{-bs}$

4. $\omega/(s^2 + \omega^2)$ **6.** $a/(s^2 - a^2)$ **8.** $1/(s - 2)$

10. $(1 - e^{-s})/s^2$ **12.** $1/(s - 1)^2$

14. $(s^2 - 1)/(s^2 + 1)^2$

16. $2(s^2 - 2s + 1)/(s - 1)(s^2 - 2s + 2)^2$

18. $(s^2 + 4)/(s^2 - 4)^2$

20. $4s/(s^2 - 4)^2$

22. $4e^{t/2} \sinh t/2$ or $2(e^t - 1)$

24. $t - \frac{1}{2} \sin 2t$

26. $-\frac{2}{3}(t - \frac{1}{3} \sinh 3t)$

28. $2e^{-t} - 1$

SECTION 3.4

2. $(s^2 - 4)/(s^2 + 4)^2$

4. $2(3s^2 + 1)/(s^2 - 1)^3$

6. $4s/(s^2 - 1)^2$

8. $2(s + 1)/(s^2 + 2s + 2)^2$

10. $(s^2 + 1)/(s^2 - 1)^2$ **12.** $\ln (s^2 - 4)/s^2$ **14.** $\ln s/(s - 2)$

16. te^{-2t}

18. $(t/4) \sinh 2t$

20. $(e^{-3t} - e^{2t})/t$

22. $2(\cos 2t - \cos t)/t$ **24.** $2(e^{-t} \cos 2t - e^{-2t} \cos t)/t$

SECTION 3.5

2. $(1 - e^{-2s} - 2se^{-2s})/(1 - e^{-2s})s^2$

4. $(1 - 2s - e^{-4s} - 2se^{-4s})/(1 - e^{-4s})s^2$

6. $(1 - e^{-2s})/(1 - e^{-4s})s$

8. $(2s - 1 + e^{-s} - se^{-s})/(1 - e^{-2s})s^2$

SECTION 3.6

2. $-5 + \frac{5}{2}e^t - 2e^{-4t} + \frac{25}{6}e^{-t}$

4. $\frac{1}{4}e^{-3t} + \frac{3}{4}e^t + te^t$

6. $\dfrac{7}{8} + \dfrac{t}{2} + \left(\dfrac{8}{3} + \dfrac{t}{3}\right)e^t + \dfrac{1}{72}e^{4t}$

8. $\frac{1}{164}\sin 20t - \frac{5}{861}\sin 21t$

10. $(\frac{2}{5}t - \frac{6}{25})e^{-t} - \frac{6}{25}\cos 2t - \frac{9}{50}\sin 2t$

12. $\frac{40}{27}(\cos 2t - \cos t) + \frac{5}{72}(\sin 2t - 2t\cos 2t) + \frac{5}{9}(\sin t - t\cos t)$

SECTION 3.7

8. te^{at}

10. $-\dfrac{1}{2\omega}\sin \omega t - (t/2)\cos \omega t$

12. $a^2 te^{at}$

SECTION 3.8

2. $2\cosh 2t$

4. $\frac{2}{3}(\cos t - \cos 2t)$

6. $2 + e^{-t}/2 - \frac{1}{2}(5\cos t - \sin t)$

8. $e^{-2t}(1 + 2t)$

10. $2 - 6e^{-3t} + 4e^{-2t}$

12. $(\frac{3}{2} + 3t)e^{-2t} - \frac{1}{2}\cos 2t$

15. (a) $(1 - \cos 6t)/36$;   (c) $5(\sin 6t - 6t\cos 6t)/72$

16. (a) $\frac{1}{36} - \frac{1}{36}(\cos 5.98t + \frac{1}{12}\sin 5.98t)e^{-t/2}$;
   (c) $-\frac{5}{6}\cos 6t + \frac{5}{6}(\cos 5.98t + \frac{1}{12}\sin 5.98t)e^{-t/2}$

17. (b) $\frac{1}{10}\sin 2t - \frac{3}{40}\cos 2t - (\frac{3}{40} - t/4)e^{-6t}$;   (f) $50te^{-6t}$

18. (c) $-\frac{1}{4}\cos 6t - \frac{1}{192}e^{-18t} + \frac{3}{64}e^{-2t}$

19. (a) $\sin 10t$;   (c) $(t/4)\sin 10t$;   (e) $10\cos 10t$

20. (e) $\frac{10}{3}(3\cos 6t - 4\sin 6t)e^{-8t}$

21. (b) $\frac{1}{25}(8\sin 20t - 6\cos 20t + 6e^{-10t} - 100te^{-10t})$;
   (d) $10te^{-10t} - 10(t - 2\pi)e^{-10(t-2\pi)}u_{2\pi}(t)$

22. (e) $\frac{10}{3}(4e^{-20t} - e^{-5t})$;   (f) $-\frac{5}{19}e^{-t} - \frac{80}{57}e^{-20t} + \frac{5}{3}e^{-5t}$

23. (b) $\frac{1}{25}(1 - \cos 5t)u_{2\pi}(t)$;   (e) $\frac{1}{25}(1 - \cos 5t)[1 + u_{4\pi}(t)]$;   (f) $\sin 5t$

24. (a) $\frac{1}{10}\sin 10t[u_{2\pi}(t) - u_{4\pi}(t)]$;   (b) $\frac{1}{10}u_{2\pi}(t)\sin 10t$;
   (e) $\frac{1}{10}\sin 10t[1 + u_{4\pi}(t)]$;   (g) $5\cos 10t + \frac{1}{2}u_{2\pi}(t)\sin 10t$;

25. (b) $\frac{1}{50}[1 - e^{-25(t-2\pi)}]u_{2\pi}(t)$;   (f) $\frac{5}{2}e^{-25t}$

26. (e) $\frac{1}{50}[1 + u_{4\pi}(t) - e^{-25t} - e^{-25(t-4\pi)}u_{4\pi}(t)]$

SECTION 4.2

2. $\begin{bmatrix} 2 & 3 & 4 & 5 \\ 3 & 4 & 5 & 6 \end{bmatrix}$

4. $\begin{bmatrix} 1 & 2 & 3 \\ 1 & 2 & 3 \\ 1 & 2 & 3 \end{bmatrix}$

6. $[1 \quad 1 \quad 1]$, $[1 \quad 1 \quad 1 \quad 0]$

8. $\begin{bmatrix} 1 & 0 & 0 \\ 0 & 1 & 0 \\ 0 & 0 & 1 \end{bmatrix}$ $\begin{bmatrix} 1 & 0 & 0 & 0 \\ 0 & 1 & 0 & -1 \\ 0 & 0 & 1 & 1 \end{bmatrix}$

10. (a);   (b);   (d);   (g);   (i);   (j);   (k)

SECTION 4.3

2. $\begin{bmatrix} 1 & 0 & 0 \\ 0 & -1 & 0 \\ 0 & 0 & 1 \end{bmatrix}$ **4.** $\begin{bmatrix} c & d \\ 0 & -b \end{bmatrix}$

6. Upper triangular system with no solutions, $x_2 = 0$, $x_2 = 1$
8. $x_1 = -x_2 = k$, $x_3 = -1$, for all k
10. $x = y = -3$ **12.** $x = \frac{43}{23}$, $y = \frac{8}{23}$
14. $x = -2$, $y = -\frac{1}{3}$, $z = -1$
16. $\begin{bmatrix} 1 \\ -1 \end{bmatrix}$ **18.** $\begin{bmatrix} 6 \\ -3 \\ -2 \end{bmatrix}$

SECTION 4.4

2. (c); (d); (f); (g); (h); (i) if $* = 0$ or 1; (j); (l)

SECTION 4.5

10.
$A - B = \begin{bmatrix} 1 & 0 & -1 \\ 1 & -1 & -2 \\ 2 & 1 & 3 \end{bmatrix}$ **12.** $\begin{bmatrix} 12 & 8 & 4 \\ 4 & -4 & -8 \\ 24 & 12 & -12 \end{bmatrix}$

14. $\begin{bmatrix} 2 & 1 & 4 \\ 1 & -1 & 2 \\ 0 & -2 & 0 \end{bmatrix} \begin{bmatrix} 1 & 0 & 2 \\ 1 & 0 & 1 \\ 1 & 0 & -3 \end{bmatrix}$

16. $(A - A^T)^T = A^T - A = -(A - A^T)$
18. $\begin{bmatrix} 0 & -5 & 4 \\ -5 & 0 & -1 \\ 4 & -1 & 4 \end{bmatrix} \begin{bmatrix} 0 & -3 & 2 \\ 3 & 0 & 3 \\ -2 & -3 & 0 \end{bmatrix}$

SECTION 4.6

2. $\begin{bmatrix} 1 & 0 & 4 \\ 0 & 2 & 4 \\ 0 & 0 & 1 \end{bmatrix}$ **4.** $a^2 + b^2 + c^2$ **6.** $\begin{bmatrix} 1 & 0 \\ 0 & -1 \end{bmatrix}$

10. $\begin{bmatrix} 1 & n \\ 0 & 1 \end{bmatrix}$ **12.** $\begin{bmatrix} 0 & -6 \\ 0 & 0 \end{bmatrix}$

18. $x_1^2 + x_2^2 - x_3^2 - x_1x_2 + 2x_2x_3$

22. $\begin{bmatrix} 1 & 1 & 1 \\ 1 & 1 & 1 \\ 1 & 1 & 1 \end{bmatrix} \begin{bmatrix} 1 & 0 & 0 \\ 0 & 0 & 0 \\ 0 & 0 & 0 \end{bmatrix} = \begin{bmatrix} 1 & 1 & 1 \\ 1 & 1 & 1 \\ 1 & 1 & 1 \end{bmatrix} \begin{bmatrix} 0 & 0 & 0 \\ 1 & 0 & 0 \\ 0 & 0 & 0 \end{bmatrix}$

28. $A = \begin{bmatrix} 1 & 1 \\ 0 & 1 \end{bmatrix}$; $B = \begin{bmatrix} 1 & 0 \\ 1 & 0 \end{bmatrix}$ **30.** $(A^T A)^T = A^T (A^T)^T = A^T A$

32. -4

34. $\begin{bmatrix} -3 & 4 & -3 \\ 3 & -4 & 3 \\ -6 & 8 & -6 \end{bmatrix}$

36. $\begin{bmatrix} 1 & 3 & 7 \\ 0 & 1 & -3 \\ 1 & -1 & 1 \end{bmatrix}$

38. $\begin{bmatrix} 3 \\ 1 \\ 1 \end{bmatrix}$

40. $\begin{bmatrix} 1 \\ -5 \\ 5 \end{bmatrix}$

42. $[-3 \quad 11 \quad 3]$

44. $\begin{bmatrix} 1 & -2 & 0 \\ -2 & 13 & 3 \\ 0 & 3 & 2 \end{bmatrix}$ $\begin{bmatrix} 10 & 6 & 1 \\ 6 & 5 & 0 \\ 0 & 0 & 1 \end{bmatrix}$

46. $\begin{bmatrix} -3 & 6 & 1 \\ -2 & 1 & -1 \\ 0 & 0 & 1 \end{bmatrix}$ $\begin{bmatrix} -6 & 3 & -2 \\ -1 & -4 & -3 \\ 0 & 0 & 1 \end{bmatrix}$

48. Not defined   **50.** Not defined   **52.** $\begin{bmatrix} 4 \\ -1 \\ -3 \end{bmatrix}$

SECTION 4.7

2. $\mathbf{A}^{-1}$ is $n \times n$ and $\mathbf{B}$ is $n \times m$. No, unless $\mathbf{C}$ is $n \times n$.
4. $(\mathbf{A}^{-1})^T (\mathbf{A}^T) = (\mathbf{A}\mathbf{A}^{-1}) = \mathbf{I}$ **6.** Set $\mathbf{B} = -\mathbf{A} = \mathbf{I}$
8. Note that $\mathbf{JB}$ has the same entries in each position in the first column. But then $\mathbf{JB} = \mathbf{I}$ is impossible.
10. By induction, using $\mathbf{A}^n (\mathbf{A}^{-1})^n = \mathbf{A}^{n-1} (\mathbf{A}\mathbf{A}^{-1})(\mathbf{A}^{-1})^{n-1} = \mathbf{A}^{n-1} (\mathbf{A}^{-1})^{n-1}$

SECTION 4.8

2. $\begin{bmatrix} \frac{1}{2} & 0 & -\frac{1}{4} \\ 0 & \frac{1}{3} & -\frac{4}{21} \\ 0 & 0 & \frac{1}{7} \end{bmatrix}$

4. $\begin{bmatrix} 0 & 0 & 1 \\ 0 & 1 & 0 \\ 1 & 0 & 0 \end{bmatrix}$

6. $\begin{bmatrix} 1 & 0 & 0 \\ 4 & -2 & 0 \\ 4 & -2 & -2 \end{bmatrix}$ $\frac{1}{2}$

14. $\frac{1}{2}\begin{bmatrix} -1 & 3 & 1 \\ 3 & -7 & -1 \\ 1 & -1 & -1 \end{bmatrix}$

16. $\frac{1}{3}\begin{bmatrix} 2 & -1 & 2 & -2 \\ -1 & 2 & -1 & 1 \\ 2 & -1 & -4 & 1 \\ -2 & 1 & 1 & 2 \end{bmatrix}$

18. Singular   **20.** Singular

22. $\frac{1}{2}\begin{bmatrix} 0 & 2 & -2 \\ -1 & -1 & 3 \\ 1 & -1 & 1 \end{bmatrix}$

24. $\frac{1}{2}\begin{bmatrix} 1 & 1 & -3 \\ 1 & 3 & -5 \\ -1 & -3 & 7 \end{bmatrix}$

SECTION 4.9

2. By induction, using Eq. 4.9.12.
4. The number of permutations of the integers 1 through n is $n!$
10. 6   **12.** 0   **14.** 36   **22.** -2
24. 0   **26.** 50   **28.** -257

SECTION 4.9.1

2. -36 **4.** -36 **6.** -276 **8.** -276
10. 0 **12.** 2 **14.** 2 **16.** 1

SECTION 4.9.2

2. $\mathbf{A}^+ = \begin{bmatrix} 3 & -6 \\ -1 & 2 \end{bmatrix}$, $\mathbf{A}$ is singular **4.** $\mathbf{A}^+ = \begin{bmatrix} 0 & -2 \\ 0 & 1 \end{bmatrix}$, $\mathbf{A}$ is singular

6.
$$\mathbf{A}^+ = \begin{bmatrix} 1 & 1 & -3 \\ 1 & 3 & -5 \\ -1 & -3 & 7 \end{bmatrix}, \quad \mathbf{A}^{-1} = \tfrac{1}{2}\mathbf{A}^+$$

8.
$$\mathbf{A}^+ = \begin{bmatrix} 1 & 1 & -3 \\ 1 & 3 & -5 \\ -1 & -3 & 7 \end{bmatrix}, \quad \mathbf{A}^{-1} = \tfrac{1}{2}\mathbf{A}^+$$

10. $x = 3, y = -3$ **12.** $x = \frac{43}{23}, y = \frac{8}{23}$
14. $x = -2, y = -13, z = -1$

SECTION 4.10

2. Linearly independent **4.** The only real value is
$k = -1$.
6. Not necessarily, for suppose $x_1 = x_2$.
8. $-2[-1 \ \ 0 \ \ 0 \ \ 1] - [2 \ \ -1 \ \ 1 \ \ 1] + [0 \ \ -1 \ \ 1 \ \ 3] = [0 \ \ 0 \ \ 0 \ \ 0]$
10. $[1 \ \ 1 \ \ 0 \ \ 1] - 2[1 \ \ 0 \ \ 0 \ \ 1] + [1 \ \ -1 \ \ 0 \ \ 1] = [0 \ \ 0 \ \ 0 \ \ 0]$

SECTION 4.11

2. From Problem 1 above

8. $\mathbf{x}_1 = \begin{bmatrix} -1 \\ 0 \\ 1 \\ 0 \end{bmatrix}$, $\mathbf{x}_2 = \begin{bmatrix} 1 \\ 1 \\ 0 \\ 0 \end{bmatrix}$, $\mathbf{x}_g = \alpha_1\mathbf{x}_1 + \alpha_2\mathbf{x}_2$

10. No free variables and therefore no basic solutions. The general solution is
$$\mathbf{x}_g = \mathbf{0}.$$

12. $\mathbf{x}_1 = \begin{bmatrix} 0 \\ -1 \\ 1 \end{bmatrix}$, $\mathbf{x}_g = \alpha_1\mathbf{x}_1$ **14.** $\mathbf{x}_1 = \begin{bmatrix} 0 \\ -1 \\ 1 \end{bmatrix}$, $\mathbf{x}_g = \alpha_1\mathbf{x}_1$

16. $\mathbf{x}_1 = \begin{bmatrix} 1 \\ -1 \\ 0 \\ \vdots \\ 0 \end{bmatrix}$, $\mathbf{x}_2 = \begin{bmatrix} 1 \\ 0 \\ -1 \\ \vdots \\ 0 \end{bmatrix}$, $\cdots$ $\mathbf{x}_{n-1} = \begin{bmatrix} 1 \\ 0 \\ 0 \\ \vdots \\ -1 \end{bmatrix}$, $\mathbf{x}_g = \sum_{1}^{n-1} \alpha_k\mathbf{x}_k$

18.

$$\mathbf{x}_1 = \begin{bmatrix} u_2 \\ -u_1 \\ 0 \\ \vdots \\ 0 \end{bmatrix}, \quad \mathbf{x}_2 = \begin{bmatrix} u_3 \\ 0 \\ -u_1 \\ \vdots \\ 0 \end{bmatrix}, \quad \dots \quad \mathbf{x}_{n-1} = \begin{bmatrix} u_{n-1} \\ 0 \\ 0 \\ \vdots \\ -u_1 \end{bmatrix}, \quad \mathbf{x}_g = \sum_{1}^{n-1} \alpha_k \mathbf{x}_k$$

20.

$$\mathbf{x}_1 = \begin{bmatrix} v_2 \\ -v_1 \\ 0 \\ \vdots \\ 0 \end{bmatrix}, \quad \mathbf{x}_2 = \begin{bmatrix} v_3 \\ 0 \\ -v_1 \\ \vdots \\ 0 \end{bmatrix}, \quad \dots \quad \mathbf{x}_{n-1} = \begin{bmatrix} v_{n-1} \\ 0 \\ 0 \\ \vdots \\ -v_1 \end{bmatrix}, \quad \mathbf{x}_g = \sum_{1}^{n-1} \alpha_k \mathbf{x}_k$$

SECTION 4.12

2. $x = 1, y = z = 0$ **4.** $x = 0, y = z = -\frac{1}{2}$

6. $x = -3, y = t = 0, z = 2$

8.

$$\mathbf{x}_g = \begin{bmatrix} 1 \\ 0 \\ 0 \\ 0 \end{bmatrix} + \alpha_1 \begin{bmatrix} -1 \\ 0 \\ 0 \\ 0 \end{bmatrix} + \alpha_2 \begin{bmatrix} 1 \\ 0 \\ 1 \\ 0 \end{bmatrix} + \alpha_3 \begin{bmatrix} -1 \\ 1 \\ 0 \\ 0 \end{bmatrix}$$

10.

$$\mathbf{x}_g = \begin{bmatrix} -2 \\ 0 \\ 2 \\ 0 \end{bmatrix} + \alpha_1 \begin{bmatrix} -3 \\ 0 \\ 2 \\ 1 \end{bmatrix} + \alpha_2 \begin{bmatrix} 1 \\ 1 \\ 0 \\ 0 \end{bmatrix}$$

12.

$$\mathbf{x}_g = \begin{bmatrix} 1 \\ 0 \\ 0 \end{bmatrix} + \alpha_1 \begin{bmatrix} -1 \\ 2 \\ 3 \end{bmatrix}$$

14.

$$\mathbf{x}_g = 1/n \begin{bmatrix} 1 \\ 1 \\ 1 \\ \vdots \\ 1 \end{bmatrix} + \alpha_1 \begin{bmatrix} 1 \\ -1 \\ 0 \\ \vdots \\ 0 \end{bmatrix} + \alpha_2 \begin{bmatrix} 1 \\ 0 \\ -1 \\ \vdots \\ 0 \end{bmatrix} + \cdots + \alpha_{n-1} \begin{bmatrix} 1 \\ 0 \\ 0 \\ \vdots \\ -1 \end{bmatrix}$$

SECTION 5.2

6. 0   **8.** 1   **10.** 0   **12.** 0

14. $x_1^2 + x_2^2 + \cdots + x_n^2$

16. 0   **18.** $2\sqrt{xy}$

20. No.   Let $\mathbf{z} = \mathbf{x}$.   **22.** Yes. $\langle \mathbf{x}, \mathbf{y} + \mathbf{z} \rangle = \langle \mathbf{x}, \mathbf{y} \rangle + \langle \mathbf{x}, \mathbf{z} \rangle$

24. $(\mathbf{u}\mathbf{u}^T)(\mathbf{u}\mathbf{u}^T) = \mathbf{u}(\mathbf{u}^T\mathbf{u})\mathbf{u}^T = \|\mathbf{u}\|^2\mathbf{u}\mathbf{u}^T = \mathbf{u}\mathbf{u}^T$

26. $\langle \mathbf{u}, \Sigma a_i \mathbf{x}_i \rangle = \Sigma a_i \langle \mathbf{u}, \mathbf{x}_i \rangle = 0$

SECTION 5.3

2.

$$\sqrt{\tfrac{1}{3}} \begin{bmatrix} 1 \\ 1 \\ 1 \end{bmatrix}, \quad \sqrt{\tfrac{1}{6}} \begin{bmatrix} 1 \\ 1 \\ -2 \end{bmatrix}$$

4.

$$\begin{bmatrix} 1 \\ 0 \\ 0 \end{bmatrix}, \quad \begin{bmatrix} 0 \\ 1 \\ 0 \end{bmatrix}, \quad \begin{bmatrix} 0 \\ 0 \\ 1 \end{bmatrix}$$

6. $\begin{bmatrix} 1 \\ 0 \\ 0 \end{bmatrix}$, $\sqrt{\frac{1}{2}}\begin{bmatrix} 0 \\ 1 \\ 1 \end{bmatrix}$, $\sqrt{\frac{1}{2}}\begin{bmatrix} 0 \\ 1 \\ -1 \end{bmatrix}$

8. $\begin{bmatrix} \sqrt{\frac{1}{2}} & 0 \\ \sqrt{\frac{1}{2}} & 0 \\ 0 & 1 \end{bmatrix}\begin{bmatrix} \sqrt{2} & 2\sqrt{2} \\ 0 & 1 \end{bmatrix}$

10. $\begin{bmatrix} \sqrt{\frac{1}{2}} & \sqrt{\frac{1}{2}} & 0 \\ 0 & 0 & 1 \\ \sqrt{\frac{1}{2}} & -\sqrt{\frac{1}{2}} & 0 \end{bmatrix}\begin{bmatrix} \sqrt{2} & 0 & \sqrt{2} \\ 0 & \sqrt{2} & 0 \\ 0 & 0 & 1 \end{bmatrix}$

12. Suppose $\Sigma c_i\mathbf{q}_i = \mathbf{0}$. Then $0 = \langle\mathbf{q}_j, \Sigma c_i\mathbf{q}_i\rangle = \Sigma c_i\langle\mathbf{q}_j, \mathbf{q}_i\rangle = c_j$

16. Use the hint and the fact that $\mathbf{A}\mathbf{A}^T$ is $k \times k$.

18. No. $\mathbf{P}_1\mathbf{P}_2\mathbf{P}_1\mathbf{P}_2 = \mathbf{P}_1\mathbf{P}_1\mathbf{P}_2\mathbf{P}_2$

20. $\mathbf{u} = \frac{1}{n}[1, 1, \ldots, 1]$

22. $\mathbf{I}$

24. $\begin{bmatrix} 2 & 1 & -1 \\ 0 & 1 & 1 \\ 0 & 0 & 2 \end{bmatrix}\mathbf{x} = \begin{bmatrix} 1 \\ 1 \\ -2 \end{bmatrix}$

SECTION 5.4

4. $\begin{bmatrix} n & \Sigma x_i & \Sigma x_i^2 \\ \Sigma x_i & \Sigma x_i^2 & \Sigma x_i^3 \\ \Sigma x_i^2 & \Sigma x_i^3 & \Sigma x_i^4 \end{bmatrix}\begin{bmatrix} \Sigma y_i \\ \Sigma x_iy_i \\ \Sigma x_i^2y_i \end{bmatrix}$

SECTION 5.5

2. $\lambda^2 - 1$; $\lambda = \pm 2$

4. $\lambda^2 - 8\lambda - 9$; $\lambda = 9$, $\lambda = -1$

6. $\lambda^2 - 4\lambda - 21$; $\lambda = 7$, $\lambda = -3$

8. $1 - \lambda$; $\lambda = 1$

10. $(1 - \lambda)^n$; $\lambda_1 = \lambda_2 \ldots = \lambda_n = 1$

12. $(6 - \lambda)(\lambda - 8)(\lambda + 2)$; $\lambda = 6$, $\lambda = 8$, $\lambda = -\frac{1}{2}$

14. $\lambda = \frac{1}{4}$, $\lambda = -\frac{1}{2}$, $\lambda = \frac{1}{6}$, $\lambda = \frac{1}{8}$, $\lambda = -\frac{1}{2}$ **18.** $(1 - \lambda)^3$; $\lambda = 1, 1, 1$

20. $(2 - \lambda)(3 - \lambda)(-1 - \lambda)$; $\lambda_1 = -1$, $\lambda_2 = 3$, $\lambda_3 = 2$

22. $(1 - \lambda)(2 - \lambda)(-1 - \lambda)\lambda$; $\lambda_1 = 0$, $\lambda_2 = -1$, $\lambda_3 = 2$, $\lambda_4 = 1$

24. $\lambda^2 + a\lambda + b$; $\lambda_{1,2} = \frac{1}{2}(-a \pm \sqrt{a^2 - 4b})$

26. $(\cos\theta - \lambda)^2 - \sin^2\theta$; $\lambda_{1,2} = \cos\theta \pm \sin\theta$

28. $\begin{bmatrix} 1 \\ 0 \end{bmatrix}$, $\begin{bmatrix} 1 \\ 0 \end{bmatrix}$

30. $\begin{bmatrix} -1 \\ i \end{bmatrix}$, $\begin{bmatrix} -1 \\ -i \end{bmatrix}$ where $i = \sqrt{-1}$

32. $\begin{bmatrix} 1 \\ 1 \end{bmatrix}$, $\begin{bmatrix} -1 \\ 5 \end{bmatrix}$

34. $\begin{bmatrix} 1 \\ 1 \\ 1 \end{bmatrix}$, $\begin{bmatrix} 1 \\ -1 \\ 1 \end{bmatrix}$, $\begin{bmatrix} -2 \\ 1 \\ 1 \end{bmatrix}$

36. $\begin{bmatrix} 1 \\ 0 \\ 1 \end{bmatrix}$, $\begin{bmatrix} -1 \\ 0 \\ 5 \end{bmatrix}$, $\begin{bmatrix} 0 \\ 1 \\ 0 \end{bmatrix}$

38. $\begin{bmatrix} 1 \\ 0 \\ 0 \\ 0 \end{bmatrix}$, $\begin{bmatrix} 1 \\ 1 \\ 0 \\ 0 \end{bmatrix}$, $\begin{bmatrix} 1 \\ 0 \\ 2 \\ 0 \end{bmatrix}$, $\begin{bmatrix} -1 \\ -1 \\ 2 \\ 2 \end{bmatrix}$

44. Let μ be an eigenvalue of $\mathbf{B}$. Then $|\mathbf{B} - \mu\mathbf{I}| = |\mathbf{A} - (k + \lambda)\mathbf{I}| = 0$.

48. Let μ be an eigenvalue of $\mathbf{A}^{-1}$. Then $\mu \neq 0$ and
$$0 = |\mathbf{A}^{-1} - \mu\mathbf{I}| = \mu|\mathbf{A}^{-1}||\mu^{-1}\mathbf{I} - \mathbf{A}| = |\mathbf{A} - \lambda\mathbf{I}|.$$

58. $\begin{bmatrix} 0 & 1 \\ -1 & 1 \end{bmatrix}$

60. $\begin{bmatrix} 0 & 1 & 0 & 0 \\ 0 & 0 & 1 & 0 \\ 0 & 0 & 0 & 1 \\ 1 & 0 & 0 & 0 \end{bmatrix}$

62. $\begin{bmatrix} 0 & 1 & 0 \\ 0 & 0 & 1 \\ \alpha & \beta & \gamma \end{bmatrix}$ where $\alpha = \lambda_1 \lambda_2 \lambda_3$, $\beta = -\lambda_1 \lambda_2 - \lambda_1 \lambda_3 - \lambda_2 \lambda_3$,
$\gamma = \lambda_1 + \lambda_2 + \lambda_3$

SECTION 5.6

2. $2; 0$

10.
They are linearly independent because $\begin{vmatrix} -1 & -1 & 1 \\ 1 & -1 & 1 \\ 0 & 2 & 3 \end{vmatrix} = 10$

30. No solutions

SECTION 5.7

2. $\alpha_1 e^{2t} \begin{bmatrix} 1 \\ 1 \end{bmatrix} + \alpha_2 e^{-2t} \begin{bmatrix} 1 \\ -3 \end{bmatrix}$

4. $\alpha_1 e^t \begin{bmatrix} 1 \\ -1 \end{bmatrix} + \alpha_2 e^{4t} \begin{bmatrix} 1 \\ 2 \end{bmatrix}$

6. $\alpha_1 e^{2t} \begin{bmatrix} 1 \\ 1 \\ 0 \end{bmatrix} + \alpha_2 e^t \begin{bmatrix} 1 \\ 0 \\ 0 \end{bmatrix} + \alpha_3 e^{3t} \begin{bmatrix} 1 \\ 1 \\ 1 \end{bmatrix}$

8. $\alpha_1 e^{-t} \begin{bmatrix} 1 \\ -1 \\ 1 \end{bmatrix} + \alpha_2 e^t \begin{bmatrix} 1 \\ 0 \\ 0 \end{bmatrix} + \alpha_3 e^{-2t} \begin{bmatrix} 1 \\ -1 \\ 2 \end{bmatrix}$

10. $\frac{1}{4} e^{2t} \begin{bmatrix} 3 \\ 1 \end{bmatrix} + \frac{3}{4} e^{-2t} \begin{bmatrix} 1 \\ -1 \end{bmatrix}$

12. $-2e^{-t} \begin{bmatrix} 1 \\ -1 \\ 1 \end{bmatrix} + e^t \begin{bmatrix} 1 \\ 0 \\ 0 \end{bmatrix} + e^{-2t} \begin{bmatrix} 1 \\ -1 \\ 2 \end{bmatrix}$

14. $-e^{-t} \begin{bmatrix} 1 \\ -1 \\ 1 \end{bmatrix} + e^{-2t} \begin{bmatrix} 1 \\ -1 \\ 2 \end{bmatrix}$

16. $i_1'' = \dfrac{C_1 - C_2}{L_1 C_1 C_2} i_1 + \dfrac{1}{L_1 C_2} i_2,$ $i_2'' = \dfrac{1}{L_2 C_2} i_1 + \dfrac{1}{L_1 C_2} i_2$

18. $\mathbf{Ax} = m^2 \mathbf{x}$ **20.** $120.7, -20.7;$ $\begin{bmatrix} 0.816 \\ 0.577 \end{bmatrix}$ $\begin{bmatrix} 0.816 \\ -0.577 \end{bmatrix}$

22. $-12, -2; (\sqrt{5}/5) \begin{bmatrix} -2 \\ 1 \end{bmatrix}$ $(\sqrt{5}/5) \begin{bmatrix} 1 \\ 2 \end{bmatrix}$

24. $y_1(t) = (-2/\sqrt{3}) \sin \sqrt{12}\,t + 2\sqrt{2} \sin \sqrt{2}\,t$
$y_2(t) = (1/\sqrt{3}) \sin \sqrt{12}\,t + 4\sqrt{2} \sin \sqrt{2}\,t$
26. $y_1(t) = 0.219 \cos 1.73t + 1.78 \cos 8.19t$
$y_2(t) = 0.312 \,(\cos 1.73t - \cos 8.19t)$

SECTION 5.8

2. $\dfrac{e^t}{2}\begin{bmatrix} t - \frac{1}{2} \\ t + \frac{1}{2} \end{bmatrix}$
4. $-\begin{bmatrix} 1 \\ 1 \end{bmatrix} - \dfrac{e^t}{2}\begin{bmatrix} t - \frac{1}{2} \\ t + \frac{1}{2} \end{bmatrix}$
6. $\begin{bmatrix} 1 \\ 0 \end{bmatrix} + e^t\begin{bmatrix} 2t - 1 \\ 2t + 1 \end{bmatrix}$

8. $e^t\begin{bmatrix} 1 - t \\ -t \end{bmatrix} + e^{-t}\begin{bmatrix} -1 \\ 0 \end{bmatrix}$

16. $\begin{bmatrix} -1 \\ -\frac{2}{3} \end{bmatrix}$
18. $-\frac{1}{2}\begin{bmatrix} \sin t + \cos t \\ \cos t \end{bmatrix}$
20. $\begin{bmatrix} 2 \sin - \cos t \\ \cos t + \sin t \end{bmatrix}$

SECTION 6.2

2. 23.7, 17.76°;   10.62, 318.3°
4. 11.18, 333.4°;   11.18, 26.57°
6. $-10\mathbf{i} + 14.14\mathbf{j} + 10\mathbf{k}$; $-0.5\mathbf{i} + 0.707\mathbf{j} + 0.5\mathbf{k}$
8. $6\mathbf{i} + 3\mathbf{j} - 8\mathbf{k}$
10. $3\mathbf{i} + 3\mathbf{j} - 4\mathbf{k}$
12. -4
14. 53.85
16. 0
18. $100\mathbf{i} + 154\mathbf{j} - 104\mathbf{k}$
20. 8
22. 107.3
28. $\frac{5}{9}$
30. -2.941
32. $-0.391\mathbf{i} - 0.557\mathbf{j} - 0.743\mathbf{k}$
34. -11.93
36. 26 N·m
38. $-40\mathbf{i} + 75\mathbf{j} + 145\mathbf{k}$
40. 100 N·m, 0

SECTION 6.3

2. $2\mathbf{i} + 4\mathbf{k}$
4. -30.8
6. 74.79
8. $-210\mathbf{i} + 600\mathbf{j}$
10. 0.00257 m/s^2; 0.0239 m/s^2
12. $-15.41°\text{C/s}$

SECTION 6.4

2. $2y\mathbf{i} + 2x\mathbf{j}$
4. $e^x(\sin 2y\mathbf{i} + 2 \cos 2y\mathbf{j})$
6. $\mathbf{r}/r^2$
8. $(-y\mathbf{i} + x\mathbf{j})/(x^2 + y^2)$
10. $(2/\sqrt{5})\mathbf{i} + (1/\sqrt{5})\mathbf{j}$
12. $0.907\mathbf{i} - 0.243\mathbf{j}$
14. $0.174\mathbf{i} + 0.696\mathbf{j} - 0.696\mathbf{k}$
16. $3x + 4y = 25$
18. $y = 2$
20. 5
22. $4\sqrt{3}$
24. 0
26. 1
28. 0
30. $\nabla \cdot \mathbf{v} = 0$. No. 1
32. 0
34. $-\mathbf{j} + 2\mathbf{k}$
36. $-0.1353\mathbf{j}$
38. 7
40. $\mathbf{k}$
42. 4
44. $-5\mathbf{i} + 10\mathbf{j} - 38\mathbf{k}$
46. $10\mathbf{i} + 4\mathbf{j} + 6\mathbf{k}$
48. $14\mathbf{i} - 9\mathbf{j} + 8\mathbf{k}$
50. irrotational
52. irrotational, solenoidal
54. irrotational
56. neither

58. irrotational, solenoidal

70. $e^x \sin y + C$

68. $(x^3 + y^3 + z^3)/3 + C$

72. $x^2z + y^3/3 + C$

SECTION 6.5

2. $\cos \theta$ **4.** $\sin \phi \sin \theta$ **6.** 0 **8.** $\cos \phi$

10. $\cos \phi$

16. $\sqrt{\dfrac{x^2 + y^2 + z^2}{x^2 + y^2}}(-y\mathbf{i} + x\mathbf{j})$ **22.** $(Ar + B/r) \cos \theta + C$

SECTION 6.6

2. $-\frac{1}{2}$ **4.** 32π **6.** 256π

12. $\rho C(\partial T/\partial t) = k\nabla^2 T$

14. $-\frac{8}{3}$ **16.** 2 **18.** 0

SECTION 7.1

4. $a_n \cos \dfrac{n\pi t}{T} + b_n \sin \dfrac{n\pi t}{T} = \dfrac{1}{T}\displaystyle\int_{-T}^{T} f(s)\left[\cos \dfrac{n\pi}{T}(s - t)\right] ds$

6. $a_0 = -2; \; a_n = \dfrac{1}{T}\displaystyle\int_{-T}^{T} (-1) \cos \dfrac{n\pi t}{T} dt = 0; \quad b_n = \dfrac{1}{T}\displaystyle\int_{-T}^{T} (-1) \sin \dfrac{n\pi t}{T} dt = 0$

Hence, $f(t) = -1$ is its own expansion.

SECTION 7.3.1

2. $-\dfrac{x^n}{a} \cos (ax) + \dfrac{nx^{n-1}}{a^2} \sin (ax) + \cdots + \dfrac{n!}{a^{n+1}} \begin{cases} (-1)^r \sin (ax), & n \text{ odd} \\ (-1)^r \cos (ax), & n \text{ even} \end{cases}$

where $r = (n - 1)/2$ if n is odd; $r = n/2$ if n is even.

4. $\dfrac{x^n}{a} \cosh (ax) - \dfrac{nx^{n-1}}{a^2} \sinh (ax) + \cdots + (-1)^n \dfrac{n!}{a^{n+1}} \begin{cases} \sinh (ax), & n \text{ odd} \\ \cosh (ax), & n \text{ even} \end{cases}$

6. $\dfrac{x^n}{a} \dfrac{(ax + b)^{\alpha+1}}{(\alpha + 1)} - \dfrac{nx^{n-1}}{a^2} \dfrac{(ax + b)^{\alpha+2}}{(\alpha + 1)(\alpha + 2)} + (-1)^n \dfrac{n!}{a^{n+1}} \dfrac{(ax + b)^{\alpha+n+1}}{(\alpha + 1) \cdots (\alpha + n + 1)}$

SECTION 7.3.2

2. $\dfrac{\pi^2}{3} + 4\pi \displaystyle\sum_{n=1}^{\infty} \dfrac{(-1)^n}{n^2} \cos (nt)$ **4.** $2\pi - \displaystyle\sum_{n=1}^{\infty} \dfrac{4(-1)^n}{n} \sin \left(\dfrac{n}{2}t\right)$

6. $\dfrac{1}{4} + \displaystyle\sum_{n=1}^{\infty} \left[\dfrac{1 - (-1)^n}{n^2\pi^2} \cos (n\pi t) - \dfrac{(-1)^n}{n\pi} \sin (n\pi t)\right]$

8. $\dfrac{10}{3} + \displaystyle\sum_{n=1}^{\infty} \left[\dfrac{-8[1 - 3(-1)^n]}{n^2\pi^2} \cos \left(\dfrac{n}{2}\pi t\right) + \dfrac{16}{n^3\pi^3}[1 - (-1)^n] \sin \left(\dfrac{n}{2}\pi t\right)\right]$

10. $1 + 2 \displaystyle\sum_{n=1}^{\infty} \dfrac{(-1)^{n-1}}{n} \sin (nt)$ **12.** $\dfrac{2}{\pi} - \dfrac{4}{\pi} \displaystyle\sum_{n=2}^{\infty} \dfrac{1}{n^2 - 1} \cos 2(n - 1)t$

SECTION 7.3.3

2. Odd: $\displaystyle 8\sum_{n=1}^{\infty}\frac{(-1)^{n-1}}{n}\sin(nt)$ **Even:** $\displaystyle \frac{\pi}{2}-\frac{4}{\pi}\sum_{n=1}^{\infty}\frac{\cos(2n-1)t}{(2n-1)^2}$

4. Odd: $\sin t$ **Even:** $\displaystyle \frac{2}{\pi}-\frac{4}{\pi}\sum_{n=1}^{\infty}\frac{\cos(2nt)}{4n^2-1}$

6. Odd: $\displaystyle 2\pi^2\sum_{n=1}^{\infty}\left[\frac{(-1)^{n+1}}{n\pi}-2\frac{1-(-1)^n}{n^3\pi^3}\right]\sin(nt)$

8. Since $f(t)=f(-t)$ for all t, we have $f(0^+)=f(0^-)$ and hence $f(0)=\lim\limits_{t\to 0}f(t)$.

SECTION 7.3.4

2. $\displaystyle \frac{1}{6}-\frac{1}{\pi^2}\cos(2\pi t)-\frac{1}{4\pi^2}\cos(4\pi t)-\cdots$

   $\displaystyle +\frac{8}{\pi^3}\sin(\pi t)+\frac{8}{27\pi^3}\sin(3\pi t)+\frac{8}{125\pi^3}\sin(5\pi t)+\cdots$

SECTION 7.3.5

2. $\displaystyle \frac{\pi}{2}-\frac{4}{\pi}\sum_{n=1}^{\infty}\frac{\cos(2n-1)t}{(2n-1)^2}$ **4.** $\displaystyle \frac{1}{2}+\frac{2}{\pi}\sum_{n=1}^{\infty}\frac{\sin(2n-1)t}{2n-1}$

6. It is the odd extension of $f_1(t)$.

SECTION 7.4

2. $-\frac{1}{6}\cos 2t$ **4.** $\frac{1}{21}\cos 2t+\frac{1}{90}\sin 4t$

6. $-\frac{1}{58}\cos 2t+\frac{5}{116}\sin 2t$

8. Set $D_n=(16-n^2)^2+(12n)^2$ and $\alpha_n=-12nb_n/D_n$ and

   $\beta_n=(16-n^2)b_n/D_n$. $Y(t)=\displaystyle\sum_{n=1}^{N}\alpha_n\cos nt+b_n\sin nt$

12. $\displaystyle \sum_{1}^{\infty}A_n\cos\frac{n\pi t}{0.001}+B_n\sin\frac{n\pi t}{0.001}$. Set $\alpha_n=100-n^2\pi^2$. Then

   $A_n=\displaystyle \frac{120[1-(-1)^n]\alpha_n}{\alpha_n^2+400n^2\pi^2}$, $B_n=\displaystyle \frac{20n\pi}{\alpha_n}A_n$

SECTION 7.5.1

2. $t^2/2+t=\frac{1}{6}+(2/\pi)\displaystyle\sum_{n=1}^{\infty}\alpha_n\cos n\pi t+\beta_n\sin n\pi t$

   where $\alpha_n=\dfrac{(-1)^n}{\pi n^2}$, $\beta_n=\dfrac{(-1)^{n+1}}{n}$

8. $F(t)=0$ for $-\pi<t<0$ and is $t^2/2$ for $0<t<\pi$. Then,

   $F(t)=\pi^2/12+\displaystyle\sum\frac{(-1)^n}{n^2}\cos nt+\beta_n\sin nt;\ n=1,\,2,\,\ldots$ where

   $\beta_n=\dfrac{\pi(-1)^{n+1}}{2n}+\dfrac{(-1)^{n+1}-1}{n^3\pi}$

10. $F(t) = \dfrac{t^2}{2} + 2\pi t = \dfrac{2\pi^2}{3} + 8\pi \sum \dfrac{(-1)^n}{n^2\pi} \cos \dfrac{nt}{2} + \dfrac{(-1)^{n+1}}{n} \sin \dfrac{nt}{2}$

for $n = 1, 2, \ldots$ and $-2\pi < t < 2\pi$

SECTION 7.5.2

2. Yes, because $|\sin t|$ is continuous and all its derivatives are either continuous or sectionally continuous. Moreover, $f(-\pi) = f(\pi)$.

$$\frac{8}{\pi} \sum_1^\infty \frac{n \sin (2nt)}{4n^2 - 1}$$

SECTION 7.5.3

2. $e^{\cos t} \cos (\sin t)$ 　　　　　　　　**4.** $\sin (\cos t) \cosh (\sin t)$

6. $\dfrac{2}{a^2 - 1}\left[\dfrac{1}{2} + \sum_1^\infty a^{-n} \cos (nt)\right]$

8. $a\dfrac{(a - \cos t)^2 - \sin^2 t}{(a^2 - 2a \cos t + 1)^2} = \sum_1^\infty \dfrac{(n + 1) \cos (nt)}{a^{n+1}}$

$2\dfrac{(a - \cos t) \sin t}{(a^2 - 2a \cos t + 1)^2} = \sum_1^\infty \dfrac{(n + 1) \sin (nt)}{a^{n+1}}$

10. $e^{-\cos t} \cos (\sin t) = 1 - \cos t + \dfrac{\cos (2t)}{2!} - \dfrac{\cos (3t)}{3!} + \cdots$

$e^{-\cos t} \sin (\sin t) = \sin t - \dfrac{\sin (2t)}{2!} + \dfrac{\sin (3t)}{3!} + \cdots$

12. $\cosh t \cos (\sin t) = 1 + \dfrac{\cos (2t)}{2!} + \dfrac{\cos (3t)}{4!} + \cdots$

$\cosh t \cos (\sin t) = \sin t + \dfrac{\sin (3t)}{3!} + \dfrac{\sin (5t)}{5!} + \cdots$

SECTION 8.1

2. Parabolic, homogeneous, linear 　　　**4.** Elliptic, non-homogeneous, linear

SECTION 8.3

2. $\dfrac{\partial T}{\partial t} = k\dfrac{\partial^2 T}{\partial x^2} + \phi$ 　　　　　**4.** $\dfrac{\partial T}{\partial t} = k\dfrac{\partial^2 T}{\partial x^2}, 100x$

SECTION 8.6

2. $(c_1 x + c_2)(c_3 t + c_4)$ 　　　　　　**4.** $A = (c_1 - c_2)i, \quad B = c_1 + c_2$

6. a) $0.1 \sin \dfrac{\pi x}{2} \cos \dfrac{\pi a t}{2}$ 　　　　　**8.** $0.1, x = \pi/2, \quad t = \pi/80$

10. $0.2 \sin \dfrac{\pi x}{4} \cos \dfrac{\pi a t}{4}$ 　　　　　**12.** 0.334 at $x = 2$

14. $2k/\pi, 2k/\pi, 2k/3\pi$

16. $\sum A_n \sin nx \sin 60 \, nt, \quad A_n = \dfrac{2}{3\pi n^2}\left(\cos \dfrac{n\pi}{4} - \cos \dfrac{3n\pi}{4}\right)$

18. Yes. For $n = 3$.

SECTION 8.7

2. $100 \sin \dfrac{\pi x}{4} e^{-n^2 \pi^2 kt/16} + 50x$ **4.** $200(1 + \sin \pi x e^{-n^2 \pi^2 kt})$

6. $\sum A_n \sin \dfrac{n\pi x}{2} e^{-n^2 \pi^2 kt/4} + 50x, \quad A_n = \dfrac{200}{n\pi}$

8. $\sum A_n \sin \dfrac{n\pi x}{2} e^{-n^2 \pi^2 kt/4} \quad A_n = \displaystyle\int_0^2 100(2x - x^2) \sin \dfrac{n\pi x}{2} dx$

10. 37.27°   **12.** 50°   **14.** 52.4°   **16.** 702 s
18. 2145 s   **22.** 3231 W   **24.** 5027 W   **26.** 335 W
32. $100 \cos x e^{-kt}$

34. $100 \sin \dfrac{x}{2} e^{-kt/4}$ **36.** $\sum A_n \sin \dfrac{2n-1}{2} x e^{-kt(2n-1)2/4}$

38. 31.2°   **40.** $5.556(x - x^2/2)$
42. $11.11(x - x^2/4) + 100$

44. $\sum A_n \sin \dfrac{2n-1}{4} \pi x e^{-k(2n-1)2 \pi^2 t/16} + g(x), \quad g(x) = 11.11(x - x^2/4)$

46. $\dfrac{100}{e^\pi - e^{-\pi}} \sin \pi x (e^{\pi y} - e^{-\pi y})$ **48.** $\sum A_n \sin n\pi x (e^{n\pi y} - e^{-n\pi y})$

50. $100 + \sum A_n \sin n\pi y (e^{n\pi x} - e^{-n\pi x})$

SECTION 8.8

2. $50 \sum r^n P_n(x)(2n + 1) \displaystyle\int_{-1}^1 x P_n(x) \, dx$

SECTION 8.9

2. $\sum e^{-k\mu_n^2 t} A_n J_0(\mu_n r)$

4. $\sum A_n e^{-k\mu_n^2 t} J_0(\mu_n r), \quad A_1 = 200/3, \quad A_2 = 1230 \displaystyle\int_0^1 r^2 J_0(3.83r) \, dr$

SECTION 9.2

6. $\delta + \dfrac{\delta^2}{2} + \dfrac{\delta^3}{8} - \dfrac{\delta^5}{128} + \cdots$

SECTION 9.4

12. $\dfrac{1}{6h}\left[2f_{i+3} - 9f_{i+2} + 14f_{i+1} - 11f_i\right]$

14. $\dfrac{1}{4h^3}\left[-7f_{i-5} + 41f_{i-4} - 98f_{i-3} + 118f_{i-2} - 71f_{i-1} + 17f_i\right]$

16. 0.0958   **18.** 0.0871   **20.** −0.405   **22.** −0.505

SECTION 9.5

2. (a) 73;   (b) 72   **4.** (a) 1.537; (b) 1.426   **6.** 10.04

SECTION 9.6

2. 0.29266   **4.** 0.29267 **6.** 0.29302, 0.123%
10. 0.995724   **12.** 0.96728 **14.** 0.55932

SECTION 9.7

2. 1.57   **4.** 1.46   **6.** 1.73   **8.** 0.394
10. 7.93

SECTION 9.8

2. 0.964   **4.** 2.57   **6.** 2.58
8. 2, 1.84, 1.44, 0.98, 0.62, 0.42
10. 2, 1.84, 1.36, 0.20, −2.91, −12.3
12. 0.38, 0.72   **14.** 0.38, 0.72
16. $y_{i+1} = y_i + \dfrac{h}{12}(23\dot{y}_i - 16\dot{y}_{i-1} + 5\dot{y}_{i-2})$

SECTION 9.9

2. 2.43, 3.04, 4.00   **3.** 0, 1.6, 2.94, 3.80, 4.04, 3.61
4. 0, 1.6, 2.94, 3.80, 4.04, 3.61   **6.** 0, 1.6, 2.94, 3.82, 4.08, 3.69
8. 0, 0.16, 0.17, 0.11, −0.01, −0.15

SECTION 9.12

2. 22 ks   **4.** 23 ks   **6.** 32 ks
8. 200, 125, 75, 25, 12.5, 12.5   **10.** 200, 162, 137, 112, 106, 106
12. 0, −0.1, 0, −0.2, −0.1, 0, 0   **14.** 0, −0.02, −0.04, −0.04, −0.02, 0
16. 0, −0.02, −0.04, −0.04, −0.02, 0

SECTION 10.2

2. 143.1°, 2.498   **4.** 216.9°, 3.785   **6.** 25   **8.** 1.265
10. 25   **12.** −527 + 336i
14. −0.3641 + 1.671i,   −1.265 − 1.151i,   1.629 − 0.5202i
16. −2 + 11i,   2 − 11i
18. $\sqrt{2}(1 + i)$, $\sqrt{2}(-1 - i)$, $\sqrt{2}(-1 + i)$, $\sqrt{2}(1 - i)$
20. 3, −3   **22.** $(x - 2)^2 + y^2 = 4$ **24.** $15x^2 + 34x + 15y^2 + 15 = 0$

SECTION 10.3

2. $2e^{\pi i}$ **4.** $2e^{3\pi i/2}$ **6.** $13e^{5.107i}$ **8.** $13e^{1.966i}$
12. −0.416 + 0.909i
14. 1   **16.** 0.26(1 − i)

18. $0.276 + 1.39i$, $-1.39 + 0.276i$, $-0.276 - 1.39i$, $1.39 - 0.276i$
20. $2 + 11i$ **22.** $-1.455 + 0.3431i$ **24.** 1.324
26. $1.63 - 1.77i$ **28.** 1.324 **30.** $(\pi/2)i$
32. $1.609 + 5.64i$ **34.** $1 + (\pi/2)i$ **36.** i
38. $9.807 - 7.958i$ **40.** $0.2739 + 0.5837i$ **42.** $\pi/2 \pm 1.317i$
44. $1.099 + \pi i$ **46.** $\pi \pm 1.317i$ **50.** $-1.964 - 0.1734i$

SECTION 10.4

2. $1, -1$ **4.** $2(x - 1) + 2yi$ **6.** $e^x \cos y + ie^x \sin y$

10. $\theta + C$ **12.** $2xy$ **14.** $2 \tan^{-1} \dfrac{y}{x}$

SECTION 10.5

2. $x = a \cos t, \ y = b \sin t$ $0 \le t \le 2\pi$
4. -2 **6.** $\frac{16}{3}$ **8.** 0
10. $\frac{8}{3}i, \ -\frac{16}{3} + 8i$ **12.** $-\frac{8}{3}i$ **14.** Yes   **16.** No
18. Yes   **20.** Yes

SECTION 10.6

2. 0 **4.** 0 **6.** 0 **8.** 0
10. 0 **12.** $2\pi i$

SECTION 10.7

2. $2\pi ei$ **4.** $(2\pi/5)i$ **6.** $-2\pi i$ **8.** $\pi(1 + i)$
10. 0 **12.** $\pi(1 + i)$ **14.** $2\pi i$ **16.** $-5.287i$
18. $(\pi/3)i$

SECTION 10.8

2. $1 - z + z^2 - z^3 + \cdots$ **4.** $-1 + 2z - 2z^2 + 2z^3 - \cdots$

6. $z + \dfrac{z^3}{3!} + \dfrac{z^5}{5!} + \cdots$

8. $-z - (z - 1)^2 - (z - 1)^3 - \cdots, \quad R = 1$

10. $-\dfrac{1}{9}\left[4 + z + \dfrac{(z + 1)^2}{3} + \dfrac{(z + 1)^3}{9} + \cdots\right], \quad R = 3$

12. $-\dfrac{1 - i}{4}\left[1 + \dfrac{1 - i}{4}(z + 2i) - \dfrac{i}{8}(z + 2i)^2 - \dfrac{1 + i}{32}(z + 2i)^3 + \cdots\right],$
$R = 2\sqrt{2}$

14. $-1 + z + z^3 - z^4 - z^6 + z^7 + \cdots$

16. $-\dfrac{1}{4} + \dfrac{3}{16}z - \dfrac{13}{64}z^2 + \dfrac{51}{256}z^3 + \cdots$

18. $e^2\left(1 - z + \dfrac{z^2}{2!} - \dfrac{z^3}{3!} + \cdots\right)$ **20.** $z^2 - \dfrac{z^6}{3!} + \dfrac{z^{10}}{3!} + \cdots$

22. $1 + z - \dfrac{z^3}{3} + \dfrac{z^4}{12} + \cdots$

24. $z + z^2 + \dfrac{z^3}{3} - \dfrac{z^4}{6} + \cdots$

26. $\dfrac{z^3}{3} - \dfrac{z^7}{42} + \dfrac{z^{11}}{1320} + \cdots$

28. $z - \dfrac{z^5}{10} + \dfrac{z^9}{216} + \cdots$

30. $e\left[z + \dfrac{(z-1)^2}{2!} + \dfrac{(z-1)^3}{3!} + \cdots \right]$

32. $1 - \dfrac{1}{2}\left(z - \dfrac{\pi}{2} \right)^2 + \dfrac{1}{12}\left(z - \dfrac{\pi}{2} \right)^4 + \cdots$

34. $1 + 2(1 - z) + 3(1 - z)^2 + \cdots$

SECTION 10.9

2. $-\dfrac{1}{2z} - \dfrac{1}{4} - \dfrac{z}{8} - \dfrac{z^2}{16} - \cdots, \quad R = 2$

4. $\dfrac{1}{e}\left(\dfrac{1}{z} + 1 + \dfrac{z}{2} + \dfrac{z^2}{6} + \cdots \right), \quad R = \infty$

6. $1 + \dfrac{1}{z} + \dfrac{1}{2z^2} + \dfrac{1}{6z^3} + \cdots, \quad |z| > 0$

8. $\dfrac{1}{2}\left[1 + \dfrac{z+1}{2} + \dfrac{(z+1)^2}{4} + \cdots \right], \quad 0 \le |z+1| < 2$

10. $-\dfrac{1}{z} - 1 - z - z^2 - \cdots, \quad 0 < |z| < 1$

12. $-\dfrac{i}{2}\left[\dfrac{1}{z-i} + \dfrac{i}{2} - \dfrac{z-i}{4} - \dfrac{(z-i)^2}{8}i + \cdots \right], \quad 0 < |z-i| < 2$

$\dfrac{1}{(z-i)^2} - \dfrac{2i}{(z-i)^3} - \dfrac{4}{(z-i)^4} + \dfrac{8}{(z-i)^5} + \cdots, \quad 2 < |z-i|$

14. $\dfrac{1}{(z+1)^2} + \dfrac{3}{(z+1)^3} + \dfrac{9}{(z+1)^4} + \cdots, \quad 3 < |z+1|$

$-\dfrac{1}{3(z+1)} - \dfrac{1}{9} - \dfrac{z+1}{27} - \dfrac{(z+1)^2}{81} - \cdots, \quad 0 < |z+1| < 3$

SECTION 10.10

2. $\frac{1}{2}$ at $-2i$, $\frac{1}{2}$ at $2i$ **4.** e at 1 **6.** $\frac{5}{9}$ at 2, $\frac{4}{9}$ at -1

8. 0 **10.** $2\pi(1+i)$ **12.** 0 **14.** $-i$

16. 0 **18.** $4\pi i$ **20.** 2.43 **22.** 0

24. $3.93i$ **26.** $\pi/4$ **28.** 1.81 **30.** $\pi/6$

32. π/e **34.** 0.773 **36.** 0.072

Index